# Lecture Notes in Computer Science          9710

Commenced Publication in 1973
Founding and Former Series Editors:
Gerhard Goos, Juris Hartmanis, and Jan van Leeuwen

More information about this series at http://www.springer.com/series/7407

Nadia Creignou · Daniel Le Berre (Eds.)

# Theory and Applications of Satisfiability Testing – SAT 2016

19th International Conference
Bordeaux, France, July 5–8, 2016
Proceedings

 Springer

*Editors*
Nadia Creignou
Aix-Marseille Université
Marseille
France

Daniel Le Berre
Université d'Artois
Lens
France

ISSN 0302-9743        ISSN 1611-3349   (electronic)
Lecture Notes in Computer Science
ISBN 978-3-319-40969-6      ISBN 978-3-319-40970-2   (eBook)
DOI 10.1007/978-3-319-40970-2

Library of Congress Control Number: 2016941614

LNCS Sublibrary: SL1 – Theoretical Computer Science and General Issues

Printed on acid-free paper

This Springer imprint is published by Springer Nature
The registered company is Springer International Publishing AG Switzerland

# Preface

This volume contains the papers presented at the 19th International Conference on Theory and Applications of Satisfiability Testing (SAT 2016) held during July 5–8, 2016, in Bordeaux, France. SAT 2016 was hosted by the Computer Science Laboratory of Bordeaux (LaBRI).

The International Conference on Theory and Applications of Satisfiability Testing (SAT) is the premier annual meeting for researchers focusing on the theory and applications of the propositional satisfiability problem, broadly construed. Aside from plain propositional satisfiability, the scope of the meeting includes Boolean optimization (including MaxSAT and pseudo-Boolean (PB) constraints), quantified Boolean formulas (QBF), satisfiability modulo theories (SMT), and constraint programming (CP) for problems with clear connections to Boolean-level reasoning. Many hard combinatorial problems can be tackled using SAT-based techniques, including problems that arise in formal verification, artificial intelligence, operations research, computational biology, cryptology, data mining, machine learning, mathematics, etc. Indeed, the theoretical and practical advances in SAT research over the past 20 years have contributed to making SAT technology an indispensable tool in a variety of domains.

SAT 2016 welcomed scientific contributions addressing different aspects of SAT interpreted in a broad sense, including (but not restricted to) theoretical advances (including exact algorithms, proof complexity, and other complexity issues), practical search algorithms, knowledge compilation, implementation-level details of SAT solvers and SAT-based systems, problem encodings and reformulations, applications (including both novel applications domains and improvements to existing approaches), as well as case studies and reports on findings based on rigorous experimentation.

A total of 70 papers were submitted this year distributed into 48 long papers, 13 short papers, and nine tool papers. The papers were reviewed by the Program Committee (33 members), with the help of 65 additional reviewers. Only one regular paper was found by the Program Committee to be out of the scope for the conference. Each of the remaining submissions was reviewed by at least three different reviewers. A rebuttal period allowed the authors to provide a feedback to the reviewers. After that, the discussion among the Program Committee took place. External reviewers supporting the Program Committee were also invited to participate directly in the discussions for the papers they reviewed. This year, the authors received a meta-review, summarizing the discussion that occurred after the rebuttal and the reasons of the final recommendation. The final recommendation was to accept 31 submissions (22 long, four short, and five tool papers) and to accept conditionally five additional papers. The latter (four long and one short) eventually satisfied the conditions for acceptance.

In addition to presentations on the accepted papers, the scientific program of SAT 2016 included three invited talks:

- Phokion Kolaitis (University of California Santa Cruz, IBM, USA) "Coping with Inconsistent Databases: Semantics, Algorithms, and Complexity"
- David Monniaux (VERIMAG University of Grenoble, CNRS, France) "Satisfiability Testing, a Disruptive Technology in Program Verification"
- Torsten Schaub (University of Potsdam, Germany, EurAI sponsored) "From SAT to ASP and Back!?"

As in previous years, SAT 2016 hosted various associated events, including four workshops on July 4:

- 6th International Workshop on the Cross-Fertilization Between CSP and SAT (CSPSAT 2016) organized by Yael Ben-Haim, Valentin Mayer-Eichberger, and Yehuda Naveh
- "Graph Structure and Satisfiability Testing" organized by Simone Bova and Stefan Mengel
- 7th Pragmatics of SAT International Workshop (PoS 2016) organized by Olivier Roussel and Allen Van Gelder
- 4th International Workshop on Quantified Boolean Formulas (QBF 2016) organized by Florian Lonsing and Martina Seidl

There were also four competitive events, which ran before the conference and whose results were disclosed during the conference:

- MAXSAT evaluation organized by Josep Argelich, Chu Min Li, Felip Manyà and Jordi Planes
- PB competition organized by Olivier Roussel
- QBF evaluation organized by Luca Pulina
- SAT competition organized by Marijn Heule, Matti Jarvisalo, and Tomas Baylo

Moreover, this year a full day of tutorials — "How to Solve My Problem with SAT?" — was organized right after the conference, on July 9.

March 2016 was a terrible month for the SAT community. On March 12, Helmut Veith, our esteemed colleague from TU Vienna, passed away at the age of 45. His work on counter example guided abstraction refinement is widely used in the SAT community, especially in recent years to tackle QBF problems: A specific session on that topic was organized during the conference. On March 13, Hilary Putnam, one of the authors of the seminal "Davis and Putnam" procedure, central in current SAT research, passed away at the age of 90. The first session on SAT solving was dedicated to his memory. Our thoughts are with their families during this difficult time.

We would like to thank everyone who contributed to making SAT 2016 a success. First and foremost we would like to thank the members of the Program Committee and the additional reviewers for their careful and thorough work, without which it would not have been possible for us to put together such an outstanding conference. We also wish to thank all the authors who submitted their work for our consideration. We thank the SAT Association chair Armin Biere, vice chair John Franco, and treasurer Hans Kleine Büning for their help and advice in organizational matters. The EasyChair conference systems provided invaluable assistance in coordinating the submission and review process, in organizing the program, as well as in the assembly of these proceedings. We also thank the local organization team for their efforts with practical aspects of local organization.

Finally, we gratefully thank the University of Bordeaux, Bordeaux INP, the Computer Science Laboratory of Bordeaux (LaBRI), the GIS Albatros (Bordeaux), the CNRS, the Laboratory of Fundamental Computer Science of Marseilles (LIF), the Lens Computer Science Research Laboratory (CRIL), the European Association for Artifical Intelligence (EurAI), the SAT association, the French-Speaking Constraints Association (AFPC), Intel, RATP and Safe-River for financial and organizational support for SAT 2016.

April 2016

Daniel Le Berre
Nadia Creignou

# Organization

## Program Committee

| | |
|---|---|
| Fahiem Bacchus | University of Toronto, Canada |
| Yael Ben-Haim | IBM Research, Israel |
| Olaf Beyersdorff | University of Leeds, UK |
| Armin Biere | Johannes Kepler University, Austria |
| Nikolaj Bjorner | Microsoft Research, USA |
| Maria Luisa Bonet | Universitat Politecnica de Catalunya, Spain |
| Sam Buss | UCSD, USA |
| Nadia Creignou | Aix-Marseille Université, LIF-CNRS, France |
| Uwe Egly | TU Wien, Austria |
| John Franco | University of Cincinnati, USA |
| Djamal Habet | Aix-Marseille Université, LSIS-CNRS, France |
| Marijn Heule | The University of Texas at Austin, USA |
| Holger Hoos | University of British Columbia, Canada |
| Frank Hutter | University of Freiburg, Germany |
| Mikolas Janota | Microsoft Research, UK |
| Matti Järvisalo | University of Helsinki, Finland |
| Hans Kleine Büning | University of Paderborn, Germany |
| Daniel Le Berre | Université d'Artois, CRIL-CNRS, France |
| Ines Lynce | INESC-ID/IST, University of Lisbon, Portugal |
| Marco Maratea | DIBRIS, University of Genoa, Italy |
| Joao Marques-Silva | Faculty of Science, University of Lisbon, Portugal |
| Stefan Mengel | CRIL-CNRS, France |
| Alexander Nadel | Intel, Israel |
| Nina Narodytska | Samsung Research America, USA |
| Jakob Nordström | KTH Royal Institute of Technology, Sweden |
| Albert Oliveras | Technical University of Catalonia, Spain |
| Roberto Sebastiani | DISI, University of Trento, Italy |
| Martina Seidl | Johannes Kepler University Linz, Austria |
| Yuping Shen | Institute of Logic and Cognition, Sun Yat-sen University, China |
| Laurent Simon | Labri, Bordeaux Institute of Technology, France |
| Takehide Soh | Information Science and Technology Center, Kobe University, Japan |
| Stefan Szeider | TU Wien, Austria |
| Allen Van Gelder | University of California, Santa Cruz, USA |

# Additional Reviewers

Abramé, André
Aleksandrowicz, Gadi
Audemard, Gilles
Banbara, Mutsunori
Baud-Berthier, Guillaume
Bayless, Sam
Berkholz, Christoph
Blinkhorn, Joshua
Bofill, Miquel
Bonacina, Ilario
Cameron, Chris
Cao, Weiwei
Chew, Leroy
de Rezende, Susanna F.
Demri, Stéphane
Dodaro, Carmine
Eggensperger, Katharina
Elffers, Jan
Falkner, Stefan
Feng, Shiguang
Fichte, Johannes Klaus
Fröhlich, Andreas
Galesi, Nicola
Ganesh, Vijay
Gange, Graeme
Gaspers, Serge
Griggio, Alberto
Guthmann, Ofer
Hermo, Montserrat
Hinde, Luke
Ibanez-Garcia, Yazmin Angelica
Ignatiev, Alexey
Ivrii, Alexander

Kolaitis, Phokion
Kotthoff, Lars
Kovásznai, Gergely
Lauria, Massimo
Lettmann, Theodor
Lindauer, Marius
Lonsing, Florian
Malod, Guillaume
Manquinho, Vasco
Marquis, Pierre
Martins, Ruben
Meel, Kuldeep S.
Miksa, Mladen
Nabeshima, Hidetomo
Naveh, Yehuda
Neves, Miguel
Oetsch, Johannes
Panagiotou, Konstantinos
Pich, Ján
Prcovic, Nicolas
Preiner, Mathias
Rozier, Kristin Yvonne
Ryvchin, Vadim
Slivovsky, Friedrich
Strichman, Ofer
Trentin, Patrick
Tveretina, Olga
Vinyals, Marc
Wang, Lingli
Widl, Magdalena
Wintersteiger, Christoph M.
Yue, Weiya

# Invited Talks

# Coping with Inconsistent Databases: Semantics, Algorithms, and Complexity

Phokion G. Kolaitis[1,2]

[1] University of California Santa Cruz, Santa Cruz, USA
[2] IBM Research – Almaden, San Jose, USA
kolaitis@cs.ucsc.edu

**Abstract.** Managing inconsistency in databases is a long-standing challenge. The framework of database repairs provides a principled approach towards coping with inconsistency in databases. Intuitively, a repair of an inconsistent database is a consistent database that differs from the given inconsistent database in a minimal way. Repair checking and consistent query answering are two fundamental algorithmic problems arising in this context. The first of these two problems asks whether, given two databases, one is a repair of the other. The second asks whether, a query is true on every repair of a given inconsistent database.

The aim of this talk is to give an overview of a body of results in this area with emphasis on the computational complexity of repair checking and consistent query answering, including the quest for dichotomy theorems. In addition to presenting open problems, the last part of the talk will include a discussion of the potential use of solvers in developing practical systems for consistent query answering.

**Keywords:** Inconsistent databases · Database dependencies · Database repairs · Consistent query answering · Computational complexity

## References

1. Afrati, F.N., Kolaitis, P.G.: Repair checking in inconsistent databases: algorithms and complexity. In: 12th International Conference on Database Theory. ICDT 2009, St. Petersburg, Russia, March 23–25, 2009, Proceedings, pp. 31–41 (2009)
2. Arenas, M., Bertossi, L.E., Chomicki, J.: Consistent query answers in inconsistent databases. In: Proceedings of the Eighteenth ACM SIGACT-SIGMOD-SIGART Symposium on Principles of Database Systems, May 31 – June 2, 1999, Philadelphia, Pennsylvania, USA, pp. 68–79 (1999)
3. Arming, S., Pichler, R., Sallinger, E.: Complexity of repair checking and consistent query answering. In: 19th International Conference on Database Theory. ICDT 2016, Bordeaux, France, March 15–18, 2016, pp. 21:1–21:18 (2016)
4. Bertossi, L.E.: Database Repairing and Consistent Query Answering. Synthesis Lectures on Data Management. Morgan & Claypool Publishers (2011)
5. ten Cate, B., Fontaine, G., Kolaitis, P.G.: On the data complexity of consistent query answering. Theory Comput. Syst. **57**(4), 843–891 (2015)

6. ten Cate, B., Halpert, R.L., Kolaitis, P.G.: Exchange-repairs: managing inconsistency in data exchange. In: Kontchakov, R., Mugnier, M.L. (eds.) RR 2014. LNCS, vol. 8741, pp. 140–156. Springer, Switzerland (2014)

7. ten Cate, B., Halpert, R.L., Kolaitis, P.G.: Practical query answering in data exchange under inconsistency-tolerant semantics. In: Proceedings of the 19th International Conference on Extending Database Technology. EDBT 2016, Bordeaux, France, March 15–16, 2016, pp. 233–244 (2016)

8. Chomicki, J.: Consistent query answering: five easy pieces. In: 11th International Conference on Database Theory - ICDT 2007, Barcelona, Spain, January 10–12, 2007, Proceedings, pp. 1–17 (2007)

9. Chomicki, J., Marcinkowski, J.: Minimal-change integrity maintenance using tuple deletions. Inf. Comput. **197**(1–2), 90–121 (2005)

10. Chomicki, J., Marcinkowski, J.: Staworko, S.: Hippo: a system for computing consistent answers to a class of SQL queries. In: Advances in Database Technology - EDBT 2004, 9th International Conference on Extending Database Technology, Heraklion, Crete, Greece, March 14–18, 2004, Proceedings, pp. 841–844 (2004)

11. Fagin, R., Kimelfeld, B., Kolaitis, P.G.: Dichotomies in the complexity of preferred repairs. In: Proceedings of the 34th ACM Symposium on Principles of Database Systems. PODS 2015, Melbourne, Victoria, Australia, May 31 – June 4, 2015, pp. 3–15 (2015)

12. Fontaine, G.: Why is it hard to obtain a dichotomy for consistent query answering? ACM Trans. Comput. Log. **16**(1), 7:1–7:24 (2015)

13. Fuxman, A., Fazli, E., Miller, R.J.: ConQuer: Efficient management of inconsistent databases. In: Proceedings of the ACM SIGMOD International Conference on Management of Data, Baltimore, Maryland, USA, June 14–16, 2005, pp. 155–166 (2005)

14. Fuxman, A., Fuxman, D., Miller, R.J.: ConQuer: a system for efficient querying over inconsistent databases. In: Proceedings of the 31st International Conference on Very Large Data Bases, Trondheim, Norway, August 30 – September 2, 2005, pp. 1354–1357 (2005)

15. Kolaitis, P.G., Pema, E.: A dichotomy in the complexity of consistent query answering for queries with two atoms. Inf. Process. Lett. **112**(3), 77–85 (2012)

16. Kolaitis, P.G., Pema, E., Tan, W.: Efficient querying of inconsistent databases with binary integer programming. PVLDB **6**(6), 397–408 (2013)

17. Koutris, P., Wijsen, J.: The data complexity of consistent query answering for self-join-free conjunctive queries under primary key constraints. In: Proceedings of the 34th ACM Symposium on Principles of Database Systems. PODS 2015, Melbourne, Victoria, Australia, May 31 – June 4, 2015, pp. 17–29 (2015)

18. Marileo, M.C., Bertossi, L.E.: The consistency extractor system: answer set programs for consistent query answering in databases. Data Knowl. Eng. **69**(6), 545–572 (2010)

19. Staworko, S., Chomicki, J., Marcinkowski, J.: Prioritized repairing and consistent query answering in relational databases. Ann. Math. Artif. Intell. **64**(2–3), 209–246 (2012)

20. Wijsen, J.: Certain conjunctive query answering in first-order logic. ACM Trans. Database Syst. **37**(2), 9 (2012)

21. Wijsen, J.: A survey of the data complexity of consistent query answering under key constraints. In: 8th International Symposium on Foundations of Information and Knowledge Systems. FoIKS 2014, Bordeaux, France, March 3–7, 2014. Proceedings, pp. 62–78 (2014)

# From SAT to ASP and Back!?

Torsten Schaub[1,2]

[1] University of Potsdam, Potsdam, Germany
[2] INRIA Rennes, Rennes, France
torsten@cs.uni-potsdam.de

Answer Set Programming (ASP; [1–4]) provides an approach to declarative problem solving that combines a rich yet simple modeling language with effective Boolean constraint solving capacities. This makes ASP a model, ground, and solve paradigm, in which a problem is expressed as a set of first-order rules, which are subsequently turned into a propositional format by systematically replacing all variables, before finally the models of the resulting propositional rules are computed. ASP is particularly suited for modeling problems in the area of Knowledge Representation and Reasoning involving incomplete, inconsistent, and changing information due to its nonmonotonic semantic foundations. As such, it offers, in addition to satisfiability testing, various reasoning modes, including different forms of model enumeration, intersection or unioning, as well as multi-objective optimization. From a formal perspective, ASP allows for solving all search problems in $NP$ (and $NP^{NP}$) in a uniform way, that is, by separating problem encodings and instances. Hence, ASP is well-suited for solving hard combinatorial search (and optimization) problems. Interesting applications of ASP include decision support systems for NASA shuttle controllers [5], industrial team-building [6], music composition [7], natural language processing [8], package configuration [9], phylogeneticics [10], robotics [11, 12], systems biology [13–15], timetabling [16], and many more. The versatility of ASP is nicely reflected by the ASP solver *clasp* [17], winning first places at various solver competitions, including ASP, MISC, PB, and SAT. In fact, *clasp* is at the heart of the open source platform potassco.-sourceforge.net. *Potassco* stands for the "Potsdam Answer Set Solving Collection" [18] and has seen more than 145000 downloads world-wide since its inception at the end of 2008.

The talk will start with a gentle introduction to ASP, while focusing on the commonalities and differences to SAT. It will discuss the different semantic foundations and describe the impact of a modelling language along with off-the-shelf grounding systems. Finally, it will highlight some resulting techniques, like meta-programming, preference handling, heuristic constructs, and theory reasoning.

## References

1. Gelfond, M., Lifschitz, V.: The stable model semantics for logic programming. In: Kowalski, R., Bowen, K. (eds.) Proceedings of the Fifth International Conference and Symposium of Logic Programming. ICLP1988, pp. 1070–1080. MIT Press (1988)

2. Baral, C.: Knowledge Representation, Reasoning and Declarative Problem Solving. Cambridge University Press (2003)
3. Gebser, M., Kaminski, R., Kaufmann, B., Schaub, T.: Answer Set solving in practice. Synthesis Lectures on Artificial Intelligence and Machine Learning. Morgan and Claypool Publishers (2012)
4. Gelfond, M., Kahl, Y.: Knowledge Representation, Reasoning, and the Design of Intelligent Agents: The Answer-Set Programming Approach. Cambridge University Press (2014)
5. Nogueira, M., Balduccini, M., Gelfond, M., Watson, R., Barry, M.: An A-prolog decision support system for the space shuttle. In: Ramakrishnan, I. (ed.) PADL 2001. LNCS, vol. 1990, pp. 169–183. Springer, Berlin (2001)
6. Grasso, G., Iiritano, S., Leone, N., Lio, V., Ricca, F., Scalise, F.: An ASP-based system for team-building in the Gioia-Tauro seaport. In: Carro, M., Peña, R. (eds.) PADL 2010. LNCS, vol. 5937, pp. 40–42. Springer, Berlin (2010)
7. Boenn, G., Brain, M., de Vos, M., Fitch, J.: Automatic composition of melodic and harmonic music by answer set programming. In: Garcia de la Banda, M., Pontelli, E. (eds.) ICLP 2008. LNCS, vol. 5366, pp. 160–174. Springer, Berlin (2008)
8. Schwitter, R.: The jobs puzzle: taking on the challenge via controlled natural language processing. Theory Pract. Logic Program. 13(4–5), 487–501 (2013)
9. Gebser, M., Kaminski, R., Schaub, T.: aspcud: A Linux package configuration tool based on answer set programming. In: Drescher, C., Lynce, I., Treinen, R., (eds.) Proceedings of the Second International Workshop on Logics for Component Configuration. LoCoCo 2011. Electronic Proceedings in Theoretical Computer Science (EPTCS), vol. 65, pp. 12–25 (2011)
10. Brooks, D., Erdem, E., Erdogan, S., Minett, J., Ringe, D.: Inferring phylogenetic trees using answer set programming. J. Autom. Reason. 39(4), 471–511 (2007)
11. Chen, X., Ji, J., Jiang, J., Jin, G., Wang, F., Xie, J.: Developing high-level cognitive functions for service robots. In: van der Hoek, W., Kaminka, G., Lespérance, Y., Luck, M., Sen, S. (eds.) Proceedings of the Ninth International Conference on Autonomous Agents and Multiagent Systems. AAMAS 2010, IFAAMAS, pp. 989–996 (2010)
12. Erdem, E., Haspalamutgil, K., Palaz, C., Patoglu, V., Uras, T.: Combining high-level causal reasoning with low-level geometric reasoning and motion planning for robotic manipulation. In: Proceedings of the IEEE International Conference on Robotics and Automation. ICRA 2011, pp. 4575–4581. IEEE (2011)
13. Erdem, E., Türe, F.: Efficient haplotype inference with answer set programming. In: Fox, D., Gomes, C. (eds.) Proceedings of the Twenty-Third National Conference on Artificial Intelligence. AAAI 2008, pp. 436–441. AAAI Press (2008)
14. Gebser, M., Schaub, T., Thiele, S., Veber, P.: Detecting inconsistencies in large biological networks with answer set programming. Theory Pract. Logic Program. 11(2–3), 323–360 (2011)
15. Gebser, M., Guziolowski, C., Ivanchev, M., Schaub, T., Siegel, A., Thiele, S., Veber, P.: Repair and prediction (under inconsistency) in large biological networks with answer set programming. In: Lin, F., Sattler, U. (eds.) Proceedings of the Twelfth International Conference on Principles of Knowledge Representation and Reasoning. KR 2010, pp.497–507. AAAI Press (2010)
16. Banbara, M., Soh, T., Tamura, N., Inoue, K., Schaub, T.: Answer set programming as a modeling language for course timetabling. Theory Pract. Logic Program. 13(4–5), 783–798 (2013)
17. Gebser, M., Kaufmann, B., Schaub, T.: Conflict-driven answer set solving: from theory to practice. Artif. Intell. 187–188, 52–89 (2012)
18. Gebser, M., Kaminski, R., Kaufmann, B., Ostrowski, M., Schaub, T., Schneider, M.: Potassco: the Potsdam answer set solving collection. AI Commun. 24(2), 107–124 (2011)

# Satisfiability Testing, a Disruptive Technology in Program Verification?

David Monniaux[1,2]

[1] Université Grenoble Alpes, VERIMAG, 38000 Grenoble, France
[2] CNRS, Verimag, 38000 Grenoble, France
David.Monniaux@imag.fr

**Abstract.** In the 2000s, progress in satisfiability testing shook automated and assisted program verification. The advent of efficient satisfiability modulo theory (SMT) solvers allowed new approaches: efficient testing and symbolic execution, new methods for generating inductive invariants, and more automated assisted proof.

Program verification consists in proving properties of software, be them safety ("whatever the program execution, some property is always satisfied") or liveness ("some action will always eventually happen"). Improvements in satisfiability testing have allowed exciting combinations of exact decision procedures [12], often based on Boolean satisfiability [2], with existing approaches.

## 1  Program Verification Before SMT

Traditionally, program verification (i) either relied heavily on the user tediously providing *inductive invariants, ranking functions* as well as proofs (ii) either simplified the problems through abstractions that sometimes were sufficient to prove the property, sometimes were not. Excessively coarse abstractions (e.g. one single interval of variation per program variable per location) could sometimes be refined by explicit partitioning [16], but its cost is exponential and thus it must be quickly limited.

Program verification was most often, and still is, considered too costly in human and algorithmic terms, and thus in most practical cases, it was replaced by *testing*, with test cases chosen by hand or through test case generation techniques. *Fuzzing* is a kind of testing where variant of input files or protocol exchanges are randomly modified so as to trigger bugs in parsers, which could be exploited as security vulnerabilities.

Both program verification and testing were transformed by the advent of a disruptive technology: *satisfiability modulo theory* (SMT), that is, efficient algorithms for checking that a first-order logic formula over a given theory is satisfiable — e.g. $(x < y \lor x \geq y + 3) \land (x \geq 0 \land x + y \leq 5)$ is a first-order formula over linear real or integer arithmetic.

The research leading to these results has received funding from the European Research Council under the European Union's Seventh Framework Programme (FP/2007-2013)/ERC Grant Agreement nr. 306595 "STATOR".

## 2  Testing

In *bounded model checking* [1], the program is unrolled up to a finite depth, and a first-order formula is generated, whose solutions are the program traces that violate the desired property before the given depth. In *symbolic execution*, sequences of program statements are translated into a first-order formula, and the feasibility of tests into new branches is checked by satisfiability testing; this approach has been successfully applied to *fuzzing*, that is, searching for inputs that trigger security violations in file or protocol parsers [8]. If symbolic execution is too costly, certain unknowns may be chosen to have concrete values, while others are left symbolic, leading to *concolic execution*.

## 3  Automatic Verification

Bounded model checking and symbolic execution cannot prove the safety of programs, except in the rare case where there is a small constant bound on execution lengths. Inductive invariant inference was also greatly transformed by the advent of satisfiability modulo theory solvers. First, inference approaches designed to operate over a control-flow graph and produce an invariant per control location were modified to traverse loop-free program fragments (*large blocks*) encoded into first order formulas, whose satisfiability is checked by SMT [7, 15]. Second, *counterexample-guided abstraction refinement* approaches mine proofs of unreachability of errors through finite unrollings for arguments that could become inductive, in particular by extraction of *Craig interpolants* [14].

Research in SMT solving has strived to extend the class of formulas handled by solvers [12]: from quantifier-free linear real arithmetic, solved by a combination of *constraint-driven clause learning* (CDCL) SAT-solving [13] and exact-precision simplex algorithm [5], and *uninterpreted functions*, solvers were extended to linear integer arithmetic, nonlinear (polynomial) arithmetic, arrays, bitvector arithmetic, character strings, data structures, and quantified formulas. Some of these combinations are undecidable or have high lower bounds on their worst-case complexity [6], yet this is not considered a major hindrance; practical efficiency is paramount.

## 4  Assisted Verification

In assisted proof, using tools such as Coq or Isabelle, the user traditionally has to provide detailed arguments why the claimed theorem is true. Automation is traditionally limited. When the theorem, or a part of it, fits within a class decidable by SMT, it is tempting to check it using SMT . It may however be unwise to blindly trust a SMT solver, a complex piece of software likely to have bugs; and some assistants require all proofs to be broken down into basic steps checked by a *proof checker*. It is therefore desirable that the solver provides an independently checkable *proof witness*; one challenge is to keep the witness small enough to be manageable while keeping the checker small [3, 11].

## 5 Challenges

Most extant SMT solvers are based on the DPLL(T) framework: a combination of a CDCL SAT solver and a decision procedure for conjunctions. This framework has known weaknesses: for instance, some industrially relevant families of formulas induce exponential behavior [9]. There have been multiple proposals for solvers based on other approaches [4, 10], but they are less mature. Quantifiers are still often difficult to handle. Verification approaches based on Craig interpolation are often brittle; more generally, reliance on random number generators sometimes results in unpredictable behavior. Better collaboration between verification and SAT experts is needed to overcome these challenges.

## References

1. Biere, A.: Bounded model checking. In: Biere et al. [2], vol. 185, pp. 455–481 (2009)
2. Biere, A., Heule, M.J.H., van Maaren, H., Walsh, T. (eds.): Handbook of Satisfiability, Frontiers in Artificial Intelligence and Applications, vol. 185. IOS Press (2009)
3. Böhme, S., Weber, T.: Fast LCF-style proof reconstruction for Z3. In: Kaufmann, M., Paulson, L.C. (eds.) ITP 2010. LNCS, vol. 6172, pp. 179–194. Springer, Berlin (2010)
4. Brain, M., D'Silva, V., Griggio, A., Haller, L., Kroening, D.: Deciding floating-point logic with abstract conflict driven clause learning. Formal Methods Syst. Des. **45**(2), 213–245 (2014)
5. Dutertre, B., de Moura, L.M.: Integrating simplex with DPLL(T). Sri-csl-06-01, SRI International, Computer Science Laboratory (2006)
6. Fischer, M.J., Rabin, M.O.: Super-exponential complexity of presburger arithmetic. In: Karp, R. (ed.) Complexity of Computation. SIAM–AMS Proceedings, pp. 27–42, no. 7. American Mathematical Society (1974). citeseer.ist.psu.edu/fischer74superexponential.html
7. Gawlitza, T., Monniaux, D.: Invariant generation through strategy iteration in succinctly represented control flow graphs. Logical Methods in Computer Science (2012)
8. Godefroid, P., Levin, M.Y., Molnar, D.: SAGE: Whitebox fuzzing for security testing. Queue **10**(1), 20:20–20:27 (2012)
9. Henry, J., Asavoae, M., Monniaux, D., Maiza, C.: How to compute worst-case execution time by optimization modulo theory and a clever encoding of program semantics. In: Zhang, Y., Kulkarni, P. (eds.) Languages, Compilers, Tools and Theory for Embedded Systems (LCTES), pp. 43–52. ACM (2014)
10. Jovanović, D., de Moura, L.: Solving non-linear arithmetic. In: IJCAR (2012)
11. Keller, C.: Extended resolution as certificates for propositional logic. In: Blanchette, J.C., Urban, J. (eds.) Proof Exchange for Theorem Proving (PxTP). EPiC Series, vol. 14, pp. 96–109. EasyChair (2013). http://www.easychair.org/publications/?page=117514525

12. Kroening, D., Strichman, O.: Decision Procedures. Springer (2008)
13. Marques-Silva, J.P., Lynce, I., Malik, S.: Conflict-driven clause learning SAT solvers. In: Biere et al. [2], vol. 185, pp. 131–153 (2009)
14. McMillan, K.L.: Lazy abstraction with interpolants. In: Ball, T., Jones, R.B. (eds.) CAV 2006. LNCS, vol. 4144, pp. 123–136. Springer, Berlin (2006)
15. Monniaux, D., Gonnord, L.: Using bounded model checking to focus fixpoint iterations. In: Yahav, E. (ed.) SAS 2011. LNCS, vol. 6887, pp. 369–385. Springer, Berlin (2011)
16. Rival, X., Mauborgne, L.: The trace partitioning abstract domain. ACM Trans. Program. Lang. Syst. **29**(5) (2007)

# Contents

## Satisfiability Applications

## Satisfiability Modulo Theory

## Beyond SAT

# Complexity

# Parameterized Compilation Lower Bounds for Restricted CNF-Formulas

Stefan Mengel$^{(\boxtimes)}$

CNRS, CRIL UMR 8188, Lens, France
mengel@cril.fr

**Abstract.** We show unconditional parameterized lower bounds in the area of knowledge compilation, more specifically on the size of circuits in decomposable negation normal form (DNNF) that encode CNF-formulas restricted by several graph width measures. In particular, we show that
- there are CNF formulas of size $n$ and modular incidence treewidth $k$ whose smallest DNNF-encoding has size $n^{\Omega(k)}$, and
- there are CNF formulas of size $n$ and incidence neighborhood diversity $k$ whose smallest DNNF-encoding has size $n^{\Omega(\sqrt{k})}$.

These results complement recent upper bounds for compiling CNF into DNNF and strengthen—quantitatively and qualitatively—known conditional lower bounds for cliquewidth. Moreover, they show that, unlike for many graph problems, the parameters considered here behave significantly differently from treewidth.

## 1  Introduction

Knowledge compilation is a preprocessing regime that aims to translate or "compile" knowledge bases, generally encoded as CNF formulas, into different representations more convenient for a task at hand. The idea is that many queries one would like to answer on the knowledge base, say clause entailment queries, are intractable in CNF encoding, but tractable for other representations. When there are many queries on the same knowledge base, as for example in product configuration, it makes sense to invest into a costly preprocessing to change the representation once in order to then speed up the queries and thus amortize the time spent on the preprocessing.

One critical question when following this approach is the choice of the representation that the knowledge is encoded into. In general, there is a trade-off between the usefulness of a representation (which queries does it support efficiently?) and succinctness (what is the size of the encoded knowledge base?). This trade-off has been studied systematically [8], leading to a fine understanding of the different representations. In particular, circuits in decomposable negation normal form (short DNNF) [6] have been identified as a representation that is more succinct than nearly all other representations while still allowing useful queries. Consequently, DNNFs play a central role in knowledge compilation.

This paper should be seen as complementing the findings of [1]: In that paper, algorithms compiling CNF formulas with restricted underlying graph structure

© Springer International Publishing Switzerland 2016
N. Creignou and D. Le Berre (Eds.): SAT 2016, LNCS 9710, pp. 3–12, 2016.
DOI: 10.1007/978-3-319-40970-2_1

were presented, showing that popular graph width measures like treewidth and cliquewidth can be used in knowledge compilation. More specifically, every CNF formula of incidence *treewidth* $k$ and size $n$ can be compiled into a DNNF of size $2^{O(k)}n$. Moreover, if $k$ is the incidence *cliquewidth*, the size bound on the encoding becomes $n^{O(k)}$. As has long been observed, $2^{O(k)}n$ is of course far preferable to $n^{O(k)}$ for nontrivial sizes of $n$—in fact, this is the main premise of the field of parameterized complexity theory, see e.g. [13]. Consequently, the results of [1] leave open the question if the algorithm for clique-width based compilation of CNF formulas can be improved.

In fact, the paper [1] already gives a partial answer to this question, proving that there is no compilation algorithm achieving fixed-parameter compilability, i.e., a size bound of $f(k)p(|F|)$ for a function $f$ and a polynomial $p$. But unfortunately this result is based on the plausible but rather non-standard complexity assumption that not all problem in W[1] have FPT-size circuits. The result of this paper is that this assumption is not necessary. We prove a lower bound of $|F|^{\Omega(k)}$ for formulas of *modular incidence treewidth* $k$ where modular treewidth is a restriction of cliquewidth proposed in [19]. It follows that the result in [1] is essentially tight. Moreover, we show a lower bound of $|F|^{\Omega(\sqrt{k})}$ for formulas of neighborhood diversity $k$ [16]. This intuitively shows that all graph width measures that are stable under adding modules, i.e., adding a new vertex that has exactly the same neighborhood as an existing vertex, behave qualitatively worse than treewidth for compilation into DNNFs.

*Related work.* Parameterized knowledge compilation was first introduced by Chen [4] and has seen some recent renewed interest, see e.g. [5,9] for work on conditional lower bounds. Unconditional lower bounds based on treewidth can e.g. be found in [20,21], but they are only for different versions of branching programs that are known to be less succinct than DNNF. Moreover, these lower bounds fail for DNNFs as witnessed by the upper bounds of [1].

There is a long line of research using graph and hypergraph width measures for problems related to propositional satisfiability, see e.g. the extensive discussion in [3]. The paper [18] gave the first parameterized lower bounds on SAT with respect to graph width measures, in particular cliquewidth. This result was later improved to modular treewidth to complement an upper bound for model counting [19] and very recently to neighborhood diversity [10], a width measure introduced in [16]. We remark that the latter result could be turned into a conditional parameterized lower bound similar to that in [1] discussed above.

Our lower bounds strongly rely on the framework for DNNF lower bounds proposed in [2] and communication theory lower bounds from [12], for more details see Sect. 3.

## 2   Preliminaries

In the scope of this paper, a *linear code* $C$ is the solution of a system of linear equations $A\bar{x} = 0$ over the boolean field $\mathbb{F}_2$. The matrix $A$ is called the *parity-*

*check matrix* of $C$. The characteristic function $f_C$ is the boolean function that, given a boolean string $e$, evaluates to 1 if and only if $e$ is in $C$.

We use the notation $[n] := \{1, \ldots, n\}$ and $[n_1, n_2] := \{n_1, n_1 + 1, \ldots, n_2\}$ to denote integer intervals. We use standard notations from graph theory and assume the reader to have a basic background in in the area [11]. By $N(v)$ we denote the open neighborhood of a vertex in a graph.

We say that two vertices $u$, $v$ in a graph $G = (V, E)$ have the same neighborhood type if and only if $N(u) \setminus \{v\} = N(v) \setminus \{u\}$. It can be shown that having the same neighborhood type is an equivalence relation on $V$ [16]. The neighborhood diversity of $G$ is defined to be the number of equivalence classes of $V$ with respect to neighborhood types.

A generalization of neighborhood diversity is *modular treewidth* which is defined as follows: From a graph $G$ we construct a new graph $G'$ by contracting all vertices sharing a neighborhood type, i.e., from every equivalence class we delete all vertices but one. The modular treewidth of $G$ is then defined to be the treewidth of $G'^1$. Modular pathwith is defined in the obvious analogous way.

We assume basic familiarity with propositional logic and in particular CNF formulas. We define the size of a CNF formula to be the overall number of occurrences of literals, i.e., the sum of the sizes of the clauses where the size of a clause is the number of literals in the clause. The incidence graph of a CNF formula $F$ has as vertices the variables and clauses of $F$ and an edge between every clause and the vertices contained in it. The projection of an assignment $a : X \rightarrow \{0, 1\}$ to a set $Z$ is the restriction of $a$ to the variable set $Z$. This definition generalizes to sets of assignments in the obvious way. Moreover, the projection of a boolean function $f$ on $X$ to $Z$ is defined as the boolean function on $Z$ whose satisfying assignments are those of $f$ projected to $Z$. For the width measures introduced above, we define the with of a formula to be that of its incidence graph.

## 3  Statement of the Main Results and Preparation of the Proof

We now state our main results. The first theorem shows that modular pathwith— and thus also more general parameters like cliquewidth and modular treewidth— do not allow fixed-parameter compilation to DNNF.

**Theorem 1.** *For every $k$ and for every $n$ big enough there is a CNF formula $F$ of size at most $n$ and modular pathwidth $k$ such that any DNNF computing the same function as $F$ must have size $n^{\Omega(k)}$.*

We also show lower bounds for neighborhood diversity that are nearly as strong as those for modular pathwidth.

---

[1] Note that the definition in [19] differs from the one we give here, but can easily be seen to be equivalent for bipartite graphs and thus incidence graphs of CNF formulas. We keep our definition to be more consistent with the definition of neighborhood diversity.

**Theorem 2.** *For every $k$ and for every $n$ big enough there is a CNF formula $F$ of size polynomial in $n$ and with neighborhood diversity $k$ such that any DNNF computing the same function as $F$ must have size $n^{\Omega(\sqrt{k})}$.*

At this point, the attentive reader may be a little concerned because we promise to prove lower bounds for DNNF which we have not even defined in the preliminaries. In fact, it is the main strength of the approach in [2] that the definition and properties of DNNF are not necessary to show our lower bounds, because we can reduce showing lower bounds on DNNF to a problem in communication complexity. Since we will not use any properties of DNNF, we have decided to leave out the definition for space reasons and refer to e.g. [8]. Here we will only use the following result.

**Theorem 3** [2]. Let $f$ be a function computed by a DNNF of size $s$. Then $f$ has a multi-partition rectangle cover of size $s$.

Now the reader might be a little puzzled about what multi-partition rectangle covers of a function are. Since we will also only use them as a black box in our proofs and do not rely on any of their properties, we have opted to leave out their definition and refer to [12]

We will use a powerful theorem which follows directly from the results in [12].

**Theorem 4** [12]. For every $n' \in \mathbb{N}$ and every $m' \leq n'/32$ there is a linear code $C$ with a $m' \times n'$ parity check matrix such that every multi-partition rectangle cover of the characteristic function $f_C$ has size at least $\frac{1}{4}2^{m'}$.

# 4    Accepting Codes by CNF Formulas

In this section we will construct CNF formulas to accept linear codes. We will first start with a naive encoding that will turn out to be of unbounded modular treewidth and thus not directly helpful to us. We will then show how to change the encoding in such a way that the modular treewidth and even the neighborhood diversity are small and the size of the resulting CNF is small enough to show meaningful lower bounds for encodings in DNNF with Theorem 4.

## 4.1   The Naive Approach

In this subsection, we show how we can check $m$ linear equations on variables $x_1, \ldots, x_n$ efficiently by CNF. The idea is to simply consider one variable after the other and remember the parity for the equations at hand. To this end, fix an $m \times n$ matrix $A = (a_{ij})$. We denote the resulting equations of the system $A\bar{x} = 0$ by $E_1, \ldots, E_m$. For each equation $E_i$ we introduce variables $z_{ij}$ for $j \in [n]$ which intuitively remembers the parity of $E_i$ up to seeing the variable $x_j$.

We encode the computations for each $E_i$ individually: Introduce constraints

$$a_{i,1}x_1 = z_{i,1}, \tag{1}$$

$$z_{i,j-1} + a_{ij}x_j = z_{ij}. \tag{2}$$

Note that $z_{i,n}$ yields the parity for equation $E_i$ which can then be checked for 0. This yields a system whose accepted inputs projected to the $x_i$ are the code words of the considered code. The constraints have all at most 3 variables, so we can encode them into CNF easily.

Unfortunately, the resulting CNF can be shown to have high modular tree-width, so it is not useful for our considerations. We will see how to reduce the modular treewidth and the neighborhood diversity of the system without blowing up the size of the resulting CNF-encoding too much.

## 4.2  Bounding Modular Treewidth

The idea for decreasing the modular treewidth is to not encode all constraints on the parities individually but combine them into larger constraints. So fix $n$ and $k$ and set $m := k\log(n)$. For each $j$, we will combine the constraints from (1) and (2) for blocks of $\log(n)$ values of $i$ into one. The resulting constraints are

$$R_1^\ell(x_1, z_{\ell\log(n)+1,1}, \ldots, z_{(\ell+1)\log(n),1}) := \{(d_1, t_{\ell\log(n)+1,1}, \ldots, t_{(\ell+1)\log(n),1}) \mid$$
$$a_{i,1}d_1 = t_{i,1}, i = \ell\log(n) + 1, \ldots, (\ell+1)\log(n)\}$$

and

$$R_j^\ell(x_i, z_{\ell\log(n)+1,j-1}, \ldots, z_{(\ell+1)\log(n),j-1}, z_{\ell\log(n)+1,j-1}, \ldots, z_{(\ell+1)\log(n),j-1}) :=$$
$$\{(d_i, t_{\ell\log(n)+1,j-1}, \ldots, t_{(\ell+1)\log(n),j-1}, t_{\ell\log(n)+1,j}, \ldots, t_{(\ell+1)\log(n),j}) \mid$$
$$t_{i,j-1} + a_{ij}d_j = t_{i,j}, i = \ell\log(n) + 1, \ldots, (\ell+1)\log(n)\}$$

for $\ell = 0, \ldots, k - 1$.

Note that the constraints $R_j^\ell$ have at most $2\log(n) + 1$ boolean variables, so we can encode them into CNF of quadratic size where every clause contains all variables of $R_j^\ell$. Moreover, the $R_j^\ell$ encode all previous constraints from (1) and (2), so the assignments satisfying all $R_j^\ell$ projected to the $x_i$ still are exactly the code words of the code we consider. Call the resulting CNF $F$.

*Claim.* $F$ has modular pathwidth at most $2k - 1$.

*Proof.* Note that the clauses introduced when translating the constraint $R_j^\ell$ into CNF have by construction all the same set of variables. Thus these clauses have the same neighborhood type, and we can for modular treewidth restrict to an instance just having one clause for each $R_j^\ell$. We call the resulting vertex in the incidence graph $r_{\ell,j}$. Next, observe that the variables $z_{\ell\log(n)+i,j}$ and $z_{\ell\log(n)+i',j}$ for $i, i' \in [\log(n)]$ appear in exactly the same clauses. Thus these variables have the same neighborhood type as well, so we can delete all but one of them, say $z_{\ell\log(n),j}$ for $\ell = 1, \ldots, k$. Call the resulting vertices in the incidence graph $s_{\ell,j}$.

The resulting graph $G = (V, E)$ has

$$V = \{x_j, s_{\ell,j}, r_{\ell,j} \mid j \in [n], \ell \in [k]\}.$$
$$E = \{x_j r_{\ell,j} \mid j \in [n], \ell \in [k]\} \cup \{s_{\ell,j-1} r_{\ell,j} \mid j \in [2,n], \ell \in [k]\}$$
$$\cup \{s_{\ell,j} r_{\ell,j} \mid j \in [n], \ell \in [k]\}$$

We construct a path decomposition of $G$ as follows: The bags are the sets

$$B_2 := \{x_1 r_{1,1}, \ldots, r_{\ell,1}\},$$
$$B_3 := \{s_{1,1}, \ldots, s_{\ell,1}, r_{1,1}, \ldots, r_{\ell,1}\}$$

and for $j = 2, \ldots n$

$$B_{3j-2} := \{s_{1,j-1}, \ldots, s_{\ell,j-1}, r_{1,j}, \ldots, r_{\ell,j}\},$$
$$B_{3j-1} := \{x_j, r_{1,j}, \ldots, r_{\ell,j}\}, \text{ and }$$
$$B_{3j} := \{s_{1,j}, \ldots, s_{\ell,j}, r_{1,j}, \ldots, r_{\ell,j}\}.$$

Ordering the bags $B_j$ with respect to their index yields a path decomposition of $G$ of width $2k - 1$.

Let us collect the results of the section into one statement.

**Lemma 1.** *For every linear code $C$ with a $k \log(n) \times n$ parity check matrix there is a CNF formula $F$ in variable sets $X$ and $Z$ such that*

- *the solution set of $F$ projected to $X$ is exactly $C$,*
- *$F$ has size $O(kn^3 \log(n)^2)$, and*
- *$F$ has modular pathwidth at most $2k - 1$.*

*Proof.* It remains only to show the size bound on $F$. Note that we have $n$ variables $x_j$ and $kn \log(n)$ variables $z_{i,j}$. Moreover, we have $kn \log(n)$ constraints $R_i^\ell$. Each of those has $2 \log(n) + 1$ variables, so it can be encoded by $O(n^2)$ clauses with $O(\log(n))$ variables each. This yields $O(kn^3 \log(n))$ clauses with $O(\log(n))$ variables. Consequently, the overall size of $F$ is $O(kn^3 \log(n)^2)$.

### 4.3 Bounding Neighborhood Diversity

Fix now two positive integers $N$ and $k$ and let $n := 32k \log(N)$ and $m := k \log(N)$ and consider $A$ with these parameters as before. We want to encode the code of $A$ by a CNF with neighborhood diversity $O(k^2)$.

To do so, we split the variables $z_{ij}$ into $O(k^2)$ sets of $\log(N)^2$ variables $S_{rs} := \{z_{ij} \mid i \in [r \log(N)+1, (r+1) \log(N)], j \in [s \log(N)+1, (s+1) \log(N)-1]\}$ for $r \in [k]$ and $s \in [32k]$. Now create for all $r \in [k]$, $s \in [32k-1]$ a constraint $R_{rs}$ in the variables $X := \{x_1, \ldots, x_n\} \cup S_{rs} \cup S_{r+1,s}$ that accepts all assignments to its variables that satisfy all constraints from 1 and 2 whose variables are variables of $R_{rs}$. Note that the resulting constraints cover all constraints of Sect. 4.1, so we still accept the code defined by $A$ after projection to $X$.

The problem now is that, since we have $\Theta(\log(N)^2)$ boolean variables in each constraint, the resulting encoding into CNF could be superpolynomial in $N$ and thus to big for our purposes. This is easily repaired by the observation that fixing the values of $x_i$, $z_{i,s \log(N)}$ and $z_{i,s \log(N)+1}$ determines the values of the other $z_{ij}$ in all satisfying assignments. Consequently, we can project out these variables of the individual constraints without changing the accepted assignments to $X$.

Call the resulting constraints $R'_{rs}$. It is easy to see that every constraint $R'_{rs}$ has only $O(\log(N))$ variables, so the CNF encoding in which every variable of $R'_{r,s}$ appears in every clause has polynomial size in $N$.

We claim that the resulting CNF $F$ has neighborhood diversity $O(k^2)$. To see this, note that the clauses introduced in the encoding of a fixed $R'_{rs}$ all have the same variables. It follows that the clause vertices in the incidence graph have $O(k^2)$ neighborhood types. The variables in $X$ all appear in all clauses, so they are all of the same neighborhood type. Finally, the vertices in each $S_{rs}$ appearing in an $R'_{rs}$ all appear in the same clauses, so they have $O(k^2)$ neighborhood types as well. This show that the incidence graph of the CNF formula $F$ has neighborhood diversity $O(k^2)$.

Let us again combine the results of this section into one summary statement.

**Lemma 2.** *For every linear code $C$ with a $k\log(N) \times 32k\log(N)$ parity check matrix there is a CNF formula $F$ in variable sets $X$ and $Z$ such that*

- *the solution set of $F$ projected to $X$ is exactly $C$,*
- *$F$ has size polynomial in $N$ and $k$, and*
- *$F$ has neighborhood diversity $O(k^2)$.*

## 5   Completing the Proof

We now combine the results of the previous sections to get our main results.

*Proof (of Theorem 1).* Let $C$ be a linear code as in Theorem 4 with parameters $n' = n^{\frac{1}{4}}$ and $m' := \log(n)\frac{k}{2} = k\log(n^{\frac{1}{4}})$. Then by Theorem 4 we know that every rectangle cover of the characteristic function $f_C$ has size at least $\frac{1}{4}2^{m'} = n^{\Omega(k)}$.

Now apply Lemma 1 to $C$ to get a CNF-formula $F$ of size less than $n$ and modular pathwidth less than $k$. Let $D$ be a DNNF representation of $F$ of minimal size $s$. Since DNNFs allow projection to a subset of variables without any increase of size [6], this yields a DNNF of size $s$ computing $f_C$. But then by Theorem 3, we get that $s \geq n^{\Omega(k)}$.

With the same proof but other parameters we get Theorem 2 from Lemma 2.

## 6   Connections to Model Counting and Affine Decision Trees

In this section we discuss connections of the findings of this paper to practical model counting. It has been shown that there is a tight connection between compilation and model counting, as runs of exhaustive DPLL-based model counting algorithms can be translated into (restricted) DNNFs [14]. Here the size of the resulting DNNF corresponds to the runtime of the model counter. Since state of the art solvers like Cachet [23] and sharpSAT [24] use exhaustive DPLL, the lower bounds in this paper can be seen as lower bounds for these programs: model counting for CNF formulas of size $n$ and, modular treewidth will take time at least $n^{\Omega(k)}$ when solved with these state-of-the-art solvers even with perfect caching and optimal branching variable choices. Note that in the light of the general conditional hardness result of [19] this is not surprising, but here we get concrete and unconditional lower bounds for a large class of algorithms used in practice. Naturally, we also directly get lower bounds for approaches that are based on compilation into DNNF as those in [7,17], so we have lower bounds for most practical approaches to model counting.

One further interesting aspect to observe is that, while the instances that we consider are in a certain sense hard for practical model counting algorithms, in fact counting their models is extremely easy. Since we just want to count the number of solutions of a system of linear equations, basic linear algebra will do the job. A similar reasoning translated to compilation is the background for the definition of affine decision trees (ADT) [15], a compilation language that intu-itively has checking an affine equation as a built-in primitive. Consequently, it is very easy to see that ADTs allow a very succinct compilation of the CNF for-mulas we consider in this paper. It follows, by setting the right parameters, that there are formulas where ADTs are exponentially more succinct than DNNF. We remark that this superior succinctness can also be observed in experiments when compiling the formulas of Sect. 4.1 with the compiler from [15].

## 7 Conclusion

We have shown that parameters like cliquewidth, modular treewidth and even neighborhood diversity behave significantly differently from treewidth for compi-lation into DNNF by giving lower bounds complementing the results of [1]. These unconditional lower bounds confirm conditional ones that had been known for some time already and improve them quantitatively. Our proofs heavily relied on the framework proposed in [2] thus witnessing the strength of this approach. We have also discussed implications for practical model counting.

One consequence of our results is that most graph width measures that allow dense incidence graphs for the input CNF—like modular treewidth or cliquewidth and unlike treewidth which forces a small number of edges—do not allow fixed-parameter compilation into DNNF. A priori, there is no reason why many edges in the incidence graphs, which translates into many big clauses, should necessarily make compilation hard. Thus it would be interesting to see if there are any width measures that allow dense graphs and fixed-parameter compilation at the same time. One width measure that might be worthwhile analyzing is the recently defined measure sm-width [22].

**Acknowledgments.** The author would like to thank Florent Capelli for helpful dis-cussions. Moreover, he thanks Jean-Marie Lagniez for helpful discussions and for exper-iments with the compiler from [15].

# References

1. Bova, S., Capelli, F., Mengel, S., Slivovsky, F.: On compiling CNFs into structured deterministic DNNFs. In: Heule, M., Weaver, S. (eds.) SAT 2015. LNCS, vol. 9340, pp. 199–214. Springer, Heidelberg (2015). doi:10.1007/978-3-319-24318-4_15
2. Bova, S., Capelli, F., Mengel, S., Slivovsky, F.: Knowledge compilation meets communication complexity. In: Proceedings of the 25th International Joint Conference on Artificial Intelligence, IJCAI 2016. IJCAI/AAAI (to appear, 2016)
3. Brault-Baron, J., Capelli, F., Mengel, S.: Understanding model counting for beta-acyclic CNF-formulas. In: Mayr, E.W., Ollinger, N. (eds.) 32nd International Symposium on Theoretical Aspects of Computer Science, STACS 2015, Garching, Germany, 4–7 March 2015. LIPIcs, vol. 30, pp. 143–156. Schloss Dagstuhl - Leibniz-Zentrum fuer Informatik (2015)
4. Chen, H.: Parameterized compilability. In: Kaelbling, L.P., Saffiotti, A. (eds.) Proceedings of the Nineteenth International Joint Conference on Artificial Intelligence, IJCAI 2005, Edinburgh, Scotland, UK, 30 July– 5 August 2005, pp. 412–417. Professional Book Center (2005)
5. Chen, H.: Parameter compilation. In: Husfeldt, T., Kanj, I.A. (eds.) 10th International Symposium on Parameterized and Exact Computation, IPEC 2015, Patras, Greece, 16–18 September 2015. LIPIcs, vol. 43, pp. 127–137. Schloss Dagstuhl - Leibniz-Zentrum fuer Informatik (2015)
6. Darwiche, A.: Decomposable negation normal form. J. ACM **48**(4), 608–647 (2001)
7. Darwiche, A.: New advances in compiling CNF into decomposable negation normal form. In López de Mántaras, R., Saitta, L. (eds.) Proceedings of the 16th Eureopean Conference on Artificial Intelligence, ECAI 2004, including Prestigious Applicants of Intelligent Systems, PAIS 2004, Valencia, Spain, 22–27 August 2004, pp. 328–332. IOS Press (2004)
8. Darwiche, A., Marquis, P.: A knowledge compilation map. J. Artif. Intell. Res. (JAIR) **17**, 229–264 (2002)
9. de Haan, R.: An overview of non-uniform parameterized complexity. Electron. Colloquium Comput. Complex. (ECCC) **22**, 130 (2015)
10. Dell, H., Kim, E.J., Lampis, M., Mitsou, V., Mömke, T.: Complexity and approximability of parameterized MAX-CSPs. In: Husfeldt, T., Kanj, I.A. (eds.) 10th International Symposium on Parameterized and Exact Computation, IPEC 2015, Patras, Greece, 16–18 September 2015. LIPIcs, vol. 43, pp. 294–306. Schloss Dagstuhl - Leibniz-Zentrum fuer Informatik (2015)
11. Diestel, R.: Graph Theory. Graduate Texts in Mathematics, vol. 173, 4th edn. Springer, Heidelberg (2012)
12. Duris, P., Hromkovic, J., Jukna, S., Sauerhoff, M., Schnitger, G.: On multi-partition communication complexity. Inf. Comput. **194**(1), 49–75 (2004)
13. Flum, J., Grohe, M.: Parameterized Complexity Theory. Texts in Theoretical Computer Science. An EATCS Series. Springer, Berlin (2006)
14. Huang, J., Darwiche, A.: DPLL with a trace: from SAT to knowledge compilation. In: Kaelbling, L.P., Saffiotti, A. (eds.) Proceedings of the Nineteenth International Joint Conference on Artificial Intelligence, IJCAI 2005, Edinburgh, Scotland, UK, 30 July– 5 August 2005, pp. 156–162. Professional Book Center (2005)
15. Koriche, F., Lagniez, J.-M., Marquis, P., Thomas, S.: Compilation, knowledge compilation for model counting : affine decision trees. In Rossi, R. (ed.) Proceedings of the 23rd International Joint Conference on Artificial Intelligence, IJCAI 2013, Beijing, China, 3–9 August 2013. IJCAI/AAAI (2013)

16. Lampis, M.: Algorithmic meta-theorems for restrictions of treewidth. In: de Berg, M., Meyer, U. (eds.) ESA 2010, Part I. LNCS, vol. 6346, pp. 549–560. Springer, Heidelberg (2010)
17. Muise, C., McIlraith, S.A., Beck, J.C., Hsu, E.I.: DSHARP: Fast d-DNNF Compilation with sharpSAT. In: Kosseim, L., Inkpen, D. (eds.) Canadian AI 2012. LNCS, vol. 7310, pp. 356–361. Springer, Heidelberg (2012)
18. Ordyniak, S., Paulusma, D., Szeider, S.: Satisfiability of acyclic and almost acyclic CNF formulas. Theor. Comput. Sci. **481**, 85–99 (2013)
19. Paulusma, D., Slivovsky, F., Szeider, S.: Model counting for CNF formulas of bounded modular treewidth. In: Portier, N., Wilke, T. (eds.) 30th International Symposium on Theoretical Aspects of Computer Science, STACS 2013, Kiel, Germany, 27 February–2 March 2013. LIPIcs, vol. 20, pp. 55–66. Schloss Dagstuhl - Leibniz-Zentrum fuer Informatik (2013)
20. Razgon, I.: No small nondeterministic read-once branching programs for CNFs of bounded treewidth. In: Cygan, M., Heggernes, P. (eds.) IPEC 2014. LNCS, vol. 8894, pp. 319–331. Springer, Heidelberg (2014)
21. Razgon, I.: On OBDDs for CNFs of bounded treewidth. In: Baral, C., De Giacomo, G., Eiter, T. (eds.) Principles of Knowledge Representation and Reasoning: Proceedings of the Fourteenth International Conference, KR 2014, Vienna, Austria, 20–24 July 2014. AAAI Press (2014)
22. Sæther, S.H., Telle, J.A.: Between treewidth and clique-width. In: Kratsch, D., Todinca, I. (eds.) WG 2014. LNCS, vol. 8747, pp. 396–407. Springer, Heidelberg (2014)
23. Sang, T., Bacchus, F., Beame, P., Kautz, H.A., Pitassi, T.: Combining component caching and clause learning for effective model counting. In: Proceedings of the Seventh International Conference on Theory and Applications of Satisfiability Testing, SAT 2004, Vancouver, BC, Canada, 10–13 May 2004
24. Thurley, M.: sharpSAT – counting models with advanced component caching and implicit BCP. In: Biere, A., Gomes, C.P. (eds.) SAT 2006. LNCS, vol. 4121, pp. 424–429. Springer, Heidelberg (2006)

# Satisfiability via Smooth Pictures

Mateus de Oliveira Oliveira[✉]

Institute of Mathematics - Czech Academy of Sciences,
Žitná 25, 115 67 Praha 1, Czech Republic
mateus.oliveira@math.cas.cz

**Abstract.** A picture over a finite alphabet $\Gamma$ is a matrix whose entries are drawn from $\Gamma$. Let $\pi : \Sigma \to \Gamma$ be a function between finite alphabets $\Sigma$ and $\Gamma$, and let $V, H \subseteq \Sigma \times \Sigma$ be binary relations over $\Sigma$. Given a picture $N$ over $\Gamma$, the picture satisfiability problem consists in determining whether there exists a picture $M$ over $\Sigma$ such that $\pi(M_{ij}) = N_{ij}$, and such that the constraints imposed by $V$ and $H$ on adjacent vertical and horizontal positions of $M$ are respectively satisfied. This problem can be easily shown to be NP-complete. In this work we introduce the notion of $s$-smooth picture. Our main result states the satisfiability problem for $s$-smooth pictures can be solved in time polynomial on $s$ and on the size of the input picture. With each picture $N$, one can naturally associate a CNF formula $F(N)$ which is satisfiable if and only if $N$ is satisfiable. In our second result, we define an infinite family of unsatisfiable pictures which intuitively encodes the pigeonhole principle. We show that this family of pictures is polynomially smooth. In contrast we show that the formulas which naturally arise from these pictures are hard for bounded-depth Frege proof systems. This shows that there are families of pictures for which our algorithm for the satisfiability for smooth pictures performs exponentially better than certain classical variants of SAT solvers based on the technique of conflict-driven clause-learning (CDCL).

**Keywords:** Smooth pictures · Bounded frege proof systems · Pigeonhole principle

## 1 Introduction

A picture over an alphabet $\Gamma$ is a matrix whose elements are drawn from $\Gamma$. Let $\pi : \Sigma \to \Gamma$ be a function between finite alphabets $\Sigma$ and $\Gamma$, and let $V, H \subseteq \Sigma \times \Sigma$ be binary relations over $\Sigma$. In the picture satisfiability problem we are given an $m \times n$ picture $N$ over $\Gamma$, and the goal is to determine whether there exists an $m \times n$ picture $M$ over $\Sigma$ such that the following conditions are satisfied. First, $N_{i,j} = \pi(M_{i,j})$ for each $i \in \{1, ..., m\}$ and $j \in \{1, ..., n\}$; second, each two consecutive vertical entries of $M$ belong to $V$; and third, each two consecutive horizontal entries of $M$ belong to $H$. If such a picture $M$ exists, we say that $M$ is a $(\pi, V, H)$-solution for $N$. Variations of the picture satisfiability problem have been studied since the seventies in the context of pattern recognition [21,23], image

© Springer International Publishing Switzerland 2016
N. Creignou and D. Le Berre (Eds.): SAT 2016, LNCS 9710, pp. 13–28, 2016.
DOI: 10.1007/978-3-319-40970-2_2

processing [5,21], tiling systems [14,22] and formal language theory [8,11,15,21]. In this work, we introduce the notion of $s$-smooth picture. Our main result states that one can determine whether an $s$-smooth picture $N$ has a $(\pi, V, H)$-solution in time $O(|\Sigma|^{e(\pi,V)} \cdot s^{e(\pi,V)} \cdot m \cdot n)$. Here, $e(\pi, V) \leq |\Sigma \times \Sigma|$ is a parameter that does not depend on the size of the picture. As an implication, we have that if $\mathcal{F}$ is a family of pictures such that each $m \times n$ picture in $\mathcal{F}$ is $poly(m, n)$-smooth, then the picture satisfiability problem for this family can be solved in polynomial time.

The pigeonhole principle states that if $m$ pigeons are placed into $m - 1$ holes, then at least one hole contains two pigeons. In a influential work, Haken showed that a family of propositional formulas $H_m$ encoding the pigeonhole principle requires resolution refutations of exponential size [2,9]. Following Haken's work, the pigeonhole principle and many of its variants have played a central hole in propositional proof complexity theory [3,19]. In particular, it has been shown that refutations of the formulas $H_m$ in constant-depth proof systems must have exponential size [12,13,18]. In our second result, we define an infinite family of pictures $P^{m,m-1}$ encoding the pigeonhole principle. Subsequently, we show that this family of pictures is $poly(m)$-smooth. This implies that our algorithm for smooth pictures is able to detect the unsatisfiability of the pictures $P^{m,m-1}$ in polynomial time.

For each fixed triple $(\pi, V, H)$ and each picture $N$ one can derive a natural constant-width CNF formula $F(N)$ which is satisfiable if and only if $N$ has a $(\pi, V, H)$-solution. Our third result states that the family of formulas $F(P^{m,m-1})$ derived from the pigeonhole pictures is still hard for constant depth Frege proof systems. The proof of this result follows by application of routine techniques to show that small refutations of $F(P^{m,m-1})$ imply small refutations of the formulas $H_m$. This last result establishes a point of comparison between our algorithm for the satisfiability of smooth pictures, and SAT solvers based on the technique of conflict-driven clause-learning (CDCL) [7,16]. Indeed, it has been shown that certain variants of CDCL-based SAT solvers, such as those introduced in [7,16], are equivalent in power to resolution-based proof systems [1,4,10,17]. Since bounded-depth Frege is stronger than the resolution proof system, our third result implies that the formulas $F(P^{m,m-1})$ derived from the pigeonhole pictures are hard for such variants of CDCL SAT solvers.

The remainder of the paper is organized as follows. Next, in Sect. 2 we introduce some notation and some basic results concerning leveled finite automata. Subsequently, in Sect. 3 we formally define the picture satisfiability problem and introduce the notion of $s$-smooth picture. In Sect. 4 we state and prove our main theorem, namely, that the satisfiability problem for pictures can be solved in time polynomial on its smoothness. In Sect. 5 we define the family of pigeonhole pictures and show that this family is polynomially smooth. In Sect. 6 we define a natural translation from pictures to constant-width CNF formulas, and in Sect. 7 we show that CNF formulas derived from the pigeonhole pictures according to our translation require exponential bounded-depth Frege proofs.

## 2  Preliminaries

A *leveled nondeterministic finite automaton* (LNFA) over an alphabet $\Sigma$ is a triple $\mathcal{A} = (Q, \Sigma, \mathfrak{R})$ where $Q$ is a set of states partitioned into subsets $Q_0, ..., Q_n$ and $\mathfrak{R} \subseteq \bigcup_{i=1}^{n} Q_{i-1} \times \Sigma \times Q_i$ is a transition relation. The states in $Q_0$ are the initial states of $\mathcal{A}$, while $Q_n$ is the set of final states of $\mathcal{A}$. For each $i \in \{0, ..., n\}$, we say that $Q_i$ is the $i$-th level of $\mathcal{A}$. The size of $\mathcal{A}$ is defined as $|\mathcal{A}| = |Q| + |\mathfrak{R}|$. We say that a string $w \in \Sigma^n$ is accepted by $\mathcal{A}$ if there exists a sequence of transitions

$$q_0 \xrightarrow{w_1} q_1 \xrightarrow{w_2} ... \xrightarrow{w_n} q_n$$

such that $q_i \in Q_i$ for each $i$ in $\{0, 1, ..., n\}$, and $(q_{i-1}, w_i, q_i) \in \mathfrak{R}_i$ for each $i$ in $\{1, ..., n\}$. We denote by $\mathcal{L}(\mathcal{A})$ the set of all strings accepted by $\mathcal{A}$. We note that all strings accepted by $\mathcal{A}$ have size $n$, i.e., $\mathcal{L}(\mathcal{A}) \subseteq \Sigma^n$. We say that $\mathcal{A}$ is a *leveled deterministic finite automaton* (LDFA) if $Q_0$ has a unique state $q_0$, and for each state $q \in Q$, and each symbol $a \in \Sigma$ there exists at most one state $q' \in Q$ such that $(q, a, q') \in \mathfrak{R}$. Let $\pi : \Sigma \to \Gamma$ be a function and let $w = w_1 w_2 ... w_n$ be a string in $\Sigma^*$. We denote by $\pi(w) = \pi(w_1)\pi(w_2)...\pi(w_n)$ the image of $w$ under $\pi$.

**Lemma 2.1 (Synchronized Product of Automata).** *Let $\mathcal{A}$ and $\mathcal{A}'$ be LNFA over $\Sigma$ accepting strings of size $n$. Let $V \subseteq \Sigma \times \Sigma$ be a binary relation over $\Sigma$. Then one can construct in time $|\mathcal{A}| \cdot |\mathcal{A}'|$ an LNFA $\mathcal{A} \otimes_V \mathcal{A}'$ accepting the following language over $\Sigma \times \Sigma$.*

$$\mathcal{L}(\mathcal{A} \otimes_V \mathcal{A}') = \{(w_1, w_1')(w_2, w_2')...(w_n, w_n') \mid w \in \mathcal{L}(\mathcal{A}),\ w' \in \mathcal{L}(\mathcal{A}'), (w_i, w_i') \in V\}.$$

## 3  Pictures

An $(m, n)$-picture over a finite set of symbols $\Sigma$ is an $m \times n$ matrix whose entries are drawn from $\Sigma$. Let $\pi : \Sigma \to \Gamma$ be a function between finite sets of symbols $\Sigma$ and $\Gamma$, and let $V, H \subseteq \Sigma \times \Sigma$ be binary relations over $\Sigma$. Finally, let $N$ be an $(m, n)$-picture over $\Gamma$. We say that an $(m, n)$-picture $M$ over $\Sigma$ is a $(\pi, V, H)$-solution for $N$ if the following conditions are satisfied.

1. $N_{i,j} = \pi(M_{i,j})$ for each $i \in \{1, ..., m\}$ and each $j \in \{1, ..., n\}$.
2. $(M_{i,j}, M_{i,j+1})$ belongs to $H$ for each $i \in \{1, ..., m\}$ and $j \in \{1, ..., n-1\}$.
3. $(M_{i,j}, M_{i+1,j})$ belongs to $V$ for each $i \in \{1, ..., m-1\}$ and $j \in \{1, ..., n\}$.

Intuitively, the symbols in $\Sigma$ may be regarded as colored versions of symbols in $\Gamma$. For each symbol $a \in \Gamma$, the set $\pi^{-1}(a) \subseteq \Sigma$ is the set of colored versions of $a$. Thus $M$ is a $(\pi, V, H)$-solution for $N$ if $M$ is a colored version of $N$ and the entries in $M$ respect the vertical and horizontal constraints imposed by $V$ and $H$ respectively. If $N$ admits a $(\pi, V, H)$-solution, then we say that $N$ is *satisfiable* (with respect to $(\pi, V, H)$). Otherwise, we say that $N$ is unsatisfiable.

**Definition 3.1 (Picture Satisfiability Problem).** *Let $\pi : \Sigma \to \Gamma$ be a function and $V, H \subseteq \Sigma \times \Sigma$ be binary relations over $\Sigma$. Given an $(m, n)$-picture $N$ over $\Gamma$, is $N$ satisfiable with respect to $(\pi, V, H)$?*

## 3.1   Smooth Pictures

Let $[n] = \{1,...,n\}$. We assume that the set $[m] \times [n] = \{(i,j) \mid i \in [m], j \in [n]\}$ is endowed with a lexicographic ordering $<$, which sets $(i,j) < (i',j')$ if either $i < i'$, or $i = i'$ and $j < j'$. We write $(i,j) \leq (i',j')$ to denote that $(i,j) = (i',j')$ or $(i,j) < (i',j')$. For each $(i,j) \in [m] \times [n]$, we let

$$S(m,n,i,j) = \{(i',j') \mid (i',j') \leq (i,j)\}$$

be the set of all positions in $[m] \times [n]$ that are (lexicographically) smaller than or equal to $(i,j)$.

Let $\pi : \Sigma \to \Gamma$ be a function, and $V, H \subseteq \Sigma \times \Sigma$ be a binary relation over $\Sigma$. We say that a function $M : S(m,n,i,j) \to \Sigma$ is an $(i,j)$-partial $(\pi, V, H)$-solution for $N$ if the following conditions are satisfied.

1. $(M_{i',j'}, M_{i',j'+1}) \in H$ for each $(i',j')$, $(i',j'+1)$ in $S(m,n,i,j)$.
2. $(M_{i',j'}, M_{i'+1,j'}) \in V$ for each $(i',j')$, $(i'+1,j')$ in $S(m,n,i,j)$.

Note that for simplicity we write $M_{i,j}$ in place of $M(i,j)$ to designate an entry of $M$. Intuitively, an $(i,j)$-partial $(\pi, V, H)$-solution for $N$ is a function that colors the positions of $N$ up to the entry $(i,j)$ with elements from $\Sigma$ in such a way that the vertical and horizontal constraints imposed by $V$ and $H$ respectively are respected. If $(i,j_1)$ and $(i,j_2)$ are positions in $S(m,n,i,j)$ with $j_1 < j_2$, then we let $M_{i,[j_1,j_2]} = M_{i,j_1}...M_{i,j_2}$ be the string formed by all entries at the $i$-th row of $M$ between positions $j_1$ and $j_2$. Now let $(i,j) \in S(m,n,i,j)$ with $(i,j) \geq (1,n)$. The $(i,j)$-boundary of $M$ is defined as follows.

$$\partial_{i,j}(M) = \begin{cases} M_{i,[1,n]} & \text{if } j = n. \\ \\ M_{i,[1,j]} \cdot M_{i-1,[j+1,n]} & \text{if } j < n. \end{cases} \quad (1)$$

In other words, if $j = n$, then $\partial(M)$ is the string consisting of all entries in the $i$-th row of $M$. On the other hand, if $j < n$, then $\partial(M)$ is obtained by concatenating the string corresponding to the first $j$ entries of row $i$ with the last $(n-j)$ entries of row $(i-1)$. The notion of boundary of a partial solution is illustrated in Fig. 1.

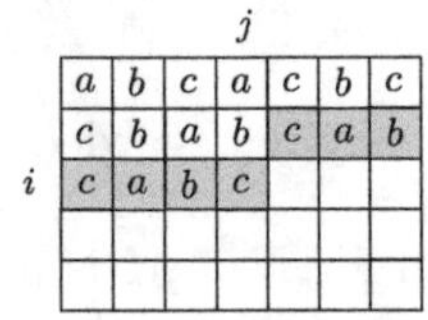

**Fig. 1.** An $(i,j)$-partial solution $M$ where $i = 3$ and $j = 4$. The grey entries form the boundary of $M$. Therefore $\partial_{i,j}(M) = cabccab$.

Below, we define the $(i,j)$-feasibility boundary of a picture $N$ over $\Gamma$ as the set of $(i,j)$-boundaries of partial solutions of $N$.

**Definition 3.2 (Feasibility Boundary).** *Let* $\pi : \Sigma \to \Gamma$ *be a function,* $V, H \subseteq \Sigma \times \Sigma$, *and* $N$ *be a* $(m, n)$-*picture over* $\Gamma$. *The* $(i, j)$-*feasibility boundary of* $N$ *with respect to* $(\pi, V, H)$, *denoted by* $\partial_{i,j}(N, \pi, V, H)$, *is defined as follows.*

$$\partial_{i,j}(N, \pi, V, H) = \{\partial_{i,j}(M) \mid M \text{ is an } (i, j)\text{-partial } (\pi, V, H)\text{-solution for } N\}. \tag{2}$$

Note that $N$ has a $(\pi, V, H)$-solution if and only if its $(m, n)$-feasibility boundary $\partial_{m,n}(N, \pi, V, H)$ is non-empty. Below we define the notion of *smooth picture*.

**Definition 3.3 (Smooth Picture).** *Let* $\pi : \Sigma \to \Gamma$ *be a function, and* $V, H \subseteq \Sigma \times \Sigma$ *be binary relations over* $\Sigma$. *We say that an* $(m, n)$-*picture* $N$ *over* $\Gamma$ *is* s-*smooth if for each* $(i, j) \geq (1, n)$, *the set* $\partial_{i,j}(N, \pi, V, H)$ *can be represented by an LDFA of size at most* $s$.

Intuitively, a picture $N$ is smooth if each of its feasibility boundaries can be efficiently represented. The main goal of this work is to show that the automata representing the boundaries of feasibility of a picture $N$ can actually be constructed in time polynomial on the smoothness parameter $s$ and on the size of $N$. Additionally, once these automata are constructed, one can proceed to actually construct a solution for $N$ if such a solution exists.

# 4   Satisfiability of Smooth Pictures in Polynomial Time

In this section we show that the satisfiability problem for smooth pictures can be solved in time polynomial on the size of the picture and on its smoothness parameter. Let $\pi : \Sigma \to \Gamma$ be a function and $V \subseteq \Sigma \times \Sigma$ be a binary relation over $\Sigma$. We let $e(\pi, V) = \max_{a \in \Gamma, b \in \Sigma} |\{c \in \pi^{-1}(a) \mid (b, c) \in V\}|$ be the extension number of $V$. Below we state our main theorem.

**Theorem 4.1 (Main Theorem).** *Let* $\pi : \Sigma \to \Gamma$ *be a function and* $V, H \subseteq \Sigma$ *be binary relations over* $\Sigma$. *Let* $N$ *be an* $(m, n)$-*picture over* $\Gamma$. *There is an algorithm that works in time* $O(|\Sigma|^{e(\pi,V)} \cdot s^{e(\pi,V)} \cdot m \cdot n)$ *and either constructs a* $(\pi, V, H)$-*solution for* $N$, *or correctly determines that no such solution exists.*

Note that as an application, we have that if $\mathcal{F}$ is a family of pictures such that for each $(m, n)$-picture $N$ in $\mathcal{F}$, $N$ is $poly(m, n)$-smooth, then the picture satisfiability problem for $\mathcal{F}$ can be solved in polynomial time.

We dedicate this section to the proof of Theorem 4.1. For $(i, j) \geq (1, n)$, let $\mathcal{A}_{ij}(N, \pi, V, H)$ be the LDFA with minimum number of states accepting the set of strings $\partial_{i,j}(N, \pi, V, H)$. The next Lemma will be used in the construction of the automaton $\mathcal{A}_{1,n}(N, \pi, V, H)$. Note that the language accepted by this automaton is simply the set of all colored versions of the first row of $N$ which satisfy constraints imposed by the horizontal relation $H$.

**Lemma 4.2.** *Let* $\pi : \Sigma \to \Gamma$ *be a function, and let* $H \subseteq \Sigma \times \Sigma$ *be a binary relation over* $\Sigma$. *Let* $w = w_1 w_2 ... w_n$ *be a string in* $\Gamma^n$. *Then one can construct* in time $O(|\Sigma|^2 \cdot n)$ an LDFA $\mathcal{A}(w, H)$ of size at most $O(|\Sigma|^2 \cdot n)$ accepting the following language.

$$\mathcal{L}(\mathcal{A}(w, H)) = \{u \in \Sigma^n \mid \pi(u) = w, \ (u_i, u_{i+1}) \in H \ for \ i \in \{1, ..., n-1\}\}.$$

*Proof.* First, we construct an automaton $\mathcal{A} = (Q, \Sigma, \mathfrak{R})$ as follows. We set $Q = Q_0 \dot{\cup} Q_1 \dot{\cup} ... \dot{\cup} Q_n$ and $\mathfrak{R} = \bigcup_{i=1}^n \mathfrak{R}_i$ where $Q_0 = \{q_0\}$, $Q_i = \{q_{i,a} \mid a \in \Sigma\}$ for each $i \in \{1, ..., n\}$, $\mathfrak{R}_1 = \{(q_0, a, q_{1,a}) \mid \pi(a) = w_1\}$, and for each $i \in \{1, ..., n-1\}$, $\mathfrak{R}_i = \{(q_{i,a}, b, q_{i+1,b}) \mid (a, b) \in H, \ \pi(b) = w_i\}$. Note that $Q$ has $|\Sigma| \cdot n + 1$ states and at most $|\Sigma|^2 \cdot n$ transitions. Now it is straightforward to check that for each string $u \in \Sigma^n$ there exists an accepting path $q_0 \xrightarrow{u_1} q_{1,u_1} \xrightarrow{u_2} ... \xrightarrow{u_n} q_{n,u_n}$ in $\mathcal{A}$ if and only if $\pi(u_1)\pi(u_2)...\pi(u_n) = w_1 w_2...w_n$ and for each $i \in \{1, ..., n-1\}$, $(u_i, u_{i+1}) \in H$. Finally, we let $\mathcal{A}(w, H) = \mathrm{Min}(\mathcal{A})$ be the minimum LDFA which accepts the same language as $\mathcal{A}$. Since $\mathcal{A}$ is acyclic, minimization can be performed in time linear on the size of $\mathcal{A}$ [20]. So the overall time to construct $\mathcal{A}(w, H)$ is still $O(|\Sigma|^2 \cdot n)$. □

As a corollary of Lemma 4.2, the LDFA $\mathcal{A}_{1,n}(N, \pi, V, H)$ can be constructed in time $O(|\Sigma|^2 \cdot n)$.

**Corollary 4.3.** *Let $\pi : \Sigma \to \Gamma$ be a function and $V, H \subseteq \Sigma \times \Sigma$ be binary relations over $\Sigma$. Then the LDFA $\mathcal{A}_{1,n}(N, \pi, V, H)$ has size $O(|\Sigma|^2 \cdot n)$ and can be constructed in time $O(|\Sigma|^2 \cdot n)$.*

*Proof.* Set $\mathcal{A}_{i,j}(N, \pi, V, H) = \mathcal{A}(w, H)$, where $w = N_{1,[1,n]}$ and $\mathcal{A}(w, H)$ is the automaton of Lemma 4.2. □

For each $(i, j) \in [m] \times [n]$ with $(i, j) < (m, n)$, let $Suc(i, j) = (i, j+1)$ if $j < n$ and $Suc(i, j) = (i+1, 1)$ if $j = n$. In other words, $Suc(i, j)$ is the smallest pair in $[m] \times [n]$ which is greater than $(i, j)$ according to the lexicographical ordering $<$. Our next step consists in showing that if $(i', j') = Suc(i, j)$, then the automaton $\mathcal{A}_{i',j'}(N, \pi, V, H)$ can be constructed in time polynomial on the size of $\mathcal{A}_{i,j}(N, \pi, V, H)$. Towards this construction we will need to introduce some notation concerning leveled finite automata.

Let $\mathcal{A} = (Q, \Sigma, \mathfrak{R})$ be a leveled finite automaton over $\Sigma$ and let $q$ be a state in $Q$. The *non-deterministic degree* of $q$, denoted $d(q)$, is defined as the maximum number of states that can be reached from $q$ when reading some fixed symbol $a \in \Sigma$.

$$d(q) = \max_{a \in \Sigma} |\{q' \mid (q, a, q') \in \mathfrak{R}\}| \qquad (3)$$

We say that $\mathcal{A}$ has non-deterministic degree $d$ if the following conditions are satisfied.

1. All states of $\mathcal{A}$ have non-deterministic degree at most $d$.
2. All states of $\mathcal{A}$ of non-deterministic degree greater than one belong to the same level.

The following lemma states that a leveled nondeterministic finite automaton $\mathcal{A}$ of non-deterministic degree $d$ can be transformed into a leveled deterministic finite automaton $det(\mathcal{A})$ of size at most $|\mathcal{A}|^{O(d)}$ which accepts the same language as $\mathcal{A}$.

**Lemma 4.4.** *Let $\mathcal{A}$ be an LNFA of non-deterministic degree $d$. Let $det(\mathcal{A})$ be the minimum LDFA accepting $\mathcal{L}(\mathcal{A})$. Then $det(\mathcal{A})$ has size at most $|\mathcal{A}|^d$ and can be constructed in time $O(|\mathcal{A}|^d)$.*

*Proof.* Let $\mathcal{A} = (Q, \Sigma, \mathfrak{R})$ where $Q = Q_0 \,\dot{\cup}\, Q_1 \,\dot{\cup}\, ... \,\dot{\cup}\, Q_n$. Assume that all states of $\mathcal{A}$ with non-deterministic degree greater than one belong to level $Q_i$. Then the sub-automaton of $\mathcal{A}$ induced by the states $Q_0 \,\dot{\cup}\, ... \,\dot{\cup}\, Q_{i-1}$ is deterministic, in the sense that all of its states have non-deterministic degree at most one. Therefore we just need to determinize the sub-automaton of $\mathcal{A}$ induced by the states $Q_i \,\dot{\cup}\, ... \,\dot{\cup}\, Q_n$. This determinization process will be achieved by an adaptation of the traditional subset construction, but with the caveat that only subsets of states of size at most $d$ need to be considered.

Let $S \subseteq Q$ be a subset of states of $\mathcal{A}$, and let $a \in \Sigma$. Then, we define $Q(S, a) = \{q' \mid \exists q \in S, (q, a, q') \in \mathfrak{R}\}$ as the set of all states reachable from some state in $S$ through a transition labeled with $a$. We note that for each $j \in \{0, ..., n-1\}$, if $S \subseteq Q_j$ then $Q(S, a) \subseteq Q_{j+1}$. We note that since each state $q \in Q_i$ has non-deterministic degree at most $d$, we have that $|Q(\{q\}, a)| \le d$ for each symbol $a \in \Sigma$. Additionally, for each $j \in \{i+1, ..., n-1\}$, the non-deterministic degree of each state in $Q_j$ is at most one. Therefore, for each subset $S \subseteq Q_j$, we have that $|Q(S, a)| \le |S|$. Therefore, to construct a deterministic version of $\mathcal{A}$ we only need to consider sets of states of size at most $d$. The construction is as follows. Consider the automaton $\mathcal{A} = (Q', \Sigma, \mathfrak{R}')$ where $Q' = Q'_0 \cup Q'_1 \cup ... \cup Q'_n$ and $\mathfrak{R} = \mathfrak{R}_1 \dot{\cup} ... \dot{\cup} \mathfrak{R}_n$. For each $j \in \{0, ..., i\}$ we let $Q'_j = \{\mathfrak{q}_{j, \{q\}} \mid q \in Q_j\}$. In other words, $Q'_j$ has one state $\mathfrak{q}_{j, \{q\}}$ for each state $q \in Q_j$. Now for $j \in \{i+1, ..., n\}$, $Q'_j = \{\mathfrak{q}_{j, S} \mid S \subseteq Q_j, |S| \le d\}$. In other words, $Q'_j$ has one state $\mathfrak{q}_{j, S}$ for each subset of $Q_j$ of size at most $d$. Now for each $j \in \{1, ..., n\}$, we set $\mathfrak{R}'_j = \{(\mathfrak{q}_{j-1, S}, a, \mathfrak{q}_{j, S'}) \mid \mathfrak{q}_{j-1, S} \in Q'_{j-1}, \mathfrak{q}_{j, S'} \in Q'_j, S' = Q(S, a)\}$. Clearly, the automaton $\mathcal{A}'$ is deterministic, and has size at most $|\mathcal{A}|^d$. Additionally we have that $w_1 w_2 ... w_n$ is accepted by $\mathcal{A}$ if and only if there exists an accepting sequence

$$\mathfrak{q}_{0, S_0} \xrightarrow{w_1} \mathfrak{q}_{1, S_1} \xrightarrow{w_2} ... \xrightarrow{w_n} \mathfrak{q}_{n, S_n}$$

in $\mathcal{A}'$ where $S_0 = \{q_0\}$ and for each $j \in \{0, ..., n-1\}$, $S_{j+1} = Q(S_j, w_{j+1})$. This implies that $\mathcal{L}(\mathcal{A}') = \mathcal{L}(\mathcal{A})$. Since $\mathcal{A}'$ is deterministic and acyclic, the minimum leveled deterministic finite automaton $det(\mathcal{A})$ accepting $\mathcal{L}(\mathcal{A}) = \mathcal{L}(\mathcal{A}')$ can be constructed in time linear on the size of $\mathcal{A}'$ [20], i.e., in time $O(|\mathcal{A}|^d)$. □

Let $(i', j') = Suc(i, j)$. The following proposition, whose proof is immediate, establishes a way of defining the boundary set $\partial_{i', j'}(N, \pi, V, H)$ in terms of $\partial_{i, j}(N, \pi, V, H)$.

**Proposition 4.5.** *Let $\pi : \Sigma \to \Gamma$ be a function, $V, H \subseteq \Sigma \times \Sigma$ be binary relations over $\Sigma$, and $N$ be an $(m \times n)$-picture over $\Gamma$. Let $\partial_{i,j}(N, \pi, V, H)$ be the set of $(i,j)$-boundaries of $N$ where $(i,j) < (m,n)$.*

*1. If $j = n$, then*

$$\partial_{i+1,1}(N, \pi, V, H) = \{aw_2...w_n \mid a \in \Sigma, \, (w_1, a) \in V\}.$$

*2. If $j < n$, then*

$$\partial_{i,j+1}(N, \pi, V, H) = \{w_1...w_j aw_{j+2}...w_n \mid a \in \Sigma, \, (w_j, a) \in H, (w_{j+1}, a) \in V\}.$$

Using Proposition 4.5, we will prove the following theorem.

**Theorem 4.6.** *Let $\pi : \Sigma \to \Gamma$ be a function, $V, H \subseteq \Sigma \times \Sigma$ be binary relations over $\Sigma$, and $N$ be a $(m,n)$-picture over $\Gamma$. Let $(i',j') = Suc(i,j)$. Then $\mathcal{A}_{i',j'}(N, \pi, V, H)$ can be constructed in time*

$$O( |\Sigma|^{e(\pi,V)} \cdot |\mathcal{A}_{i,j}(N, \pi, V, H)|^{e(\pi,V)} ).$$

*Proof.* Let $\mathcal{A} = \mathcal{A}_{ij}(N, \pi, V, H) = (Q, \Sigma, \mathfrak{R})$ be the minimum LDFA accepting the set of boundaries $\partial_{ij}(N, \pi, V, H)$ where $Q = Q_0 \,\dot{\cup}\, Q_1 \,\dot{\cup}\, ... \,\dot{\cup}\, Q_n$ and $\mathfrak{R} = \mathfrak{R}_1 \,\dot{\cup}\, ... \,\dot{\cup}\, \mathfrak{R}_n$. Now let $\mathcal{A}' = (Q', \Sigma, \mathfrak{R}')$ be the LDFA obtained from $\mathcal{A}$ as follows. First set $Q' = Q'_0 \,\dot{\cup}\, Q'_1 \,\dot{\cup}\, ... \,\dot{\cup}\, Q'_n$ where $Q'_0 = Q_0$ and $Q'_i = \{q^a \mid q \in Q_i, \, a \in \Sigma\}$. Subsequently set $\mathfrak{R}' = \mathfrak{R}_1 \,\dot{\cup}\, ... \,\dot{\cup}\, \mathfrak{R}_n$ where $\mathfrak{R}'_1 = \{(q_0, a, q^a) \mid (q_0, a, q) \in \mathfrak{R}_1\}$ and for each $i \in \{2, ..., n\}$, $\mathfrak{R}'_i = \{(q^a, b, r^b) \mid (q, b, r) \in \mathfrak{R}_i, \, a \in \Sigma \}$. Then $\mathcal{A}'$ is still deterministic, and $q_0 \xrightarrow{w_1} q_1 \xrightarrow{w_2} ... \xrightarrow{w_n} q_n$ is an accepting path in $\mathcal{A}$ if and only if $q_0 \xrightarrow{w_1} q_1^{w_1} \xrightarrow{w_2} ... \xrightarrow{w_n} q_n^{w_n}$ is an accepting path in $\mathcal{A}'$. In other words, $\mathcal{A}'$ accepts precisely the same language as $\mathcal{A}$. Note that $|\mathcal{A}'| \leq |\Sigma| \cdot |\mathcal{A}|$. Now using $\mathcal{A}'$ we will construct a non-deterministic automaton $\mathcal{A}'' = (Q', \Sigma, \mathfrak{R}'')$ of non-deterministic degree at most $e(\pi, V)$ which accepts the language $\partial_{i',j'}(N, \pi, V, H)$ where $(i',j') = Suc(i,j)$. Note that the set of states of $\mathcal{A}''$ is a copy of the set of states of $\mathcal{A}'$. Now to define the set of transitions $\mathfrak{R}$ of $\mathcal{A}''$ we need to consider two cases.

1. In the first case, $j = n$, and therefore $i' = i+1$ and $j' = 1$. In this case we set $\mathfrak{R}''_k = \mathfrak{R}'_k$ for each $k \in \{2, ..., n\}$, and

$$\mathfrak{R}''_1 = \{(q_0, a, r^b) \mid a \in \pi^{-1}(N_{i+1,1}), \, (q_0, b, r^b) \in \mathfrak{R}'_1, \, (b, a) \in V\}.$$

Therefore, we have that $q_0 \xrightarrow{w_1} q_1^{w_1} \xrightarrow{w_2} q_2^{w_2} ... \xrightarrow{w_n} q_n^{w_n}$ is an accepting path of $\mathcal{A}'$ if and only if for each $a$ with $(w_1, a) \in V$, $q_0 \xrightarrow{a} q_1^{w_1} \xrightarrow{w_2} q_2^{w_2} ... \xrightarrow{w_n} q_n$ is an accepting path of $\mathcal{A}''$. Stated otherwise,

$$\mathcal{L}(\mathcal{A}'') = \{aw_2...w_n \mid a \in \Sigma, \, (w_1, a) \in V, w \in \mathcal{L}(\mathcal{A}')\}.$$

By Proposition 4.5, we have that $\mathcal{L}(\mathcal{A}'') = \partial_{i',j'}(N, \pi, V, H)$.

2. In the second case, $j < n$, and therefore $i' = i$ and $j' = j + 1$. In this case, we set $\mathfrak{R}_k = \mathfrak{R}'_k$ for $k \neq j + 1$. We define $\mathfrak{R}''_{j+1}$ as follows.

$$\mathfrak{R}''_{j+1} = \{(q^c, a, r^b) \mid (q^c, b, r^b) \in \mathfrak{R}_k, \ a \in \pi^{-1}(N_{i,j+1}), \ (c, a) \in H, \ (b, a) \in V\}.$$

Then we have that $q_0 \xrightarrow{w_1} q_1^{w_1} \xrightarrow{w_2} q_2^{w_2} \dots \xrightarrow{w_n} q_n^{w_n}$ is an accepting path of $\mathcal{A}$ if and only if for each $a \in \Sigma$ with $(w_j, a) \in H$ and $(w_{j+1}, a) \in V$, the path

$$q_0 \xrightarrow{w_1} \dots \xrightarrow{w_j} q_j^{w_j} \xrightarrow{a} q_{j+1}^{w_{j+1}} \dots \xrightarrow{w_n} q_n^{w_n}$$

is an accepting path of $\mathcal{A}''$. In other words,

$$\mathcal{L}(\mathcal{A}'') = \{w_1 \dots w_j a w_{j+2} \dots w_n \mid w \in \mathcal{L}(\mathcal{A}'), \ a \in \Sigma, \ (w_j, a) \in H, \ (w_{j+1}, a) \in V\}.$$

Since $\mathcal{L}(\mathcal{A}') = \partial_{i,j}(N, \pi, V, H)$, by Proposition 4.5 we have that $\mathcal{L}(\mathcal{A}'') = \partial_{i',j'}(N, \pi, V, H)$.

In both cases considered above, $\mathcal{A}''$ has as many states and transitions as $\mathcal{A}'$. Nevertheless, contrary to $\mathcal{A}'$, $\mathcal{A}''$ is non-deterministic. Fortunately, the non-deterministic degree of $\mathcal{A}''$ is upper bounded by $e(\pi, V)$. This implies that by Lemma 4.4, the minimum leveled deterministic finite automaton accepting the same language as $\mathcal{A}''$ can be constructed in time

$$O(|\mathcal{A}''|^{e(\pi,V)}) = O(|\Sigma|^{e(\pi,V)} \cdot |\mathcal{A}_{i,j}(N, \pi, V, H)|^{e(\pi,V)}).$$

$\square$

Now, for each $(i, j) \geq (1, n)$ we know how to construct the automaton $\mathcal{A}_{i,j}(N, \pi, V, H)$ accepting the set $\partial_{ij}(N, \pi, V, H)$. Note that since by assumption, the picture $N$ is $s$-smooth, we have that the collection of all automata $\mathcal{A}_{i,j}(N, \pi, V, H)$ can be constructed in time $O(|\Sigma|^{e(\pi,V)} \cdot s^{e(\pi,V)} \cdot m \cdot n)$. If the last automaton $\mathcal{A}_{m,n}(N, \pi, V, H)$ accepts the empty language, then the picture $N$ has no $(\pi, V, H)$-solution. On the other hand, if this language is not empty, then we still need to construct a solution. Let $\mathcal{A}_i = \mathcal{A}_{i,n}(N, \pi, V, H)$ be the automaton accepting the boundary set $\partial_{i,n}(N, \pi, V, H)$. Let $\gamma : \Sigma \times \Sigma \to \Sigma$ be a projection which sets $\gamma(a, b) = a$ for each pair $(a, b) \in \Sigma \times \Sigma$. In other words, $\gamma$ erases the second coordinate of each pair $(a, b) \in \Sigma \times \Sigma$. For a string $u = (a_1, b_1)(a_2, b_2) \dots (a_n, b_n) \in (\Sigma \times \Sigma)^n$, we let $\gamma(u) = a_1 a_2 \dots a_n$. Also, for a string $w \in \Sigma^n$, let $\mathcal{A}(w)$ be the minimum LDFA that accepts $w$, and no other string. We will construct a $(\pi, V, H)$-solution $M$ for $N$ row by row, starting from the last row of $M$ and finishing at the first. The construction is inductive. More precisely, we construct a sequence $w^n, w^{n-1}, \dots, w^1$ of strings in $\Sigma^n$ as follows.

1. Let $w^n$ be an arbitrary string in $\mathcal{L}(\mathcal{A}_n)$.
2. For $i = n - 1, \dots, 1$:
   (a) Let $w^i$ be an arbitrary string in $\mathcal{L}(\gamma(\mathcal{A}_i \otimes_V \mathcal{A}(w^{i+1})))$.
3. Set $M$ to be the $(m, n)$-picture over $\Sigma$ such that $w^i$ is the $i$-th row of $M$.

We note that the string $w^i$ selected from $\mathcal{L}(\gamma(\mathcal{A}_i \otimes_V \mathcal{A}(w^{i+1})))$ is simply a string in the boundary $\partial_{i,n}(N, \pi, V, H)$ satisfying the property that $(w_k^i, w_k^{i+1}) \in V$ for each $k \in \{1, ..., n\}$. In other words, $w^i$ is vertically compatible with $w^{i+1}$. Additionally, since $w^i$ by definition belongs to $\partial_{i,n}(N, \pi, V, H)$ we have that for each $k \in \{1, ..., n-1\}$, $(w_k^i, w_{k+1}^i) \in H$. In other words, the horizontal constraints imposed by the relation $H$ are satisfied by each $w^i$. This shows that the picture $M$ obtained by setting, for each $i \in \{1, ..., m\}$, $w^i$ as the $i$-th row of $M$ is a valid $(\pi, V, H)$-solution for $N$. We note that the time to select each string $w^i$ is of the order of $O(|\mathcal{A}_i|)$. Therefore, the total algorithm still runs in the time $O(|\Sigma|^{e(\pi, V)} \cdot s^{e(\pi, V)} \cdot m \cdot n)$. This proves Theorem 4.1. $\square$

## 5    Pigeonhole Pictures

In this section we define a family of pictures formalizing the pigeonhole principle. We call these pictures the pigeonhole pictures. Subsequently, we will show that this family is polynomially smooth. This implies that our algorithm for the satisfiability of smooth pictures is able to determine whether a given element of this family has a solution, and in the case it has, the algorithm is able to construct it. On the other hand, we will show in Sect. 7 that CNF formulas derived from the pigeonhole pictures require constant-depth Frege proofs of exponential size.

Our primary alphabet $\Gamma = \{\oplus, \circ, \odot\}$ has a left symbol $\oplus$, used to fill all entries of the first column of the picture, a right symbol $\odot$, used to fill all entries of the last column of the picture, and a middle symbol $\circ$, used to fill all entries in between the first and last columns. Our colored alphabet is defined as $\Sigma = \{b, g, bb, bg, gb, gg, rr\}$. Finally, the projection $\pi : \Sigma \to \Gamma$ is such that $\pi(\{bb, bg, gb, gg, rr\}) = \{\circ\}$, $\pi(b) = \oplus$ and $\pi(g) = \odot$. Intuitively, the letters $r, b, g$ stand for "red", "blue" and "green" respectively. The symbols in $\{bb, bg, gb, gg, rr\}$ are called double colors. The left symbol $\oplus$ can only be colored with $b$, then right symbol $\odot$ can only be colored with $g$, and the middle symbol can be colored with any double color in $\{bb, bg, gb, gg, rr\}$. The color red serves to indicate presence of a pigeon. The colors blue and green serve to mark the "footprint" of a pigeon. The vertical and horizontal relations are defined as follows:

$$V = \{ (xb, yb), (xb, rr), (rr, xg), (xg, yg) \mid x, y \in \{b, g\} \} \cup \{(b, b), (g, g)\}$$

$$H = \{ (bx, by), (bx, rr), (rr, gx), (gx, gy) \mid x, y \in \{b, g\} \}$$
$$\cup \tag{4}$$
$$\{ (b, bz), (b, rr), (rr, g), (gz, g) \mid z \in \{b, g\} \}$$

We define the $(m, n)$-pigeonhole picture, $P^{m,n}$, as the $m \times (n+2)$ picture, where all entries in the first column are filled with the left symbol $\oplus$, all entries in the last column are filled with the right symbol $\odot$, and all other entries are filled with the middle symbol $\circ$.

**Definition 5.1 (Pigeonhole Picture).** *The $(m, n)$-pigeonhole picture $P^{m,n}$ is the $m \times (n + 2)$ picture over $\Gamma$ defined as follows.*

$$P_{ij}^{m,n} = \begin{cases} \oplus & \text{if } j = 1. \\ \odot & \text{if } j = n + 2. \\ \circ & \text{otherwise.} \end{cases} \tag{5}$$

Intuitively, the rows of $P^{m,n}$ correspond to pigeons, while the middle columns correspond to holes. Pigeons are not allowed to occupy the first nor the last columns. If $M$ is a $(\pi, V, H)$-solution for the pigeonhole picture then the fact that $M_{i,j} = rr$ indicates that the $i$-th pigeon is placed at the hole represented by column $j$. The fact that $M_{i,j} = bx$ for some $x \in \{g, b\}$ indicates that the $i$-th pigeon is placed in some column greater than $j$, while the fact that $M_{i,j} = gx$ for some $x \in \{b, g\}$ indicates that the $i$-th pigeon is placed in some column smaller than $j$. Analogously, if $M_{i,j} = xb$ for some $x \in \{b, g\}$, then the pigeon that is placed at the $j$-th hole is greater than $i$, while if $M_{i,j} = xg$, then the pigeon that is placed at the $j$-th hole is smaller than $i$.

Note also that the horizontal relation guarantees that a pigeon must occur in each row of a satisfying assignment. This is because there is no allowed pair $(xy, x'y')$ where $x$ is blue and $x'$ is green. Therefore in a satisfying assignment, any row must have at least one position colored with $rr$. Now for the columns we note that if some pigeon occurs in a position $(i, j)$ then the second color in each entry below $(i, j)$ must be green, while the second color of each entry above $(i, j)$ must be blue. Therefore no two pigeons are allowed to appear on the same column of a satisfying assignment.

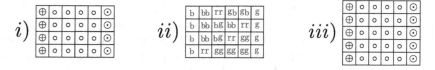

**Fig. 2.** $(i)$ The pigeonhole picture $P^{4,4}$. $(ii)$ A solution for $P^{4,4}$. $(iii)$ The pigeonhole picture $P^{5,4}$ is unsatisfiable. (Color figure online)

We note that if $m > n$, then pigeonhole picture $P^{m,n}$ is unsatisfiable. The following theorem says that the family of pigeonhole pictures is polynomially smooth. This implies that our algorithm for the satisfiability of smooth pictures can be used to decide the unsatisfiability of the pigeonhole pictures in polynomial time.

**Theorem 5.2.** *Let $P^{m,n}$ be the $(m, n)$-pigeonhole picture. Then for each $i, j \in \{1, .., m\} \times \{1, ..., n+1\}$, the boundary $\partial_{ij}(P^{m,n}, \pi, V, H)$ is accepted by an LDFA of size at most $O(m \cdot n)$.*

*Proof.* Let $M$ be a solution for $P^{m,n}$. If $M_{i,j} = y_1 y_2$ where $i \in \{2, ..., n+1\}$, then we say that $y_1$ is the first coordinate of $M_{i,j}$ while $y_2$ is the second coordinate of $M_{i,j}$.

We sketch the proof in the case in which $j = n + 2$. In other words we will show how to construct the boundaries corresponding to full rows. The generalization for smaller values of $j$ is straightforward. Thus let $M$ be an $(i, n + 2)$ partial solution. Then the border of $M$ is simply its $i$-th row. Since $M$ is a partial solution, there exists a unique $k$ such that $M_{i,k} = rr$. Additionally, if $k' < k$ then the first coordinate of $M_{i,k'}$ is necessarily blue, indicating that a pigeon has not occurred in that row up to position $k'$. On the other hand, if $k' > k$, then the first coordinate of $M_{i,k'}$ is green, indicating that a pigeon has already occurred at row $i$. Now, we also have that precisely $i$ pigeons must be present in $M$, one for each row. Therefore, there must exist precisely $i - 1$ entries of row $i$ whose second coordinate is green. Indeed, for an entry $M_{i,k'}$ with $k' \neq k$, if the second coordinate of $M_{i,k'}$ is green, then we know that a pigeon has already occurred at column $k'$. On the other hand, if $M_{i,k'}$ is blue then we know that no pigeon occurred yet at column $k'$. It turns out that the converse also holds. Namely, if each two consecutive entries of row $i$ satisfy the horizontal constraints imposed by the horizontal relation $H$ and there exists a unique entry which is equal to $rr$ and precisely $i - 1$ entries whose second coordinate is green, then we known that all other entries of $M$ can be filled in such a way that it is an $(i, n + 2)$ solution. In other words, the strings $w \in \Sigma^{n+2}$ belonging to the border of an $(i, n + 2)$ partial solution are characterized by the following properties.

1. The first entry of $w$ is $b$ and the last entry is $g$.
2. $(w_k, w_{k+1}) \in H$ for each $k \in \{1, ..., n + 1\}$
3. There exists a unique $k$ such that $w_k = rr$
4. $w$ has precisely $i - 1$ entries whose second coordinate is green.

Now one can implement a leveled deterministic finite automaton with $O(i \cdot n)$ transitions and levels $Q_0, Q_1, ..., Q_{n+1}$ which accepts a string $w \in \Sigma^{n+2}$ if and only if the conditions above are satisfied. Note that the three first conditions are immediate to verify using such an automaton. The fourth condition can be implemented by considering that each level $Q_k$ is split into subsets of states $Q_k^r$ for $r \in \{1, ..., i\}$, where the states in $Q_k^r$ indicate that from the $k$ first read symbols of $w$, $r$ of them have the second coordinate green. $\qquad\square$

**Corollary 5.3.** *Let $P^{m,n}$ be the pigeonhole picture. Then the algorithm devised in the proof of Theorem 4.1 determines in time $O(m^4 \cdot n^4)$ whether $P^{m,n}$ has a $(\pi, V, H)$-solution. In case such a solution exists the algorithm constructs it.*

*Proof.* By Theorem 5.2, $P^{m,n}$ is $O(m \cdot n)$-smooth. Additionally, the extension number of $(\pi, V)$ is $e(\pi, V) = 3$. Therefore from Theorem 4.6, we can construct each automaton $\mathcal{A}_{i,j}(P^{m,n}, \pi, V, H)$ in time at most $O(m^3 \cdot n^3)$. Since there are $m \cdot n$ such automata, the whole algorithm takes time at most $O(m^4 \cdot n^4)$. $\qquad\square$

# 6    From Pictures to Constant Width CNF Formulas

Let $\pi : \Sigma \to \Gamma$ be a function, $V, H \subseteq \Sigma \times \Sigma$ be binary relations over $\Sigma$, and $M$ be an $m \times n$-picture over $\Gamma$. Next we define a constant width CNF formula $F(M)$ that is satisfiable if and only if $M$ is $(\pi, V, H)$-satisfiable.

Let $S_{ij} = \pi^{-1}(M_{ij})$ be the set of colored versions of the symbol $M_{ij}$. The formula $F(M)$ has a variable $x_{ija}$ for each $(i,j) \in [m] \times [n]$, and each symbol $a \in S_{ij}$. Intuitively, the variable $x_{ija}$ is true if the position $(i,j)$ of a solution picture is set to $a$. The following set of clauses specifies that in a satisfying assignment, precisely one symbol of $S_{ij}$ occupies the position $(i,j)$.

$$\text{OneSymbol}(M, i, j) \equiv \bigvee_{s \in S_{ij}} x_{ijs} \ \wedge \bigwedge_{s,s' \in S_{ij}, s \neq s'} (\overline{x}_{ijs} \vee \overline{x}_{ijs'}) \tag{6}$$

The next set of clauses expresses the fact that no pair of symbols $(a, a') \notin H$ occur in consecutive horizontal positions at row $i$.

$$\text{Horizontal}(M, i) \equiv \bigwedge_{(a,a') \notin H, j \in \{1,...,n-1\}} (\overline{x}_{ija} \vee \overline{x}_{i(j+1)a'}) \tag{7}$$

Similarly, the following set of of clauses expresses the fact that that no pair of symbols $(a, a') \notin V$ occurs in consecutive vertical positions at column $j$.

$$\text{Vertical}(M, j) \equiv \bigwedge_{(a,a') \notin V, i \in \{1,...,m-1\}} (\overline{x}_{ija} \vee \overline{x}_{(i+1)ja'}) \tag{8}$$

Finally, we set the formula $F(M)$ as follows.

$$F(M) \equiv \bigwedge_{i=1}^{m} \text{Horizontal}(M, i) \wedge \bigwedge_{j=1}^{n} \text{Vertical}(M, j) \wedge \bigwedge_{ij} \text{OneSymbol}(M, i, j) \tag{9}$$

## 7    Lower Bound for Bounded Depth Frege Proofs

Let $P^{m,m-1}$ be the pigeonhole pictures as defined in Sect. 5. In this section we will show that bounded-depth Frege refutations of the family of formulas $\{F(P^{m,m-1})\}_{m \in \mathbb{N}}$ require exponential size. Recall that a Frege system is specified by a finite set of rules of the form

$$\frac{\varphi_0(q_1, ..., q_m)}{\varphi_1(q_1, ..., q_m), ..., \varphi_r(q_1, ..., q_m)} \tag{10}$$

where $q_1, ..., q_m$ are variables and $\varphi_0, \varphi_1, ..., \varphi_r$ are formulas in the language $\vee$, $\wedge$, $\neg$, $0$, $1$, and $q_1, ..., q_m$. The only requirement is that the rules are sound and complete [6]. An instance of the rule is obtained by substituting particular formulas $\psi_1, ..., \psi_m$ (in the language $\vee$, $\neg$, $0$, $1$, $y_{ij}$) for the variables $q_1, ..., q_m$. A rule in which $r = 0$ is called an axiom scheme. The size of a formula $F$ is the number of symbols $\{\vee, \wedge, \neg\}$ in it. The depth of a formula is the size of the longest path from the root of $F$ to one of its leaves. We say that a proof has depth $d$ if all formulas occurring in it have depth at most $d$.

Now consider the family of pigeonhole formulas $\{H_m\}_{m \in \mathbb{N}}$. For each $m \in \mathbb{N}$, $H_m$ has variables $y_{ij}$ for $i \in \{1, ..., m\}$ and $j \in \{1, ..., m-1\}$ and the following clauses.

1. $(y_{i,1} \vee ... \vee y_{i,m-1})$ for each $i \in \{1, ..., m\}$.
2. $(\neg y_{i,j} \vee \neg y_{k,j})$ for each $i, k \in \{1, ..., m\}$, $j \in \{1, ..., m-1\}$.

Intuitively, when set to true, the variable $y_{i,j}$ indicates that pigeon $i$ sits at hole $j$. Clauses of the first type specify that each pigeon has to sit in at least one hole, clauses of the second type specify that no two distinct pigeons sit in the same hole. Clearly, the formula $H_m$ is unsatisfiable for each $m \in \mathbb{N}$. The following theorem states that $H_m$ is hard for bounded depth Frege systems.

**Theorem 7.1** ([13,18]). *For each Frege proof system $F$ there exists a constant $c$ such that for each $d$, and each sufficiently large $m$, every depth-$d$ refutation of $H_m$ must have size at least $2^{m^{c^{-d}}}$.*

Now let $F(P^{m,m-1})$ be the formula derived from the pigeonhole picture $P^{m,m-1}$. This formula has the following variables[1]:

$$x_{i,j,bb}, \quad x_{i,j,bg}, \quad x_{i,j,gb}, \quad x_{i,j,gg} \quad i \in \{1, ..., m\}, j \in \{2, ..., m\} \tag{11}$$

$$x_{i,1,b}, \quad x_{i,m+1,g}, \quad \text{for } i \in \{1, ..., m\} \tag{12}$$

We will consider a substitution of variables that transforms the formula $F(P^{m,m-1})$ into a formula $F'$ in the variables $y_{ij}$ used by the formula $H_m$. Intuitively, the variable $x_{i,j,rr}$ which expresses that the pigeon $i$ is being placed at column $j$ in a solution for the picture $P^{m,m-1}$, is mapped to the variable $y_{i,j-1}$ which expresses that the pigeon $i$ is placed at hole $j-1$. Note that the $j$-th column of a solution for the picture $P^{m,m-1}$ is the $(j-1)$-th hole. The other variables are then mapped to a conjunction of disjunctions. For instance, the variable $x_{i,j,bb}$ is true in an hypothetical satisfying assignment of $F(P^{m,m-1})$ if and only if the pigeon at row $i$ occurs after the $j$-th entry of this row, and the pigeon at column $j$ appears after the $i$-th entry of this column. Analogue substitutions can be made with respect to the other variables. These substitutions are formally specified below.

1. For $i \in \{1, ..., m\}$:
   (a) $x_{i,1,b} \rightarrow (y_{i,1} \vee ... \vee y_{i,m-1})$
   (b) $x_{i,(m+1),g} \rightarrow (y_{i,1} \vee ... \vee y_{i,m-1})$
2. For $i \in \{1, ..., m\}$, $j \in \{2, ..., m\}$
   (a) $x_{i,j,rr} \rightarrow y_{i,(j-1)}$
   (b) $x_{i,j,bb} \rightarrow (y_{i,j} \vee ... \vee y_{i,m-1}) \wedge (y_{i+1,j} \vee ... \vee y_{m,j})$
   (c) $x_{i,j,bg} \rightarrow (y_{i,j} \vee ... \vee y_{i,m-1}) \wedge (y_{1,j} \vee ... \vee y_{i-1,j})$
   (d) $x_{i,j,gb} \rightarrow (y_{i,1} \vee ... \vee y_{i,j-2}) \wedge (y_{i+1,j} \vee ... \vee y_{m,j})$
   (e) $x_{i,j,gg} \rightarrow (y_{i,1} \vee ... \vee y_{i,j-2}) \wedge (y_{1,j} \vee ... \vee y_{i-1,j})$

---

[1] Recall that the first and last columns of the pigeonhole picture do not correspond to holes.

Let $F'$ be the formula that is obtained from $F(P^{m,m-1})$ by replacing its variables according to the substitutions defined above. Then the formula $F'$ has only variables $y_{ij}$. Additionally, the implication $H_m \Rightarrow F'$ can be proved by a bounded depth Frege proof of polynomial size. Now suppose that $\Pi$ is a depth-$d$ Frege refutation of the formula $F(P^{m,m-1})$ of size $S$. Then, if we replace all variables occurring in formulas of $\Pi$ according to the substitutions above, we get a depth $d+2$ Frege refutation $\Pi'$ of the formula $F'$ whose size is at most $O(m) \cdot S$. But since the implication $H_m \Rightarrow F'$ has a Frege proof of size $poly(m)$, we have that $\Pi'$ also can be used to construct a refutation of $H_m$ of size $poly(m) \cdot S$. Therefore by Theorem 7.1, the size of $S$ must be at least $2^{m^{c'-d}}$ for some constant $c'$ independent on $m$. $\qquad\qquad\square$

**Acknowledgments.** This work was supported by the European Research Council, grant number 339691, in the context of the project Feasibility, Logic and Randomness (FEALORA). The author thanks Pavel Pudlák and Neil Thapen for enlightening discussions on Frege proof systems.

# References

1. Beame, P., Kautz, H., Sabharwal, A.: Towards understanding and harnessing the potential of clause learning. J. Artif. Intell. Res. **22**, 319–351 (2004)
2. Beame, P., Pitassi, T.: Simplified and improved resolution lower bounds. In: Proceedings of the 37th Annual Symposium on Foundations of Computer Science, pp. 274–282. IEEE (1996)
3. Buss, S.R., et al.: Resolution proofs of generalized pigeonhole principles. Theoret. Comput. Sci. **62**(3), 311–317 (1988)
4. Buss, S.R., Hoffmann, J., Johannsen, J.: Resolution trees with lemmas: resolution refinements that characterize DLL algorithms with clause learning. Logical Meth. Comput. Sci. **4**, 1–18 (2008)
5. Cherubini, A., Reghizzi, S.C., Pradella, M., San, P.: Picture languages: Tiling systems versus tile rewriting grammars. Theoret. Comput. Sci. **356**(1), 90–103 (2006)
6. Cook, S.A., Reckhow, R.A.: The relative efficiency of propositional proof systems. J. Symbol. Logic **44**(01), 36–50 (1979)
7. Eén, N., Sörensson, N.: An extensible SAT-solver. In: Giunchiglia, E., Tacchella, A. (eds.) SAT 2003. LNCS, vol. 2919, pp. 502–518. Springer, Heidelberg (2004)
8. Giammarresi, D., Restivo, A.: Recognizable picture languages. Int. J. Pattern Recogn. Artif. Intell. **6**(2&3), 241–256 (1992)
9. Haken, A.: The intractability of resolution. Theoret. Comput. Sci. **39**, 297–308 (1985)
10. Hertel, P., Bacchus, F., Pitassi, T., Van Gelder, A.: Clause learning can effectively P-simulate general propositional resolution. In: Proceedings of the 23rd National Conference on Artificial Intelligence (AAAI 2008), pp. 283–290 (2008)
11. Kim, C., Sudborough, I.H.: The membership and equivalence problems for picture languages. Theoret. Comput. Sci. **52**(3), 177–191 (1987)
12. Krajíček, J.: Lower bounds to the size of constant-depth propositional proofs. J. Symbol. Logic **59**(01), 73–86 (1994)

13. Krajíček, J., Pudlák, P., Woods, A.: An exponential lower bound to the size of bounded depth frege proofs of the pigeonhole principle. Random Struct. Algorithms **7**(1), 15–39 (1995)
14. Latteux, M., Simplot, D.: Recognizable picture languages and domino tiling. Theoret. Comput. Sci. **178**(1), 275–283 (1997)
15. Maurer, H.A., Rozenberg, G., Welzl, E.: Using string languages to describe picture languages. Inf. Control **54**(3), 155–185 (1982)
16. Moskewicz, M.W., Madigan, C.F., Zhao, Y., Zhang, L., Malik, S.: Chaff: engineering an efficient SAT solver. In: Proceedings of the 38th Annual Design Automation Conference, pp. 530–535. ACM (2001)
17. Pipatsrisawat, K., Darwiche, A.: On the power of clause-learning SAT solvers as resolution engines. Artif. Intell. **175**(2), 512–525 (2011)
18. Pitassi, T., Beame, P., Impagliazzo, R.: Exponential lower bounds for the pigeonhole principle. Comput. Complex. **3**(2), 97–140 (1993)
19. Raz, R.: Resolution lower bounds for the weak pigeonhole principle. J. ACM (JACM) **51**(2), 115–138 (2004)
20. Revuz, D.: Minimisation of acyclic deterministic automata in linear time. Theoret. Comput. Sci. **92**(1), 181–189 (1992)
21. Rosenfeld, A.: Picture Languages: Formal Models for Picture Recognition. Academic Press (2014)
22. Simplot, D.: A characterization of recognizable picture languages by tilings by finite sets. Theoret. Comput. Sci. **218**(2), 297–323 (1999)
23. Stromoney, G., Siromoney, R., Krithivasan, K.: Abstract families of matrices and picture languages. Comput. Graph. Image Process. **1**(3), 284–307 (1972)

# Solution-Graphs of Boolean Formulas and Isomorphism

Patrick Scharpfenecker$^{(\boxtimes)}$ and Jacobo Torán

Institute of Theoretical Computer Science, University of Ulm, Ulm, Germany
{patrick.scharpfenecker,jacobo.toran}@uni-ulm.de

**Abstract.** The solution graph of a Boolean formula on $n$ variables is the subgraph of the hypercube $H_n$ induced by the satisfying assignments of the formula. The structure of solution graphs has been the object of much research in recent years since it is important for the performance of SAT-solving procedures based on local search. Several authors have studied connectivity problems in such graphs focusing on how the structure of the original formula might affect the complexity of the connectivity problems in the solution graph.

In this paper we study the complexity of the isomorphism problem of solution graphs of Boolean formulas and we investigate how this complexity depends on the formula type.

We observe that for general formulas the solution graph isomorphism problem can be solved in exponential time while in the cases of 2CNF formulas, as well as for CPSS formulas, the problem is in the counting complexity class C$_=$P, a subclass of PSPACE. We also prove a strong property on the structure of solution graphs of Horn formulas showing that they are just unions of partial cubes.

In addition we give a PSPACE lower bound for the problem on general Boolean functions. We prove that for 2CNF, as well as for CPSS formulas the solution graph isomorphism problem is hard for C$_=$P under polynomial time many one reductions, thus matching the given upper bound.

**Keywords:** Solution graph · Isomorphism · Counting · Partial cube

## 1 Introduction

Schaefer provided in [17] a well known dichotomy result for the complexity of the satisfiability problem on different classes of Boolean formulas. He showed that for formulas constructed from specific Boolean functions (now called Schaefer functions), satisfiability is in P while for all other classes, satisfiability is NP-complete. Surprisingly, there are no formulas of intermediate complexity.

More recently, Gopalan et al. and Schwerdtfeger [9,19] uncovered a similar behavior for connectivity problems on solution graphs of Boolean formulas.

P. Scharpfenecker—Supported by DFG grant TO 200/3-1.

N. Creignou and D. Le Berre (Eds.): SAT 2016, LNCS 9710, pp. 29–44, 2016.
DOI: 10.1007/978-3-319-40970-2_3

The solution graph of a Boolean formula on $n$ variables is the subgraph of the $n$-dimensional hypercube induced by all satisfying assignments. The study of solution graphs of Boolean formulas has been the object of important research in recent years, especially for the case of random formula instances. It has been observed both empirically and analytically that the solution space breaks in many small connected components as the ratio between variables and clauses in the considered formulas approaches a critical threshold [1,15]. This phenomenon explains the better performance on random formulas of SAT-solvers based on message passing with decimation than those based on local search or DPLL procedures (see e.g. [8]). The motivation behind the works of [9,19] was to obtain new information about the connectivity properties of the solution space for different types of Boolean formulas. Introducing some new classes of Boolean functions, they were able to prove a dichotomy result for the $st$-connectivity problem [9], as well as a trichotomy result for connectivity [19]. For different formula classes the complexity of the connectivity problem is either in P, or complete for coNP or for PSPACE while for $st$-connectivity it is either in P or PSPACE-complete.

In this paper we look further in the solution space of Boolean formulas studying the complexity of the isomorphism of their solution graphs. In other words, we consider the following natural questions: given two Boolean formulas, how hard is it to test if their solution graphs are isomorphic? Does the complexity of the problem depend on the structure of the formula? Observe that isomorphism of solution graphs is a very strong concept of equivalence between formulas, stronger than Boolean isomorphism [2] and stronger than saying that both formulas have the same number of satisfying assignments. Since the complexity of the general graph isomorphism problem, GI, is not completely settled (see [13]), one might expect that it would be hard to obtain a complete classification for solution graph isomorphism. We show in fact that for different types of Boolean formulas, the complexity of the isomorphism problem on their solution graphs varies. We also characterize completely the complexity of the problem for some types of Boolean formulas. For solution graphs of 2CNF formulas, isomorphism of a single connected component is exactly as hard as testing Graph Isomorphism. For a collection of such components (encoded by a single 2CNF formula), the isomorphism problem is complete for the complexity class $C_=P$, a complexity class defined in terms of exact counting. This means that deciding isomorphism of the solution graphs of 2CNF formulas is exactly as hard as testing if two such formulas have the same number of satisfying assignment. This result also holds for the more general class of CPSS formulas (definitions in the preliminaries section), showing that for this class of formulas isomorphism and counting have the same complexity. For the upper bound we use a recent result on the isometric dimension of partial cubes [18], the fact that GI is low for the class $C_=P$ [12], as well as the closure of this class under universal quantification [10]. The hardness property uses a result of Curticapean [7], where it is proven that $SamePM$, the problem to decide if two given graphs have the same number of perfect matchings is complete for $C_=P$. We show that this problem can be reduced to the verification of whether two 2CNF formulas have the same number of satisfying solutions, implying that this problem and even *Iso(CPSS)*, the isomorphism problem of CPSS solution graphs, are complete for $C_=P$.

For the other types of formulas used in [9,19], built from Schaefer, safely tight and general functions, we observe that the corresponding solution graph isomorphism problems can be solved in EXP, thus improving the trivial NEXP upper bound.

For classes of functions that are not safely tight, we can also improve the $C_=P$ lower bound and show that the isomorphism problem for their solution graphs is in fact hard for PSPACE.

Figure 1 summarizes the complexity results for isomorphism of solution graphs for specific classes of formulas.

| Function | Hard for | Upper bound |
|---|---|---|
| CPSS | $C_=P$ | $C_=P$ |
| Schaefer, not CPSS | $C_=P$ | EXP |
| safely tight, not Schaefer | $C_=P$ | EXP |
| not safely tight | PSPACE | EXP |

**Fig. 1.** Classification of isomorphism problems.

While we could not improve the EXP upper bound for the isomorphism of solution graphs corresponding to Horn formulas, we prove a strong new property for the structure of such graphs which might help to develop a non-trivial isomorphism algorithm. We show that the set of solutions between a locally minimal and locally maximal solution is a partial cube. Therefore a solution graph can be seen as taking a partial cube for every locally maximal solution and glueing them together.

While there is no direct connection between the isomorphism problem for solution graphs and SAT-solving methods, the study of isomorphism questions provides new insights on the structure of solution graphs and on the number of satisfying assignments for certain formula classes that might be useful in further SAT-related research.

## 2   Preliminaries

For two words $x, y \in \{0,1\}^n$, $\Delta(x,y)$ denotes the Hamming-distance between them. We associate words in $\{0,1\}^n$ with subsets of $[n] = \{1,\ldots,n\}$ in the standard way.

We mostly deal with undirected graphs without self-loops. For such a graph $G = (V, E)$ with vertex set $V = [n]$ and edge set $E \subseteq \binom{V}{2}$, its simplex graph (see e.g. [4]) is defined as $simplex(G) = (V', E')$ with $V'$ as the set of all cliques (including the empty clique) in $G$ and $E' = \{\{u,v\} \in \binom{V'}{2} \mid \Delta(u,v) = 1\}$.

So $G' = simplex(G)$ is the set of all cliques of $G$ and two cliques are connected iff they differ (considered as strings of $\{0,1\}^n$) in one element. We will only consider the simplex graph of bipartite graphs. As these graphs have only cliques of size at most 2, $|V'| = |V| + |E| + 1$. The graph $G'$ contains all original nodes $V$, a node $u = \{i,j\}$ for every edge $\{i,j\} \in G$ which is connected to $\{i\}$ and $\{j\}$ and a new node $o = \emptyset$ which is connected to all original nodes.

Two graphs $G = (V, E)$ and $H = (V', E')$ with $V = V' = [n]$ are isomorphic iff there is a bijection $\pi : V \to V'$ such that for all $u, v \in V : (u, v) \in E \Leftrightarrow (\pi(u), \pi(v)) \in E'$. If such a bijection exists we write $G \cong H$, if not, $G \not\cong H$. The graph isomorphism problem (GI) is the decision problem of whether two given graphs are isomorphic. Given a class of graphs $C$, *Iso(C)* denotes the graph isomorphism problem on graphs in $C$.

The Boolean isomorphism problem consists in deciding, given two Boolean formulas $F$ and $G$ on variables $x_1, \ldots, x_n$, whether there is a signed permutation $\pi$ of the $n$ variables such that for all $x \in \{0,1\}^n$, $F(x_1, \ldots, x_n) = G(\pi(x_1), \ldots, \pi(x_n))$.

We deal with different classes of formulas. 2CNF denotes the class of formulas in conjunctive normal form and with exactly two literals per clause. For a 2CNF formula $F(x_1, \ldots, x_n)$ we define the directed implication graph $I(F) = (V, E)$ on nodes $V = \{x_1, \ldots, x_n, \overline{x_1}, \ldots, \overline{x_n}\}$ and edges $(k, l) \in E$ with $k, l \in V$ iff there is no solution to $F$ which falsifies the clause $(k \to l)$. By replacing all variables in a cycle with a single variable we get the reduced implication graph $RI(F)$. We say that a 2CNF formula $F$ is reduced if $I(F) = RI(F)$.

We deal mostly with standard complexity classes like P, NP, EXP and NEXP. A class that might not be so familiar is the counting class $C_=P$ [22]. This consists of the class of problems $A$ for which there is a nondeterministic polynomial time Turing machine $M$ and a polynomial time computable function $f$ such that for each $x \in \{0,1\}^*$, $x \in A$ iff the number of accepting paths of $M(x)$ is exactly $f(x)$. The standard complete problem for $C_=P$ is ExactSAT: given a Boolean formula $F$ and a number $k$, does $F$ have exactly $k$ satisfying assignments?

## 2.1 Solution Graphs of Boolean Formulas

Intuitively, a solution graph for a given Boolean formula is the induced subgraph on all satisfying solution represented in a host graph. In this paper we only consider induced subgraphs of the $n$-dimensional hypercube $H_n$ which is the graph with $V = \{0,1\}^n$ and $E = \{\{u, v\} | \Delta(u, v) = 1\}$.

**Definition 1.** *Let $F(x_1, \ldots, x_n)$ be an arbitrary Boolean formula. Then the solution graph $G_F$ is the subgraph of the $n$-dimensional hypercube $H_n$ induced by all satisfying solutions $x$ of $F$.*

Note that two satisfying solutions are connected by an edge iff their Hamming distance is one. For a set of Boolean formulas $D$ (for example $D = 2CNF$), *Iso(D)* denotes the isomorphism problem on the class of solution graphs of $D$-formulas.

Given a graph $G$ and two nodes $u, v$, $d(u, v)$ is the length of the shortest path between $u$ and $v$ in $G$ or $\infty$ if there is no such path.

**Definition 2.** *An induced subgraph $G$ of $H_n$ is a partial cube iff for all $x, y \in G$, $d(x, y) = \Delta(x, y)$. We call such an induced subgraph "isometric". The isometric dimension of a graph $G$ is the smallest $n$ such that $G$ embeds isometrically into $H_n$.*

**Definition 3.** *A graph $G = (V, E)$ is a median graph iff for all nodes $u, v, w \in V$ there is a unique $b \in V$ which lies on the shortest paths between $(u, v)$, $(u, w)$ and $(v, w)$. Then $b$ is called the median of $u, v$ and $w$.*

For any Boolean function $F : \{0, 1\}^n \rightarrow \{0, 1\}$ we can represent $F$ with the subset of all its satisfying assignments in $\{0, 1\}^n$. A Boolean function $F \subseteq \{0, 1\}^n$ is closed under a ternary operation $\odot : \{0, 1\}^3 \rightarrow \{0, 1\}$ iff $\forall x, y, z \in F : \odot(x, y, z) := (\odot(x_1, y_1, z_1), \ldots, \odot(x_n, y_n, z_n)) \in F$. Note that we abuse the notation of a ternary operation to an operation on three bit-vectors by applying the operation bitwise on the three vectors. For $R$ a set of Boolean functions with arbitrary arities (for example $R = \{(\overline{x} \vee y), (x \oplus y), (x \oplus y \oplus z)\}$, we define $SAT(R)$ to be the satisfiability problem for all Boolean formulas which are conjunctions of instantiations of functions in $R$. For the given example $R$, $F(x, y, z) = (\overline{z} \vee y) \wedge (x \oplus y)$ is a formula in which every clause is an instantiation of an $R$-function. Similarly, $Conn(R)$ $(stConn(R))$ is the connectivity (reachability) problem, given a conjunction $F$ of $R$-functions (and $s, t$), is the solution graph connected (is there a path from $s$ to $t$). We mostly use $F$ for Boolean formulas/functions and $R, S$ for sets of functions.

Note that $r \in R$ can be an arbitrary Boolean function as for example $r = (x \oplus y)$ or $r = (x \vee \overline{y} \vee \overline{z}) \wedge (\overline{x} \vee z)$. With $Horn_n$ we define the set of all Horn-clauses of size up to $n$. The ternary majority function $maj : \{0, 1\}^3 \rightarrow \{0, 1\}$ is defined as $maj(a, b, c) = (a \wedge b) \vee (a \wedge c) \vee (b \wedge c)$.

In the next definitions we recall some terms introduced in [9,19].

**Definition 4.** *A Boolean function $F$ is*

- *bijunctive, iff it is closed under $maj(a, b, c)$.*
- *affine, iff it is closed under $a \oplus b \oplus c$.*
- *Horn, iff it is closed under $a \wedge b$.*
- *dual-Horn, iff it is closed under $a \vee b$.*
- *IHSB−, iff it is closed under $a \wedge (b \vee c)$.*
- *IHSB+, iff it is closed under $a \vee (b \wedge c)$.*

*A function has such a property componentwise, iff every connected component in the solution graph is closed under the corresponding operation. A function $F$ has the additional property "safely", iff the property still holds for every function $F'$ obtained by identification of variables[1].*

In the case of Horn-formulas, the usual definition (the conjunction of Horn-clauses, clauses with at most one positive literal) implies that the represented functions are Horn.

---

[1] Identifying two variables corresponds to replacing one of them with the other variable.

**Definition 5.** *A set of functions $R$ is Schaefer (CPSS) if at least one of the following conditions holds:*

- *every function in $R$ is bijunctive.*
- *every function in $R$ is Horn (and safely componentwise IHSB−).*
- *every function in $R$ is dual-Horn (and safely componentwise IHSB+).*
- *every function in $R$ is affine.*

If we have a Boolean formula $F$ which is built from a set $R$ of CPSS functions we say that $F$ is CPSS. Clearly, every CPSS formula is Schaefer. We later use a bigger class of functions called *safely tight*. This class properly contains all Schaefer sets of functions.

**Definition 6.** *A set $R$ of functions is (safely) tight if at least one of the following conditions holds:*

- *every function in $R$ is (safely) componentwise bijunctive.*
- *every function in $R$ is (safely) OR-free.*
- *every function in $R$ is (safely) NAND-free.*

*A function is OR-free if we can not derive $(x \vee y)$ by fixing variables. Similarly, a function is NAND-free if we can not derive $(\overline{x} \vee \overline{y})$ by fixing variables.*

## 3    Isomorphism for Solution Graphs

We now turn our attention to the isomorphism problem on solution graphs. In general the solution graph of a formula can have an exponential number of connected components and each component might be of exponential size (in the formula size). The NP upper bound for GI translates directly into a NEXP upper bound for the isomorphism of solution graphs.

Based on the celebrated new algorithm from Babai for Graph Isomorphism [3] running in time $n^{\log^{O(1)}}$, it is not hard to see that the isomorphism of solution graphs is in EXP: for two given Boolean formulas on $n$ variables, we can construct explicitly their solution graphs in time $O(2^n)$ and then apply Babai's algorithm on them, resulting in a $2^{n^{O(1)}}$ algorithm. But we do not need such a strong result, the algorithm of Luks for testing isomorphism of bounded degree graphs [14] suffices.

**Proposition 7.** *The problem to decide for two given Boolean formulas whether their respective solution graphs are isomorphic is in EXP.*

*Proof.* Luks [14] gave an algorithm for graph isomorphism with time-complexity $|V|^{deg(G)}$. A solution graph embedded in the hypercube $H_n$ has degree at most $n - 1$. The running time of Luks algorithm on such graphs is bounded by $2^{n^2}$. □

By restricting the encoding formula, we can get better upper bounds. Theorem 13 will show that the isomorphism problem for $CPSS$ encoding formulas is in $C_{=}P$, a subclass of PSPACE. For this, we need the following two results.

**Theorem 8** [18]. *Given a CPSS function $F(x_1, \ldots, x_n)$, every connected component of $F$ is a partial cube of isometric dimension at most $n$.*

**Theorem 9** [16], **Theorem 5.72.** *For any two finite isomorphic partial cubes $G_1$ and $G_2$ on a set $X$, there is an automorphism of the cube $H(X)$ that maps one of the partial cubes onto the other. Moreover, for any isomorphism $\alpha : G_1 \to G_2$, there is a Boolean automorphism $\sigma : H(X) \to H(X)$ such that $\sigma$ on $G_1$ is exactly $\alpha$.*

We note that isomorphism of (explicitly given) partial cubes is already GI-complete. The hardness follows from the observation that for any graph, its simplex is a median graph. The other two facts we need is that median graphs are partial cubes (see e.g. [16], Theorem 5.75), and the fact that a given pair of graphs $G, H$, can be transformed in logarithmic space into a pair of bipartite graphs $G', H'$ so that $simplex(G') \cong simplex(H')$ iff $G \cong H$.

To see this we first suppose that for two given general graphs $G, H$ we know that $|E| \neq |V|$. This could easily be enforced in an isomorphism-preserving logspace reduction. In a next step, we replace each edge $(u, v)$ in both graphs with the gadget $(u, z_{u,v}), (z_{u,v}, v)$ where $z_{u,v}$ is a new vertex. This yields two new bipartite graphs $G', H'$ which are isomorphic iff $G$ and $H$ were isomorphic. But then $simplex(G') \cong simplex(H')$ iff $G \cong H$. This implies the following Lemma.

**Lemma 10.** *Isomorphism for median graphs is GI-complete under logarithmic space many-one reductions.*

Note that it is known that median graphs can be exactly embedded as a solution graph of a reduced 2CNF formula (see e.g. [5]). Lemma 10 gives therefore an alternative reduction to the one given in [6] between Boolean isomorphism for 2CNF formulas and GI. With Theorem 9 we get:

**Corollary 11.** *The Isomorphism Problem for reduced 2CNF solution graphs is GI-complete under logarithmic space many-one reductions.*

*Proof.* The hardness part follows from the observation given above. By Theorem 9 two partial cubes are isomorphic iff there is an automorphism of the whole hypercube mapping one partial cube to the other. But such an automorphism is just a Boolean automorphism of the Boolean function. For general boolean formulas this problem is hard for NP and in $\Sigma_2$ [2], but for Schaefer-formulas, which contain 2CNF formulas, this problem can be reduced in polynomial time to GI (see [6]) by creating a unique normal form and looking for a syntactic isomorphism of the formulas.                                                                    □

This basically tells us that even if we look at two exponentially sized, isomorphic partial cubes embedded in the hypercube $H_n$, finding an isomorphism is as easy as finding a Boolean isomorphism. The problem is more complex when the solution graphs might have more than one connected component. We face the additional problem that single connected components may not have a single formula representing just this component. For the isomorphism of solution graphs of CPSS functions we will show an upper bound of $C_=P$. For this we need the following Lemma showing that the problem of testing if there is an isomorphism between two connected components which maps a given solution to another given solution, can be reduced to GI.

**Lemma 12.** *The following problem is reducible to GI: given CPSS functions $F$ and $G$ and two satisfying solutions $s$ and $t$, decide whether there is an isomorphism $\pi$ between the connected components containing $s$ and $t$ with $\pi(s) = t$.*

*Proof.* We know by Theorem 9 that if two partial cubes are isomorphic, then there is always a Boolean isomorphism[2]. One could easily guess a candidate permutation of variables for the isomorphism. But it is not clear how to verify that this permutation is in fact an isomorphism. To reduce this problem to GI we would have to extract a single connected component and create a formula which contains only this subgraph. In general, this is not possible. We use the construction depicted in Fig. 2 to achieve such an extraction which is enough in the case of isomorphism.

We describe this construction which basically performs a walk on the solution graph beginning at a given node $s$. We use several blocks of variables. Given the original variables $x = (x_1, \ldots, x_n)$, we create new blocks of variables $x^i$ and $x^{i,j}$ for $i \in \{0, \ldots, w\}$ and $j \in \{1, \ldots, n\}$ ($w$ to be fixed later), each containing $n$ variables. For example $x^0 = (x_1^0, \ldots, x_n^0)$ and $x^{0,4} = (x_1^{0,4}, \ldots, x_n^{0,4})$. We fix the first block of variables $x^0$ to $s \in \{0,1\}^n$. We then add $n$ new blocks $x^{0,1}, \ldots, x^{0,n}$ such that every $x^{0,j}$ may only differ from $x^0$ in bit $j$. If $x_j^0 = 0$ we add the clause $(x_j^0 \to x_j^{0,j})$, if $x_j^0 = 1$ we add the clause $(\overline{x_j^0} \to \overline{x_j^{0,j}})$. This will not be relevant in the first step as these clauses are obviously satisfied but this ensures that a walk never returns to $s$: if we add these clauses for all later steps and for example for the case $x_j^0 = 0$ there is an $i$ such that $x_j^i = 1$, then for all $i' > i$, $x_j^{i'} \neq 0$.

All other variables have to be equivalent to the variables in the previous block (or the previous variables could get reused in $x_j^{0,j}$, except $x_j^0$). In addition we add for every $j$ the clauses $F(x^{0,j})$ to ensure that every following block of $x^0$ satisfies $F$. Obviously, every $x^{0,j}$ has distance at most 1 from $x^0 = s$ and is a node in the solution graph. We then add a new block $x^1$ such that $x_j^1 = x_j^{0,j}$. This performs all steps of the previous branching-step in parallel and we require $x^1$ to satisfy $F$.

Although all nodes visited in the branching-step have distance at most 1 from $s$, the nodes described in $x^1$ have distance $\sum_{j \in [n]} d(x^0, x^{0,j})$. Therefore $x^1$ may not be in the same connected component as $x^0$. We now show that, in the case of CPSS functions, this can never happen. Let us assume w.l.o.g. that exactly the first $k$ blocks have distance 1.

*Claim.* Let $S$ be a CPSS function with satisfying solution $x$. If $x^i$ for $1 \leq i \leq k \leq n$ is equal to $x$ with the $i$-th bit flipped, $x' = x_1^1, \ldots, x_k^k, x_{k+1}, \ldots, x_n$ and $x^1, \ldots, x^k, x'$ all satisfy $S$, then there is a path from $x$ to $x'$.

---

[2] Boolean isomorphisms are signed permutations: they may map variables to variables and may flip variables.

Proof: Obviously, $x^i$ is connected to $x$ as $d(x, x^i) = 1$. We show by induction on $j$ that for all $j$ with $1 \leq j \leq k$, $y^j = x_1^1, \ldots, x_j^j, x_{k+1}, \ldots, x_n$ satisfies $S$. As their consecutive distances are 1 the statement follows. For $j = 1$ we know that $x^1 = y^1$. Now let $y^j$ satisfy $S$. If $S$ is componentwise bijunctive, then $maj(x', y^j, x^{j+1}) = y^{j+1}$ satisfies $S$ by its closure property. If $S$ is not componentwise bijunctive it is Horn and componentwise IHSB- (or dual). Again the closure property gives $x' \wedge (y^j \vee x^{j+1}) = y^{j+1}$. ∎

Note that the given construction on $(F, x)$ creates a formula $F'$ such that every satisfying solution is a walk on $F$ of length $w := n$ starting at $x$ (we use $n$ branch and reduce blocks). Therefore, the set of all satisfying solutions to $F'$ is the set of all walks on $F$ where every step is the traversal of a complete subhypercube and in every step the walk may refuse to take a step and remain at the previous node. Obviously, if there is an isomorphism $\pi$ mapping the two components onto each other such that $\pi(x) = y$, then there is an isomorphism mapping the sets of paths onto each other. This isomorphism just has to use $\pi$ for every block and has to exchange the parallel steps according to $\pi$.

We can now reduce the Boolean isomorphism question between $F'$ and $G'$ to $GI$ (again, using [6]) implementing the additional properties with graph gadgets. We therefore force all blocks to be mapped internally in the same way and we force the $n$ parallel blocks to be mapped exactly as each block is mapped internally. The result are two graphs which are isomorphic iff there is an isomorphism mapping the components rooted at $x$ and $y$ in $F$ and $G$ onto each other so that $x$ gets mapped to $y$. □

**Theorem 13.** *Iso(CPSS)* $\in C_=P$.

*Proof.* The proof uses the fact that GI is low for the class $C_=P$ [12]. This means that a nondeterministic polynomial time algorithm with a $C_=P$ acceptance mechanism and having access to an oracle for GI, can be simulated by an algorithm of the same kind, but without the oracle. In symbols $C_=P^{GI} = C_=P$.

We already know that the solution graphs of CPSS functions consist of at most an exponential number of connected components and every such component is a partial cube. For two solution graphs $F$ and $G$ to be isomorphic there has to be a bijection mapping each connected component of $F$ onto an isomorphic component of $G$.

One way to check the existence of such a bijection is by looking at each possible partial cube and counting the number of connected components isomorphic to it in both graphs. If the numbers match for all partial cubes, the graphs are isomorphic. Instead of checking all possible partial cubes, which would be too many, one only has to check the ones which occur in the graphs. For $x \in \{0, 1\}^n$ let $A_x$ and $B_x$ be the sets

$$A_x = \{y \in \{0, 1\}^n \mid F(y) = 1 \wedge F_x \cong F_y \text{with an isomorphism mapping } x \text{ to } y\}$$

$$B_x = \{y \in \{0, 1\}^n \mid G(y) = 1 \wedge F_x \cong G_y \text{with an isomorphism mapping } x \text{ to } y\}$$

The existence of an isomorphism between $F_x$ and $F_y$ (or $G_y$) mapping $x$ to $y$ can be checked with a GI oracle (as proven in Lemma 12). Our algorithm checks for every $x \in \{0,1\}^n$ satisfying $F$, whether $||A_x|| = ||B_x||$. The same test is performed for all $x$ satisfying $G$. Both tests are successful iff the graphs are isomorphic. Clearly the graphs are isomorphic iff both tests succeed.

This procedure shows that the problem is in the class $\forall C_= P^{GI}$.[3] Using the mentioned fact that GI is low for $C_= P$, this class coincides with $\forall C_= P$. In addition, Green showed [10] that $C_= P$ is closed under universal quantification, i.e. $\forall C_= P = C_= P$. We conclude that $Iso(CPSS) \in C_= P$.  □

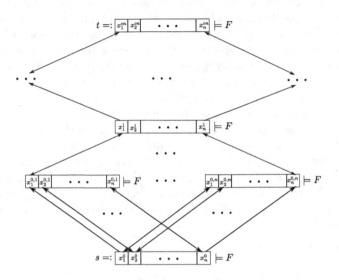

**Fig. 2.** A walk on solution graphs.

In Theorem 13 we exploited the fact that CPSS functions consist of partial cubes of small isometric dimension. But for general Schaefer functions this property does not hold. The solution graph might have an exponential isometric dimension or the connected subgraphs might even not be partial cubes. Therefore it seems improbable that the $C_= P$-algorithm can be adapted for general Schaefer solution graphs. These graphs should admit a better lower bound. Unfortunately, we can only provide such a lower bound for the more powerful class of Boolean functions that are not safely tight.

**Theorem 14.** *Let $S$ be a set of functions which is not safely tight. Then $Iso(S)$ is hard for PSPACE under logarithmic-space reductions.*

*Proof.* The proof is based on the reduction from $s, t$-connectivity to GI from [11]. We know that the $s, t$-connectivity problem for functions that are not safely tight is PSPACE-complete [9]. We give a construction of solution graphs that have

---

[3] We use the same quantifier notation which is common for the classes in the polynomial time hierarchy.

colored vertices as a way to distinguish some vertices. Later we show how the formulas can be modified to produce the colors in their solution graphs. Given a formula $F$ built on functions from $S$, as well as satisfying assignments $s$ and $t$, we create two copies of $G_F$ (which is the solution graph defined by $F$) and color vertex $s$ in one of the copies with color white and with black in the second copy. Let $G_{F'}$ be the disjoint union of the two copies. Now we consider two copies $G_{F_1}$ and $G_{F_2}$ of $G_{F'}$. We color in $G_{F_1}$ one of the copies of vertex $t$ with the grey color while in $G_{F_2}$, the second copy of $t$ is colored grey. All other nodes have no color. There is a path from $s$ to $t$ in $G_F$ iff $G_{F_1}$ and $G_{F_2}$ are not isomorphic.

This construction can easily be performed with solution graphs. Given the formula $F(x_1, \ldots, x_n)$, two disjoint copies of the encoded graph are defined by the formula

$$F'(a, b, x_1, \ldots, x_n) = (a \leftrightarrow b) \wedge F(x_1, \ldots, x_n)$$

using two new variables $a$ and $b$. Coloring a vertex by attaching a gadget to it can be done with the following construction. We assume w.l.o.g. that the node we want to color is $0^n$ in the solution graph of an arbitrary formula $G$. We add to $0^n$ the graph $H_m$ with $m > n$ as neighbor. Then $G'(x_1, \ldots, x_n, y_1, \ldots, y_m) = G(x_1, \ldots, x_n) \wedge \bigwedge_{i \leq n, j \leq m} (x_i \rightarrow \overline{y_j})$. The new graph can be described as the old solution graph of $\tilde{G}$ but $0^n$ now is the minimal node of a new, complete hypercube on $m$ variables. Note that $0^n$ is the only node of the original solution graph which is part of a hypercube of dimension $m$. In addition, it is the only node of the hypercube of dimension $m$ which is connected to some of the old nodes. This completes the reduction and shows that $\overline{Iso(S)}$ is hard for PSPACE and therefore $Iso(S)$ is hard for coPSPACE = PSPACE.    □

The given construction uses new clauses which are Horn and 2CNF and can even be applied to simpler classes of formulas. The following statements use the hardness results of [18] with the reduction in Theorem 14.

**Corollary 15.**  *Iso(2CNF) is hard for NL and Iso(Horn$_3$) is hard for P under logspace reductions.*

Note that the resulting solution graphs in this corollary can not have more than two connected components. The isomorphism for these $2CNF$ graphs is therefore polynomial time reducible to GI.

## 4   Structure of Solution Graphs of Horn Formulas

While [18] showed that CPSS formulas contain only partial cubes of small isometric dimension as connected components, Horn formulas may encode partial cubes of exponential isometric dimension or graphs which are not even partial cubes. So for the isomorphism question, things seem to get more complicated. We give an interesting property for Horn solution graphs which suggests that *Iso(Horn)* might be easier than general solution graph isomorphism.

Let $d_m(a, b)$ denote the monotone distance between $a$ and $b$. So $d_m(a, b) < \infty$ iff there is a strictly monotone increasing path from $a$ to $b$ or vice versa.

In [9] it is shown that in OR-free formulas there is a unique minimal satisfying assignment in every connected component. As Horn-formulas are OR-free, given an assignment $y$ satisfying a Horn-formula $F$, the connected component of $y$ contains a unique minimal satisfying assignment. For the next result we will assume w.l.o.g. that this minimal satisfying assignment in the connected component of $y$ is $0^n$. If this is not true, we could modify $F$ setting all variables to 1 which are 1 in $y$ and get a formula $F'$ on less variables where $0^{n'}$ is the required minimal satisfying assignment. The resulting formula satisfies this property and still contains the connected component corresponding to $y$ in $F$. With $[y]_F := \{a \in \{0,1\}^n \mid d_m(a,y) < \infty\}$ we denote the set of all nodes $a$ lying between $0^n$ and $y$ for which there is a monotone increasing path from $a$ to $y$.

**Theorem 16.** *For every solution $y$ to a Horn-formula $F$, $[y]_F$ is a partial cube.*

*Proof.* Let $a, b \in [y]_F$ be two arbitrary nodes. We show that $d(a,b) = \Delta(a,b)$. In case the two monotone increasing paths $a = a_1, \ldots, a_k = y$ from $a$ to $y$ and $b = b_1, \ldots, b_l = y$ from $b$ to $y$ are already of total length $\Delta(a,b)$, then we are done. Otherwise, suppose that there is at least one variable $x_i$ which gets increased to 1 in both paths. The positions in the path where such variables are increased may differ. Every variable can be classified as either not changed in any of the paths, changed in only one path (and therefore contributing to $\Delta(a,b)$), or changed in both paths. We can now construct the shorter path from $a_1 \wedge y = a_1$ over $a_1 \wedge b_{l-1}$ and $a_1 \wedge b_1 = a \wedge b = b_1 \wedge a_1$ back to $b_1 \wedge a_2$ and $b_1 \wedge a_k = b_1$. Figure 3 illustrates in the first row the original path and in the second row the new path.

| $a_1$ | $a_2$ | $\ldots$ | $a_k = b_l = y$ | $b_{l-1}$ | $\ldots$ | $b_2$ | $b_1$ |
|---|---|---|---|---|---|---|---|
| $a_1 \wedge b_l = a_1$ | $a_1 \wedge b_{l-1}$ | $\ldots$ | $a_1 \wedge b_1 = a \wedge b$ | $b_1 \wedge a_2$ | $\ldots$ | $b_1 \wedge a_{k-1}$ | $b_1 \wedge a_k = b_1$ |

**Fig. 3.** Original and shorted paths from $a_1$ to $b_1$ over $y = b_l = a_k$.

Note that all these nodes are in $G_F$ as Horn-formulas are closed under conjunction and the overall sum of nodes in this sequence is the same as in the original path. But as the first half is the conjunction of $a_1$ with every node in the second half, every variable which gets increased in both halves (0 in $a_1$) will lead to two identical consecutive nodes in the first half. By symmetry, the same happens in the new second half. This path is now two nodes shorter for every variable which was changed in both paths. All remaining flips are still present.     □

Figure 4 gives a minimal example (with repeated $y$ node in the middle) which illustrates how an increasing/decreasing path can be transformed to a shortest path of the same length as the Hamming distance between the source and target nodes. The original path has length 6 with one common variable in both halves while the shortcut has length 4, which is optimal.

| Long path | $a = 11000$ | 11010 | 11011 | 11111 | 11111 | 01111 | 00111 | 00011 $= b$ |
|-----------|-------------|-------|-------|-------|-------|-------|-------|---------------|
| Optimal path | $a = 11000$ | 01000 | 00000 | 00000 | 00000 | 00010 | 00011 | 00011 $= b$ |

**Fig. 4.** Finding shortcuts in Horn solution graphs.

This result shows that Horn solution graphs encode for every locally maximal solution $y$ a partial cube $[y]_F$ and every intersection of two such partial cubes $[y]_F \cap [y']_F = [z]_F$ is also a partial cube. We point out that a similar statement holds for dual-Horn-formulas.

## 5    *Iso(2CNF)* and the number of perfect matchings

We showed in Theorem 13 that $Iso(2CNF) \in \mathrm{C}_=\mathrm{P}$. In this section we show that $Iso(2CNF)$ is also hard for $\mathrm{C}_=\mathrm{P}$. For this we will consider several reductions involving the following decision problems:

*Same2SAT*: Given two 2CNF-formulas $F$ and $F'$, does the number of satisfying assignments for $F$ and $F'$ coincide?

*SamePM*: Given two graphs, does the number of perfect matchings in each of the graphs coincide?

Curticapean [7] showed recently that *SamePM* is $\mathrm{C}_=\mathrm{P}$-complete. In a series of reductions, Valiant [20, 21] proved that the (functional) problem of computing the permanent can be Turing reduced to computing the number of satisfying assignments of a 2CNF formula. This reduction queries a polynomial number of $\#SAT$ instances and uses the answers (which are numbers of polynomial length) in a Chinese-remainder fashion to compute the original number of perfect matchings. This argument does not work in the context of many-one reductions and decision problems. We take ideas from these reductions to show that *SamePM* is many-one reducible to *Same2SAT* and to *Iso(2CNF)*.

**Theorem 17.** *SamePM is polynomial time many one reducible to Same2SAT.*

*Proof.* Valiant [21] gave a way to Turing reduce the problem of counting perfect matchings to the problem of counting satisfying assignments of a 2CNF formula by counting all matchings as an intermediate step.

Reducing the number of matchings (perfect or not) of a given graph $B$ to the number of satisfying solutions of a formula is easy. We define a variable $x_e$ for each edge $e$ in $B$ and for each pair of edges $e, e'$ with a common vertex we create a clause $(\overline{x_e} \vee \overline{x_{e'}})$. If $F_B$ is the conjunction of all these clauses, the set of satisfying assignments for $F_B$ coincides with the set of matchings in $B$.

The number of perfect matchings of a graph $B$ with $n$ vertices can be computed from the number of all matchings in $B$ and some derived graphs $B_k$. For this, let $b_i$ be the number of matchings with exactly $i$ unmatched nodes. Then $b_0$ is the number of perfect matchings that we want to compute, while $b_{n-2}$ is the number of edges in $B$. Let us define a modification $B_k$ of $B$ ($1 \leq k \leq n$) consisting of a copy of $B$ and for every node $u$ in $B$, $k$ otherwise isolated nodes

$u_1, \ldots, u_k$ with a single edge connecting each of them to $u$. Now each matching in $B$ can be extended in $B_k$ by matching each non-matched node of $B$ to one of its $k$ new neighbors. Each original matching of $B$ with $i$ unmatched nodes corresponds to $(k+1)^i$ matchings in $B_k$. The total number of matchings $c_k$ in $B_k$ is $\sum_{i=0}^{n} b_i \cdot (k+1)^i$. The following equation system describes the relation between matchings in $B_k$ graphs and in $B$.

$$\begin{pmatrix} 1 & 1 & 1 & \cdots & 1 \\ 1 & 2 & 4 & \cdots & 2^n \\ \vdots & \vdots & \vdots & \ddots & \vdots \\ 1 & (n+1) & (n+1)^2 & \cdots & (n+1)^n \end{pmatrix} \times \begin{pmatrix} b_0 \\ b_1 \\ \vdots \\ b_n \end{pmatrix} = \begin{pmatrix} c_0 \\ c_1 \\ \vdots \\ c_n \end{pmatrix}$$

The $(n+1) \times (n+1)$ matrix $V$ is a Vandermonde-matrix and can therefore be inverted in polynomial time. The $c$ coefficients are numbers of matchings, that can be reduced to numbers of satisfying assignments of 2CNF formulas. The first entry of $V^{-1} \times (c_0, \ldots, c_n)^T$ is $b_0$, the number of perfect matchings in $B$ that we want to compute. Given $V^{-1}$ and 2CNF formulas $F_0, \ldots, F_n$ having respectively $c_0, \ldots, c_n$ satisfying assignments (the formulas can be created from $B_0, \ldots, B_n$ with the aforementioned reduction), $b_0$ can be computed as the sum and difference of $c_i$'s multiplied by coefficients defined by $V^{-1}$.

If we are given two graphs $B_1$ and $B_2$, on $n$ vertices by doing the same construction we get two sets of coefficients ($c^1$ and $c^2$) and the number of perfect matchings in $B_1$ and $B_2$ coincide iff the following statement holds:

$$(V_{1,1}^{-1}, \ldots, V_{1,n+1}^{-1}) \times (c_0^1, \ldots, c_n^1)^T = (V_{1,1}^{-1}, \ldots, V_{1,n+1}^{-1}) \times (c_0^2, \ldots, c_n^2)^T$$

The $c$ coefficients in the equation can be expressed as numbers of solutions of 2CNF formulas, while the other numbers are rational numbers. Inverting the Vandermonde matrix leads to rational numbers of length at most polynomial in $n$. Therefore, using an appropriate factor, we can multiply both sides of this equation by the same factor and reduce every rational number to an integer of polynomial length. This equation can be transformed so that both sides contain only additions and multiplications of positive numbers. These can be implemented as numbers of satisfying assignments of 2CNF formulas using the following gadgets. Note that input formulas are all anti-monotone and therefore have the satisfying solution $0^n$ and we maintain this solution through all constructions.

Multiplying the number of satisfying assignments of 2CNF-formulas can be achieved by the conjunction of both formulas (with disjoint sets of variables).

The sum of the solution sets is again a conjunction of both formulas (with disjoint sets of variables) with the following modification: For two fixed satisfying assignments $0^n$ of $F$ and $0^m$ of $F'$, we add the clauses $\bigwedge_{i \in [n], j \in [m]} (x_i \to \overline{y_j})$. So for every solution $v' \neq 0^n$ in $F$, the variables of $F'$ get fixed to $0^m$. By symmetry the same holds for all $v' \neq 0^m$ satisfying $F'$. This corresponds to the disjoint union of the solution sets except for $0^{n+m}$ which occurs only once. So we add a new variable $b$ and add the clauses $\bigwedge_{i \in [n]} (x_i \to \overline{b}) \wedge \bigwedge_{j \in [m]} (y_j \to \overline{b})$ in the same way as before. This duplicates $0^{n+m}$ as $b$ is allowed to be 1 or 0 but if we deviate from this assignment, we fix $b$ to 0. The number of satisfying solutions is therefore the sum of $F$ and $F'$ and $0^{n+m+1}$ is still a satisfying solution.

For encoding the coefficients of the inverse Vandermonde matrix we need a way to transform a positive integer $k$ into a 2CNF-formula $G$ with exactly $k$ satisfying solutions. This can be achieved by looking at the binary encoding of $k = (k_1, \ldots, k_l)_2$. For every $i$ with $k_i = 1$ we create the 2CNF formula $G_i = \bigwedge_{j \in [i]} (x_i \vee \overline{x_i})$ on $i$ variables and take the sum of all these formulas (as described before) where every formula has its own set of variables and contains $0^i$ as satisfying solution. This new formula $G$ has exactly $k$ satisfying solutions.

We form for both sides of the equation 2CNF-formulas implementing these computations, and get two formulas $F, F'$ that have the same number of satisfying assignments iff $B_1$ and $B_2$ have the same number of perfect matchings.     □

**Theorem 18.** *Same2SAT is polynomial time many-one reducible to Iso(2CNF).*

*Proof.* Two formulas having only isolated satisfying assignments have the same number of solutions iff their solution graphs are isomorphic. A formula $F$ can be transformed into another one $F'$ with the same number of solutions but having only isolated satisfying assignments. This can be done by duplicating each occurring variable $x$ with a new variable $x'$ and adding the restriction $(x \leftrightarrow x')$. The Hamming distance between two solutions in $F'$ is then at least two.     □

These reductions plus Theorem 13 imply:

**Corollary 19.**   *Same2SAT and Iso(2CNF) are $C_=P$-complete.*

**Corollary 20.** *Iso(Horn) and Iso(safely tight) are hard for $C_=P$.*

This last result follows from the observation that all constructed 2CNF formulas, those for counting matchings, as well as those for multiplication and summation constructions are also Horn.

# References

1. Achlioptas, D., Coja-Oghlan, A., Ricci-Tersenghi, F.: On the solution-space geometry of random constraint satisfaction problems. Random Struct. Algorithms **38**(3), 251–268 (2011)
2. Agrawal, M., Thierauf, T.: The Boolean isomorphism problem. In: Proceedings of 37th Conference on Foundations of Computer Science, pp. 422–430. IEEE Computer Society Press (1996)
3. Babai, L.: Graph isomorphism in quasipolynomial time. In: Proceedings of 48th Annual Symposium on the Theory of Computing, STOC (2016)
4. Bandelt, H.-J., van de Vel, M.: Embedding topological median algebras in products of dendrons. Proc. London Math. Soc. **3**(58), 439–453 (1989)
5. Bandelt, H.J., Chepoi, V.: Metric graph theory and geometry: a survey. Contemp. Math. **453**, 49–86 (2008)

6. Böhler, E., Hemaspaandra, E., Reith, S., Vollmer, H.: Equivalence and isomorphism for boolean constraint satisfaction. In: Bradfield, J.C. (ed.) CSL 2002 and EACSL 2002. LNCS, vol. 2471, pp. 412–426. Springer, Heidelberg (2002)

7. Curticapean, R.: Parity separation: a scientifically proven method for permanent weight loss. arXiv preprint. arXiv:1511.07480 (2015)

8. Gableske, O.: SAT Solving with Message Passing. Ph.D. thesis, University of Ulm (2016)

9. Gopalan, P., Kolaitis, P.G., Maneva, E., Papadimitriou, C.H.: The connectivity of boolean satisfiability: computational and structural dichotomies. SIAM J. Comput. **38**(6), 2330–2355 (2009)

10. Green, F.: On the power of deterministic reductions to C=P. Math. Syst. Theor. **26**(2), 215–233 (1993)

11. Jenner, B., Köbler, J., McKenzie, P., Torán, J.: Completeness results for graph isomorphism. J. Comput. Syst. Sci. **66**(3), 549–566 (2003)

12. Köbler, J., Schöning, U., Torán, J.: Graph isomorphism is low for PP. Comput. Complex. **2**(4), 301–330 (1992)

13. Köbler, J., Schöning, U., Torán, J.: The Graph Isomorphism Problem: its Structural Complexity. Birkhauser, Boston (1993)

14. Luks, E.M.: Isomorphism of graphs of bounded valence can be tested in polynomial time. J. Comput. Syst. Sci. **25**(1), 42–65 (1982)

15. Mézard, M., Mora, T., Zecchina, R.: Clustering of solutions in the random satisfiability problem. Phys. Rev. Lett. **94**(19), 197205 (2005)

16. Ovchinnikov, S.: Graphs and Cubes. Universitext. Springer, New York (2011)

17. Schaefer, T.J.: The complexity of satisfiability problems. In: Proceedings of the Tenth Annual ACM Symposium on Theory of Computing - STOC 1978, pp. 216–226. ACM Press, New York (1978)

18. Scharpfenecker, P.: On the structure of solution-graphs for boolean formulas. In: Kosowski, A., Walukiewicz, I. (eds.) FCT 2015. LNCS, vol. 9210, pp. 118–130. Springer, Heidelberg (2015)

19. Schwerdtfeger, K.W.: A computational trichotomy for connectivity of boolean satisfiability. JSAT **8**(3/4), 173–195 (2014)

20. Valiant, L.G.: The complexity of computing the permanent. Theor. Comput. Sci. **8**(2), 189–201 (1979)

21. Valiant, L.G.: The complexity of enumeration and reliability problems. SIAM J. Comput. **8**, 410–421 (1979)

22. Wagner, K.W.: The complexity of combinatorial problems with succinct input representation. Acta Informatica **23**(3), 325–356 (1986)

# Strong Backdoors for Default Logic

Johannes K. Fichte[1]([✉]), Arne Meier[2], and Irina Schindler[2]

[1] Technische Universität Wien, Wien, Austria
jfichte@dbai.tuwien.ac.at
[2] Leibniz Universität Hannover, Hannover, Germany
{meier,schindler}@thi.uni-hannover.de

**Abstract.** In this paper, we introduce a notion of backdoors to Reiter's propositional default logic and study structural properties of it. Also we consider the problems of backdoor detection (parameterised by the solution size) as well as backdoor evaluation (parameterised by the size of the given backdoor), for various kinds of target classes (CNF, HORN, KROM, MONOTONE, POSITIVE-UNIT). We show that backdoor detection is fixed-parameter tractable for the considered target classes, and backdoor evaluation is either fixed-parameter tractable, in para-$\Delta_2^P$, or in para-NP, depending on the target class.

## 1 Introduction

In the area of non-monotonic logic one aims to find formalisms that model human-sense reasoning. It turned out that this kind of reasoning is quite different from classical deductive reasoning as in the classical approach the addition of information always leads to an increase of derivable knowledge. Yet, intuitively, human-sense reasoning does not work in that way: the addition of further facts might violate previous assumptions and can therefore significantly decrease the amount of derivable conclusions. Hence, in contrast to the classical process the behaviour of human-sense reasoning is non-monotonic. In the 1980s, several kinds of formalisms have been introduced, most notably, circumscription [26], default logic [33], autoepistemic logic [29], and non-monotonic logic [27]. A good overview of this field is given by Marek and Truszczyński [25].

In this paper, we focus on Reiter's *default logic (DL)*, which has been introduced in 1980 [33] and is one of the most fundamental formalism to model human-sense reasoning. DL extends the usual logical derivations by rules of default assumptions (*default rules*). Informally, default rules follow the format "in the absence of contrary information, assume ...". Technically, these patterns are taken up in triples of formulas $\frac{\alpha:\beta}{\gamma}$, which express "if prerequisite $\alpha$ can be deduced and justification $\beta$ is never violated then assume conclusion $\gamma$". Default rules can be used to enrich calculi of different kinds of logics. Here, we consider a variant of propositional formulas, namely, formulas in conjunctive normal form (CNF). A key concept of DL is that an application of default rules must not lead to an inconsistency if conflicting rules are present, instead such

N. Creignou and D. Le Berre (Eds.): SAT 2016, LNCS 9710, pp. 45–59, 2016.
DOI: 10.1007/978-3-319-40970-2_4

rules should be avoided if possible. This concept results in the notion of *stable extensions*, which can be seen as a maximally consistent view of an agent with respect to his knowledge base together in combination with its set of default rules. The corresponding decision problem, i.e., the extension existence problem, then asks whether a given default theory has a *consistent stable extension*, and is the problem of our interest. The computationally hard part of this problem lies in the detection of the order and "applicability" of default rules, which is a quite challenging task as witnessed by its $\Sigma_2^p$-completeness. In 1992, Gottlob showed that many important decision problems, beyond the extension existence problem, of non-monotonic logics are complete for the second level of the polynomial hierarchy [21] and thus are of high intractability.

A prominent approach to understand the intractability of a problem is to use the framework of parameterised complexity, which was introduced by Downey and Fellows [10,11]. The main idea of parameterised complexity is to fix a certain structural property (the *parameter*) of a problem instance and to consider the computational complexity of the problem in dependency of the parameter. Then ideally, the complexity drops and the problem becomes solvable in polynomial time when the parameter is fixed. Such problems are called *fixed-parameter tractable* and the corresponding parameterised complexity class, which contains all fixed-parameter tractable problems, is called FPT. For instance, for the propositional satisfiability problem (SAT) one (naïve) parameter is the number of variables of the given formula. Then, for a given formula $\varphi$ of size $n$ and $k$ variables its satisfiability can be decided in time $O(n \cdot 2^k)$, i.e., polynomial (even linear) runtime in $n$ if $k$ is assumed to be fixed.

The invention of new parameters can be quite challenging, however, SAT has so far been considered under many different parameters [4,30,35,40]. A concept that provides a parameter and has been widely used in theoretical investigations of propositional satisfiability are backdoors [20,24,41]. The size of a backdoor can be seen as a parameter with which one tries to exploit a small distance of a formula from being tractable. More detailed, given a class $\mathcal{F}$ of formulas and a formula $\varphi$, a subset $B$ of its variables is a *strong $\mathcal{F}$-backdoor* if the formula $\varphi$ under every truth assignment over $B$ yields a formula that belongs to the class $\mathcal{F}$. Using backdoors usually consists of two phases: (i) finding a backdoor (backdoor detection) and (ii) using the backdoor to solve the problem (backdoor evaluation). If $\mathcal{F}$ is a class where SAT is tractable and backdoor detection is fixed-parameter tractable for this class, like the class of all Horn or Krom formulas, we can immediately conclude that SAT is fixed-parameter tractable when parameterised by the size of a smallest strong $\mathcal{F}$-backdoor.

*Related Work.* Backdoors for propositional satisfiability have been introduced by Williams, Gomes, and Selman [41,42]. The concept of backdoors has recently been lifted to some non-monotonic formalisms as abduction [32], answer set programming [16,17], and argumentation [12]. Beyond the classification of Gottlob [21], the complexity of fragments, in the sense of Post's lattice, has been considered by Beyersdorff et al. extensively for default logic [1], and for autoepistemic logic by Creignou et al. [8]. Also parameterised analyses of non-monotonic logics in the spirit of Courcelle's theorem [6,7] have recently been considered by Meier et al. [28]. Further, Gottlob et al. studied treewidth as a parameter for various non-monotonic logics [22] and also considered a more CSP focused non-monotonic context within the parameterised complexity setting [23].

*Contribution.* In this paper, we introduce a notion of backdoors to propositional default logic and study structural properties therein. Then we investigate the parameterised complexity of the problems of backdoor detection (parameterised by the solution size) and evaluation (parameterised by the size of the given backdoor), with respect to the most important classes of CNF formulas, e.g., CNF, KROM, HORN, MONOTONE, and POSITIVE-UNIT. Informally, given a formula $\varphi$ and an integer $k$, the detection problem asks whether there exists a backdoor of size $k$ for $\varphi$. Backdoor evaluation then exploits the distance $k$ for a target formula class to solve the problem for the starting formula class with a "simpler" complexity. Our classification shows that detection is fixed-parameter tractable for all considered target classes. However, for backdoor evaluation starting at CNF the parameterised complexity depends, as expected, on the target class: the parameterised complexity then varies between para-$\Delta_2^p$ (MONOTONE), para-NP (KROM, HORN), and FPT (POSITIVE-UNIT).

## 2   Preliminaries

We assume familiarity with standard notions in computational complexity, the complexity classes P and NP as well as the polynomial hierarchy. For more detailed information, we refer to other standard sources [11,19,31].

*Parameterised Complexity.* We follow the notion by Flum and Grohe [18]. A *parameterised (decision) problem* $L$ is a subset of $\Sigma^* \times \mathbb{N}$ for some finite alphabet $\Sigma$. Let $C$ be a classical complexity class, then para-$C$ consists of all parameterised problems $L \subseteq \Sigma^* \times \mathbb{N}$, for which there exists an alphabet $\Sigma'$, a computable function $f\colon \mathbb{N} \to \Sigma'^*$, and a (classical) problem $L' \subseteq \Sigma^* \times \Sigma'^*$ such that (i) $L' \in C$, and (ii) for all instances $(x, k) \in \Sigma^* \times \mathbb{N}$ of $L$ we have $(x, k) \in L$ if and only if $(x, f(k)) \in L'$. For the complexity class P, we write FPT instead of para-P. We call a problem in FPT *fixed-parameter tractable* and the runtime $f(k) \cdot |x|^{O(1)}$ also *fpt-time*. Additionally, the parameterised counterparts of NP and $\Delta_2^p = P^{NP}$, which are denoted by para-NP and para-$\Delta_2^p$, are relevant in this paper.

*Propositional Logic.* Next, we provide some notions from propositional logic. We consider a finite set of propositional variables and use the symbols $\top$ and $\bot$ in the standard way. A *literal* is a variable $x$ (positive literal) or its negation $\neg x$ (negative literal). A *clause* is a finite set of literals, interpreted as the disjunction of these literals. A propositional formula in *conjunctive normal form (CNF)* is a finite set of clauses, interpreted as the conjunction of its clauses. We denote the class of all CNF formulas by CNF. A clause is *Horn* if it contains at most one positive literal, *Krom* if it contains two literals, *monotone* if it contains only positive literals, and *positive-unit* if it contains at most one positive literal. We say that a CNF formula has a certain property if all its clauses have the property. We consider several classes of formulas in this paper. Table 1 gives an overview on these classes and defines clause forms for these classes.

**Table 1.** Considered normal forms. In the last row, $\ell_i^+$ denote positive, and $\ell_i^-$ negative literals; and $n$ and $m$ are integers such that $n, m \geq 0$.

| Class | Clause description | Clause forms |
|---|---|---|
| CNF | no restrictions | $\{\ell_1^+, \ldots, \ell_n^+, \ell_1^-, \ldots, \ell_m^-)$ |
| HORN | at most one positive literal | $\{\ell^+, \ell_1^-, \ldots, \ell_n^-\}, \{\ell_1^-, \ldots, \ell_m^-\}$ |
| KROM | binary clauses | $\{\ell_1^+, \ell_2^+\}, \{\ell^+, \ell^-\}, \{\ell_1^-, \ell_2^-\}$ |
| MONOTONE | no negation, just positive literals | $\{\ell_1^+, \ldots, \ell_n^+\}$ |
| POSITIVE-UNIT | only positive unit clauses | $\{\ell^+\}$ |

A formula $\varphi'$ is a *subformula* of a CNF formula $\varphi$ (in symbols $\varphi' \subseteq \varphi$) if for each clause $C' \in \varphi'$ there is some clause $C \in \varphi$ such that $C' \subseteq C$. We call a class $\mathcal{F}$ of CNF formulas *clause-induced* if whenever $F \in \mathcal{F}$, all subformulas $F' \subseteq F$ belong to $\mathcal{F}$. Note that all considered target classes in this paper are clause-induced.

Given a formula $\varphi \in$ CNF, and a subset $X \subseteq \mathrm{Vars}(\varphi)$, then a *(truth) assignment* is a mapping $\theta \colon X \to \{0,1\}$. The *truth (evaluation)* of propositional formulas is defined in the standard way, in particular, $\theta(\bot) = 0$ and $\theta(\top) = 1$. We extend $\theta$ to literals by setting $\theta(\neg x) = 1 - \theta(x)$ for $x \in X$. By $\mathbb{A}(X)$ we denote the set of all assignments $\theta \colon X \to \{0,1\}$. For simplicity of presentation, we sometimes identify the set of all assignments by its corresponding literals, i.e., $\mathbb{A}(X) = \{ \{\ell_1, \ldots, \ell_{|X|}\} \mid x \in X, \ell_i \in \{x, \neg x\} \}$. We write $\varphi[\theta]$ for the *reduct* of $\varphi$ where every literal $\ell \in X$ is replaced by $\top$ if $\theta(\ell) = 1$, then all clauses that contain a literal $\ell$ with $\theta(\ell) = 1$ are removed and from the remaining clauses all literals $\ell'$ with $\theta(\ell') = 0$ are removed. We say $\theta$ *satisfies* $\varphi$ if $\varphi[\theta] \equiv \top$, $\varphi$ is *satisfiable* if there exists an assignment that satisfies $\varphi$, and $\varphi$ is *tautological* if all assignments $\theta \in \mathbb{A}(X)$ satisfy $\varphi$. Let $\varphi, \psi \in$ CNF and $X = \mathrm{Vars}(\varphi) \cup \mathrm{Vars}(\psi)$. We write $\varphi \models \psi$ if and only if for all assignments $\theta \in \mathbb{A}(X)$ it holds that all assignments $\theta$ that satisfy $\varphi$ also satisfy $\psi$. Further, we define the *deductive closure* of $\varphi$ as $\mathrm{Th}(\varphi) := \{ \psi \in$ CNF $\mid \varphi \models \psi \}$.

Note that any assignment $\theta \colon \mathrm{Vars}(\varphi) \to \{0,1\}$ can also be represented by the CNF formula $\bigwedge_{\theta(x)=1} x \wedge \bigwedge_{\theta(x)=0} \neg x$. Therefore, we often write $\theta \models \varphi$ if $\varphi[\theta] \equiv \top$ holds.

We denote with $\mathrm{SAT}(\mathcal{F})$ the problem, given a propositional formula $\varphi \in \mathcal{F}$ asking whether $\varphi$ is satisfiable. The problem $\mathrm{TAUT}(\mathcal{F})$ is defined over a given formula $\varphi \in \mathcal{F}$ asking whether $\varphi$ tautological.

## 2.1   Default Logic

We follow notions by Reiter [33] and define a *default rule* $\delta$ as a triple $\frac{\alpha:\beta}{\gamma}$; $\alpha$ is called the *prerequisite*, $\beta$ is called the *justification*, and $\gamma$ is called the *conclusion*; we set $prereq(\delta) := \alpha$, $just(\delta) := \beta$, and $concl(\delta) := \gamma$. If $\mathcal{F}$ is a class of formulas, then $\frac{\alpha:\beta}{\gamma}$ is an $\mathcal{F}$-default rule if $\alpha, \beta, \gamma \in \mathcal{F}$. An $\mathcal{F}$-*default theory* $\langle W, D \rangle$ consists of a set of propositional formulas $W \subseteq \mathcal{F}$ and a set $D$ of $\mathcal{F}$-default rules. We sometimes call $W$ the *knowledge base* of $\langle W, D \rangle$. Whenever we do not explicitly state the class $\mathcal{F}$, we assume it to be CNF.

**Definition 1 (Fixed point semantics, [33]).** *Let $\langle W, D \rangle$ be a default theory and $E$ be a set of formulas. Then $\Gamma(E)$ is the smallest set of formulas such that: (1) $W \subseteq \Gamma(E)$, (2) $\Gamma(E) = \mathrm{Th}(\Gamma(E))$, and (3) for each $\frac{\alpha:\beta}{\gamma} \in D$ with $\alpha \in \Gamma(E)$ and $\neg\beta \notin E$, it holds that $\gamma \in \Gamma(E)$. $E$ is a stable extension of $\langle W, D \rangle$, if $E = \Gamma(E)$. An extension is* inconsistent *if it contains $\bot$, otherwise it is called* consistent.

A definition of stable extensions beyond fixed point semantics, which has been introduced by Reiter [33] as well, uses the principle of a stage construction.

**Proposition 2 (Stage construction, [33]).** *Let $\langle W, D \rangle$ be a default theory and $E$ be a set of formulas. Then define $E_0 := W$ and*

$$E_{i+1} := \mathrm{Th}(E_i) \cup \left\{ \gamma \;\middle|\; \frac{\alpha:\beta}{\gamma} \in D, \alpha \in E_i \text{ and } \neg\beta \notin E \right\}.$$

*$E$ is a stable extension of $\langle W, D \rangle$ if and only if $E = \bigcup_{i \in \mathbb{N}} E_i$. The set*

$$G = \left\{ \frac{\alpha:\beta}{\gamma} \in D \;\middle|\; \alpha \in E \wedge \neg\beta \notin E \right\}$$

*is called the set of* generating defaults. *If $E$ is a stable extension of $\langle W, D \rangle$, then $E = \mathrm{Th}(W \cup \{ concl(\delta) \mid \delta \in G \})$.*

*Example 3.* Let $W = \emptyset$, $W' = \{x\}$, $D_1 = \{\frac{x:y}{\neg y}, \frac{\neg x:y}{\neg y}\}$, and $D_2 = \{\frac{x:z}{\neg y}, \frac{x:y}{\neg z}\}$. The default theory $\langle W, D_1 \rangle$ has only the stable extension $\mathrm{Th}(W)$. The default theory $\langle \{x\}, D_1 \rangle$ has no stable extension. The default theory $\langle \{x\}, D_2 \rangle$ has the stable extensions $\mathrm{Th}(\{x, \neg y\})$ and $\mathrm{Th}(\{x, \neg z\})$.

The following example illustrates that a default theory might contain "contradicting" default rules that cannot be avoided in the process of determining extension existence. Informally, such default rules prohibit stable extensions. Note that there are also less obvious situations where "chains" of such default rules interact with each other.

*Example 4.* Consider $W'$ and $D_2$ from Example 3 and let $D_2' = D_2 \cup \{\frac{\top:\beta}{\neg\beta}\}$ for some formula $\beta$. The default theory $\langle W', D_2' \rangle$ has no stable extension $\mathrm{Th}(W)$ unless $W \cup \{\neg y\} \models \neg\beta$ or $W \cup \{\neg z\} \models \neg\beta$.

Technically, the definition of stable extensions allows inconsistent stable extensions. However, Marek and Truszczyński showed that inconsistent extensions only occur if the set $W$ is already inconsistent where $\langle W, D \rangle$ is the theory of interest [25, Corollary 3.60]. An immediate consequence of this result explains the interplay between consistency and stability of extensions more subtle: (i) if $W$ is consistent, then every stable extension of $\langle W, D \rangle$ is consistent, and (ii) if $W$ is inconsistent, then $\langle W, D \rangle$ has a stable extension. In Case (ii) the stable extension consists of all formulas $\mathcal{L}$. Hence, it makes sense to consider only consistent stable extensions as the relevant ones. Moreover, we refer by $\mathrm{SE}(\langle W, D \rangle)$ to the set of all consistent stable extensions of $\langle W, D \rangle$.

A main computational problem for DL is the *extension existence problem*, defined as follows where $\mathcal{F}$ is a class of propositional formulas: The extension existence problem, $\mathrm{EXT}(\mathcal{F})$, asks, given an $\mathcal{F}$-default theory $\langle W, D \rangle$, whether $\langle W, D \rangle$ has a consistent stable extension.

The following proposition summarises relevant results for the extension existence problem for certain classes of formulas.

**Proposition 5.** *(1)* $\mathrm{EXT}(\mathrm{CNF})$ *is* $\Sigma_2^p$*-complete [21], (2)* $\mathrm{EXT}(\mathrm{HORN})$ *is* NP-*complete [38, 39], and (3)* $\mathrm{EXT}(\mathrm{POSITIVE\text{-}UNIT}) \in \mathrm{P}$ *[1].*

## 2.2 The Implication Problem

The implication problem is an important (sub-)problem when reasoning with default theories. In the following, we first formally introduce the implication problem for classes of propositional formulas, and then state its (classical) computational complexity for the classes HORN and KROM.

The implication problem $\mathrm{IMP}(\mathcal{F})$ asks, given a set $\Phi$ of $\mathcal{F}$-formulas and a formula $\psi \in \mathcal{F}$, whether $\Phi \models \psi$ holds.

Beyersdorff et al. [1] have considered all Boolean fragments of $\mathrm{IMP}(\mathcal{F})$ and completely classified its computational complexity concerning the framework of Post's lattice. However, Post's lattice talks only about restrictions on allowed Boolean functions. Since several subclasses of CNF, like HORN or KROM, use the Boolean functions "$\wedge$","$\neg$", and "$\vee$", such classes are unrestricted from the perspective of Post's lattice. Still, efficient algorithms are known for such classes from propositional satisfiability. The next results state a similar behaviour for the implication problem. The proof can be found in an extended version [15].

**Lemma 6.** $\mathrm{IMP}(\mathrm{KROM}) \in \mathrm{P}$.

Similar to the proof of Lemma 6 one can show the same complexity for the implication problem of HORN formulas. However, its complexity is already known from the work by Stillman [38].

**Proposition 7 ([38, Lemma 2.3]).** $\mathrm{IMP}(\mathrm{HORN}) \in \mathrm{P}$.

# 3   Strong Backdoors

In this section, we lift the concept of backdoors to the world of default logic. First, we review backdoors from the propositional setting [41, 42], where a backdoor is a subset of the variables of a given formula. Formally, for a class $\mathcal{F}$ of formulas and a formula $\varphi$, a *strong $\mathcal{F}$-backdoor* is a set $B$ of variables such that for all assignments $\theta \in \mathbb{A}(B)$, it holds that $\varphi[\theta] \in \mathcal{F}$.

Backdoors in propositional satisfiability follow the binary character of truth assignments. Each variable of a given formula is considered to be either true or false. However, reasoning in default logic has a ternary character. When we consider consistent stable extensions of a given default theory then one of the following three cases holds for some formula $\varphi$ with respect to an extension $E$: (i) $\varphi$ is contained in $E$, (ii) the negation $\neg\varphi$ is contained in $E$, or (iii) neither $\varphi$ nor $\neg\varphi$ is contained in $E$ (e.g., for the theory $\langle\{x\}, D_2\rangle$, from Example 3, neither $b$ nor $\neg b$ is contained in any of the two stable extensions, and $b$ is a variable). Since we need to weave this trichotomous point of view into a backdoor definition for default logic, the original definition of backdoors cannot immediately be transferred (from the SAT setting) to the scene of default logic. The first step is a notion of *extended literals* and *reducts*. The latter step can be seen as a generalisation of assignment functions to our setting.

**Definition 8 (Extended literals and reducts).** *An* extended literal *is a literal or a fresh variable $x_\varepsilon$. For convenience, we further define $\sim\ell = x$ if $\ell = \neg x$ and $\sim\ell = \neg x$ if $\ell = x$. Given a formula $\varphi$ and an extended literal $\ell$, then the* reduct $\rho_\ell(\varphi)$ *is obtained from $\varphi$ such that*

1. *if $\ell$ is a literal: then all clauses of $\varphi$ that contain $\ell$ are deleted and all literals $\sim\ell$ are deleted from all clauses of $\varphi$,*
2. *if $\ell = x_\varepsilon$: then all occurrences of literals $\neg x, x$ are deleted from all clauses of $\varphi$.*

*Let $\langle W, D\rangle$ be a default theory and $\ell$ be an extended literal, then*

$$\rho_\ell(W, D) := \left( \rho_\ell(W), \left\{ \frac{\rho_\ell(\alpha) : \rho_x(\beta)}{\rho_\ell(\gamma) \wedge y_i} \;\middle|\; \delta_i = \frac{\alpha : \beta}{\gamma} \in D \right\} \right),$$

*where $y_i$ is a fresh variable, and $\rho_\ell(W)$ is $\bigcup_{\omega \in W} \rho_\ell(\omega)$.*

Later (in the proof of Lemma 15), we will see why we need the $y_i$s.

*Example 9.* Given the default theory $\langle W, D\rangle = \{\{x\}, \{\frac{x:y}{\neg y \vee x}\}\}$ we will exemplify the notion of reductions for $x$, $\neg x$, and $x_\varepsilon$. An application of Definition 8 leads to $\rho_x(W, D) = \langle\{\top\}, \{\frac{\top:y}{y_1}\}\rangle$, $\rho_{\neg x}(W, D) = \langle\{\bot\}, \{\frac{\bot:y}{\neg y \wedge y_1}\}\rangle = \rho_{x_\varepsilon}(W, D)$.

In the next step, we incorporate the notion of extended literals into sets of assignments. Therefore, we introduce *threefold assignment sets*. Let $X$ be a set of variables, then we define

$$\mathbb{T}(X) := \{\{a_1, \ldots, a_{|X|}\} \mid x \in X \text{ and } a_i \in \{x, \neg x, x_\varepsilon\}\}.$$

Technically, $\mathbb{A}(X) \subsetneq \mathbb{T}(X)$ holds. However, $\mathbb{T}(X)$ additionally contains variables $x_\varepsilon$ that will behave as "don't care" variables encompassing the trichotomous reasoning approach explained above. For $Y \in \mathbb{T}(X)$ the reduct $\rho_Y(W, D)$ is the consecutive application of all $\rho_y(\cdot)$ for $y \in Y$ to $\langle W, D \rangle$. Observe that the order in which we apply the reducts to $\langle W, D \rangle$ is not important.

The following proposition ensures that the addition of the $y_i$s from Definition 8 does not influence negatively our reasoning process, i.e., implication of formulas is invariant under adding conjuncts of fresh variables to the premise.

**Proposition 10.** *Let* $\varphi, \psi \in$ CNF *be two formulas and* $y \notin \mathrm{Vars}(\varphi) \cup \mathrm{Vars}(\psi)$. *Then* $\varphi \models \psi$ *if and only if* $\varphi \wedge y \models \psi$.

Now we show that implication for CNF formulas that do not contain tautological clauses is invariant under the application of "deletion reducts" $\rho_{x_\varepsilon}(\cdot)$. The proof of the following results can be found in an extended version [15].

**Lemma 11.** *Let* $\psi, \varphi \in$ CNF *be two formulas that do not contain tautological clauses. If* $\psi \models \varphi$, *then* $\rho_{x_\varepsilon}(\psi) \models \rho_{x_\varepsilon}(\varphi)$ *for every variable* $x \in \mathrm{Vars}(\varphi) \cup \mathrm{Vars}(\psi)$.

The next lemma shows that implication for CNF formulas is invariant under the application of reducts over $\mathbb{A}$, i.e., the usual assignments.

**Lemma 12.** *Let* $\psi, \varphi$ *be two* CNF *formulas, and* $X \subseteq \mathrm{Vars}(\psi) \cup \mathrm{Vars}(\varphi)$. *If* $\psi \models \varphi$, *then* $\rho_Y(\psi) \models \rho_Y(\varphi)$ *holds for every set* $Y \in \mathbb{A}(X)$.

We denote by BD-IMP(CNF $\to \mathcal{F}$) the parameterised version of the problem IMP(CNF) where additionally a strong $\mathcal{F}$-backdoor is given and the parameter is the size of the strong $\mathcal{F}$-backdoor.

**Corollary 13.** *Given a class* $\mathcal{F} \in \{$POSITIVE-UNIT, HORN, KROM$\}$ *of CNF formulas. Then* BD-IMP(CNF $\to \mathcal{F}$) $\in$ FPT.

A combination of Lemmas 11 and 12 now yields a generalisation for CNF formulas that do not contain tautological clauses. Note that the crucial difference is the use of $\mathbb{T}$ instead of $\mathbb{A}$ in the claim of the result.

**Corollary 14.** *Let* $\psi, \varphi$ *be two* CNF *formulas that do not contain tautological clauses, and* $X \subseteq \mathrm{Vars}(E) \cup \mathrm{Vars}(\varphi)$ *be a set of variables. If* $\psi \models \varphi$ *then for every set* $Y \in \mathbb{T}(X)$ *it holds* $\rho_Y(\psi) \models \rho_Y(\varphi)$.

The following lemma is an important cornerstone for the upcoming section. It intuitively states that we do not loose any stable extensions under the application of reducts. Before we can start with the lemma we need to introduce a bit of notion. For a set $D = \{\delta_1, \ldots, \delta_n\}$ of default rules and a set $E$ of formulas we define y-concl$(D, E) := \{ concl(\delta_i) \mid 1 \le i \le n, \delta_i \in D, E \models y_i \}$, that is, the set of conclusions of default rules $\delta_i$ such that $y_i$ is implied by all formulas in $E$. Further, for a set $X$ of variables, we will extend the notion for SE$(\cdot)$ as follows:

$$\mathrm{SE}(\langle W, D \rangle, X) := \bigcup_{Y \in \mathbb{T}(X)} \{ \mathrm{Th}(W \cup \text{y-concl}(D, E)) \mid E \in \mathrm{SE}(\rho_Y(W, D)) \}.$$

**Lemma 15.** *Let $\langle W, D\rangle$ be a CNF default theory with formulas that do not contain tautological clauses, and $X$ be a set of variables from $\mathrm{Vars}(W, D)$. Then, $\mathrm{SE}(\langle W, D\rangle) \subseteq \mathrm{SE}(\langle W, D\rangle, X)$.*

*Proof.* Let $\langle W, D\rangle$ be the given default theory, $X \subseteq \mathrm{Vars}(W, D)$, and $E \in \mathrm{SE}(\langle W, D\rangle)$ be a consistent stable extension of $\langle W, D\rangle$.

Now suppose for contradiction that $E \notin \mathrm{SE}(\langle W, D\rangle, X)$. Further, let $G$ be the set of generating defaults of $E$ by Proposition 2, and w.l.o.g. let $G := \{\delta_1, \ldots, \delta_k\}$ also denote the order in which these defaults are applied. Thus it holds that $E = \mathrm{Th}(W \cup \{concl(\delta) \mid \delta \in G\})$. Hence, $W \models prereq(\delta_1)$ holds and further fix a $Y \in \mathbb{T}(X)$ which agrees with $E$ on the implied literals from $\mathrm{Vars}(W, D)$, i.e., $x \in Y$ if $E \models x$ for $x \in \mathrm{Vars}(W, D)$, $\neg x \in Y$ if $\models \neg x$, and $x_\varepsilon \in Y$ otherwise. Then, by Corollary 14 we know that also $\bigwedge_{\omega \in W} \rho_Y(\omega) \models \rho_Y(prereq(\delta_1))$ is true. Furthermore, we get that

$$\bigwedge_{\omega \in W} \rho_Y(\omega) \wedge \bigwedge_{1 \leq j \leq i} \rho_Y(concl(\delta_j)) \models \rho_Y(prereq(\delta_{i+1}))$$

holds for $i < k$. Thus, by definition of $\rho_Y(W, D)$, the reducts of the knowledge base $W$ and the derived conclusions together trivially imply the $y_i$s, i.e., it holds that

$$\bigwedge_{\omega \in W} \rho_Y(\omega) \wedge \bigwedge_{1 \leq i \leq k} \rho_Y(concl(\delta_i)) \models \bigwedge_{1 \leq i \leq k} y_i.$$

As neither $E \models prereq(\delta)$ holds for some $\delta \in D \setminus G$, nor $E \cup \{concl(\delta) \mid \delta \in G\} \models \delta'$ is true for some $\delta' \in D \setminus G$, $E$ is a consistent set, and $Y$ agrees with $E$ on the implied variables from $\mathrm{Vars}(W, D)$, we get that no further default rule $\delta$ is triggered by $\rho_Y(W)$ or $\rho_Y(W \cup \{concl(\delta) \mid \delta \in D \setminus G\})$.

Further, it holds that no justification is violated as $E \models \neg\beta$ for some $\beta \in \bigcup_{\delta \in G} just(\delta)$ would imply that $\rho_Y(E) \models \neg\rho_Y(\beta)$ also holds by Corollary 14. Thus, eventually $E' = \mathrm{Th}(\rho_Y(W) \cup \{\rho_Y(concl(\delta)) \mid \delta \in G\})$ is a stable extension with respect to $\rho_Y(W, D)$. But, the set of conclusions of $G$ coincides with y-concl$(D, E')$ therefore

$$E = \mathrm{Th}(W \cup \{concl(\delta) \mid \delta \in G\})$$
$$= \mathrm{Th}(W \cup \text{y-concl}(D, E')) \in \mathrm{SE}(\langle W, D\rangle, X)$$

holds, which contradicts our assumption. Thus, the lemma applies. □

We have seen that it is important to disallow tautological clauses. However, the detection of this kind of clauses is possible in polynomial time. Therefore, we assume in the following that a given theory contains no tautological clauses. This is not a very weak restriction as (i) $\varphi \wedge C \equiv \varphi$ for any tautological clause $C$, and (ii) $C \equiv \top$ for any tautological clause $C$.

The following example illustrates how reducts maintain existence of stable extensions.

*Example 16.* The default theory $\langle W, D \rangle = \{\{x\}, \{\frac{x:y}{\neg y \lor x}\}\}$ has the extension $E := \text{Th}(x, \neg y \lor x)$ and yields the following cases for the set $B = \{x\}$: $\rho_x(W, D) = \langle \{\top\}, \{\frac{\top:y}{y_1}\} \rangle$, yields $\text{SE}(\rho_x(W, D)) = \{\text{Th}(y_1)\}$, and, both, $\rho_{\neg x}(W, D)$ and $\rho_{x_\varepsilon}(W, D)$ yield an empty set of stable extensions. Thus, with y-concl$(D, \text{Th}(y_1)) = \{\neg y \lor x\}$ we get $\text{Th}(\{\neg y \lor x\} \cup \{x\})$, which is equivalent to the extension $E$ of $\langle W, D \rangle$.

Now, we are in the position to present a definition of strong backdoors for default logic.

**Definition 17 (Strong Backdoors for Default Logic).** *Given a* CNF *default theory* $\langle W, D \rangle$, *a set* $B \subseteq \text{Vars}(W, D)$ *of variables, and a class* $\mathcal{F}$ *of formulas. We say that* $B$ *is a* strong $\mathcal{F}$-backdoor *if for each* $Y \in \mathbb{T}(B)$ *the reduct* $\rho_Y(W, D)$ *is an* $\mathcal{F}$-default theory.

# 4 Backdoor Evaluation

In this section, we investigate the evaluation of strong backdoors for the extension existence problem in default logic with respect to different classes of CNF formulas. Formally, the problem of strong backdoor evaluation $\text{EVALEXT}(\mathcal{F} \to \mathcal{F}')$ for extension existence is defined as follows. Given an $\mathcal{F}$-default theory $\langle W, D \rangle$ and a strong $\mathcal{F}'$-backdoor $B \subseteq \text{Vars}(W) \cup \text{Vars}(D)$, asking does $\langle W, D \rangle$ have a stable extension?

First, we study the complexity of the "extension checking problem", which is a main task we need to accomplish when using backdoors as our approach following Lemma 15 yields only "stable extension candidates". Formally, given a default theory $\langle W, D \rangle$ and a finite set $\Phi$ of formulas, the problem EC asks whether $\text{Th}(\Phi) \in \text{SE}(\langle W, D \rangle)$ holds.

Rosati [34] classified the extension checking problem as complete for the complexity class $\Theta_2^P = \Delta_2^P[\log]$, which allows only logarithmic many oracle questions to an NP oracle. We will later see that a simpler version suffices for our complexity analysis. Therefore, we state in Algorithm 1 an adaption of Rosatis algorithm [34, Figure 1] to our notation showing containment (only) in $\Delta_2^P$.

**Proposition 18 ([34, Figure 1, Theorem 4]).** EC $\in \Delta_2^P$.

In a way, extension checking can be compared to model checking in logic. In default logic the complexity of the extension existence problem EXT is twofold: using the approach of Proposition 2 (i) one has to non-deterministically guess the set (and ordering) of the generating defaults, and (ii) one has to verify whether the generating defaults lead to an extension. For (ii), one needs to answer quadratic many implication questions. Hence, the problem is in $\text{NP}^{\text{NP}}$. Thus, a straightforward approach for EC omits the non-determinism in (i) and achieves the result in $\text{P}^{\text{NP}}$.

---

**Algorithm 1.** Extension checking algorithm [34, Theorem 4]

---

**Input**: Set $E$ of formulas and a default theory $\langle W, D \rangle$
**Output**: True if and only if $E$ is a stable extension of $\langle W, D \rangle$
1   $D' := \emptyset$
2   **forall the** $\frac{\alpha:\beta}{\gamma} \in D$ **do** // (1) Classify unviolated justifications.
3     **if** $E \not\models \neg\beta$ **then**   $D' := D' \cup \{\frac{\alpha:}{\gamma}\}$
    // (2) Compute extension candidate of justification-free theory.
4   $E' := W$
5   **while** $E'$ *did change in the last iteration* **do**
6     **forall the** $\frac{\alpha:}{\gamma} \in D'$ **do**
7       **if** $E' \models \alpha$ **then**   $E' := E' \wedge \gamma$
    // (3) Does the candidate match the extension?
8   **if** $E \models E'$ **and** $E' \models E$ **then return** true **else return** false

---

**Theorem 19.** EVALEXT(CNF $\rightarrow$ HORN) $\in$ para-NP.

*Proof.* Let $\langle W, D \rangle$ be a given CNF default theory and $B \subseteq \mathrm{Vars}(W, D)$ be the given backdoor. In order to evaluate the backdoor we have to consider the $|\mathbb{T}(X)| = 3^{|B|}$ many different reducts to HORN default theories. For each of them we have to non-deterministically guess a set of generating defaults $G$. Then, we use Algorithm 1 to verify whether $W \wedge \bigwedge_{g \in G} g$ is a stable extension (extensions can be represented by generating defaults; see Proposition 2). IMP(HORN) $\in$ P by Proposition 7. Hence, stable extension checking is in P for HORN formulas. Then, after finding an extension $E$ with respect to the reduct default theory $\rho_Y(W, D)$, we need to compute the corresponding extension $E'$ with respect to the original default theory. Here we just need to verify simple implication questions of the form $E \models y_i$ for $1 \leq i \leq |D|$. Next, we need to verify whether $E'$ is a valid extension for $\langle W, D \rangle$ using Algorithm 1. Note that Corollary 13 shows that the implication problem of propositional HORN formulas parameterised by the size of the backdoor is in FPT, hence we can compute the implication questions inline. As the length of the used formulas is bounded by the input size and the relevant parameter is the same as for the input this step runs in fpt-time. Together this yields a para-NP algorithm.     $\square$

The following corollary summarises the results for the remaining classes. Detailed proofs can be found in an extended version [15].

**Corollary 20.** *1.* EVALEXT(CNF $\rightarrow$ MONOTONE) $\in$ para-$\Delta_2^p$,
*2.* EVALEXT(CNF $\rightarrow$ KROM) $\in$ para-NP, *and*
*3.* EVALEXT(CNF $\rightarrow$ POSITIVE-UNIT) $\in$ FPT.

## 5   Backdoor Detection

In this section, we study the problem of finding backdoors, formalised in terms of the following parameterised problem. The problem BDDETECT(CNF $\rightarrow \mathcal{F}$) asks, given a CNF default theory $T$ and an integer $k$, parameterised by $k$, if $T$ has a strong $\mathcal{F}$-backdoor of size at most $k$. If the target class $\mathcal{F}'$ is clause-induced, we can use a decision algorithm for BDDETECT($\mathcal{F} \to \mathcal{F}'$) to find the backdoor using self-reduction [11,37]. The proof can be found in an extended version [15].

**Lemma 21.** *Let $\mathcal{F}$ be a clause-induced class of* CNF *formulas. If* BDDETECT *(*CNF $\to \mathcal{F}$*) is fixed-parameter tractable, then also computing a strong $\mathcal{F}$-backdoor of size at most $k$ of a given default theory $T$ is fixed-parameter tractable (for parameter $k$).*

The following theorem provides interesting target classes, where we can determining backdoors in fpt-time.

**Theorem 22.** *Let $\mathcal{C} \in \{$HORN, POSITIVE-UNIT, KROM, MONOTONE$\}$, then the problem* BDDETECT(CNF $\to \mathcal{C}$) $\in$ FPT.

*Proof.* Let $\langle W, D \rangle$ be a CNF default theory and $\mathsf{F} := W \cup \{ \mathit{prereq}(\delta),$ $\mathit{just}(\delta), \mathit{concl}(\delta) \mid \delta \in D \}$. Since each class $\mathcal{C} \in \{$HORN, POSITIVE-UNIT, KROM, MONOTONE$\}$ is clause-induced and then obviously $\rho_Z(\varphi) \subseteq \rho_Y(\varphi)$ holds for any $Z \in \mathbb{T}(X)$, we have to consider only the case $Y = \{ x_\varepsilon \mid x \in X \}$ to construct a strong $\mathcal{C}$-backdoor of $\langle W, D \rangle$.

Case $\mathcal{C} =$ MONOTONE: A CNF formula $\varphi$ is monotone if every literal appears only positively in any clause $C \in \varphi$ where $\varphi \in \mathsf{F}$. We can trivially construct a smallest strong MONOTONE-backdoor by taking all negative literals of clauses in formulas of $\mathsf{F}$ in linear time. Hence, the claim holds.

For $\mathcal{C} \in \{$HORN, POSITIVE-UNIT, KROM$\}$ we follow known constructions from the propositional setting [35]. Therefore, we consider certain (hyper-)graph representations of the given theory and establish that a set $B \subseteq \mathrm{Vars}(\mathsf{F})$ is a strong $\mathcal{C}$-backdoor of $\langle W, D \rangle$ if and only if $B$ is a $d$- hitting set of the respective (hyper-)graph representation of $\langle W, D \rangle$, where $d$ depends on the class of formulas, i.e., $d = 2$ for HORN and POSITIVE-UNIT and $d = 3$ for KROM. A 2- hitting set (*vertex cover*) of a graph $G = (V, E)$ is a set $S \subseteq V$ such that for every edge $uv \in E$ we have $\{u, v\} \cap S \neq \emptyset$. A *3-hitting set* of a hypergraph $H = (\mathsf{V}, \mathsf{E})$, where f.a. $E \in \mathsf{E}$ it holds $|E| \leq 3$, is a set $S \subseteq \mathsf{V}$ such that for every hyperedge $E \in \mathsf{E}$ we have $E \cap S \neq \emptyset$. Then, a vertex cover of size at most $k$, if it exists, can be found in time $O(1.2738^k + kn)$ [5] and a 3-hitting set of size at most $k$, if it exists, can be found in time $O(2.179^k + n^3)$ [14], which gives us then a strong $\mathcal{C}$-backdoor of $\langle W, D \rangle$. It remains to define the specific graph representations and to establish the connection to strong $\mathcal{C}$-backdoors.

Definition of the various (hyper-)graphs: For $\mathcal{C} =$ HORN we define a graph $G_T^+$ on the set of variables of $\mathsf{F}$, where two distinct variables $x$ and $y$ are joined by an edge if there is a formula $\varphi \in \mathsf{F}$ and some clause $C \in \varphi$ with $x, y \in C$. For $\mathcal{C} =$ POSITIVE-UNIT we define a graph $G_T$ on the set of variables of $\mathsf{F}$, where two distinct variables $x$ and $y$ are joined by an edge if there is a formula $\varphi \in \mathsf{F}$ and some clause $C \in \varphi$ with $l_x, l_y \in C$ where $l_x \in \{x, \neg x\}$ and $l_y \in \{y, \neg y\}$. For $\mathcal{C} =$ KROM we define a hypergraph $H_T$ on the variables $\mathrm{Vars}(\mathsf{F})$ where distinct variables $x, y, z$ are joined by a hyperedge if there is a formula $\varphi \in \mathsf{F}$ and some clause $C \in \varphi$ with $\{x, y, z\} \subseteq \text{Vars}(C)$. The correctness proof is shown in the report [15].     □

Now, we can use Theorem 22 to strengthen the results of Theorem 19 and Corollary 20 by dropping the assumption that the backdoor is given.

**Corollary 23.** *The problem* EVALEXT(CNF → $\mathcal{C}$) *parameterised by the size of a smallest strong $\mathcal{C}$-backdoor of the given theory is (1) in* para-$\Delta_2^P$ *if* $\mathcal{C} =$ MONOTONE, *(2) in* para-NP *if* $\mathcal{C} \in \{\text{HORN}, \text{KROM}\}$, *and (3) in* FPT *if* $\mathcal{C} =$ POSITIVE-UNIT.

## 6   Conclusion

We have introduced a notion of strong backdoors for propositional default logic. In particular, we investigated on the parameterised decision problems backdoor detection and backdoor evaluation. We have established that backdoor detection for the classes CNF, HORN, KROM, MONOTONE, and POSITIVE-UNIT are fixed-parameter tractable whereas for evaluation the classification is more complex. If CNF is the starting class and HORN or KROM is the target class, then backdoor evaluation is in para-NP. If MONOTONE is the target class, then backdoor evaluation is in para-$\Delta_2^P$, which thus can be solved by an fpt-algorithm that can query a SAT solver multiple times [9]. For POSITIVE-UNIT as target class backdoor evaluation is fixed-parameter tractable.

An interesting task for future research is to consider the remaining Schaefer classes [36], e.g., dual-Horn, 1- and 0-valid, as well as the classes renamable-Horn and QHorn [2,3], and to investigate whether we can generalise the concept of Theorem 19. We have established for backdoor evaluation the upper bounds para-NP and para-$\Delta_2^P$, respectively. We think that it would also be interesting to establish corresponding lower bounds. Finally, a direct application of quantified Boolean formulas in the context of propositional default logic, for instance, via the work of Egly et al. [13] or exploiting backdoors similar to results by Fichte and Szeider [16], might yield new insights.

**Acknowledgements.** The first author gratefully acknowledges support by the Austrian Science Fund (FWF), Grant Y698. He is also affiliated with the Institute of Computer Science and Computational Science at University of Potsdam, Germany. The second and third author gratefully acknowledge support by the German Research Foundation (DFG), Grant ME 4279/1-1. The authors thank Jonni Virtema for pointing out Lemma 6 and Sebastian Ordyniak for discussions on Lemma 11. The authors acknowledge the helpful comments of the anonymous reviewers.

## References

1. Beyersdorff, O., Meier, A., Thomas, M., Vollmer, H.: The complexity of reasoning for fragments of default logic. J. Logic Comput. **22**(3), 587–604 (2012)
2. Boros, E., Crama, Y., Hammer, P.L.: Polynomial-time inference of all valid implications for horn and related formulae. Ann. Math. Artif. Intell. **1**(1–4), 21–32 (1990)

3. Boros, E., Hammer, P.L., Sun, X.: Recognition of q-Horn formulae in linear time. Discrete Appl. Math. **55**(1), 1–13 (1994)
4. Chen, J., Chor, B., Fellows, M.R., Huang, X., Juedes, D.W., Kanji, I.A., Xia, G.: Tight lower bounds for certain parameterized NP-hard problems. Inf. Comput. **201**(2), 216–231 (2005)
5. Chen, J., Kanj, I.A., Xia, G.: Improved upper bounds for vertex cover. Theor. Comput. Sci. **411**(40–42), 3736–3756 (2010)
6. Courcelle, B.: Graph rewriting: an algebraic and logic approach. In: van Leeuwen, J. (ed.) Handbook of Theoretical Computer Science. Volume Formal Models and Semantics, pp. 193–242. Elsevier Science Publishers, North-Holland (1990)
7. Courcelle, B., Engelfriet, J.: Graph Structure and Monadic Second-order Logic: A Language Theoretic Approach. Cambridge University Press, Cambridge (2012)
8. Creignou, N., Meier, A., Thomas, M., Vollmer, H.: The complexity of reasoning for fragments of autoepistemic logic. ACM Trans. Comput. Logic **13**(2), 1–22 (2012)
9. de Haan, R., Szeider, S.: Fixed-parameter tractable reductions to SAT. In: Sinz, C., Egly, U. (eds.) SAT 2014. LNCS, vol. 8561, pp. 85–102. Springer, Heidelberg (2014)
10. Downey, R.G., Fellows, M.R.: Parameterized Complexity. Monographs in Computer Science. Springer, New York (1999)
11. Downey, R.G., Fellows, M.R.: Fundamentals of Parameterized Complexity. Texts in Computer Science. Springer, London (2013)
12. Dvořák, W., Ordyniak, S., Szeider, S.: Augmenting tractable fragments of abstract argumentation. Artif. Intell. **186**, 157–173 (2012)
13. Egly, U., Eiter, T., Tompits, H., Woltran, S.: Solving advanced reasoning tasks using quantified boolean formulas. In: Kautz, H., Porter, B. (eds.) Proceedings of the 17th Conference on Artificial Intelligence (AAAI 2000), Austin, TX, USA, pp. 417–422. The AAAI Press, July 2000
14. Fernau, H.: A top-down approach to search-trees: improved algorithmics for 3-hitting set. Algorithmica **57**(1), 97–118 (2010)
15. Fichte, J., Meier, A., Schindler, I.: Strong Backdoors for Default Logic. Technical report. CoRR: abs/1483200, arXiv (2016)
16. Fichte, J.K., Szeider, S.: Backdoors to normality for disjunctive logic programs. ACM Trans. Comput. Logic **17**(1), 7 (2015)
17. Fichte, J.K., Szeider, S.: Backdoors to tractable answer-set programming. Artif. Intell. **220**, 64–103 (2015)
18. Flum, J., Grohe, M.: Describing parameterized complexity classes. Inf. Comput. **187**(2), 291–319 (2003)
19. Flum, J., Grohe, M.: Parameterized Complexity Theory. Theoretical Computer Science, vol. 14. Springer, Heidelberg (2006)
20. Gaspers, S., Szeider, S.: Backdoors to satisfaction. In: Bodlaender, H.L., Downey, R., Fomin, F.V., Marx, D. (eds.) Fellows Festschrift 2012. LNCS, vol. 7370, pp. 287–317. Springer, Heidelberg (2012)
21. Gottlob, G.: Complexity results for nonmonotonic logics. J. Logic Comput. **2**(3), 397–425 (1992)
22. Gottlob, G., Pichler, R., Wei, F.: Bounded treewidth as a key to tractability of knowledge representation and reasoning. Artif. Intell. **174**(1), 105–132 (2010)
23. Gottlob, G., Scarcello, F., Sideri, M.: Fixed-parameter complexity in AI and nonmonotonic reasoning. Artif. Intell. **138**(1–2), 55–86 (2002)
24. Gottlob, G., Szeider, S.: Fixed-parameter algorithms for artificial intelligence, constraint satisfaction, and database problems. Comput. J. **51**(3), 303–325 (2006). Survey paper

25. Marek, V.W., Truszczyński, M.: Nonmonotonic Logic: Context-Dependent Reasoning. Artificial Intelligence. Springer, Heidelberg (1993)
26. McCarthy, J.: Circumscription – a form of non-monotonic reasoning. Artif. Intell. **13**, 27–39 (1980)
27. McDermott, D., Doyle, J.: Non-montonic logic I. Artif. Intell. **13**, 41–72 (1980)
28. Meier, A., Schindler, I., Schmidt, J., Thomas, M., Vollmer, H.: On the parameterized complexity of non-monotonic logics. Arch. Math. Logic **54**(5–6), 685–710 (2015)
29. Moore, R.C.: Semantical considerations on modal logic. Artif. Intell. **25**, 75–94 (1985)
30. Ordyniak, S., Paulusma, D., Szeider, S.: Satisfiability of acyclic and almost acyclic CNF formulas. Theor. Comput. Sci. **481**, 85–99 (2013)
31. Papadimitriou, C.H.: Computational Complexity. Addison-Wesley, Reading (1994)
32. Pfandler, A., Rümmele, S., Szeider, S.: Backdoors to abduction. In: Rossi, F. (ed.) Proceedings of the 23rd International Joint Conference on Artificial Intelligence (IJCAI 2013), Beijing, China, pp. 1046–1052. The AAAI Press, August 2013
33. Reiter, R.: A logic for default reasoning. Artif. Intell. **13**, 81–132 (1980)
34. Rosati, R.: Model checking for nonmonotonic logics: algorithms and complexity. In: Dean, T. (ed.) Proceedings of the 16th International Joint Conference on Artificial Intelligence (ICJAI 1999), Stockholm, Sweden. The AAAI Press, July 1999
35. Samer, M., Szeider, S.: Fixed-parameter tractability. In: Biere, A., Heule, M., van Maaren, H., Walsh, T. (eds.) Handbook of Satisfiability, pp. 425–454. IOS Press, Amsterdam (2009)
36. Schaefer, T.J.: The complexity of satisfiability problems. In: Lipton, R.J., Burkhard, W.A., Savitch, W.J., Friedman, E.P., Aho, A.V. (eds.) Proceedings of the 10th Annual ACM Symposium on Theory of Computing (STOC 1978), San Diego, CA, USA, pp. 216–226. Association for Computing Machinery, New York (1978)
37. Schnorr, C.-P.: On self-transformable combinatorial problems. In: König, H., Korte, B., Ritter, K. (eds.) Mathematical Programming at Oberwolfach. Mathematical Programming Studies, vol. 14, pp. 225–243. Springer, Heidelberg (1981)
38. Stillman, J.P.: It's not my default: the complexity of membership problems in restricted propositional default logics. In: Dietterich, T., Swartout, W. (eds.) Proceedings of the 8th National Conference on Artificial Intelligence (AAAI 1990), Boston, MA, USA, vol. 1, pp. 571–578. The AAAI Press, July 1990
39. Stillman, J.P.: The complexity of horn theories with normal unary defaults. In: Proceedings of the 8th Canadian Artificial Intelligence Conference (AI 1990) (1990)
40. Szeider, S.: On fixed-parameter tractable parameterizations of SAT. In: Giunchiglia, E., Tacchella, A. (eds.) SAT 2003. LNCS, vol. 2919, pp. 188–202. Springer, Heidelberg (2004)
41. Williams, R., Gomes, C., Selman, B.: Backdoors to typical case complexity. In: Gottlob, G., Walsh, T. (eds.) Proceedings of the 18th International Joint Conference on Artificial Intelligence (IJCAI 2003), Acapulco, Mexico, pp. 1173–1178. Morgan Kaufmann, August 2003
42. Williams, R., Gomes, C., Selman, B.: On the connections between backdoors, restarts, and heavy-tailedness in combinatorial search. In: Informal Proceedings of the 6th International Conference on Theory and Applications of Satisfiability Testing (SAT 2003), Portofino, Italy, pp. 222–230, May 2003

# The Normalized Autocorrelation Length of Random Max $r$-Sat Converges in Probability to $(1 - 1/2^r)/r$

Daniel Berend and Yochai Twitto[✉]

Ben-Gurion University, Beer Sheva 84105, Israel
{berend,twittoy}@cs.bgu.ac.il

**Abstract.** In this paper we show that the so-called normalized autocorrelation length of random Max $r$-Sat converges in probability to $(1 - 1/2^r)/r$, where $r$ is the number of literals in a clause. We also show that the correlation between the numbers of clauses satisfied by a random pair of assignments of distance $d = cn$, $0 \leq c \leq 1$, converges in probability to $((1 - c)^r - 1/2^r)/(1 - 1/2^r)$. The former quantity is of interest in the area of landscape analysis as a way to better understand problems and assess their hardness for local search heuristics. In [34], it has been shown that it may be calculated in polynomial time for any instance, and its mean value over all instances was discussed. Our results are based on a study of the variance of the number of clauses satisfied by a random assignment, and the covariance of the numbers of clauses satisfied by a random pair of assignments of an arbitrary distance. As part of this study, closed-form formulas for the expected value and variance of the latter two quantities are provided. Note that all results are relevant to random $r$-Sat as well.

**Keywords:** Combinatorial optimization · Max Sat · Fitness landscapes · Autocorrelation length · Local search

## 1   Introduction

In the Maximum Satisfiability (Max Sat) problem, we are given a multiset of clauses over some boolean variables. Each clause is a disjunction of literals (a variable or its negation) over different variables. We seek a truth (true/false) assignment for the variables, maximizing the number of satisfied (made true) clauses. In the Max $r$-Sat problem, each clause is restricted to consist of at most $r$ literals. Here we restrict our attention to instances with clauses consisting of exactly $r$ literals each. This restricted problem is also known as Max E$r$-Sat.

Let $n$ be the number of variables. Denote the variables by $v_1, v_2, \ldots, v_n$. The number of clauses is denoted by $m$, and the clauses by $C_1, C_2, \ldots, C_m$. We denote the clause-to-variable ratio by $\alpha = m/n$. We use the terms "positive variable" and "negative variable" to refer to a variable and to its negation, respectively.

© Springer International Publishing Switzerland 2016
N. Creignou and D. Le Berre (Eds.): SAT 2016, LNCS 9710, pp. 60–76, 2016.
DOI: 10.1007/978-3-319-40970-2_5

Whenever we find it convenient, we consider the truth values `true` and `false` as binary 1 and 0, respectively.

As Max $r$-Sat (for $r \geq 2$) is NP-hard [7, pp. 455–456], large-sized instances cannot be exactly solved in an efficient manner (unless $P = NP$), and one must resort to approximation algorithms and heuristics. Numerous methods have been suggested for solving Max $r$-Sat, e.g. [5,9,10,15,22,24,28,31,32], and an annual competition of solvers has been held since 2006 [6]. Satisfiability related questions attracted a lot of attention from the scientific community. As an example, one may consider the well-studied satisfiability threshold question [1,13,14,16,18, 26]. For a comprehensive overview of the whole domain of satisfiability we refer to [8].

Using Walsh analysis [20], an efficient way of calculating moments of the number of satisfied clauses of a given instance of Max $r$-Sat was suggested in [21]. Simulation results for the variance and higher moments of the number of clauses satisfied by a random assignment over the ensemble of all instances were provided as well. We provide closed-form formula, asymptotics, and convergence proof for the variance.

An interesting study of Max 3-Sat is provided in [29]. The authors claimed that many instances share similar statistical properties and provided empirical evidence for it. Simulation results on the autocorrelation of a random walk on the assignments space were provided for several instances, as well as extrapolation for the typical instance. Finally, a novel heuristic was introduced, ALGH, which exploits long-range correlations found in the problem's landscape. This heuristic outperformed GSAT [32] and WSAT [31]. A slightly better version of this heuristic, based on clustering instead of averaging, is provided in another paper [30] of the same authors. This version turned out to outperform all the heuristics implemented at that time in the Sat solver framework UBCSAT [35]. Our convergence in probability proofs mathematically validate their simulative results regarding similarity for several statistical properties, including the long-range correlation they used for their heuristics.

In [23], the authors analyze how the way random instances are generated affects the autocorrelation and fitness-distance correlation. These quantities are considered fundamental to understanding the hardness of instances for local search algorithms. They raised the question of similarity of the landscape of different instances. In [3], the autocorrelation coefficient of several problems was calculated, and problem hardness was classified accordingly. We contribute one more result for this classification.

Elaboration on correlations and on the way of harnessing them to design well-performing local search heuristics and memetic algorithms is provided in [27]. The importance of selecting an appropriate neighborhood operator for producing the smoothest possible landscape was emphasized. For some landscapes, the autocorrelation length is shown to be associated with the average distance between local optima. This may be used to facilitate the design of operators that lead memetic algorithms out of the basin of attraction of a local optimum they reached.

In [34], it is shown how to use Walsh decomposition [20] to efficiently calculate the exact autocorrelation function and autocorrelation length of any given instance of Max $r$-Sat. Furthermore, this decomposition is used to approximate the expectation of these quantities over the ensemble of all instances. The approximation is based on mean-field approximation [36] under some assumption on the statistical fluctuation of the approximated quantity. Formulas for these expectations are provided only in terms of Walsh coefficients, and thus give less insight as to their actual values. We substantially improve the result regarding the autocorrelation length, by showing its normalized version converges in probability to an explicit constant.

This paper deals with the variance of the number of clauses satisfied by a random assignment, and the covariance of the numbers of clauses satisfied by a random pair of assignments at an arbitrary distance. We obtain explicit formulas for the expected value and the variance of these quantities. Asymptotics of these expressions are provided as well. From the asymptotics we conclude that the variance of the number of clauses satisfied by a random assignment is usually quite close to the expected value of this variance. Based on these results, we show that the correlation between the numbers of clauses satisfied by a random pair of assignments of distance $d = cn$, $0 \leq c \leq 1$, converges in probability to $((1-c)^r - 1/2^r)/(1 - 1/2^r)$. Our main result is that the so-called normalized autocorrelation length [19] of (random) Max $r$-Sat converges in probability to $(1 - 1/2^r)/r$.

The latter quantity, which is closely related to the ruggedness of landscapes, is of interest in the area of landscape analysis [3,4,11,17,23,25,34]. It is fundamental to the theory and design of local search heuristics [12,27]. According to the autocorrelation length conjecture [33], in many landscapes, the number of local optima can be estimated using an expression based on this quantity. Our result reveals the normalized autocorrelation length of (random) Max $r$-Sat, improving a former result [34] that expressed it only in terms of Walsh coefficients [20].

In Sect. 2 we present our main results, and in Sect. 3 the proofs. Some elaboration and discussion are provided in Sect. 4. All our results immediately apply to random $r$-Sat, as both random Max $r$-Sat and random $r$-Sat deal with the same collection of random instances – the collection of random $r$-CNF formulas. We choose to present our results in the context of Max $r$-Sat, and to omit the prefix "random", assuming this is the default when not mentioned otherwise.

## 2    Main Results

Throughout the paper we deal with three probability spaces. The first consists of all instances with $m$ clauses of length $r$ over $n$ variables. As any $r$ of the variables may appear in a clause, and each may be positive or negative, the number of instances is $\left(\binom{n}{r}2^r\right)^m$. All instances are equally likely, namely each has a probability of $1/\left(\binom{n}{r}2^r\right)^m$. The second probability space consists of all $2^n$ equally likely truth assignments. The third consists of all the $2^n\binom{n}{d}$ equally likely pairs of truth assignments of distance $d$. The distance between two assignments is the so-called Hamming distance, i.e., the number of variables they assign differently.

We use the subscripts $\mathcal{I}$, $A$, and $d$ to specify that a certain quantity is associated with the first, second, or third probability space, respectively. We use $I$, $a$, and $(a,b)$ to denote a random instance, a random assignment, and a random pair of assignments at distance $d$ from each other, respectively. Let the random variable $S(I,a)$ ($R(I,a)$, resp.) be the number of clauses of $I$ satisfied (unsatisfied, resp.) by the assignment $a$.

For a given instance $I$, let $\rho(d) = \mathrm{Corr}_d(S(I,a),S(I,b))$ be the correlation (coefficient) between the numbers of clauses satisfied by a random pair of assignments at distance $d$ from each other. The autocorrelation length [19], given by $l = -1/\ln(|\rho(1)|)$, is a one-number summary of the ruggedness of the landscape of the instance. The higher its value, the smoother is the landscape. The *normalized* autocorrelation length is simply $l/n$. A slightly different quantity for summarizing ruggedness is the autocorrelation coefficient [2], defined by $\xi = 1/(1-\rho(1))$. Similarly, the *normalized* autocorrelation coefficient is $\xi/n$. These two measures, $l/n$ and $\xi/n$, are asymptotically the same. I.e., their quotient approaches 1 as $n$ grows larger. We arbitrarily choose to work with the latter.

This quantity converges in probability to a constant independent of the clause-to-variable ratio $\alpha = m/n$, as stated in our main theorem, which improves the result of [34]. There, this quantity is provided only in terms of Walsh coefficients [20], along with a mean-field approximation [36]. In our results regarding convergence in probability, here and afterward, the random variables are always defined on $\mathcal{I}$. Namely, they are defined on the probability space consisting of all (equally likely) instances with $m$ clauses of length $r$ over $n$ variables.

**Theorem 1.**    *For Max r-Sat:*

$$\frac{\xi}{n} \xrightarrow[n \to \infty]{\mathcal{P}} \frac{1 - 1/2^r}{r}.$$

To prove Theorem 1, we will first formulate and prove two more theorems. Besides being building blocks for the proof of the main theorem, each of these is of independent interest. The first summarizes our results regarding the variance of the number of clauses satisfied by a random assignment. In the provided asymptotics, we assume that $n \to \infty$, $m = \alpha n$ for some constant $\alpha > 0$, and $r$ is constant.

**Theorem 2.** *For Max r-Sat, the expected value and the variance (over all instances) of the variance of the number of clauses satisfied by a random assignment are given by:*

$$E_{\mathcal{I}}(V_A(S(I,a))) = \frac{m}{2^r}\left(1 - \frac{1}{2^r}\right), \qquad (2.A)$$

$$V_{\mathcal{I}}(V_A(S(I,a))) = \frac{2m(m-1)}{2^{4r}} \left( \sum_{t=0}^{r} \frac{\binom{r}{t}\binom{n-r}{r-t}}{\binom{n}{r}} \cdot 2^t - 1 \right) \tag{2.B.1}$$

$$= \frac{\alpha^2 r^2}{2^{4r-1}} n + O(1). \tag{2.B.2}$$

*In particular,*

$$\frac{V_A(S(I,a))}{n} \xrightarrow[n\to\infty]{\mathcal{P}} \frac{\alpha}{2^r} \left( 1 - \frac{1}{2^r} \right). \tag{2.C}$$

The second theorem generalizes the results further, and summarizes our results regarding the covariance of the numbers of clauses satisfied by a random pair of assignments at an arbitrary distance from each other. Here, in the asymptotics we also assume that $d = cn$, $0 \le c \le 1$.

**Theorem 3.** *For Max r-Sat, the expected value and the variance (over all instances) of the covariance of the numbers of clauses satisfied by a random pair of assignments of distance d are given by:*

$$E_{\mathcal{I}}(\mathrm{Cov}_d(S(I,a), S(I,b))) = \frac{m}{2^r} \left( \frac{\binom{n-r}{d}}{\binom{n}{d}} - \frac{1}{2^r} \right) \tag{3.A.1}$$

$$= \frac{\alpha}{2^r} \left( (1-c)^r - \frac{1}{2^r} \right) n + O(1), \tag{3.A.2}$$

$$V_{\mathcal{I}}(\mathrm{Cov}_d(S(I,a), S(I,b)))$$
$$= \frac{2m(m-1)}{2^{4r}} \left( \sum_{t=0}^{r} \left( \frac{\binom{r}{t}\binom{n-r}{r-t}}{\binom{n}{r}} \cdot 2^t \cdot \sum_{s=0}^{t} \frac{\binom{t}{s}\binom{n-t}{d-s}}{\binom{n}{d}} \cdot \frac{\binom{n-t}{d-s}}{\binom{n}{d}} \right) - 1 \right) \tag{3.B.1}$$

$$= \frac{\alpha^2 r^2 (2c-1)^2}{2^{4r-1}} n + O(1). \tag{3.B.2}$$

*In particular,*

$$\frac{\mathrm{Cov}_d(S(I,a), S(I,b))}{n} \xrightarrow[n\to\infty]{\mathcal{P}} \frac{\alpha}{2^r} \left( (1-c)^r - \frac{1}{2^r} \right). \tag{3.C}$$

Finally, notice that (2.C) and (3.C) together lead immediately to the following corollary regarding the convergence in probability of the correlation of the numbers of clauses satisfied by a random pair of assignments at an arbitrary distance from each other.

**Corollary 1.** *For Max r-Sat:*

$$\rho(d) \xrightarrow[n\to\infty]{\mathcal{P}} \frac{(1-c)^r - 1/2^r}{1 - 1/2^r}.$$

## 3    Proofs

Theorem 2 follows immediately from Theorem 3, by applying the latter with $d = 0$. We provide the proof of Theorem 3, and then the proof of Theorem 1, which rely heavily on the former theorems.

*Proof (Theorem 3).* As $R(I, a) = m - S(I, a)$, we may work with the covariance of the numbers of *unsatisfied* clauses, instead of that of the numbers of satisfied clauses. Define the following random variable:

$$R_i(I, a) = \begin{cases} 1, & \text{the assignment } a \text{ does not satisfy the clause } C_i, \\ 0, & \text{otherwise.} \end{cases}$$

We start with a single clause. For the sake of readability we write $R_i(a)$ instead of $R_i(I, a)$.

$$\begin{aligned}
\text{Cov}_d(R_i(a), R_i(b)) &= E_d(R_i(a) \cdot R_i(b)) - E_d(R_i(a)) \cdot E_d(R_i(b)) \\
&= P_d(R_i(a) = R_i(b) = 1) - P_d(R_i(a) = 1) \cdot P_d(R_i(b) = 1) \\
&= P_d(R_i(a) = 1) \cdot P_d(R_i(b) = 1 \mid R_i(a) = 1) - \frac{1}{2^{2r}} \\
&= \frac{1}{2^r} \left( \frac{\binom{n-r}{d}}{\binom{n}{d}} - \frac{1}{2^r} \right).
\end{aligned}$$

Next, we calculate the covariance for any specific instance. Again, we use the shorthand $R(a)$ instead of $R(I, a)$.

$$\begin{aligned}
\text{Cov}_d(R(a), R(b)) &= \text{Cov}_d \left( \sum_{i=1}^{m} R_i(a), \sum_{j=1}^{m} R_j(b) \right) \\
&= \sum_{i=1}^{m} \sum_{j=1}^{m} \text{Cov}_d(R_i(a), R_j(b)) \\
&= \sum_{i=1}^{m} \text{Cov}_d(R_i(a), R_i(b)) + 2 \sum_{1 \le i < j \le m} \text{Cov}_d(R_i(a), R_j(b)) \\
&= \frac{m}{2^r} \left( \frac{\binom{n-r}{d}}{\binom{n}{d}} - \frac{1}{2^r} \right) + 2 \sum_{1 \le i < j \le m} \text{Cov}_d(R_i(a), R_j(b)).
\end{aligned} \tag{1}$$

For $1 \le i < j \le m$ we have:

$$\begin{aligned}
E_\mathcal{I}(\text{Cov}_d(R_i(a), R_j(b))) &= E_\mathcal{I}(E_d(R_i(a) \cdot R_j(b))) - E_\mathcal{I}(E_d(R_i(a)) \cdot E_d(R_j(b))) \\
&= E_d(E_\mathcal{I}(R_i(a) \cdot R_j(b))) - \frac{1}{2^{2r}} \\
&= E_d(E_\mathcal{I}(R_i(a)) \cdot E_\mathcal{I}(R_j(b))) - \frac{1}{2^{2r}} \\
&= 0.
\end{aligned} \tag{2}$$

Here, the penultimate transition stems from the fact that, for a given pair of assignments $a, b$, the variables $R_i(a), R_j(b)$ are independent, as their associated clauses are selected uniformly at random and with repetitions from all possible clauses. Overall, the second addend on the right-hand side of (1) vanishes, which proves (3.A.1). The asymptotics provided in (3.A.2) is immediate.

Now, let us prove (3.B.1). We have:

$$V_{\mathcal{I}}(\mathrm{Cov}_d(R(a), R(b))) = V_{\mathcal{I}}\left(\frac{m}{2^r}\left(\frac{\binom{n-r}{d}}{\binom{n}{d}} - \frac{1}{2^r}\right) + 2\sum_{1\leq i<j\leq m}\mathrm{Cov}_d(R_i(a), R_j(b))\right)$$

$$= 4\sum_{1\leq i<j\leq m}V_{\mathcal{I}}(\mathrm{Cov}_d(R_i(a), R_j(b)))$$

$$+ 8\sum_{\substack{1\leq i<j\leq m \\ 1\leq k<l\leq m \\ (i<k)\vee(i=k\wedge j<l)}}\mathrm{Cov}_{\mathcal{I}}(\mathrm{Cov}_d(R_i(a), R_j(b)), \mathrm{Cov}_d(R_k(a), R_l(b)))$$

$$= 4g + 8\sum_{\substack{1\leq i<j\leq m \\ 1\leq k<l\leq m \\ (i<k)\vee(i=k\wedge j<l)}}h_{ijkl}$$

$$= 4g + 8h.$$

Next, we calculate $g$ as follows:

$$g = \sum_{1\leq i<j\leq m}V_{\mathcal{I}}(\mathrm{Cov}_d(R_i(a), R_j(b)))$$

$$= \sum_{1\leq i<j\leq m}\left(E_{\mathcal{I}}\left(\mathrm{Cov}_d^2\left(R_i(a), R_j(b)\right)\right) - E_{\mathcal{I}}^2\left(\mathrm{Cov}_d\left(R_i(a), R_j(b)\right)\right)\right)$$

$$= \sum_{1\leq i<j\leq m}\left(E_{\mathcal{I}}\left(\left(E_d(R_i(a)\cdot R_j(b)) - E_d(R_i(a))\cdot E_d(R_j(b))\right)^2\right) - 0\right) \quad // \text{ by (2)}$$

$$= \sum_{1\leq i<j\leq m}E_{\mathcal{I}}\left(\left(E_d(R_i(a)\cdot R_j(b)) - \frac{1}{2^{2r}}\right)^2\right)$$

$$= \sum_{1\leq i<j\leq m}E_{\mathcal{I}}\left(E_d^2(R_i(a)\cdot R_j(b)) - \frac{E_d(R_i(a)\cdot R_j(b))}{2^{2r-1}} + \frac{1}{2^{4r}}\right)$$

$$= \sum_{1\leq i<j\leq m}\left(E_{\mathcal{I}}\left(E_d^2(R_i(a)\cdot R_j(b))\right) - \frac{E_{\mathcal{I}}(E_d(R_i(a)\cdot R_j(b)))}{2^{2r-1}} + \frac{1}{2^{4r}}\right)$$

$$= \sum_{1\leq i<j\leq m}\left(E_{\mathcal{I}}\left(E_d^2(R_i(a)\cdot R_j(b))\right) - \frac{1}{2^{2r}2^{2r-1}} + \frac{1}{2^{4r}}\right) \quad // \text{ as done in (2)}$$

$$= \sum_{1\leq i<j\leq m}\left(E_{\mathcal{I}}\left(P_d^2(R_i(a) = R_j(b) = 1)\right) - \frac{1}{2^{4r}}\right).$$

To continue the calculation, we consider the clauses $C_i$ and $C_j$. We denote by $t$ the number of variables shared by $C_i$ and $C_j$. Let $s$ be the number of variables, out of the $t$ common ones, whose sign (as literals) is different in the two clauses. The remaining $t - s$ shared variables have the same sign in the two clauses.

The probability that $C_i$ and $C_j$ share exactly $t$ variables, from which exactly $s$ are of different sign, is

$$\frac{\binom{r}{t}\binom{n-r}{r-t}}{\binom{n}{r}} \cdot \frac{\binom{t}{s}}{2^t}.$$

Given $t$ and $s$, we have

$$P_d(R_i(a) = R_j(b) = 1)) = \frac{1}{2^r} \cdot \frac{\binom{n-t}{d-s}}{\binom{n}{d}} \cdot \frac{1}{2^{r-t}}.$$

Thus,

$$E_\mathcal{I}\left(P_d^2(R_i(a) = R_j(b) = 1)\right) = \sum_{t=0}^{r}\sum_{s=0}^{t} \frac{\binom{r}{t}\binom{n-r}{r-t}}{\binom{n}{r}} \cdot \frac{\binom{t}{s}}{2^t} \cdot \left(\frac{2^t}{2^{2r}} \cdot \frac{\binom{n-t}{d-s}}{\binom{n}{d}}\right)^2$$

$$= \frac{1}{2^{4r}} \sum_{t=0}^{r} \left(\frac{\binom{r}{t}\binom{n-r}{r-t}}{\binom{n}{r}} \cdot 2^t \cdot \sum_{s=0}^{t} \frac{\binom{t}{s}\binom{n-t}{d-s}}{\binom{n}{d}} \cdot \frac{\binom{n-t}{d-s}}{\binom{n}{d}}\right).$$

Plugging the last expression into the expression for $g$, we arrive at the final form of $g$:

$$g = \frac{m(m-1)}{2} \cdot \frac{1}{2^{4r}} \cdot \left(\sum_{t=0}^{r}\left(\frac{\binom{r}{t}\binom{n-r}{n-t}}{\binom{n}{r}} \cdot 2^t \cdot \sum_{s=0}^{t} \frac{\binom{t}{s}\binom{n-t}{d-s}}{\binom{n}{d}} \cdot \frac{\binom{n-t}{d-s}}{\binom{n}{d}}\right) - 1\right).$$

The expression for $4g$ is exactly the right-hand side of (3.B.1), so to conclude the proof it suffices to show that $h = 0$. In fact, we will see that every single term $h_{ijkl}$ in the sum appearing in the expression for $h$ vanishes. Let us expand this term.

$$h_{ijkl} = \mathrm{Cov}_\mathcal{I}\left(\mathrm{Cov}_d\left(R_i(a), R_j(b)\right), \mathrm{Cov}_d\left(R_k(a), R_l(b)\right)\right)$$

$$= \mathrm{Cov}_\mathcal{I}\left(E_d\left(R_i(a) \cdot R_j(b)\right) - \frac{1}{2^{2r}}, E_d\left(R_k(a) \cdot R_l(b)\right) - \frac{1}{2^{2r}}\right)$$

$$= \mathrm{Cov}_\mathcal{I}\left(P_d(R_i(a) = R_j(b) = 1), P_d(R_k(a) = R_l(b) = 1)\right).$$

To prove that $h_{ijkl} = 0$, we will show that $P_d(R_i(a) = R_j(b) = 1)$ and $P_d(R_k(a) = R_l(b) = 1)$ are independent. Observe that the addends in the sum can be classified to two categories: addends for which $|\{i, j, k, l\}| = 4$, and those for which $|\{i, j, k, l\}| = 3$. Either way, we have

$$P_d(R_i(a) = R_j(b) = 1) = \frac{2^{T_1}}{2^{2r}} \cdot \frac{\binom{n-T_1}{d-S_1}}{\binom{n}{d}},$$

$$P_d(R_k(a) = R_l(b) = 1) = \frac{2^{T_2}}{2^{2r}} \cdot \frac{\binom{n-T_2}{d-S_2}}{\binom{n}{d}},$$

where $T_1$ is the number of variables shared by $C_i$ and $C_j$, $S_1$ is the number of common variables whose sign (as literals) is different in $C_i$ and $C_j$, and $T_2$ and $S_2$ are defined similarly with respect to $C_k$ and $C_l$.

The variables $T_1$ and $S_1$ are determined solely by the way $C_j$ selected in relation to $C_i$. Similarly, $T_2$ and $S_2$ are determined solely by the way $C_l$ selected in relation to $C_k$. This holds even if, for example, $i = k$. Thus, the variables $(T_1, S_1)$, and the variables $(T_2, S_2)$, are independent. Consequently, $P_d(R_i(a) = R_j(b) = 1)$ and $P_d(R_k(a) = R_l(b) = 1)$ are independent as well. This means that $h_{ijkl} = 0$, which proves (3.B.1).

Regarding (3.B.2), a routine calculation shows that, regardless the value of $0 \le c \le 1$, the sum appearing in (3.B.1) is

$$1 + \frac{r^2(2c-1)^2}{n} + O\left(\frac{1}{n^2}\right).$$

Plugging this approximation into (3.B.1), and using the fact that $m = \alpha n$, we arrive at (3.B.2).

Finally, to prove (3.C), denote $X_n = \mathrm{Cov}_d(R(a), R(b))/n$. It suffices to show that the expected value of $X_n$ converges to the right-hand side of (3.C), and that its variance converges to 0. Those two convergences follow from (3.A.2) and (3.B.2) directly:

$$E_{\mathcal{I}}(X_n) = \frac{1}{n} \cdot E_{\mathcal{I}}(\mathrm{Cov}_d(R(a), R(b)))$$

$$= \frac{\alpha}{2^r}\left((1-c)^r - \frac{1}{2^r}\right) + O\left(\frac{1}{n}\right),$$

$$V_{\mathcal{I}}(X_n) = \frac{1}{n^2} \cdot V_{\mathcal{I}}(\mathrm{Cov}_d(R(a), R(b)))$$

$$= \frac{\alpha^2 r^2 (2c-1)^2}{2^{4r-1}} \cdot \frac{1}{n} + O\left(\frac{1}{n^2}\right).$$

This proves (3.C), which concludes the proof of the theorem.

*Proof (Theorem 1).* Denote

$$Y_n = V_A(R(a))/n,$$
$$Z_n = V_A(R(a)) - \mathrm{Cov}_1(R(a), R(b)).$$

By (2.C), the random variable $Y_n$ converges in probability to $\alpha(1 - 1/2^r)/2^r$. In the following, we will show that $Z_n$ converges in probability to $\alpha r/2^r$. As $\xi/n = Y_n/Z_n$, this will imply the theorem. To this end, it suffices to show that

$$E_{\mathcal{I}}(Z_n) \xrightarrow[n\to\infty]{} \frac{\alpha r}{2^r}, \tag{3}$$

$$V_{\mathcal{I}}(Z_n) \xrightarrow[n\to\infty]{} 0. \tag{4}$$

Using (2.A) and (3.A.1), we get:

$$E_{\mathcal{I}}(Z_n) = E_{\mathcal{I}}(V_A(R(a))) - E_{\mathcal{I}}(\mathrm{Cov}_1(R(a), R(b)))$$

$$= \frac{m}{2^r}\left(1 - \frac{1}{2^r}\right) - \frac{m}{2^r}\left(\frac{\binom{n-r}{1}}{\binom{n}{1}} - \frac{1}{2^r}\right)$$

$$= \frac{\alpha n}{2^r} \cdot \frac{r}{n} = \frac{\alpha r}{2^r}.$$

To prove (4), denote:

$$g = V_A(R(a)) \cdot \mathrm{Cov}_1(R(a), R(b)).$$

Then:

$$
\begin{aligned}
V_{\mathcal{I}}(Z_n) &= V_{\mathcal{I}}(V_A(R(a))) + V_{\mathcal{I}}(\mathrm{Cov}_1(R(a), R(b))) \\
&\quad - 2\mathrm{Cov}_{\mathcal{I}}(V_A(R(a)), \mathrm{Cov}_1(R(a), R(b))) \\
&= V_{\mathcal{I}}(V_A(R(a))) + V_{\mathcal{I}}(\mathrm{Cov}_1(R(a), R(b))) \\
&\quad - 2E_{\mathcal{I}}(g) + 2E_{\mathcal{I}}(V_A(R(a)))E_{\mathcal{I}}(\mathrm{Cov}_1(R(a), R(b))).
\end{aligned}
\tag{5}
$$

Theorems 2 and 3 provide an explicit form for each of the terms on the right-hand side, with the exception of $E_{\mathcal{I}}(g)$.

By (1),

$$V_A(R(a)) = \frac{m}{2^r}\left(1 - \frac{1}{2^r}\right) + 2\sum_{1 \le i < j \le m} \mathrm{Cov}_0(R_i(a), R_j(b)),$$

and

$$\mathrm{Cov}_1(R(a), R(b)) = \frac{m}{2^r}\left(1 - \frac{1}{2^r} - \frac{r}{n}\right) + 2\sum_{1 \le k < l \le m} \mathrm{Cov}_1(R_k(a), R_l(b)).$$

Thus,

$$
\begin{aligned}
E_{\mathcal{I}}(g) &= E_{\mathcal{I}}(V_A(R(a)) \cdot \mathrm{Cov}_1(R(a), R(b))) \\
&= \frac{m^2}{2^{2r}}\left(1 - \frac{1}{2^r}\right)\left(1 - \frac{1}{2^r} - \frac{r}{n}\right) \\
&\quad + \frac{m}{2^r}\left(1 - \frac{1}{2^r} - \frac{r}{n}\right) \cdot 2\sum_{1 \le i < j \le m} E_{\mathcal{I}}(\mathrm{Cov}_0(R_i(a), R_j(b))) \\
&\quad + \frac{m}{2^r}\left(1 - \frac{1}{2^r}\right) \cdot 2\sum_{1 \le k < l \le m} E_{\mathcal{I}}(\mathrm{Cov}_1(R_k(a), R_l(b))) \\
&\quad + 4\sum_{\substack{1 \le i < j \le m \\ 1 \le k < l \le m}} E_{\mathcal{I}}(\mathrm{Cov}_0(R_i(a), R_j(b)) \cdot \mathrm{Cov}_1(R_k(a), R_l(b))).
\end{aligned}
\tag{6}
$$

By (2), the second and third addends on the right-hand side of (6) both vanish. Moreover, following an argument similar to that applied to $h_{ijkl}$ in the proof of Theorem 3, we conclude that $\mathrm{Cov}_0(R_i(a), R_j(b))$ and $\mathrm{Cov}_1(R_k(a), R_l(b))$ are independent whenever $3 \leq |\{i, j, k, l\}| \leq 4$. Thus, the last sum on the right-hand side of (6) reduces to the sum over terms with $i = k$ and $j = l$. Applying those insights to $E_\mathcal{I}(g)$, we arrive at:

$$E_\mathcal{I}(g) = \frac{m^2}{2^{2r}} \left(1 - \frac{1}{2^r}\right) \left(1 - \frac{1}{2^r} - \frac{r}{n}\right)$$
$$+ 4 \sum_{1 \leq i < j \leq m} E_\mathcal{I}(\mathrm{Cov}_0(R_i(a), R_j(b)) \cdot \mathrm{Cov}_1(R_i(a), R_j(b))). \tag{7}$$

Now, let us convert the expectation appearing in the last expression to a simpler form, which will allow us to calculate it directly:

$$E_\mathcal{I}(\mathrm{Cov}_0(R_i(a), R_j(b)) \cdot \mathrm{Cov}_1(R_i(a), R_j(b)))$$
$$= E_\mathcal{I}((E_0(R_i(a)R_j(b)) - 1/2^{2r}) \cdot (E_1(R_i(a)R_j(b)) - 1/2^{2r}))$$
$$= E_\mathcal{I}(E_0(R_i(a)R_j(b)) \cdot E_1(R_i(a)R_j(b)))$$
$$\quad - E_\mathcal{I}(E_0(R_i(a)R_j(b)))/2^{2r} - E_\mathcal{I}(E_1(R_i(a)R_j(b)))/2^{2r} + 1/2^{4r}$$
$$= E_\mathcal{I}(E_0(R_i(a)R_j(b)) \cdot E_1(R_i(a)R_j(b)))$$
$$\quad - 1/2^{4r} - 1/2^{4r} + 1/2^{4r} \quad // \text{ as done in (2)}$$
$$= E_\mathcal{I}(P_0(R_i(a) = R_j(b) = 1) \cdot P_1(R_i(a) = R_j(b) = 1)) - 1/2^{4r}.$$

As in the proof of Theorem 3, we now consider the clauses $C_i$ and $C_j$. We denote by $t$ the number of variables shared by $C_i$ and $C_j$. Let $s$ be the number of variables, out of the $t$ common ones, whose sign (as literals) is different in the two clauses. The remaining $t - s$ shared variables have the same sign in the two clauses. A direct calculation, as in that proof, with $d = 0$ and $d = 1$, yields:

$$E_\mathcal{I}(P_0(R_i(a) = R_j(b) = 1) \cdot P_1(R_i(a) = R_j(b) = 1))$$
$$= \sum_{t=0}^{r} \sum_{s=0}^{t} \frac{\binom{r}{t}\binom{n-r}{r-t}}{\binom{n}{r}} \cdot \frac{\binom{t}{s}}{2^t} \cdot \frac{\binom{n-t}{-s}}{2^{2r-t}} \cdot \frac{\binom{n-t}{1-s}}{2^{2r-t}n}$$
$$= \frac{1}{2^{4r}} \sum_{t=0}^{r} \frac{\binom{r}{t}\binom{n-r}{r-t}}{\binom{n}{r}} \cdot 2^t \cdot \left(1 - \frac{t}{n}\right).$$

The last equality follows by a routine simplification, after observing that the terms for which $s > 0$ vanish. Plugging these values into (7), we arrive at the final form of $E_\mathcal{I}(g)$:

$$E_\mathcal{I}(g) = \frac{m^2}{2^{2r}} \left(1 - \frac{1}{2^r}\right) \left(1 - \frac{1}{2^r} - \frac{r}{n}\right)$$
$$+ \frac{2m(m-1)}{2^{4r}} \left(\sum_{t=0}^{r} \frac{\binom{r}{t}\binom{n-r}{r-t}}{\binom{n}{r}} \cdot 2^t \cdot \left(1 - \frac{t}{n}\right) - 1\right).$$

Now that we have an explicit form for all the terms on the right-hand side of (5), we can obtain an explicit expression for $V(Z_n)$, which after a routine simplification turns out to be:

$$V_{\mathcal{I}}(Z_n) = \frac{2m(m-1)}{2^{4r}} \cdot \frac{1}{n^2} \cdot \sum_{t=0}^{r} \frac{\binom{r}{t}\binom{n-r}{r-t}}{\binom{n}{r}} \cdot 2^t \cdot t(t+1).$$

Recall that $m = \alpha n$ for some $\alpha > 0$. Thus, the product of the two factors outside the sum (in the last expression) is $\Theta(1)$. The leading addend in the sum is obtained for $t = 1$, and it is $\Theta(1/n)$. Thus, we conclude that $V_{\mathcal{I}}(Z_n) = \Theta(1/n)$, which completes the proof.

## 4 Discussion

In this paper we have provided some results characterizing the ensemble of all (equally likely) $r$-CNF formulas. These results apply to both random Max $r$-Sat and random $r$-Sat. Along the paper we chose to present them in the context of random Max $r$-Sat instances.

In this section we discuss how our results directly apply and characterize random Max $r$-Sat instances by giving some examples and interpretations of our results. We give some general motivation and implications of our results in the context of local search and landscape theory. Finally, we elaborate on the results of [34], compare them with ours, and clarify our relative contribution.

Consider, for example, the variance of the number of clauses satisfied by a random assignment, explored in Theorem 2. Let $\varepsilon > 0$, and consider the proportion of instances for which the absolute difference between $V_A(S(I, a))/n$ and $\alpha(1 - 1/2^r)/2^r$ exceeds $\varepsilon$. By (2.C), this proportion tends to 0 as $n$ grows, which gives a very accurate understanding of this variance over the ensemble of all (equally likely) instances. Similar statements can be made regarding the covariance, correlation coefficient, and normalized autocorrelation length, using the results in Theorem 3, Corollary 1, and Theorem 1, respectively.

Formulas (2.B.1) and (2.B.2), as well as (3.B.1) and (3.B.2), give closed expressions and clear asymptotics for useful quantities, usually simulated or approximated (e.g., [21, 29, 30]).

In the local search community, and especially among researchers in the area of landscape analysis, it is a common belief that the ruggedness of the landscape of an instance is one of the main factors for its hardness. The commonly used measures to summarize ruggedness are the autocorrelation length (or alternatively the autocorrelation coefficient) and the fitness-distance correlation. These measures are believed to assess the hardness of combinatorial optimization problems for local search heuristics. These beliefs have been empirically confirmed in various studies [2–4], and a work aiming at better understanding the landscape of combinatorial optimization problems and classifying their hardness for local search is continuously conducted [11, 12, 23, 25, 29, 30, 33, 34].

Our results provide insights on the structure of the landscapes of instances of Max $r$-Sat, and their similarity. They are also a step toward a richer classification of hardness of combinatorial optimization problems, in the context of local search. We provide a simple expression for the autocorrelation length of Max $r$-Sat, and other interesting statistical measures for this problem as well. Results in a similar vein, regarding the autocorrelation length of problems like TSP, QAP, GC, GBP, etc., are summarized in [3, Table 1].

In several problems, the autocorrelation length has been calculated with respect to various neighborhood operators, and a clear suggestion for the designers of local search heuristics came out: use the neighborhood operator with the largest autocorrelation length [2,4,27]. Such operators induce smoother landscape, which leads to better performance of local search heuristics. Some researchers even suggest performing amendments to the problem in a way that leads to an equivalent problem with a larger autocorrelation length [3]. Our result regarding the autocorrelation length may serve as a baseline for assessing the relevance of any future neighborhood operator or amendment for Max $r$-Sat.

In many heuristics, after a local optimum is reached, the whole search is repeated from a different, randomly selected starting point. This is a simple, clean way to restart. Yet, as the heuristic may have already yielded a quite good assignment, one may prompt for further exploration of the landscape in the vicinity of this assignment. In such cases, one may want to perform a jump from the local optimum that is not too large, so as to stay in the vicinity of the local optimum. On the other hand, a too small jump would fall in the basin of attraction of the local optimum, which will lead to the same local optimum again. To this end, the autocorrelation length may provide assisting information, as it hints on the average distance between local optima [27]. The specific formulation of using it to calculate the size of the jump is a subject for further research.

Research efforts to formulate the autocorrelation length of Max $r$-Sat done in the last decades culminated in [34]. In both our paper and [34], there is interest in the correlation between the number of clauses satisfied by a random assignment, and an assignment obtained from it due to some random changes. In our paper, we measure the change according to the Hamming distance between the two assignments. In [34], one starts from the initial assignment, makes a number of random steps (bit flips), and then compares the initial assignment with the final one. If the distance in the first version, and the number of steps in the second, are the same, then we may expect to be closer to the initial assignment in the second version, as some of the steps may well bring us closer to it. However, as the autocorrelation length only depends on two assignments at a distance of 1 from one another, the two versions are equivalent in this respect.

The first result of [34] is that the exact autocorrelation length of any instance can be computed in polynomial time. The computation starts by noticing that the space of real-valued functions on the space of all assignments has a very simple basis. The space of all assignments is clearly equivalent to the discrete $n$-dimensional cube. It is of size $2^n$, and thus the space of real-valued functions on it is a $2^n$-dimensional linear space. The basis with which [34] works is that of

Walsh functions. We will not define the basis here. Suffice it to say that the basis consists of the functions $\psi_{\mathbf{i}}$, where $\mathbf{i}$ goes over all binary vectors of length $n$. An especially convenient property of this basis is that it is orthonormal. Also, while the basis is of exponential size, the fitness function for any instance of Max $r$-Sat, which gives the number of clauses satisfied by any assignment, involves only a linear number of $\psi_{\mathbf{i}}$'s (as a function of the number of clauses). The key to the algorithm is the fact that one can write the fitness function $f$ explicitly in the form

$$f(x) = \sum_{\mathbf{i}} w_{\mathbf{i}} \psi_{\mathbf{i}}(x),$$

where the summation is over a set of size $O(m)$ and the weights $w_{\mathbf{i}}$ can be calculated explicitly. Moreover, the $\psi_{\mathbf{i}}$'s are the eignefunctions of the matrix which gives the connection between the value of the fitness function at an assignment and at a random neighbor of that assignment.

Now denote, for each integer $p$ between 1 and $r$:

$$W^{(p)} = \sum_{\langle \mathbf{i}, \mathbf{i} \rangle} w_{\mathbf{i}}^2 / \sum_{\mathbf{i}} w_{\mathbf{i}}^2.$$

Using these notations, one can give a formula for the following version of the correlation $\rho(d)$. Recall that $\rho(d)$ is the correlation (coefficient) between the numbers of clauses satisfied by a random pair of assignments at Hamming distance $d$ from each other. In [34], instead of changing the initial random assignment for $d$ of the variables, there is some prescribed number of 1-variable changes. The correlation (coefficient) between the numbers of clauses satisfied by a pair of random assignments, the second of which is obtained from the first by a sequence of $s$ random changes of a single variable, is denoted by $r(s)$. Note that for a single change, there is no difference between the two versions: $r(1) = \rho(1)$. In [34, Prop.1] it is shown that

$$r(s) = \sum_{p \neq 0} W^{(p)} (1 - 2p/n)^s.$$

From this, the autocorrelation length is immediately found for any instance.

Next, [34] continues to estimate the mean value of $r(s)$. To this end, we start by finding the mean value of each $w_{\mathbf{i}}^2$. Let $p = \langle \mathbf{i}, \mathbf{i} \rangle$ and put:

$$\pi' = \binom{r}{p} / \binom{n}{p}.$$

It is shown that the mean of $w_{\mathbf{i}}^2$ is

$$\frac{2^n}{4^r} \sum_{z=-m}^{m} z^2 b_z,$$

where

$$b_z = \sum_{\substack{z_1, z_2 \geq 0 \\ z_1 - z_2 = z}} \frac{m! (\pi'/2)^{z_1 + z_2} (1 - \pi')^{m - (z_1 + z_2)}}{z_1! z_2! (m - (z_1 + z_2))!}.$$

This enables one to calculate, for each value of the parameters, the mean value of any $w_i^2$ or sum of several ones. Assuming that the average of the quotient may be approximated by the quotient of the averages, one can estimate the average value of $r(1)$.

In our approach, we: (i) find an *explicit* expression that approximates $r(1)$, and thereby an expression approximating the normalized autocorrelation length, and (ii) prove *rigorously* that the normalized autocorrelation length converges in probability to the latter expression as the number of variables grows.

Finally, we note that the nature of industrial instances, for example, is subtly different from random ones. Their underlying probability model, if any, is different, and they should be addressed separately. We hope our work will encourage a concise analysis of modeled practical instances ensembles.

**Acknowledgment.** The work of the second author was partially supported by the Lynne and William Frankel Center for Computer Science.

# References

1. Achlioptas, D., Peres, Y.: The threshold for random $k$-SAT is $2^k \log 2 - O(k)$. J. Am. Math. Soc. **17**(4), 947–973 (2004)
2. Angel, E., Zissimopoulos, V.: Autocorrelation coefficient for the graph bipartitioning problem. Theoret. Comput. Sci. **191**(1), 229–243 (1998)
3. Angel, E., Zissimopoulos, V.: On the classification of NP-complete problems in terms of their correlation coefficient. Discrete Appl. Math. **99**(1), 261–277 (2000)
4. Angel, E., Zissimopoulos, V.: On the landscape ruggedness of the quadratic assignment problem. Theor. Comput. Sci. **263**(1), 159–172 (2001)
5. Ansótegui, C., Bonet, M.L., Levy, J.: SAT-based MaxSAT algorithms. Artif. Intell. **196**, 77–105 (2013)
6. Argelich, J., Li, C.M., Manyá, F., Planes, J.: MaxSat Evaluations. http://www.maxsat.udl.cat/
7. Ausiello, G., Crescenzi, P., Gambosi, G., Kann, V., Marchetti-Spaccamela, A., Protasi, M.: Complexity and Approximation: Combinatorial Optimization Problems and Their Approximability Properties, 2nd edn. Springer-Verlag (2003)
8. Biere, A., Heule, M., van Maaren, H.: Handbook of Satisfiability, vol. 185. IOS press, Amsterdam (2009)
9. de Boer, P.-T., Kroese, D.P., Mannor, S., Rubinstein, R.Y.: A tutorial on the cross-entropy method. Ann. Oper. Res. **134**(1), 19–67 (2005)
10. Chen, R., Santhanam, R.: Improved algorithms for sparse MAX-SAT and MAX-$k$-CSP. In: Heule, M., Weaver, S. (eds.) SAT 2015. LNCS, vol. 9340, pp. 33–45. Springer, Heidelberg (2015). doi:10.1007/978-3-319-24318-4_4
11. Chicano, F., Luque, G., Alba, E.: Autocorrelation measures for the quadratic assignment problem. Appl. Math. Lett. **25**(4), 698–705 (2012)
12. Chicano, F., Luque, G., Alba, E.: Problem understanding through landscape theory. In: Proceedings of the 15th Annual Conference Companion on Genetic and Evolutionary Computation, pp. 1055–1062. ACM (2013)
13. Chvátal, V., Reed, B.: Mick gets some (the odds are on his side) [satisfiability]. In: Proceedings of the 33rd Annual Symposium on Foundations of Computer Science, pp. 620–627. IEEE (1992)

14. Coja-Oghlan, A.: The asymptotic $k$-sat threshold. In: Proceedings of the 46th Annual ACM Symposium on Theory of Computing, pp. 804–813. ACM (2014)
15. Davies, J., Bacchus, F.: Solving MAXSAT by solving a sequence of simpler SAT instances. In: Lee, J. (ed.) CP 2011. LNCS, vol. 6876, pp. 225–239. Springer, Heidelberg (2011)
16. Ding, J., Sly, A., Sun, N.: Proof of the satisfiability conjecture for large k. arXiv preprint (2014). arXiv:1411.0650
17. Fontana, W., Stadler, P.F., Bornberg-Bauer, E.G., Griesmacher, T., Hofacker, I.L., Tacker, M., Tarazona, P., Weinberger, E.D., Schuster, P.: RNA folding and combinatory landscapes. Phy. Rev. E **47**(3), 2083–2099 (1993)
18. Friedgut, E., Bourgain, J.: Sharp thresholds of graph properties, and the $k$-sat problem. J. Am. Math. Soc. **12**(4), 1017–1054 (1999)
19. García-Pelayo, R., Stadler, P.F.: Correlation length, isotropy and meta-stable states. Physica D Nonlinear Phenom. **107**(2), 240–254 (1997)
20. Goldberg, D.E.: Genetic Algorithms and Walsh Functions: A Gentle Introduction. Clearinghouse for Genetic Algorithms, Department of Mechanical Engineering, University of Alabama (1988)
21. Heckendorn, R.B., Rana, S., Whitley, D.: Polynomial time summary statistics for a generalization of MAXSAT. In: Proceedings of the 11th Annual Conference on Genetic and Evolutionary Computation, pp. 281–288. Morgan Kaufmann (1999)
22. Heras, F., Larrosa, J., Oliveras, A.: MiniMaxSAT: An efficientWeighted Max-SAT solver. J. Artif. Intell. Res. (JAIR) **31**, 1–32 (2008)
23. H. Hoos, H., Smyth, K., Stützle, T.: Search space features underlying the performance of stochastic local search algorithms for MAX-SAT. In: Yao, X., et al. (eds.) PPSN 2004. LNCS, vol. 3242, pp. 51–60. Springer, Heidelberg (2004)
24. Luo, C., Cai, S., Wu, W., Jie, Z., Su, K.-W.: CCLS: An efficient local search algorithm for weighted maximum satisfiability. IEEE Trans. Comput. **64**(7), 1830–1843 (2014)
25. Malan, K.M., Engelbrecht, A.P.: A survey of techniques for characterising fitness landscapes and some possible ways forward. Inf. Sci. **241**, 148–163 (2013)
26. Mertens, S., Mézard, M., Zecchina, R.: Threshold values of random $k$-SAT from the cavity method. Random Struct. Algorithms **28**(3), 340–373 (2006)
27. Merz, P., Freisleben, B.: Fitness landscapes and memetic algorithm design. In: Corne, D., Dorigo, M., Glover, F., Dasgupta, D., Moscato, P., Poli, R., Price, K.V. (eds.) New Ideas in Optimization, pp. 245–260. McGraw-Hill, New York (1999)
28. Narodytska, N., Bacchus, F.: Maximum satisfiability using core-guided MaxSAT resolution. In: Proceedings of the Twenty-Eighth AAAI Conference on Artificial Intelligence, pp. 2717–2723. AAAI Press (2014)
29. Prügel-Bennett, A., Tayarani-Najaran, M.-H.: Maximum satisfiability: Anatomy of the fitness landscape for a hard combinatorial optimization problem. IEEE Trans. Evol. Comput. **16**(3), 319–338 (2012)
30. Qasem, M., Prügel-Bennett, A.: Learning the large-scale structure of the MAX-SAT landscape using populations. IEEE Trans. Evol. Comput. **14**(4), 518–529 (2010)
31. Selman, B., Kautz, H., Cohen, B.: Local search strategies for satisfiability testing. Cliques Coloring Satisfiability Second DIMACS Implement. Chall. **26**, 521–532 (1993)
32. Selman, B., Levesque, H., Mitchell, D.: A new method for solving hard satisfiability problems. In: Proceedings of the Tenth National Conference on Artificial Intelligence, pp. 440–446. AAAI Press (1992)

33. Stadler, P.: Fitness landscapes. In: Lässig, M., Valleriani, A. (eds.) Biological Evolution and Statistical Physics, pp. 183–204. Springer, Heidelberg (2002)
34. Sutton, A.M., Whitley, L.D., Howe, A.E.: A polynomial time computation of the exact correlation structure of $k$-satisfiability landscapes. In: Proceedings of the 11th Annual Conference on Genetic and Evolutionary Computation, pp. 365–372. ACM (2009)
35. Tompkins, D.A.D., Hoos, H.H.: UBCSAT: An implementation and experimentation environment for SLS algorithms for SAT and MAX-SAT. In: Hoos, H.H., Mitchell, D.G. (eds.) SAT 2004. LNCS, vol. 3542, pp. 306–320. Springer, Heidelberg (2005)
36. Williams, C.P., Hogg, T.: Exploiting the deep structure of constraint problems. Artif. Intell. **70**(1), 73–117 (1994)

# Tight Upper Bound on Splitting by Linear Combinations for Pigeonhole Principle

Vsevolod Oparin[1,2]($\boxtimes$)

[1] St. Petersburg Academic University, 8/3 Khlopina, St.Petersburg 194021, Russia
[2] Steklov Institute of Mathematics at St. Petersburg,
27 Fontanka, St.Petersburg 191023, Russia
oparin.vsevolod@gmail.com

**Abstract.** The usual DPLL algorithm uses splittings (branchings) on single Boolean variables. We consider an extension to allow splitting on linear combinations mod 2, which yields a search tree called a linear splitting tree. We prove that the pigeonhole principle has linear splitting trees of size $2^{O(n)}$. This is near-optimal since Itsykson and Sokolov [1] proved a $2^{\Omega(n)}$ lower bound. It improves on the size $2^{\Theta(n \log n)}$ for splitting on single variables; thus the pigeonhole principle has a gap between linear splitting and the usual splitting on single variables. This is of particular interest since the pigeonhole principle is not based on linear constraints. We further prove that the perfect matching principle has splitting trees of size $2^{O(n)}$.

## 1 Introduction

Splitting is a well known method for solving NP-hard problems. In the case of the satisfiability problem for a Boolean CNF formula, the highly successful DPLL algorithms are based on splitting [2,3]. DPLL works in the following way to search for satisfying assignments for a CNF formula $\phi$. The algorithm chooses a variable $x$ and a first value $\alpha$ to substitute. The algorithm substitutes $x = \alpha$ and runs recursively on the simplified formula $\phi|_{x=\alpha}$. If this does not succeed, it tries $\phi|_{x=1\oplus\alpha}$. If there is no success again, the algorithm returns FAIL. Otherwise, it returns a satisfying assignment.

There is extensive research on hard examples for DPLL algorithms. It is well known that systems of linear equations mod 2, such as the Tseitin tautologies, are hard for DPLL and resolution [1,4–6]. However, they can be quickly solved by splitting on linear combinations mod 2. Thus it is natural to consider generalizing DPLL to use linear splitting.

A linear splitting algorithm maintains a system of linear equations over $\mathbb{F}_2$. Initially, the system is empty. Instead of choosing a single variable, the algorithm chooses a linear form $\sum_i \alpha_i \cdot x_i$ and a first value $\beta$. The algorithm adds the equation $\sum_i \alpha_i \cdot x_i = \beta$ to the system and runs itself recursively. The second call runs with the equation $\sum_i \alpha_i \cdot x_i = 1 + \beta$.

Research is partially supported by the Government of the Russian Federation under Grant 14.Z50.31.0030.

© Springer International Publishing Switzerland 2016
N. Creignou and D. Le Berre (Eds.): SAT 2016, LNCS 9710, pp. 77–84, 2016.
DOI: 10.1007/978-3-319-40970-2_6

On every step of the recursion the algorithm checks three conditions before splitting again.

- If the system is inconsistent, the algorithm backtracks. This condition can be checked in polynomial time by Gaussian elimination.
- If the system violates any single clause, the algorithm backtracks. The system violates a clause $C = l_1 \vee l_2 \vee \cdots \vee l_k$, if for every $i \in [k]$ the system plus the equation $l_i = 1$ is inconsistent. All clauses can be checked in polynomial time.
- If the system has exactly one solution, the algorithm returns a satisfying assignment. This is also can be checked by Gaussian elimination.

Otherwise, the algorithm selects a linear form for the next splitting. The entire execution tree is a binary tree called a linear splitting tree.

Several prior works combine linear substitutions and splitting. For example, the recent algorithm of Seto and Tamaki [7] solves the satisfiability problem for an arbitrary formula of linear size $c \cdot n$ in $2^{(1-\mu_c) \cdot n}$ steps using splitting by linear forms. Kulikov and Demenkov [8] use a formula which is non-trivial up to the linear number of linear substitutions to show a lower bound of $3n - o(n)$ for the circuit complexity over the full binary basis.

Itsykson and Sokolov [1] provide a family of hard formulas for linear splitting. In particular, they prove that the pigeonhole principle, where $m$ pigeons fly to $n$ holes, has no linear splitting tree of size less than $2^{\Omega(n)}$. On the other hand, it is known that splitting by single Boolean variables gives a tree of size $2^{O(n \log n)}$ and this bound is tight [9]. So it is natural to ask whether the bound for the pigeonhole principle with linear splitting is tight.

We answer this by showing that the pigeonhole principle has a linear splitting tree of size $2^{O(n)}$. The pigeonhole principle is not based on linear constraints, so this gap between splitting by single variables and by linear forms is interesting.

We also consider the perfect matching principle built on arbitrary graphs. Any graph of odd cardinality has a polynomial-size splitting tree [1], but nothing is known about graphs of even cardinality. For an arbitrary $n$, we prove a $2^{O(n)}$ upper bound on splitting trees for graphs on $n$ vertices.

## 2   Preliminaries

We will use the following notation: $[n] = \{1, 2, \ldots, n\}$. Let $X = \{x_1, \ldots, x_n\}$ be a set of variables that take values from $\mathbb{F}_2$. A linear form is a polynomial $\sum_{i=1}^{n} \alpha_i \cdot x_i$ over $\mathbb{F}_2$.

Consider a binary tree $T$ with edges labeled by linear equalities. For every vertex $v$ of $T$ we denote by $\Phi_v^T$ the system of all equalities that are written along the path from the root to vertex $v$.

A *linear splitting tree* for a CNF formula $\varphi$ is a binary tree $T$ with the following properties. Every internal node is labeled by a linear form that depends on variables from $\varphi$. For every internal node labeled by a linear form $f$, one of the incident edges going to the children is labeled by $f = 0$, and the other one is labeled by $f = 1$.

For every leaf $v$ of the tree exactly one of the following conditions holds: (1) The system $\Phi_v^T$ has no solution. We call such leaf degenerate. (2) The system $\Phi_v^T$ is satisfiable but violates a clause $C$ of the formula $\varphi$. We say that such leaf violates clause $C$. (3) The system $\Phi_v^T$ has exactly one solution and the solution satisfies the formula $\varphi$. We call such leaf satisfying.

A linear splitting tree may be viewed as a tree of recursive calls for the algorithm solving SAT for the CNF formula $\varphi$. The algorithm maintains a system of linear equations $\Phi$ and starts with a given formula $\varphi$ and $\Phi = \texttt{True}$. Given a formula $\varphi$ and a system of linear equations $\Phi$, the algorithm looks for a satisfying assignment of $\varphi \wedge \Phi$. At every step the algorithm chooses a linear form $f$ and a value $\alpha \in \mathbb{F}_2$ and makes two recursive calls: on the input $(\varphi, \Phi \wedge (f = \alpha))$ and on the input $(\varphi, \Phi \wedge (f = 1 + \alpha))$.

The algorithm backtracks in one of the three cases: (1) The system $\Phi$ has no solution; (2) The system $\Phi$ contradicts a clause $C$ of the formula $\varphi$. (A system $\Psi$ contradicts a clause $(l_1 \vee l_2 \vee \cdots \vee l_k)$ iff for all $i \in [k]$ the system $\Psi \wedge (l_i = 1)$ is unsatisfiable.) (3) The system $\Phi$ has a unique solution that satisfies $\varphi$. All three cases can be checked in polynomial time.

Note that if it is enough to find merely one satisfying assignment, the algorithm may stop at the first satisfying leaf. In the case of unsatisfiable formulas, the algorithm must traverse the whole splitting tree.

**Proposition 1 [1].** *For every linear splitting tree $T$ for a formula $\varphi$ it is possible to construct a splitting tree without degenerate leaves. The number of vertices in the new tree is at most the number of vertices in $T$.*

## 3   Upper Bound for the Pigeonhole Principle

Let we have $m$ pigeons and $n$ holes. Every pigeon should fly to at least one hole. The pigeonhole principle states that if $m > n$, there exists a hole with at least two pigeons inside.

We encode the reverse statement into an unsatisfiable CNF formula. For $i \in [m]$ and $j \in [n]$ let $x_{i,j}$ be a variable such that the $i$-th pigeon flies to the $j$-th hole iff $x_{i,j} = 1$.

We encode the fact that the $i$-th pigeon flies somewhere by the clause

$$\bigvee_{j \in [n]} x_{i,j}.$$

Also we encode, that the $j$-th hole accepts at most one pigeon by the set of clauses

$$\neg x_{i_1,j} \vee \neg x_{i_2,j}$$

for every $i_1 \neq i_2 \in [m]$

We denote the conjunction of all these clauses by $\mathrm{PHP}_n^m$. Obviously, the formula $\mathrm{PHP}_n^m$ is unsatisfiable if $m > n$.

**Theorem 1.** *For all $m > n$ there exists a linear splitting tree for $PHP_n^m$ of size* $2^{O(n)}$.

*Proof.* The formula $PHP_n^{n+1}$ is a subformula of $PHP_n^m$. So it is enough to build a tree for $PHP_n^{n+1}$ only.

We construct the tree by induction on $n$. The base $n = 1$ is trivial.

For $n > 1$, we reduce $PHP_n^{n+1}$ to multiple copies of $PHP_{n/2}^{n/2+1}$. This is done by building a linear splitting tree $T$ of size $2^{O(n)}$. Every leaf of $T$ either will violate one of the clauses or will correspond to an instance of $PHP_{n/2}^{n/2+1}$. (The second kind of leaves will become the root of a tree for $PHP_{n/2}^{n/2+1}$.)

The logarithm $LG(n)$ of the size of the whole splitting tree for $PHP_n^{n+1}$ can be expressed by the inequality

$$LG(n) \leq \log(2^{O(n)} \cdot 2^{LG(\lfloor n/2 \rfloor)}) = O(n) + LG(\lfloor n/2 \rfloor).$$

Hence $LG(n) = O(n)$. So the size of the tree is $2^{O(n)}$.

We split the pigeons into two almost equal parts $L = [1, \lfloor (n + 1)/2 \rfloor]$ and $R = [\lfloor (n + 1)/2 \rfloor + 1, n + 1]$. We refer to these parts as "left" and "right", respectively. For every hole $j$ we define two linear forms:

$$\begin{aligned}
\text{LEFT}(j) &= \bigoplus_{i \in L} x_{i,j}, \\
\text{RIGHT}(j) &= \bigoplus_{i \in R} x_{i,j}.
\end{aligned}$$

The tree $T$ starts with a full binary tree $T_Q$ of height $2n$. Every branch in $T_Q$ queries the values $\text{LEFT}(j)$ and $\text{RIGHT}(j)$ for every hole $j$. So $T_Q$ has $2^{2n}$ leaves. $T$ will be defined from $T_Q$ by replacing each leaf $\ell$ of $T_Q$ with a polynomial-size subtree $T_\ell$. In each $T_\ell$, all but possibly one of its leaves will be labeled by violated clauses (see Fig. 1).

Fix a leaf $\ell$ of tree $T_Q$. For each hole $j$, we have fixed values of $\text{LEFT}(j)$ and $\text{RIGHT}(j)$. There are four cases.

1. $\text{LEFT}(j) = 1$, $\text{RIGHT}(j) = 1$.
2. $\text{LEFT}(j) = 0$, $\text{RIGHT}(j) = 1$.
3. $\text{LEFT}(j) = 1$, $\text{RIGHT}(j) = 0$.
4. $\text{LEFT}(j) = 0$, $\text{RIGHT}(j) = 0$.

1. If Case 1 holds for any hole $j$, the splitting tree $T_\ell$ has size $O(n^2)$ and finds a violated clause. $T_\ell$ can be described as a tree of recursive calls of the following algorithm. First, we go through all pigeons in the left part and split by $x_{i,j}$ for $i \in L$. Once $x_{i,j} = 1$ is found, we go through the pigeons of the right part and do the same until we find $x_{i',j} = 1$. Both variables exist since $\text{LEFT}(j) = \text{RIGHT}(j) = 1$. Once two non-zero variables are found, we return a violated clause.

   Otherwise, we form $T_\ell$ by chaining together splitting trees $T_j$, one for each hole $j$. $T_j$ is formed depending on which of the Cases 2–4 holds.

2. Suppose Case 2 holds, so $\text{LEFT}(j) = 0$. The leaves of the tree $T_j$ either will violate an injectivity clause for hole $j$ or will ensure that no left pigeon flies to hole $j$. The tree $T_j$ has the following structure. For every left pigeon $i$ we split by the variable $x_{i,j}$. If $x_{i,j} = 1$, we can find a violated clause. Since $\text{LEFT}(j) = 0$, there must be another $x_{i',j} = 1$ for $i' \in L$. We split by $x_{i',j}$ for every pigeon $i' \in L\backslash\{i\}$ and find a violated clause.

   Otherwise, the values for all left pigeons $i \in L$ are zero. In this case, we come to a leaf at which we know no left pigeon flies to the $j$-th hole.
3. Suppose Case 3 holds, so $\text{RIGHT}(j) = 0$. The tree $T_j$ is formed dually as above, and each leaf of $T_j$ either will violate an injectivity clause for hole $j$ or will ensure that no right pigeon flies to hole $j$.
4. Suppose Case 4 holds, so both $\text{LEFT}(j) = 0$ and $\text{RIGHT}(j) = 0$. $T_j$ is formed as in the previous two cases, but now we split on $x_{i,j}$ for all pigeons $i$. Each leaf of $T_j$ either will violate an injectivity clause or will ensure that no pigeon flies to hole $j$.

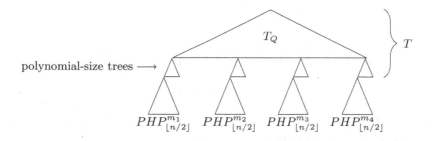

**Fig. 1.** Tree structure for $\text{PHP}_n^m$. The small polynomial-size trees contain trees $T_j$ chained together.

By design every tree $T_j$ has exactly one leaf not labeled by a violated clause. We call such a leaf *free*. We connect all trees $T_j$ each to the next one using free leaves, forming a *chain of trees*. The chain of trees forms the tree $T_\ell$ and has size $O(n^3)$. We attach chain $T_\ell$ to the leaf $\ell$.

The last tree in the chain has exactly one free leaf. At this leaf if any hole $j$ has $\text{LEFT}(j) = 0$, then no pigeon flies there from the left part $L$. Likewise, if any hole $j$ has $\text{RIGHT}(j) = 0$, then no pigeon flies there from the right part $R$. We separate holes into two disjoint parts: the first part has the holes $j$ with $\text{LEFT}(j) = 1$, the second part has the holes $j$ with $\text{RIGHT}(j) = 1$. The pigeons in $L$ can fly only to the first part, the pigeons in $R$ can fly only to the second part.

We show that at least one part of holes is less than the number of pigeons that fly there. Let $h_l$ and $h_r$ be the number of holes with $\text{LEFT}(j) = 1$ and $\text{RIGHT}(j) = 1$, respectively. We prove that either $h_l < |L|$ or $h_r < |R|$ by contradiction. Suppose $h_l \geq |L|$ and $h_r \geq |R|$. Since the sets of holes of the subformulas are distinct, $h_l + h_r \leq n$.

So

$$n \geq h_l + h_r \geq |L| + |R| = n + 1,$$

which is impossible.

Since $L$ and $R$ are less than $\lceil n/2 \rceil$, we can take a set of holes and pigeons that form a formula $PHP_{n/2}^{n/2+1}$ and attach a tree for this formula to the free leaf of the chain (see Fig. 1).

## 4   Upper Bound on the Perfect Matching Principle

In terms of CNF encoding, the perfect matching principle is similar to the pigeonhole principle. The formula $PMP_G$, built on an arbitrary graph $G = \langle V, E \rangle$, encodes that every vertex has exactly one edge, taken into the matching. Formally, we provide a variable $x_e$ for each edge $e \in E$. For every vertex $v$ we encode that there exists at least one edge taken into the matching:

$$\bigvee_{u \in V : (u,v) \in E} x_{(u,v)}.$$

Also for every pair of edges $(u, v)$ and $(w, v)$ with a common endpoint $v$ we encode, that they can not be both taken into the matching:

$$\neg x_{(u,v)} \vee \neg x_{(w,v)}.$$

The formula $PMP_G$ is the conjunction of all these clauses. Obviously, if the graph $G$ has no perfect matching, the formula is unsatisfiable.

Itsykson and Sokolov proved the following proposition.

**Proposition 2** [1]. *Let $G$ be a graph on an odd number of vertices. Then the formula $PMP_G$ has a splitting tree of a polynomial size.*

Using Theorem 1 and Proposition 2, we prove the following theorem. Note that $n$ can be even.

**Theorem 2.** *Let $G = \langle V, E \rangle$ be a graph on $n$ vertices, which has no perfect matching. Then the formula $PMP_G$ has a splitting tree of the size $2^{O(n)}$.*

*Proof.* We use Tutte's criterion to prove the theorem.

**Criterion 1 (Tutte, 1947).** *A graph $G$ has a perfect matching iff for any set $S \subseteq V$ the following statement holds: $o(G - S) \leq |S|$, where $G - S$ denotes the graph $G$ without vertices of the set $S$ and $o(G - S)$ denotes the number of connected components with odd cardinality in the obtained graph.*

We reduce the problem to the pigeonhole principle. Suppose the graph $G$ has no perfect matching. Let $S \subseteq V$ be a set such that $|S| < o(G - S)$.

Let $v_1, v_2, \cdots, v_n$ be the vertices of the set $S$ and $C_1, C_2, \cdots, C_m$ be the odd-cardinality connected components of the graph $G - S$. For every vertex $v_j$ and connected component $C_i$ we introduce a variable

$$y_{i,j} = \bigoplus_{(u,v_j) \in E, u \in C_j} x_{u,v_j}.$$

By the criterion $m > n$. Let us construct a formula $\text{PHP}_n^m$, built on $y$'s, and build a splitting tree $T_y$ of size $2^{O(n)}$ as it was done in Theorem 1. Every node in the tree $T_y$ has a linear form on $y$'s.

We build a tree $T_x$ using the structure of $T_y$. We expand all $y$'s into the xor of $x$'s. At some nodes of $T_x$ we may have empty linear forms: no edge connects a vertex of $S$ and a connected component of $G - S$. In this case, one of the outgoing edges is labeled by the equation $0 = 1$. We truncate the corresponding subtree since the system becomes inconsistent.

We replace all leaves of $T_y$ by polynomial-size trees that finds violated clauses of $\text{PMP}_G$. Fix a leaf $\ell$ of $T_y$ labeled by a clause $C_\ell$. We replace corresponding leaf of $T_x$ by a tree $T_\ell$. The structure of $T_\ell$ depends on clause $C_\ell$. There are two possible cases.

1. Clause $C_\ell$ is of type $\neg y_{i_1,j} \vee \neg y_{i_2,j}$. Then there exist two connected components $C_{i_1}$ and $C_{i_2}$ and vertex $v_j \in S$ s.t. $y_{i_1,j} = 1$ and $y_{i_2,j} = 1$.
2. Clause $C_\ell$ is of type $\bigvee_j y_{i,j}$. Then there exists connected component $C_i$ s.t. $y_{i,j} = 0$ for every vertex $v_j \in S$.

1. We have at least two edges in the matching coming to the vertex $v_j$. Tree $T_\ell$ corresponds to the recursive tree of the following algorithm that finds these edges. Check every edge $e$ between $v_j$ and $C_{i_1}$. Once the edge $e_1$ with $x_{e_1} = 1$ is found, switch to the second component $C_{i_2}$ and repeat the search. Once the second edge $e_2$ with $x_{e_2} = 1$ is found, return falsified clause $\neg x_{e_1} \vee \neg x_{e_2}$. Both edges exist since $y_{i_1,j} = \bigoplus_{(u,v_j) \in E, u \in C_{i_1}} x_{(u,v_j)} = 1$ and $y_{i_2,j} = \bigoplus_{(u,v_j) \in E, u \in C_{i_2}} x_{(u,v_j)} = 1$. Tree $T_\ell$ has size $O(n^2)$.

2. We have $y_{i,j} = 0$ for every $v_j \in S$. It means that either $x_{u,v} = 0$ for every $u \in C_i$ and $v \in S$ or there are at least two $x_{(u,v_j)} = 1$ for a fixed vertex $v_j$.

First, we ensure that the variables $x_{u,v} = 0$. Tree $T_\ell$ begins with a splitting tree $T_i$ that corresponds to the following algorithm. The algorithm goes through all variables $x_e$ for all edges between $S$ and $C_i$. Once, the algorithm finds $x_{(u,v_j)} = 1$ for a vertex $v_j$, it starts to look for the second $x_{(u',v_j)} = 1$ for all $u' \in C_i \setminus \{u\}$. There must exist such a variable since $y_{i,j} = \bigoplus_{(u,v_j) \in E, u \in C_i} x_{(u,v_j)} = 0$. Once the variable is found, the algorithm returns a violated clause.

If all variables are zero, we end up at a free leaf of $T_i$ where $C_i$ has no outgoing edge taken into the matching. We consider $C_i$ as a graph of the odd-cardinality and use Proposition 2 to get a polynomial-size splitting tree $T_{C_i}$. We attach $T_{C_i}$ to the free leaf of tree $T_i$ forming $T_\ell$.

# 5   Open Question

Tight bounds on splitting trees for perfect matching is still an open question. Itsykson and Sokolov provided polynomial-size splitting trees for graphs on odd number of vertices. We have just proved, that formula built on an arbitrary graph has a splitting tree of size $2^{O(n)}$. It is an interesting question if the formula $PMP_G$ has exponential lower bounds for arbitrary graphs or even such case can be solved with polynomial-size splitting trees.

**Acknowledgements.** The work is performed according to the Russian Government Program of Competitive Growth of Kazan Federal University. The author is grateful to Dmitry Itsykson for fruitful discussions and to Sam Buss for valuable advices that improve readability of the paper.

# References

1. Itsykson, D., Sokolov, D.: Lower bounds for splittings by linear combinations. In: Csuhaj-Varjú, E., Dietzfelbinger, M., Ésik, Z. (eds.) MFCS 2014, Part II. LNCS, vol. 8635, pp. 372–383. Springer, Heidelberg (2014)
2. Davis, M., Logemann, G., Loveland, D.: A machine program for theorem-proving. Commun. ACM **5**, 394–397 (1962)
3. Davis, M., Putnam, H.: A computing procedure for quantification theory. J. ACM **7**, 201–215 (1960)
4. Tseitin, G.S.: On the complexity of derivation in the propositional calculus. Zapiski nauchnykh seminarov LOMI **8**, 234–259 (1968). English translation of this volume: Consultants Bureau, N.Y., 1970, pp. 115–125
5. Urquhart, A.: Hard examples for resolution. JACM **34**(1), 209–219 (1987)
6. Alekhnovich, M., Hirsch, E.A., Itsykson, D.: Exponential lower bounds for the running time of DPLL algorithms on satisfiable formulas. J. Autom. Reasoning **35**(1–3), 51–72 (2005)
7. Seto, K., Tamaki, S.: A satisfiability algorithm and average-case hardness for formulas over the full binary basis. In: 2012 IEEE 27th Annual Conference on Computational Complexity (CCC), pp. 107–116, June 2012
8. Demenkov, E., Kulikov, A.S.: An elementary proof of a 3n-o(n) lower bound on the circuit complexity of affine dispersers. In: Murlak, F., Sankowski, P. (eds.) MFCS 2011. LNCS, vol. 6907, pp. 256–265. Springer, Heidelberg (2011)
9. Dantchev, S., Riis, S.: Tree resolution proofs of the weak pigeon-hole principle. In: 16th Annual IEEE Conference on Computational Complexity, pp. 69–75 (2001)

# Satisfiability Solving

# Extreme Cases in SAT Problems

Gilles Audemard[1] and Laurent Simon[2(✉)]

[1] CRIL, Artois University, Arras, France
audemard@cril.fr
[2] LaBRI, University Bordeaux, Bordeaux, France
lsimon@labri.fr

**Abstract.** With the increasing performance of SAT solvers, a lot of distinct problems, coming from very disparate fields, are added to the pool of *Application* problems, regularly used to rank solvers. These problems are also widely used to measure the positive impact of any new idea. We show in this paper that many of them have extreme behaviors that any SAT solvers must cope with. We show that, by adding a few, simple, human-readable, indicators, we can let Glucose choose between four strategies to show important improvements on the set of the hardest problems from all the competitions between 2002 and 2013 included. Moreover, once the SAT solver has been specialized, we show that a new restart polarity policy can improve even more the results. Without the first specialization step mentioned above, this new and effective policy would have been jugged inefficient. Our final Glucose is capable of solving 20 % more problems than the original one, while speeding up also UNSAT answers.

## 1 Introduction

In the late 90's, benchmarks used for SAT solvers evaluations were often gathered under the same pool of problems. With the introduction of CDCL solvers [11, 17,18] a few years later (and their specialized data structure for huge problems) appeared the need to partition them into random problems, "hand crafted" problems (designed to give solvers a hard time) and "industrial" problems. It was clear that techniques developed in each category had to be fundamentally distinct [7,10,13]. And so, the first SAT competition (in 2002) proposed to use this partitioning to award solvers in a satisfying way [6]. This high-granularity partitioning remains a standard in the community.

However, with the increasing interest in SAT technologies, a number of new problems were reduced to SAT, coming from distinct areas (*e.g.* biology, cryptography, graph problems, factorization, hardware/software verifications,...). All non random, non handcrafted problems were considered under the same pool of "Application" problems [5], even if the pool offer a high discrepancy in problems structures. This would not be problematic if the practical SAT field was not so much driven by empirical results: any new idea must be validated by an increasing number of solved instances. Moreover, each competition or race takes place

© Springer International Publishing Switzerland 2016
N. Creignou and D. Le Berre (Eds.): SAT 2016, LNCS 9710, pp. 87–103, 2016.
DOI: 10.1007/978-3-319-40970-2_7

every year, with its own benchmarks selection rules. And thus, a new idea can be very appealing one year, and uninteresting one year before or one year after, depending on the selected problems.

In this paper, we show that the initial partitioning into "applications" problems can be counter-productive. It is partially based on our experience with Glucose [2–4] and in the last developments of SAT solvers [8,9,19,23] where a number of different techniques are embedded into single core SAT engines. Without any benchmark specialization technique, we were almost unable to improve Glucose performances on the whole set of distinct problems. In most of the cases, what was gained on one side was lost on the other. Moreover, once we were able to specialize Glucose for only 4 typical use cases, we were able to develop an additional technique to increase its performances. In total, by mixing the two new techniques presented in this paper, we were able to solve 20 % more problems on all the hardest problems found in the literature from 2002 to 2013, included, which is a one-year unseen improvement in the history of Glucose.

The contributions of this paper are twofold. Firstly, we show that a lot of series of problems found in the last competitions show *unusual* SAT engine regimes. We ran Glucose on a large number of difficult benchmarks and gathered many data during the search, such as the number of decisions between each conflict, the depth when a conflict occurs, the number of glue clauses learnt... We show that many *families* of benchmarks (benchmarks from the same generator/problem encodings) exhibit extreme behaviors, sometimes well beyond any initial guess. By focusing on only 3 extreme cases, we were able to adapt Glucose search parameters and techniques. Secondly, once adaptation is included in the general Glucose engine, it opens rooms for further improvements: the second contribution of this paper is a specialized phase cache mechanism [21] used only when restarting. This new phase cache scheme is motivated by a number of experimentations showing that the phase can be partitioned into two forces (one for SAT and one for UNSAT), casting new lights on a fundamental mechanism of CDCL SAT solvers. What is striking in our results is that this mechanism seems uninteresting on the original Glucose solver but shows a clear improvement on the new *adaptative* Glucose.

In the first part of the paper, we show, Sect. 2 that a number of problems have extreme behaviors. We also introduce Glucose adaptation policy in this section. In the Sect. 3, we present the new phase cache scheme, motivated by empirical evidences studying the role of the phase on SAT problems. The Sect. 4 presents the empirical evaluations of the new techniques proposed in this paper, pointing out that different benchmarks selections can be misleading. Before concluding, Sect. 6, we discuss our results Sect. 5 in front of recent trends in SAT solving.

## 2 SAT Benchmarks: On Dealing with Many Extreme Cases

As we will show in this section, the set of benchmarks we consider in SAT contests is full of outliers. These *special* problems are not exceptions by themselves. They belong to particular series of outliers. It was already pointed out in [5] that problems can be partitioned into buckets, depending on their origin. However, these buckets were proposed w.r.t. the origin of problems (crypto, hardware verification, ... ) and not with respect to the solver's behavior on them. These series of outliers may be problematic: a simple serie containing many benchmarks can completely change the results of a contest, or, more importantly, give a wrong feedback of a new idea to be tested against traditional testbeds. This will be discussed in detail Sect. 5.

In order to properly detect outliers, we gathered all industrial/application instances from all previous known competitions from 2002 to 2013, included and done our best to remove duplicate problems. We obtained a first set of 2,632 problems. In the first part of this paper, we kept only problems not solved in 100,000 conflicts, after preprocessing (we used Minisat Satelite build-in pre-processing) and gathered a set of statistics at this point. Using preprocessing allowed to solve many problems by itself but we think that CDCL solvers are typically launched on pre-processed SAT formulas and thus we need to study their behavior on simplified formulas. We obtained a set of 1,164 not trivial problems to study (66 % of the original problems are trivially solved). The current section is based on the study of these 1,164 problems[1].

## 2.1 Outliers Everywhere

In the following, we report the distribution of some characteristic measures we chose as witnesses of the underlying mechanisms operating during the SAT solver run. These measures are the number of decisions levels, the number of decisions between two conflicts, the number of successive conflicts, the number of non binary glue clauses and the number of unit propagations performed during the search.

**Decision Levels.** The first indicator we propose to observe is the decision level at which conflicts are found. We averaged them for each of the 1164 formulas (after 100,000 conflicts) and reported the distribution of these values Fig. 1. 10 % of problems have conflicts occurring with more than 400 decisions and 10 % with less than 30. Clearly enough, we cannot expect a CDCL solver to have the same behavior on these two kinds of problems. We rather could expect that a technique developed for one extreme will not work for the other (and *vice-versa*). For instance, it is questionable to systematically use Clause Learning when decision levels are smaller than a few dozens. More interestingly, most of the outliers are in fact families of problems: the set of benchmarks with the highest decision levels are hwmcc12miters (from competition 2013), li-test4 (2003), 9dlx-vliw (2004). With the lowest decision levels, we find families longmult (2008), desgen-gss (2009), kummling-grosmann-ctl (2013), traffic (2011), eq.atree.braun (2007). It is very interesting to notice that some hard problems may exhibit an

---

[1] The list of problems with additional informations are available at http://www.labri.fr/perso/lsimon/sat16.

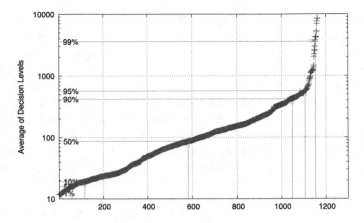

**Fig. 1.** Average Decision Levels at which conflicts were reached (over all the problems taking more than 100,000 conflicts to solve). Note that the Y-axis is a log axis.

average number of decision of less than 20 levels, showing a more "combinatoric" structure.

**Number of Decisions per Conflicts.** Another very simple measure is the total number of decisions made during the first 100,000 conflicts. This measure approximates the number of decisions made between two successive conflicts (if we forget about restarts). A large number should identifies problems where conflicts are somehow far away from each others. Figure 2 plots the distribution of these values. It is striking to notice that if half of the benchmarks have less than 3 decisions per conflicts to make, 10 % of the benchmarks have more than 20 decisions between two successive conflicts, which is surprisingly large. Also, a large number of problems have only one decision per conflict, on average. Clearly enough, this shows a large discrepancy in the intrinsic structure of problems. Let us notice here that this measure may be a little bit biased at the start of the search, where a lot of decisions are based on uninitialized heuristics values.

Let us focus on families of outliers instances. The sets of instances bivium (2013), hitag (2013), par32 (2002), desgen-gss (2009) have less than 1.1 decisions per conflicts on average. Those benchmarks have a clear "combinatoric" behavior, which is consistent with their origin. In any cases, we cannot expect the conflict analysis to give very good insights in terms of backjump information. At the opposite, we find the hwmcc12miters (2013), li-test4 (2003), 9dlx-vliw (2004) with 30 to 3000 decisions between conflicts, which is really above any expectation.

**Number of Successive Conflicts.** Figure 3 reports the number of successive conflicts, *i.e.* conflict directly occurring after another conflict, without decision. A high number of successive conflicts may point out problems on which resolution/FUIP is not powerful enough to analyze conflicts. This figure is less

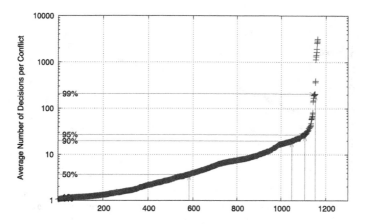

**Fig. 2.** Average Number of Decisions per conflicts over all the problems taking more than 100,000 conflicts to solve (total number of decisions divided by 100,000). The Y-axis is a log axis.

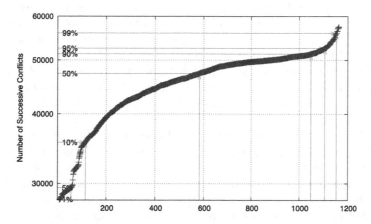

**Fig. 3.** A successive conflict is a conflict directly occurring after another conflict, without decision. The Y-axis is a log axis.

pathological than the first two ones we reported above but the 10 first and 10 last percents clearly exhibit extreme behaviors. The low values for this number are all the nossum-sha1 (2013) problems (around 30,000), followed by the very hard MD5 (2014) problems (around 32,000). At the opposite, the highest vales are typically observed for vmpc (2006) and bio-rpcl-xits (2009). Beside the 20 % of outliers we identified here, it was surprising to see that, for all problems, at least 30 % of conflicts are obtained without further decision[2], which is a lot.

---

[2] This observation was originally pointed out by Lakhdar Saïs (CRIL, Lens) during one of the National ANR-Funded project "UNLOC" in 2010.

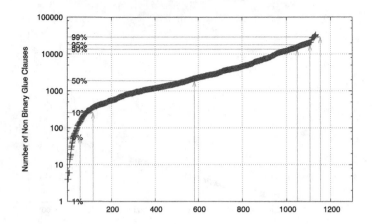

**Fig. 4.** Number of Non-binary Glues Clauses. The Y-axis is a log axis.

**Non-binary Glue Clauses.** Glue clauses are one of the key ingredients of Glucose. However, some problems have a lot of binary learnt clauses (that are trivially also glue clauses). Measuring the number of non binary glue clauses may give additional hints on the structure of the problem. Moreover, the strategy of Glucose forces the solver to keep all glue clauses forever, whereas for a typical CDCL solver, only binary clauses are kept forever. Thus, we could expect problems with a lot of non binary glue clauses to be well suited for Glucose. Figure 4 reports the cumulative distribution of this measure. The Homer (2002), stable (2014), ndhf-xits (2009) have 0 pure glue clauses in the first 100,000 conflicts. The bivium (2013) have no more than 128 pure glue clauses and the hitag (2013) at most 280 (those two families of problems are typically hard for Glucose). We can also report a lot of kummling-grosmann-ctl (2013) and carseq-pb (2013) ones. At the opposite, the ibm-SAT-dat-k (2002–2012) problems have a very large number of true glue clauses (around 20,000). We typically find here BMC problems from the industry (goldberg-bmc). Note that ibm-SAT-dat-k problems are also the problems that produced the more binary clauses. Interestingly, the goldberg-bmc2-cnt10 problem (2002) produced 45,122 glue clauses (with 15,926 binary clauses). It's almost half of the learnt clauses that are glues.

**Unit Propagations.** A well known measure for the performance of SAT solvers is their ability to handle unit propagations. However, what we report here is the high discrepancy of this number, depending on series of problems. Figure 5 reports the distribution of the number of propagations done during the first 100,000 conflicts. Here again, many discrepancies can be reported. The smallest values are from Homer (2002), stable (2014), aes (2011). The largest values are for atco (2014), arcfour (2013), hwmcc12miters (2013), ibm-sat-dat, ACG (2009) and gss (2009). On these extreme values we have mores than 10,000 propagations per conflicts, which is, once again, beyond any initial guess we could have. To solve this kind of problems, we should probably use a more

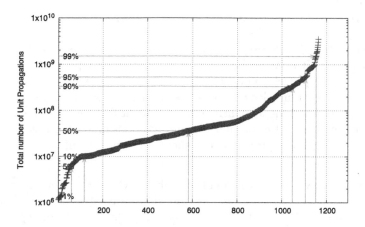

**Fig. 5.** Total Unit Propagations fired during the first 100,000 conflicts to solve. The Y-axis is a log axis.

sophisticated preprocessing, taking into account equivalences between literals for instance, or use failed literal probings [25].

## 2.2   A Few Other Measures

We report Table 1 a selection of other measures we tested, to emphasize that on most of mesures we tried extreme behaviors can be observed. BJ means Backjump level (number of decision levels cancelled after conflict analysis), DL is the decision level depth, where conflict occurred, LBD is the average LBD value of learnt clauses, Size their size, BR the number of Glucose blocked restart, FUIPisDEC is the percentage of conflicts where the FUIP literal was the last decision. As we can read on the table, a number of these measures have impressive skewness and kurtosis values, which typically witnesses the existence of extreme values.

**Table 1.** Summary of some other measures performed on the 1,164 instances.

| Measure | BJ | DL | LBD | Size | BR | FUIPisDec |
|---|---|---|---|---|---|---|
| Min | 1.05 | 10.81 | 4.17 | 5.28 | 0 | 24 % |
| Max | 3154.9 | 8559.46 | 83.81 | 3395.46 | 1730 | 91 % |
| Median | 3.3 | 86.69 | 10.45 | 25.51 | 298.5 | 61 % |
| Mean | 24.57 | 218.29 | 14.28 | 74.02 | 344.54 | 63 % |
| St. Dev. | 194.73 | 623.29 | 11.61 | 242.35 | 339.80 | 12 % |
| Skewness | 12.69 | 8.98 | 3.20 | 9.01 | 1.51 | 0.34 |
| Kurtosis | 174.34 | 99.02 | 14.70 | 100.68 | 6.04 | 2.33 |

### 2.3 Synthesis: Choosing the Proper Strategy

We showed above the high discrepancy of CDCL behaviors on so called "Application" benchmarks. It may not be realistic, or even counter-productive, to expect a general CDCL approach to be efficient on all problems with the same parameters/techniques. We thus decided to focus the tuning of Glucose on each set of extreme problems, independently. Once we identified sets of typical outliers, we tried an important set of parameters and methods on each set, separately. We report here the final configuration that showed the best results.

**Low Decision Levels.** For this kind of problem, Chanseok [19] strategies have were the best ones. We thus kept all the clauses until LBD 4 and used his reduction strategies. In this strategy, all clauses with LBD<4 are kept and only a few clauses are kept at each reduction, based on their activity in conflict analysis (original Minisat strategy). By using this, we were able to solve 20 (over 30) Bivium/Hitag additional problems (the original Glucose was only able to solve 1 of these problems).

Let us point out here the differences between our approach and the one proposed in [19]. First, we don't focus on the SAT vs UNSAT supposed status of the formula. Second, we don't launch 2 separate method on the same instance. The original method in [19] alternates between 2 regimes. Here, we finally have only one strategy, once decided.

**Successive Conflicts.** If this number is low, this is typically nossum and MD5 problems. In this case, it has been also reported [19] that a Luby restart [15] and a very high VSIDS value (0.999 instead of 0.95) is efficient. We were not able to solve any addition MD5 benchmarks (they are all very hard) but our parameters allowed to solve 43 nossum instead of 18 in the original Glucose (+25 benchs).

If this number is high, then we experimented a set of parameters to target VMPC problems. Those cryptographic problems are very hard to solve. We generated a set of easier problems to experiment with. On our set, we add a few additional options. We implemented a strategy in the idea of minisat-blbd where restarting is considered as a way to diversify the search, by forcing the first descent after a restart to be randomized. This method only seems to give good results on this last set of problems.

**Pure Glue Clauses.** On the set of ibm problems, identified by a very high number of pure glue clauses, we observed that a very low VSIDS constant of 0.91 is the best choice. The reason for this is still unknown.

## 3   Playing with the Phase

The phase saving mechanism is one of the essential components of SAT solvers [21], especially when used with very fast restarts. However, there are still open questions about it. *is the phase trying to search for a model (SAT) or a contradiction (UNSAT)?* The phase is updated during conflict analysis (which would suggest memorizing where contradictions occurred) but also during the *backjump* (which would suggest memorizing unharmful assignments, *i.e.* independant ones). In the seminal paper [21] the ability of the phase to memorize partial solutions of independant problems was pointed out, suggesting that the phase was more a witness of partial solution found so far.

In this section, we investigate the question of an optimized phase caching scheme. As we will show, all our attempts to use the phase to help searching for a model (SAT) failed. However, we were able to specialize this simple cache scheme for UNSAT search. In any cases, we think our experimental results will cast a clearer view of the phase mechanisms.

## 3.1   A Phase for SAT, a Phase for UNSAT

The phase is a very simple cache scheme to choose a literal polarity when branching. When branching, the phase is only responsible for the polarity of the chosen variable. If finding the good phase for SAT instances is as hard as solving it, for UNSAT problems, finding a good phase is not as intuitive. We propose to split the phase updates into two typical cases: (1) when the assignments were responsible for a conflict and (2) when they were independent of the conflict. The case (1) occurs during conflict analysis, on the last decision level, only on variables seen during conflict analysis. All other assignments made at the last decision level, not seen during conflict analysis, are variables propagated but not implied in the conflict. Those last assignments can be considered as unharmful and belong to case (2). The other set of unharmful assignments are the variables assigned at a decision level between the last decision level and the backjump level (strictly). Those variables were conflict independent. Let's write phaseUNSAT the specialized phase for the cases (1) and phaseSAT the phase for cases (2). We will call polarity the standard phase cache scheme (phase updated at every unassignment).

As an example, we illustrate the above 2 cases Fig. 6. We represent a formula, and a sequence of decisions/propagations leading to a conflict (level 4). Conflict analysis involves variables $x_9, x_{13}, x_7$ from last decision level and $x_3$ and $x_8$ that belong to the asserting clause. All these literals update their phaseUNSAT to their respecting values when backjumping. Conversely, literals $x_{10}, x_{12}$ from last decision level and $x_2, x_{11}$ are unassigned after backjumping and do not participate to conflict analysis. Those variables update their phaseSAT. All the above variables update their polarity when backjumping. A last note: we also refined our phaseSAT/phaseUNSAT updates to take into account the frequent restarts. Obviously, when restarting, we don't have any information about the harmfulness of any assignment. Thus, when restarting, no phaseSAT or phaseUNSAT updates are performed (which is not the case for polarity).

$c_1 = x_1 \vee x_4$
$c_2 = x_1 \vee \overline{x_3} \vee \overline{x_8}$
$c_3 = x_1 \vee x_8 \vee x_{12}$
$c_4 = x_2 \vee x_{11}$
$c_5 = \overline{x_3} \vee \overline{x_7} \vee x_{13}$
$c_6 = \overline{x_3} \vee \overline{x_7} \vee \overline{x_{13}} \vee x_9$
$c_7 = \overline{x_8} \vee \overline{x_7} \vee \overline{x_9}$
$c_8 = \overline{x_7} \vee x_{10}$
$c_9 = \overline{x_7} \vee x_1 \vee x_{12}$

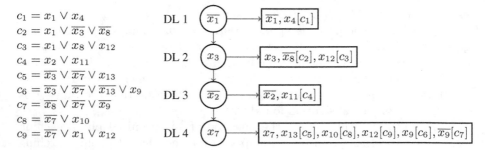

Resolution Steps:
$\beta_1 = res(x_9, \phi_7, c_6) = \overline{x_3} \vee x_8 \vee \overline{x_7} \vee \overline{x_{13}}$
$\beta = res(x_{13}, \beta_1, c_5) = \overline{x_7} \vee \overline{x_3} \vee x_8$

**Fig. 6.** Sequence of decisions (decision literals are in circle nodes) followed by propagations (in rectangle). Each clause reason of a propagation is in brackets. A conflict occurs at decision level 4. Resolutions steps are also displayed. The backjump level is equal to 2. The asserting literal is $\overline{x_7}$.

## 3.2  Comparing Phase Cache Values with the Final Models

We computed the phaseSAT/phaseUNSAT values for 573 hard problems (see Sect. 4 for a description of those benchmarks) with a timeout of 10,000s. We only obtained 123 SAT answers to be analyzed. For each of them, we compare the final solution with the two above caching schemes. On Fig. 7 we report for each formula, the number of phaseSAT values agreeing with the solution versus the number of phaseUNSAT values also agreeing with the solution (points below the line $y = x$). We also report the number of phaseSAT values that disagree with the solution versus the number of phaseUNSAT values that disagree with the solution (points above the line ($y = x$). It is striking to notice how clear this figure is. The phaseSAT is always much more close to the solution than the phaseUNSAT. Moreover, the phaseUNSAT much more disagrees with the solution than the phaseSAT.

Let us notice here that some values for the phaseSAT/phaseUNSAT arrays may not be initialized (if a variable is never seen during conflict analysis, its phaseUNSAT value will be unknown). In this experiment, we only count values for initialized values. Thus, phaseUNSAT values are the last polarity of a variable seen during conflict analysis (if seen), and we may have counted much less initialized values in phaseUNSAT than in phaseSAT (or *vice-versa*). This could give the wrong feeling that one array is closer to the solution that the other (by considering many more values). This is why we had to report the two scatter plots Fig. 7 (one scatter plot cannot be deduced from the other one).

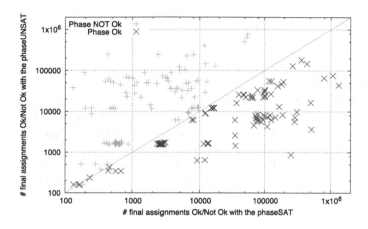

**Fig. 7.** Comparison of the number of assignments in the final solution that agree/disagree with `phaseSAT` or `phaseUNSAT`

### 3.3    A First Attempt for SAT...that Failed

We *played* with the `phaseSAT` polarity in order to improve `Glucose` performances on SAT instances. We implemented the following two techniques:

– During the first descent after a restart, assign decision variables with respect to the `phaseSAT` polarity: try to first get closer to a solution.
– Use the `phaseSAT` polarity when `Glucose` decides to postpone a restart, when approaching a full assignment [4]: the number of decision levels is suddenly increasing, meaning that a solution may be found, and may be using a more SAT-related polarity would help to complete the current partial assignment.

These two techniques were unsatisfactory. The number of solved instances was approximatively the same for both of them. Of course, on some satisfiable instances the gain was substantial, but as it is often the case, what was gain on one side, was lost on the other. However, it is important to notice that this technique did not change the overall performances of `Glucose` on unsatisfiable instances.

### 3.4    Refining the Phase for UNSAT, When Restarting

We first tried to replace the `polarity` scheme by the `phaseUNSAT` polarity only. The results were clearly bad, showing that keeping the `phaseSAT` polarity was important. However, the `phaseUNSAT` has a clear meaning: it indicates where conflicts are. Thus, we only used this polarity during the first descent after a restart, in the hope to reach conflicts even faster. The results were, this time, clearly good. We report the detailed results of this new technique in the next section (Sect. 4). As we will see in the following, however, results are highly dependent of the selection of considered problems and, more intriguibly, the time out.

## 4   Experimentations

In this section we compare different combinations of Glucose, with and without the adaptive tuning (denoted by A), and, with and without the phaseUNSAT scheme (denoted by P). We add in this experiment the last versions of lingeling that participate to previous SAT competitions. Our results may look a little bit disparate (we report results on 3 sets of benchmarks, with different timeouts) but, as we will explain in the later, we had to deal with surprising results that needed further investigations. Before starting to analyze the results, let us note the lingeling-baq version we used is the Machine Learning version, using auto-tuning and learning on the set of problems we used (see Sect. 5).

| Solver | SAT | UNS | Total |
|---|---|---|---|
| Glucose | 81 | 151 | 232 |
| Glucose A | 98 | 162 | 260 |
| Glucose P | 83 | 150 | 233 |
| Glucose A+P | **107** | 168 | **275** |
| lingeling-ayv | 93 | 158 | 251 |
| lingeling-baq | 77 | **185** | 262 |

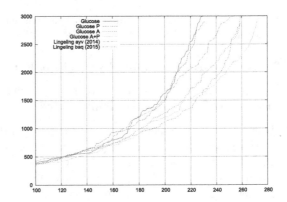

**Fig. 8.** Comparing several versions of Glucose and lingeling on the set of 573 hard instances for Glucose. The table provides the number of SAT/UNSAT instances solved, right part gives the classical cactus plot. Time limit is set to 3,000 s.

     The first experiments is done on a subset of formulas coming from competitions 2002 to 2013 included. Here, we selected 573 problems among the 2,632 ones and took only ones that cannot be solved in less than one minute with GLUCOSE-SYRUP working on 12 cores[3]. Our goal here is to have hard, but not so hard problems, suitable for exhaustive testing on a cluster of CPUs. We simply wanted to limit the number of short jobs requests. We limited the CPU time to 3,000 s in this first experiment, which should be largely sufficient for our purposes. Figure 8 shows the results. If we start to compare Glucose with and without the phaseUNSAT scheme, we can see approximatively the same results. Conversely, the adaptive tuning of Glucose seems promising, it can solve more satisfiable and unsatisfiable instances than the classical version of Glucose. When both components (A+P) are activated, we obtain an even stronger solver, which is able to solve 43 more instances than the basic version of Glucose. It

---

[3] More informations on this set of problems available at http://www.labri.fr/perso/lsimon/sat16.

is however surprising to see that the UNSAT phase component seems inefficient. We do not have yet a satisfactory explanation for this.

The comparison with `lingeling` is also surprizing. The last version is amazingly powerful on unsatisfiable instances, but performances fall down on satisfiable ones. This last version uses an adaptive tuning trained on SAT'2014 benchmarks, which contains problems we used too in this experiment. It may be possible that the adaptive behavior fails to classify satisfiable instances. On this set of benchmarks, our last version of `Glucose` solves the most instances.

| Solver | SAT | UNS | Total |
|---|---|---|---|
| Glucose | 104 | 127 | 231 |
| Glucose A | **115** | 128 | 243 |
| Glucose P | 106 | 127 | 233 |
| Glucose A+P | 112 | 129 | 241 |
| lingeling-ayv | 99 | 143 | 243 |
| lingeling-baq | 106 | **146** | **252** |

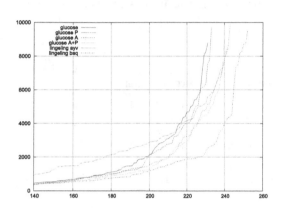

**Fig. 9.** Comparing several versions of `Glucose` and `lingeling` on SAT 2014 application benchmarks. The table provides the number of SAT/UNSAT instances solved, right part gives the classical cactus plot. Time limit is set to 10,000 s.

We were surprized by the very good results of our last version of `Glucose` in comparison to `lingeling-baq`, especially because this version of lingeling was trained using Machine Learning. We thus decided to investigate this. The first thing to investigate is the increasing of the time out. The second thing was of course to change the set of problems we considered.

Let us now focus on SAT 2014 competition benchmarks with a 10,000 s time out, reported Fig. 9. Here again, playing with the phase seems useless without the adaptive strategy of `Glucose`. However, the adaptive strategy always produces strong results which is a very good result: our adaptative strategy was based on benchmarks from 2002 to 2013, and it pays for this new set of problems.

On these set of benchmarks, around 210 have *normal* behaviors and around 40 were solved before the limit of 100,000 conflicts. Then, the adjustment is done on a small set of benchmarks. Twenty of them have very low decision levels, and `Glucose` adjusts its strategy in consequence (see Sect. 2.3). For thirty instances, the tuning was due to low successive conflicts and for 6 of them, to the high number of successive conflicts. Finally, none of the benchmarks exhibit a large number of non binary glue clauses. Nevertheless, the adaptive tuning allows to solve 12 more instances than the original version.

There is however a negative point to discuss: the activation of both components (P and A) does not seem as effective as what we observed on the first set of benchmarks, with a small time out. Then, our first promising results are a little counteracted here and, once again, we may just speculate some explanation (see below). This is however partly due to the time limit, since with a cutoff of 3,000 s, Glucose-A solves 217 instances and Glucose-A+P, 220. Of course, our first set of benchmarks contains many different families of instances and may be some of them are not represented here. It is quite visible that the adaptive tuning of lingeling was adjusted on this set of benchmarks leading to a very robust solver. But, here again, the results of lingeling on satisfiable instances are not as impressive as on unsatisfiable ones (Fig. 10).

| Solver | SAT | UNS | Total |
|---|---|---|---|
| Glucose | 148 | 108 | 256 |
| Glucose A | **154** | 108 | **262** |
| Glucose P | 151 | 108 | 259 |
| Glucose A+P | 150 | 109 | 259 |
| lingeling-ayv | 146 | 111 | 257 |
| lingeling-baq | 146 | **115** | 261 |

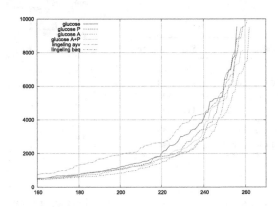

**Fig. 10.** Comparing several versions of Glucose and lingeling on SAT 2015 application benchmarks. Table provides the number of SAT/UNSAT instances solved, right part gives the classical cactus plot. Time limit is set to 10,000 s.

The last set of benchmarks comes from the SAT-RACE 2015, used with a 10,000s cutoff. Results are quite comparable with the above one: the adaptive framework allows to solve few more instances and the combination of both components seems, one more time, useless. Here again, the lack of diversity of benchmarks may explain it but, more importantly, as reported by the shape of the cactus plot, our strategy A+P pays for small time out. Our guess is that by intensificating the search on UNSAT, our solver may be more greedy on the search for an UNSAT proof. However, the learning mechanism could also "repair" the phase s.t. after a while it reduce the differences between the two approaches.

The tuning of lingeling provides less spectacular results than for the 2014 SAT competition one. Classification does not help so much the basis solver in term of solved instances, but, it is much more faster with it. However, we still experiment an important gain on small time outs (less than 3,000 s).

## 5    Discussion

Portfolio approaches exploit the fact that different solvers are better at solving different kinds of problems. The exist for CSP (Constraint Satisfaction Problems) [1,20], SAT [16,24] or either ASP (Answer Set Programming) [12]. After an intensive offline training phase, these solvers use Machine Learning techniques to choose the best solver to launch. Recently, most of the progresses observed in the SAT field were partially due to benchmarks adaptations. For instance, inprocessing techniques [14], uses a number of different techniques that can be turned off on the fly if proved inadequate. Moreover, the recent success of [9] is certainly due to solver specialization for different typical cases (however, this last solver was not made publicly available by the author). More importantly, in the SAT RACE 2015, at least two other solvers used benchmark recognition/adaptation. The first one is LINGELING solver [8]. This version uses a k-nearest neighbor (KNN) algorithm to classify a given benchmark (based on the number of variables/clauses) inside one of the buckets instances of the SAT Competition 2014. If the classifier agrees, then parameters are tuned in order to try to use the best strategy with respect to the given bucket. The other solver that used tuning mechanism is CRYPTOMINISAT [23]. This solver also used machine learning techniques [22] to choose, after 160,000 conflicts, one among the 13 different setup of the solver. The training phase was done on benchmarks from previous competitions (2009, 2011, 2013).

Our work is related to these approaches but our final goal is distinct, written in a more long-term view of SAT solver developments. We try to understand the different SAT solver regimes and try to extract only human understandable ones. By comparison, we use five measures to separate benchmarks in families, where SATZILLA uses more than fifty ones. Once those extreme cases were identified, we proposed to work on each "bucket" of instance. We tried to propose new techniques, designed to fit the observed behaviors (not all tried techniques are reported in this paper but, for instance, we tried a number of specialized branching heuristics when we observed a small number of FUIP decision variables, as reported Table 1, without success). This is not the same thing as trying as many parameters as possible, blindly. A portfolio approach in the late 90's will not have found a parameter to change any DPLL solvers into a CDCL one. Moreover, our approach is guided by the willing of understanding benchmarks characteristics. Even if Machine Learning can outputs a decision tree for parameter tuning, we think it is much more simple and understandable to talk about problems with low decision levels, or large number of successive conflicts. We even think that adapting CDCL solvers for this special case is an important direction for future research.

# 6    Conclusion

It is very hard to find consistent ranking of solvers on different set of benchmarks, even when considering only well established set of benchmarks used in different SAT competition and SAT races. This is very problematic since, in the SAT community, experimental evidences are often required for any new idea. In this paper, we showed that a better partitioning of benchmarks is required. We showed that the set of "application" problems is full of outliers, that must be specially targeted: more than just adapting current methods by playing on parameters, we need to develop new ideas to solve them efficiently. We thus propose to use a partition method based on SAT solver regimes rather than the origin of problems. We think that problems that, for instance, need less than 30 decisions before conflicts should be attacked with the same, new, methods, whatever they initially represent.

Based on the above observations, we proposed a very simple version of Glucose, using only 3 special additional regimes that are human understandable. This version of Glucose (called adapt) is really strong on a set of hard problems containing all the problems found in the literature from 2002 to 2013. We also showed how efficient it is in the last sets of benchmarks (2015 in particular). Moreover, we were able to add a new phase policy on top of this version of Glucose, which is very efficient on the hardest benchmarks from 2002 to 2013. On this exhaustive set of interesting problems, our new version of Glucose is capable of solving 20 % more problems, which is a one year unseen improvement of Glucose.

**Acknowledgments.** Computer time was provided (1) the MCIA (Mésocentre de Calcul Intensif Aquitain) of the Univ. of Bordeaux and of the Univ. of Pau and des Pays de l'Adour and (2) the CRIL (Univ. d'Artois). The authors were partially supported by the ANR-2016 "SATAS" ANR-15-CE40-0017 National French Project.

# References

1. Amadini, R., Gabbrielli, M., Mauro, J.: Sunny: a lazy portfolio approach for constraint solving. arXiv preprint arXiv:1311.3353 (2013)
2. Audemard, G., Simon, L.: Predicting learnt clauses quality in modern SAT solvers. In: Proceedings of IJCAI, pp. 399–404 (2009)
3. Audemard, G., Simon, L.: Glucose 2.1: Aggressive, but reactive, clause database management, dynamic restarts (system description). In: Pragmatics of SAT 2012 (POS 2012), dans le cadre de SAT 2012, June 2012
4. Audemard, G., Simon, L.: Refining restarts strategies for SAT and UNSAT. In: Milano, M. (ed.) CP 2012. LNCS, vol. 7514, pp. 118–126. Springer, Heidelberg (2012)
5. Belov, A., Heule, M.J., Diepold, D., Järvisalo, M.: The application and the hard combinatorial benchmarks in sat competition 2014. SAT COMPETITION 2014, p. 81 (2014)
6. Le Berre, D., Simon, L.: The essentials of the SAT 2003 competition. In: Giunchiglia, E., Tacchella, A. (eds.) SAT 2003. LNCS, vol. 2919, pp. 452–467. Springer, Heidelberg (2004)
7. Biere, A.: (p)lingeling. http://fmv.jku.at/lingeling
8. Biere, A.: Lingeling and friends entering the sat race 2015 (2015)
9. Chen, J.: Minisat bcd and abcdsat: Solvers based on blocked clause decomposition. In: SAT RACE 2015 solvers description (2015)
10. Dubois, O., Dequen, G.: A backbone-search heuristic for efficient solving of hard 3-sat formulae. In: Proceedings of the Seventeenth International Joint Conference on Artificial Intelligence, IJCAI, pp. 248–253 (2001)

11. Eén, N., Sörensson, N.: An extensible SAT-solver. In: Giunchiglia, E., Tacchella, A. (eds.) SAT 2003. LNCS, vol. 2919, pp. 502–518. Springer, Heidelberg (2004)
12. Gebser, M., Kaminski, R., Kaufmann, B., Schaub, T., Schneider, M.T., Ziller, S.: A portfolio solver for answer set programming: preliminary report. In: Delgrande, J.P., Faber, W. (eds.) LPNMR 2011. LNCS, vol. 6645, pp. 352–357. Springer, Heidelberg (2011)
13. Heule, M., van Maaren, H.: March_dl: Adding adaptive heuristics and a new branching strategy. JSAT **2**(1–4), 47–59 (2006)
14. Järvisalo, M., Heule, M.J.H., Biere, A.: Inprocessing rules. In: Gramlich, B., Miller, D., Sattler, U. (eds.) IJCAR 2012. LNCS, vol. 7364, pp. 355–370. Springer, Heidelberg (2012)
15. Luby, M., Sinclair, A., Zuckerman, D.: Optimal speedup of Las Vegas algorithms. In: Proceedings of ISTCS, pp. 128–133 (1993)
16. Malitsky, Y., Sabharwal, A., Samulowitz, H., Sellmann, M.: Algorithm portfolios based on cost-sensitive hierarchical clustering. In: Proceedings of the 23rd International Joint Conference on Artificial Intelligence (IJCAI) (2013)
17. Marques-Silva, J., Sakallah, K.: GRASP - a new search algorithm for satisfiability. In: ICCAD 1996, pp. 220–227 (1996)
18. Moskewicz, M., Madigan, C., Zhao, Y., Zhang, L., Malik, S.: Chaff: Engineering an efficient SAT solver. In: Proceedings of DAC, pp. 530–535 (2001)
19. Chanseok, O.: Patching minisat to deliver performance of modern sat solvers (2015)
20. O'Mahony, E., Hebrard, E., Holland, A., Nugent, C., O'Sullivan, B.: Using case-based reasoning in an algorithm portfolio for constraint solving. In: Irish Conference on Artificial Intelligence and Cognitive Science, pp. 210–216 (2008)
21. Pipatsrisawat, K., Darwiche, A.: A lightweight component caching scheme for satisfiability solvers. In: Marques-Silva, J., Sakallah, K.A. (eds.) SAT 2007. LNCS, vol. 4501, pp. 294–299. Springer, Heidelberg (2007)
22. Ross Quinlan, J.: C4. 5: programs for machine learning. Elsevier (2014)
23. Soos, M., Lindauer, M.: The cryptominisat-4.4 set of solvers at the sat race 2015 (2015)
24. Xu, L., Hutter, F., Hoos, H., Leyton-Brown, K.: Satzilla: Portfolio-based algorithm selection for sat. J. Artif. Intell. Res. **32**, 565–606 (2008)
25. LeBerre, D.: Exploiting the real power of unit propagation lookahead. In: Proceedings of SAT (2001)

# Improved Static Symmetry Breaking for SAT

Jo Devriendt[1(✉)], Bart Bogaerts[1,2], Maurice Bruynooghe[1],
and Marc Denecker[1]

[1] KU Leuven – University of Leuven, Celestijnenlaan 300A, Leuven, Belgium
{jo.devriendt,bart.bogaerts,maurice.bruynooghe,
marc.denecker}@cs.kuleuven.be
[2] Helsinki Institute for Information Technology HIIT,
Department of Computer Science, Aalto University, 00076 Aalto, Finland

**Abstract.** An effective SAT preprocessing technique is the construction of symmetry breaking formulas: auxiliary clauses that guide a SAT solver away from needless exploration of symmetric subproblems. However, during the past decade, state-of-the-art SAT solvers rarely incorporated symmetry breaking. This suggests that the reduction of the search space does not outweigh the overhead incurred by detecting symmetry and constructing symmetry breaking formulas. We investigate three methods to construct more effective symmetry breaking formulas. The first method simply improves the encoding of symmetry breaking formulas. The second detects special symmetry subgroups, for which complete symmetry breaking formulas exist. The third infers binary symmetry breaking clauses for a symmetry group as a whole rather than longer clauses for individual symmetries. We implement these methods in a symmetry breaking preprocessor, and verify their effectiveness on both hand-picked problems as well as the 2014 SAT competition benchmark set. Our experiments indicate that our symmetry breaking preprocessor improves the current state-of-the-art in static symmetry breaking for SAT and has a sufficiently low overhead to improve the performance of modern SAT solvers on hard combinatorial instances.

## 1 Introduction

Hard combinatorial problems often exhibit symmetry. Eliminating symmetry potentially boosts solver performance as it prevents a solver from needlessly exploring isomorphic parts of a search space. One common method to eliminate symmetries is to add symmetry breaking formulas to the problem specification [1,6], which is called *static symmetry breaking*. For the Boolean satisfiability problem (SAT), the state-of-the-art tool SHATTER [3] implements this technique; it functions as a preprocessor that can be used with any SAT solver. *Dynamic symmetry breaking*, on the other hand, interferes in the search process by adding symmetric versions of learned clauses [8,23] or by avoiding symmetric choices [21].

Symmetry properties of a SAT problem are typically detected by first converting the problem to a colored graph such that the graph's automorphism

© Springer International Publishing Switzerland 2016
N. Creignou and D. Le Berre (Eds.): SAT 2016, LNCS 9710, pp. 104–122, 2016.
DOI: 10.1007/978-3-319-40970-2_8

group corresponds to a symmetry group of the SAT problem. Subsequently, this automorphism group is detected by tools such as NAUTY [19], SAUCY [15] or BLISS [14].

Static symmetry breaking proceeds by adding formulas that exclude symmetric assignments. A symmetry breaking formula is *sound* if it preserves at least one assignment from each class of symmetric assignments and *complete* if it preserves at most one assignment from each class. One sound symmetry breaking constraint is the *lex-leader* constraint, which holds exactly for those assignments lexicographically smaller than their symmetric image [6]. The conjunction of lex-leader constraints for every symmetry in a symmetry group constitutes a complete symmetry breaking constraint for that group. However, symmetry groups tend to be too large to enforce lex-leader for each symmetry. Instead, *partial* symmetry breaking adds lex-leader constraints only for a set of generators of the group [18]. While effective for many symmetric problems, the pruning power of partial symmetry breaking depends heavily on the chosen set of generators and whether or not compositions of these generators are eliminated as well [16].

In this paper, we present BREAKID,[1] a symmetry breaking preprocessor for SAT that follows in SHATTER's footsteps. BREAKID sports several improvements compared to SHATTER, ranging from small "hacks" to avoid known problems, as well as novel ideas to exploit a symmetry group's structure. Three of these improvements are investigated in-depth in this paper. Firstly, we evaluate a *compact CNF encoding* of the lex-leader constraint. Secondly, we show how to detect *row interchangeability symmetry* subgroups, for which a small set of generators exists such that their lex-leader constraints do break the subgroup completely. Thirdly, we show how to generate *binary symmetry breaking clauses* not based on individual generators, but on algebraic properties of the entire symmetry group. A common theme in the second and third improvement is to simultaneously adjust the set of generators and the variable ordering, both needed to construct lex-leader constraints.

We evaluate the proposed symmetry breaking improvements individually, and verify the effectiveness of both SHATTER and BREAKID on 2014's SAT competition instances. From our experiments, we conclude that (i) more compact CNF encodings have a small impact on runtime and memory consumption, (ii) row-interchangeability detection improves performance on several problems, (iii) group-based binary symmetry breaking clauses are effective for a particular class of problems, (iv) BREAKID outperforms SHATTER on most benchmarks, and (v) symmetry breaking leads to significant performance gains on the hard combinatorial instances of 2014's SAT competition.

After some preliminaries in Sect. 2, we present the improvements in Sects. 3, 4 and 5 respectively. Section 6 describes how these ideas are combined in BREAKID and Sect. 7 contains the experimental evaluation. We conclude in Sect. 8.

---

[1] Pronounced "Break it!".

## 2   Preliminaries

*Satisfiability problem.* Let $\Sigma$ be a set of Boolean variables and $\mathbb{B} = \{\mathbf{t}, \mathbf{f}\}$ the set of Boolean values. For each $x \in \Sigma$, there exist two *literals*; the *positive* literal denoted by $x$ and the *negative* literal denoted by $\neg x$. The set of all literals over $\Sigma$ is denoted $\overline{\Sigma}$. A *clause* is a finite disjunction of literals, and a *formula* is a finite conjunction of clauses (as usual, we assume formulas are in conjunctive normal form (CNF)). An *assignment* $\alpha$ is a mapping $\Sigma \to \mathbb{B}$. We extend $\alpha$ to literals as $\alpha(\neg x) = \neg \alpha(x)$, where $\neg \mathbf{t} = \mathbf{f}$ and $\neg \mathbf{f} = \mathbf{t}$. An assignment satisfies a formula iff at least one literal from each clause is mapped to $\mathbf{t}$ by $\alpha$. The Boolean Satisfiability (SAT) problem consists of deciding whether there exists an assignment that satisfies a propositional formula.

*Group theoretical concepts.* A *permutation* is a bijection from a set to itself. We write permutations in *cycle notation*: $(abc)(de)$ is the permutation that maps $a$ to $b$, $b$ to $c$, $c$ to $a$, swaps $d$ with $e$, and maps all other elements to themselves. A *swap* is a non-trivial permutation that is its own inverse. Permutations form algebraic groups under the composition relation ($\circ$). A set of permutations $P$ is a set of *generators* for a permutation group $\Pi$ if each permutation in $\Pi$ is a composition of permutations from $P$. The group $Grp(P)$ is the permutation group *generated* by all compositions of permutations in $P$. The orbit $Orb_\Pi(x)$ of an element $x$ under a permutation group $\Pi$ is the set $\{\pi(x) \mid \pi \in \Pi\}$. The *support* $Supp(\pi)$ of a permutation $\pi$ is the set of elements $\{x \mid \pi(x) \neq x\}$. The support $Supp(\Pi)$ of a permutation group $\Pi$ is the union of the supports of permutations in $\Pi$. A group $\Pi$ *stabilizes* an element $x$ if $x \notin Supp(\Pi)$. The *stabilizer subgroup* $Stab_\Pi(x)$ of a permutation group $\Pi$ for an element $x$ is the group $\{\pi \in \Pi \mid \pi(x) = x\}$, or equivalently, the largest subgroup of $\Pi$ that stabilizes $x$.

*Symmetry in SAT.* Let $\pi$ be a permutation of $\overline{\Sigma}$. We extend $\pi$ to clauses: $\pi(l_1 \vee \ldots \vee l_n) = \pi(l_1) \vee \ldots \vee \pi(l_n)$; to formulas: $\pi(c_1 \wedge \ldots \wedge c_n) = \pi(c_1) \wedge \ldots \wedge \pi(c_n)$; to assignments: $\pi(\alpha)(l) = \alpha(\pi(l))$. A *symmetry* of a formula $\phi$ is a permutation $\pi$ of $\overline{\Sigma}$ that *commutes with negation* (i.e., $\pi(\neg l) = \neg \pi(l)$) and that *preserves satisfaction* to $\phi$ (i.e., $\pi(\alpha)$ satisfies $\phi$ iff $\alpha$ satisfies $\phi$). A permutation of literals $\pi$ that commutes with negation and for which $\pi(\phi) = \phi$ is a *syntactical* symmetry of $\phi$. Typically, only syntactical symmetry is exploited, since this type of symmetry can be detected with relative ease. The practical techniques presented in this paper are no exception, though all presented theorems hold for non-syntactical symmetry as well.

*Symmetry breaking.* Symmetry *breaking* aims at eliminating symmetry, either by *statically* posting symmetry breaking constraints that invalidate symmetric assignments, or by altering the search space *dynamically* to avoid symmetric search paths. A (static) symmetry breaking formula for SAT is presented in Sect. 3. If $\Pi$ is a symmetry group, then a symmetry breaking formula $\psi$ is *sound* if for each assignment $\alpha$ there exists at least one symmetry $\pi \in \Pi$ such that

$\pi(\alpha)$ satisfies $\psi$; $\psi$ is *complete* if for each assignment $\alpha$ there exists at most one symmetry $\pi \in \Pi$ such that $\pi(\alpha)$ satisfies $\psi$ [27].

# 3   Compact CNF Encodings of the Lex-Leader Constraint

A classic approach to static symmetry breaking is to construct *lex-leader constraints*.

**Definition 1 (Lex-leader constraint [6]).** *Let $\phi$ be a formula over $\Sigma$, $\pi$ a symmetry of $\phi$, $\preceq_x$ an order on $\Sigma$ and $\preceq_\alpha$ the lexicographic order induced by $\preceq_x$ on the set of $\Sigma$-assignments. A formula $LL_\pi$ over $\Sigma' \supseteq \Sigma$ is a lex-leader constraint for $\pi$ if for each $\Sigma$-assignment $\alpha$, there exists a $\Sigma'$-extension of $\alpha$ that satisfies $LL_\pi$ iff $\alpha \preceq_\alpha \pi(\alpha)$.*

In other words, each assignment whose symmetric image under $\pi$ is smaller, is eliminated by $LL_\pi$. It is easy to see that the conjunction of $LL_\pi$ for all $\pi$ in some $\Pi' \subseteq \Pi$ is a sound (but not necessarily complete) symmetry breaking constraint for $\Pi$.

An efficient encoding of the lex-leader constraint $LL_\pi$ as a conjunction of clauses is given by Aloul et al. [1], where each variable in $Supp(\pi)$ leads to 2 clauses of size 3 and 2 clauses of size 4. Below, we give a derivation of a more compact encoding of $LL_\pi$ as a conjunction of 3 clauses of size 3 for each variable in $Supp(\pi)$, which is more compact and hence reduces the overhead introduced by posting it. A similar encoding is presented by Sakallah [22] but has not been experimentally evaluated before.

**Theorem 1 (Compact encoding of lex-leader constraint).** *Let $\pi$ be a symmetry, let $Supp(\pi) = \{x_1, \ldots, x_n\}$ be ordered such that $x_i \preceq_x x_j$ iff $i \leq j$ and let $\{y_0, \ldots, y_{n-1}\}$ be a set of auxiliary variables disjoint from $Supp(\pi)$. The following set of clauses is a lex-leader constraint for $\pi$:*

$$
\begin{array}{c|c}
y_0 & y_j \vee \neg y_{j-1} \vee \neg x_j \quad 1 \leq j < n \\
\neg y_{i-1} \vee \neg x_i \vee \pi(x_i) \quad 1 \leq i \leq n & y_j \vee \neg y_{j-1} \vee \pi(x_j) \quad 1 \leq j < n
\end{array}
$$

*Proof.* Crawford et al. proposed the following lex-leader constraint [6]:

$$\forall i : (\forall j < i : x_j \Leftrightarrow \pi(x_j)) \Rightarrow \neg x_i \vee \pi(x_i) \tag{1}$$

Assuming $\mathbf{f} < \mathbf{t}$, this constraint expresses that the value of a variable $x_i$ must be less than or equal to the value of $\pi(x_i)$ if for all smaller variables $x_j$, $x_j$ has the same value as $\pi(x_j)$. As such, it encodes a valid lex-leader constraint.

Aloul et al. [1] noticed that the antecedent $(\forall j < i : x_j \Leftrightarrow \pi(x_j))$ is recursively reified by introducing auxiliary variables $y_j$:

$$y_j \Leftrightarrow (y_{j-1} \wedge (x_j \Leftrightarrow \pi(x_j))) \tag{2}$$

where the base case $y_0$ is fixed to be true. In essence, $y_j$ holds iff $x_k$ and $\pi(x_k)$ have the same truth value for $1 \leq k \leq j$. Eq. (1) then translates to:

$$y_0 \tag{3}$$
$$y_j \Leftrightarrow (y_{j-1} \wedge (x_j \Leftrightarrow \pi(x_j))) \qquad\qquad 1 \leq j < n \tag{4}$$
$$y_{i-1} \Rightarrow \neg x_i \vee \pi(x_i) \qquad\qquad 1 \leq i \leq n \tag{5}$$

Note that by (5), if $y_{j-1}$ holds, then $\neg x_j \vee \pi(x_j)$ holds. Hence, $y_{j-1} \wedge (x_j \Leftrightarrow \pi(x_j))$ simplifies to $y_{j-1} \wedge (x_j \vee \neg\pi(x_j))$, and (4) simplifies to:

$$y_j \Leftrightarrow (y_{j-1} \wedge (x_j \vee \neg\pi(x_j))) \qquad\qquad 1 \leq j < n \tag{4'}$$

Lastly, we observe that $y_j$ only occurs negatively in formula (5). Thus, only one implication in the definition of $y_j$ (4') is needed to enforce (5). Relaxing the constraints when $y_j$ must be false leads to:

$$y_j \Leftarrow (y_{j-1} \wedge (x_j \vee \neg\pi(x_j))) \qquad\qquad 1 \leq j < n \tag{4''}$$

Working out the implications and applying distributivity of $\wedge$ and $\vee$ in Eq. (3, 4'', 5) leads to the CNF formula in this theorem and hence, concludes our proof.                                                                    □

The relaxation introduced by step (4'') does not weaken the symmetry breaking capacity of our encoding, as it only weakens the constraints on auxiliary variables not permuted by any original symmetry. However, (4'') still represents weaker constraints than (4'). In Sect. 7 we give experimental results that compare the presented compact encoding with an "unrelaxed" clausal encoding based on (4'), having an extra binary and ternary clause. It turns out that the relaxation of (4'') leads to slightly less overhead.

Note that the condition $y_j$ is satisfied in fewer assignments as $j$ increases, so the marginal effect of posting the above constraints decreases as $j$ increases. Still, the marginal cost is stable at three clauses and one auxiliary variable regardless of $j$. Because of this, the size of lex-leader constraints is often limited by putting an upper bound $k$ on the number of auxiliary variables to be introduced [2], resulting in a shorter lex-leader constraint $LL_\pi^k$. BREAKID also employs this limit on the size of its lex-leader constraints, by default posting a conservative $LL_\pi^{50}$ for each generator symmetry $\pi$.

## 4    Exploiting Row Interchangeability

An important type of symmetry is *row interchangeability*, which is present when a subset of variables can be structured as a two-dimensional matrix such that each permutation of the rows induces a symmetry. This form of symmetry is common; often it stems from an interchangeable set of objects in the original problem domain, with each row of variables expressing certain properties of one particular object. Examples are intercheangeability of pigeons or holes in a pigeonhole problem, interchangeability of nurses in a nurse scheduling problem, fleets of interchangeable trucks in a delivery system etc. Exploiting this type of matrix symmetry with adjusted symmetry breaking techniques can significantly improve SAT performance [8,17]. In this section, we present a novel way of automatically dealing with row interchangeability in SAT.

*Example 1 (Row interchangeability in graph coloring).* Let $\phi$ be a CNF formula expressing the graph coloring constraint that two directly connected vertices cannot have the same color. Let $\Sigma = \{x_{11}, \ldots, x_{nm}\}$ be the set of variables, with intended interpretation that $x_{ij}$ holds iff vertex $j$ has color $i$. Given the nature of the graph coloring problem, all colors are interchangeable, so each color permutation $\rho$ induces a symmetry $\pi_\rho$ of $\phi$. More formally, the color interchangeability symmetry group consists of all symmetries

$$\pi_\rho : \overline{\Sigma} \to \overline{\Sigma} : x_{ij} \mapsto x_{\rho(i)j}, \neg x_{ij} \mapsto \neg x_{\rho(i)j}$$

If we structure $\Sigma$ as a matrix where $x_{ij}$ occurs on row $i$ and column $j$, then each permutation of rows corresponds to a permutation of colors, and hence a symmetry.

**Definition 2 (Row interchangeability in SAT [8]).** *A variable matrix $M$ is a bijection $M : Ro \times Co \to \Sigma'$ from two sets of indices $Ro$ and $Co$ to a set of variables $\Sigma' \subseteq \Sigma$. We refer to $M(r, c)$ as $x_{rc}$. The $r$'th row of $M$ is the sequence of variables $[x_{r1}, \ldots, x_{rm}]$, the $c$'th column is the sequence $[x_{1c}, \ldots, x_{nc}]$. A formula $\phi$ exhibits* row interchangeability *symmetry if there exists a variable matrix $M$ such that for each permutation $\rho : Ro \to Ro$*

$$\pi_\rho^M : \overline{\Sigma}' \to \overline{\Sigma}' : x_{rc} \mapsto x_{\rho(r)c}, \neg x_{rc} \mapsto \neg x_{\rho(r)c}$$

*is a symmetry of $\phi$. The row interchangeability symmetry group of a matrix $M$ is denoted as $R_M$.*

A useful property of row interchangeability is that it is broken completely by only a linearly sized symmetry breaking formula [10,25].

**Theorem 2 (Complete symmetry breaking for row interchangeabil-ity [8]).** *Let $\phi$ be a formula and $R_M$ a row interchangeability symmetry group of $\phi$ with $Ro = \{1, \ldots, n\}$ and $Co = \{1, \ldots, m\}$. If the total variable order $\preceq_x$ on $\Sigma$ satisfies $x_{ij} \preceq_x x_{i'j'}$ iff $i < i'$ or $(i = i'$ and $j \leq j')$, then the conjunction of lex-leader constraints for $\pi_{(k\ k+1)}^M$ with $1 \leq k < n$ breaks $M_R$ completely.*

In words, Theorem 2 guarantees that for a natural ordering of the variable matrix, the lex-leader constraints for the swap of each two subsequent rows form a complete symmetry breaking formula for row interchangeability. The condition that the order "matches" the variable matrix is important: the theorem no longer holds without it.

If we are able to detect that a formula exhibits row interchangeability, we can break it completely by choosing the right order and posting the right lex-leader constraints. In practice, symmetry detection tools for SAT only present us with a set of generators for the symmetry group, which contains no information on the *structure* of this group. The challenge is to derive row interchangeability subgroups from these generators.

## 4.1   Row Interchangeability Detection Algorithm

Given a set of generators $P$ for a symmetry group $\Pi$ of a formula $\phi$, the task at hand is to detect a variable matrix $M$ that represents a row interchangeability subgroup $R_M \subseteq \Pi$. We present an algorithm that is sound, but incomplete in the sense that it does not guarantee that all row interchangeability subgroups present are detected.

The first step is to find an initial row interchangeable variable matrix $M$ consisting of three rows. This is done by selecting two swap symmetries $\pi_1$ and $\pi_2$ that represent two swaps of three rows. More formally, suppose $\pi_1$ and $\pi_2$ are such that (with $r = Supp(\pi_1) \cap Supp(\pi_2)$) the following three conditions hold (i) $\pi_1 = \pi_1^{-1}$ and $\pi_2 = \pi_2^{-1}$, (ii) $r$, $\pi_1(r)$ and $\pi_2(r)$ are pairwise disjoint, and (iii) $Supp(\pi_1) = r \cup \pi_1(r)$ and $Supp(\pi_2) = r \cup \pi_2(r)$. In this case, $r$, $\pi_1(r)$ and $\pi_2(r)$ form three rows of a row interchangeable variable matrix, and $\pi_1$ and $\pi_2$ are swaps of those rows.

If, after inspecting all pairs of swaps in $P$, no initial three-rowed matrix is found, the algorithm stops, in which case we do not know whether a row interchangeability subgroup exists. However, our experiments indicate that for many problems, an initial three-rowed matrix can be derived from a detected set of generator symmetries.

The second step maximally extends the initial variable matrix $M$ with new rows. The idea is that for each symmetry $\pi \in P$ and each row $r$ of $M$, $\pi(r)$ is a candidate row to add to $M$. This is the case if $\pi(r)$'s literals are disjoint from $M$'s literals and swapping $\pi(r)$ with $r$ is a syntactical symmetry of $\phi$.

Pseudocode is given in Algorithm 1. This algorithm terminates since both $P$ and the number of rows in any row interchangeability matrix are finite. The algorithm is sound: each time a row is added, it is interchangeable with at least one previously added row and hence, by induction, with all rows in $M$. If $k$ is the largest support size of a symmetry in $P$, then finding an initial row interchangeable matrix based on two row swap symmetries in $P$ takes $O(|P|^2 k)$ time. With an optimized implementation that avoids duplicate combinations of generators and rows, extending the initial matrix with extra interchangeable rows has a complexity of $O(|P||Ro||\phi|k)$, with $Ro$ the set of row indices of $M$. Algorithm 1 then has a complexity of $O(|P|^2 k + |P||Ro||\phi|k)$.

As mentioned before, the algorithm is not complete: it might not be possible to construct an initial matrix, or even given an initial matrix, there is no guarantee to detect all possible row extensions, as only the set of generators instead of the whole symmetry group is used to calculate a new candidate row.

It is straightforward to extend Algorithm 1 to detect multiple row interchangeability subgroups. After detecting a first row interchangeability subgroup $R_M$, remove any generators from $P$ that also belong to $R_M$. This can be done by

input  : $P, \phi$
output: $M$
1  identify two swaps $\pi_1, \pi_2 \in P$ that induce an initial variable matrix $M$ with 3 rows;
2  **repeat**
3      **foreach** *permutation* $\pi$ *in* $P$ **do**
4          **foreach** *row* $r$ *in* $M$ **do**
5              **if** $\pi(r)$ *is disjoint from* $M$ *and swapping* $r$ *and* $\pi(r)$ *is a symmetry of* $\phi$ **then**
6                 add $\pi(r)$ as a new row to $M$;
7              **end**
8          **end**
9      **end**
10  **until** *no extra rows are added to* $M$;
11  **return** $M$;

**Algorithm 1.** Row interchangeability detection

standard algebraic group membership tests, which are efficient for interchangeability groups [24]. Then, repeat Algorithm 1 with the reduced set of generator symmetries until no more row interchangeability subgroups are detected.

*Example 2 (Example 1 continued.).* Let $\varphi$ and the $x_{ij}$ be as in Example 1. Suppose we have five colors and three vertices. Vertex 1 is connected to vertex 2 and 3; vertices 2 and 3 are not connected. This problem has a symmetry group $\Pi$ induced by the interchangeability of the colors and by a swap on vertex 2 and 3. A set of generators for $\Pi$ is $\{\pi_{(12)}, \pi_{(23)}, \pi_{(34)}, \pi_{(45)}, \nu_{23}\}$, where $\pi_{(ij)}$ is the symmetry that swaps colors $i$ and $j$ (as in the previous example), and $\nu_{23}$ is the symmetry obtained by swapping vertices 2 and 3, i.e.,[2]

$$\nu_{23} = (x_{12}\ x_{13})(x_{22}\ x_{23})(x_{32}\ x_{33})(x_{42}\ x_{43})(x_{52}\ x_{53})$$

For these generators, it is obvious that the swaps $\pi_{(ij)}$ generate a row interchangeability symmetry group. However, a symmetry detection tool might return the alternative set of symmetry generators $P = \{\pi_{(12)}, \pi_{(23)}, \sigma_1, \sigma_2, \nu_{23}\}$ with

$$\sigma_1 = \pi_{(35)} \circ \pi_{(13)} = (x_{11}\ x_{31}\ x_{51})(x_{12}\ x_{32}\ x_{52})(x_{13}\ x_{33}\ x_{53})$$
$$\sigma_2 = \pi_{(34)} \circ \nu_{23} = (x_{12}\ x_{13})(x_{22}\ x_{23})(x_{31}\ x_{41})(x_{32}\ x_{43})(x_{42}\ x_{33})(x_{52}\ x_{53})$$

The challenge is to detect the color interchangeability subgroup starting from $P$.

The first step of Algorithm 1 searches for two swaps in $P$ that combine to a 3-rowed variable matrix. $\pi_{(12)}$ and $\pi_{(23)}$ fit the bill, resulting in a variable matrix $M$ with rows:

$$[x_{11}, x_{12}, x_{13}] \quad [x_{21}, x_{22}, x_{23}] \quad [x_{31}, x_{32}, x_{33}]$$

---

[2] We omit negative literals from the cycle notation, noting that a symmetry always commutes with negation.

Applying $\sigma_1$ on the third row results in:

$$[x_{51}, x_{52}, x_{53}]$$

which after a syntactical check on $\phi$ is confirmed to be a new row to add to $M$.

Unfortunately, the missing row $[x_{41}, x_{42}, x_{42}]$ is not derivable by applying any generator in $P$ on rows in $M$, so the algorithm terminates.

The failure of detecting the missing row in Example 2 stems from the fact that the generators $\sigma_1$ and $\sigma_2$ are obtained by complex combinations of symmetries in the interchangeability subgroup and the symmetry $\nu_{23}$. This inspires a small extension of Algorithm 1. As soon as the algorithm reaches a fixpoint, we call the original symmetry detection tool to search for a set of generators of the subgroup that stabilizes all but one rows of the matrix $M$ found so far. This results in "simpler" generators that do not permute the literals of the excluded rows. Tests on CNF instances show that this simple extension, although giving no theoretical guarantees, often manages to find new generators that, when applied on the current set of rows, construct new rows. After detecting row stabilizing symmetries, Algorithm 1 resumes from line 2, aiming to extend the matrix further by applying the extended set of generators. This process ends when even the new generators can no longer derive new rows.

*Example 3 (Example 2 continued.).* The only symmetry of the problem that stabilizes the variables $\{x_{11}, x_{12}, x_{13}, x_{21}, x_{22}, x_{23}, x_{31}, x_{32}, x_{33}\}$ is $\pi_{(45)}$, which has the missing row $[x_{41}, x_{42}, x_{42}]$ as image of the fifth row.

The matrix, which now contains all variables, allows one to completely break the color interchangeability. The symmetry between vertices 2 and 3 is not expressed in the matrix, but can still be broken by a lex-leader constraint, as described in Sect. 6.

## 5   Generating Binary Symmetry Breaking Clauses

Partial symmetry breaking, where lex-leader constraints are posted only for a set of generators of a symmetry group $\Pi$, is motivated by the infeasibility of posting lex-leader constraints for all symmetries in $\Pi$. An alternative we explore here is to post only a very short lex-leader constraint, namely $LL_\pi^1$, but do this for a large number of $\pi \in \Pi$. As already mentioned in Sect. 3, the first parts of the lex-leader constraint breaks comparatively more symmetry than later parts, so in that sense, posting $LL_\pi^1$ is the most cost-effective way of breaking $\pi$.

Note that $LL_\pi^1$ is equivalent to the binary clause $\neg x \vee \pi(x)$ where $x$ is the smallest variable in $Supp(\pi)$ according to $\preceq_x$. To construct as many of these binary clauses as possible without enumerating the whole symmetry group $\Pi$, we use a greedy approach that starts from the generators of $\Pi$ and exploits the freedom to choose the variable order $\preceq_x$ as well as the fact that one can easily compute the orbit of a literal in $\Pi$.

**Theorem 3 (Binary symmetry breaking clauses).** *Let $\Pi$ be a non-trivial symmetry group of $\phi$, $\preceq_x$ an ordering of $\Sigma$, and $x^*$ the $\preceq_x$-smallest variable in $Supp(\Pi)$. For each $x \in Orb_\Pi(x^*)$, the binary clause $\neg x^* \vee x$ is entailed by $LL_\pi$ for some $\pi \in \Pi$.*

*Proof.* If $x = x^*$, the theorem is trivially true. If $x \neq x^*$, there exists a $\pi \in \Pi$ with $\pi(x^*) = x$ since $x \in Orb_\Pi(x^*)$. Since $x^*$ is the smallest variable in $Supp(\Pi)$, it is also the smallest in $Supp(\pi)$. Theorem 1 shows that $y_0$ and $\neg y_0 \vee \neg x^* \vee \pi(x^*)$ are two clauses in $LL_\pi$. Resolution on $y_0$ leads to $\neg x^* \vee \pi(x^*)$, where $\pi(x^*) = x$.

Theorem 3 allows to construct small lex-leader clauses for $\Pi$ without enumerating individual members of $\Pi$; it suffices to compute the orbit of the smallest variable in $Supp(\Pi)$ to derive a set of binary symmetry breaking clauses. Theorem 3 holds for all symmetry groups, so also for any subgroup $\Pi'$ of $\Pi$. In particular, if $\Pi'$ stabilizes the smallest variable in $Supp(\Pi)$, applying Theorem 3 to $\Pi'$ results in different clauses than applying it to $\Pi$, as $\Pi'$ has a different smallest variable in its support.

*Example 4.* Let $P = \{(ab)(cdef)\}$ and $\Pi = Grp(P) = \{(ab)(cdef), (ab)(cfed), (ce)(df)\}$.[3] With order $a \preceq_x b \preceq_x c \preceq_x d \preceq_x e \preceq_x f$, $a$ is the $\preceq_x$-smallest variable of $Supp(\Pi)$. Theorem 3 guarantees that $\neg a \vee b$ is a consequence of the lex-leader constraints for $\Pi$. Let $\Pi' = Stab_\Pi(a) = \{(ce)(df)\}$, then $c$ is the $\preceq_x$-smallest variable of $Supp(\Pi')$, hence also $\neg c \vee e$ is entailed by the lex-leader constraints for $\Pi$.

If we assume a different order $\preceq_x'$, different binary clauses are obtained. For instance, let $c$ be the $\preceq_x'$-smallest variable of $Supp(\Pi)$. Then Theorem 3 allows us to post the clauses $\neg c \vee d$, $\neg c \vee e$ and $\neg c \vee f$ as symmetry breaking clauses. The stabilizer subgroup $Stab_\Pi(c)$ is empty, so no further binary clauses can be derived for this order.

A *stabilizer chain* is a sequence of stabilizer subgroups starting with the full group $\Pi$ and ending with the trivial group containing only the identity, where each next subgroup in the chain stabilizes an extra element. Given a variable order $\preceq_x$, applying Theorem 3 to each subgroup in a stabilizer chain stabilizing literals according to $\preceq_x$ for a symmetry group $\Pi$, is equivalent to constructing all $LL_\pi^1$ for $\pi \in \Pi$ under $\preceq_x$ [13]. This stabilizer chain idea was also used by Puget to efficiently break all-different constraints in a constraint programming context [20].

However, as shown by Example 4, the variable order influences the number of binary symmetry clauses derivable by a stabilizer chain of $\Pi$. We present an algorithm that, given a set of generator symmetries $P$ for symmetry group $\Pi$, decides a total order on a subset of variables, and constructs binary symmetry breaking clauses for those variables based on a simultaneously constructed sequence of subgroups stabilizing those variables. The constructed sequence of subgroups stabilizing the literals is no actual stabilizer chain, as each of the subgroups equals $Grp(P')$ for some subset $P' \subseteq P$. The advantage of this approach

---

[3] We again omit negative literals in cycle notation.

```
   input  : P
   output: LL^{bin}, Ord
1  initialize Q = P and initialize Ord as an empty list;
2  while Q ≠ ∅ do
3  │   O is a largest orbit of Grp(Q);
4  │   x* is a variable in O for which {π ∈ Q | π(x*) ≠ x*)} is minimal;
5  │   add x* to Ord as last variable;
6  │   foreach x ∈ O do
7  │   │   add ¬x* ∨ x to LL^{bin};
8  │   end
9  │   Q = {π ∈ Q | π(x*) = x*};
10 end
11 return LL^{bin}, Ord;
```
**Algorithm 2.** Binary symmetry breaking clause generation

is simplicity of the algorithm and low computational complexity, although it would be interesting future work to compute an actual stabilizer chain using for instance the *Schreier-Sims algorithm* [24].

In detail, our algorithm starts with an empty variable order $Ord$ and a copy $Q$ of the given set of generators $P$. It iteratively chooses a suitable variable $x^*$ as next in the variable order, constructs binary clauses based on $Grp(Q)$, and removes any permutations $\pi \in Q$ for which $x \in Supp(\pi)$. As a result, at each iteration, $Grp(Q)$ stabilizes all variables in $Ord$ except the last variable $x^*$, allowing the construction of binary symmetry breaking clauses $\neg x^* \vee x$ for each $x \in Orb_{Grp(Q)}(x^*)$, as per Theorem 3.

A suitable next variable $x^*$ is one that induces a high number of binary symmetry breaking clauses, but removes few symmetries from $Q$ so that the following iterations of the algorithm still have a reasonably sized symmetry group to work with. One way to satisfy these requirements is to pick $x^*$ such that $Orb_{Grp(Q)}(x^*)$ is maximal, and $\{\pi \in Q \mid \pi(x^*) \neq x^*)\}$ is minimal compared to other literals of $x^*$'s orbit.

Pseudocode is given in Algorithm 2. This algorithm terminates, as while $Q \neq \emptyset$, $x^*$ belongs to a largest orbit of $Grp(Q)$, so $x^* \neq \pi(x^*)$ for at least one $\pi \in Q$. As a result, $Q$ shrinks in size during each iteration, eventually becoming the empty set. The complexity of Algorithm 2 is dominated by finding the largest orbit of $Grp(Q)$, which is $O(|Q||\Sigma|)$, resulting in a total complexity of $O(|P|^2|\Sigma|)$.

Worst case, $O(|Supp(\Pi)|^2)$ binary clauses are constructed by Algorithm 2. In particular, if some subgroup $\Pi'$ of $\Pi$ represents an interchangeable set of $n$ variables, $n(n-1)/2$ binary clauses are derived. However, in this case $\Pi'$ also represents a row interchangeability symmetry group, which is completely broken by techniques from Sect. 4. Performing row interchangeability detection and breaking before binary clause generation can avoid quadratic sets of binary clauses. Section 6 shows this is indeed the order by which BREAKID performs its symmetry breaking.

# 6    Putting it all Together as BreakID

This section describes how the improvements presented in the previous section combine with eachother and with standard symmetry breaking techniques in the symmetry breaking preprocessor BREAKID.

BREAKID has been around since 2013, when a preliminary version obtained the gold medal in the hard combinatorial sat+unsat track of 2013's SAT competition [5]. This early version incorporated all of SHATTER's symmetry breaking techniques and used a primitive row interchangeability detection algorithm that enumerated symmetries to detect as many row swap symmetries as possible [8]. We developed BREAKID2 in 2015, using the ideas presented in this paper. BREAKID2 entered the main track of 2015's SAT race in combination with GLUCOSE 4.0, placing 10th, ahead of all other GLUCOSE variants. The experiments in the next section are run with a slightly updated version – BREAKID2.1– which has more usability features and reduced memory overhead. For the remainder of this paper, we use BREAKID to refer to the particular implementation BREAKID2.1. BREAKID's source code is published online [7].

## 6.1    BREAKID's High Level Algorithm

Preprocessing a formula $\phi$ by symmetry breaking in BREAKID starts with removing duplicate clauses and duplicate literals in clauses from $\phi$, as SAUCY cannot handle duplicate edges. Then, a call to SAUCY constructs an initial generator set $P$ of the syntactical symmetry group of $\phi$.

Thirdly, BREAKID detects row interchangeability subgroups $R_M$ of $Grp(P)$ by Algorithm 1. The program incorporates the variables of the support of all $R_M$ in a global variable order $Ord$ such that the conjunction of $LL_{\pi_\rho^M}$ under $Ord$ for all subsequent row swaps $\pi_\rho^M$ forms a complete symmetry breaking formula for $R_M$.[4] After adding the complete symmetry breaking formula of each $R_M$ to an initial set of symmetry breaking clauses $\psi$, we also remove all symmetries in $P$ that belong to some $R_M$, since these symmetries are broken completely already.

Next, using the pruned $P$, binary clauses for $Grp(P)$ are constructed by Algorithm 2, which simultaneously decides a set of variables to be smallest under $Ord$.[5]

Finally, $Ord$ is supplemented with missing variables until it is total, and limited lex-leader constraints $LL_\pi^{50}$ are constructed for each $\pi$ left in $P$. These lex-leader constraints incorporate two extra refinements also used by SHATTER; one for *phase-shifted* variables and one for the *largest variable in a symmetry cycle* [1].

Algorithm 3 gives pseudocode for BREAKID's high-level routine described above.

---

[4] In case two detected row interchangeability matrices overlap, it is not always possible to choose the order on the variables so that both are broken completely. In this case, one of the row interchangeability groups will only be broken partially.

[5] A small adaptation to Algorithm 2 ensures BREAKID only selects smallest variables that are not permuted by a previously detected row interchangeability group.

**input**  : $\phi$
**output**: $\psi$
1  remove duplicate clauses from $\phi$ and duplicate literals from clauses in $\phi$;
2  run **Saucy**  to detect a set of symmetry generators $P$;
3  initialize $\psi$ as the empty formula and $Ord$ as an empty sequence of ordered variables;
4  detect **row interchangeability** subgroups $R_M$;
5  **foreach** *row interchangeability $R_M$ subgroup of $Grp(P)$* **do**
6  |    remove $P \cap R_M$ from $P$;
7  |    add complete symmetry breaking clauses for $R_M$ to $\psi$;
8  |    add $Supp(R_M)$ to the back of $Ord$ accordingly;
9  **end**
10  add **binary clauses** for $Grp(P)$ to $\psi$, add corresponding variables to the front of $Ord$;
11  add missing variables to the middle of $Ord$;
12  **foreach** $\pi \in P$ **do**
13  |    add $LL_\pi^{50}$ to $\psi$, utilizing **Shatter's optimizations**;
14  **end**
15  **return** $\psi$;

Algorithm 3. Symmetry breaking by BREAKID

# 7    Experiments

In this section, we verify the effectiveness of the proposed techniques separately, and investigate the feasibility of using BREAKID in the application and hard-combinatorial track of 2014's SAT competition We use eight benchmark sets:

- **app14:** the application track of 2014's SAT competition (300 instances)
- **app14sym:** subset of app14 for which SAUCY detected symmetry (164 instances)
- **hard14:** the hard-combinatorial track of 2014's SAT competition (300 instances)
- **hard14sym:** subset of hard14 for which SAUCY detected symmetry (159 instances)
- **hole:** 8 unsatisfiable pigeonhole instances
- **urquhart:** 6 unsatisfiable Urquhart instances
- **channel:** 10 unsatisfiable channel routing instances
- **color:** 10 unsatisfiable graph coloring instances

Pigeonhole and Urquhart problems are provably hard for purely resolution-based SAT solvers, in the sense that even for very small instances astronomical running time is needed to decide satisfiability of the problem [12,26]. The employed channel routing and graph coloring instances are highly symmetric, exhibiting strong row interchangeability. They are taken from SYMCHAFF's benchmark set [21]. The graph coloring instances were also used in 2005 and 2007's SAT competitions.

As SAT-solver, we use GLUCOSE 4.0 [4], which is based on MINISAT [9]. We include the symmetry breaking preprocessor SHATTER [3] bundled with

SAUCY 3.0 [15] in our experiments. The resources available to each experiment were 10GB of memory and 3600 s on an Intel(R) Xeon(R) E3-1225 cpu. The operating system was Ubuntu 14.04 with Linux kernel 3.13. Unless noted otherwise, all results *include* any preprocessing step, such as deduplicating the input CNF, symmetry detection by SAUCY and symmetry breaking clause generation by SHATTER or BREAKID. Detailed experimental results are available online [7].

## 7.1  Compact Symmetry Breaking Clauses

We first investigate the influence of the compact lex-leader encoding presented in Sect. 3. The experiment consists of running BREAKID with the *standard* encoding used in SHATTER (four clauses for each variable in a symmetry's support), with BREAKID's default *compact* encoding (three clauses), and with an *unrelaxed* encoding that does not relax the constraints on the auxiliary variables (five clauses). To focus on the difference between the encodings, in this experiment, BREAKID does not exploit row interchangeability, does not generate binary clauses, and does not limit the size of the lex-leader formulas. The benchmark sets employed are app14sym, hard14sym, hole, urquhart, channel and color. The results are presented in Table 1.

**Table 1.** Number of solved instances for standard, unrelaxed and compact lex-leader encoding, as well as average runtime and memory consumption of GLUCOSE (excluding BREAKID's preprocessing) for solved instances.

| | app14sym | | | hard14sym | | | hole | urquhart | channel | color |
|---|---|---|---|---|---|---|---|---|---|---|
| | avg mem | avg time | solved | avg mem | avg time | solved | solved | solved | solved | solved |
| standard | 334 MB | 597.6 s | **113** | 323 MB | 662.9 s | 106 | 4 | 3 | 2 | 3 |
| unrelaxed | 349 MB | 611.1 s | **113** | 336 MB | 708.8 s | 107 | 4 | 3 | 2 | 3 |
| compact | **329 MB** | **589.6 s** | 112 | **305 MB** | **638.0 s** | **108** | 4 | 3 | 2 | 3 |

The theoretical advantage of having a more compact encoding is not translated into a significant increase in the number of solved instances. We do observe average runtime and memory consumption correlating with the size of the encoding, being lowest for the compact encoding and highest for the unrelaxed encoding. We conclude that none of the clausal encodings strongly outperforms the others. That said, the compact encoding enjoys a small runtime and memory advantage over both other encodings.

## 7.2  Row Interchangeability and Binary Clauses

To assess the influence of exploiting row interchangeability and binary clauses, we set up an experiment with four versions of BREAKID:

- BREAKID(): without both row interchangeability and binary clauses
- BREAKID(r): with row interchangeability and without binary clauses

– BREAKID(b): without row interchangeability and with binary clauses
– BREAKID(r,b): with both row interchangeability and binary clauses

Each of these versions uses the compact encoding, and limits the lex-leader formulas to 50 auxiliary variables, symmetries used to completely break row interchangeability excepted. The results are summarized in Table 2.

**Table 2.** Number of solved instances for BREAKID configurations with and without (r)ow-interchangeability and (b)inary clauses. Also includes average number of corresponding symmetry breaking clauses introduced.

| | BREAKID() | BREAKID(b) | | BREAKID(r) | | BREAKID(r,b) | | |
|---|---|---|---|---|---|---|---|---|
| | solved | (b) clauses | solved | (r) clauses | solved | (b) clauses | (r) clauses | solved |
| app14sym | 113 | 37552 | 111 | 10245 | **114** | 190 | 10245 | **114** |
| hard14sym | 108 | 207719 | 105 | 2926 | **112** | 308 | 2926 | 110 |
| hole | 4 | 427 | 3 | 1627 | **8** | 0 | 1627 | **8** |
| urquhart | 3 | 99 | **6** | 0 | 3 | 99 | 0 | **6** |
| channel | 2 | 9893 | 2 | 15421 | **10** | 0 | 15421 | **10** |
| color | 3 | 1469 | 4 | 1481 | 5 | 656 | 1481 | **6** |

A first observation is that the binary clause improvement shows mixed results. It performs very well on urquhart, allowing all instances to be solved in less than a second, but struggles with app14sym and hard14sym instances. The main reason is the huge amount of binary clauses derived, amounting over 5 million on some instances. However, activating row interchangeability fixes this problem by not allowing symmetries from row interchangeability groups to be used to construct binary clauses.

The row interchangeability improvement is more successful, improving performance on all benchmark sets except urquhart. Focusing on hole, full row interchangeability is detected for all instances, so each instance became polynomially solvable given the presence of symmetry breaking clauses. This is a significant improvement to the preliminary version of BREAKID [8]. A similar effect is seen for channel, where activating row interchangeability allows deciding all instances in less than a minute. For the benchmark set as a whole, row interchangeability was detected in 54 % of the instances for which SAUCY could detect symmetry.

We conclude that row interchangeability exploitation is a significant improvement, while binary clauses have the potential to improve performance on certain types of problems. Furthermore, row interchangeability compensates for weaknesses of the binary clause approach, and the combination of the two yields the best overall performance.

### 7.3   Comparison to Shatter and Performance on the 2014 SAT Competition

This experiment compares BREAKID to state-of-the-art solving configurations. We use app14, hard14, hole, urquhart, channel and color as benchmark sets.

We effectively run all application and hard-combinatorial instances of 2014's SAT competition. The solving configurations used are (with GLUCOSE as SAT engine):

- GLUCOSE: pure GLUCOSE without symmetry breaking.
- SHATTER: SHATTER is run after first deduplicating the input CNF.
- BREAKID: compact encoding, row interchangeability and binary clauses activated.
- BREAKID(100s): same as BREAKID but SAUCY is forced to stop detecting symmetry after 100 s of preprocessing have elapsed.

We present the number of instances solved within resource limits, as well as the average time needed to detect symmetry and generate symmetry breaking clauses in Table 3.

**Table 3.** Number of solved instances for GLUCOSE, SHATTER, BREAKID and BREAKID limiting SAUCY to 100 s. Also includes average preprocessing time in seconds.

|  | GLUCOSE | SHATTER | | BREAKID | | BREAKID(100s) | |
| --- | --- | --- | --- | --- | --- | --- | --- |
|  | solved | pre-time | solved | pre-time | solved | pre-time | solved |
| hole | 2 | 0.0s | 3 | 0.1s | **8** | 0.1s | **8** |
| urquhart | 2 | 0.1s | 2 | 0.2s | **6** | 0.2s | **6** |
| channel | 2 | 3.2s | 2 | 9.7s | **10** | 9.7s | **10** |
| color | 3 | 2.9s | 2 | 4.1s | **6** | 4.1s | **6** |
| app14 | **214** | 6.3s | 210 | 74.9s | 209 | 14.7s | 211 |
| hard14 | 164 | 159.2s | 178 | 181.3s | 183 | 14.8s | **187** |

First, the two BREAKID variants are the only configurations that handle hole, urquhart and channel efficiently, as SHATTER constructs lex-leader constraints for the wrong set of symmetry generators, and GLUCOSE gets lost in the symmetrical search space for non-trivial instances. A similar observation is made for color instances, though even BREAKID remains unable to solve 4 instances.

On hard14, SHATTER outperforms GLUCOSE, while both BREAKID approaches outperform SHATTER. So for these instances, symmetry detection and breaking is worth the incurred overhead. The preprocessing time needed by SHATTER is almost completely due to SAUCY's symmetry detection, which exceeds 3600 s for 9 instances. BREAKID(100s) solves this problem by limiting the time consumed by SAUCY to 100s, resulting in the best performance on hard14, adding 23 solved instances compared to plain GLUCOSE. Of course, both BREAKID approaches increase the preprocessing overhead by detecting row interchangeability and constructing binary clauses.

As far as app14 is concerned, the benefit of a smaller search space does not outweigh the overhead of detecting symmetry and introducing symmetry breaking clauses.

# 8    Conclusion

In this paper, we presented novel improvements to state-of-the-art symmetry breaking for SAT. Common themes were to adapt the variable order and the set of generator symmetries by which to construct lex-leader constraints. BREAKID implements these ideas and functions as a symmetry breaking preprocessor in the spirit of SHATTER. Our experiments with BREAKID show the potential for these techniques separately and combined. We observed that BREAKID outperforms SHATTER, and is a particularly effective preprocessor for hard-combinatorial SAT-problems.

The algorithms presented are effective, but also incomplete, e.g., not all row interchangeability is detected, no maximal set of binary symmetry breaking clauses is derived etc. Coupling BREAKID to a computational group algebra system such as GAP [11] has the potential to alleviate these issues.

Alternatively, it might be interesting to compare different methods of graph automorphism detection, and investigate how hard it is to adjust their internal search algorithms to put out more useful symmetry generators, stabilizer chains for binary clauses, or even row interchangeability symmetry groups. Jefferson & Petrie already started this research in a constraint programming context [13].

**Acknowledgement.** This research was supported by the project GOA 13/010 Research Fund KU Leuven and projects G.0489.10, G.0357.12 and G.0922.13 of FWO (Research Foundation - Flanders). Bart Bogaerts is supported by the Finnish Center of Excellence in Computational Inference Research (COIN) funded by the Academy of Finland (grant #251170).

# References

1. Aloul, F., Ramani, A., Markov, I., Sakallah, K.: Solving difficult SAT instances in the presence of symmetry. In: Proceedings of the 39th Design Automation Conference, 2002, pp. 731–736 (2002)
2. Aloul, F.A., Markov, I.L., Sakallah, K.A.: Shatter: efficient symmetry-breaking for boolean satisfiability. In: Design Automation Conference, pp. 836–839 (2003)
3. Aloul, F.A., Sakallah, K.A., Markov, I.L.: Efficient symmetry breaking for Boolean satisfiability. IEEE Trans. Comput. **55**(5), 549–558 (2006)
4. Audemard, G., Simon, L.: Predicting learnt clauses quality in modern SAT solvers. In: Boutilier, C. (ed.) IJCAI, pp. 399–404 (2009)
5. Balint, A., Belov, A., Heule, M.J., Järvisalo, M.: The 2013 international SAT competition (2013). satcompetition.org/2013
6. Crawford, J.M., Ginsberg, M.L., Luks, E.M., Roy, A.: Symmetry-breaking predicates for search problems. In: Principles of Knowledge Representation and Reasoning, pp. 148–159. Morgan Kaufmann (1996)
7. Devriendt, J., Bogaerts, B.: BreakID, a symmetry breaking preprocessor for SAT solvers (2015). bitbucket.org/krr/breakid
8. Devriendt, J., Bogaerts, B., Bruynooghe, M.: BreakIDGlucose: On the importance of row symmetry. In: Proceedings of the Fourth International Workshop on the Cross-Fertilization Between CSP and SAT (CSPSAT) (2014). http://lirias. kuleuven.be/handle/123456789/456639

9. Eén, N., Sörensson, N.: An extensible SAT-solver. In: Giunchiglia, E., Tacchella, A. (eds.) SAT 2003. LNCS, vol. 2919, pp. 502–518. Springer, Heidelberg (2004)

10. Flener, P., Frisch, A.M., Hnich, B., Kiziltan, Z., Miguel, I., Pearson, J., Walsh, T.: Breaking row and column symmetries in matrix models. In: Van Hentenryck, P. (ed.) CP 2002. LNCS, vol. 2470, pp. 462–477. Springer, Heidelberg (2002). http://dx.doi.org/10.1007/3-540-46135-3_31

11. The GAP Group: GAP - Groups, Algorithms, and Programming, Version 4.7.9 (2015). www.gap-system.org

12. Haken, A.: The intractability of resolution. Theor. Comput. Sci. **39**, 297–308 (1985). http://www.sciencedirect.com/science/article/pii/0304397585901446. Third Conference on Foundations of Software Technology and Theoretical Computer Science

13. Jefferson, C., Petrie, K.E.: Automatic generation of constraints for partial symmetry breaking. In: Lee, J. (ed.) CP 2011. LNCS, vol. 6876, pp. 729–743. Springer, Heidelberg (2011). http://dx.doi.org/10.1007/978-3-642-23786-7_55

14. Junttila, T., Kaski, P.: Engineering an efficient canonical labeling tool for large and sparse graphs. In: Applegate, D., Brodal, G.S., Panario, D., Sedgewick, R. (eds.) Proceedings of the Ninth Workshop on Algorithm Engineering and Experiments and the Fourth Workshop on Analytic Algorithms and Combinatorics, pp. 135–149. SIAM (2007)

15. Katebi, H., Sakallah, K.A., Markov, I.L.: Symmetry and satisfiability: an update. In: Strichman, O., Szeider, S. (eds.) SAT 2010. LNCS, vol. 6175, pp. 113–127. Springer, Heidelberg (2010)

16. Lee, J.H.M., Li, J.: Increasing symmetry breaking by preserving target symmetries. In: Milano, M. (ed.) CP 2012. LNCS, vol. 7514, pp. 422–438. Springer, Heidelberg (2012). http://dx.doi.org/10.1007/978-3-642-33558-7_32

17. Lynce, I., Marques-Silva, J.: Breaking symmetries in SAT matrix models. In: Marques-Silva, J., Sakallah, K.A. (eds.) SAT 2007. LNCS, vol. 4501, pp. 22–27. Springer, Heidelberg (2007). http://dx.doi.org/10.1007/978-3-540-72788-0_6

18. McDonald, I., Smith, B.: Partial symmetry breaking. In: Van Hentenryck, P. (ed.) CP 2002. LNCS, vol. 2470, pp. 431–445. Springer, Heidelberg (2002). http://dx.doi.org/10.1007/3-540-46135-3_29

19. McKay, B.D., Piperno, A.: Practical graph isomorphism. J. Symbolic Comput. **60**, 94–112 (2014). http://www.sciencedirect.com/science/article/pii/S0747717113001193

20. Puget, J.F.: Breaking symmetries in all-different problems. In: Proceedings of the 19th International Joint Conference on Artificial Intelligence, IJCAI 2005, pp. 272–277 (2005)

21. Sabharwal, A.: Symchaff: exploiting symmetry in a structure-aware satisfiability solver. Constraints **14**(4), 478–505 (2009). http://dx.doi.org/10.1007/s10601-008-9060-1

22. Sakallah, K.A.: Symmetry and satisfiability. In: Biere, A., Heule, M.J.H., van Maaren, H., Walsh, T. (eds.) Handbook of Satisfiability, vol. 185, pp. 289–338. IOS Press, Amsterdam (2009)

23. Schaafsma, B., Heule, M.J.H., van Maaren, H.: Dynamic symmetry breaking by simulating Zykov contraction. In: Kullmann, O. (ed.) SAT 2009. LNCS, vol. 5584, pp. 223–236. Springer, Heidelberg (2009).http://dx.doi.org/10.1007/978-3-642-02777-2_22

24. Seress, Á.: Permutation Group Algorithms. Cambridge University Press (2003). cambridgeBooksOnline. http://dx.doi.org/10.1017/CBO9780511546549

25. Shlyakhter, I.: Generating effective symmetry-breaking predicates for search problems. Discrete Appl. Math. **155**(12), 1539–1548 (2007). http://dx.doi.org/10.1016/j.dam.2005.10.018
26. Urquhart, A.: Hard examples for resolution. J. ACM 34(1), 209–219. http://doi.acm.org/10.1145/7531.8928
27. Walsh, T.: Symmetry breaking constraints: Recent results. CoRR abs/1204.3348 (2012)

# Learning Rate Based Branching Heuristic for SAT Solvers

Jia Hui Liang[(✉)], Vijay Ganesh, Pascal Poupart, and Krzysztof Czarnecki

University of Waterloo, Waterloo, Canada
jliang@gsd.uwaterloo.ca

**Abstract.** In this paper, we propose a framework for viewing solver branching heuristics as optimization algorithms where the objective is to maximize the *learning rate*, defined as *the propensity for variables to generate learnt clauses*. By viewing online variable selection in SAT solvers as an optimization problem, we can leverage a wide variety of optimization algorithms, especially from machine learning, to design effective branching heuristics. In particular, we model the variable selection optimization problem as an online multi-armed bandit, a special-case of *reinforcement learning*, to learn branching variables such that the learning rate of the solver is maximized. We develop a branching heuristic that we call *learning rate branching* or LRB, based on a well-known multi-armed bandit algorithm called *exponential recency weighted average* and implement it as part of MiniSat and CryptoMiniSat. We upgrade the LRB technique with two additional novel ideas to improve the learning rate by accounting for *reason side rate* and exploiting *locality*. The resulting LRB branching heuristic is shown to be faster than the VSIDS and conflict history-based (CHB) branching heuristics on 1975 application and hard combinatorial instances from 2009 to 2014 SAT Competitions. We also show that CryptoMiniSat with LRB solves more instances than the one with VSIDS. These experiments show that LRB improves on state-of-the-art.

## 1 Introduction

Modern Boolean SAT solvers are a critical component of many innovative techniques in security, software engineering, hardware verification, and AI such as solver-based automated testing with symbolic execution [9], bounded model checking [11] for software and hardware verification, and planning in AI [27] respectively. Conflict-driven clause-learning (CDCL) SAT solvers [4,6,12,23, 24,29] in particular have made these techniques feasible as a consequence of their surprising efficacy at solving large classes of real-world Boolean formulas. The development of various heuristics, notably the Variable State Independent Decaying Sum (VSIDS) [24] branching heuristic (and its variants) and conflict analysis techniques [23], have dramatically pushed the limits of CDCL solver performance. The VSIDS heuristic is used in the most competitive CDCL SAT solvers such as Glucose [4], Lingeling [6], and CryptoMiniSat [29]. Since its

© Springer International Publishing Switzerland 2016
N. Creignou and D. Le Berre (Eds.): SAT 2016, LNCS 9710, pp. 123–140, 2016.
DOI: 10.1007/978-3-319-40970-2_9

introduction in 2001, VSIDS has remained one of the most effective and dominant branching heuristic despite intensive efforts by many researchers to replace it [7,15,16,28]. In early 2016, we provided the first branching heuristic that is more effective than VSIDS called the conflict history-based (CHB) branching heuristic [19]. The branching heuristic introduced in this paper, which we refer to as *learning rate branching* (LRB), significantly outperforms CHB and VSIDS.

In this paper, we introduce a general principle for designing branching heuristics wherein online variable selection in SAT solvers is viewed as an optimization problem. The objective to be maximized is called the *learning rate* (LR), a numerical characterization of a variable's propensity to generate learnt clauses. The goal of the branching heuristic, given this perspective, is to select branching variables that will maximize the cumulative LR during the run of the solver. Intuitively, achieving a perfect LR of 1 implies the assigned variable is responsible for every learnt clause generated during its lifetime on the assignment trail.

We put this principle into practice in this paper. Although there are many algorithms for solving optimization problems, we show that *multi-armed bandit learning* (MAB) [31], a special-case of *reinforcement learning*, is particularly effective in our context of selecting branching variables. In MAB, an agent selects from a set of actions to receive a reward. The goal of the agent is to maximize the cumulative rewards received through the selection of actions. As we will describe in more details later, we abstract the branching heuristic as the agent, the available branching variables are abstracted as the actions, and LR is defined to be the reward. Abstracting online variable selection as a MAB problem provides the bridge to apply MAB algorithms from the literature directly as branching heuristics. In our experiments, we show that the MAB algorithm called *exponential recency weighted average* (ERWA) [31] in our abstraction surpasses the VSIDS and CHB branching heuristics at solving the benchmarks from the 4 most recent SAT Competitions in an apple-to-apple comparison. Additionally, we provide two extensions to ERWA that increases its ability to maximize LR and its performance as a branching heuristic. The final branching heuristic, called *learning rate branching* (LRB), is shown to dramatically outperform CryptoMiniSat [29] with VSIDS.

## 1.1    Contributions

**Contribution I:** We define a principle for designing branching heuristics, that is, a branching heuristic should maximize the *learning rate* (LR). We show that this principle yields highly competitive branching heuristics in practice.

**Contribution II:** We show how to abstract online variable selection in the *multi-armed bandit* (MAB) framework. This abstraction provides an interface for applying MAB algorithms directly as branching heuristics. Previously, we developed the *conflict history-based* (CHB) branching heuristic [19], also inspired by MAB. The key difference between this paper and CHB is that in the case of CHB the rewards are known a priori, and there is no metric being optimized. Whereas in this work, the learning rate is being maximized and is unknown a priori, which requires a bona fide machine learning algorithm to optimize under uncertainty.

**Contribution III:** We use the MAB abstraction to develop a new branching heuristic called *learning rate branching* (LRB). The heuristic is built on a well-known MAB algorithm called *exponential recency weighted average* (ERWA). Given our domain knowledge of SAT solving, we extend ERWA to take advantage of *reason side rate* and *locality* [20] to further maximize the learning rate objective. We show in comprehensive apple-to-apple experiments that it outperforms the current state-of-the-art VSIDS [24] and CHB [19] branching heuristics on 1975 instances from four recent SAT Competition benchmarks from 2009 to 2014 on the application and hard combinatorial categories. Additionally, we show that a modified version of CryptoMiniSat with LRB outperforms Glucose, and is very close to matching Lingeling over the same set of 1975 instances.

## 2  Preliminaries

### 2.1  Simple Average and Exponential Moving Average

Given a time series of numbers $\langle r_1, r_1, r_2, ..., r_n \rangle$, the *simple average* is computed as $avg(\langle r_1, ..., r_n \rangle) = \sum_{i=1}^{n} \frac{1}{n} r_i$. Note that every $r_i$ is given the same coefficient (also called *weight*) of $\frac{1}{n}$.

In a time series however, recent data is more pertinent to the current situation than old data. For example, consider a time series of the price of a stock. The price of the stock from yesterday is more correlated with today's price than the price of the stock from a year ago. The *exponential moving average* (EMA) [8] follows this intuition by giving the recent data higher weights than past data when averaging. Incidentally, the same intuition is built into the multiplicative decay in VSIDS [5,20]. The EMA is computed as $ema_\alpha(\langle r_1, ..., r_n \rangle) = \sum_{i=1}^{n} \alpha(1-\alpha)^{n-i} r_i$ where $0 < \alpha < 1$ is called the *step-size* parameter. $\alpha$ controls the relative weights between recent and past data. EMA can be computed incrementally as $ema_\alpha(\langle r_1, ..., r_n \rangle) = (1 - \alpha) \cdot ema_\alpha(\langle r_1, ..., r_{n-1} \rangle) + \alpha r_n$, and we define the base case $ema_\alpha(\langle \rangle) = 0$.

### 2.2  Multi-Armed Bandit (MAB)

We will explain the MAB problem [31] through a classical analogy. Consider a gambler in a casino with $n$ slot machines, where the objective of the gambler is to maximize payouts received from these machines. Each slot machine has a probability distribution describing its payouts, associating a probability with every possible value of payout. This distribution is hidden from the gambler. At any given point in time, the gambler can play one of the $n$ slot machines, and hence has $n$ actions to choose from. The gambler picks an action, plays the chosen slot machine, and receives a *reward* in terms of monetary payout by sampling that slot machine's payout probability distribution. The MAB problem is to decide which actions to take that will maximize the cumulative payouts.

If the probability distributions of the slot machines were revealed, then the gambler would simply play the slot machine whose payout distribution has the highest mean. This will maximize expected payouts for the gambler. Since the probability distribution is hidden, a simple MAB algorithm called *sample-average* [31] estimates the true mean of each distribution by averaging the samples of observed payouts. For example, suppose there are 2 slot machines. The gambler plays the first and the second slot machine 4 times each, receiving the 4 payouts $\langle 1, 2, 3, 4 \rangle$ and $\langle 5, 4, 3, 2 \rangle$ respectively. Then the algorithm will estimate the mean of the first and second slot machines' payout distributions as $avg(\langle 1, 2, 3, 4 \rangle) = 2.5$ and $avg(\langle 5, 4, 3, 2 \rangle) = 3.5$ respectively. Since the second slot machine has a higher estimated mean, the choice is to play the second slot machine. This choice is called *greedy*, that is, it chose the action it estimates to be the best given extant observations. On the other hand, choosing a non-greedy action is called *exploration* [31].

The sample-average algorithm is applicable if the hidden probability distributions are fixed. If the distributions change over time, then the problem is called *nonstationary*, and requires different algorithms. For example, suppose a slot machine gives smaller and smaller payouts the more it has been played. The older the observed payout, the bigger the difference between the current probability distribution and the distribution from which the payout was sampled. Hence, older observed payouts should have smaller weights. This gives rise to the *exponential recency weighted average* [31] (ERWA) algorithm. Instead of computing the simple average of the observed payouts, use EMA to give higher weights to recent observations relative to distant observations. Continuing the prior example, ERWA estimates the mean payout of the first and second slot machines as $ema_\alpha(\langle 1, 2, 3, 4 \rangle) = 3.0625$ and $ema_\alpha(\langle 5, 4, 3, 2 \rangle) = 2.5625$ respectively where $\alpha = 0.5$. Therefore ERWA estimates the first slot machine to have a higher mean, and hence the greedy action is to play the first slot machine.

## 2.3  Clause Learning

The defining feature of CDCL solvers is to analyze every conflict it encounters to learn new clauses to block the same conflicts, and up to exponentially similar conflicts, from re-occurring. The solver maintains an implication graph, a directed acyclic graph where the vertices are assigned variables and edges record the propagations between variables induced by Boolean constraint propagation. A clause is falsified when all of its literals are assigned to false, and in this circumstance, the solver can no longer proceed with the current assignment. The solver analyzes the implication graph and *cuts* the graph into two sides: the *conflict side* and the *reason side*. The conflict side must contain all the variables from the falsified clause and the reason side must contain all the decision variables. A learnt clause is generated on the variables from the reason side incident to the cut by negating the current assignments to those variables. In practice, the implication graph is typically cut at the first unique implication point [33]. Upon learning a clause, the solver backtracks to an earlier state where no clauses are falsified and proceeds from there.

## 2.4 The VSIDS Branching Heuristic

VSIDS can be seen as a ranking function that maintains a floating point number for each Boolean variable in the input formula, often called activity. The activities are modified in two interweaving operations called the bump and the multiplicative decay. Bump increases the activity of a variable additively by 1 whenever it appears in either a newly learnt clause or the conflict side of the implication graph. Decay periodically decreases the activity of every variable by multiplying all activities by the decay factor $\delta$ where $0 < \delta < 1$. Decay typically occurs after every conflict. VSIDS ranks variables in decreasing order of activity, and selects the unassigned variable with the highest activity to branch on next. This variable is called the *decision* variable. A separate heuristic, typically phase-saving [26], will select the polarity to assign the decision variable.

# 3 Contribution I: Branching Heuristic as Learning Rate (LR) Optimization

The branching heuristic is responsible for assigning variables through decisions that the SAT solver makes during a run. Although most of the assignments will eventually revert due to backtracking and restarts, the solver guarantees progress due to the production of learnt clauses. It is well-known that branching heuristics play a significant role in the performance of SAT solvers. To frame branching as an optimization problem, we need a metric to quantify the degree of contribution from an assigned variable to the progress of the solver, to serve as an objective to maximize. Since producing learnt clauses is a direct indication of progress, we define our metric to be the variable's propensity to produce learnt clauses. We will now define this formally.

Clauses are learnt via conflict analysis on the implication graph that the solver constructs during solving. A variable $v$ *participates* in generating a learnt clause $l$ if either $v$ appears in $l$ or $v$ is resolved during the conflict analysis that produces $l$ (i.e., appears in the conflict side of the implication graph induced by the cut that generates $l$). In other words, $v$ is required for the learning of $l$ from the encountered conflict. Note that only assigned variables can participate in generating learnt clauses. We define $I$ as the interval of time between the assignment of $v$ until $v$ transitions back to being unassigned. Let $P(v, I)$ be the number learnt clauses in which $v$ participates during interval $I$ and let $L(I)$ be the number of learnt clauses generated in interval $I$. The *learning rate* (LR) of variable $v$ at interval $I$ is defined as $\frac{P(v,I)}{L(I)}$. For example, suppose variable $v$ is assigned by the branching heuristic after 100 learnt clauses are produced. It participates in producing the 101-st and 104-th learnt clause. Then $v$ is unassigned after the 105-th learnt clause is produced. In this case, $P(v, I) = 2$ and $L(I) = 5$ and hence the LR of variable $v$ is $\frac{2}{5}$.

The exact LR of a variable is usually unknown during branching. In the previous example, variable $v$ was picked by the branching heuristic after 100 learnt clauses are produced, but the LR is not known until after the 105-th learnt clause is produced. Therefore optimizing LR involves a degree of uncertainty, which makes the problem well-suited for learning algorithms. In addition, the LR of a variable changes over time due to modifications to the learnt clause database, stored phases, and assignment trail. As such, estimating LR requires nonstationary algorithms to deal with changes in the underlying environment.

## 4   Contribution II: Abstracting Online Variable Selection as a Multi-Armed Bandit (MAB) Problem

Given $n$ Boolean variables, we will abstract online variable selection as an $n$-armed bandit optimization problem. A branching heuristic has $n$ actions to choose from, corresponding to branching on any of the $n$ Boolean variables. The expressions *assigning a variable* and *playing an action* will be used interchangeably. When a variable $v$ is assigned, then $v$ can begin to participate in generating learnt clauses. When $v$ becomes unassigned, the LR $r$ is computed and returned as the reward for playing the action $v$. The terms *reward* and *LR* will be used interchangeably. The MAB algorithm uses the reward to update its internal estimates of the action that will maximize the rewards.

The MAB algorithm is limited to picking actions corresponding to unassigned variables, as the branching heuristic can only branch on unassigned variables. This limitation forces some exploration, as the MAB algorithm cannot select the same action again until the corresponding variable is unassigned due to backtracking or restarting. Although the branching heuristic is only assigning one variable at a time, it indirectly assigns many other variables through propagation. We include the propagated variables, along with the branched variables, as plays in the MAB framework. That is, branched and propagated variables will all receive their own individual rewards corresponding to their LR, and the MAB algorithm will use all these rewards to update its internal estimates. This also forces some exploration since a variable ranked poorly by the MAB algorithm can still be played through propagation.

## 5   Contribution III: Learning Rate Branching (LRB) Heuristic

Given the MAB abstraction, we first use the well-known ERWA bandit algorithm as a branching heuristic. We will upgrade ERWA with two novel extensions to arrive at the final branching heuristic called the *learning rate branching* (LRB) heuristic. We will justify these extensions experimentally through the lens of MAB, that is, these extensions are better at maximizing the LR rewards. We will demonstrate empirically the effectiveness of LRB at solving the benchmarks from the 4 previous SAT Competitions.

## 5.1 Exponential Recency Weighted Average (ERWA)

We will explain how to apply ERWA as a branching heuristic through the MAB abstraction. First we will provide a conceptual explanation, that is easier to comprehend. Then we will provide a complementary explanation from the implementation's perspective, which is equivalent to the conceptual explanation, but provides more details.

Conceptually, each variable $v$ maintains its own time series $ts_v$ containing the observed rewards for $v$. Whenever a variable $v$ transitions from assigned to unassigned, ERWA will calculate the LR $r$ for $v$ (see Sect. 3) and append the reward $r$ to the time series by updating $ts_v \leftarrow append(ts_v, r)$. When the solver requests the next branching variable, ERWA will select the variable $v^*$ where $v^* = argmax_{v \in U}(ema_\alpha(ts_v))$ and $U$ is the set of currently unassigned variables.

The actual implementation takes advantage of the incrementality of EMA to avoid storing the time series $ts$, see Algorithm 1 for pseudocode of the implementation. Alternative to the above description, each variable $v$ maintains a floating point number $Q_v$ representing $ema_\alpha(ts_v)$. When $v$ receives reward $r$, then the implementation updates $Q_v$ using the incrementality of EMA, that is, $Q_v \leftarrow (1 - \alpha) \cdot Q_v + \alpha \cdot r$ (see line 24 of Algorithm 1). When the solver requests the next branching variable, the implementation will select the variable $v^*$ where $v^* = argmax_{v \in U} Q_v$ and $U$ is the set of currently unassigned variables (see line 28 of Algorithm 1). Note that $Q_v$ can be stored in a priority queue for all unassigned variables $v$, hence finding the maximum will take logarithmic time in the worst-case. The implementation is equivalent to the prior conceptual description, but significantly more efficient in both memory and time.

For our experiments, we initialize the step-size $\alpha = 0.4$. We follow the convention of typical ERWA to decrease the step-size over time [31]. After each conflict, the step-size is decreased by $10^{-6}$ until it reaches 0.06 (see line 14 in Algorithm 1), and remains at 0.06 for the remainder of the run. This step-size management is equivalent to the one in CHB [19] and is similar to how the Glucose solver manages the VSIDS decay factor by increasing it over time [4].

## 5.2 Extension: Reason Side Rate (RSR)

Recall that LR measures the participation rate of variables in generating learnt clauses. That is, variables with high LR are the ones that frequently appear in the generated learnt clause and/or the conflict side of the implication graph. If a variable appears on the reason side near the learnt clause, then these variables just missed the mark. We show that accounting for these close proximity variables, in conjunction with the ERWA heuristic, optimizes the LR further.

More precisely, if a variable $v$ appears in a reason clause of a variable in a learnt clause $l$, but does not occur in $l$, then we say that $v$ *reasons* in generating the learnt clause $l$. We define $I$ as the interval of time between the assignment of $v$ until $v$ transitions back to being unassigned. Let $A(v, I)$ be the number of learnt clauses which $v$ reasons in generating in interval $I$ and let $L(I)$ be the

**Algorithm 1.** Pseudocode for ERWA as a branching heuristic using our MAB abstraction for maximizing LR.

```
1:  procedure INITIALIZE                                    ▷ Called once at the start of the solver.
2:      α ← 0.4                                                                 ▷ The step-size.
3:      LearntCounter ← 0                          ▷ The number of learnt clauses generated by the solver.
4:      for v ∈ Vars do                 ▷ Vars is the set of Boolean variables in the input CNF.
5:          Qᵥ ← 0                                                        ▷ The EMA estimate of v.
6:          Assignedᵥ ← 0                                              ▷ When v was last assigned.
7:          Participatedᵥ ← 0                   ▷ The number of learnt clauses v participated in
                                                        generating since Assignedᵥ.
8:
9:  procedure AFTERCONFLICTANALYSIS(learntClauseVars ⊆ Vars, conflictSide ⊆ Vars)    ▷
        Called after a learnt clause is generated from
        conflict analysis.
10:      LearntCounter ← LearntCounter + 1
11:      for v ∈ conflictSide ∪ learntClauseVars do
12:          Participatedᵥ ← Participatedᵥ + 1
13:      if α > 0.06 then
14:          α ← α − 10⁻⁶
15:
16:  procedure ONASSIGN(v ∈ Vars)                       ▷ Called when v is assigned by branching or prop-
                                                           agation.
17:      Assignedᵥ ← LearntCounter
18:      Participatedᵥ ← 0
19:
20:  procedure ONUNASSIGN(v ∈ Vars)                     ▷ Called when v is unassigned by backtracking or
                                                           restart.
21:      Interval ← LearntCounter − Assignedᵥ
22:      if Interval > 0 then                              ▷ Interval = 0 is possible due to restarts.
23:          r ← Participatedᵥ/Interval.                                        ▷ r is the LR.
24:          Qᵥ = (1 − α) · Qᵥ + α · r                            ▷ Update the EMA incrementally.
25:
26:  function PICKBRANCHLIT                             ▷ Called when the solver requests the next branch-
                                                           ing variable.
27:      U ← {v ∈ Vars | isUnassigned(v)}
28:      return argmaxᵥ∈U Qᵥ                           ▷ Use a priority queue for better performance.
```

number of learnt clauses generated in interval $I$. The *reason side rate* (RSR) of variable $v$ at interval $I$ is defined as $\frac{A(v,I)}{L(I)}$.

Recall that in ERWA, the estimates are updated incrementally as $Q_v \leftarrow (1-\alpha) \cdot Q_v + \alpha \cdot r$ where $r$ is the LR of $v$. This extension modifies the update to $Q_v \leftarrow (1-\alpha) \cdot Q_v + \alpha \cdot (r + \frac{A(v,I)}{L(I)})$ where $\frac{A(v,I)}{L(I)}$ is the RSR of $v$ (see line 20 in Algorithm 2). Note that we did not change the definition of the reward. The extension simply encourages the algorithm to select variables with high RSR when deciding to branch. We hypothesize that variables observed to have high RSR are likely to have high LR as well.

## 5.3   Extension: Locality

Recent research shows that VSIDS exhibits locality [20], defined with respect to the community structure of the input CNF instance [1,20,25]. Intuitively, if the solver is currently working within a community, it is best to continue focusing on the same community rather than exploring another. We hypothesize that high LR variables also exhibit locality, that is, the branching heuristic can achieve higher LR by restricting exploration.

**Algorithm 2.** Pseudocode for ERWA as a branching heuristic with the RSR extension. The pseudocode $Algorithm1.method(...)$ is calling out to the code in Algorithm 1. The procedure $PickBranchLit$ is unchanged.

```
 1: procedure INITIALIZE
 2:     Algorithm1.Initialize()
 3:     for v ∈ Vars do                          ▷ Vars is the set of Boolean variables in the input CNF.
 4:         Reasonedᵥ ← 0                         ▷ The number of learnt clauses v reasoned in gen-
                                                     erating since Assignedᵥ.
 5:
 6: procedure AFTERCONFLICTANALYSIS(learntClauseVars ⊆ Vars, conflictSide ⊆ Vars)
 7:     Algorithm1.AfterConflictAnalysis(learntClauseVars, conflictSide)
 8:     for v ∈ (⋃_{u∈learntClauseVars} reason(u)) ∖ learntClauseVars do
 9:         Reasonedᵥ ← Reasonedᵥ + 1
10:
11: procedure ONASSIGN(v ∈ Vars)
12:     Algorithm1.OnAssign()
13:     Reasonedᵥ ← 0
14:
15: procedure ONUNASSIGN(v ∈ Vars)
16:     Interval ← LearntCounter − Assignedᵥ
17:     if Interval > 0 then                      ▷ Interval = 0 is possible due to restarts.
18:         r ← Participatedᵥ/Interval.           ▷ r is the LR.
19:         rsr ← Reasonedᵥ/Interval.             ▷ rsr is the RSR.
20:         Qᵥ = (1 − α) · Qᵥ + α · (r + rsr)     ▷ Update the EMA incrementally.
```

**Algorithm 3.** Pseudocode for ERWA as a branching heuristic with the locality extension. $AfterConflictAnalysis$ is the only procedure modified.

```
 1: procedure AFTERCONFLICTANALYSIS(learntClauseVars ⊆ Vars, conflictSide ⊆ Vars)
 2:     Algorithm2.AfterConflictAnalysis(learntClauseVars, conflictSide)
 3:     U ← {v ∈ Vars | isUnassigned(v)}
 4:     for v ∈ U do
 5:         Qᵥ ← 0.95 × Qᵥ.
```

Inspired by the VSIDS decay, this extension multiplies the $Q_v$ of every unassigned variable $v$ by 0.95 after each conflict (see line 5 in Algorithm 3). Again, we did not change the definition of the reward. The extension simply discourages the algorithm from exploring inactive variables. This extension is similar to the decay reinforcement model [13,32] where unplayed arms are penalized by a multiplicative decay. The implementation is optimized to do the multiplications in batch. For example, suppose variable $v$ is unassigned for $k$ conflicts. Rather than executing $k$ updates of $Q_v \leftarrow 0.95 \times Q_v$, the implementation simply updates once using $Q_v \leftarrow 0.95^k \times Q_v$.

### 5.4 Putting it all Together to Obtain the Learning Rate Branching (LRB) Heuristic

The *learning rate branching* (LRB) heuristic refers to ERWA in the MAB abstraction with the RSR and locality extensions. We show that LRB is better at optimizing LR than the other branching heuristics considered, and subsequently has the best overall performance of the bunch.

# 6   Experimental Results

In this section, we discuss the detailed and comprehensive experiments we performed to evaluate LRB. First, we justify the extensions of LRB by demonstrating their performance vis-a-vis improvements in learning rate. Second, we show that LRB outperforms the state-of-the-art VSIDS and CHB branching heuristic. Third, we show that LRB achieves higher rewards/LR than VSIDS, CHB, and LRB sans the extensions. Fourth, we show the effectiveness of LRB within a state-of-the-art CDCL solver, namely, CryptoMiniSat [29]. To better gauge the results of these experiments, we quote two leading SAT solver developers, Professors Audemard and Simon [3]:

> "We must also say, as a preliminary, that improving SAT solvers is often a cruel world. To give an idea, improving a solver by solving at least ten more instances (on a fixed set of benchmarks of a competition) is generally showing a critical new feature. In general, the winner of a competition is decided based on a couple of additional solved benchmarks."

## 6.1   Setup

The experiments are performed by running CDCL solvers with various branching heuristics on StarExec [30], a platform designed for evaluating logic solvers. The StarExec platform uses the Intel Xeon CPU E5-2609 at 2.40 GHz with 10240 KB cache and 24 GB of main memory, running on Red Hat Enterprise Linux Workstation release 6.3, and Linux kernel 2.6.32-431.1.2.el6.x86_64. The benchmarks for the experiments consist of all the instances from the previous 4 SAT Competitions (2014, 2013, 2011, and 2009), in both the application and hard combinatorial categories. For each instance, the solver is given 5000 seconds of CPU time and 7.5 GB of RAM, abiding by the SAT Competition 2013 limits.

Our experiments test different branching heuristics on a *base* CDCL solver, where the only modification is to the branching heuristic to give a fair apple-to-apple comparison. Our base solver is MiniSat version 2.2.0 [12] (simp version) with one modification to use the popular *aggressive LBD-based clause deletion* proposed by the authors of the Glucose solver in 2009 [2]. Since MiniSat is a relatively simple solver with very few features, it is ideal for our base solver to better isolate the effects swapping branching heuristics in our experiments. Additionally, MiniSat is the basis of many competitive solvers and aggressive LBD-based clause deletion is almost universally implemented, hence we believe the results of our experiments will generalize to other solver implementations.

## 6.2   Experiment: Efficacy of Extensions to ERWA

In this experiment, we demonstrate the effectiveness of the extensions we proposed for LRB. We modified the base solver by replacing the VSIDS branching heuristic with ERWA. We then created two additional solvers, one with the RSR extension and another with both the RSR and locality extensions. We ran these

**Table 1.** Comparison of our extensions on the base CDCL solver (MiniSat 2.2 with aggressive LBD-based clause deletion). The entries show the number of instances solved for the given solver and benchmark, the higher the better. Green is best, red is worst.

| Benchmark | Status | ERWA | ERWA+ RSR | ERWA+ RSR + Locality (LRB) |
|---|---|---|---|---|
| 2009 Application | SAT | 85 | 84 | 85 |
| | UNSAT | 122 | 120 | 121 |
| | BOTH | 207 | 204 | 206 |
| 2009 Hard Combinatorial | SAT | 98 | 99 | 101 |
| | UNSAT | 65 | 68 | 69 |
| | BOTH | 163 | 167 | 170 |
| 2011 Application | SAT | 105 | 105 | 103 |
| | UNSAT | 98 | 101 | 98 |
| | BOTH | 203 | 206 | 201 |
| 2011 Hard Combinatorial | SAT | 95 | 88 | 93 |
| | UNSAT | 45 | 61 | 65 |
| | BOTH | 140 | 149 | 158 |
| 2013 Application | SAT | 125 | 133 | 132 |
| | UNSAT | 89 | 95 | 95 |
| | BOTH | 214 | 228 | 227 |
| 2013 Hard Combinatorial | SAT | 113 | 110 | 116 |
| | UNSAT | 97 | 108 | 110 |
| | BOTH | 210 | 218 | 226 |
| 2014 Application | SAT | 111 | 108 | 116 |
| | UNSAT | 82 | 77 | 77 |
| | BOTH | 193 | 185 | 193 |
| 2014 Hard Combinatorial | SAT | 87 | 92 | 91 |
| | UNSAT | 73 | 87 | 89 |
| | BOTH | 160 | 179 | 180 |
| **TOTAL (excluding duplicates)** | **SAT** | **638** | **632** | 654 |
| | **UNSAT** | **574** | **619** | 625 |
| | **BOTH** | **1212** | **1251** | 1279 |

3 solvers over the entire benchmark and report the number of instances solved by these solvers within the time limit in Table 1. ERWA solves a total of 1212 instances, ERWA with the RSR extension solves a total of 1251 instances, and ERWA with the RSR and locality extensions (i.e., LRB) solves a total of 1279 instances. See Fig. 1 for a cactus plot of the solving times.

## 6.3    Experiment: LRB vs VSIDS vs CHB

In this experiment, we compare LRB with VSIDS [24] and CHB [19]. Our base solver is MiniSat 2.2 which already implements VSIDS. We then replaced VSIDS in the base solver with LRB and CHB to derive 3 solvers in total, with the only difference being the branching heuristic. We ran these 3 solvers on the entire benchmark and present the results in Table 2. LRB solves a total of 1279 instances, VSIDS solves a total of 1179 instances, and CHB solves a total of 1235 instances. See Fig. 1 for a cactus plot of the solving times.

**Table 2.** Apple-to-apple comparison between branching heuristics (LRB, CHB, and VSIDS) in a version of MiniSat 2.2 with aggressive LBD-based clause deletion. The entries show the number of instances in the benchmark the given branching heuristic solves, the higher the better. Green is best, red is worst. The LRB version (we dub as MapleSAT), outperforms the others.

| Benchmark | Status | LRB | VSIDS | CHB |
|---|---|---|---|---|
| 2009 Application | SAT | 85 | 83 | 89 |
| | UNSAT | 121 | 125 | 119 |
| | BOTH | 206 | 208 | 208 |
| 2009 Hard Combinatorial | SAT | 101 | 100 | 103 |
| | UNSAT | 69 | 66 | 67 |
| | BOTH | 170 | 166 | 170 |
| 2011 Application | SAT | 103 | 95 | 106 |
| | UNSAT | 98 | 99 | 96 |
| | BOTH | 201 | 194 | 202 |
| 2011 Hard Combinatorial | SAT | 93 | 88 | 102 |
| | UNSAT | 65 | 48 | 47 |
| | BOTH | 158 | 136 | 149 |
| 2013 Application | SAT | 132 | 127 | 137 |
| | UNSAT | 95 | 86 | 79 |
| | BOTH | 227 | 213 | 216 |
| 2013 Hard Combinatorial | SAT | 116 | 115 | 122 |
| | UNSAT | 110 | 73 | 96 |
| | BOTH | 226 | 188 | 218 |
| 2014 Application | SAT | 116 | 105 | 115 |
| | UNSAT | 77 | 94 | 73 |
| | BOTH | 193 | 199 | 188 |
| 2014 Hard Combinatorial | SAT | 91 | 91 | 90 |
| | UNSAT | 89 | 59 | 76 |
| | BOTH | 180 | 150 | 166 |
| **TOTAL (excluding duplicates)** | **SAT** | **654** | **626** | 673 |
| | **UNSAT** | 625 | **553** | **562** |
| | **BOTH** | 1279 | 1179 | 1235 |

## 6.4    Experiment: LRB and Learning Rate

In this experiment, we measure the efficacy of the 5 branching heuristics from Tables 1 and 2 at maximizing the LR. For each instance in the benchmark, we solve the instance 5 times with the 5 branching heuristics implemented in the base solver. For each branching heuristic, we track all the observed rewards (i.e., LR) and record the mean observed reward at the end of the run, regardless if the solver solves the instance or not. We then rank the 5 branching heuristics by their mean observed reward for that instance. A branching heuristic gets a rank of 1 (resp. 5) if it has the highest (resp. lowest) mean observed reward for that instance. For each branching heuristic, we then average its ranks over the entire benchmark and report these numbers in Table 3. The experiment shows that LRB is the best heuristic in terms of maximizing the reward LR (corresponding to a rank closest to 1) in almost every category. In addition, the experiment shows that the RSR and locality extensions increase the observed rewards relative to vanilla ERWA. Somewhat surprisingly, VSIDS and CHB on average observe higher rewards (i.e., LR) than ERWA, despite the fact that VSIDS and CHB are designed without LR as an explicit objective. This perhaps partly explains the effectiveness of those two heuristics.

**Table 3.** The average ranking of observed rewards compared between different branching heuristics in MiniSat 2.2 with aggressive LBD-based clause deletion. The lower the reported number, the better the heuristic is at maximizing the observed reward relative to the others. Green is best, red is worst.

| Benchmark | Status | LRB | ERWA | ERWA + RSR | VSIDS | CHB |
|---|---|---|---|---|---|---|
| 2009 Application | SAT | 2.41 | 3.79 | 3.42 | 2.51 | 2.87 |
| | UNSAT | 2.13 | 4.16 | 3.32 | 2.90 | 2.49 |
| | BOTH | 2.25 | 4.01 | 3.36 | 2.74 | 2.65 |
| 2009 Hard Combinatorial | SAT | 2.43 | 3.30 | 3.03 | 3.29 | 2.95 |
| | UNSAT | 2.18 | 4.18 | 3.48 | 3.22 | 1.94 |
| | BOTH | 2.33 | 3.66 | 3.21 | 3.26 | 2.53 |
| 2011 Application | SAT | 2.25 | 3.61 | 3.02 | 2.77 | 3.35 |
| | UNSAT | 2.14 | 3.82 | 3.22 | 3.49 | 2.33 |
| | BOTH | 2.20 | 3.72 | 3.12 | 3.13 | 2.85 |
| 2011 Hard Combinatorial | SAT | 2.57 | 3.47 | 2.98 | 3.46 | 2.53 |
| | UNSAT | 2.57 | 3.72 | 3.32 | 3.54 | 1.85 |
| | BOTH | 2.57 | 3.56 | 3.11 | 3.49 | 2.27 |
| 2013 Application | SAT | 2.33 | 3.60 | 3.16 | 2.49 | 3.41 |
| | UNSAT | 2.02 | 4.16 | 3.07 | 3.39 | 2.37 |
| | BOTH | 2.19 | 3.85 | 3.12 | 2.89 | 2.95 |
| 2013 Hard Combinatorial | SAT | 2.51 | 3.57 | 2.91 | 3.03 | 2.98 |
| | UNSAT | 1.99 | 3.92 | 2.65 | 4.26 | 2.18 |
| | BOTH | 2.24 | 3.75 | 2.78 | 3.65 | 2.58 |
| 2014 Application | SAT | 2.27 | 3.68 | 3.21 | 2.50 | 3.35 |
| | UNSAT | 2.24 | 4.34 | 3.20 | 2.82 | 2.40 |
| | BOTH | 2.25 | 4.01 | 3.21 | 2.66 | 2.88 |
| 2014 Hard Combinatorial | SAT | 2.43 | 3.51 | 3.03 | 2.78 | 3.26 |
| | UNSAT | 1.81 | 4.38 | 2.69 | 3.82 | 2.30 |
| | BOTH | 2.11 | 3.96 | 2.85 | 3.31 | 2.76 |
| **TOTAL (excluding duplicates)** | SAT | 2.45 | 3.53 | 3.10 | 2.72 | 3.20 |
| | UNSAT | 2.12 | 4.08 | 3.10 | 3.41 | 2.30 |
| | BOTH | 2.28 | 3.81 | 3.10 | 3.07 | 2.74 |

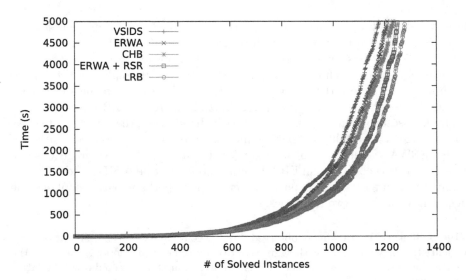

**Fig. 1.** A cactus plot of the 5 branching heuristics in MiniSat 2.2 with aggressive LBD-based clause deletion. The benchmark consists of the 4 most recent SAT Competition benchmarks (2014, 2013, 2011, 2009) including both the application and hard combinatorial categories, excluding duplicate instances. A point $(x, y)$ on the plot is interpretted as: $y$ instances in the benchmark took less than $x$ seconds to solve for the branching heuristic. The further right and further down, the better.

### 6.5   Experiment: LRB vs State-of-the-Art CDCL

In this experiment, we test how LRB-enchanced CryptoMiniSat competes against the state-of-the-art solvers CryptoMiniSat [29], Glucose [4], and Lingeling [6] which all implement VSIDS. We modified CryptoMiniSat 4.5.3 [29] by replacing VSIDS with LRB, leaving everything else unmodified. We ran unmodified CryptoMiniSat, Glucose, and Lingeling, along with the LRB-enchanced CryptoMiniSat on the benchmark and report the results in Table 4. LRB improved CryptoMiniSat on 6 of the 8 benchmarks and solves 59 more instances overall.

## 7   Related Work

The Chaff solver introduced the VSIDS branching heuristic in 2001 [24]. Although many branching heuristics have been proposed [7, 15–17, 22, 28], VSIDS and its variants remain as the dominant branching heuristic employed in modern CDCL SAT solvers. Carvalho and Marques-Silva used rewards based on learnt clause length and backjump size to improve VSIDS [10]. More precisely, the bump value of VSIDS is increased for short learnt clauses and/or long backjumps. Their usage of rewards is unrelated to the definition of rewards in the reinforcement learning and multi-armed bandits context. Lagoudakis and Littman used reinforcement learning to dynamically switch between a fixed set of 7 well-known

**Table 4.** Apple-to-apple comparison between four state-of-art solvers: CryptoMiniSat (CMS) with LRB heuristic, CMS with VSIDS, Glucose, and Lingeling. The table shows the number of instances solved per SAT Competition benchmark, categorized as SAT or UNSAT instances. CMS with LRB (we dub as MapleCMS) outperforms CMS with VSIDS on most benchmarks.

| Benchmark | Status | CMS with LRB | CMS with VSIDS | Glucose | Lingeling |
|---|---|---|---|---|---|
| 2009 Application | SAT | 85 | 87 | 83 | 80 |
| | UNSAT | 140 | 143 | 138 | 141 |
| | BOTH | 225 | 230 | 221 | 221 |
| 2009 Hard Combinatorial | SAT | 102 | 95 | 90 | 98 |
| | UNSAT | 71 | 65 | 70 | 83 |
| | BOTH | 173 | 160 | 160 | 181 |
| 2011 Application | SAT | 106 | 97 | 94 | 94 |
| | UNSAT | 122 | 129 | 127 | 134 |
| | BOTH | 228 | 226 | 221 | 228 |
| 2011 Hard Combinatorial | SAT | 86 | 86 | 80 | 88 |
| | UNSAT | 57 | 49 | 44 | 66 |
| | BOTH | 143 | 135 | 124 | 154 |
| 2013 Application | SAT | 115 | 109 | 104 | 100 |
| | UNSAT | 120 | 115 | 111 | 122 |
| | BOTH | 235 | 224 | 215 | 222 |
| 2013 Hard Combinatorial | SAT | 116 | 114 | 115 | 114 |
| | UNSAT | 114 | 101 | 106 | 117 |
| | BOTH | 230 | 215 | 221 | 231 |
| 2014 Application | SAT | 107 | 102 | 99 | 101 |
| | UNSAT | 118 | 127 | 120 | 141 |
| | BOTH | 225 | 229 | 219 | 242 |
| 2014 Hard Combinatorial | SAT | 89 | 85 | 79 | 89 |
| | UNSAT | 122 | 100 | 93 | 119 |
| | BOTH | 211 | 185 | 172 | 208 |
| **TOTAL (excluding duplicates)** | **SAT** | **619** | **598** | **575** | **589** |
| | **UNSAT** | **738** | **700** | **685** | **782** |
| | **BOTH** | **1357** | **1298** | **1260** | **1371** |

SAT branching heuristics [18]. Their technique requires offline training on a class of similar instances. Loth et al. used multi-armed bandits for directing the growth of the search tree for Monte-Carlo Tree Search [21]. The rewards are computed based on the relative depth failure of the tree walk. Fröhlich et al. used the UCB algorithm from multi-armed bandits to select the candidate variables to define the neighborhood of a stochastic local search for the theory of bitvectors [14]. The rewards they are optimizing is to minimize the number of unsatisfied clauses. Liang et al. also applied ERWA as a branching heuristic called CHB [19]. As stated earlier, CHB is neither an optimization nor learning algorithm since the rewards are computed on past events.

# 8   Conclusions and Future Work

In this paper, we provide three main contributions, and each has potential for further enhancements.

**Contribution I:** We define LR as a metric for the branching heuristic to optimize. LR captures the intuition that the branching heuristic should assign variables which are likely to generate a high quantity of learnt clauses with no regards to the "quality" of those clauses [2]. A new metric that captures quality should encourage better clause learning. Or perhaps branching heuristics can be stated as a multi-objective optimization problem where a good heuristic would balance the tradeoff between quality and quantity of learnt clauses.

Additionally, we would like to stress that the starting point for this research was a model of CDCL SAT solvers as an interplay between branching heuristic and clause learning. The branching heuristic guides the search, and has great impact on the clauses that will be learnt during the run of the solver. In the reverse direction, clause learning provides feedback to guide branching heuristics like VSIDS, CHB, and LRB. We plan to explore a mathematical model where the branching heuristic is an inductive engine (machine learning), and the conflict analysis is a deductive feedback mechanism. Such a model could enable us to prove complexity theoretic results that at long last might explain why CDCL SAT solvers are so efficient for industrial instances.

**Contribution II:** We chose MAB as the optimization method in this paper, but many other optimization techniques can be applied to optimize LR. The most natural extension to our work here is to incorporate the internal state of the solver and apply stateful reinforcement learning. The state, for example, could be the current community the solver is focused on and exploiting this information could improve the locality of the branching heuristic [20].

**Contribution III:** We based LRB on one MAB algorithm, ERWA, due to its low computational overhead. The literature of multi-armed bandits is very rich, and provides many alternative algorithms with a wide spectrum of characteristics and assumptions. It is fruitful to explore the MAB literature to determine the best algorithm for branching in CDCL SAT solvers.

Finally, as our experimental results suggest, the line of research we have just started exploring, namely, branching heuristics as machine learning algorithms (and branching as an optimization problem) has already shown considerable improvement over previous state-of-the-art branching heuristics such as VSIDS and CHB, and affords a rich design space of heuristics to explore in the future.

# References

1. Ansótegui, C., Giráldez-Cru, J., Levy, J.: The community structure of SAT formulas. In: Cimatti, A., Sebastiani, R. (eds.) SAT 2012. LNCS, vol. 7317, pp. 410–423. Springer, Heidelberg (2012)
2. Audemard, G., Simon, L.: Predicting learnt clauses quality in modern SAT solvers. In: Proceedings of the 21st International Jont Conference on Artifical Intelligence, IJCAI 2009, pp. 399–404. Morgan Kaufmann Publishers Inc., San Francisco (2009)
3. Audemard, G., Simon, L.: Refining restarts strategies for SAT and UNSAT. In: Milano, M. (ed.) CP 2012. LNCS, vol. 7514, pp. 118–126. Springer, Heidelberg (2012)

4. Audemard, G., Simon, L.: Glucose 2.3 in the SAT 2013 Competition. In: Proceedings of SAT Competition 2013, pp. 42–43 (2013)
5. Biere, A.: Adaptive restart strategies for conflict driven SAT solvers. In: Kleine Büning, H., Zhao, X. (eds.) SAT 2008. LNCS, vol. 4996, pp. 28–33. Springer, Heidelberg (2008)
6. Biere, A.: Lingeling, Plingeling, PicoSAT and PrecoSAT at SAT Race 2010. FMV Report Series Technical report 10(1) (2010)
7. Biere, A., Fröhlich, A.: Evaluating CDCL variable scoring schemes. In: Heule, M., Weaver, S. (eds.) SAT 2015. LNCS, vol. 9340, pp. 405–422. Springer, Heidelberg (2015). doi:10.1007/978-3-319-24318-4_29
8. Brown, R.G.: Exponential smoothing for predicting demand. Oper. Res. **5**, 145–145 (1957)
9. Cadar, C., Ganesh, V., Pawlowski, P.M., Dill, D.L., Engler, D.R.: EXE: automatically generating inputs of death. In: Proceedings of the 13th ACM Conference on Computer and Communications Security, CCS 2006, pp. 322–335. ACM, New York (2006)
10. Carvalho, E., Marques-Silva, J.P.: Using rewarding mechanisms for improving branching heuristics. In: Proceedings of the Seventh International Conference on Theory and Applications of Satisfiability Testing (2004)
11. Clarke, E., Biere, A., Raimi, R., Zhu, Y.: Bounded model checking using satisfiability solving. Form. Methods Syst. Des. **19**(1), 7–34 (2001)
12. Eén, N., Sörensson, N.: An extensible SAT-solver. In: Giunchiglia, E., Tacchella, A. (eds.) SAT 2003. LNCS, vol. 2919, pp. 502–518. Springer, Heidelberg (2004)
13. Erev, I., Roth, A.E.: Predicting how people play games: reinforcement learning in experimental games with unique, mixed strategy equilibria. Am. Econ. Rev. **88**(4), 848–881 (1998)
14. Fröhlich, A., Biere, A., Wintersteiger, C., Hamadi, Y.: Stochastic local search for satisfiability modulo theories. In: Proceedings of the Twenty-Ninth AAAI Conference on Artificial Intelligence, AAAI 2015, pp. 1136–1143. AAAI Press (2015)
15. Gershman, R., Strichman, O.: HAIFASAT: a new robust SAT solver. In: Ur, S., Bin, E., Wolfsthal, Y. (eds.) HVC 2005. LNCS, vol. 3875, pp. 76–89. Springer, Heidelberg (2006)
16. Goldberg, E., Novikov, Y.: BerkMin: a fast and robust sat-solver. Discrete Appl. Math. **155**(12), 1549–1561 (2007)
17. Jeroslow, R.G., Wang, J.: Solving propositional satisfiability problems. Ann. Math. Artif. Intell. **1**(1–4), 167–187 (1990)
18. Lagoudakis, M.G., Littman, M.L.: Learning to select branching rules in the DPLL procedure for satisfiability. Electron. Notes Discrete Math. **9**, 344–359 (2001)
19. Liang, J.H., Ganesh, V., Poupart, P., Czarnecki, K.: Exponential recency weighted average branching heuristic for SAT solvers. In: Proceedings of AAAI 2016 (2016)
20. Liang, J.H., Ganesh, V., Zulkoski, E., Zaman, A., Czarnecki, K.: Understanding VSIDS branching heuristics in conflict-driven clause-learning SAT solvers. In: Liang, J.H., Ganesh, V., Zulkoski, E., Zaman, A., Czarnecki, K. (eds.) HVC 2015. LNCS, vol. 9434, pp. 225–241. Springer, Heidelberg (2015). doi:10.1007/978-3-319-26287-1_14
21. Loth, M., Sebag, M., Hamadi, Y., Schoenauer, M.: Bandit-based search for constraint programming. In: Schulte, C. (ed.) CP 2013. LNCS, vol. 8124, pp. 464–480. Springer, Heidelberg (2013)
22. Marques-Silva, J.: The impact of branching heuristics in propositional satisfiability algorithms. In: Barahona, P., Alferes, J.J. (eds.) EPIA 1999. LNCS (LNAI), vol. 1695, pp. 62–74. Springer, Heidelberg (1999)

23. Marques-Silva, J.P., Sakallah, K.A.: GRASP-a new search algorithm for satisfiability. In: Proceedings of the 1996 IEEE/ACM International Conference on Computer-aided Design, ICCAD 1996, pp. 220–227. IEEE Computer Society, Washington, DC (1996)
24. Moskewicz, M.W., Madigan, C.F., Zhao, Y., Zhang, L., Malik, S.: Chaff: engineering an efficient SAT solver. In: Proceedings of the 38th Annual Design Automation Conference, DAC 2001, pp. 530–535. ACM, New York (2001)
25. Newsham, Z., Ganesh, V., Fischmeister, S., Audemard, G., Simon, L.: Impact of community structure on SAT solver performance. In: Sinz, C., Egly, U. (eds.) SAT 2014. LNCS, vol. 8561, pp. 252–268. Springer, Heidelberg (2014)
26. Pipatsrisawat, K., Darwiche, A.: A lightweight component caching scheme for satisfiability solvers. In: Marques-Silva, J., Sakallah, K.A. (eds.) SAT 2007. LNCS, vol. 4501, pp. 294–299. Springer, Heidelberg (2007)
27. Rintanen, J.: Planning and SAT. In: Biere, A., Heule, M., van Maaren, H., Walsh, T. (eds.) Handbook of Satisfiability, vol. 185, pp. 483–504. IOS Press, Amsterdam (2009)
28. Ryan, L.: Efficient Algorithms for Clause-Learning SAT Solvers. Master's thesis, Simon Fraser University (2004)
29. Soos, M.: CryptoMiniSat v4. In: SAT Competition, p. 23 (2014)
30. Stump, A., Sutcliffe, G., Tinelli, C.: StarExec: a cross-community infrastructure for logic solving. In: Demri, S., Kapur, D., Weidenbach, C. (eds.) IJCAR 2014. LNCS, vol. 8562, pp. 367–373. Springer, Heidelberg (2014)
31. Sutton, R.S., Barto, A.G.: Reinforcement Learning: An Introduction, vol. 1. MIT Press Cambridge, Massachusetts (1998)
32. Yechiam, E., Busemeyer, J.R.: Comparison of basic assumptions embedded in learning models for experience-based decision making. Psychon. Bull. Rev. **12**(3), 387–402 (2005)
33. Zhang, L., Madigan, C.F., Moskewicz, M.H., Malik, S.: Efficient conflict driven learning in a boolean satisfiability solver. In: Proceedings of the 2001 IEEE/ACM International Conference on Computer-aided Design, ICCAD 2001, pp. 279–285. IEEE Press, Piscataway (2001)

# On the Hardness of SAT with Community Structure

Nathan Mull[(✉)], Daniel J. Fremont[(✉)], and Sanjit A. Seshia[(✉)]

University of California, Berkeley, USA
{nathan.mull,dfremont,sseshia}@berkeley.edu

**Abstract.** Recent attempts to explain the effectiveness of Boolean satisfiability (SAT) solvers based on conflict-driven clause learning (CDCL) on large industrial benchmarks have focused on the concept of community structure. Specifically, industrial benchmarks have been empirically found to have good community structure, and experiments seem to show a correlation between such structure and the efficiency of CDCL. However, in this paper we establish hardness results suggesting that community structure is not sufficient to explain the success of CDCL in practice. First, we formally characterize a property shared by a wide class of metrics capturing community structure, including "modularity". Next, we show that the SAT instances with good community structure according to any metric with this property are still NP-hard. Finally, we also prove that with high probability, random unsatisfiable modular instances generated from the "pseudo-industrial" *community attachment* model of Giráldez-Cru and Levy have exponentially long resolution proofs. Such instances are therefore hard for CDCL on average, indicating that actual industrial instances easily solved by CDCL may have some other relevant structure not captured by this model.

## 1  Introduction

Over the last 20 years Boolean satisfiability (SAT) solvers have become widely used tools for solving problems in many domains [1,2]. This is largely the result of the *conflict-driven clause learning* (CDCL) paradigm, introduced in the mid-1990s [3–5] and much developed since then. This success of SAT solving in practice is perhaps surprising in light of the NP-hardness of SAT, which is widely interpreted to mean that the problem admits no efficient algorithms. This has led to a line of research trying to answer the basic question: *why does CDCL perform so well in practice?* In other words, what is it about industrial SAT instances that allows them to seemingly avoid the worst-case behavior of CDCL?

One possible explanation is that SAT is significantly easier *on average* than in the worst case. For algorithms like CDCL that are based on resolution (which we will discuss in more detail below), this was ruled out by the discovery that random instances require exponentially-long resolution proofs [6]. Of course, industrial instances are generally highly *non-random*, so another possibility is that such instances tend to fall into a tractable class of problems. For example,

© Springer International Publishing Switzerland 2016
N. Creignou and D. Le Berre (Eds.): SAT 2016, LNCS 9710, pp. 141–159, 2016.
DOI: 10.1007/978-3-319-40970-2_10

SAT is known to be fixed-parameter tractable with respect to various natural parameters such as treewidth and clique-width [7]. However, it is unclear whether these parameters are always small in practice. Moreover, if the goal is to analyze the success of CDCL, the existence of *different* algorithms that take advantage of small (say) treewidth is not relevant: what matters is whether it correlates with CDCL performance, and in fact there is evidence against this [8].

Parameters more relevant to CDCL are the sizes of *backdoors* [9] and *backbones* [10]. In essence, a backdoor is a set of variables which if assigned cause the instance to become solvable by simplification with no further search, while the backbone is the set of variables which can only be assigned one way in every satisfying assignment. Correlations between the sizes of backdoors and backbones and the performance of CDCL have been observed empirically, and some "structured" instances do seem to have small backdoors [11, 12]. Unfortunately, these experiments have all been limited by the computational difficulty of estimating backdoor sizes, and it is unclear whether they are representative of a majority of large industrial benchmarks.

None of these ideas have adequately covered the whole variety of industrial instances, leaving a significant gap between our theoretical understanding of when SAT is easy and the reality of CDCL's effectiveness in practice. Of course, despite the intuition that industrial instances have some common underlying structure that explains why CDCL is so effective on them, it is probable that no single explanation suffices. There are particular types of industrial instances that are easy for a specific, known reason that does not apply to all other industrial instances [13]. However, it is still worthwhile to seek general explanations covering as many different types of instances as possible.

One recent approach has focused on the concept of *community structure*, as measured by *modularity* [14]. The variables of an instance with "good community structure" (high modularity) can be partitioned into relatively small sets such that few clauses span multiple sets. It has been found that industrial instances exhibit significantly better community structure than random instances [15], and that community structure does empirically correlate with CDCL performance [16, 17]. This makes community structure a plausible candidate for the "hidden structure" underlying the effectiveness of CDCL on industrial benchmarks. In fact, community structure has been used as the basis for a model of random "pseudo-industrial" instances, the *community attachment* model [17], which is designed to reflect the properties of industrial benchmarks.

However, as yet there has been little theoretical analysis connecting community structure to CDCL performance. The only relevant work we are aware of is that of Ganian and Szeider [18], who observe that SAT remains NP-hard for highly modular instances. They also give a tractability result for a parameter "h-modularity" inspired by community structure. However, this parameter is significantly different from the usual modularity, there is no evidence that it is small in industrial benchmarks, and the tractability result is via an algorithm completely different from CDCL.

In this paper, we extend the connection between community structure and worst-case complexity, and establish the first theoretical result on the average-case performance of CDCL on modular instances. Specifically, we:

- Define the *polynomial clique metrics* (PCMs), a broad class of graph metrics that includes modularity and other popular measures of graph clustering (Sect. 3.1).
- Show that the set of SAT instances which have "good community structure" according to any PCM is still NP-hard (Sect. 3.2).
- Prove that on random unsatisfiable instances from the community attachment model with fewer than $\Theta(n^{1/10})$ communities, CDCL takes exponential time with high probability (Sect. 4).

Based on these results, we suggest that community structure by itself may not be an adequate explanation for the effectiveness of CDCL in practice. We begin in Sect. 2 with background on SAT, CDCL, and community structure both generally and as recently applied to SAT, and conclude in Sect. 5 with a discussion of our results and some directions for future work.

## 2 Background

### 2.1 SAT

The *Boolean satisfiability* or SAT problem is to decide, given a Boolean formula $\varphi(\boldsymbol{x})$ over a vector of variables $\boldsymbol{x}$, whether or not there is a *satisfying assignment* to $\boldsymbol{x}$ that makes the formula true. In this paper, we make the common assumption that the formula $\varphi$ is in *conjunctive normal form* (CNF): it is a conjunction $\psi_1 \wedge \cdots \wedge \psi_m$ of *clauses*, where each clause $\psi_i$ is a disjunction $\ell_{i1} \vee \cdots \vee \ell_{ik}$. Here each $\ell_{ij}$ is a *literal*: either a variable from $\boldsymbol{x}$ or the negation of such a variable. We also assume that every clause has the same length $k$. A formula $\varphi$ satisfying these conditions is called a $k$-SAT formula.

Given a partial assignment $\rho$ to some of the variables $\boldsymbol{x}$, the *restriction* $\varphi\lceil_\rho$ of $\varphi(\boldsymbol{x})$ to $\rho$ is the formula obtained from $\varphi$ by removing all clauses satisfied by $\rho$ and all literals falsified by $\rho$. We can apply a restriction to any list of clauses analogously. The *size* of the restriction is the number of variables assigned by $\rho$.

### 2.2 Resolution and CDCL

*Resolution* [19] is a fundamental proof system that underlies modern SAT solving algorithms. It consists of a single rule stating that from clauses $(v \vee \boldsymbol{w})$ and $(\neg v \vee \boldsymbol{u})$ that have occurrences of $v$ with opposite polarities, we may infer the clause $(\boldsymbol{w} \vee \boldsymbol{u})$. As we will see in a moment, the importance of resolution for our purposes is that in order to establish that a formula $\varphi$ is unsatisfiable, SAT solvers based on CDCL implicitly construct a *resolution refutation* of $\varphi$: a derivation of a contradiction (the empty clause) from $\varphi$ using the resolution rule. This effectively means that the runtime of such a solver cannot be shorter than the length of the shortest such refutation.

To make this precise we need to define what we mean by CDCL. *Conflict-driven clause learning* [5] describes a class of algorithms that extend the Davis–Putnam–Logemann–Loveland (DPLL) algorithm [20]. DPLL is a classical search algorithm that assigns each variable in turn, backtracking if a clause is falsified by the current assignment. If at any point there is a clause with only a single unassigned variable, then that variable can immediately be given the assignment which satisfies the clause — a rule called *unit propagation*. If we eventually assign every variable, then we have found a satisfying assignment; otherwise, the search will backtrack all the way to the top level, every possible assignment will have been tried, and the formula is unsatisfiable.

CDCL-type algorithms augment this procedure by *learning* at every backtrack point a new clause that summarizes the reason why the current partial assignment falsifies the formula [3,4]. This *conflict clause* $C$ is derived by resolving the falsified clause $F$ with one or more other clauses that were used to assign variables in $F$ by unit propagation. As a result $C$ is always derivable from the original formula $\varphi$ by resolution, and writing out all clauses learned by CDCL when $\varphi$ is unsatisfiable gives a resolution refutation of $\varphi$ [21]. So the shortest such refutation gives a lower bound for the runtime of CDCL. This is true regardless of the heuristics used by the particular CDCL variant to decide which variable to assign and its polarity, how exactly to derive the conflict clause, and when to restart search from the beginning (see [21] for a more precise statement). We also note that pre- or inprocessing techniques that add no clauses (e.g. blocked clause elimination [22]) or add only clauses derived via resolution (e.g. variable elimination [23]) will not affect the lower bound.

### 2.3   Random SAT Instances

To study the performance of CDCL on "typical" instances, we use the framework of *average-case complexity*, which analyzes the efficiency of algorithms on *random* instances drawn from a particular distribution. We will be interested in complexity lower bounds that hold for almost all sufficiently large instances:

**Definition 1.** *An event $X$ occurs **with high probability** in terms of $n$ if* $\Pr[X] \to 1$ *as* $n \to \infty$.

For example, if flipping $n$ fair coins, with high probability at least 49 % will be heads.

Perhaps the simplest distribution over SAT instances arises from fixing the numbers of variables, clauses, and variables per clause, and then sampling uniformly:

**Definition 2.** $F_k(n, m)$ *is the uniform distribution over $k$-CNF formulas with $n$ variables and $m$ clauses.*

This *random $k$-SAT model* has been widely studied, and is known to be difficult on average for CDCL (for clause-variable ratios in a certain range) by the resolution lower bound discussed above: with high probability, a random unsatisfiable instance has only exponentially long resolution refutations [24]. As we will discuss shortly, our work extends this result to a more recent random SAT model that favors instances that are "pseudo-industrial" in the sense of having good community structure.

## 2.4    Community Structure

The notion of community structure has a long history in many fields [25]. The essential idea is that graphs with "good community structure" can be broken into relatively small pieces, *communities*, that are densely connected internally but only sparsely connected to each other. There are a number of metrics which have been proposed to make this notion formal, of which one of the most popular is *modularity* [14]. We consider unweighted graphs as weighted graphs with all weights 1.

**Definition 3.** *Let $G = (V, E)$ and let $\delta = \{C_1, \ldots, C_n\}$ be a vertex partition. Let $\deg v$ be the degree of $v$, and $w(x, y)$ be the weight of the edge $(x, y)$ or zero if there is no such edge. The **modularity** (or Q-value) of $G$ is*

$$Q = \max_{\delta} \sum_{C \in \delta} \left[ \frac{\sum_{x,y \in C} w(x, y)}{\sum_{x,y \in V} w(x, y)} - \left( \frac{\sum_{x \in C} \deg x}{\sum_{x \in V} \deg x} \right)^2 \right].$$

While work on community structure in SAT instances has focused on modularity, there are several competing metrics that have been used to measure community structure in other domains. In Appendix A of this paper,[1] we consider four: silhouette index, conductance, coverage, and performance [26].

Finally, we introduce notation for two graphs that will be useful in this paper: $K_n$, the complete graph on $n$ vertices, and $K_n^m$, consisting of $m$ disjoint copies of $K_n$.

## 2.5    SAT and Community Structure

Recent work on the community structure of SAT instances begins by associating to each instance its *variable incidence graph* (also known as the *primal graph*).

**Definition 4.** *Let $\varphi$ be a CNF formula. The **variable incidence graph (VIG)** of $\varphi$ is the graph $G_\varphi = (V, E)$ where $V$ is the set of all variables occurring in $\varphi$ and $E$ is the set $\{(v_1, v_2) : v_1, v_2 \in V$ and they appear together in some clause of $\varphi\}$.*

Some works use a *weighted* version of this graph with $w(v_1, v_2) = \sum_{cl} \left[ 1 / \binom{|cl|}{2} \right]$, where the sum is over all clauses in which both $v_1$ and $v_2$ appear [15,17]. This ensures that each clause contributes an equal amount to the total weight of

---

[1] Available in the full version [27].

the graph regardless of its length. Our results apply to both the weighted and unweighted versions.

Obviously, the graph $G_\varphi$ does not preserve all information about the instance $\varphi$. In particular, the polarities of the literals are ignored. But the graph does capture significant structural information: for example, if the graph has two connected components on variables $x$ and $y$ then the formula $\varphi(x, y)$ can be split into $\psi(x) \wedge \chi(y)$ and each subformula solved independently. In practice a perfect decomposition is rare, but one into *almost* independent parts is more plausible. This is exactly the idea of community structure, and leads us naturally to consider applying modularity to SAT instances.

**Definition 5.** *The **modularity** of a formula $\varphi$ is the modularity of $G_\varphi$.*

As was mentioned earlier, it has been found empirically that modularity correlates with CDCL performance [16]. This is a claim about the *average* behavior of CDCL over a wide variety of industrial benchmarks, not about its behavior on any specific instance. Thus it is naturally formalized in the average-case complexity framework discussed above, by giving a distribution that favors instances that are "industrial" in character. One such proposal, based on the idea that the key commonality of industrial instances is their good community structure, is the *community attachment* model of Giráldez-Cru and Levy [17]. In addition to the numbers of variables and clauses, this model has parameters controlling the number of communities and the (expected) fraction of clauses that lie within a single community instead of spanning multiple communities.

**Definition 6.** *Let $N$ be a set of $n$ variables. A **partition of $N$ into $c$ communities** is a partition $S = \{S_1, \ldots, S_c\}$ of $N$ such that $|S_i| = n/c$. A clause is **within a community** if it contains only variables from a single $S_i$. A **bridge clause** is a clause whose variables are all in different communities.*

**Definition 7 [17]    (Community Attachment Model).** *Let $n, m, c, k \in \mathbb{N}$ and $p \in [0, 1]$ such that $c$ divides $n$ and $2 \le k \le c \le n/k$. Then $F_k(n, m, c, p)$ is the distribution over $k$-CNF formulas with $n$ variables and $m$ clauses given by the following procedure: first, choose a random partition of $n$ variables into $c$ communities. With probability $p$, choose a clause uniformly among clauses within a community, and otherwise choose uniformly among bridge clauses. Generate $m$ clauses independently in this way.*

*Remark 1.* We define bridge clauses in a way that matches the community attachment model, but as we will discuss below our results also hold for a modified model where a bridge clause is any clause not within a single community.

Like the random $k$-SAT model $F_k(n, m)$, the model $F_k(n, m, c, p)$ ranges over $k$-CNF formulas with $n$ variables and $m$ clauses, and each clause is chosen independently of the others. However, in this model the clauses are of two different types: those lying entirely within a community, and those spread across $k$ different communities. The probability $p$ controls how likely a clause is to be of the first type versus the second.

The idea behind this model is that by picking $c$ and $p$ appropriately, one is likely to obtain instances that decompose into loosely-connected communities, as has been observed in actual industrial instances. More precisely, the expected modularity of an instance drawn from $F_k(n, m, c, p)$ is lower bounded by $p-(1/c)$, so that for nontrivial $c$ highly modular instances can be generated by setting $p$ large enough [17]. Furthermore, Giráldez-Cru and Levy find experimentally that high-modularity instances generated with this model are solved more quickly by CDCL than by look-ahead solvers, and the reverse is true for low-modularity instances [17]. This parallels the same observation for industrial instances versus random instances. Thus, they conclude, $F_k(n, m, c, p)$ is a more realistic model of industrial instances than the random $k$-SAT model $F_k(n, m)$.

## 3    Worst-Case Hardness

In this section, we propose a simple class of graph metrics that we argue should include most metrics quantifying community structure. We show that modularity is in fact within the class, as are several other popular graph clustering metrics. However, we demonstrate that the set of SAT instances that have "good community structure" according to any metric in the class is NP-hard. Therefore, no such metric can be a guaranteed indicator of the difficulty of a SAT instance.

### 3.1    A Class of "Modularity-Like" Graph Metrics

We begin by formalizing what we mean by a graph metric.

**Definition 8.** *A **graph metric** is a function $m$ from weighted graphs to $[0, 1]$. Given $m$ and any $\epsilon \in [0, 1]$, $\mathsf{SAT}_{m,\epsilon}$ is the class of all SAT instances $\varphi$ such that $m(G_\varphi) \geq 1 - \epsilon$.*

For example, if $m$ is modularity then $\mathsf{SAT}_{m,\epsilon}$ consists of the "high modularity" formulas, where "high" means any modularity above $1 - \epsilon$.

In general we are interested in graph metrics that represent a notion of community structure, assigning larger values to graphs which have such a structure than those that do not. For such a metric $m$, consider the following property:

**Definition 9.** *A graph metric $m$ is a **polynomial clique metric (PCM)** if for all $\epsilon > 0$, there is a poly-time computable function $c : \mathbb{N} \to \mathbb{N}$ with at most polynomial growth and some $n_0 \in \mathbb{N}$ such that for all $n \geq n_0$, if $K$ is $K_n$ with any positive edge weights then $m(K^{c(n)}) \geq 1 - \epsilon$.*

*Remark 2.* If using the unweighted version of the variable incidence graph, our proofs will work using a relaxed definition that applies only to $K = K_n$ with unit weights.

In essence, the definition states that for any (sufficiently large) size $n$, at most a polynomial number of copies of $K_n$ are needed to produce a graph that $m$ considers to have "good community structure". This is a natural property for modularity-like metrics to have, since copies of $K_n$ are in some sense ideal communities: internally connected as much as possible, with no external edges. Of course we would not consider a single copy of $K_n$ to have good community structure, so the definition of a PCM only requires that such structure be obtained for *some* number of copies at most polynomial in $n$.

Next we demonstrate that the PCMs are a large class including modularity and several other popular clustering metrics. While the other metrics have not been experimentally evaluated in the context of SAT, this still supports our claim that the PCM property is a natural one for metrics of community structure to have. For lack of space, we defer the definitions and analysis of the metrics other than modularity to Appendix A [27].

**Theorem 1.** *Modularity is a PCM.*

*Proof.* Fix any $\epsilon > 0$ and $n \geq 2$. Let $K$ be $K_n$ with arbitrary positive edge weights, and let $G = K^c$. Let $\delta$ be the vertex partition that groups two vertices iff they are in the same copy of $K$. Then since each community is identical, and there are $c$ communities, $\sum_{x,y \in C} w(x,y) / \sum_{x,y \in V} w(x,y) = 1/c$ and $\sum_{x \in C} \deg x / \sum_{x \in V} \deg x = 1/c$ for any $C \in \delta$. Therefore, $Q(G) \geq c(1/c - (1/c)^2) = 1 - 1/c$. Putting $c = 1/\epsilon$, we have $Q(G) \geq 1 - \epsilon$. Since $c$ is $O(1)$ with respect to $n$, $Q$ is a PCM. $\quad\square$

**Theorem 2.** *Silhouette index, conductance, coverage, and performance are PCMs.*

## 3.2    Hardness of PCM-Modular Instances

Now we show that the SAT instances which have "good community structure" according to a PCM are no easier in the worst case than any other instance. The PCMs thus form a wide class of metrics which cannot be used as a guaranteed indicator of the difficulty of a SAT instance. Our reduction can be viewed as a variation of that suggested by Ganian and Szeider [18] to show NP-hardness in the specific case of modularity.

**Theorem 3.** *For any PCM $m$, the class $\mathsf{SAT}_{m,\epsilon}$ is NP-hard for all $\epsilon > 0$.*

*Proof.* Given a SAT instance $\phi$, we will convert it into an equisatisfiable instance of $\mathsf{SAT}_{m,\epsilon}$ in polynomial time. Let $V$ be the set of all variables occurring in $\phi$, along with new variables as necessary so that $|V| \geq n_0$. Fixing a variable $x$ not in $V$, let $\psi$ be the formula obtained by adding to $\phi$ all clauses of the form $x \vee y \vee z$ with $y, z \in V$. Clearly, the VIG of $\psi$ is $K_n$ with $n = |V| + 1 \geq n_0$. Furthermore, $\phi$ and $\psi$ are equisatisfiable, since we can simply assert $x$ to satisfy all the new clauses. Now letting $\chi$ be the conjunction of $c(n)$ disjoint copies of $\psi$ (i.e. copies with variables renamed so none are common), the variable incidence graph $G$ of $\chi$ is $K_n^{c(n)}$ (with some positive weights). By the PCM property, we have $m(G) \geq 1 - \epsilon$, so $\chi \in \mathsf{SAT}_{m,\epsilon}$. Since $\chi$ and $\psi$ are clearly equisatisfiable, so are $\chi$ and $\phi$, and thus this procedure gives a reduction from SAT to $\mathsf{SAT}_{m,\epsilon}$. Finally, the procedure is polynomial-time since $c(n)$ has at most polynomial growth and can be computed in polynomial time.      □

## 4   Average-Case Hardness

In contrast to the previous section, we now consider the difficulty of modular instances for a particular class of algorithms, namely those like CDCL which prove unsatisfiability by effectively constructing a resolution refutation. While these results are therefore more specific, they are also much more powerful: they show that modular instances are difficult not just in the worst case but also on average.

Our argument is largely based on the resolution lower bound of Beame and Pitassi [24], which can be used to establish the hardness of instances from the random $k$-SAT model. In order to use that result, we need to show that most instances from the community attachment model have certain *sparsity* properties used by the proof. So our main steps, detailed in Sects. 4.1–4.5 below, are as follows:

1. Define a new distribution $\overline{F}_k(n, m, c, p; m')$ over $k$-CNF formulas that works by taking a *random subformula* of an instance from the random $k$-SAT model $F_k(n, m')$.
2. Show that this new distribution is in fact identical to the community attachment model $F_k(n, m, c, p)$.
3. Observe that the sparsity properties are inherited by subformulas, so the sparsity result in [24] for the random $k$-SAT model $F_k(n, m')$ transfers to the community attachment model $F_k(n, m, c, p)$.
4. Adapt the Beame–Pitassi argument [24] to obtain an exponential lower bound on the resolution refutation length.
5. Conclude that CDCL takes exponential time on unsatisfiable formulas from the community attachment model $F_k(n, m, c, p)$ with high probability.

### 4.1   Defining the New Distribution

We begin by defining our new distribution $\overline{F}_k(n, m, c, p; m')$, which takes an additional parameter $m'$ that we will specify in Sect. 4.3.

**Definition 10.** *Let $n, m, c, k, m' \in \mathbb{N}$ and $p \in [0, 1]$ such that $2 \leq k \leq c \leq n/k$. Then $\overline{F}_k(n, m, c, p; m')$ is the distribution over $k$-CNF formulas with $n$ variables and $m$ clauses defined by Algorithm 1 (which is such a distribution by Lemma 1 below).*

---

**Algorithm 1.** defining the distribution $\overline{F}_k(n, m, c, p; m')$

---
1: choose $\phi$ from $F_k(n, m')$
2: choose a uniformly random partition of the $n$ variables into $c$ communities
3: $h \leftarrow c\binom{n/c}{k}/\binom{n}{k}$
4: $b \leftarrow (n/c)^k \binom{c}{k}/\binom{n}{k}$
5: $\psi \leftarrow$ the empty formula on $n$ variables
6: **for all** clauses $C$ of $\phi$ **do**
7:     **with probability** $p$ **do**
8:         **if** $C$ is within a community **then**
9:             add $C$ to $\psi$
10:    **otherwise do**
11:        **with probability** $h/b$ **do**
12:            **if** $C$ is a bridge clause **then**
13:                add $C$ to $\psi$
14:    **if** $|\psi| = m$ **then return** $\psi$                     ▷ the algorithm "succeeds"
15: choose a fresh $\psi$ from $F_k(n, m, c, p)$
16: **return** $\psi$                                              ▷ the algorithm "fails"

---

**Lemma 1.** *For all parameters satisfying the conditions of Definition 10, Algorithm 1 defines a probability distribution over $k$-CNF formulas with $n$ variables and $m$ clauses.*

*Proof.* First we must check that $h/b \leq 1$ so that the algorithm is well-defined. We have

$$\frac{h}{b} = \frac{c\binom{n/c}{k}}{\left(\frac{n}{c}\right)^k \binom{c}{k}} \leq \frac{c\left(\frac{n}{c}\right)^k}{k! \left(\frac{n}{c}\right)^k \left(\frac{c}{k}\right)^k} = \frac{k^k}{k! \, c^{k-1}} \leq \frac{1}{(k-1)!} \leq 1,$$

since we assume $c \geq k$. Algorithm 1 always terminates, returning a formula $\psi$ from either line 14 or 16. In the first case, $\psi$ is a subset of $\phi$, which is drawn from $F_k(n, m')$ and so has $k$-CNF clauses over $n$ variables. Furthermore, the algorithm does not return from line 14 unless $\psi$ has $m$ clauses. In the second case, $\psi$ is drawn from $F_k(n, m, c, p)$, and so again is a $k$-CNF formula with $n$ variables and $m$ clauses.                                                          □

## 4.2  Comparing the Distribution to the Community Attachment Model

Next we prove that our definition via Algorithm 1 is equivalent to the usual community attachment definition. Since the algorithm adds each clause independently, in essence this amounts to showing that each clause is within a community with probability $p$.

**Lemma 2.** *For any $m' \in \mathbb{N}$, the distribution $\overline{F}_k(n, m, c, p; m')$ is identical to the distribution $F_k(n, m, c, p)$.*

*Proof.* When Algorithm 1 returns a formula $\psi$ from line 16, $\psi$ is drawn from $F_k(n, m, c, p)$, and so the two distributions are trivially identical. So we need only consider the case when the algorithm returns from line 14. Because the algorithm handles each clause of $\phi$ independently (until $m$ clauses are added), it suffices to show that when a clause is added to $\psi$, it is within a community with probability $p$ and is otherwise a bridge clause. Starting from line 7 of Algorithm 1, let $C_{\text{comm}}$ be the event that the clause $C$ is within a community, $C_{\text{bridge}}$ the event that $C$ is a bridge clause, and $C_{\text{added}}$ the event that $C$ is added to $\psi$. Let $A$ be the event that the algorithm takes the random branch on line 7 instead of the branch on line 10. Then we have

$$\Pr[C_{\text{comm}}|C_{\text{added}}] = \Pr[C_{\text{comm}}|A, C_{\text{added}}]\Pr[A|C_{\text{added}}]$$
$$+ \Pr[C_{\text{comm}}|\overline{A}, C_{\text{added}}]\Pr[\overline{A}|C_{\text{added}}].$$

The second term is zero because $\overline{A}$ means the algorithm takes the branch on line 10 and thus only adds the clause if it is a bridge clause. Likewise, $\Pr[C_{\text{comm}}|A, C_{\text{added}}] = 1$ because the branch on line 7 only adds the clause if it is within a community. So

$$\Pr[C_{\text{comm}}|C_{\text{added}}] = \Pr[A|C_{\text{added}}] = \Pr[C_{\text{added}}|A]\Pr[A] \ / \ \Pr[C_{\text{added}}].$$

By straightforward counting arguments, $\Pr[C_{\text{comm}}] = c\binom{n/c}{k}/\binom{n}{k} = h$ and $\Pr[C_{\text{bridge}}] = (n/c)^k\binom{c}{k}/\binom{n}{k} = b$. Since the coin flips on lines 7 and 11 are independent of $C$, we have $\Pr[C_{\text{added}}|A] = \Pr[C_{\text{comm}}] = h$ and $\Pr[C_{\text{added}}|\overline{A}] = (h/b)\Pr[C_{\text{bridge}}] = h$. Also $\Pr[A] = p$, so

$$\Pr[C_{\text{added}}] = \Pr[C_{\text{added}}|A]\Pr[A] + \Pr[C_{\text{added}}|\overline{A}]\Pr[\overline{A}] = hp + h(1 - p) = h.$$

Plugging these into the expression above we obtain $\Pr[C_{\text{comm}}|C_{\text{added}}] = p$. So each clause added to $\psi$ is within a community with probability $p$, and otherwise by construction it must be a bridge clause. Therefore when Algorithm 1 returns from line 14, it is equivalent to generating $m$ clauses independently, each of which is a uniformly random clause within a community with probability $p$, and otherwise a uniformly random bridge clause. So $\overline{F}_k(n, m, c, p; m')$ is identical to $F_k(n, m, c, p)$. □

## 4.3    Transferring Subformula-Inherited Properties

Algorithm 1 can "fail" by adding fewer than the desired number of clauses $m$ to $\psi$, then falling back on the community attachment model as a backup. Otherwise, the algorithm "succeeds", returning on line 14 a formula that was built up from clauses of $\phi$ and is therefore a subformula of it. Since our goal is to have the formulas from this distribution inherit properties from $\phi$, we need to ensure that Algorithm 1 succeeds with high probability. We can do this by taking $m'$, the number of clauses in $\phi$, to be large enough: then even if a given clause is only added to $\psi$ with a small probability, overall we are likely to add $m$ of them. As we will see in the proof, the probability of adding a clause is roughly $1/c^{k-1}$, so taking $m'$ to be slightly larger than $c^{k-1}m$ will suffice. We use the following standard tail bound.

**Lemma 3.** *If $B(n, p)$ is the number of successes in $n$ Bernoulli trials each with success probability $p$, then for $k < pn$ we have*

$$\Pr[B(n, p) \leq k] \leq \exp\left(\frac{-(pn - k)^2}{2pn}\right).$$

**Lemma 4.** *Suppose that $c$ is $o(n)$, $m \to \infty$ as $n \to \infty$, and $m' = (1 + \epsilon)c^{k-1}m$ for some $\epsilon > 0$. Then Algorithm 1 returns from line 14 with high probability.*

*Proof.* As shown in Lemma 2, the probability that starting from line 7 the clause $C$ will be added to $\psi$ is $h$. So the probability that Algorithm 1 returns from line 14 is $\Pr[B(m', h) \geq m] = 1 - \Pr[B(m', h) \leq m - 1]$. Now observe that

$$hc^{k-1} = \frac{c^k \binom{n/c}{k}}{\binom{n}{k}} = \frac{n(n - c) \cdots (n - c(k + 1))}{n(n - 1) \cdots (n - k + 1)} \leq 1.$$

Furthermore, we have

$$\lim_{n \to \infty} hc^{k-1} = \lim_{n \to \infty} \frac{c^k \binom{n/c}{k}}{\binom{n}{k}} = \lim_{n \to \infty} \left[\frac{\binom{n/c}{k}}{\frac{(n/c)^k}{k!}} \cdot \frac{\frac{n^k}{k!}}{\binom{n}{k}}\right]$$

$$= \left[\lim_{n \to \infty} \frac{\binom{n/c}{k}}{\frac{(n/c)^k}{k!}}\right]\left[\lim_{n \to \infty} \frac{\frac{n^k}{k!}}{\binom{n}{k}}\right] = 1,$$

where in evaluating the second-to-last limit we use the fact that $c$ is $o(n)$ and so $\lim_{n \to \infty}(n/c) = \infty$. So for sufficiently large $n$ we have $hc^{k-1} \geq 1 - \epsilon/2(1 + \epsilon)$, and therefore

$$hm' = h(1 + \epsilon)c^{k-1}m \geq \left(1 - \frac{\epsilon}{2(1 + \epsilon)}\right)(1 + \epsilon)m = (1 + \epsilon/2)m.$$

Applying Lemma 3, we have

$$\Pr[B(m', h) \leq m - 1] \leq \exp\left(\frac{-[hm' - (m - 1)]^2}{2hm'}\right) \leq \exp\left(\frac{-[(1 + \epsilon/2)m - m]^2}{2h(1 + \epsilon)c^{k-1}m}\right)$$

$$= \exp\left(\frac{-m(\epsilon/2)^2}{2(1 + \epsilon) \cdot hc^{k-1}}\right) \leq \exp!\left(\frac{-m\epsilon^2}{8(1 + \epsilon)}\right),$$

which goes to zero as $m \to \infty$, and therefore as $n \to \infty$. So with high probability, Algorithm 1 will return from line 14.                                                               □

Now it is simple to show that subformula-inherited properties are indeed passed down from random $k$-SAT instances to instances drawn from our distribution. Here "subformula-inherited" simply means that if $\varphi$ has the property, then any formula made up of a subset of the clauses of $\varphi$ also has the property. For example, being satisfiable is subformula-inherited, but being unsatisfiable is not.

**Lemma 5.** *Suppose that $c$ is $o(n)$, $m \to \infty$ as $n \to \infty$, $m' = (1 + \epsilon)c^{k-1}m$ for some $\epsilon > 0$, and $P$ is a subformula-inherited property. Then if a formula drawn from $F_k(n, m')$ has property $P$ with high probability, a formula drawn from $\overline{F}_k(n, m, c, p; m')$ has property $P$ with high probability.*

*Proof.* Run Algorithm 1 to sample from $\overline{F}_k(n, m, c, p; m')$. Let $P_\psi$ and $P_\phi$ respectively be the events that the returned formula $\psi$ and the formula $\phi$ from line 1 have property $P$. Also let $R$ be the event that the algorithm returns from line 14. When the algorithm returns from line 14, $\psi$ is a subformula of $\phi$, and since $P$ is inherited by subformulas we have $\Pr[P_\psi|R] \geq \Pr[P_\phi]$. Now as $\phi$ is drawn from $F_k(n, m')$, the event $P_\phi$ occurs with high probability, and so $\Pr[P_\psi|R] \to 1$ as $n \to \infty$. By Lemma 4, the event $R$ also happens with high probability, so $\Pr[P_\psi] \geq \Pr[P_\psi \wedge R] = \Pr[P_\psi|R] \cdot \Pr[R] \to 1$ as $n \to \infty$. Therefore $\psi$ has property $P$ with high probability. $\qquad\square$

Together, Lemmas 2 and 5 show that subformula-inherited properties of random $k$-SAT instances are also possessed (with high probability) by instances from the community attachment model.

## 4.4   Proving the Resolution Lower Bounds

Now we transition to adapting the argument of Beame and Pitassi [24]. The proof uses two types of sparsity conditions. Both view a clause $C$ as a set of variables, so that another set of variables $X$ "contains" $C$ if and only if every variable in $C$ is in $X$.

**Definition 11.** *A formula is $n'$-**sparse** if every set of $s \leq n'$ variables contains at most $s$ clauses.*

**Definition 12.** *Let $n' < n''$. A formula is $(n', n'', y)$-**sparse** if every set of $s$ variables with $n' < s \leq n''$ contains at most $ys$ clauses.*

These are both clearly subformula-inherited.

The Beame–Pitassi argument [24] is broken into three major lemmas, each of which we will use without change. The last lemma establishes the sparsity properties above for the random $k$-SAT model.

**Lemma 6 [24].** *Let $n' \leq n$ and $F$ be an unsatisfiable CNF formula in $n$ variables with clauses of size at most $k$ that is both $n'$-sparse and $(n'(k+\epsilon)/4, n'(k+\epsilon)/2, 2/(k+\epsilon))$-sparse. Then any resolution proof $P$ of the unsatisfiability of $F$ must include a clause of length at least $\epsilon n'/2$.*

**Lemma 7 [24].** *Let $P$ be a resolution refutation of $F$ of size $S$. Given $\beta > 0$, say the **large clauses** of $P$ are those clauses mentioning more than $\beta n$ distinct variables. Then with probability at least $1 - 2^{1-\beta t/4}S$, a random restriction of size $t$ sets all large clauses in $P$ to 1.*

**Lemma 8 [24].** *Let* $x > 0$, $1 \geq y > 1/(k-1)$, *and* $z \geq 4$. *Fix a restriction* $\rho$ *on* $t \leq \min\{xn/2, x^{1-1/y(k-1)}n^{1-1/(k-1)}/z\}$ *variables. Drawing* $F$ *from* $F_k(n, m)$ *with*

$$m \leq \frac{y}{e^{1+1/y}2^{k+1/y}}x^{1/y-(k-1)}n,$$

*then with probability at least* $1 - 2^{-t} - (2^k + 1)/z^{k-1}$, $F\lceil_\rho$ *is both* $(xn/2, xn, y)$- *sparse and* $xn$-*sparse.*

We can now combine these to prove the analog of the main theorem of Beame and Pitassi for modular instances. Our argument is almost identical to theirs: the only difference is that we apply Lemma 8 to larger instances from the random $k$-SAT model,[2] so that our results above will give us sparsity for modular instances of the correct size embedded in them as subformulas.

**Theorem 4.** *Let* $k \geq 3$, $0 < \epsilon < 1$, *and* $x$, $t$, $z$, $c$ *be functions of* $n$ *such that* $x > 0$, $t$ *and* $z$ *are* $\omega(1)$, $c$ *is* $o(n)$, *and* $t$ *satisfies the conditions of Lemma 8 for all sufficiently large* $n$. *Then with high probability, an unsatisfiable formula drawn from* $F_k(n, m, c, p)$ *with*

$$m \leq \frac{1}{2^{7k/2}(1+\epsilon)c^{k-1}}x^{-(k-2-\epsilon)/2}n$$

*does not have a resolution refutation of size* $\leq 2^{\frac{\epsilon}{4(k+\epsilon)}xt}/8$.

*Proof.* Let $S = 2^{\frac{\epsilon}{4(k+\epsilon)}xt}/8$ and let $U$ be the set of unsatisfiable $k$-CNF formulas with $n$ variables and $m$ clauses. For each $\varphi \in U$ fix a shortest resolution refutation $P_\varphi$, and let $W \subseteq U$ be the set of $\varphi$ such that $|P_\varphi| \leq S$. Let $R$ be the set of all restrictions of size $t$, and for any formula $\varphi$ and $\rho \in R$ let $L(\varphi, \rho)$ be the indicator function for the event that either $\varphi$ is satisfiable or $P_\varphi\lceil_\rho$ contains a clause of length at least $\epsilon xn/(k + \epsilon)$. Now for any $\varphi \in W$, by Lemma 7 with $\beta = \epsilon x/(k + \epsilon)$ we have

$$\sum_\rho \frac{L(\varphi, \rho)}{|R|} \leq 2^{1-\frac{\epsilon}{4(k+\epsilon)}xt}S = 2^{1-\frac{\epsilon}{4(k+\epsilon)}xt}(2^{\frac{\epsilon}{4(k+\epsilon)}xt}/8) = 1/4.$$

Let $X$ be a random variable defined over a restriction $\rho$ and equal to $\Pr_\varphi[L(\varphi, \rho)|\varphi \in W]$, where $\varphi$ is distributed as $F_k(n, m, c, p)$. Putting a uniform distribution on $\rho$ and writing $q(\psi)$ for the conditional distribution $\Pr_\varphi[\varphi = \psi|\varphi \in W]$,

$$\mathbb{E}_\rho[X] = \sum_\rho \frac{1}{|R|}\Pr_\varphi[L(\varphi, \rho)|\varphi \in W] = \sum_{\psi \in W} q(\psi)\left[\sum_\rho \frac{L(\psi, \rho)}{|R|}\right] \leq \sum_{\psi \in W} q(\psi)\frac{1}{4} = \frac{1}{4}.$$

---

[2] Note that as required by its statement, we are applying Lemma 8 to formulas drawn from $F_k(n, m)$, *not* to formulas drawn from $\overline{F}_k(n, m, c, p; m')$. Lemmas 6 and 7 work for any formula, so we may use all three lemmas precisely as proved in [24].

So by Markov's inequality,

$$\Pr_{\rho}[X \geq 1/2] \leq \frac{\mathbb{E}_{\rho}[X]}{1/2} \leq 1/2,$$

and therefore there is some $\rho'$ such that $\Pr_{\varphi}[L(\varphi, \rho') | \varphi \in W] \leq 1/2$. In other words, there is a restriction that eliminates large clauses from a random $\varphi \in W$ with probability at least $1/2$.

Now let $y = 2/(k + \epsilon)$. Since $k \geq 3$ and $\epsilon < 1$ we have $y \geq 1/(k-1)$ and

$$\frac{y}{e^{1+1/y} 2^{k+1/y}} = 2\left[(k+\epsilon)e^{1+\frac{k+\epsilon}{2}} 2^{k+\frac{k+\epsilon}{2}}\right]^{-1} \geq 2(k+\epsilon)^{-1} e^{-\frac{k}{2}-\frac{3}{2}} 2^{-\frac{3k}{2}-\frac{1}{2}}$$

$$= 2(k+\epsilon)^{-1} e^{-3/2} 2^{-1/2} 2^{-k(3+\log_2 e)/2}$$

$$\geq 2(k+1)^{-1} e^{-3/2} 2^{-1/2} 2^{-2.23k} \geq 2^{-1.23k} 2^{-2.23k} \geq 2^{-7k/2}.$$

By our assumption on $m$,

$$(1+\epsilon)c^{k-1} m \leq 2^{-7k/2} x^{-(k-2-\epsilon)/2} n = 2^{-7k/2} x^{1/y - (k-1)} n$$

$$\leq \frac{y}{e^{1+1/y} 2^{k+1/y}} x^{1/y - (k-1)} n.$$

Finally, since $z$ is $\omega(1)$ we have $z \geq 4$ for sufficiently large $n$, and then all the conditions of Lemma 8 are satisfied by $y$, $z$, $t$, and $m' = (1+\epsilon)c^{k-1} m$. Therefore for a formula $\varphi$ drawn from $F_k(n, m')$, $\varphi\lceil_{\rho'}$ is simultaneously $(xn/2, xn, 2/(k+\epsilon))$-sparse and $xn$-sparse with probability at least $1 - 2^{-t} - (2^k + 1)/z^{k-1}$. Since $t$ and $z$ are $\omega(1)$, $\varphi$ has this property with high probability. Furthermore, the property is inherited by subformulas, so by Lemma 5 it also holds with high probability for formulas drawn from $\overline{F}_k(n, m, c, p; m')$. Then by Lemma 2 the same is true for formulas drawn from $F_k(n, m, c, p)$.

Now let $n' = 2xn/(k+\epsilon)$. Since $k + \epsilon \geq 3$, we have $n' \leq xn$ and so $xn$-sparsity implies $n'$-sparsity. Also note that

$$\frac{xn}{2} = \frac{2xn(k+\epsilon)}{4(k+\epsilon)} = \frac{n'(k+\epsilon)}{4} \text{ and } xn = \frac{n'(k+\epsilon)}{2}.$$

So by Lemma 6, when drawing an unsatisfiable formula $\varphi$ from $F_k(n, m, c, p)$, with high probability every resolution refutation of $\varphi\lceil_{\rho'}$ has a clause of length at least $\epsilon n'/2 = \epsilon xn/(k+\epsilon)$. That is, $\Pr_{\varphi}[L(\varphi, \rho') \mid \varphi \in U] \to 1$ as $n \to \infty$. So

$$\Pr_{\varphi}[\varphi \in W | \varphi \in U] = \frac{\Pr_{\varphi}[\varphi \in W \wedge \overline{L(\varphi, \rho')} \mid \varphi \in U]}{\Pr_{\varphi}[\overline{L(\varphi, \rho')} \mid \varphi \in W]} \leq \frac{\Pr_{\varphi}[\overline{L(\varphi, \rho')} \mid \varphi \in U]}{1/2} \to 0$$

as $n \to \infty$. Therefore with high probability, an unsatisfiable instance drawn from $F_k(n, m, c, p)$ does not have a resolution refutation of size $\leq S$. $\qquad\square$

Next we instantiate this general result to obtain exponential lower bounds for the refutation length when the number of communities is not too large. We use slightly different arguments for $k \geq 4$ and $k = 3$, again following Beame and Pitassi [24]. As the computations are uninteresting, we defer the proofs to Appendix B [27]. The basic idea is to let $x$ go to zero fast enough that the bound on $m$ required by Theorem 4 is satisfied when $m = O(n)$, but slowly enough that the length bound is of the form $2^{O(n^\lambda)}$.

**Theorem 5.** *Suppose that $k \geq 4$, $m = O(n)$, and $c = O(n^\alpha)$ for some $\alpha < \frac{k-2}{4(k-1)}$. Then there is some $\lambda > 0$ so that with high probability, an unsatisfiable formula drawn from $F_k(n, m, c, p)$ does not have a resolution refutation of size $2^{O(n^\lambda)}$.*

**Theorem 6.** *Suppose $m = O(n)$ and $c = O(n^\alpha)$ for some $\alpha < 1/10$. Then there is some $\lambda > 0$ so that with high probability, an unsatisfiable formula drawn from $F_3(n, m, c, p)$ does not have a resolution refutation of size $2^{O(n^\lambda)}$.*

### 4.5   Deducing a Lower Bound on CDCL Runtime

Finally, we can conclude that unsatisfiable random instances from $F_k(n, m, c, p)$ with sufficiently few communities usually take exponential time for CDCL to solve.

**Theorem 7.** *If $m = O(n)$ and $c = O(n^\alpha)$ for any $\alpha < 1/10$, the runtime of CDCL on an unsatisfiable formula $\varphi$ from $F_k(n, m, c, p)$ is exponential with high probability.*

*Proof.* If $\varphi$ is unsatisfiable, the runtime of CDCL on $\varphi$ is lower bounded (up to a polynomial factor) by the length of the shortest resolution refutation of $\varphi$ [21]. If $k = 3$, then the shortest refutation of $\varphi$ is exponentially long with high probability by Theorem 6. If instead $k \geq 4$, the same is true by Theorem 5, since $1/10 < \frac{k-2}{4(k-1)}$. Therefore with high probability, CDCL will take exponential time to prove $\varphi$ unsatisfiable.                                                             □

*Remark 3.* By picking a sufficiently high clause-variable ratio, we can ensure $\varphi$ is unsatisfiable with high probability, so that CDCL takes exponential time on average for formulas drawn from $F_k(n, m, c, p)$ (not just the unsatisfiable ones).

We also note that our proof technique is not sensitive to the details of how the community attachment model is defined. For example, changing the definition of a bridge clause so that the variables do not all have to be in different communities requires only minor changes to the proof (detailed in Appendix B) [27].

**Theorem 8.** *Let $\widetilde{F}_k(n, m, c, p)$ be the community attachment model modified so that any clause that is not within a single community counts as a bridge clause. Then if $m = O(n)$ and $c = O(n^\alpha)$ for any $\alpha < 1/10$, the runtime of CDCL on an unsatisfiable formula $\varphi$ from $\widetilde{F}_k(n, m, c, p)$ is exponential with high probability.*

Thus we have showed that similarly to unsatisfiable random $k$-SAT instances, unsatisfiable random *modular* instances (as formalized by the community attachment model) are hard on average for CDCL as long as they do not have too many communities.

# 5    Discussion

We have introduced a broad class of "modularity-like" graph metrics, the polynomial clique metrics, and showed that no PCM can be a guaranteed indicator of whether a SAT instance is easy (unless P = NP). This is perhaps not too surprising in light of the fact that the VIG throws away the Boolean information in the formula. While the VIG has received the most attention in recent work on community structure, it would be worthwhile to investigate other graph encodings that preserve more information. Regardless, our result does indicate that it may be difficult to define a tractable class of SAT instances based purely on modularity or its variants. Furthermore, by setting up a concrete barrier (the PCM property) that must be avoided to obtain a tractable class, our result can help guide future attempts to find a graph metric that does work.

Our result on the community attachment model $F_k(n, m, c, p)$ is more interesting, as it shows that instances from this model are exponentially hard for CDCL even on average (when $c$ is small enough). An important point is that the result is actually nontrivial when $p < 1$, unlike for $p = 1$. In the latter case there are no bridge clauses, so the instances consist of $c$ independent problems of size $n/c$, and since we assume $c = O(n^{1/10})$ each problem has size $\Omega(n^{9/10})$. So by the old results on random $k$-SAT CDCL would take exponential time to solve even the easiest problem, and so likewise for the original instance (with a slightly smaller exponent on $n$). On the other hand, when $p < 1$ it is conceivable that the bridge clauses could actually make the instances easier, by adding some extra propagation power or easier-to-find contradictions that would make the whole instance easier to solve than any individual community. Our result effectively says that this happens with vanishing probability as $n \to \infty$.

The case $p = 1$ also brings out an important caveat when interpreting our result as evidence that community structure doesn't explain CDCL's effectiveness on industrial instances. Our result shows that such structure isn't enough to bring random formulas down from exponential-time-on-average to polynomial-time-on-average. However, it could decrease the time from (say) $2^{n^{1/2}}$ to $2^{n^{1/4}}$, which could be the difference between intractability and tractability if $n$ is small enough. On the other hand, given the enormous size of many industrial instances it isn't clear whether this is really all that is happening. It would be interesting to do experiments on parametrized families of industrial instances to see whether CDCL actually avoids exponential behavior, or if the point of blow-up is just pushed out far enough that we tend not to encounter it in practice.

Another important aspect of our result is the limit on the number of communities. It does not apply when communities have logarithmic size, for example, so that $c = \Theta(n/\log n)$. In fact it is easy to see that the result cannot hold in this case: if one of the communities is unsatisfiable then it will have a polynomial-length resolution refutation, and as $c \to \infty$ the probability that at least one community is unsatisfiable by itself goes to 1. So with high probability the entire instance has a short refutation, and CDCL could in theory solve it in polynomial time. A clear direction for future work is to see whether improved proof techniques can extend our results to larger numbers of communities, closing the gap between $O(n^{1/10})$ and $\Omega(n/\log n)$. This is also another way our results can inform future experiments: it would be interesting to explore a variety of growth rates for $c$ above $n^{1/10}$ and see how the performance of CDCL changes. Our results indicate that high modularity alone may not be adequate to ensure good performance even on average, but that it could be rewarding to investigate more refined notions of "good community structure" that somehow restrict the number of communities.

**Acknowledgments.** The authors thank Vijay Ganesh, Holger Hoos, Zack Newsham, Markus Rabe, Stefan Szeider, and several anonymous reviewers for helpful discussions and comments. This work is supported in part by the National Science Foundation Graduate Research Fellowship Program under Grant No. DGE-1106400, by the Hellman Family Faculty Fund, by gifts from Microsoft and Toyota, and by TerraSwarm, one of six centers of STARnet, a Semiconductor Research Corporation program sponsored by MARCO and DARPA.

# References

1. Marques-Silva, J.: Practical applications of Boolean satisfiability. In: Proceedings of the 9th International Workshop on Discrete Event Systems, pp. 74–80 (2008)
2. Vizel, Y., Weissenbacher, G., Malik, S.: Boolean satisfiability solvers and their applications in model checking. Proc. IEEE **103**(11), 2021–2035 (2015)
3. Marques-Silva, J.P., Sakallah, K.A.: GRASP - a new search algorithm for satisfiability. In: ICCAD, pp. 220–227 (1996)
4. Bayardo Jr., R.J., Schrag, R.: Using CSP look-back techniques to solve real-world SAT instances. In: Proceedings of the Fourteenth National Conference on Artificial Intelligence and Ninth Innovative Applications of Artificial Intelligence Conference, pp. 203–208 (1997)
5. Marques-Silva, J., Lynce, I., Malik, S.: Conflict-driven clause learning SAT solvers. In: Biere, A., Heule, M., van Maaren, H., Walsh, T. (eds.) Handbook of Satisfiability. Frontiers in Artificial Intelligence and Applications, vol. 185. IOS Press (2009)
6. Chvátal, V., Szemerédi, E.: Many hard examples for resolution. J. ACM **35**(4), 759–768 (1988)
7. Samer, M., Szeider, S.: Fixed-parameter tractability. In: Biere, A., Heule, M., van Maaren, H., Walsh, T. (eds.) Handbook of Satisfiability. Frontiers in Artificial Intelligence and Applications, vol. 185. IOS Press (2009)
8. Mateescu, R.: Treewidth in industrial SAT benchmarks. Technical report MSR-TR-2011-22, Microsoft Research, February 2011
9. Williams, R., Gomes, C.P., Selman, B.: Backdoors to typical case complexity. In: Proceedings of the Eighteenth International Joint Conference on Artificial Intelligence, IJCAI 2003, pp. 1173–1178 (2003)
10. Monasson, R., Zecchina, R., Kirkpatrick, S., Selman, B., Troyansky, L.: Determining computational complexity from characteristic phase transitions. Nature **400**(6740), 133–137 (1999)
11. Kilby, P., Slaney, J.K., Thiébaux, S., Walsh, T.: Backbones and backdoors in satisfiability. In: Proceedings of the Twentieth National Conference on Artificial Intelligence and the Seventeenth Innovative Applications of Artificial Intelligence Conference, pp. 1368–1373 (2005)

12. Gregory, P., Fox, M., Long, D.: A new empirical study of weak backdoors. In: Stuckey, P.J. (ed.) CP 2008. LNCS, vol. 5202, pp. 618–623. Springer, Heidelberg (2008)

13. Liang, J.H., Ganesh, V., Czarnecki, K., Raman, V.: SAT-based analysis of large real-world feature models is easy. In: Proceedings of the 19th International Software Product Line Conference, SPLC, pp. 91–100 (2015)

14. Newman, M.E., Girvan, M.: Finding and evaluating community structure in networks. Phys. Rev. E **69**(2), 026113 (2004)

15. Ansótegui, C., Giráldez-Cru, J., Levy, J.: The community structure of SAT formulas. In: Cimatti, A., Sebastiani, R. (eds.) SAT 2012. LNCS, vol. 7317, pp. 410–423. Springer, Heidelberg (2012)

16. Newsham, Z., Ganesh, V., Fischmeister, S., Audemard, G., Simon, L.: Impact of community structure on SAT solver performance. In: Sinz, C., Egly, U. (eds.) SAT 2014. LNCS, vol. 8561, pp. 252–268. Springer, Heidelberg (2014)

17. Giráldez-Cru, J., Levy, J.: A modularity-based random SAT instances generator. In: 24th International Joint Conference on Artificial Intelligence, IJCAI 2015 (2015)

18. Ganian, R., Szeider, S.: Community structure inspired algorithms for SAT and #SAT. In: Heule, M., Weaver, S. (eds.) SAT 2015. LNCS, vol. 9340, pp. 223–237. Springer, Heidelberg (2015). doi:10.1007/978-3-319-24318-4_17

19. Robinson, J.A.: A machine-oriented logic based on the resolution principle. J. ACM (JACM) **12**(1), 23–41 (1965)

20. Davis, M., Logemann, G., Loveland, D.: A machine program for theorem-proving. Commun. ACM **5**(7), 394–397 (1962)

21. Beame, P., Kautz, H.A., Sabharwal, A.: Understanding the power of clause learning. In: Proceedings of the Eighteenth International Joint Conference on Artificial Intelligence, IJCAI 2003, pp. 1194–1201 (2003)

22. Järvisalo, M., Biere, A., Heule, M.: Blocked clause elimination. In: Esparza, J., Majumdar, R. (eds.) TACAS 2010. LNCS, vol. 6015, pp. 129–144. Springer, Heidelberg (2010)

23. Eén, N., Biere, A.: Effective preprocessing in SAT through variable and clause elimination. In: Bacchus, F., Walsh, T. (eds.) SAT 2005. LNCS, vol. 3569, pp. 61–75. Springer, Heidelberg (2005)

24. Beame, P., Pitassi, T.: Simplified and improved resolution lower bounds. In: Proceedings of the 37th Annual Symposium on Foundations of Computer Science, pp. 274–282. IEEE (1996)

25. Fortunato, S.: Community detection in graphs. Phys. Rep. **486**(3–5), 75–174 (2010)

26. Almeida, H., Guedes, D., Meira Jr., W., Zaki, M.J.: Is there a best quality metric for graph clusters? In: Gunopulos, D., Hofmann, T., Malerba, D., Vazirgiannis, M. (eds.) ECML PKDD 2011, Part I. LNCS, vol. 6911, pp. 44–59. Springer, Heidelberg (2011)

27. Mull, N., Fremont, D.J., Seshia, S.A.: On the hardness of SAT with community structure. ArXiv e-prints (2016). http://arxiv.org/abs/1602.08620

# Trade-offs Between Time and Memory in a Tighter Model of CDCL SAT Solvers

Jan Elffers[1], Jan Johannsen[2], Massimo Lauria[3], Thomas Magnard[4], Jakob Nordström[1(✉)], and Marc Vinyals[1]

[1] KTH Royal Institute of Technology, Stockholm, Sweden
{elffers,jakobn,vinyals}@kth.se
[2] Ludwig-Maximilians-Universität München, Munich, Germany
Jan.Johannsen@ifi.lmu.de
[3] Universitat Politècnica de Catalunya, Barcelona, Spain
lauria@cs.upc.edu
[4] École Normale Supérieure, Paris, France
magnard@clipper.ens.fr

**Abstract.** A long line of research has studied the power of conflict-driven clause learning (CDCL) and how it compares to the resolution proof system in which it searches for proofs. It has been shown that CDCL can polynomially simulate resolution even with an adversarially chosen learning scheme as long as it is asserting. However, the simulation only works under the assumption that no learned clauses are ever forgotten, and the polynomial blow-up is significant. Moreover, the simulation requires very frequent restarts, whereas the power of CDCL with less frequent or entirely without restarts remains poorly understood. With a view towards obtaining results with tighter relations between CDCL and resolution, we introduce a more fine-grained model of CDCL that captures not only time but also memory usage and number of restarts. We show how previously established strong size-space trade-offs for resolution can be transformed into equally strong trade-offs between time and memory usage for CDCL, where the upper bounds hold for CDCL without any restarts using the standard 1UIP clause learning scheme, and the (in some cases tightly matching) lower bounds hold for arbitrarily frequent restarts and arbitrary clause learning schemes.

## 1 Introduction

For two decades the dominant strategy for solving the Boolean satisfiability problem (SAT) in practice has been *conflict-driven clause learning (CDCL)* [5,27,28]. Although SAT is an NP-complete problem, and is hence widely believed to be intractable in the worst case, CDCL SAT solvers have turned out to be immensely successful over a wide range of application areas. An important problem is to understand how such SAT solvers can be so efficient and what theoretical limits exist on their performance.

***Previous Work.*** At the core, CDCL searches for proofs in the proof system *resolution* [14]. While pre- and inprocessing techniques can, and sometimes do, go

© Springer International Publishing Switzerland 2016
N. Creignou and D. Le Berre (Eds.): SAT 2016, LNCS 9710, pp. 160–176, 2016.
DOI: 10.1007/978-3-319-40970-2_11

significantly beyond resolution (incorporating, e.g., solving of linear equations mod 2 and reasoning with cardinality constraints), understanding the power of even just the fundamental CDCL algorithm seems like an interesting and challenging problem in its own right. Three crucial aspects of CDCL solvers, which are the focus of our work, are running time, memory usage, and restart policy.

In resolution, time is modelled by the size/length complexity measure, in that lower bounds on proof size yield lower bounds on the running time of CDCL solvers. Resolution proof size is a well-studied measure. It is not hard to show that it need never be larger than exponential in the formula size, and such exponential lower bounds were shown already in, e.g., [20,25,31].

Another more recently studied measure is *(clause) space*, measured as the number of clauses needed in memory while verifying a proof.[1] We remark that although the study of space was originally motivated by SAT solving concerns, it is not a priori clear to what extent this abstract space measure corresponds to CDCL memory usage. Space need never be more than linear in the worst case [24], even though such proofs might have exponential size, and optimal linear lower bounds on space were obtained in [1,10,24].

More interesting than such space bounds is perhaps what can be said regarding simultaneous optimization of time and space, which is the setting in which SAT solvers operate. There are strong trade-offs [6,9,11] showing that this is not possible in general. What this means is that one can find formulas for which (a) there are short proofs and (b) also space-efficient proofs but (c) no proof can get close to being simultaneously both size- and space-efficient.

Regarding restarts, such a concept does not quite make sense for resolution proofs and so has not been studied in that context as far as we are aware.

It is natural to ask to what extent upper and lower bounds for resolution apply to CDCL. By comparison, it is well understood that the *DPLL* method [22,23] searches for proofs in *tree-like resolution*, which incurs an exponential loss in performance as compared to general resolution. There has been a long line of research investigating how CDCL compares to general resolution, e.g., [7,18,26,32], culminating in the result by [30] that CDCL viewed as a proof system polynomially simulates resolution with respect to size/time. The non-constructive part of this result is that variable decisions are not done according to some concrete heuristic but are provided as helpful advice to the solver. This limitation is probably inherent, since a fully algorithmic result would have unexpected implications in complexity theory [2]. It is worth noting, however, that in independent work [3] showed that for resolution proofs where all clauses have constant size, using a *random* variable selection heuristic will yield a constructive polynomial-time simulation.

---

[1] We mention for completeness that there is also a *total space* measure counting the number of literals in memory, which has been studied in, e.g., [1,13,15,16], but for our purposes clause space seems like a more relevant measure to focus on.

One strength of [3, 30] is that the results hold for any learning scheme as long as it is *asserting* (an assumption that anyway lies at the heart of the CDCL algorithm). The results also have a few less desirable aspects, however:

- The simulations require very frequent restarts. Only the first conflict after each restart is useful, and after that one has to wait for the next restart to make any further progress.
- There is also a large polynomial blow-up in the simulations, which means that for practical purposes these simulations are far too inefficient to yield really concrete insights into CDCL performance as compared to resolution.
- Finally, and most seriously, the results crucially rely on the assumption that no learned clause is ever forgotten. This is unrealistic, as typically around 90–95 % of learned clauses are erased during CDCL search and this is absolutely essential for performance.

It would be desirable to obtain results relating CDCL and resolution that also take the above aspects into account.

Addressing one of these concerns, a more fine-grained study of the power of CDCL without restarts has been conducted in, e.g., [8, 17–19]. One problematic aspect here is that the models studied appear to be quite far from actual CDCL behaviour. Some papers assume non-standard and rather artificial preprocessing steps. Others study CDCL models that do not enforce that unit clauses are propagated or that do not trigger conflict analysis as soon as a clause is falsified. In the latter case, as a result one gets very limited restrictions on what the clause learning schemes are, and it is hard even to talk about what "conflict analysis" is supposed to mean in this context. This is not an issue for results establishing lower bounds limiting what CDCL can do—here a stronger model of CDCL only makes the results stronger—but for upper bounds the results become too optimistic, indicating that the theoretical CDCL model can do much better than what seems possible in practice. As a case in point, there are currently no known separations between general resolution and CDCL without restarts, but part of the reason for this appears to be that the models of CDCL without restarts are clearly too strong to be realistic.

We are not aware of any work on models measuring not only time but also memory consumption in a proof system formalizing CDCL. As discussed above, one can define a space measure for resolution proofs, but it is not clear what relation, if any, there is between this space measure and the size of the clause database during CDCL execution.

***Our Contributions.*** In this work, we present a proof system that tightly models running time, memory usage, and restarts in CDCL. The model draws heavily on [3, 30], combined with ideas from [18] to capture memory and restarts. Indeed, we do not claim any key new technical insights for this part of our work, but rather it is more a matter of carefully studying previous models and painstakingly putting the pieces together to get as clean and simple a proof system as possible that is nevertheless significantly "closer to the metal" than in previous papers.

Our CDCL proof system enforces unit propagation and triggers conflict analysis directly at a conflict. It can incorporate any asserting learning scheme (as long as it is based on resolution derivations from the current conflict and reason clauses), and this scheme is specified explicitly as a parameter. Right from the definitions one obtains natural measures of time, memory usage, and restarts. Variable decisions are still provided externally, just as in [3,30], but in principle one could also plug in, say, the most commonly used *VSIDS (variable state independent decaying sum)* decision scheme with *phase saving* and analyse what proofs can be generated using these heuristics (though this is not the focus of our current work). Since we are now managing the database of learned clauses explicitly, we also have to specify a clause database reduction policy. In this paper, the decisions about which clauses to delete are also provided to the solver, but the model allows to plug in a concrete reduction policy as well.

We argue that the proof system we present faithfully models possible execution traces during CDCL search. Some interesting questions to study in this model are as follows:

1. Do upper and lower bounds on resolution size and space transfer to this CDCL proof system?
2. How does CDCL compare to general resolution if we want efficient simulations with respect to *both* time and space, and in addition aim for at most constant-factor blow-ups rather than arbitrary polynomial blow-ups?
3. What is the power of CDCL without restarts compared to the subsystems of tree-like resolution or so-called *regular resolution*? (Briefly, regularity is the somewhat SAT solver-like restriction on resolution that along each path in the proof any variable is branched over only once.)

The worst-case upper bounds on size and space in resolution carry over to time and memory usage in CDCL, and it turns out that this can in fact be read off from [29], although that paper uses quite a different language. More interestingly, we show that there is a straightforward translation from CDCL to resolution that preserves both time and space, and so we obtain that all size and space lower bounds previously established for resolution apply also to CDCL (which, in particular, was not at all obvious for space).

This means that the lower bounds on time-space trade-offs in [6,9,11] also hold for CDCL. But this does not yet yield true trade-offs, since for such results we also want *upper bounds*. That is, we want to show that CDCL can find time- or space-efficient proofs optimizing just one of these measures in isolation. It is known how to construct such proofs in resolution, but these proofs are not obviously CDCL-like. Since SAT solving was mentioned as a motivation for [6,9,11] it is a relevant question whether the size-space trade-offs shown in these papers correspond to anything one could expect to see in practice, or whether the size- and space-efficient proofs have such peculiar structure that nothing similar can be found by CDCL proof search.

The main contribution of our work is to address the question of whether true time-space trade-offs can be established for CDCL. Finding an answer turns out to be surprisingly technically challenging, and we are not able to prove the known trade-offs for exactly the same formulas as in [6,9,11] However, for many of the formulas it is possible to modify them slightly to obtain CDCL trade-offs with essentially the same parameters. An additional feature of these trade-offs is that all our upper bounds hold for CDCL *without any restarts* using the standard *1UIP (first unique implication point)* learning scheme, while the (often tightly matching) lower bounds hold for arbitrarily frequent restarts and arbitrarily chosen clause learning schemes (even non-asserting ones).

We leave as open problems whether CDCL with 1UIP clause learning and with or without restarts can simulate or be separated from general or regular resolution, respectively. While those problems still look quite challenging, we hope and believe that it should be possible to make progress by investigating them in a model that more closely resembles what happens during CDCL proof search in practice, such as the model presented in this paper.

***Organization of This Paper.*** In Sect. 2 we describe our proof system modelling CDCL. Section 3 gives an overview of our time-space trade-off results. Since the proofs are quite long and technical, however, we have to defer essentially all of them to the full-length version of the paper, and in this extended abstract we only sketch the proof of a simpler (but still nontrivial) trade-off for CDCL *with* restarts. We make some concluding remarks in Sect. 4.

## 2   Modelling CDCL as a Proof System

We start by describing our model of CDCL and how it is formalized as a proof system. As already mentioned, this is very much inspired by [3,30], but with ideas added from [18]. We want to remark right away that we describe the model at a level of detail that might seem excessive to SAT practitioners familiar with CDCL. We do so precisely because a serious issue with many contributions on the theoretical side has been that they fail to get crucial details of the model right, as discussed in the introduction.

***Preliminaries.*** Let us first fix some standard notation and terminology. A *literal* $a$ over a Boolean variable $x$ is either $x$ itself or its negation $\overline{x}$ (a *positive* or *negative* literal, respectively). A *clause* $C = a_1 \vee \cdots \vee a_k$ is a disjunction of literals, where the clause is *unit* if it contains only one literal. A *CNF formula* $F$ is a conjunction of clauses $F = C_1 \wedge \cdots \wedge C_m$. We think of clauses and formulas as sets, so that the order of elements is irrelevant and there are no repetitions.

A *resolution derivation* of $C$ from $F$ is a sequence of clauses $(C_1, C_2, \ldots, C_\tau)$ such that $C_\tau = C$ and every $C_i$ is either a clause in $F$ (an *axiom*) or is derived from clauses $C_j, C_k$ with $j, k < i$, by the *resolution rule*

$$\frac{C \vee x \qquad D \vee \overline{x}}{C \vee D}, \tag{1}$$

where we say that $C \vee x$ and $D \vee \overline{x}$ are *resolved* over $x$. A derivation is *trivial* if all variables resolved over are distinct and each $C_i$ either is an axiom or is derived from a resolution rule application where one of the resolved clauses is an axiom.

A *resolution refutation* of, or *resolution proof* for, an unsatisfiable formula $F$ is a derivation of the empty clause $\perp$ (containing no literals) from the axioms in $F$. The *length* or *size* of a proof is the number of clauses in it counted with repetitions. The space of a proof at step $t$ is the number of clauses at steps $\leq t$ that are used in applications of the resolution rule at steps $\geq t$. The space of a proof is obtained by measuring the space at each step and taking the maximum.

**A Formal Description of CDCL.** A CDCL solver running on a formula $F$ decides variable assignments and propagates values that follow from such assignments until a clause is falsified, at which point a learned clause is added to the clause database $\mathcal{D}$ (where we always have $F \subseteq \mathcal{D}$) and the search backtracks. A key concept is the current partial assignment maintained by the solver together with some book-keeping why variables were set this way, which we refer to as the *trail*. This is a sequence $s = (x_1 = b_1/*, x_2 = b_2/*, \ldots, x_\ell = b_\ell/*)$ where all variables are distinct and where $* = \mathsf{d}$ indicates that the assignment is a decision and $* = C$ that it was propagated by the clause $C$. We write $s_{\leq j}$ and $s_{<j}$ to denote the subsequences that are the prefixes of length $j$ and $j - 1$ of $s$, respectively. We denote the empty trail by $\epsilon$.

The *decision level* of an assignment $x_j = b_j/*$ is the number of decision assignments in $s_{\leq j}$. The decision level of a (non-empty) trail is that of its last assignment. Identifying a trail $s$ with the partial assignment it defines, we write $C\restriction_s$ to denote the clause $C$ *restricted by* $s$, which is the trivially true clause if $s$ satisfies $C$ and otherwise $C$ with all literals falsified by $s$ removed, and this notation is extended to sets of clauses by taking unions. If a trail $s$ falsifies a clause $C$, we say that $C$ is *asserting* if it has a unique variable at the maximum decision level of $s$. If so, the second largest decision level represented in $C$ is the *assertion level* of $C$.

A trail $s = (x_1 = b_1/*, \ldots, x_\ell = b_\ell/*)$ is *legal* with respect to a formula $F$ and clause database $\mathcal{D} \supseteq F$ if the following holds:

- $\mathcal{D}\restriction_{s_{<\ell}}$ does not contain the empty clause;
- if the $j$th element of $s$ is $x_j = b_j/\mathsf{d}$, then $\mathcal{D}\restriction_{s_{<j}}$ does not contain a unit clause;
- if the $j$th element of $s$ is $x_j = b_j/C$, then $C$ is contained in $\mathcal{D}$ and has the property that $C\restriction_{s_{<j}}$ is unit and is satisfied by setting $x_j = b_j$.

This captures properties that must hold during CDCL search, and so in what follows trails are implicitly required to be legal unless otherwise specified.

At each point in time, the solver is in a *CDCL state* $(F, \mathcal{D}, s)$, where at the beginning $\mathcal{D} = F$ and $s = \epsilon$. It is convenient to describe the solver as being in one of the four modes **Default** (where it starts), **Unit**, **Conflict**, or **Decision**, where transitions are performed as described below (guided by plug-in components that specify the detailed behaviour; also to be discussed in what follows):

**Default.** If all variables in $F$ have been assigned, the solver halts and outputs SAT together with the assignment $s$. If $s$ falsifies a clause in $\mathcal{D}$, the solver moves to **Conflict**. Otherwise, if $\mathcal{D}\restriction_s$ contains a unit clause, the solver transits to **Unit** mode. If none of the above cases apply, the solver uses its *restart* *policy* to decide whether to set $s = \epsilon$, and its *clause database reduction policy* to decide whether to shrink $\mathcal{D}$ to $\mathcal{D}' \subsetneq \mathcal{D}$, where $\mathcal{D}'$ must still contain $F$ and all clauses mentioned in the current trail $s$, after which it moves to **Decision**.

**Conflict.** If $s$ has decision level 0, the solver outputs `UNSAT`. Otherwise it applies the *learning scheme* to derive an asserting clause $C$ and then *backjumps* by updating the state to $(F, \mathcal{D} \cup \{C\}, s')$ (where $s'$ is the prefix of $s$ that contains all assignments with decision level less than or equal to the assertion level of $C$), and shifts to **Unit** mode.

**Unit.** The solver uses the *unit propagation scheme* to pick a clause $C$ in $F \cup \mathcal{D}$ such that $C{\restriction}_s$ is unit, extends $s$ with the assignment $x = b/C$ that satisfies $C{\restriction}_s$, and moves to **Default** mode.

**Decision.** The solver uses the *decision scheme* to determine an assignment $x = b/\mathrm{d}$ with which to extend the trail and moves to **Default** mode.

We say that a CDCL state $(F, \mathcal{D}, s)$ is *stable* if it makes the solver move from **Default** mode to **Decision** mode and a *conflict state* if it causes a move from **Default** to **Conflict**. We remark that CDCL solvers typically apply restarts and database reductions only in the first stable state after a conflict. However, it is not hard to see that from a proof complexity point of view the solver does not get any stronger by allowing these steps to be performed at any stable state, and since this simplifies the description we have done so above.

In order to obtain a concrete CDCL implementation, one needs to instantiate the components referred to above. Let us briefly discuss how this can be done.

For the *clause learning scheme* the assumption is that the clause is derivable in resolution from the clause falsified (the *conflict clause*) and the clauses causing unit propagations (the *reason clauses*) and that the learned clause is always asserting. For our upper bounds we use the 1UIP learning scheme from [33], which is simply a trivial resolution derivation from the conflict clause and the reason clauses processed in reverse order up to the first point when there remains only one variable of maximal decision level in the clause.[2]

The *restart policy* determines when the solver should clear the trail and start over from the beginning (but keeping the rest of the solver state). From a theoretical point of view adding more frequent restarts can only make the solver more powerful. Hence, in order to obtain the strongest possible result we want to prove our upper bounds on CDCL with a strict no-restarts policy and our lower bounds in a setting with no restrictions on restarts.

If there is more than one unit clause that can propagate in **Unit** mode, the *unit propagation scheme* determines in which order the clauses are chosen. Typically this will depend somewhat randomly on low-level implementation details, and therefore we try to prove our upper and lower bounds for the settings when the order chosen is maximally unhelpful and maximally helpful, respectively.

The *decision scheme* is used to choose the next variable to assign when there are no unit propagations. The dominant heuristic in practice is VSIDS [28], but

---

[2] In fact, our results hold for any UIP scheme, but for simplicity we focus on 1UIP, which is anyway dominant in practice.

for our theoretical analysis we follow [3,30] by allowing the decisions to be chosen externally by a helpful oracle and fed to the solver.

The *database reduction policy*, finally, regulates when and how to forget learned clauses. Making this aspect explicit is the main difference between our work and [3,30]—the latter papers crucially need the unrealistic assumption that no learned clauses may ever be erased. In principle, here one could plug in, say, the *literal block distance (LBD)* heuristic in [4] to decide which clauses to throw away or keep, but in this work we will let this, too, be part of the external input provided to construct a CDCL proof.

***Formalizing CDCL as a Proof System.*** In order to construct a proof system corresponding to CDCL, we will simply let the proofs be execution traces that contain enough information to allow efficient verification that they are consistent with the detailed description of the CDCL model above. More formally, we say that a *CDCL trace* $\pi$ is an ordered sequence of the following types of elements:

- decisions $x_i = b/\mathsf{d}$;
- unit propagations $x_i = b/C$ (with reason $C$);
- learned clauses $\mathsf{add}\, C/\sigma_C$ (with conflict analysis $\sigma_C$);
- deletions of clauses $\mathsf{del}\, C$;
- restarts $\mathsf{R}$.

Given a CDCL model with components as above partially or fully specified, a trace $\pi$ is *legal*, or is a *CDCL proof*, if it is consistent with an execution of the CDCL model as described above.

We say that a CDCL trace is a *CDCL proof of unsatisfiablity* or *CDCL refutation* of $F$ if it is legal and makes the CDCL solver output UNSAT, and that it is a *CDCL proof of satisfiablity* if the output is SAT. It should be clear that if the components specified are efficiently computable, then CDCL traces are efficiently verifiable and constitute a proof system in the sense of [21] (and since all traces we construct will be *legal*, we will sometimes use the words "trace" and "proof" as synonyms).

The *time* of a CDCL proof $\pi$ is the number of elements in the sequence plus the sum of the length of all conflict analysis resolution derivations $\sigma_C$, i.e., the total number of variable decisions, propagations, and steps in conflict analysis. The *space* of the proof at a given point in time is the number of learned clauses $|\mathcal{D} \setminus F|$, i.e., the number of statements $\mathsf{add}\, C/\sigma_C$ minus the number of statements $\mathsf{del}\, C$ up to that point, and the space of a proof is obtained by taking the maximum over all time steps in it.

These measures are intended to capture the execution time and memory usage of a CDCL solver execution described by the trace $\pi$, and in addition we want them to translate to length and space bounds for resolution. This is indeed the case, as we state in the next theorem (the proof of which is provided in the full-length version of this paper).

**Theorem 1.** *If there is a CDCL proof with some learning scheme using trivial resolution (in particular, 1UIP) refuting a CNF formula $F$ in time $\tau$ and space $s$, then $F$ has a resolution refutation in length at most $\tau$ and space $s + \mathrm{O}(1)$.*

The meaning of this theorem is that all lower bounds on length and space in resolution automatically carry over to impossibility results for conflict-driven clause learning. These results hold even for very general models of CDCL, with arbitrarily frequent restarts and arbitrarily smart decision and database reduction heuristics, as long as the clause learning scheme is realistic. In order to prove upper bounds for CDCL, however—i.e., showing that proof search can (at least sometimes) be performed in a time- and space-efficient manner compared to the best-case resolution proof scenario—we have to work harder.

## 3   Overview of Time-Space Trade-Off Results

In this section we survey the kind of CDCL time-space trade-offs obtained in this paper, and discuss some of the challenges that have to be overcome when establishing such results.

***Statement of Trade-off Theorems.*** Our first set of trade-off results are for formulas defined in terms of pebble games as described in [12]. Given a directed acyclic graph (DAG) $G$ with source vertices $S$ and a unique sink vertex $z$, and with all non-sources having fan-in 2, we identify vertices with variables and define the *pebbling formula* $Peb_G$ to consist of the following clauses:

– for all $s \in S$, the unit clause $s$ (*source axioms*),
– for all $w$ with predecessors $u, v$, the clause $\overline{u} \vee \overline{v} \vee w$ (*pebbling axioms*),
– for the sink $z$, the unit clause $\overline{z}$ (*sink axiom*).

These formulas are not too interesting, since it is easy to see that they are solved immediately by unit propagation, but if we replace each variable by an exclusive or of two new, fresh variables, and then expand out to CNF we obtain a *XORified pebbling formula* $Peb_G^{\oplus}$ as in Fig. 1b. Given the right kind of graphs, [11] showed that such formulas have strong trade-offs between length and space in resolution, and we are able to lift most of these results to CDCL. We give two examples of such results below.

**Theorem 2 (Robust trade-offs (informal)).** *There are XORified pebbling formulas $F_n$ of size $\Theta(n)$ such that:*

– *CDCL with 1UIP learning and no restarts can refute $F_n$ in time $O(n)$ and space $O(n/\log n)$ simultaneously.*
– *CDCL with 1UIP learning and no restarts can refute $F_n$ in space $O\big((\log n)^2\big)$ and time $n^{O(\log n)}$ simultaneously.*
– *Any CDCL refutation of $F_n$ in space $o(n/\log n)$ requires time at least $n^{\Omega(\log \log n)}$ regardless of learning scheme and restart policy.*

**Theorem 3 (Exponential trade-offs (informal)).** *There are XORified pebbling formulas $F_n$ of size $\Theta(n)$ such that:*

– *CDCL with 1UIP learning and no restarts can refute $F_n$ in time $O(n)$ and space $O\big(\sqrt[4]{n}\big)$ simultaneously.*

$$(u_1 \vee u_2)$$
$$\wedge\ (\overline{u}_1 \vee \overline{u}_2)$$
$$\wedge\ (v_1 \vee v_2)$$
$$\wedge\ (\overline{v}_1 \vee \overline{v}_2)$$
$$\wedge\ (w_1 \vee w_2)$$
$$\wedge\ (\overline{w}_1 \vee \overline{w}_2)$$
$$\wedge\ (v_1 \vee \overline{v}_2 \vee \overline{w}_1 \vee w_2 \vee y_1 \vee y_2)$$
$$\wedge\ (v_1 \vee \overline{v}_2 \vee \overline{w}_1 \vee w_2 \vee \overline{y}_1 \vee \overline{y}_2)$$
$$\wedge\ (\overline{v}_1 \vee v_2 \vee w_1 \vee \overline{w}_2 \vee y_1 \vee y_2)$$
$$\wedge\ (\overline{v}_1 \vee v_2 \vee w_1 \vee \overline{w}_2 \vee \overline{y}_1 \vee \overline{y}_2)$$
$$\wedge\ (\overline{v}_1 \vee v_2 \vee \overline{w}_1 \vee w_2 \vee y_1 \vee y_2)$$
$$\wedge\ (\overline{v}_1 \vee v_2 \vee \overline{w}_1 \vee w_2 \vee \overline{y}_1 \vee \overline{y}_2)$$

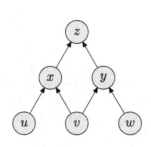

**(a)** Pyramid graph $\Pi_2$ of height 2.

$$\wedge\ (u_1 \vee \overline{u}_2 \vee v_1 \vee \overline{v}_2 \vee x_1 \vee x_2)$$
$$\wedge\ (u_1 \vee \overline{u}_2 \vee v_1 \vee \overline{v}_2 \vee \overline{x}_1 \vee \overline{x}_2)$$
$$\wedge\ (u_1 \vee \overline{u}_2 \vee \overline{v}_1 \vee v_2 \vee x_1 \vee x_2)$$
$$\wedge\ (u_1 \vee \overline{u}_2 \vee \overline{v}_1 \vee v_2 \vee \overline{x}_1 \vee \overline{x}_2)$$
$$\wedge\ (\overline{u}_1 \vee u_2 \vee v_1 \vee \overline{v}_2 \vee x_1 \vee x_2)$$
$$\wedge\ (\overline{u}_1 \vee u_2 \vee v_1 \vee \overline{v}_2 \vee \overline{x}_1 \vee \overline{x}_2)$$
$$\wedge\ (\overline{u}_1 \vee u_2 \vee \overline{v}_1 \vee v_2 \vee x_1 \vee x_2)$$
$$\wedge\ (\overline{u}_1 \vee u_2 \vee \overline{v}_1 \vee v_2 \vee \overline{x}_1 \vee \overline{x}_2)$$
$$\wedge\ (v_1 \vee \overline{v}_2 \vee w_1 \vee \overline{w}_2 \vee y_1 \vee y_2)$$
$$\wedge\ (v_1 \vee \overline{v}_2 \vee w_1 \vee \overline{w}_2 \vee \overline{y}_1 \vee \overline{y}_2)$$

$$\wedge\ (x_1 \vee \overline{x}_2 \vee y_1 \vee \overline{y}_2 \vee z_1 \vee z_2)$$
$$\wedge\ (x_1 \vee \overline{x}_2 \vee y_1 \vee \overline{y}_2 \vee \overline{z}_1 \vee \overline{z}_2)$$
$$\wedge\ (x_1 \vee \overline{x}_2 \vee \overline{y}_1 \vee y_2 \vee z_1 \vee z_2)$$
$$\wedge\ (x_1 \vee \overline{x}_2 \vee \overline{y}_1 \vee y_2 \vee \overline{z}_1 \vee \overline{z}_2)$$
$$\wedge\ (\overline{x}_1 \vee x_2 \vee y_1 \vee \overline{y}_2 \vee z_1 \vee z_2)$$
$$\wedge\ (\overline{x}_1 \vee x_2 \vee y_1 \vee \overline{y}_2 \vee \overline{z}_1 \vee \overline{z}_2)$$
$$\wedge\ (\overline{x}_1 \vee x_2 \vee \overline{y}_1 \vee y_2 \vee z_1 \vee z_2)$$
$$\wedge\ (\overline{x}_1 \vee x_2 \vee \overline{y}_1 \vee y_2 \vee \overline{z}_1 \vee \overline{z}_2)$$
$$\wedge\ (z_1 \vee \overline{z}_2)$$
$$\wedge\ (\overline{z}_1 \vee z_2)$$

**(b)** XORified pebbling formula $Peb_{\Pi_2}^{\oplus}$.

**Fig. 1.** Example pebbling formula for the pyramid of height 2.

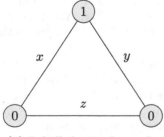

**(a)** Labelled triangle graph.

$$(x \vee y)$$
$$\wedge\ (\overline{x} \vee \overline{y})$$
$$\wedge\ (x \vee \overline{z})$$
$$\wedge\ (\overline{x} \vee z)$$
$$\wedge\ (y \vee \overline{z})$$
$$\wedge\ (\overline{y} \vee z)$$

**(b)** Corresponding Tseitin formula.

**Fig. 2.** Example Tseitin formula.

- *CDCL with 1UIP learning and no restarts can refute $F_n$ in space $\mathrm{O}\left(\sqrt[8]{n}\right)$ and time $n^{\mathrm{O}\left(\sqrt[8]{n}\right)}$ simultaneously.*
- *Any CDCL refutation of $F_n$ in space $\mathrm{O}\left(n^{1/4-\epsilon}\right)$ for $\epsilon > 0$ requires exponential time regardless of learning scheme and restart policy.*

The other formula family considered in this paper are *Tseitin formulas*, which are defined in terms of undirected graphs with vertices labelled 0/1 in such a way that the total sum of all vertex labels is odd. The variables of the formula are the edges of the graph. For every vertex we add a constraint saying that the parity of the number of true edges incident to the vertex is equal to the vertex label. Summing over all vertices, each edge is counted exactly twice and hence the total number of true edges must be even. But this contradicts that the sum of the labels is odd, and thus the formulas are unsatisfiable. Figure 2b gives an example Tseitin formula generated from the labelled graph in Fig. 2a.

Using Tseitin formulas over long, skinny grids, we can build on [9] to obtain the following trade-off, which applies even for superlinear space.

**Theorem 4 (Superlinear space trade-offs (informal)).** *For a Tseitin formula $F_{w,\ell}$ over a grid graph with $w$ rows and $\ell$ columns, $1 \leq w \leq \ell^{1/4}$, and with double edges between every two vertices at horizontal distance one or vertical distance one, it holds that*

- *CDCL with 1UIP learning and no restarts can refute $F_{w,\ell}$ in time $\mathrm{O}(2^{5w}\ell)$ and space $\mathrm{O}(2^{2w})$.*
- *CDCL with 1UIP learning and no restarts can refute $F_{w,\ell}$ in space $\mathrm{O}(w\log(\ell))$ and time $\mathrm{O}\left(\ell^{\mathrm{O}(w)}\right)$.*
- *For any CDCL refutation in time $\tau$ and space $s$, regardless of learning scheme and restart policy, it holds that*

$$\tau = \left(\frac{2^{\Omega(w)}}{s}\right)^{\Omega\left(\frac{\log\log\ell}{\log\log\log\ell}\right)}.$$

***Proof Techniques and Technical Challenges.*** All the trade-offs stated in Theorems 2, 3, and 4 are known to hold for resolution, and so by Theorem 1 we immediately obtain that the lower bounds carry over to CDCL. What we need to show in order to establish these theorems is that CDCL can find proofs that match the upper bounds in resolution.

The general idea how we would like to do this is clear: given a resolution proof $\pi = (C_1, C_2, \ldots, C_\tau)$, we should force the CDCL solver to efficiently learn the clauses $C_i$ one by one, making sure at all times that the clause database size is comparable to the space complexity of the resolution proof. This seems hard to do, however, and somewhat ironically what causes trouble for us are the unit propagations that otherwise make CDCL so efficient. To illustrate the problem, suppose that we have learned $C \vee x$ and $D \vee \overline{x}$ and now want to learn their resolvent $C \vee D$. It would be nice to decide on all literals in $C \vee D$ being false, after which we could get a conflict on $x$. But there might be other clauses in the database that propagate literals to "wrong values" before we manage to falsify all of $C \vee D$, and if so the CDCL search will veer off in another direction and we will not be able to learn this resolvent.

This highlights two technical difficulties that we need to be able to deal with:

- Not only do we have to decide on variables in the right order, but we have to make sure that no other unexpected (and unwanted) propagations occur.
- In contrast to resolution, where having more clauses at your disposal never hurts, keeping too many learned clauses in the clause database can actually hinder the CDCL search. This is also a striking contrast to [3,30], where a key technical lemma is precisely that having more clauses in the database can only be helpful.

We do not know how to simulate general resolution efficiently with respect to length and space simultaneously, even using ever so frequent restarts. And an additional problem is that we want to know—in order to better understand basic CDCL reasoning with unit propagation and conflict analysis—whether for the pebbling and Tseitin formulas presented above CDCL can find efficient proofs *even without restarts*. This makes our task substantially more complicated.

If we allow suitably frequent restarts, however, it is not too hard to show that CDCL can efficiently simulate the "canonical" resolution proofs for these formulas. To give at least some flavour of the technical arguments needed to reason about the CDCL proof system, we conclude this section with a description of how this result can be proven for pebbling formulas.

***Pebbling Formula Upper Bound for CDCL with Restarts.*** A pebbling formula encodes the *black pebble game* played on a DAG $G$, where we start with $G$ being empty and want to finish with a pebble on the sink $z$. A vertex can be pebbled if its predecessors have pebbles (vacuously true for sources), and pebbles can always be removed. The *time* of a pebbling is the number of moves before $z$ is reached and the *space* is the maximum number of pebbles on vertices of $G$ at any point.

Resolution can simulate such pebblings by deriving, whenever a vertex $w$ is pebbled, the two *pebble clauses* $w_1 \vee w_2$ and $\overline{w}_1 \vee \overline{w}_2$ saying that the exclusive or $w_1 \oplus w_2$ is true, and by erasing these clauses whenever a pebble is removed. For a source vertex the pebble clauses are already available as source axioms in the formula (see Figure 1b), and it is not hard to show that resolution can efficiently propagate exclusive ors from predecessors to successors. Once pebble clauses have been derived for the sink $z$, contradiction immediately follows from the sink axioms.

We want to mimic this in CDCL as described in the algorithm Pebble in Fig. 3 producing a CDCL trace. For a pebble placement, we want to learn first $w_1 \vee w_2$, corresponding to "half of the pebble" on $w$, and then $\overline{w}_1 \vee \overline{w}_2$. How to do this is described in the procedure HalfPebble in Fig. 4, where the notation $x^b$ for $b \in \{0, 1\}$ is used as a compact way of denoting $x^1 = x$ and $x^0 = \overline{x}$.

When a pebble is placed on $w$ in the pebbling, we let the CDCL solver make the decisions $(w_1 = 0/d, w_2 = 0/d)$ with the goal of learning $w_1 \vee w_2$. Then we decide values for the variables of the predecessors $u$ and $v$ of $w$, and since there are clauses in memory encoding $u_1 \oplus u_2$ and $v_1 \oplus v_2$ this will provoke repeated conflicts until finally the clause $w_1 \vee w_2$ is learned. Since this only involves a constant number of variables, the time and space required for this is constant, and our goal can be achieved, e.g., as described in FindConflicts in Fig. 5.

But now we run into problems. At this point the CDCL solver will backjump to the decision $w_1 = 0$, where the learned clause $w_1 \vee w_2$ asserts $w_2 = 1$. As the next step, we want to generate conflicts that lead to the "second half" of the pebble $\overline{w}_1 \vee \overline{w}_2$ being learned, but there is no way this can happen since the decision $w_1 = 0$ is on the trail and the clause $\overline{w}_1 \vee \overline{w}_2$ is thus satisfied. Moreover, if the solver is not allowed to restart, then this satisfying assignment is fixed on the trail, and no new conflict could possibly cause a backjump to before this assignment. Therefore, the solver is forced to continue the proof search elsewhere. This turns out to be a major obstacle, which we are able to circumvent only by substantial extra work involving reordering the pebbling and using a different algorithm. Unfortunately the technical arguments are rather intricate and the algorithm too long to describe; we refer to the full-length version for the details.

If we instead give the solver the option to restart at this point, it can clear the trail and also forget all unnecessary clauses. This means that the decisions $(w_1 = 1/\mathsf{d}, w_2 = 1/\mathsf{d})$ can be made, after which the clause $\overline{w}_1 \vee \overline{w}_2$ is learned in

---

**Input**: a black pebbling $\mathcal{P}$
1 **foreach** *(move,w) in $\mathcal{P}$ where $w$ is not a source or the sink* **do**
2     **if** *move is Add* **then**
3        HalfPebble $(w, 0)$
4        HalfPebble $(w, 1)$
5     **else**
6        del $w_1 \vee w_2$
7        del $\overline{w}_1 \vee \overline{w}_2$
8 PebbleSink

---

**Fig. 3.** Procedure Pebble $(\mathcal{P})$

---

**Input**: A vertex $w$, a Boolean $b$
1 Decide $w_1 = b/\mathsf{d}$
2 Decide $w_2 = b/\mathsf{d}$
3 FindConflicts $(w, b)$
4 Learn $w_1^{1-b} \vee w_2^{1-b}$ and assert $w_2 = 1 - b/\mathsf{u}$
5 Restart R
6 **foreach** *clause $C \in \mathcal{D} \setminus F$ such that $|C| > 2$* **do**
7    del $C$

---

**Fig. 4.** Procedure HalfPebble $(w, b)$

the same way as above. To conclude, we again trigger a restart and erase all auxiliary clauses that are no longer needed.

Pebble removals are very straightforward to simulate: the only condition that could stop us from erasing the clauses $w_1 \vee w_2$ and $\overline{w}_1 \vee \overline{w}_2$ is if they are reasons for propagated literals on the trail, but since we have just made a restart the trail is empty. Formalizing the arguments above, we obtain the following lemma.

---

**Input**: A vertex $w$ with predecessors $u$ and $v$, a Boolean $b$
1 Decide $u_1 = 0/d$
2 Propagate $u_2 = 1/u_1 \vee u_2$
3 Decide $v_1 = 0/d$
4 Learn $w_1^{1-b} \vee w_2^{1-b} \vee u_1 \vee \overline{u}_2 \vee v_1$
5 Assert $v_1 = 1/w_1^{1-b} \vee w_2^{1-b} \vee u_1 \vee \overline{u}_2 \vee v_1$
6 Learn $w_1^{1-b} \vee w_2^{1-b} \vee u_1$
7 Assert $u_1 = 1/w_1^{1-b} \vee w_2^{1-b} \vee u_1$
8 Propagate $u_2 = 0/\overline{u}_1 \vee \overline{u}_2$
9 Decide $v_1 = 0/d$
10 Learn $w_1^{1-b} \vee w_2^{1-b} \vee \overline{u}_1 \vee u_2 \vee v_1$
11 Assert $v_1 = 1/w_1^{1-b} \vee w_2^{1-b} \vee \overline{u}_1 \vee u_2 \vee v_1$

---

**Fig. 5.** Procedure `FindConflicts` $(w, b)$

**Lemma 5.** *If $\mathcal{P}$ is a black pebbling of $G$ in space $s$ and time $\tau$, then there is a CDCL proof of $Peb_G^{\oplus}$ with restarts using the 1UIP learning scheme and any unit propagation scheme in space at most $2s + 3$ and time $O(\tau)$.*

*Proof.* Given a pebbling $\mathcal{P}$ we generate a trace as described by the procedure `Pebble`. Note that this procedure maintains the invariant that the pebble clauses for a non-source vertex $w$ are in the clause database if and only if there is a pebble on $w$. No other clauses are in memory. The space bound follows from this invariant, and the time bound holds by construction.

It remains to check that the trace thus generated is legal. Observe that the clauses in memory only propagate if at least one variable from each vertex they mention is set. Since the decision sequence mentions at most three vertices at the same time, we only need to reason about clauses that mention these vertices.

The correctness of `FindConflicts` is straightforward to verify, since the order of unit propagations can be seen to be uniquely determined. At the end of `FindConflicts`, the assignments to $w_1$ and $w_2$ are decisions and all predecessor variables $u_1, u_2, v_1, v_2$ are set by unit propagation. Since one of the conflicting clauses, a pebbling axiom, contains the decision variables $w_1$ and $w_2$, they have to appear in any conflict clause. The remaining variables involved in the conflict have maximal decision level, so they cannot appear in an asserting clause because $w_2$ already appears. Therefore, we learn the clause $w_1^{1-b} \vee w_2^{1-b}$ and assert $w_2 = 1 - b/u$. We only erase clauses after a restart, so no erased clauses can be reasons for unit propagations. This concludes the proof.

# 4   Concluding Remarks

In this paper, we present a proof system that closely models conflict-driven clause learning (CDCL) and yields natural measures not only of running time but also of memory usage and number of restarts. To the best of our knowledge, previous papers considered either zero restarts or very frequent restarts, and none of the models captured space. We show that lower bounds on proof size and space in resolution carry over to this CDCL proof system. Furthermore, we establish that currently known trade-offs between size and space in resolution can be transformed into essentially equally strong trade-offs between time and memory usage for CDCL, where the upper bounds are achieved by CDCL without any restarts using the standard 1UIP clause learning scheme, and the lower bounds apply even for arbitrarily frequent restarts and arbitrary clause learning schemes.

The focus of our work is theoretical, namely to see if CDCL proof search is in principle subject to the kind of trade-offs shown previously for the resolution proof system in which it searches for proofs. Since the answer turns out to be yes, an interesting direction for future work would be to investigate experimentally whether anything like these time-space trade-offs show up also in practice.

Two other interesting problems are whether CDCL with 1UIP can simulate general resolution efficiently with respect to both time and space (measuring time only, a polynomial simulation follows from [30]), and whether CDCL with 1UIP and without restarts can simulate or be separated from regular resolution. If one believes that a separation should be more likely, a first step could be to revisit the formulas in [17,19] and study them in our model, which is much closer to actual CDCL search and where proving lower bounds might therefore be easier. It should be said, though, that both of these problems still look like formidable challenges.

A more specialized question along the same lines, but still quite intriguing, is what can be said if VSIDS and phase saving is plugged into our CDCL model. The VSIDS heuristic seems like an important part of what makes CDCL SAT solvers so successful in practice, and yet there are also theoretical combinatorial formulas where it seems to be less useful. It would be interesting if one could find explicit examples of formulas where VSIDS in combination with phase saving goes provably wrong compared to the best possible resolution proof, causing a large polynomial or even superpolynomial blow-up in proof size.

**Acknowledgements.** We are grateful to the anonymous *SAT* conference reviewers for detailed comments that helped improve the exposition in this paper.

The third author performed this work while at KTH Royal Institute of Technology, and most of the work of the second and fourth author was done while visiting KTH. The first, third, fifth, and sixth author were funded by the European Research Council under the European Union's Seventh Framework Programme (FP7/2007–2013) / ERC grant agreement no. 279611 as well as by Swedish Research Council grant 621-2012-5645. The third author was also supported by the European Research Council under the European Union's Horizon 2020 Research and Innovation Programme/ERC grant agreement no. 648276.

# References

1. Alekhnovich, M., Ben-Sasson, E., Razborov, A.A., Wigderson, A.: Space complexity in propositional calculus. SIAM J. Comput. **31**(4), 1184–1211 (2002). Preliminary version in STOC 2000
2. Alekhnovich, M., Razborov, A.A.: Resolution is not automatizable unless W[P] is tractable. SIAM J. Comput. **38**(4), 1347–1363 (2008). Preliminary version in FOCS 2001
3. Atserias, A., Fichte, J.K., Thurley, M.: Clause-learning algorithms with many restarts and bounded-width resolution. J. Artif. Intell. Res. **40**, 353–373 (2011). Preliminary version in SAT 2009
4. Audemard, G., Simon, L.: Predicting learnt clauses quality in modern SAT solvers. In: Proceedings of the 21st International Joint Conference on Artificial Intelligence (IJCAI 2009), pp. 399–404, July 2009
5. Bayardo Jr., R.J., Schrag, R.: Using CSP look-back techniques to solve real-world SAT instances. In: Proceedings of the 14th National Conference on Artificial Intelligence (AAAI 1997), pp. 203–208, July 1997
6. Beame, P., Beck, C., Impagliazzo, R.: Time-space tradeoffs in resolution: superpolynomial lower bounds for superlinear space. In: Proceedings of the 44th Annual ACM Symposium on Theory of Computing (STOC 2012), pp. 213–232, May 2012
7. Beame, P., Kautz, H., Sabharwal, A.: Towards understanding and harnessing the potential of clause learning. J. Artif. Intell. Res. **22**, 319–351 (2004). Preliminary version in IJCAI 2003
8. Beame, P., Sabharwal, A.: Non-restarting SAT solvers with simple preprocessing can efficiently simulate resolution. In: Proceedings of the 28th National Conference on Artificial Intelligence (AAAI 2014), pp. 2608–2615. AAAI Press, July 2014
9. Beck, C., Nordström, J., Tang, B.: Some trade-off results for polynomial calculus. In: Proceedings of the 45th Annual ACM Symposium on Theory of Computing (STOC 2013), pp. 813–822, May 2013
10. Ben-Sasson, E., Galesi, N.: Space complexity of random formulae in resolution. Random Struct. Algorithms **23**(1), 92–109 (2003). Preliminary version in CCC 2001
11. Ben-Sasson, E., Nordström, J.: Understanding space in proof complexity: separations and trade-offs via substitutions. In: Proceedings of the 2nd Symposium on Innovations in Computer Science (ICS 2011). pp. 401–416, January 2011
12. Ben-Sasson, E., Wigderson, A.: Short proofs are narrow–resolution madesimple. J. ACM **48**(2), 149–169 (2001). Preliminary version in STOC 1999
13. Bennett, P., Bonacina, I., Galesi, N., Huynh, T., Molloy, M., Wollan, P.: Space proof complexity for random 3-CNFs. Technical report arXiv.org:1503.01613, April 2015
14. Blake, A.: Canonical Expressions in Boolean Algebra. Ph.D. thesis, University of Chicago (1937)
15. Bonacina, I.: Total space in resolution is at least width squared. In: Proceedings of the 43rd International Colloquium on Automata, Languages and Programming (ICALP 2016), (to appear, July 2016)
16. Bonacina, I., Galesi, N., Thapen, N.: Total space in resolution. In: Proceedings of the 55th Annual IEEE Symposium on Foundations of Computer Science (FOCS 2014), pp. 641–650, October 2014
17. Bonet, M.L., Buss, S., Johannsen, J.: Improved separations of regular resolution from clause learning proof systems. J. Artif. Intell. Res. **49**, 669–703 (2014)

18. Buss, S.R., Hoffmann, J., Johannsen, J.: Resolution trees with lemmas: resolution refinements that characterize DLL-algorithms with clause learning. Logical Meth. Comput. Sci. **4**(4:13), 1–28 (2008)
19. Buss, S.R., Kołodziejczyk, L.: Small stone in pool. Logical Meth. Comput. Sci. **10**, 1–22 (2014)
20. Chvátal, V., Szemerédi, E.: Many hard examples for resolution. J. ACM **35**(4), 759–768 (1988)
21. Cook, S.A., Reckhow, R.: The relative efficiency of propositional proof systems. J. Symbolic Logic **44**(1), 36–50 (1979)
22. Davis, M., Logemann, G., Loveland, D.: A machine program for theorem proving. Commun. ACM **5**(7), 394–397 (1962)
23. Davis, M., Putnam, H.: A computing procedure for quantification theory. J. ACM **7**(3), 201–215 (1960)
24. Esteban, J.L., Torán, J.: Space bounds for resolution. Inf. Comput. **171**(1), 84–97 (2001). Preliminary versions of these results appeared in STACS 1999 and CSL 1999
25. Haken, A.: The intractability of resolution. Theor. Comput. Sci. **39**(2–3), 297–308 (1985)
26. Hertel, P., Bacchus, F., Pitassi, T., Van Gelder, A.: Clause learning can effectively P-simulate general propositional resolution. In: Proceedings of the 23rd National Conference on Artificial Intelligence (AAAI 2008), pp. 283–290, July 2008
27. Marques-Silva, J.P., Sakallah, K.A.: GRASP: a search algorithm for propositional satisfiability. IEEE Trans. Comput. **48**(5), 506–521 (1999). Preliminary version in ICCAD 1996
28. Moskewicz, M.W., Madigan, C.F., Zhao, Y., Zhang, L., Malik, S.: Chaff: engineering an efficient SAT solver. In: Proceedings of the 38th Design Automation Conference (DAC 2001). pp. 530–535, June 2001
29. Nieuwenhuis, R., Oliveras, A., Tinelli, C.: Solving SAT and SAT modulo theories: from an abstract Davis-Putnam-Logemann-Loveland procedure to DPLL(T). J. ACM **53**(6), 937–977 (2006)
30. Pipatsrisawat, K., Darwiche, A.: On the power of clause-learning SAT solvers as resolution engines. Artif. Intell. **175**, 512–525 (2011). Preliminary version in CP 2009
31. Urquhart, A.: Hard examples for resolution. J. ACM **34**(1), 209–219 (1987)
32. Van Gelder, A.: Pool resolution and its relation to regular resolution and DPLL with clause learning. In: Sutcliffe, G., Voronkov, A. (eds.) LPAR 2005. LNCS (LNAI), vol. 3835, pp. 580–594. Springer, Heidelberg (2005)
33. Zhang, L., Madigan, C.F., Moskewicz, M.W., Malik, S.: Efficient conflict driven learning in boolean satisfiability solver. In: Proceedings of the IEEE/ACM International Conference on Computer-Aided Design (ICCAD 2001), pp. 279–285, November 2001

# Satisfiability Applications

# A SAT Approach to Branchwidth

Neha Lodha, Sebastian Ordyniak, and Stefan Szeider$^{(\boxtimes)}$

Algorithms and Complexity Group, TU Wien, Vienna, Austria
{neha,ordyniak,sz}@ac.tuwien.ac.at

**Abstract.** Branch decomposition is a prominent method for structurally decomposing a graph, hypergraph or CNF formula. The width of a branch decomposition provides a measure of how well the object is decomposed. For many applications it is crucial to compute a branch decomposition whose width is as small as possible. We propose a SAT approach to finding branch decompositions of small width. The core of our approach is an efficient SAT encoding which determines with a single SAT-call whether a given hypergraph admits a branch decomposition of certain width. For our encoding we developed a novel partition-based characterization of branch decomposition. The encoding size imposes a limit on the size of the given hypergraph. In order to break through this barrier and to scale the SAT approach to larger instances, we developed a new heuristic approach where the SAT encoding is used to locally improve a given candidate decomposition until a fixed-point is reached. This new method scales now to instances with several thousands of vertices and edges.

## 1 Introduction

*Background.* Branch decomposition is a prominent method for structurally decomposing a graph or hypergraph. This decomposition method was originally introduced by Robertson and Seymour [17] in their Graph Minors Project and has become a key notion in discrete mathematics and combinatorial optimization. Branch decompositions can be used to decompose other combinatorial objects such as matroids, integer-valued symmetric submodular functions, and propositional CNF formulas (after dropping of negations, clauses can be considered as (hyper-)edges). The width of a branch decomposition provides a measure of how well it decomposes the given object; the smallest width over its branch decompositions denotes the *branchwidth* of an object. Many hard computational problems can be solved efficiently by means of dynamic programming along a branch decomposition of small width. Prominent examples include the traveling salesman problem [6], the #P-complete problem of propositional model counting [3], and the generation of resolution refutations for unsatisfiable CNF formulas [2]. In fact, all decision problems on graphs that can be expressed in monadic second order logic can be solved in linear time on graphs that admit a branch decomposition of bounded width [10].

---

Dedicated to the memory of Helmuth Veith (1971–2016).

© Springer International Publishing Switzerland 2016
N. Creignou and D. Le Berre (Eds.): SAT 2016, LNCS 9710, pp. 179–195, 2016.
DOI: 10.1007/978-3-319-40970-2_12

A bottleneck for all these algorithmic applications is the space requirement of dynamic programming, which is typically single or double exponential in the width of the given branch decomposition. Hence it is crucial to compute first a branch decomposition whose width is as small as possible. This is very similar to the situation in the context of treewidth, where the following was noted about inference on probabilistic networks of bounded treewidth [15]:

> [...] since inference is exponential in the tree-width, a small reduction in tree-width (say by even by 1 or 2) can amount to one or two orders of magnitude reduction in inference time.

Hence small improvements in the width can change a dynamic programming approach from unfeasible to feasible. The boundary between unfeasible and feasible width values strongly depends on the considered problem and the currently available hardware. For instance, Cook and Seymour [6] mention a threshold of 20 for the Traveling Salesman Problem in 2003. Today one might consider a higher threshold. Computing an optimal branch decomposition is NP-hard [19].

*Contribution.* In this paper we propose a practical SAT-based approach to finding a branch decompositions of small width. At the core of our approach is an efficient SAT encoding which takes a hypergraph $H$ and an integer $w$ as input and produces a propositional CNF formula which is satisfiable if and only if $H$ admits a branch decomposition of width $w$. By multiple calls of the solver with various values of $w$ we can determine the smallest $w$ for which the formula is satisfiable (i.e., the branchwidth of $H$), and we can transform the satisfying assignment into an optimal branch decomposition. Our encoding is based on a novel *partition-based* characterization of branch decompositions in terms of certain sequences of partitions of the set of edges. This characterization together with clauses that express cardinality constraints allow for an efficient SAT encoding that scales up to instances with about hundred edges. The computationally most expensive part in this procedure is to determine the optimality of $w$ by checking that the formula corresponding to a width of $w - 1$ is unsatisfiable. If we do not insist on optimality and aim at good upper bounds, we can scale the approach to larger hypergraphs with over two hundred edges.

The number of clauses in the formula is polynomial in the size of the hypergraph and the given width $w$, but the order of the polynomial can be quintic, hence there is a firm barrier to the scalability of the approach to larger hypergraphs. In order to break through this barrier, we developed a new *SAT-based local improvement* approach where the encoding is not applied to the entire hypergraph but to certain smaller hypergraphs that represent local parts of a current candidate branch decomposition. The overall procedure thus starts with a branch decomposition obtained by a heuristic method and then tries to improve it locally by multiple SAT-calls until a fixed-point (or timeout) is reached. This method scales now to instances with several thousands of vertices and edges and branchwidth upper bounds well over hundred. We believe that a similar approach using a SAT-based local improvement could also be developed for other (hyper)graph width measures.

*Related Work.* Previously, SAT techniques have been proposed for other graph width measures: Samer and Veith [18] proposed a SAT encoding for *treewidth*, based on a characterization of treewidth in terms of elimination orderings (that is, the encoding entails variables whose truth values determine a permutation of the vertices, and the width of a corresponding decomposition is then bounded by cardinality constraints). This approach was later improved by Berg and Järvisalo [4] who empirically evaluated various SAT and MaxSAT strategies for treewidth encodings based on elimination orderings. Heule and Szeider [11] developed a SAT approach for computing the *clique-width* of graphs. For this purpose they developed a novel partition-based characterization of clique-width. Our encoding of branchwidth was inspired by this. However, the two encodings are different as clique-width and branchwidth are entirely different notions.

For finding branch decompositions of smallest width, Robertson and Seymour [17] suggested an exponential-time algorithm which was later implemented by Hicks [12]. Further exponential-time algorithms have been proposed (see, for instance [9,14]) but there seem to be no implementations. Ulusal [20] proposed an encoding to integer programming (CPLEX). One could also find suboptimal branch decompositions based on the related notion of tree decompositions; however, finding an optimal tree decomposition is again NP-hard, and by transforming it into a branch decomposition one introduces an approximation error factor of up to 50 % [17] which makes this approach prohibitive in practice. For practical purposes one therefore mainly resorts to heuristic methods that compute suboptimal branch decompositions [6,13,16].

Due to the space restrictions several proofs have been omitted.

## 2   Preliminaries

*Formulas and Satisfiability.* We consider propositional formulas in Conjunctive Normal Form (*CNF formulas*, for short), which are conjunctions of clauses, where a clause is a disjunction of literals, and a literal is a propositional variable or a negated propositional variables. A CNF formula is *satisfiable* if its variables can be assigned true or false, such that each clause contains either a variable set to true or a negated variable set to false. The satisfiability problem (SAT) asks whether a given formula is satisfiable.

*Graphs and Branchwidth.* We consider finite hypergraphs and undirected graphs. For basic terminology on graphs we refer to a standard text book [8]. For a (hyper-)graph $H$ we denote by $V(H)$ the vertex set of $H$ and by $E(H)$ the edge set of $H$. If $E \subseteq E(H)$, we denote by $H \setminus E$ the hypergraph with vertices $V(H)$ and edges $E(H) \setminus E$. Let $G$ be a simple undirected graph. The *radius* of $G$, denoted by $\mathrm{rad}(G)$, is the minimum integer $r$ such that $G$ has a vertex from which all other vertices are reachable via a path of length at most $\mathrm{rad}(G)$. The *center* of $G$ is the set of vertices such that all other vertices of $G$ can be reached via a path of length at most $\mathrm{rad}(G)$. We will often consider various forms of trees, i.e., connected acyclic graphs, as they form the backbone of branch decompositions.

Let $T$ be an undirected tree. We will always assume that $T$ is rooted (in some arbitrary vertex $r$) and hence the parent and child relationships between its vertices are well-defined. We say that $T$ is ternary if every non-leaf vertex of $T$ has degree exactly three. We will write $p_T(t)$ (or just $p(t)$ if $T$ is clear from the context) to denote the parent of $t \in V(T)$ in $T$. We also write $T_t$ to denote the subtree of $T$ rooted in $t$, i.e., the component of $T \setminus \{\{t, p_T(t)\}\}$ containing $t$. For a tree $T$, we denote by $h(T)$, the *height* of $T$, i.e., the length of a longest path between the root and any leaf of $T$ plus one. It is well-known that every tree has at most two center vertices, moreover, if it has two center vertices then they form the endpoints of an edge in the tree.

Let $H$ be a hypergraph. Every subset $E$ of $E(H)$ defines a *cut* of $H$, i.e., the pair $(E, E(H) \setminus E)$. We denote by $\delta_H(E)$ (or just $\delta(E)$ if $H$ is clear form the context) the set of *cut vertices* of $E$ in $H$, i.e., $\delta(E)$ contains all vertices incident to both an edge in $E$ and an edge in $E(H) \setminus E$. Note that $\delta(E) = \delta(E(H) \setminus E)$.

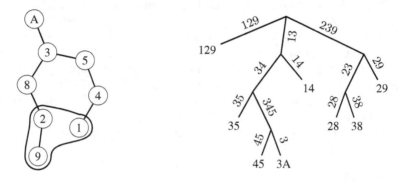

**Fig. 1.** A hypergraph $H$ (left) and an optimal branch decomposition $(T, \gamma)$ of $H$ (right). The labels of the leaves of $T$ are the edges assigned to them by $\gamma$ and the labels of the edges of $T$ are the cut vertices of that edge.

A *branch decomposition* $\mathcal{B}(H)$ of $H$ is a pair $(T, \gamma)$, where $T$ is a ternary tree and $\gamma : L(T) \to E(H)$ is a bijection between the edges of $H$ and the leaves of $T$ (denoted by $L(T)$). For simplicity, we write $\gamma(L)$ to denote the set $\{\gamma(l) \mid l \in L\}$ for a set of leaves $L$ of $T$ and we also write $\delta(T')$ instead of $\delta(\gamma(L(T')))$ for a subtree $T$ of $T'$. For an edge $e$ of $T$, we denote by $\delta_\mathcal{B}(e)$ (or simply $\delta(e)$ if $\mathcal{B}$ is clear from the context), the set of *cut vertices* of $e$, i.e., the set $\delta(T')$, where $T'$ is any of the two components of $T \setminus \{e\}$. Observe that $\delta_\mathcal{B}(e)$ consists of the set of all vertices $v$ such that there are two leaves $l_1$ and $l_2$ of $T$ in distinct components of $T \setminus \{e\}$ such that $v \in \gamma(l_1) \cap \gamma(l_2)$. The *width* of an edge $e$ of $T$ is the number of cut vertices of $e$, i.e., $|\delta_\mathcal{B}(e)|$ and the *width* of $\mathcal{B}$ is the maximum width of any edge of $T$. The *branchwidth* $\mathrm{bw}(H)$ of $H$ is the minimum width over all branch decompositions of $H$ (or 0 if $|E(G)| = 0$ and $H$ has no branch decomposition). We also define the *depth* of $\mathcal{B}$ as the radius of $T$. Figure 1 illustrates a branch decomposition of a small hypergraph. In the figure and in the remainder of the paper we will often denote a set $\{1, 2, 3, A\}$ of vertices as $123A$. We will use the following property of branch decompositions.

**Proposition 1.** *Let $\mathcal{B} := (T, \gamma)$ and $\mathcal{B}' := (T', \gamma')$ be two branch decompositions of the same hypergraph $H$. Then there is bijection $\alpha : V(T) \to V(T')$ between the vertices of $T$ such that $l \in L(T)$ if and only if $\alpha(l) \in L(T')$ and moreover $\gamma(l) = \gamma'(\alpha(l))$ for every $l \in L(T)$. In other words w.l.o.g. one can assume that $\mathcal{B}$ and $\mathcal{B}'$ differ only in terms of the edges of $T$ and $T'$.*

*Partitions.* As partitions play an important role in our reformulation of branch-width, we recall some basic terminology. A *partition* of a set $S$ is a set $P$ of nonempty subsets of $S$ such that any two sets in $P$ are disjoint and $S$ is the union of all sets in $P$. The elements of $P$ are called *equivalence classes*. Let $P, P'$ be partitions of $S$. Then $P'$ is a *refinement* of $P$ if for any two elements $x, y \in S$ that are in the same equivalence class of $P'$ are also in the same equivalence class of $P$ (this entails the case $P = P'$). Moreover, we say that $P'$ is a *$k$-ary refinement* of $P$ if additionally it holds that for every $p \in P$ there are $p_1, \dots, p_k$ in $P'$ such that $p = \bigcup_{1 \leq i \leq k} p_i$.

# 3   Partition-Based Reformulation of Branchwidth

One might be tempted to think that the original characterization of branch decompositions as ternary trees leads to a very natural and efficient SAT encoding for the existence of a branch decomposition of a certain width. In particular, in the light of Proposition 1 one could encode the branch decomposition as a formlula by fixing all vertices of the tree (as well as the bijection on the leaves) and then employing variables to guess the children for each inner vertex of the tree. We have tried this approach, however, to our surprise the performance of the encoding based on this characterization of branch decomposition was very poor. We therefore opted to develop a different encoding based on a new partition-based characterization of branch decomposition which we will introduce next. Compared to this, the original encoding was clearly inferior, resulting in an encoding size that was always at least twice as large and overall solving times that where longer by a factor of 3–10, even after several rounds of fine-tuning and experimenting with natural variants.

Let $H$ be a hypergraph. A *derivation* $\mathcal{P}$ of $H$ of *length* $l$ is a sequence $(P_1, \dots, P_l)$ of partitions of $E(G)$ such that:

D1 $P_1 = \{\, \{e\} \mid e \in E(H) \,\}$ and $P_l = \{E(H)\}$ and
D2 for every $i \in \{1, \dots, l-1\}$, $P_i$ is a 2-ary refinement of $P_{i+1}$ and
D3 $P_{l-1}$ is a 3-ary refinement of $P_l$.

The *width* of $\mathcal{P}$ is the maximum size of $\delta_H(E)$ over all sets $E \in \bigcup_{1 \leq i < l} P_i$. We will refer to $P_i$ as the *$i$-th level* of the derivation $\mathcal{P}$ and we will refer to elements in $\bigcup_{1 \leq i \leq l} P_i$ as *sets* of the derivation. We will show that any branch decomposition can be transformed into a derivation of the same width and also the other way around. The following example illustrates the close connection between branch decompositions and derivations.

*Example 1.* Consider the branch decomposition $\mathcal{B}$ given in Fig. 1. Then $\mathcal{B}$ can, e.g., be translated into the derivation $\mathcal{P} = (P_1, \ldots, P_5)$ defined by:

$$P_1 = \Big\{ \{129\}, \{35\}, \{45\}, \{3A\}, \{14\}, \{28\}, \{38\}, \{29\} \Big\}$$

$$P_2 = \Big\{ \{129\}, \{35\}, \{45, 3A\}, \{14\}, \{28\}, \{38\}, \{29\} \Big\}$$

$$P_3 = \Big\{ \{129\}, \{35, 45, 3A\}, \{14\}, \{28, 38\}, \{29\} \Big\}$$

$$P_4 = \Big\{ \{129\}, \{35, 45, 3A, 14\}, \{28, 38, 29\} \Big\}$$

$$P_5 = \Big\{ \{129, 35, 45, 3A, 14, 28, 38, 29\} \Big\}$$

The width of $\mathcal{B}$ is equal to the width of $\mathcal{P}$.

The following theorem shows that derivations provide an alternative characterization of branch decompositions.

**Theorem 1.** *Let $H$ be a hypergraph and $w$ and $d$ two integers. $H$ has a branch decomposition of width at most $w$ and depth at most $d$ if and only if $H$ has a derivation of width at most $w$ and length at most $d$.*

One important parameter influencing the size of the encoding for the existence of a derivation is the length of the derivation. The next theorem shows a tight upper bound on the length of any derivation required to rule out the existence of a branch decomposition. Observe that a simple caterpillar (i.e., a path where each inner vertex has one additional "pending" neighbor) shows that the bound given below is tight. The main observations behind the following theorem are that every branch decomposition has depth at most $\lfloor |E(H)|/2 \rfloor$ and moreover for certain branch decompositions one can further reduce its depth by replacing subtrees at the bottom of the branch decomposition containing at most $\lceil w/e \rceil$ leaves with complete binary subtrees of height at most $\lceil \log \lfloor w/e \rfloor \rceil$.

**Theorem 2.** *Let $H$ be a hypergraph, $e$ the maximum size over all edges of $H$, and $w$ an integer. Then the branchwidth of $H$ is at most $w$ if and only if $H$ has a derivation of width at most $w$ and length at most $\lfloor |E(H)|/2 \rfloor - \lceil w/e \rceil + \lceil \log \lfloor w/e \rfloor \rceil$.*

## 4   Encoding

Let $H$ be a hypergraph with $m$ edges and $n$ vertices, and let $w$ and $d$ be positive integers. We will assume that the vertices of $H$ are represented by the numbers from 1 to $n$ and the edges of $H$ by the numbers from 1 to $m$. The aim of this section is to construct a formula $F(H, w, d)$ that is satisfiable if and only if $H$ has derivation of width at most $w$ and length at most $d$. Because of Theorem 2 (after setting $d$ to the value specified in the theorem) it holds that $F(H, w, d)$ is satisfiable if and only if $H$ has branchwidth at most $w$. To achieve this aim we first construct a formula $F(H, d)$ that is satisfiable if and only if $H$ has a derivation of length at most $d$ and then we extend this formula by adding constrains that restrict the width of the derivation to $w$.

## 4.1   Encoding of a Derivation of a Hypergraph

The formula $F(H, d)$ uses the following variables. A *set variable* $s(e, f, i)$, for every $e, f \in E(H)$ with $e < f$ and every $i$ with $0 \leq i \leq d$. Informally, $s(e, f, i)$ is true whenever $e$ and $f$ are contained in the same set at level $i$ of the derivation. A *leader variable* $l(e, i)$, for every $e \in E(H)$ and every $i$ with $0 \leq i \leq d$. Informally, the leader variables will be used to uniquely identify the sets at each level of a derivation, i.e., $l(e, i)$ is true whenever $e$ is the smallest edge in a set at level $i$ of the derivation.

We now describe the clauses of the formula. The following clauses ensure (D1) and that the derivation is a sequence of refinements.

$$(\neg s(e, f, 0)) \wedge (s(e, f, d)) \wedge (\neg s(e, f, i) \vee s(e, f, i + 1))$$
$$\text{for } e, f \in E(H), \ e < f, \ 1 \leq i < d$$

The following clauses ensure that the relation of being in the same set is transitive.

$$(\neg s(e, f, i) \vee \neg s(e, g, i) \vee s(f, g, i))$$
$$\wedge (\neg s(e, f, i) \vee \neg s(f, g, i) \vee s(e, g, i))$$
$$\wedge (\neg s(e, g, i) \vee \neg s(f, g, i) \vee s(e, f, i)) \text{ for } e, f, g \in E(H), \ e < f < g, \ 1 \leq i \leq d$$

The following clauses ensure that $l(e, i)$ is true if and only if $e$ is the smallest edge contained in some set at level $i$ of a derivation.

$$\underbrace{\left(l(e, i) \vee \bigvee_{f \in E(H), f < e} s(f, e, i)\right)}_{A} \wedge \underbrace{\bigwedge_{f \in E(H), f < e} (\neg l(e, i) \vee \neg s(f, e, i))}_{B}$$
$$\text{for } e \in E(H), \ 1 \leq i \leq d$$

Part $A$ ensures that $e$ is a leader or it is in a set with an edge which is smaller than $e$ and the part $B$ ensures that if $e$ is not in same set with any smaller edge then it is a leader. The following clauses ensure that at most two sets in the partition at level $i$ can be combined into a set in the partition at level $i + 1$, i.e., together with the clauses above it ensures (D2).

$$\neg l(e, i) \vee \neg l(f, i) \vee \neg s(e, f, i + 1) \vee l(e, i + 1) \vee l(f, i + 1)$$
$$\text{for } e, f \in E(H), \ e < f, \ 1 \leq i < d - 1$$

The following clauses ensure that at most three sets in the partition at level $d - 1$ can be combined into a set in the partition at level $d$, i.e., together with the clauses above it ensures (D3).

$$\neg l(e, d - 1) \vee \neg l(f, d - 1) \vee \neg l(g, d - 1) \vee \neg s(e, f, d) \vee \neg s(e, g, d)$$
$$\vee l(e, d) \vee l(f, d) \vee l(g, d) \qquad \text{for } e, f, g \in E(H), \ e < f < g$$

All of the above clauses together ensure (D1), (D2), and (D3). We also add the following redundant clauses.

$$l(e, i) \vee \neg l(e, i + 1) \qquad\qquad\qquad \text{for } e \in E(H), \ 1 \leq i < d$$

These clauses use the observation that if an edge is not a leader at level $i$ then it cannot be a leader at level $i + 1$. The formula $F(H, d)$ contains at most $\mathcal{O}(m^2 d)$ variables and $\mathcal{O}(m^3 d)$ clauses.

## 4.2   Encoding of a Derivation of Bounded Width

Next we describe how $F(H, d)$ can be extended to restrict the width of the derivation. The main idea is to first identify the set of cut vertices for the sets in the derivation and then restrict their sizes. To this end we first need to introduce new variables (and later clauses), which allow us to identify cut vertices of edge sets in the derivation. In particular, we introduce a *cut variable* $c(e, u, i)$ for every $e \in E(H)$, $u \in V(H)$ and $i$ with $1 \leq i \leq d$. Informally, $c(e, u, i)$ is true if $u$ is a cut vertex of the set containing $e$ at level $i$ of the derivation. In order to restrict the size of the sets of cut vertices later on we do not need the reverse direction of the previous statement. Recall that a vertex $u$ is a cut vertex for some set $p$ of the derivation if there are two distinct edges incident to $u$ such that one of them is contained in $p$ and the other one is not.

*Defining the Cut Vertices.* In the following we will present an encoding that has turned out to give the best results in our case. The main idea behind the encoding is to only define the variables $c(e, u, i)$ for the leading edges $e$ in the current derivation.

The following clauses ensure that whenever two edges incident to a vertex are not in the same set at level $i$ of the derivation, then the vertex is a cut vertex for every leading edge of the sets containing the incident edges.

$$\neg l(e, i) \lor c(e, u, i) \lor s(\min\{e, f\}, \max\{e, f\}, i) \lor \neg s(\min\{e, g\}, \max\{e, g\}, i)$$
$$\text{for } e, f, g \in E(H),\ e \neq f,\ e \neq g,\ u \in V(H),\ u \in f,\ u \in g,\ 1 \leq i \leq d$$

$$\neg l(e, i) \lor s(\min\{e, f\}, \max\{e, f\}, i) \lor c(e, u, i)$$
$$\text{for } e, f \in E(H),\ e \neq f,\ u \in V(H),\ u \in e,\ u \in f,\ 1 \leq i \leq d$$

Additionally, we add the following redundant clauses that ensure the "monotonicity" of the cut vertices, i.e., if $u$ is a cut vertex for a set at level $i$ and for the corresponding set at level $i + 2$, then it also has to be a cut vertex at level $i + 1$.

**Table 1.** An illustration of the behavior of the sequential counter for the case that $H$ has six vertices (labeled from 1 to 6) and $w = 4$. The first column identifies the vertex $u$, the second column gives the value of the variable $c(e, u, i)$ for a fixed edge $e$ and a fixed level $i$ and the last four columns give the values of the variables $\#(e, u, i, j)$.

| $u$ | $c(e, u, i)$ | $j$ 1 | 2 | 3 | 4 |
|---|---|---|---|---|---|
| 1 | 0 | 0 | 0 | 0 | 0 |
| 2 | 1 | 1 | 0 | 0 | 0 |
| 3 | 1 | 1 | 1 | 0 | 0 |
| 4 | 0 | 1 | 1 | 0 | 0 |
| 5 | 1 | 1 | 1 | 1 | 0 |
| 6 | 0 | 1 | 1 | 1 | 0 |

$$\neg l(e, i) \vee \neg l(e, i+1) \vee \neg l(e, i+2) \vee \neg c(e, u, i) \vee \neg c(e, u, i+2) \vee c(e, u, i+1))$$

$$\text{for } e \in E(H), u \in V(H), 1 \leq i \leq d-2$$

The definition of cut vertices adds at most $\mathcal{O}(mnd)$ variables and at most $\mathcal{O}(m^3nd)$ clauses.

*Restricting the Size of the Cuts.* Next we describe how to restrict the size of all sets of cut vertices to $w$ and thereby complete the encoding of $F(H, w, d)$. In particular, our aim is to restrict the number of vertices $u \in V(H)$ for which a variable $c(e, u, i)$ is true for some $e \in E(H)$ and $1 \leq i \leq d$. In this paper we will only present the *sequential counter* approach [18] since this approach has turned out to provide the best results in our setting. We also considered the *order encoding* [11] with less promising results. For the sequential counter, we will introduce a *counter variable* $\#(e, u, i, j)$ for every $e \in E(H)$, $u \in V(H)$, $1 \leq i \leq d$, $1 \leq j \leq w$.

The idea of the sequential counter is illustrated in Table 1. Informally, $\#(e, u, i, j)$ is true if $u$ is the lexicographically $j$-th cut vertex of the edge $e$. We need the following clauses.

$$(\neg\#(e, u-1, i, j) \vee \#(e, u, i, j)) \wedge (\neg c(e, u, i) \vee \neg\#(e, u-1, i, j-1))$$
$$\vee\#(e, u, i, j)) \wedge (\neg c(e, u, i) \vee \neg\#(e, u-1, i, w))$$

$$\text{for } e \in E(H), 2 \leq u \leq |V(H)|, 1 \leq i \leq d, 1 \leq j \leq w$$

$$\neg c(e, u, i) \vee \#(e, u, i, 1) \qquad\qquad \text{for } e \in E(H), 1 \leq u \leq |V(H)|, 1 \leq i \leq d$$

This completes the construction of the formula $F(H, w, d)$. In total $F(H, w, d)$ has at most $\mathcal{O}(m^2d + mndw) \subseteq \mathcal{O}(m^3 + m^2n^2)$ variables and at most $\mathcal{O}(m^3nd + mndw) \subseteq \mathcal{O}(m^4n + m^2n^2)$ clauses. By construction, $F(H, w, d)$ is satisfiable if and only if $H$ has a derivation of width at most $w$ and length at most $d$. Because of Theorem 1, we obtain:

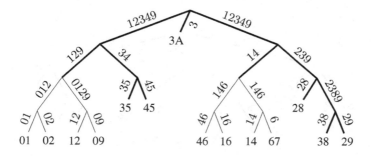

**Fig. 2.** A branch decomposition $\mathcal{B}$ of the graph $H$ given in Fig. 3 together with an example of a local branch decomposition $\mathcal{B}_L$ (highlighted by thicker edges) chosen by our algorithm.

**Theorem 3.** *The formula $F(H, w, d)$ is satisfiable if and only if $H$ has a branch decomposition of width at most $w$ and depth at most $d$. Moreover, a corresponding branch decomposition can be constructed from a satisfying assignment of $F(H, w, d)$ in linear time in terms of the number of variables of $F(H, w, d)$.*

## 5  Local Improvement

The encoding presented in the previous section allows us to compute the exact branchwidth of hypergraphs up to a certain size. Due to the instinct difficulty of the problem one can hardly hope to go much further beyond this size barrier with an exact method. In this section we therefore propose a local improvement approach that employs our SAT encoding to improve small parts of an heuristically obtained branch decomposition. Our local improvement procedure can be seen as a kind of local search procedure that at each step tries to replace a part of the branch decomposition with an better one found

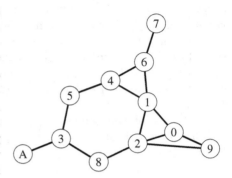

**Fig. 3.** The graph $H$ used to illustrate the main idea behind our local improvement procedure.

by means of the SAT encoding and repeats this process until a fixed-point (or timeout) is reached.

Let $H$ be a hypergraph and $\mathcal{B} := (T, \gamma)$ a branch decomposition of $H$. For a connected ternary subtree $T_L$ of $T$ we define the *local branch decomposition* $\mathcal{B}_L := (T_L, \gamma_L)$ of $\mathcal{B}$ by setting $\gamma_L(l) = \delta_{\mathcal{B}}(e)$ for every leaf $l \in L(T_L)$, where $e$ is the (unique) edge incident to $l$ in $T_L$. We also define the hypergraph $H(T_L)$ as the hypergraph that has one (hyper-)edge $\gamma_L(l)$ for every leaf $l$ of $T_L$ and whose vertices are defined as the union of all these edges. We observe that $\mathcal{B}_L$ is a branch decomposition of $H(T_L)$. The main idea behind our approach, which

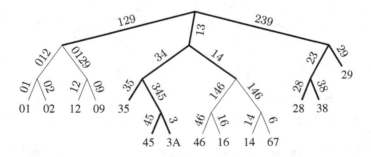

**Fig. 4.** The improved branch decomposition $\mathcal{B}'$ obtained from $\mathcal{B}$ after replacing the local branch decomposition $\mathcal{B}_L$ of $H(T_L)$ with an optimal branch decomposition $\mathcal{B}'_L$ of $H(T_L)$ obtained from our SAT encoding. See Fig. 2 for an illustration of $\mathcal{B}$ and $\mathcal{B}_L$.

we will formalize below, is that we can obtain a new branch decomposition of $H$ by replacing the part of $\mathcal{B}$ formed by $\mathcal{B}_L$ with any branch decomposition of $H(T_L)$. In particular, by replacing $\mathcal{B}_L$ with a branch decomposition of $H(T_L)$ of lower width, we will potentially improve the branch decomposition $\mathcal{B}$. This idea is illustrated in Figs. 2 and 4.

input  : A hypergraph $H$
output: A branch decomposition of $H$

$\mathcal{B} \leftarrow$ BDHeuristic($H$) // ($\mathcal{B} := (T, \gamma)$)
improved $\leftarrow$ *true*
while improved do
    $M \leftarrow$ "the set of edges $e$ of $\mathcal{B}$ whose width
      ($|\delta_{\mathcal{B}}(e)|$) is maximum"
    $\mathcal{C} \leftarrow$ "the set of components of $T[M]$"
    improved $\leftarrow$ *false*
    for $C \in \mathcal{C}$ do
      $\mathcal{B}_L \leftarrow$ LocalBD $(\mathcal{B}, C)$
      $\mathcal{B}'_L \leftarrow$ ImproveLD $(\mathcal{B}_L)$
      if $\mathcal{B}'_L \neq$ NULL then
        $\mathcal{B} \leftarrow$ Replace $(\mathcal{B}, \mathcal{B}_L, \mathcal{B}'_L)$
        improved $\leftarrow$ *true*
      else
        break

**Algorithm 1.** Local Improvement

input  : A branch decomposition
      $\mathcal{B} := (T, \gamma)$ of $H$ and a
      branch decomposition
      $\mathcal{B}_L := (T_L, \gamma_L)$ of $H(T_L)$
output: An "improved" branch
      decomposition of $H(T_L)$

if $|T_L| >$globalbudget then
    return *NULL*
$w \leftarrow$ "the width of $\mathcal{B}_L$"
repeat
    $\mathcal{B}_D \leftarrow$ SATSolve($H(T_L), w$)
    if $\mathcal{B}_D \neq$ NULL then
      $\mathcal{B}'_L \leftarrow \mathcal{B}_D$
    $w \leftarrow w - 1$
until $\mathcal{B}_D ==$ NULL
return $\mathcal{B}'_L$

**Algorithm 2.** ImproveLD

A general outline of our algorithm is given in Algorithm 1. The algorithm uses two global parameters: globalbudget gives an upper bound on the size of the local branch decomposition and the function length(H,w), which is only used by the function SATSolve explained below, provides an upper bound on the length of a derivation which will be considered by our SAT encoding.

Given a hypergraph $H$, the algorithm first computes a (not necessarily optimal) branch decomposition $\mathcal{B} := (T, \gamma)$ of $H$ using, e.g., the heuristics from [6,13]. The algorithm then computes the set $M$ of maximum cut edges of $T$, i.e., the set of edges $e$ of $T$ with $|\delta(e)| = w$, where $w$ is the width of $\mathcal{B}$. It then computes the set $\mathcal{C}$ of components of $T[M]$ and for every component $C \in \mathcal{C}$ it calls the function LocalBD to obtain a local branch decomposition $\mathcal{B}_L := (T_L, \gamma_L)$ of $\mathcal{B}$, which contains (at least) all the edges of $C$. The function LocalBD is given in Algorithm 3 and will be described later. Given $\mathcal{B}_L$ the algorithm tries to compute a branch decomposition $\mathcal{B}'_L := (T'_L, \gamma'_L)$ of $H(T_L)$ with smaller width than $\mathcal{B}_L$ using the function ImproveLD, which is described later. If successful, the algorithm updates $\mathcal{B}$ by replacing the part of $\mathcal{B}$ represented by $T_L$ with $\mathcal{B}'_L$ according to Theorem 4 and proceeds with line 4. If on the other hand $\mathcal{B}_L$ cannot be improved, the algorithm proceeds with the next component $C$ of $T[M]$. This process is repeated until none of the components $C$ of $T[M]$ lead to an improvement.

The function LocalBD, which is given in Algorithm 3, computes a local branch decomposition $\mathcal{B}_L := (T_L, \gamma_L)$ of $\mathcal{B}$ that contains at least all edges in the component $C$ and which should be small enough to ensure solvability found by our

**input** : A branch decomposition $\mathcal{B} := (T, \gamma)$ of $H$ and a component $C$
of $T$
**output**: A local branch decomposition of $\mathcal{B}$

1  $w \leftarrow$ "the width of $\mathcal{B}$"
2  $T_L \leftarrow C$
3  **for** $c \in V(C)$ *with* $\deg_C(c) = 2$ **do**
4  | "add the unique third neighbor and its edge incident to $c$ to $T_L$"
5  $Q \leftarrow$ "the set of leaves of $T_L$"
6  **while** $Q \neq \emptyset$ *and* $|T_L| \leq$ **globalbudget** $-2$ **do**
7  | $l \leftarrow Q.\text{pop}()$
8  | **if** "$l$ *is not a leaf of* $T$" **then**
9  | | $c, c' \leftarrow$ "the two neighbors of $l$ in $T$ which are not neighbors of $l$
   | | in $T_L$"
10 | | **if** $\delta_{\mathcal{B}}(\{l, c\}) < w$ *and* $\delta_{\mathcal{B}}(\{l, c'\}) < w$ **then**
11 | | | "add $c$ and $c'$ together with their edges incident to $l$ to $T_L$"
12 | | | $Q.\text{push}(c)$
13 | | | $Q.\text{push}(c')$
14 **return** "the local branch decomposition of $\mathcal{B}$ represented by $T_L$"

**Algorithm 3.** Local Selection (LocalBD)

SAT encoding as follows. In the beginning $T_L$ is set to the connected ternary subtree of $T$ obtained from $T[C]$ after adding the (unique) third neighbor of any vertex $v$ of $C$ that has degree exactly two in $T[C]$. It then proceeds by processing the (current) leaves of $T_L$ in a breadth first search manner, i.e., in the beginning all the leaves of $T_L$ are put in a first-in first-out queue $Q$. If $l$ is the current leaf of $T_L$, which is not a leaf of $T$, the algorithm adds the two additional neighbors of $l$ in $T$ to $T_L$ and adds them to $Q$. It proceeds in this manner until the number of edges in $T_L$ does exceed the global budget.

The function ImproveLD tries to compute a branch decomposition of $H(T_L)$ with lower width than $\mathcal{B}_L$ using our SAT encoding. In particular, if the size of $T_L$ does not exceed the global budget (in which case it would be highly unlikely that a lower width branch decomposition can be found using our SAT encoding), the function calls the function SATSolve with decreasing widths $w$ until SATSolve does not return a branch decomposition any more. Here, the function SATSolve uses the formula $F(H(T_L), w, d)$ from Theorem 3 with $d$ set to length(H,w) to test whether $H(T_L)$ has a branch decomposition of width at most $w$ and depth at most $d$. If so (and if the SAT-solver solves the formula within a predefined timeout) SATSolve returns the corresponding branch decomposition; otherwise it returns NULL.

Last but not least the function Replace replaces the part of $\mathcal{B}$ represented by $\mathcal{B}_L$ with the new branch decomposition $\mathcal{B}'_L$ according to Theorem 4.

Let $H$ be a hypergraph, $\mathcal{B} := (T, \gamma)$ a branch decomposition of $H$, $T_L$ a connected ternary subtree of $T$, $\mathcal{B}_L := (T_L, \gamma_L)$ be the local branch decomposition of $\mathcal{B}$ corresponding to $T_L$, and let $\mathcal{B}'_L := (T'_L, \gamma')$ be any branch decomposition of $H(T_L)$. Note that because $\mathcal{B}_L$ and $\mathcal{B}'_L$ are branch decompositions of the same hypergraph $H(T_L)$, we obtain from Proposition 1 that we can assume that $V(T_L) = V(T'_L)$ and $\gamma = \gamma'$. We define the *locally improved* branch decomposition, denoted by $\mathcal{B}(\frac{\mathcal{B}_L}{\mathcal{B}'_L})$, to be the branch decomposition obtained from $\mathcal{B}$ by replacing the part corresponding to $\mathcal{B}_L$ with $\mathcal{B}'_L$, i.e., the tree of $\mathcal{B}'$ is obtained from $T$ by removing all edges of $T_L$ from $T$ and replacing them with the edges of $T'_L$ and the bijection of $\mathcal{B}'$ is equal to $\gamma$.

**Theorem 4.** $\mathcal{B}(\frac{\mathcal{B}_L}{\mathcal{B}'_L})$ *is a branch decomposition of $H$, whose width is the maximum the width of $\mathcal{B}'_L$ and maximum width over all edges $e \in E(T) \setminus E(T_L)$ in $\mathcal{B}$.*

# 6  Experimental Results

We have implemented the single SAT encoding and the SAT-based local improvement method and tested them on various benchmark instances, including famous named graphs from the literature [21], graphs from TreewidthLIB [5] which origin from a broad range of applications, and a series of circular clusters [7] which are hypergraphs denoted $C_v^e$ with $v$ vertices and $v$ edges of size $e$. Throughout we used the SAT-solver Glucose 4.0 (with standard parameter setting) as it performed best in our initial tests compared to other solvers such as GlueMiniSat 2.2.8, Lingeling, and Riss 4.27. We run the experiments on a 4-core Intel Xeon CPU E5649, 2.35 GHz, 72 GB RAM machine with Ubuntu 14.04 with each process having access to at most 8 GB RAM.

## 6.1  Single SAT Encoding

To determine the branchwidth of a graph or hypergraph with our encoding, one could either start from $w = 1$ and increase $w$ until the formula becomes satisfiable, or by setting $w$ to an upper bound on the branchwidth obtained by a heuristic method, and decrease it until the formula becomes unsatisfiable. For both approaches the solving time at the threshold (i.e., for the largest $w$ for which the formula is unsatisfiable) is, as one would expect, by far the longest. Table 2 shows this behavior on some typical instances. Hence whether we determine the branchwidth from below or from above does not matter much. A more elaborate binary search strategy could save some time, but overall the expected gain is little compared to the solving time at the threshold. The size of the encoding is manageable for graphs and hypergraphs for up to about 100 edges. The solving time varies and depends on the structure of the (hyper)graph. We could determine the exact branchwidth of many famous graphs known from the literature, see Table 3. For many of the graphs the exact branchwidth has not been known before. We also tested the encoding on circular cluster hypergraphs $C_{2i-1}^i$. We were able to solve instances up to the hypergraph $C_{51}^{26}$, for which we established a branchwidth of 42.

**Table 2.** Distribution of solving time in seconds for various values of $w$ for some famous named graphs of branchwidth 6.

| $w$ | 2 | 3 | 4 | 5 | 6 | 7 | 8 | 9 | 10 |
|---|---|---|---|---|---|---|---|---|---|
| Graph | unsat | unsat | unsat | unsat | sat | sat | sat | sat | sat |
| Kittell | 19.5 | 87.6 | 400.8 | 204.7 | 103.3 | 40.4 | 22.5 | 18.5 | 11.2 |
| Errera | 5.7 | 22.7 | 79.4 | 1530.9 | 12.0 | 7.3 | 6.7 | 5.4 | 6.1 |
| Folkman | 3.4 | 13.7 | 98.6 | 2747.0 | 6.1 | 5.3 | 3.7 | 3.8 | 5.2 |
| Poussin | 3.3 | 9.2 | 68.7 | 941.2 | 4.5 | 3.5 | 3.9 | 2.9 | 3.4 |

**Table 3.** Exact branchwidth $w$ of famous named graphs known from the literature. Column $d$ indicates the smallest depth of a branch decomposition of width $w$ for the considered graph.

| Graph | $|V|$ | $|E|$ | $w$ | $d$ | Graph | $|V|$ | $|E|$ | $w$ | $d$ | Graph | $|V|$ | $|E|$ | $w$ | $d$ |
|---|---|---|---|---|---|---|---|---|---|---|---|---|---|---|
| Watsin | 50 | 75 | 6 | 8 | McGee | 24 | 36 | 7 | 7 | Dürer | 12 | 18 | 4 | 6 |
| Kittell | 23 | 63 | 6 | 8 | Nauru | 24 | 36 | 6 | 7 | Franklin | 12 | 18 | 4 | 6 |
| Holt | 27 | 54 | 9 | 9 | Hoffman | 16 | 32 | 6 | 6 | Frucht | 12 | 18 | 3 | 6 |
| Shrikhande | 16 | 48 | 8 | 7 | Desargues | 20 | 30 | 6 | 6 | Herschel | 11 | 18 | 4 | 6 |
| Errera | 17 | 45 | 6 | 7 | Dodecahedron | 20 | 30 | 6 | 6 | Tietze | 12 | 18 | 4 | 6 |
| Brinkmann | 21 | 42 | 8 | 7 | Flower Snark | 20 | 30 | 6 | 6 | Petersen | 10 | 15 | 4 | 6 |
| Clebsch | 16 | 40 | 8 | 7 | Goldner-Harary | 11 | 27 | 4 | 6 | Pmin | 9 | 12 | 3 | 5 |
| Folkman | 20 | 40 | 6 | 7 | Pappus | 18 | 27 | 6 | 6 | Wagner | 8 | 12 | 4 | 5 |
| Paley13 | 13 | 39 | 7 | 7 | Sousselier | 16 | 27 | 5 | 6 | Moser spindle | 7 | 11 | 3 | 6 |
| Poussin | 15 | 39 | 6 | 7 | Chvátal | 12 | 24 | 6 | 6 | Prism | 6 | 9 | 3 | 5 |
| Robertson | 19 | 38 | 8 | 7 | Grötzsch | 11 | 20 | 5 | 6 | Butterfly | 5 | 6 | 2 | 3 |

## 6.2  SAT-Based Local Improvement

We tested our local improvement method on graphs with several thousands of vertices and edges and with initial branch decomposition of width over 200. In particular, we tested it on all graphs from TreewidthLIB omitting graphs that are minors from other graphs as well as small graphs with 80 or fewer edges (small graphs can be solved with the single SAT encoding). These are in total 684 graphs with up to 5934 vertices and 17770 edges. We ran our SAT-based local improvement algorithm on each graph with a timeout of 6 h, where each SAT-call had a timeout of 600 s. We used a global budget of 80 and set the depth to 0.6 times the upper bound provided by Theorem 2. We computed the initial branch decomposition by a greedy heuristic kindly provided to us by Hicks [12].

From the 684 graphs, our SAT-based local improvement algorithm could improve the width of the initial branch decomposition for 290 graphs. In some cases the improvement was significant. Table 4 shows the graphs with the best improvement.

**Table 4.** Results for SAT-based local improvement for instances from TreewidthLIB. Column $iw$ gives the width of the initial branch decomposition, $w$ the width of the branch decomposition obtained by local improvement.

| Graph | $|V|$ | $|E|$ | $iw$ | $w$ | Graph | $|V|$ | $|E|$ | $iw$ | $w$ |
|---|---|---|---|---|---|---|---|---|---|---|
| inithx.i.2-pp | 363 | 8897 | 55 | 45 | celar10-pp-002 | 76 | 421 | 20 | 16 |
| fpsol2.i.2-pp | 333 | 7910 | 48 | 39 | fpsol2.i.2 | 451 | 8691 | 53 | 49 |
| fpsol2.i.3-pp | 333 | 7907 | 48 | 39 | graph05-wpp | 94 | 397 | 28 | 24 |
| graph13 | 458 | 1877 | 141 | 134 | graph04-pp | 179 | 678 | 52 | 48 |
| fpsol2.i.3 | 425 | 8688 | 53 | 48 | nrw1379.tsp | 1379 | 4115 | 42 | 38 |
| graph13pp-pp | 374 | 1722 | 133 | 128 | nrw1379.tsp-pp | 1367 | 4081 | 42 | 38 |
| graph09-pp | 405 | 1525 | 128 | 123 | bn_36 | 1444 | 4181 | 45 | 42 |
| bn_31-pp | 1148 | 3317 | 40 | 36 | inithx.i.1-pp | 317 | 12720 | 68 | 65 |
| bn_4 | 100 | 574 | 42 | 38 | u724.tsp | 724 | 2117 | 29 | 26 |
| celar08-pp-003 | 76 | 421 | 20 | 16 | water-wpp | 22 | 96 | 11 | 8 |
| celar08pp-pp.dgf-034 | 76 | 421 | 20 | 16 | celar05-pp | 80 | 426 | 18 | 15 |
| celar09-pp-002 | 76 | 421 | 20 | 16 | mulsol.i.2-pp | 116 | 2468 | 62 | 59 |

## 6.3   Discussion

As discussed earlier, we are aware of only two implemented algorithms that determine the exact branchwidth of a graph or hypergraph: Hick's combinatorial algorithm based on tangles [12], and Ulusal's integer programming encoding [20]. Since neither of the two implementations are available to the public, we were not able to provide an up-to-date comparision with their approaches but instead performed the comparision with respect to the results stated in the papers. It should therefore be taken into account that hardware and software improved since the time their results were obtained. As Ulusal [20] reports, the integer programming encoding could not solve hypergraphs with more than 13 edges, for instance, it could only solve the circular clusters $C_{2i-1}^i$ up to $i = 7$, whereas we could go up to $i = 26$.

Because of the high branchwidth of these hypergraphs, they are also far out of reach for the tangles-based algorithm. On small graphs the tangles-based algorithm and our SAT encoding perform similarly. For very small branchwidth the tangles-based algorithm can deal with larger graphs (according to [12]) whereas our SAT encoding can deal with graphs with larger branchwidth. These differences in scalability can be explained by the space requirements of the two approaches: The tangles-based algorithm requires space that is exponential in the branchwidth, whereas our SAT encoding requires polynomial space that depends only linearly on the branchwidth.

Our experiments show that the SAT-based local improvement approach scales well to large graphs with several thousands of vertices and edges and branchwidth upper bounds well over hundred. These are instances that are by far out of reach for any known exact method, in particular, for the tangles-based algorithm which cannot handle large branchwidth. The use of our SAT encoding which scales well with the branchwidth is therefore essential for these instances.

Our results on TreewidthLIB instances show that in some cases the obtained improvement can make a difference of whether a dynamic programming algorithm that uses the obtained branch decomposition is feasible or not.

# 7    Final Remarks

We have presented a first SAT encoding for branchwidth and introduced the new method of SAT-based local improvements for branch decompositions. Both methods are based on a novel partition-based formulation of branch decompositions. Our experiments show that the single encoding outperforms a known integer programming method and performs competitively with the best known combinatorial method. Our SAT-based local improvement method provides the means for scaling the SAT-approach to much larger instances and exhibits a fruitful new application field of SAT solvers.

For both the single SAT encoding and the SAT-based local improvement we see several possibilities for further improvement. For the encoding one can try other ways for stating cardinality constraints and one could apply incremental SAT solving techniques. Further, one could consider alternative encoding techniques based on MaxSAT, which have been shown effective for related problems [4]. For the local improvement we see various directions for further research. For instance, when a local branch decomposition cannot be improved, one could use the SAT solver to obtain an alternative branch decomposition of the same width but where other parameters are optimized, e.g., the number of maximum cuts. This could propagate into adjacent local improvement steps and yield an overall branch decomposition of smaller width.

Finally we would like to mention that branch decompositions are the basis for several other (hyper)graph width measures such as rankwidth and Boolean-width [1], and we leave the investigation on how our approaches can be extended to these width measures for future research.

**Acknowledgement.** We thank Illya Hicks for providing us the code of his branch-width heuristics and acknowledge support by the Austrian Science Fund (FWF, projects W1255-N23 and P-27721).

# References

1. Adler, I., Bui-Xuan, B.-M., Rabinovich, Y., Renault, G., Telle, J.A., Vatshelle, M.: On the boolean-width of a graph: structure and applications. In: Thilikos, D.M. (ed.) WG 2010. LNCS, vol. 6410, pp. 159–170. Springer, Heidelberg (2010)
2. Alekhnovich, M., Razborov, A.A.: Satisfiability, branch-width and Tseitin tautologies. In: Proceedings of the 43rd Annual IEEE Symposium on Foundations of Computer Science (FOCS 2002), pp. 593–603 (2002)

3. Bacchus, F., Dalmao, S., Pitassi, T.: Algorithms and complexity results for #SAT and Bayesian inference. In: 44th Annual IEEE Symposium on Foundations of Computer Science (FOCS 2003), pp. 340–351 (2003)
4. Berg, J., Järvisalo, M.: SAT-based approaches to treewidth computation: an evaluation. In: 26th IEEE International Conference on Tools with Artificial Intelligence, ICTAI 2014, Limassol, Cyprus, 10–12 November 2014, pp. 328–335. IEEE Computer Society (2014)
5. Bodlander, H.: TreewidthLIB a benchmark for algorithms for treewidth and related graph problems. http://www.staff.science.uu.nl/~bodla101/treewidthlib/
6. Cook, W., Seymour, P.: Tour merging via branch-decomposition. INFORMS J. Comput. 15(3), 233–248 (2003)
7. Cornuéjols, G.: Combinatorial Optimization: Packing and Covering. Regional Conference Series in Applied Mathematics. Society for Industrial and Applied Mathematics, Carnegie Mellon University, Pittsburgh, Pennsylvania (2001)
8. Diestel, R.: Graph Theory. Graduate Texts in Mathematics, vol. 173, 2nd edn. Springer, New York (2000)
9. Fomin, F.V., Mazoit, F., Todinca, I.: Computing branchwidth via efficient triangulations and blocks. Discr. Appl. Math. 157(12), 2726–2736 (2009)
10. Grohe, M.: Logic, graphs, and algorithms. In: Flum, J., Grädel, E., Wilke, T. (eds.) Logic and Automata: History and Perspectives. Texts in Logic and Games, vol. 2, pp. 357–422. Amsterdam University Press (2008)
11. Heule, M.J.H., Szeider, S.: A SAT approach to clique-width. In: Järvisalo, M., Van Gelder, A. (eds.) SAT 2013. LNCS, vol. 7962, pp. 318–334. Springer, Heidelberg (2013)
12. Hicks, I.V.: Graphs, branchwidth, and tangles! Oh my! Networks 45(2), 55–60 (2005)
13. Hicks, I.V.: Branchwidth heuristics. Congr. Numer. 159, 31–50 (2002)
14. Hliněný, P., Oum, S.: Finding branch-decompositions and rank-decompositions. SIAM J. Comput. 38(3), 1012–1032 (2008)
15. Kask, K., Gelfand, A., Otten, L., Dechter, R.: Pushing the power of stochastic greedy ordering schemes for inference in graphical models. In: Burgard, W., Roth, D. (eds.) Proceedings of the Twenty-Fifth AAAI Conference on Artificial Intelligence, AAAI 2011, San Francisco, California, USA, 7–11 August 2011. AAAI Press (2011)
16. Overwijk, A., Penninkx, E., Bodlaender, H.L.: A local search algorithm for branchwidth. In: Černá, I., Gyimóthy, T., Hromkovič, J., Jefferey, K., Královič, R., Vukolić, M., Wolf, S. (eds.) SOFSEM 2011. LNCS, vol. 6543, pp. 444–454. Springer, Heidelberg (2011)
17. Robertson, N., Seymour, P.D.: Graph minors X. Obstructions to tree-decomposition. J. Combin. Theory Ser. B 52(2), 153–190 (1991)
18. Samer, M., Veith, H.: Encoding treewidth into SAT. In: Kullmann, O. (ed.) SAT 2009. LNCS, vol. 5584, pp. 45–50. Springer, Heidelberg (2009)
19. Seymour, P.D., Thomas, R.: Call routing and the ratcatcher. Combinatorica 14(2), 217–241 (1994)
20. Ulusal, E.: Integer Programming Models for the Branchwidth Problem. Ph.D. thesis, Texas A&M University, May 2008
21. Weisstein, E.: MathWorld online mathematics resource. http://mathworld.wolfram.com

# Computing Maximum Unavoidable Subgraphs Using SAT Solvers

C.K. Cuong[✉] and M.J.H. Heule

Department of Computer Science,
The University of Texas at Austin,
Austin, TX, USA
{ckcuong,marijn}@cs.utexas.edu

**Abstract.** Unavoidable subgraphs have been widely studied in the context of Ramsey Theory. The research in this area focuses on highly structured graphs such as cliques, cycles, paths, stars, trees, and wheels. We propose to study *maximum unavoidable subgraphs* measuring the size in the number of edges. We computed maximum unavoidable subgraphs for graphs up to order nine via SAT solving and observed that these subgraphs are less structured, although all are bipartite. Additionally, we found large unavoidable bipartite subgraphs up to order twelve. We also present the concept of multi-component unavoidable subgraphs and show that large multi-component subgraphs are unavoidable in small graphs. We envision that maximum unavoidable subgraphs can be exploited using an alternative approach to breaking graph symmetries.

**Keywords:** Satisfiability solving · Unavoidable subgraph · Combinatorics · Graph theory · Symmetry breaking

## 1 Introduction

Satisfiability (SAT) solvers have become very powerful tools to solve hard-combinatorial problems that have only *few* symmetries. Recent successes in this direction are solving Erdős discrepancy problem [10] and the Boolean Pythagorean Triples problem [9]. However, for hard-combinatorial problems with *many* symmetries, such as Ramsey numbers [7], SAT solvers may not be the strongest tools around. For example, the most impressive result regarding Ramsey numbers, solving $R(4,5)$ [12], is two decades old and cannot be reproduced with SAT solving yet. In this paper, we propose to study a special kind of hard-combinatorial problems that could be helpful to bridge this gap.

Consider the fully connected undirected graph of order $n$, in short $K_n$. A graph $G$ is called an *unavoidable subgraph* of $K_n$ if $G$ occurs as a fully red or fully blue subgraph in any red/blue edge-coloring of $K_n$. Unavoidable subgraphs have been widely studied, as can be observed in a survey paper [13] that cites over 600 papers on the subject. This research area focuses on highly structured graphs such as cliques, cycles [3], paths [6], stars [8], trees [4], and wheels [14].

© Springer International Publishing Switzerland 2016
N. Creignou and D. Le Berre (Eds.): SAT 2016, LNCS 9710, pp. 196–211, 2016.
DOI: 10.1007/978-3-319-40970-2_13

We propose to investigate less structured graphs: the *maximum* unavoidable subgraphs with the size measured in the number of edges.

We compute the maximum unavoidable subgraphs via SAT solving. We observe that the maximum unavoidable subgraphs for small graphs are all bipartite and conjecture that this is also the case for large graphs. Another interesting observation made within the experimented range is that $K_{n+1}$ always has a strictly larger unavoidable subgraph than $K_n$ for $n > 3$. The difference in size between the maximum unavoidable subgraphs of $K_n$ and $K_{n+1}$ is typically one edge and sometimes two edges. Consequently, the size of the maximum unavoidable subgraphs (measured in the number of edges) grows faster than the size of $K_n$ (measured in the number of vertices).

The conventional notion of unavoidable subgraphs considers only connected (single-component) subgraphs. We introduce the concept of *multi-component* unavoidable subgraphs: each component occurs in either red or blue in all red/blue edge-colorings of $K_n$ — although some components may occur in blue, while others occur in red. Starting with $K_6$, some interesting patterns can be observed in the largest found multi-component unavoidable subgraphs.

The state-of-the-art symmetry-breaking methods for SAT [1] or specifically for graphs [5] are not powerful enough to make some reasonably simple unavoidable subgraph problems solvable via SAT techniques. For example, consider the problem whether a star of eight edges, in short $S_8$, is an unavoidable subgraph of $K_{15}$. Using a short combinatorial argument one can solve this problem: $K_{15}$ has an odd number of edges (105), so not all vertices can have exactly seven red and seven blue edges. Hence, one vertex must have at least eight red or at least eight blue edges, or equivalently $S_8$ as a monochromatic subgraph.

We envision that knowledge about the maximum unavoidable subgraphs, both the single and the multi-component variants, could be a basis for novel symmetry-breaking techniques for SAT solvers. All edges in a component can be replaced by a single edge in the component, allowing significant simplification of graph problems.

The remainder of this paper is structured as follows. First we present some background information on unavoidable subgraphs in Sect. 2. In Sect. 3, we describe how to encode unavoidable subgraph problems into SAT. Computing single and multi-component unavoidable subgraphs is discussed in Sects. 4 and 5, respectively. Section 6 presents a method to exploit unavoidable subgraphs and we draw some conclusions in Sect. 7.

## 2    Unavoidable Subgraphs and Motivation

All graphs mentioned in the paper are undirected. Before presenting the definition of an unavoidable subgraph, we first introduce the concept of *graph isomorphism*: Two graphs $G$ and $H$ are isomorphic if there exists an edge-preserving bijection from the vertices of $G$ to the vertices of $H$. We say that two isomorphic graphs occur in the same *isomorphism class*.

**Definition 1 (Unavoidable subgraph).** *A graph G is called an unavoidable subgraph of the fully-connected graph $K_n$ if for all red/blue edge-colorings EC of $K_n$, there exists a subgraph H of $K_n$ such that*

1. *H is isomorphic to G.*
2. *H is monochromatic, either in red or blue, under the coloring EC of $K_n$.*

We want to emphasize that Definition 1 does not require that $H$ is the same graph for different red/blue edge-colorings of $K_n$. A small example of unavoidable subgraphs is shown in Fig. 1. This figure lists all red/blue edge-colorings of $K_3$. Notice that a monochromatic path of two edges occurs in all graphs. Hence a path of two edges is an unavoidable subgraph of $K_3$.

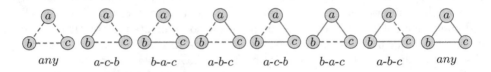

| any | a-c-b | b-a-c | a-b-c | a-c-b | b-a-c | a-b-c | any |

**Fig. 1.** All red/blue edge-colorings of $K_3$. For readability, we draw red and blue edges using solid and dashed lines, respectively. Observe that all colored graphs contain a monochromatic path of two edges (as shown below the graphs).

The two propositions below follow from the definition of unavoidable subgraphs. We will refer to them in Sect. 4:

**Proposition 1.** *If G is an unavoidable subgraph of $K_n$, then G is also an unavoidable subgraph of $K_m$ for all $m \geq n$.*

**Proposition 2.** *If G is an unavoidable subgraph of $K_n$, then all subgraphs of G are also unavoidable subgraphs of $K_n$.*

Many "nicely" structured unavoidable subgraphs have been heavily studied [13] such as cliques, cycles [3], paths [6], stars [8], trees [4], and wheels [14]. We propose to study unavoidable subgraphs somewhat differently compared to existing work. Instead of searching for graphs with a well-defined structure, we want to compute *maximum unavoidable subgraphs* (measured in the number of edges). Such maximum unavoidable subgraphs may not have a clear structure.

We argue that maximum unavoidable subgraphs are interesting, because they allow for an alternative symmetry-breaking approach for graph problems: given an avoidable subgraph, we can simplify graph problems by enforcing that all edges in the unavoidable subgraph are either all present or all absent. The larger the unavoidable subgraph, the stronger the symmetry-breaking predicate that can be derived from it. In Sect. 6 we will explain this in more detail.

# 3  SAT Encoding of Unavoidable Subgraph Problems

We employ a SAT solver to check whether a given graph $G$ of order $k$ is an unavoidable subgraph of the complete graph $K_n$ where $k \leq n$. The SAT encoding is illustrated in the following example: check whether a path of two edges is an unavoidable subgraph of $K_3$. There are three paths of two edges in $K_3$ as shown in Fig. 2. A path of two edges is an unavoidable subgraph of $K_3$ if and only if for any red/blue edge-coloring of $K_3$, at least one of the three paths in Fig. 2 is monochromatic. Let $ab$, $ac$, and $bc$ denote the Boolean variables representing the color of the edges connecting vertices $a$ and $b$, $a$ and $c$, and $b$ and $c$, respectively. If a Boolean variable has value *true*, the corresponding edge has color red, otherwise it has color blue. Then, the following Boolean formula represents the fact that at least one of the three paths in Fig. 2 is monochromatic:

$$\mathcal{F}_{G,K_3} = (ab \wedge bc) \vee (\overline{ab} \wedge \overline{bc}) \vee (ab \wedge ac) \vee (\overline{ab} \wedge \overline{ac}) \vee (ac \wedge bc) \vee (\overline{ac} \wedge \overline{bc})$$

Determining whether a path of two edges is an unavoidable subgraph of $K_3$ is equivalent to checking the validity of $\mathcal{F}_{G,K_3}$, i.e., checking if $\mathcal{F}_{G,K_3}$ holds for all truth assignments to the variables. This is then equivalent to checking if the negation of $\mathcal{F}_{G,K_3}$ is *unsatisfiable*.

$$\overline{\mathcal{F}_{G,K_3}} = (\overline{ab} \vee \overline{bc}) \wedge (ab \vee bc) \wedge (\overline{ab} \vee \overline{ac}) \wedge (ab \vee ac) \wedge (\overline{ac} \vee \overline{bc}) \wedge (ac \vee bc)$$

Since $\overline{\mathcal{F}_{G,K_3}}$ is in conjunctive normal form (CNF), SAT solvers can solve it directly. Thus, determining an unavoidable subgraph problem can be converted into a SAT problem as illustrated. The construction of $\overline{\mathcal{F}_{G,K_n}}$ is described in Algorithm 1. In particular, given the complete graph $K_n$ and its subgraph $G$ that we want to check if $G$ is unavoidable in $K_n$, for each subgraph $H$ of $K_n$ that is isomorphic to $G$, we construct two clauses: (1) disjunction of *positive* literals representing *red* color of edges in $H$, and (2) disjunction of *negative* literals representing *blue* color of edges in $H$. The formula $\overline{\mathcal{F}_{G,K_n}}$ is then the conjunction of all of these clauses.

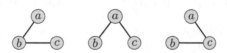

**Fig. 2.** All paths of two edges in $K_3$.

The most expensive computation in Algorithm 1 is the function call at line 2, which generates all subgraphs of $K_n$ that are isomorphic to $G$. A naive approach for generating all isomorphic graphs is to apply all permutations on every set of $k$ nodes taken from $n$ nodes, where $k$ is the number of nodes in $G$. Each permutation application will of course produce an isomorphic graph to $G$. However, duplicate graphs can be produced since different permutations can return

---

**Algorithm 1.** SAT Encoding of An Unavoidable Subgraph Problem

---
1: **function** UNAVOID-SUBGRAPH-SAT-ENCODING$(G, K_n)$
2:      $\mathcal{G} \leftarrow$ ISOMORPHIC-SUBGRAPHS-GEN$(G, K_n)$
3:      $\overline{\mathcal{F}_{G,K_n}} \leftarrow \varnothing$
4:      **for** each $H \in \mathcal{G}$ **do**
5:          $C_r \leftarrow$ disjunction of *positive* literals representing *red* color of edges in $H$
6:          $C_b \leftarrow$ disjunction of *negative* literals representing *blue* color of edges in $H$
7:          $\overline{\mathcal{F}_{G,K_n}} \leftarrow \overline{\mathcal{F}_{G,K_n}} \wedge C_r \wedge C_b$
8:      **end for**
9:      **return** $\overline{\mathcal{F}_{G,K_n}}$
10: **end function**

---

the same graph. The number of duplicate graphs can be huge in comparison with the number of non-duplicate isomorphic graphs, depending on the graph's structure and size. More importantly, the number of permutations grows rapidly as the number of nodes in $G$ increases. And the number of non-duplicate isomorphic graphs is usually very small in comparison with the number of permutations. Consequently, we will come up with spending most of the time computing duplicate graphs, which are unnecessary. We want to avoid these computations. Instead we just apply all permutations on the first set of $k$ nodes taken from $n$ nodes. After this step, we will recognize which permutations generate duplicate graphs. We can then avoid applying these permutations on all remaining sets of $k$ nodes. The details of this approach are described in Algorithm 2. The output of this algorithm is a collection of all non-duplicate subgraphs of $K_n$ that are isomorphic to the input graph $G$.

Algorithm 2 only applies the set $P_k$ of all permutations on the first set of $k$ nodes taken from $n$ nodes in the first loop (lines 7–14). This loop also constructs a *minimal* set $P'_k$ of permutations that generates all non-duplicate isomorphic graphs to $G$ for any set of $k$ nodes (recall that $k$ is the number of nodes in $G$). After this loop, the algorithm will apply $P'_k$ instead of $P_k$ for all remaining sets of $k$ nodes (lines 16–22).

## 4   Computing Unavoidable Subgraphs Using SAT Solvers

We are interested in computing unavoidable subgraphs mechanically. By exploiting the `gtools` programs in the `nauty` package [11], we are able to generate all non-isomorphic graphs of order $k$ ($\leq 10$) very quickly. For each generated graph $G$, we can mechanically check whether it is unavoidable for a given complete graph $K_n$ by converting into a SAT formula $\overline{\mathcal{F}_{G,K_n}}$ as described in the previous section, and then employing a SAT solver to check the formula's satisfiability. We use the `glucose 3.0` SAT solver [2] to do the satisfiability check.

### 4.1   Breaking Symmetries

Symmetries are the main obstacle for checking the satisfiability of $\overline{\mathcal{F}_{G,K_n}}$ efficiently. To counter this obstacle, we add symmetry-breaking predicates (SBP)

---

**Algorithm 2.** Computing Isomorphic Subgraphs

---
1: **function** ISOMORPHIC-SUBGRAPHS-GEN($G, K_n$)
2:    $P_k \leftarrow$ all permutations of $\{0, 1, 2, ..., k-1\}$, where $k$ is the order of $G$
3:    $Q_k^n \leftarrow$ all $k$-combinations of $k$ nodes taken from the vertex set of $K_n$
4:    $\mathcal{G} \leftarrow \varnothing$          ▷ Output: all subgraphs of $K_n$ that are isomorphic to $G$
5:    $P_k' \leftarrow \varnothing$    ▷ A subset of $P_k$ s.t. its application generates non-duplicate graphs

6:    Pick the first $k$-combination $Q \in Q_k^n$

7:    **for each** $\pi \in P_k$ **do**
8:        $Q' \leftarrow \pi(Q)$
9:        $G' \leftarrow Q'(G)$               ▷ Construct an isomorphic graph $G'$ of $G$
10:        **if** $G' \notin \mathcal{G}$ **then**              ▷ Check if $G'$ is not duplicate in $\mathcal{G}$
11:            $\mathcal{G} \leftarrow \mathcal{G} \cup \{G'\}$
12:            $P_k' \leftarrow P_k' \cup \{\pi\}$
13:        **end if**
14:    **end for**

15:    $Q_k^n \leftarrow Q_k^n \backslash \{Q\}$                ▷ We are done with the first combination of $Q_k^n$

16:    **for each** $Q \in Q_k^n$ **do**
17:        **for each** $\pi \in P_k'$ **do**
18:            $Q' \leftarrow \pi(Q)$
19:            $G' \leftarrow Q'(G)$              ▷ Construct an isomorphic graph $G'$ of $G$
20:            $\mathcal{G} \leftarrow \mathcal{G} \cup \{G'\}$              ▷ $G'$ is guaranteed to be non-duplicate in $\mathcal{G}$
21:        **end for**
22:    **end for**

23:    **return** $\mathcal{G}$

24: **end function**

---

to $\overline{\mathcal{F}_{G,K_n}}$ before calling a SAT solver. Unavoidable subgraph problems have two symmetries: any permutation of the vertices and swapping the edge colors. The edge color symmetry can easily be broken by selecting an edge and forcing it to a color by adding a unit clause.

Several methods have been developed to break such graph symmetries [1,5]. Existing techniques can be viewed as enforcing a lexicographic order on the rows of the adjacency matrix. Let $A_G$ be the adjacency matrix representing a graph $G$ of order $n$. The predicate $A_G[i] \preceq_{\{i,j\}} A_G[j]$ states that the binary representation of the $i$-th row of $A_G$ is less than or equal to the binary representation of the $j$-th row of $A_G$ when excluding the $i$-th and $j$-th columns of $A_G$.

**Definition 2 (Linear symmetry break).** *We define the linear symmetry-breaking predicates (L-SBP) for graph $G$ as follows:*

$$L\text{-}SBP(G) = \bigwedge_{i=1}^{n-1} A_G[i] \preceq_{\{i,i+1\}} A_G[i+1]$$

**Table 1.** The number of graphs that satisfy the symmetry-breaking predicates L-SBP and Q-SBP as compared to the number of isomorphism classes of graphs of order 5 to 12. The number of satisfying assignments were computed using `sharpSAT` [15].

| $n$ | L-SBP | Q-SBP | # isomorphism classes |
|---|---|---|---|
| 5 | 46 | 43 | 34 |
| 6 | 325 | 276 | 156 |
| 7 | 4,045 | 3,158 | 1,044 |
| 8 | 89,812 | 66,595 | 12,346 |
| 9 | 3,583,903 | 2,587,488 | 274,668 |
| 10 | 258,518,959 | 184,193,025 | 12,005,168 |
| 11 | 33,859,710,152 | 23,962,961,317 | 1,018,997,864 |
| 12 | 8,086,937,704,176 | 5,700,915,311,729 | 165,091,172,592 |

The symmetry-breaking tool `shatter` [1] adds L-SBP to graph problems such as unavoidable subgraph formulas.

Codish et al. [5] demonstrated that the comparator $\preceq_{\{i,j\}}$ is not transitive: $A_G[h] \preceq_{\{h,i\}} A_G[i] \wedge A_G[i] \preceq_{\{i,j\}} A_G[j]$ does not imply $A_G[h] \preceq_{\{h,j\}} A_G[j]$. Enforcing $A_G[i] \preceq_{\{i,j\}} A_G[j]$ for all $1 \leq i < j \leq n$ is a valid symmetry-breaking predicate for graph problems [5]. We will refer to this method as *quadratic symmetry-breaking method* since it adds a quadratic number of constraints to graph problems.

**Definition 3 (Quadratic symmetry break).** *We define the quadratic symmetry-breaking predicates (Q-SBP) for graph $G$ as follows:*

$$Q\text{-}SBP(G) = \bigwedge_{1 \leq i < j \leq n} A_G[i] \preceq_{\{i,j\}} A_G[j]$$

Table 1 shows the number of graphs / assignments that satisfy the symmetry-breaking predicates L-SBP and Q-SBP as well as the number of isomorphism classes. Notice that Q-SBP is slightly better than L-SBP. Both numbers are not close to the number of isomorphism classes meaning that many symmetries are not broken by both methods.

The impact of symmetry-breaking predicates on the time required to solve some large unavoidable subgraph problems is shown in Table 2. Notice that the runtime is reduced by orders of magnitude for the larger unavoidable subgraphs. However, there is no clear difference between the L-SBP and Q-SBP methods.

## 4.2    Enumerating Unavoidable Subgraphs

Algorithm 3 shows the pseudo-code of enumerating unavoidable subgraphs. In principle, this algorithm can be applied to compute all unavoidable subgraphs of a given complete graph $K_n$ by setting $k$ equal to $n$. Nonetheless, the number of

**Table 2.** Runtime in seconds to compute the maximum or largest found unavoidable subgraphs of $K_6$ to $K_{12}$ shown in Fig. 3 using `glucose` 3.0. The experiments were run on 3.5 GHz Intel Xeon E31280 processors with 8 MB L3 cache size. A '-' means a timeout after 24 h.

| $n$ | 6 | 7 | 8 | 9 | 10 | 11 | 12 |
|---|---|---|---|---|---|---|---|
| No SBP | 0 | 0.025 | 0.38 | 4.85 | 11,690.70 | - | - |
| L-SBP | 0 | 0 | 0.01 | 0.11 | 4.77 | 18.73 | 312.60 |
| Q-SBP | 0 | 0 | 0.01 | 0.11 | 7.98 | 19.40 | 303.40 |

---

**Algorithm 3.** Computing Unavoidable Subgraphs of $k$ Nodes for $K_n$

---

1: **function** UNAVOID-SUBGRAPHS-ORDER-K-GEN($k, K_n$)
2:  $\mathcal{G} \leftarrow$ all non-isomorphic graphs of order $k$ generated using **nauty**
3:  $\mathcal{H} \leftarrow \varnothing$          ▷ Output: all unavoidable subgraphs of order $k$ in $K_n$
4:  **for** each $G \in \mathcal{G}$ **do**
5:      $\mathcal{F}_{G,K_n} \leftarrow$ UNAVOID-SUBGRAPH-SAT-ENCODING($G, K_n$)
6:      **if** UNSAT($\mathcal{F}_{G,K_n} \wedge SBP(\mathcal{F}_{G,K_n})$) **then**
7:          $\mathcal{H} \leftarrow \mathcal{H} \cup \{G\}$
8:      **end if**
9:  **end for**
10:  **return** $\mathcal{H}$
11: **end function**

---

non-isomorphic subgraphs of $K_n$ grows rapidly as $n$ increases, so it is impractical to exhaustedly check the unavoidability of all non-isomorphic subgraphs of large complete graphs.

We reduced the number of evaluated subgraphs by ignoring several kinds of graphs that cannot be unavoidable in a given complete graph. For example, if the maximum degree of a graph exceeds some threshold, it cannot be unavoidable in a given complete graph. One may think this threshold degree is $\lfloor n/2 \rfloor$ for all $K_n$ by reasoning that there always exists a red/blue edge-coloring on any complete graph $K_n$ such that: among edges connected to each vertex $v$ in $K_n$, the number of edges of the same color is at most $\lfloor n/2 \rfloor$. As a result, subgraphs with the maximum degree higher than $\lfloor n/2 \rfloor$ cannot be unavoidable in $K_n$. However, this is not true when $n \equiv 3 \pmod 4$ as stated in Theorem 1.

**Theorem 1.** *For every $K_n$ with $n \equiv 3 \pmod 4$, there exists a graph $G$ such that $G$ is unavoidable in $K_n$ and the maximum degree of $G$ is at least $\lfloor n/2 \rfloor + 1$.*

*Proof.* We prove by contradiction. Suppose there exists a red/blue edge-coloring of $K_n$ such that among edges connected to each vertex $v$ in $K_n$, the number of edges of the same color is at most $\lfloor n/2 \rfloor$. (1)

Since $n \equiv 3 \pmod 4$, $n$ is an odd number. For each vertex $v$ in $K_n$, there are $(n - 1)$ edges connected to $v$, which is therefore an even number. (2)

From (1) and (2), we can claim that the number of red and blue edges connected to each $v$ in $K_n$ are both equal to $\lfloor n/2 \rfloor$. As a result, the number of red and blue edges in $K_n$ must be equal. (3)

Since $n \equiv 3 \pmod 4$, $K_n$ has an odd number of edges, or exactly $8l^2 + 10l + 3$ with $l = \lfloor n/4 \rfloor$. Hence, the number of red and blue edges in $K_n$ must be different, which contradicts (3). Thus there must exist a subgraph $G$ in $K_n$ with the maximum degree at least $(\lfloor n/2 \rfloor + 1)$ and $G$ is unavoidable in $K_n$.

By using our SAT solving approach, we proved that for all $n \leq 18$, the star $S_{\lfloor n/2 \rfloor + 2}$ is not an unavoidable subgraph of $K_n$ when $n \equiv 3 \pmod 4$, and $S_{\lfloor n/2 \rfloor + 1}$ is not an unavoidable subgraph of $K_n$ for other values of $n$[1]. Since the star $S_k$ (i.e., the complete bipartite graph $K_{1,k}$) is the smallest graph with the maximum degree $k$, if $S_k$ is not unavoidable in $K_n$, then no other graphs with the maximum degree at least $k$ is unavoidable in $K_n$ by the contrapositive of Proposition 2. For that reason, given a complete graph $K_n$ such that $n \leq 18$, we only need to check the unavoidability of subgraphs with the maximum degree at most $(\lfloor n/2 \rfloor + 1)$ if $n \equiv 3 \pmod 4$, and subgraphs with the maximum degree at most $\lfloor n/2 \rfloor$ for the other case of $n$.

**Table 3.** The number of graphs to be evaluated while searching for maximum unavoidable subgraphs of $K_n$ as compared to the number of isomorphism classes of graphs of order $n$ with $3 \leq n \leq 9$. Our approach imposes an upper bound on the maximum degree of graphs and a lower bound on the number of edges. We start with the fact that $K_2$ is the maximum unavoidable subgraph for itself.

| $n$ | 3 | 4 | 5 | 6 | 7 | 8 | 9 |
|---|---|---|---|---|---|---|---|
| # isomorphism classes | 4 | 11 | 34 | 156 | 1,044 | 12,346 | 274,668 |
| # checked graphs | 2 | 2 | 6 | 35 | 97 | 291 | 904 |

Since we are interested in discovering maximum unavoidable subgraphs — instead of all unavoidable subgraphs of a given complete graph $K_n$ — we skipped evaluating subgraphs that are smaller than the largest known unavoidable subgraph of $K_n$. For example, when we know that certain graph $G$ is a maximum unavoidable subgraph of $K_n$, then we only need to evaluate subgraphs with at least $(|E(G)| + 1)$ edges for $K_{n+1}$, because $G$ is also unavoidable in $K_{n+1}$ by Proposition 1. In particular, we first check subgraphs satisfying the maximum degree requirement as stated above and their number of edges equal to $(|E(G)| + 1)$. If at least one of them is unavoidable in $K_n$, we then check subgraphs with the number of edges equal to $(|E(G)| + 2)$, and so on. The process will stop if neither one of subgraphs with the number of edges $(|E(G)| + i)$ is unavoidable in $K_n$. From the contrapositive of Proposition 2, we can claim that there does not exist a subgraph $H$ in $K_n$ s.t. $|E(H)| \geq (|E(G)| + i)$ and $H$ is

---

[1] Currently, our system is not able to prove this property for $n > 18$ due to out of computational resources.

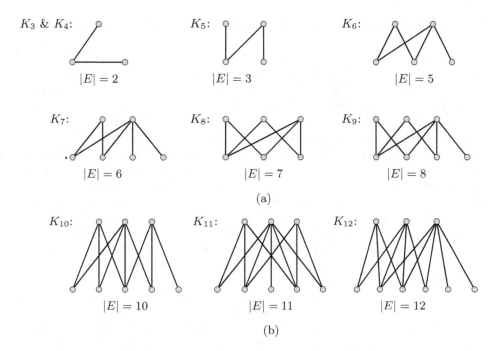

$K_3$ & $K_4$:    $|E| = 2$

$K_5$:    $|E| = 3$

$K_6$:    $|E| = 5$

$K_7$:    $|E| = 6$

$K_8$:    $|E| = 7$

$K_9$:    $|E| = 8$

(a)

$K_{10}$:    $|E| = 10$

$K_{11}$:    $|E| = 11$

$K_{12}$:    $|E| = 12$

(b)

**Fig. 3.** Some large single-component unavoidable subgraphs of $K_3$ to $K_{12}$: (a) maximum unavoidable subgraphs of $K_3$ to $K_9$, all of them are bipartite; (b) maximum bipartite unavoidable subgraphs discovered of $K_{10}$ to $K_{12}$.

unavoidable in $K_n$. In other words, a maximum unavoidable subgraph of $K_n$ must have $(|E(G)| + i - 1)$ edges. Using this approach, we are able to find maximum unavoidable subgraphs for $K_3$ to $K_9$ effectively by reducing the number of subgraphs to be checked a substantial amount (see Table 3).

### 4.3    Results on Single-Component Unavoidable Subgraphs

Figure 3 (a) shows the maximum unavoidable subgraphs for each $K_n$ with $3 \leq n \leq 9^2$. Observe that all maximum unavoidable subgraphs are *bipartite*. We used this observation to find large unavoidable subgraphs for $K_{10}$ to $K_{12}$. The bipartite restriction was required to make the number of graphs to be evaluated manageable — apart from the other restrictions of the maximum degree and the minimum number of edges. Figure 3 (b) shows the largest found unavoidable subgraphs for $K_{10}$ to $K_{12}$. We are unable to determine whether these unavoidable bipartite subgraphs are maximal: the evaluation of many graphs with one more edge required more than 24 h SAT solving time (the limit on our cluster).

---

[2] We found multiple maximum unavoidable subgraphs for $K_9$. Figure 3 (a) shows one of them.

## 5   Computing Multi-component Unavoidable Subgraphs

The concept unavoidable subgraphs as stated in Definition 1 can be generalized to multiple components, such that each component must occur monochromatic in all red/blue edge-colorings of $K_n$. We will represent a multiple-component graph $G$ by a graph with edge labels. Edges are in the same component of $G$ if and only if they have the same label. In this section, we show how to compute multi-component unavoidable subgraphs using SAT solvers. We will discuss in Sect. 6, how multi-component unavoidable subgraphs can be helpful in constructing symmetry-breaking predicates for graph problems. Given a multi-component graph $G$, we write $G_i$ to denote the subgraph of $G$ that has only the edges with label $i$. The adjacency matrix of a multi-component graphs is constructed as follows: $A_G(i,j) = A_G(j,i) = k$ if the edge connecting nodes $i$ and $j$ has label $k$ and $A_G(i,j) = A_G(j,i) = 0$ if no edge connects nodes $i$ and $j$. Figure 4 shows a 3-component graph and its adjacency matrix. In the visualizations, edges with label 1 are shown as solid lines, edges with label 2 as dashed lines, and edges with label 3 as dotted lines.

Algorithm 4 describes the SAT encoding algorithm for the case $m = 2$. Given a graph $G$ consisting of two components, first all occurrences of $G$ in $K_n$ are computed using ISOMORPHIC-SUBGRAPHS-GEN$(G, K_n)$. Notice that we had to generalize this procedure to deal with graphs that have labelled edges. For each possible presence of $G$ in $K_n$, we have 4 clauses stating that at least one component must be non-monochromatic. The algorithm can be easily generalized to $m$-component unavoidable subgraphs by adding $2^m$ clauses for each possible presence of the graph in $K_n$.

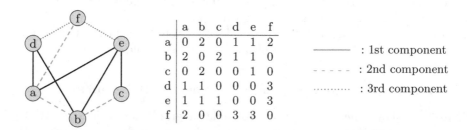

| | a | b | c | d | e | f |
|---|---|---|---|---|---|---|
| a | 0 | 2 | 0 | 1 | 1 | 2 |
| b | 2 | 0 | 2 | 1 | 1 | 0 |
| c | 0 | 2 | 0 | 0 | 1 | 0 |
| d | 1 | 1 | 0 | 0 | 0 | 3 |
| e | 1 | 1 | 1 | 0 | 0 | 3 |
| f | 2 | 0 | 0 | 3 | 3 | 0 |

———— : 1st component

- - - - - : 2nd component

············ : 3rd component

**Fig. 4.** An example illustrating the adjacency matrix of a 3-component graph.

Our mechanism for computing $m$-component unavoidable subgraphs is incremental from some $(m-1)$-component unavoidable subgraph $(m \geq 2)$. Algorithm 5 describes our mechanism for computing 2-component unavoidable subgraphs from a given single-component unavoidable subgraph. In this algorithm, we use the notation $V(G)$ to denote the vertex set of $G$, and $E_i(G)$ to denote the edges with label $i$ in $G$. Given an unavoidable subgraph $G$ of $K_n$, we generate all possible supergraphs that have one additional edge in the second component.

---

**Algorithm 4.** SAT Encoding of A 2-Component Unavoidable Subgraph Problem

---

1: **function** TWO-COMP-UNAVOID-SUBGRAPH-SAT-ENCODING$(G, K_n)$
2:     $\mathcal{G} \leftarrow$ ISOMORPHIC-SUBGRAPHS-GEN$(G, K_n)$
3:     $\overline{\mathcal{F}_{G,K_n}} \leftarrow \varnothing$
4:     **for** each $H \in \mathcal{G}$ **do**            ▷ Suppose $H$ consists of 2 components $H_1$ and $H_2$
5:         $C_{r_1} \leftarrow$ disjunction of *positive* literals representing *red* color of edges in $H_1$
6:         $C_{b_1} \leftarrow$ disjunction of *negative* literals representing *blue* color of edges in $H_1$

7:         $C_{r_2} \leftarrow$ disjunction of *positive* literals representing *red* color of edges in $H_2$
8:         $C_{b_2} \leftarrow$ disjunction of *negative* literals representing *blue* color of edges in $H_2$

9:         $\overline{\mathcal{F}_{G,K_n}} \leftarrow \overline{\mathcal{F}_{G,K_n}} \wedge (C_{r_1} \vee C_{r_2}) \wedge (C_{r_1} \vee C_{b_2}) \wedge (C_{b_1} \vee C_{r_2}) \wedge (C_{b_1} \vee C_{b_2})$
10:     **end for**
11:     **return** $\overline{\mathcal{F}_{G,K_n}}$
12: **end function**

---

**Algorithm 5.** Computing 2-Component Unavoidable Subgraphs of $K_n$ from a Given Subgraph

---

1: **function** TWO-COMP-UNAVOID-SUBGRAPHS-GEN$(G, K_n)$
2:     $\mathcal{G} \leftarrow$ all supergraphs $H$ of $G$ s.t. $|V(H)| = |V(G)|$, $|E_1(H)| = |E_1(G)|$, and $|E_2(H)| = |E_2(G)| + 1$
3:     $\mathcal{H} \leftarrow \varnothing$            ▷ Output: all 2-component graphs (extended from $G$)
                                                        ▷ that are unavoidable in $K_n$.
4:     **for** each $H \in \mathcal{G}$ **do**
5:         $\overline{\mathcal{F}_{H,K_n}} \leftarrow$ TWO-COMP-UNAVOID-SUBGRAPH-SAT-ENCODING$(H, K_n)$
6:         **if** UNSAT$(\overline{\mathcal{F}_{H,K_n}} \wedge SBP(\overline{\mathcal{F}_{H,K_n}}))$ **then**
7:             $\mathcal{H} \leftarrow \mathcal{H} \cup \{H\} \cup$ TWO-COMP-UNAVOID-SUBGRAPHS-GEN$(H, K_n)$
8:         **end if**
9:     **end for**
10:     **return** $\mathcal{H}$
11: **end function**

---

For each of those generated graphs, we check whether it is also unavoidable in $K_n$. If it is unavoidable, we recursively call the same procedure again (line 7).

The idea of this algorithm can be generalized to $m$-component unavoidable subgraph computation for all $m \geq 2$. Let $S(n)$ denote the size of maximum single-component unavoidable subgraphs of $K_n$. As heuristic to reduce the vast number of possible multi-component graphs, we start with a single component with at least $S(n) - 1$ edges. Second, we compute which one of them can be extended using a second component consisting of two edges (not necessarily connected). Notice that components consisting of a single edge can be ignored as each single edge is always monochromatic. This is repeated by trying to extend the graphs with two components to three components by adding another component with two edges. Once the starting points of unavoidable subgraphs have been determined, we try to extend them by adding edges to each component. We repeat this until no multi-component graphs can be further extended. Figure 5 displays

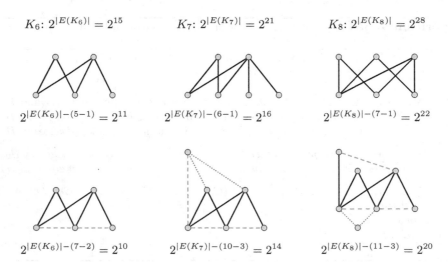

$K_6: 2^{|E(K_6)|} = 2^{15}$     $K_7: 2^{|E(K_7)|} = 2^{21}$     $K_8: 2^{|E(K_8)|} = 2^{28}$

$2^{|E(K_6)|-(5-1)} = 2^{11}$     $2^{|E(K_7)|-(6-1)} = 2^{16}$     $2^{|E(K_8)|-(7-1)} = 2^{22}$

$2^{|E(K_6)|-(7-2)} = 2^{10}$     $2^{|E(K_7)|-(10-3)} = 2^{14}$     $2^{|E(K_8)|-(11-3)} = 2^{20}$

**Fig. 5.** Comparison between maximum single-component and largest found multi-component unavoidable subgraphs of $K_6$, $K_7$, and $K_8$. The top row displays single-component graphs. The bottom row displays multi-component graphs: 2-component for $K_6$, 3-component for $K_7$ and $K_8$. For all these three cases, symmetry-breaking predicates derived from largest found multi-component unavoidable subgraphs result in smaller search spaces (measured in the number of graphs) as compared to the ones derived from maximum single-component unavoidable subgraphs.

our discovery of largest found multi-component unavoidable subgraphs for $K_6$, $K_7$, and $K_8$. One interesting observation is that the maximum single-component unavoidable subgraph of $K_6$ is an induced subgraph of the maximum single-component unavoidable subgraphs for $K_7$ and $K_8$. The same observation can also be made for the largest found multi-component unavoidable subgraphs.

The above procedure does not necessarily produce maximum multi-component unavoidable subgraphs. The used heuristics, i.e., starting with a single component of at least $S(n) - 1$ edges, may prevent us finding a maximum multi-component unavoidable subgraph. Also, we restricted the search to graphs that have at most three components. In future work, we want to compute maximum multi-component unavoidable subgraphs by applying the methods without heuristics and restrictions.

## 6     Deriving Symmetry-Breaking Predicates from Unavoidable Subgraphs

In order to compute unavoidable subgraphs efficiently, symmetry-breaking techniques were used as discussed in Sect. 4.1. Although adding symmetry-breaking predicates reduced the solving costs substantially, we also observed that these predicates are not strong enough to solve some relatively easy unavoidable subgraph problems using SAT. For example, the problem whether the star $S_8$ is unavoidable in $K_{15}$ cannot be solved even after symmetry breaking. This section describes how knowledge about some unavoidable subgraphs could be turned into an alternative approach to symmetry breaking.

The method works as follows. Given a known unavoidable subgraph $G$ of $K_n$ consisting of $m$ components, construct a predicate that forces for each component that all of its edges are equal. For example, recall that a path of two edges is an unavoidable subgraph of $K_3$. First, we pick a concrete path, say $b$-$a$-$c$, and force both edges ($a$-$b$ and $a$-$c$) to be equal, i.e., either both present or both absent in graphs of order 3. There are only four graphs of order 3 that satisfy this constraint and these are shown in Fig. 6. Moreover, these four graphs represent exactly the four isomorphism classes of graphs of order 3. This implies that the symmetry-breaking predicate derived from a path of two edges is *perfect* for graphs of order 3, i.e., exactly one graph from each isomorphism class satisfies the symmetry-breaking predicate.

**Fig. 6.** All graphs of order 3 satisfying the symmetry-breaking predicate that forces the path $b$-$a$-$c$ to be either present or absent.

More concretely, given an unavoidable subgraph $G$ of $K_n$. Let $|E_1(G)| = l$ and let $e_1, e_2,..., e_l$ denote the Boolean variables representing the edges with label 1. The symmetry-breaking predicate forcing all these variables (edges) to be equivalent is

$$e_1 \leftrightarrow e_2 \leftrightarrow e_3 \leftrightarrow \cdots \leftrightarrow e_l$$

The $(l - 1)$ equivalence relations above can be expressed using a cycle of $l$ implications, which can be represented using $l$ binary clauses as follows:

$$e_1 \leftrightarrow e_2 \leftrightarrow e_3 \leftrightarrow \cdots \leftrightarrow e_l \quad \equiv \quad e_1 \overset{\frown}{\to} e_2 \to e_3 \to \cdots \to e_l$$

$$\equiv \quad (\overline{e_1} \vee e_2) \wedge (\overline{e_2} \vee e_3) \wedge ... \wedge (\overline{e_l} \vee e_1)$$

Thus, a monochromatic component consisting of $l$ edges can be encoded with $l$ binary clauses. Applying the method for all components of $G$ results in a symmetry-breaking predicate of $|E(G)|$ binary clauses. Since any unavoidable subgraph $G$ of $K_n$ is also an unavoidable subgraph of $K_m$ where $m \geq n$, we can apply the symmetry-breaking predicate derived from $G$ in checking the unavoidability of other subgraphs of $K_m$. How useful are these symmetry-breaking predicates? Using the largest found unavoidable subgraph of $K_{11}$ as symmetry-breaking predicate to compute the largest found unavoidable subgraph of $K_{12}$ reduces the runtime to 532.48 (s). Not as strong as the L-SBP and Q-SBP predicates (see Table 2), but still reasonably well.

**Table 4.** Number of graphs of order $n$ that satisfy symmetry-breaking predicates. L-SBP-bin: the binary clauses of L-SBP; Q-SBP-bin: the binary clauses of Q-SBP; and USG-SBP: the symmetry-breaking predicates based on unavoidable subgraphs.

| $n$ | L-SBP-bin | Q-SBP-bin | USG-SBP |
|---|---|---|---|
| 6 | 5,210 | 4,672 | 1,024 |
| 7 | 196,608 | 181,248 | 16,384 |
| 8 | 14,680,064 | 13,664,256 | 1,048,576 |

In general, given an unavoidable subgraph $G$ of $K_n$ with $m$ components, the number of graphs of order $n$ that satisfy the symmetry-breaking predicate is $2^{|E(K_n)|+m-|E(G)|}$, because $|E(G)| - m$ edges will be depending on $m$ edges (the representatives of the components). Figure 5 shows how much the search space is reduced when converting the maximum or largest known unavoidable subgraphs for $K_6$, $K_7$, and $K_8$ into symmetry-breaking predicates. In all these three cases, the largest known multi-component unavoidable subgraphs are more effective than the maximum single-component unavoidable subgraphs. When comparing these numbers with Table 1, the results do not look impressive. However, the existing symmetry-breaking methods use many long clauses. If we only consider the binary clauses in symmetry-breaking predicates of existing techniques, then symmetry-breaking predicates derived from unavoidable subgraphs look much more interesting, see Table 4. The number of graphs that satisfy the predicates is much smaller.

The main question that arises is: how can we improve the unavoidable subgraph based predicates? A possible answer is finding non-binary clauses that can be added similar to the L-SBP and Q-SBP methods. Alternatively, one can search for *asymmetric unavoidable subgraphs*, such as: is a cycle of four edges occurring in red or a path of two edges occurring in blue unavoidable in $K_4$? We observed that this asymmetric subgraph is indeed unavoidable. Converting such asymmetric unavoidable subgraphs into a symmetry-breaking predicate makes it stronger. These topics will be the focus of future work.

## 7    Conclusions

We studied and computed maximum unavoidable subgraphs using SAT solvers. During our experiments we observed that all maximum unavoidable subgraphs are bipartite and conjecture that this holds in general. Also, it appears that the maximum unavoidable subgraphs of $K_{n+1}$ are strictly larger than the maximum unavoidable subgraphs of $K_n$ for $n > 3$. Symmetry breaking was crucial to obtain our results. However, we also observed that current symmetry-breaking techniques are not strong enough to compute some relatively simple unavoidable subgraph problems using SAT. We demonstrated how unavoidable subgraphs can be converted into symmetry-breaking predicates. We are hopeful that this approach helps to improve symmetry-breaking techniques for SAT.

**Acknowledgements.** The authors are supported by the National Science Foundation under grant number CCF-1526760 and acknowledge the Texas Advanced Computing Center (TACC) at The University of Texas at Austin for providing grid resources that have contributed to the research results reported within this paper.

# References

1. Aloul, F.A., Sakallah, K.A., Markov, I.L.: Efficient symmetry breaking for boolean satisfiability. In: Gottlob, G., Walsh, T. (eds.) Proceedings of the Eighteenth International Joint Conference on Artificial Intelligence, IJCAI 2003, Acapulco, Mexico, 9–15 August 2003, pp. 271–276. Morgan Kaufmann (2003)
2. Audemard, G., Simon, L.: Predicting learnt clauses quality in modern SAT solvers. In: Boutilier, C. (ed.) Proceedings of the 21st International Joint Conference on Artificial Intelligence, IJCAI 2009, Pasadena, California, USA, pp. 399–404, 11–17 July 2009
3. Bondy, J.A., Erdős, P.: Ramsey numbers for cycles in graphs. J. Combin. Theory Ser. B **14**(1), 46–54 (1973)
4. Burr, S.A., Erdős, P.: Extremal Ramsey theory for graphs. Utilitas Math. **9**, 247–258 (1976)
5. Codish, M., Miller, A., Prosser, P., Stuckey, P.J.: Breaking symmetries in graph representation. In: Proceedings of IJCAI 2013, pp. 510–516. IJCAI/AAAI (2013)
6. Gerencsér, L., Gyárfás, A.: On Ramsey-type problems. Annales Universitatis Scientiarum Budapestinensis **10**, 167–170 (1967)
7. Graham, R.L., Rothschild, B.L., Spencer, J.H.: Ramsey Theory. A Wiley-Interscience Publication. Wiley, New York (1990)
8. Harary, F.: Recent results on generalized ramsey theory for graphs. In: Alavi, Y., Lick, D.R., White, A.T. (eds.) Graph Theory and Applications. Lecture Notes in Mathematics, vol. 303, pp. 125–138. Springer, Heidelberg (1972)
9. Heule, M.J.H., Kullmann, O., Marek, V.W.: Solving and verifying the boolean pythagorean triples problem via cube-and-conquer. In: Creignou, N., Le Berre, D. (eds.) SAT 2016. LNCS, vol. 9710, pp. 228–245. Springer, Switzerland (2016)
10. Konev, B., Lisitsa, A.: A SAT attack on the Erdős discrepancy conjecture. In: Sinz, C., Egly, U. (eds.) SAT 2014. LNCS, vol. 8561, pp. 219–226. Springer, Heidelberg (2014)
11. McKay, B.D., Piperno, A.: Practical graph isomorphism, II. J. Symbol. Comput. **60**, 94–112 (2014)
12. McKay, B.D., Radziszowski, S.P.: R(4, 5) = 25. J. Graph Theory **19**(3), 309–322 (1995)
13. Radziszowski, S.P.: Small Ramsey numbers. Electron. J. Combin., #DS1 (2014)
14. Radziszowski, S.P., Jin, X.: Paths, cycles and wheels in graphs without antitriangle. Australas. J. Combin. **9**, 221–232 (1994)
15. Thurley, M.: sharpSAT – counting models with advanced component caching and implicit BCP. In: Biere, A., Gomes, C.P. (eds.) SAT 2006. LNCS, vol. 4121, pp. 424–429. Springer, Heidelberg (2006)

# Heuristic NPN Classification for Large Functions Using AIGs and LEXSAT

Mathias Soeken[1]([✉]), Alan Mishchenko[2], Ana Petkovska[1], Baruch Sterin[2], Paolo Ienne[1], Robert K. Brayton[2], and Giovanni De Micheli[1]

[1] EPFL, Lausanne, Switzerland
mathias.soeken@epfl.ch
[2] UC Berkeley, Berkeley, CA, USA

**Abstract.** Two Boolean functions are NPN equivalent if one can be obtained from the other by negating inputs, permuting inputs, or negating the output. NPN equivalence is an equivalence relation and the number of equivalence classes is significantly smaller than the number of all Boolean functions. This property has been exploited successfully to increase the efficiency of various logic synthesis algorithms. Since computing the NPN representative of a Boolean function is not scalable, heuristics have been proposed that are not guaranteed to find the representative for all functions. So far, these heuristics have been implemented using the function's truth table representation, and therefore do not scale for functions exceeding 16 variables.

In this paper, we present a symbolic heuristic NPN classification using And-Inverter Graphs and Boolean satisfiability techniques. This allows us to heuristically compute NPN representatives for functions with much larger number of variables; our experiments contain benchmarks with up to 194 variables. A key technique of the symbolic implementation is SAT-based procedure LEXSAT, which finds the lexicographically smallest satisfiable assignment. To our knowledge, LEXSAT has never been used before in logic synthesis algorithms.

## 1 Introduction

Researchers have intensively studied the classification of Boolean functions in the past. One of the frequently used classifications is based on NPN equivalence [5,7,8,11,16]. Two Boolean functions $f$ and $g$ are *Negation-Permutation-Negation (NPN) equivalent*, denoted $f =_{\text{NPN}} g$, if one can be obtained from the other by negating (i.e., complementing) inputs, permuting inputs, or negating the output. This notion of equivalence is motivated by the logic representation because NPN equivalent functions are invariant to the "shape" of an expression. As an example, $(x_1 \vee x_2) \wedge \bar{x}_3 =_{\text{NPN}} (\bar{x}_2 \vee x_3) \wedge x_1$ by replacing $x_1 \leftarrow \bar{x}_2, x_2 \leftarrow x_3$, and $\bar{x}_3 \leftarrow x_1$. This is especially true in frequently used logic representations, such as *And-Inverter Graphs (AIGs)* [9,15], in which negations are represented by complemented edges. An *equivalence class* represents a set of functions such that each two functions belonging to the class are NPN equivalent. For each

© Springer International Publishing Switzerland 2016
N. Creignou and D. Le Berre (Eds.): SAT 2016, LNCS 9710, pp. 212–227, 2016.
DOI: 10.1007/978-3-319-40970-2_14

equivalence class, one function, called *representative*, is selected uniquely among all functions of the class.

*Exact NPN classification* refers to the problem of finding, for a Boolean function $f$, its representative $r(f)$ of an NPN class. Thus, an exact NPN classification algorithm can be used to decide NPN equivalence of two functions $f$ and $g$, since $r(f) = r(g) \Leftrightarrow f =_{\text{NPN}} g$. Often $r(f)$ is chosen to be the function in the equivalence class that has the smallest truth table representation. The smallest truth table is the one with the smallest integer value, when considering the truth table as a binary number. To the best of our knowledge, there is no better way to exactly find $r(f)$ than to exhaustively enumerate all $n!$ permutations and all $2^{n+1}$ negations. To cope with this complexity, *heuristic NPN classification* has been proposed that computes $\tilde{r}(f) \geq r(f)$, i.e., it may not necessarily find the smallest truth table representation. Since $\tilde{r}(f) = \tilde{r}(g) \Rightarrow f =_{\text{NPN}} g$, heuristic NPN classification algorithms are nevertheless very helpful in many applications. For example, it has been applied to logic optimization [17,20] and technology mapping [1,4,10,16]. Finally, another option is to perform *Boolean matching* that directly checks if $f =_{\text{NPN}} g$, without first generating the representatives. Several algorithms solve this problem [1], but to the best of our knowledge, there is no good heuristics for Boolean matching without the representative.

The existing algorithms for heuristic NPN computation are implemented using truth tables [5,10] and therefore can only be efficiently applied to functions with up to 16 variables. The operations that are performed on the functional representation in heuristic NPN classification algorithms are typically: (i) negating an input, (ii) negating an output, (iii) swapping two inputs, and (iv) comparing two functions. When using truth tables as an underlying representation, all these operations can easily be implemented.

In this paper, we propose an implementation of two existing heuristic NPN classification algorithms using AIGs and LEXSAT. While the first three of the above described operations are simple to implement in AIGs, the comparison of two functions is difficult. Thus, our algorithm uses LEXSAT, which is a variant of the Boolean satisfiability (SAT) problem in which the lexicographically smallest (or greatest) assignment is returned if the problem is satisfiable. Our experimental evaluations show that, by using AIGs as function representation together with LEXSAT, heuristic NPN classification can be applied to functions with up to 194 variables while providing the same quality as the truth table based implementation at the cost of additional runtime.

The paper is structured as follows. Section 2 introduces necessary background. Section 3 reviews two existing heuristic NPN classification algorithms. Section 4 describes the truth table based implementation of the algorithms and the proposed AIG based implementation using LEXSAT. Section 5 shows the results of the experimental evaluation and Sect. 6 discusses possible performance improvements. Section 7 concludes the paper.

## 2    Preliminaries

### 2.1    Boolean Functions

We assume that the reader is familiar with Boolean functions. A *Boolean function* $f : \mathbb{B}^n \to \mathbb{B}$ evaluates bitvectors $x = x_1 \ldots x_n$ of size $n$. If $f(x) = 1$, we call $x$ a *satisfying assignment*. Given two assignments $x = x_1 \ldots x_n$ and $y = y_1 \ldots y_n$, we say that $x$ is *lexicographically smaller* than $y$, denoted $x < y$, if there exists a $k \geq 1$ such that $x_k < y_k$ and $x_i = y_i$ for all $i < k$. We use the notation $x^1$ and $x^0$ to refer to $x$ and $\bar{x}$, respectively.

The truth table of $f$ is the bitstring obtained by concatenating the function values for the assignments $0 \ldots 00, 0 \ldots 01, \ldots, 1 \ldots 10, 1 \ldots 11$ in the given order, i.e., the lexicographically smallest assignment $0 \ldots 00$ corresponds to the most significant bit. The notion of lexicographic order can be transferred to truth tables. We also use $f$ to refer to its truth table representation, if it is clear from the context. We use $\top$ and $\bot$ to refer to tautology and contradiction, respectively.

### 2.2    NPN Equivalence

**Definition 1 (NPN equivalence, [7]).** *Let $g(x_1, \ldots, x_n)$ be a Boolean function. Given a permutation $\pi \in S_n$ ($S_n$ is the symmetric group over $n$ elements) and a phase-bitvector $\varphi = (p, p_1, \ldots, p_n) \in \mathbb{B}^{n+1}$, we define*

$$apply(g, \pi, \varphi) = g^p(x_{\pi(1)}^{p_1}, \ldots, x_{\pi(n)}^{p_n}). \tag{1}$$

*We say that two Boolean functions $f(x_1, \ldots, x_n)$ and $g(x_1, \ldots, x_n)$ are Negation-Permutation-Negation (NPN) equivalent, if there exists $\pi \in S_n$ and $\varphi \in \mathbb{B}^{n+1}$ such that $f(x_1, \ldots, x_n) = apply(g, \pi, \varphi)$, i.e., $g$ can be made equivalent to $f$ by negating inputs, permuting inputs, or negating the output. We call the pair $\pi, \varphi$ an NPN signature.*

NPN equivalence is an equivalence relation that partitions the set of all Boolean functions over $n$ variables into a smaller set of NPN classes. As an example, all $2^{2^n}$ Boolean functions over $n$ variables can be partitioned into $2, 4, 14, 222, 616126$ NPN classes for $n = 1, 2, 3, 4, 5$ [8].

**Definition 2 (NPN classes and representatives).** *We refer to the NPN class of a function $f$ as $[f]$ and define $f =_{NPN} g$, if and only if $[f] = [g]$. As the representative $r(f)$ of each NPN class $[f]$, we take the function with the smallest truth table. In other words, $r(f) \leq g$ for all $g \in [f]$.*

*Example 1.* The truth table of $x_1 \vee x_2$ is 0111, and its NPN representative is $x_1 \wedge x_2 = \overline{\bar{x}_1 \vee \bar{x}_2}$ which truth table is 0001. Note that $g =_{\text{NPN}} apply(g, \pi, \varphi)$ for all $\pi \in S_n$ and all $\varphi \in \mathbb{B}^{n+1}$. For a detailed introduction into NPN classification the reader is referred to the literature [1,18].

---

**Algorithm 1.** LEXSAT implementation from [13, Ex. 7.2.2.2-109].

    **Input**    : Boolean function $f(x_1, \ldots, x_n)$
    **Output** : min $f$ if $f \neq \bot$, otherwise *unsatisfiable*
1  set $s \leftarrow \mathrm{SAT}(f)$;
2  **if** $s = $ *unsatisfiable* **then return** *unsatisfiable*;
3  set $y_1, \ldots, y_n \leftarrow s$ and $y_{n+1} \leftarrow 1$;
4  set $d \leftarrow 0$;
5  **while** *true* **do**
6    | set $d \leftarrow \min\{j > d \mid y_j = 1\}$;
7    | **if** $d > n$ **then return** $y_1, \ldots, y_n$;
8    | set $s \leftarrow \mathrm{SAT}(f, \{x_1^{y_1}, \ldots, x_{d-1}^{y_{d-1}}, \bar{x}_d\})$;
9    | **if** $s \neq$ *unsatisfiable* **then** set $y_1, \ldots, y_n \leftarrow s$;
10 **end**

---

### 2.3 Lexicographic SAT

**Definition 3 (Lexicographically smallest (greatest) assignment).** *Let* $f$ :
$\mathbb{B}^n \to \mathbb{B}$ *with* $f \neq \bot$. *Then* $x \in B^n$ *is the* lexicographically smallest assignment
*of* $f$, *if* $f(x) = 1$ *and* $f(x') = 0$ *for all* $x' < x$. *We denote this* $x$ *as* min $f$.
*Analogously,* $x \in B^n$ *is the* lexicographically greatest assignment *of* $f$, *if* $f(x) =$
1 *and* $f(x') = 0$ *for all* $x' > x$. *We denote this* $x$ *as* max $f$.

Based on this definition, *lexicographic SAT* (LEXSAT) is a decision procedure
that for a given function $f$, returns min $f$ when the problem is satisfiable, or
returns *unsatisfiable*, when $f = \bot$. LEXSAT is NP-hard and complete for the
class FP$^{\mathsf{NP}}$ [14]. Knuth [13] proposes an implementation that calls a SAT solver
several times to refine the assignment, and which is described in Algorithm. 1. In
the algorithmic description $\mathrm{SAT}(f, a)$ refers to the default SAT decision procedure for a Boolean function $f$ and optional assumptions $a$ that allows incremental
solving. SAT returns *unsatisfiable*, when $f = \bot$, or a satisfying assignment, when
the problem is satisfiable. Further, to use the SAT solver, we assume that $f$ is
translated into a CNF representation (e.g., using [6,21]). By replacing $y_j = 1$
with $y_j = 0$ in Line 6 and $\bar{x}_d$ with $x_d$ in Line 8, Algorithm 1 can find max $f$.

## 3  Heuristic NPN Classification

To the best of our knowledge there is no efficient algorithm to find the representative $r(f)$ for a given Boolean function $f : \mathbb{B}^n \to \mathbb{B}$, and an exhaustive
exploration of all functions in $[f]$ is required. More efficient algorithms can be
found when using a heuristic to *approximate* the representative. Such heuristics
do not visit all functions in $[f]$ and therefore are not guaranteed to find $r(f)$,
only a function $\tilde{r}(f) \geq r(f)$ that is locally minimal to all visited ones is returned.

    In the following subsections, we describe in detail two heuristic NPN classi-
fication algorithms from the literature for which truth table based implementations exists, e.g., in *ABC* [3]. The first heuristic [10] is called *flip-swap*, which

---

**Algorithm 2.** Flip-swap heuristic for NPN classification.

---
**Input**   : Boolean function $f \in \mathbb{B}^n \to \mathbb{B}$
**Output** : NPN signature $\pi \in S_n, \varphi \in \mathbb{B}^{n+1}$

1  set $\pi \leftarrow \pi_e$ and $\varphi \leftarrow 0 \ldots 0$;
2  **repeat**
3      set *improvement* $\leftarrow 0$;
4      **for** $i = 0, \ldots, n$ **do**
5          **if** $apply(f, \pi, \varphi \oplus 2^i) < apply(f, \pi, \varphi)$ **then**
6              set $\varphi \leftarrow \varphi \oplus 2^i$;
7              set *improvement* $\leftarrow 1$;
8          **end**
9      **end**
10     **for** $d = 1, \ldots, n-2$ **do**
11         **for** $i = 1, \ldots, n-d$ **do**
12             set $j \leftarrow i + d$;
13             **if** $apply(f, \pi \circ (i, j), \varphi) < apply(f, \pi, \varphi)$ **then**
14                 set $\pi \leftarrow \pi \circ (i, j)$;
15                 set *improvement* $\leftarrow 1$;
16             **end**
17         **end**
18     **end**
19 **until** *improvement* $= 0$;
20 **return** $\pi, \varphi$;

---

in alternating steps flips single bits and permutes pairs of indices. The second algorithm [10] is called *sifting* (inspired by BDD sifting [19]), which considers only adjacent indices for flipping and swapping.

### 3.1   Flip-Swap Heuristic

Algorithm 2 shows the flip-swap heuristic, which in alternating steps first tries to flip single bits in the phase $\varphi$ and then permutes pairs of indices in $\pi$. The phase is initialized to consist only of 0, and the permutation is initially the identity permutation $\pi_e$. The first **for**-loop, which flips bits (Line 4), also covers negating the whole function when $i = 0$. Note that all index pairs in the second **for**-loop (Line 10), which swaps inputs, are enumerated ordered by their distance such that adjacent pairs are considered first. Flipping and swapping is repeated until no more improvement, i.e., no smaller representative, can be found. The algorithm returns a permutation $\pi$ and a phase-bitvector $\varphi$, which lead to the smallest function that is considered as representative $\tilde{r}(f)$ of the input function $f$. One can obtain $\tilde{r}(f)$ by executing $apply(f, \pi, \varphi)$.

### 3.2   Sifting Heuristic

Algorithm 3 shows the sifting heuristic that was presented by Huang et al. [10]. The idea is that a *window* is shifted over adjacent pairs of input variables $x_{\pi(i)}$

---

**Algorithm 3.** Sifting heuristic for NPN classification.

**Input**  : Boolean function $f \in \mathbb{B}^n \to \mathbb{B}$
**Output** : NPN signature $\pi \in S_n, \varphi \in \mathbb{B}^{n+1}$

```
1  set π ← π_e and φ ← 0...0;
2  repeat
3  │  set improvement ← 0;
4  │  for i = 1,...,n − 1 do                              /* in alternating order */
5  │  │  set σ, σ̂ ← π_e and ψ, ψ̂ ← 0...0;
6  │  │  for j = 1,...,8 do
7  │  │  │  if 4 | j then
8  │  │  │  │  set σ ← σ ∘ (i, i + 1);
9  │  │  │  end
10 │  │  │  else if 2 | j then
11 │  │  │  │  set ψ ← ψ ⊕ 2^{i+1};
12 │  │  │  end
13 │  │  │  else
14 │  │  │  │  set ψ ← ψ ⊕ 2^i;
15 │  │  │  end
16 │  │  │  if apply(f, π ∘ σ, φ ⊕ ψ) < apply(f, π ∘ σ̂, φ ⊕ ψ̂) then
17 │  │  │  │  set σ̂ ← σ and ψ̂ ← ψ;
18 │  │  │  │  set improvement ← 1;
19 │  │  │  end
20 │  │  end
21 │  │  set π ← π ∘ σ̂ and φ ← φ ⊕ ψ̂;
22 │  end
23 until improvement = 0;
24 return π, φ;
```

---

and $x_{\pi(i+1)}$. The adjacent variables are negated and swapped (Lines 6–15) in the following way that guarantees that all eight possibilities are obtained by applying simple operations, and the initial configuration is restored in the end.

$$xy \xrightarrow[j=1]{\overline{x}y} \bar{x}y \xrightarrow[j=2]{\overline{x}y} \bar{x}\bar{y} \xrightarrow[j=3]{\overline{x}y} x\bar{y} \xrightarrow[j=4]{\leftrightarrow} y\bar{x} \xrightarrow[j=5]{\overline{x}y} \bar{y}\bar{x} \xrightarrow[j=6]{\overline{x}y} \bar{y}x \xrightarrow[j=7]{\overline{x}y} yx \xrightarrow[j=8]{\leftrightarrow} xy \tag{2}$$

Whenever $j$ is a multiple of 4 (i.e., $j = 4$ and $j = 8$), the variables are swapped. Otherwise, whenever $j$ is even (i.e., $j = 2$ and $j = 6$), the second variable is negated, and in all other cases the first variable is negated.

All eight configurations are evaluated and the best configuration is stored in $\hat{\sigma}$ and $\hat{\psi}$. This procedure is repeated as long as an improvement can be obtained. The window is moved in alternating order over the variables, i.e., first from left to right and then from right to left.

# 4   Implementations

The crucial parts in the algorithms presented in Sect. 3 are the checks in Lines 5 and 13 from Algorithm 2, and Line 16 from Algorithm 3, which derive the representative of the given function by applying the given permutation and phase. Four operations are required to perform these checks, which are (i) negating an input, (ii) negating an output, (iii) swapping two inputs, and (iv) comparing the two functions. This section describes two implementations for these four operations: (i) based on truth tables which scales to functions with up to 16 variables and (ii) based on AIGs and LEXSAT which scales for large functions.

## 4.1   Truth Table Based Implementation

A truth table is represented as the binary expansion of a nonnegative number $t \in [0, 2^n)$. The most significant bit represents the assignment $0 \ldots 0$ and the least significant bit represents the assignment $1 \ldots 1$. In this representation the truth table of the function $x_1 \wedge x_2$ is 0001 and of $x_1 \vee x_2$ is 0111. The truth table for a variable $x_{n-i}$ in a $n$-variable Boolean function is $\mu_{n,i} = \frac{2^{2^n}-1}{2^{2^i}+1}$ for $0 \le i < n$. For $n = 3$, we have $0000\,1111$, $0011\,0011$, and $0101\,0101$ for $x_1$, $x_2$, and $x_3$, respectively. The comparison of two functions given in their truth table representation is straightforward, e.g., by comparing their integer values. The other three operations can be implemented using well-known bitwise arithmetic as illustrated next. A truth table $t$ for an $n$-variable Boolean function can be negated by flipping each bit, i.e., $t \leftarrow \bar{t}$. In order to negate the polarity of a variable $x_i$ in a truth table $t$ for an $n$-variable Boolean function, we compute $t \leftarrow \big((t \,\&\, \mu_{n,i}) \ll 2^{n-i+1}\big) \mid \big((t \,\&\, \bar{\mu}_{n,i}) \gg 2^{n-i+1}\big)$. The operations '&', '|', '$\ll$', '$\gg$' are bitwise AND, bitwise OR, logical left-shift, and logical right-shift. Two variables $x_i$ and $x_j$ with $i > j$ can be swapped in a truth table $t$ by performing the two operations $t' \leftarrow (t \oplus (t \gg \delta)) \,\&\, 2^j$ and $t \leftarrow t \oplus t' \oplus (t' \ll \delta)$, where $\delta = i - j$ and '$\oplus$' is bitwise XOR.

All the described operations can be implemented very efficiently when $n$ is small. For example, a 6-variable function fits into a word that requires one memory cell on a 64-bit computer architecture. Almost all of the bitwise operations have a machine instruction counterpart that can be processed within one clock cycle. Warren, Jr. [22] and Knuth [12] describe all these bitwise manipulations and give more detailed equations.

## 4.2   AIG Based Implementation Using LEXSAT

For large functions, we propose representing the function using the AIG data structure, which represents a logic network using two-input AND gates and edges connecting them. The edges may be complemented, representing inverters over these edges. For an AIG, negating the function, negating the polarity of a single variable, and swapping two variables is trivial. This can be achieved by complementing fanout edges from the primary output and primary inputs, and by swapping two AIG nodes of the corresponding primary inputs, respectively.

Since the truth table is not explicitly represented by an AIG, the lexico-graphic comparison in Algorithm 2 cannot be performed directly. We find that LEXSAT can be used to solve this problem. The next theorem, which is the main contribution of this paper, explains how the comparison can be done.

**Theorem 1.** *Let $g : \mathbb{B}^n \to \mathbb{B}$ and $h : \mathbb{B}^n \to \mathbb{B}$ be two Boolean functions. Then $g < h$, if and only if*

$$g \neq h \quad and \quad g(\min(g \oplus h)) = 0.$$

*Proof.* '$\Leftarrow$': The condition can only hold if $g \neq h$. Then, $g \oplus h \neq \perp$ and its lexicographic smallest assignment $x = \min(g \oplus h)$ is the smallest assignment for which $g$ and $h$ differ. Hence, $x$ is the first bit-position in which the truth table representations of $g$ and $h$ differ. (Recall that the most significant bit corresponds to all variables set to 0.) If $g(x) = 0$, then $h(x) = 1$, and $g$ must be lexicographically smaller than $h$.

'$\Rightarrow$': Obviously, $g \neq h$. Let $x$ be the smallest assignment for which $g$ and $h$ differ, i.e., $x = \min(g \oplus h)$. Since $g < h$, we have $g(x) = 0$.    $\square$

Based on Theorem 1, the following steps are necessary in an AIG based imple-mentation to compute whether $g < h$.

(1) Create an AIG for the *miter* $m(x_1, \ldots, x_n) = g(x_1, \ldots, x_n) \oplus h(x_1, \ldots, x_n)$ by matching the inputs and pairing the outputs using an XOR operation [2].
(2) Encode the AIG for $m$ as a CNF.
(3) Solve LEXSAT for the variables $x_1, \ldots, x_n$ by assuming the output of $m$ to be 1.
(4) If a satisfying assignment $x$ exists and if simulating $g(x)$ returns 0, then $g < h$.

In our use of this procedure, we will obtain $g$ and $h$ from $apply(f, \pi, \varphi)$ and $apply(f, \pi', \varphi')$, respectively, and the result will determine whether the current permutation and phase-bitvectors will be updated.

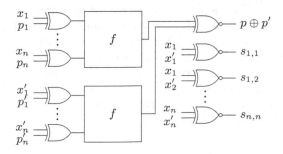

**Fig. 1.** Shared miter construction. The additional variables $p_i$ and $p_i'$ control the phase of the inputs and the output, while the additional variables $s_{i,j}$ allow permuting inputs.

**Simple Miter and Shared Miter Approach.** The lexicographic comparison has to be done several times by following the above mentioned steps that require to build and encode the miter each time. We call this the *simple miter* approach. Contrary to this approach, we propose a *shared miter* approach in which the miter is only created once and can be reconfigured. Then, steps (1) and (2) are performed only once. The shared miter is equipped with additional *inputs* $p_1, \ldots, p_n, p'_1, \ldots, p'_n$ and *outputs* $p, p', s_{1,1}, s_{1,2}, \ldots, s_{n,n}$ to reconfigure w.r.t. different permutations and phases. These inputs and outputs can be assigned using assumption literals in the LEXSAT calls in step (3).

The details are illustrated in Fig. 1. The assumption literals to control the phase for inputs and outputs are $(p, p_1, \ldots, p_n) = \varphi$ and $(p', p'_1, \ldots, p'_n) = \varphi'$. Instead of assuming the output of $m$ to be 1 in step (3), it is assumed to be $\bar{p} \oplus p'$ (note the XNOR gate at the outputs). To take input permutation into consideration, we assume $s_{i,j} = 1$, if $\pi_i = \pi'_j$. This is ensured by having a quadratic number of XNOR gates for each pair of inputs $x_i$ and $x'_j$. The lexicographically smallest assignment is determined with respect to $x_1, \ldots, x_n$. Technically, the simulation in step (4) is not required since the simulation value is contained in the satisfying assignment, however, the runtime required for simulation is negligible. As the experimental results from Sect. 5 show, the simple miter approach and the shared miter approach trade off solving time against encoding time: the LEXSAT instances in the simple miter approach are simpler to solve, however, more time is spent for encoding the AIG of the miter for each lexicographic comparison.

**Encoding the AIG.** The process of translating an AIG into a conjunctive normal form (CNF), which is a set of clauses, is called *encoding*. In order to call LEXSAT, the AIG of the miter needs to be encoded into a CNF. Several techniques for encoding exist, but the most conventional one is the *Tseytin encoding* [21] that introduces a new variable for each gate, and expresses the relation of the output to the inputs using clauses. For example, an AND gate $c = a \wedge b$ is encoded with the three clauses $(a \vee \bar{c})(b \vee \bar{c})(\bar{a} \vee \bar{b} \vee c)$. Since AIGs contain only AND gates, the SAT formula consists of clauses that have the form as the three clauses of the AND gate, with literals inverted with respect to complemented edges.

Particularly for AIGs, one can do much better by using logic synthesis techniques to derive a smaller set of clauses. We call this technique *EMS encoding* [6] due to the authors' last names. EMS encoding applies cut-based technology mapping in which the objective is not area or delay but clause count. Several applications indicate that this encoding is particularly desirable when CNFs are generated from circuits and have a positive effect on the SAT solving runtime [6]. However, due to the overhead of technology mapping, the EMS encoding requires more runtime than the Tseytin encoding.

**Table 1.** Evaluating the quality of the heuristics for small functions.

| # Variables | # Functions | # Classes / Runtime (s) | | | | | |
|---|---|---|---|---|---|---|---|
| | | Exact | | Flip-swap | | Sifting | |
| 6 | 40195 | 191 | 13941.11 | 441 | 5.65 | 368 | 24.26 |
| 8 | 81864 | 1274 | > 4 h | 2409 | 51.30 | 2251 | 127.55 |
| 10 | 19723 | 1707 | > 4 h | 2472 | 56.76 | 2360 | 81.25 |

## 5  Experiments

We have implemented all presented algorithms using $CirKit$[1] in the commands
'*npn*' and '*satnpn*'. All experiments are carried out on an Intel Xeon E5 CPU
with 2.60 GHz and 128 GB main memory running Linux 3.13. First, Sect. 5.1
shows results of an experiment that compares different heuristics against the
exact NPN classification algorithm. However, this can only be done for small
functions, for which the exact algorithm finishes in reasonable time. On the
other hand, Sect. 5.2 shows results of an experiment that evaluates scalability
by applying the heuristics to larger functions beyond the applicablity of truth
table based implementations.

### 5.1  Quality Evaluation

We applied both the truth table based implementation and the AIG based implementation of both heuristics to small Boolean functions that were harvested
using structural cut enumeration in all instances of the MCNC, ISCAS, and
ITC benchmark sets as suggested by Huang et al. [10] Based on the computed
representative, all functions are partitioned into a smaller set of classes. Table 1
shows the number of classes generated by an exact algorithm for NPN classification and by each of the two heuristic algorithms presented in Sect. 3. Although
the heuristic NPN classification algorithms can be applied to truth tables up to
16 variables, exact classification does not scale since all $n! \cdot 2^{n+1}$ permutations and
phases have to be evaluated in the worst case. Runtimes are obtained from the
truth table based methods. For example, the exact and heuristic classifications
were applied to 40195 distinct functions of six variables. When partitioning this
set based on the NPN representatives, 191 classes are obtained using exact NPN
classification. The flip-swap heuristic cannot determine all representatives correctly and partitions the 40195 functions into 441 classes. The sifting heuristics
shows better results and returns 368 classes.

The best quality is reached with exact classification, but it cannot be applied
efficiently to larger functions due to the exponential search space. Nevertheless,
the heuristic classification can have a huge contribution in some logic synthesis
algorithms. For example, some optimizations which require a large computational effort can often be limited to only representatives of each NPN class. In

---

[1] github.com/msoeken/cirkit.

**Table 2.** Benchmark properties.

| Benchmark | Inputs | Outputs | Max. inputs |
|-----------|--------|---------|-------------|
| c432 | 36 | 7 | 36 |
| c499 | 41 | 32 | 41 |
| c880 | 60 | 26 | 45 |
| c1355 | 41 | 32 | 41 |
| c1908 | 33 | 25 | 33 |
| c2670 | 233 | 140 | 119 |
| c3540 | 50 | 22 | 50 |
| c5315 | 178 | 123 | 67 |
| c7552 | 207 | 108 | 194 |

**Table 3.** Runtime (in seconds) and number of LEXSAT calls (in millions) of the heuristics for the benchmark *c7552*. In bold are the number of SAT calls (in millions).

| Heuristic | LEXSAT | Single miter | | | | Shared miter | | | |
|-----------|--------|--------------|---|---|---|--------------|---|---|---|
| | calls | Tseytin | | EMS | | Tseytin | | EMS | |
| Flip-swap | 0.8 M | 4416.10 | **57 M** | 5348.17 | **63 M** | 62357.90 | **22 M** | 29184.40 | **23 M** |
| Sifting | 1.8 M | 9450.02 | **123 M** | 12018.30 | **135 M** | 76582.10 | **49 M** | 44896.10 | **47 M** |

the case of 6-variable functions, when using the sifting heuristic, it means that such computation only needs to be performed for 368 functions, instead of 40195.

We observed that both the truth table based and the AIG based implementation generate the same results in terms of quality. However, the AIG based implementation showed significantly worse performance when applied to small functions. This is due to the extra overhead spent on encoding and SAT solving which is only amortized for larger functions for which the truth table based implementation is not scalable. For large functions, the exponential size representation of truth tables becomes the bottleneck.

## 5.2   Scalability Evaluation

In order to demonstrate scalability of the AIG based implementation of the heuristic NPN classification algorithms, as well as to evaluate the different implementation options, we have applied both heuristics to the combinational instances from the ISCAS benchmark suite. Since these benchmarks realize multiple output Boolean functions we ran the algorithm on each output cone separately. The reported results are cumulative for all outputs. Table 2 shows number of inputs and outputs of the used benchmarks, as well as the maximum number of inputs in any of the output cones. Since the runtimes for *c7552* are comparably high, we report them separately in Table 3 and allow a better scaling of the other benchmarks in the plots. As the main objective is runtime in this experi-

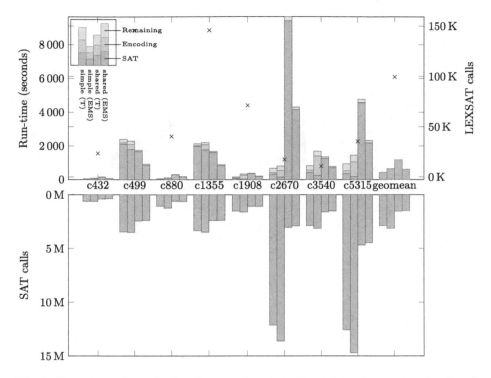

**Fig. 2.** Experimental results for flip-swap heuristic. Bars show the runtime (top) and the number of SAT calls (bottom). Cross marks show the number of LEXSAT calls. (Color figure online)

ment, there is no direct comparison of both heuristics. A qualitative comparison of both approaches has already been provided in the previous section.

The results of the experiments are presented in the plots in Fig. 2 for the flip-swap heuristic and in Fig. 3 for the sifting heuristic. Both plots are organized in the same way. We ran each benchmark in four configurations: simple miter with Tseytin encoding, simple miter with EMS encoding, shared miter with Tseytin encoding, and shared miter with EMS encoding. For each configuration, we plot the runtime in seconds in the upper axis and the number of total SAT calls in the lower axis. The runtime is separated into three parts: the lowest (blue) part shows the runtime spent on SAT solving, the middle (red) part shows the runtime spent on encoding, and the upper (brown) part shows the remaining runtime. Finally, cross marks show the number of LEXSAT calls in the upper axis. Note that the number of LEXSAT call is identical in each configuration. The last entry in the plots gives the geometric mean of the overall runtimes and SAT calls for each configuration.

The simple miter approach is often faster compared to the shared miter approach, especially for larger benchmarks. It can be seen that percentage of runtime spent on SAT solving is much smaller for the simple miter approach,

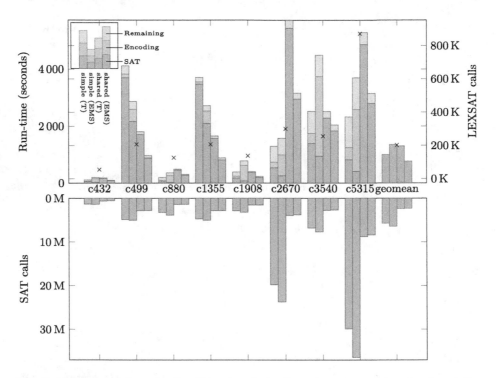

**Fig. 3.** Experimental results for sifting heuristic. Bars show the runtime (top) and the number of SAT calls (bottom), while cross marks show the number of LEXSAT calls. (Color figure online)

since more time is spent on generating and encoding the miter. In the simple miter approaches the time spent on encoding can matter. Although the EMS encoding can reduce the runtime on SAT solving significantly (see, e.g., *c2670*, *c3540*, and *c5315*) the encoding time becomes the new bottleneck and eventually results in an overall larger runtime. This effect is not evident in the shared miter approach where a very small percentage of time is spent on encoding. The EMS method, due to the improved encoding, and the resulting improvement in SAT solving, reduces the overall runtime by about half overall. Note also that the overall number of SAT calls is about two times larger in the simple miter approach compared to the shared miter approach.

**Challenges to Performance.** Symmetric or functionally independent variables can cause a problem for the algorithm, since they do not change the function after permutation and negation, respectively. For two symmetric variables $x_i$ and $x_j$ we have $apply(f, \pi, \varphi) = apply(f, \pi \circ (i, j), \varphi)$ and for a functionally independent variable $x_i$ we have $apply(f, \pi, \varphi) = apply(f, \pi, \varphi \oplus 2^i)$. Consequently, the LEXSAT calls for the comparisons in the heuristic yield UNSAT and correspond to equivalence checking of two equivalent circuits. In practice, the situation may be slightly better due to several structural similarities of both circuits. In the experimental evaluation, this problem became evident for the multiplier bench-mark *c6288* making a heuristic NPN classification based on SAT infeasible.

# 6    Possible Improvements

In this section, we discuss several improvements that are not considered in the current implementation, but could be help reduce runtime more.

**Symmetry and Functional Dependency.** One can circumvent the problem with symmetric and functional independent variables by setting a time limit to the execution of a LEXSAT call. If the timeout is reached one can try random simulation as a last resort and, if this also fails, then one can proceed while not updating the current permutation $\pi$ and phase $\varphi$. This shows positive effects on the overall runtime but can degrade the quality by increasing the number of distict equivalence classes.

**Partial EMS Encoding.** The heuristics update the current permutation and phase by swapping two variables or negating one variable. Consequently, in most of the cases only small parts of the updated circuit change. The mapping of the unchanged part of the circuit and the resulting CNF stay the same and do not need to be recomputed. Making the EMS encoding aware of changes in the circuit reduces runtime. Since in case of EMS, the encoding often is the predominant part of the overall runtime, a significant speedup can be expected.

**Avoid Miter Construction.** Note that $\max f < \max g$ implies $f < g$. The other direction is not true which can readily be seen from $f = 1010$ and $g = 1100$. This check may be faster than first constructing the miter and calling LEXSAT on it. We have tried to integrate this check as a preprocessing step before the miter construction in the simple miter approach. However, as in most cases we had $\max f = \max g$, this resulted in a higher overall runtime. The check can be better integrated using a thread, that is started at the same time of the miter construction, and is terminated when a conclusive answer is found faster.

**Using Simulation to Skip Some SAT Calls.** For simplicity, consider the computation of $\max f$ for one function, as in the above subsection, rather than for the miter of two functions, as elsewhere in the paper. Assume that we found $\max f$, which is an assignment of $n$ input variables $x$ such that $f(x) = 1$, and $f(y) = 0$ for all assignments $y > x$. Now take assignment $x$ and generate $n$ assignments, which are distance-1 from $x$, by flipping the value of one input at a time. Perform bit-parallel simulation of $f$ using these distance-1 assignments and observe the output of $f$. If $f(d) = 1$ for some distance-1 assignment $d$, we know that flipping the corresponding input cannot reduce $\max f$, and so we can skip the SAT call. On the other hand, if $f(d) = 0$, flipping the corresponding input may lead to a smaller $\max f$. The reduction in $\max f$ is possible if flipping this input does not create a new 1 for a lexicographically larger assignment, which has a 0 before (follows from the fact that the original assignment is $\max f$). As a result, in the case when $f(d) = 0$, we need a LEXSAT call to check whether flipping this input leads to an improvement in $\max f$. However, when $f(d) = 1$ the call can be skipped to reduce the runtime.

**Native LEXSAT Solver.** A considerably faster LEXSAT algorithm can be obtained by directly modifying the SAT solver (see Example 7.2.2.2–275 in [13]). To this end, instead of using the assumption interface of MiniSAT, as we did in this paper, we can modify decision heuristics of the SAT solver to always perform decisions on the input variables in the order, which is imposed by the user of the LEXSAT algorithm. Decisions on other variables can be performed in any order.

**Using Binary Search in LEXSAT.** Another possibility to improve the runtime of LEXSAT implementation is to use an approach based on binary search. Instead of trying to append to the array of assumptions one literal at a time, we can begin by putting one half of all literals, then one quarter, and so on. The binary search approach can be taken one step further: We can profile and see how often, on average, the first step of binary search leads to a SAT or UNSAT call. Based on the result, we could try to make the first step to be, say, 25 % or 75 % of the total number of assumptions, instead of 50 % (as in the naïve binary search). This way, we could have a better "success rate" of searching, which could lead to a faster LEXSAT implementation.

## 7  Conclusions

In this paper, we have shown how heuristic NPN classification algorithms can be implemented based on AIGs and SAT instead of truth tables. The new implementation can be applied to larger functions. The key aspect of the proposed approach is finding the lexicographically smallest assignment of a SAT instance using the LEXSAT algorithm. An experimental evaluation shows that using the AIG based implementation, the heuristic NPN classification algorithms can be applied to functions with hundreds of inputs. Our current implementation is preliminary and there are several possibilities to reduce the overall runtime, some of them have been outlined in Sect. 6.

**Acknowledgments.** This research was supported by H2020-ERC-2014-ADG 669354 CyberCare, by the German Academic Exchange Service (DAAD) in the PPP 57134066, and partly by the NSF/NSA grant "Enhanced equivalence checking in cryptoanalytic applications" at University of California, Berkeley.

## References

1. Benini, L., Micheli, G.D.: A survey of Boolean matching techniques for library binding. ACM Trans. Design Autom. Electr. Syst. **2**(3), 193–226 (1997)
2. Brand, D.: Verification of large synthesized designs. In: Proceedings of the International Conference on Computer Aided Design, pp. 534–537 (1993)

3. Brayton, R., Mishchenko, A.: ABC: an academic industrial-strength verification tool. In: Touili, T., Cook, B., Jackson, P. (eds.) CAV 2010. LNCS, vol. 6174, pp. 24–40. Springer, Heidelberg (2010)
4. Chatterjee, S., Mishchenko, A., Brayton, R.K., Wang, X., Kam, T.: Reducing structural bias in technology mapping. IEEE Trans. Comput. Aided Des. Integr. Circuits Syst. **25**(12), 2894–2903 (2006)
5. Debnath, D., Sasao, T.: Efficient computation of canonical form for Boolean matching in large libraries. In: Proceedings of the Asia and South Pacific Design Automation Conference, pp. 591–96 (2004)
6. Eén, N., Mishchenko, A., Sörensson, N.: Applying logic synthesis for speeding up SAT. In: Marques-Silva, J., Sakallah, K.A. (eds.) SAT 2007. LNCS, vol. 4501, pp. 272–286. Springer, Heidelberg (2007)
7. Goto, E., Takahasi, H.: Some theorems useful in threshold logic for enumerating Boolean functions. In: International Federation for Information Processing Congress, pp. 747–52 (1962)
8. Harrison, M.A.: Introduction to Switching and Automata Theory. McGraw-Hill, New York (1965)
9. Hellerman, L.: A catalog of three-variable Or-inverter and And-inverter logical circuits. IEEE Trans. Electr. Comput. **12**, 198–223 (1963)
10. Huang, Z., Wang, L., Nasikovskiy, Y., Mishchenko, A.: Fast Boolean matching based on NPN classification. In: Proceedings of the 2013 International Conference on Field Programmable Technology, pp. 310–313 (2013)
11. Katebi, H., Markov, I.L.: Large-scale Boolean matching. In: Proceedings of the Design, Automation and Test in Europe Conference and Exhibition, pp. 771–76 (2010)
12. Knuth, D.E.: The Art of Computer Programming, vol. 4A. Addison-Wesley, Reading, Massachusetts (2011)
13. Knuth, D.E.: The Art of Computer Programming, vol. 4, Fascicle 6: Satisfiability. Addison-Wesley, Reading, Massachusetts (2015)
14. Krentel, M.W.: The complexity of optimization problems. J. Comput. Syst. Sci. **36**(3), 490–509 (1988)
15. Kuehlmann, A., Paruthi, V., Krohm, F., Ganai, M.K.: Robust Boolean reasoning for equivalence checking and functional property verification. IEEE Trans. Comput. Aided Des. Integr. Circuits Syst. **21**(12), 1377–1394 (2002)
16. Mailhot, F., Micheli, G.D.: Algorithms for technology mapping based on binary decision diagrams and on Boolean operations. IEEE Trans. Comput. Aided Des. Integr. Circuits Syst. **12**(5), 599–620 (1993)
17. Mishchenko, A., Chatterjee, S., Brayton, R.: DAG-aware AIG rewriting: a fresh look at combinational logic synthesis. In: Proceedings of the 43rd Design Automation Conference, pp. 532–536 (2006)
18. Muroga, S.: Logic Design and Switching Theory. Wiley, New York (1979)
19. Rudell, R.: Dynamic variable ordering for ordered binary decision diagrams. In: Proceedings of the International Conference on Computer Aided Design, pp. 42–47 (1993)
20. Soeken, M., Amarù, L.G., Gaillardon, P., De Micheli, G.: Optimizing majority-inverter graphs with functional hashing. In: Proceedings of the Design, Automation and Test in Europe Conference and Exhibition, pp. 1030–1035 (2016)
21. Tseytin, G.S.: On the complexity of derivation in propositional calculus. In: Slisenko, A.P. (ed.) Studies in Constructive Mathematics and Mathematical Logic, Part II, Seminars in Mathematics, pp. 115–125. Springer, NewYork (1970)
22. Warren, H.S.: Hacker's Delight, 2nd edn. Addison-Wesley, Reading (2012)

# Solving and Verifying the Boolean Pythagorean Triples Problem via Cube-and-Conquer

Marijn J.H. Heule[1]([✉]), Oliver Kullmann[2], and Victor W. Marek[3]

[1] The University of Texas at Austin, Austin, USA
marijn@cs.utexas.edu
[2] Swansea University, Swansea, UK
[3] University of Kentucky, Lexington, USA

**Abstract.** The *boolean Pythagorean Triples problem* has been a long-standing open problem in Ramsey Theory: Can the set $\mathbb{N} = \{1, 2, \dots\}$ of natural numbers be divided into two parts, such that no part contains a triple $(a, b, c)$ with $a^2 + b^2 = c^2$ ? A prize for the solution was offered by Ronald Graham over two decades ago. We solve this problem, proving in fact the impossibility, by using the *Cube-and-Conquer* paradigm, a hybrid SAT method for hard problems, employing both look-ahead and CDCL solvers. An important role is played by dedicated look-ahead heuristics, which indeed allowed to solve the problem on a cluster with 800 cores in about 2 days. Due to the general interest in this mathematical problem, our result requires a formal proof. Exploiting recent progress in unsatisfiability proofs of SAT solvers, we produced and verified a proof in the DRAT format, which is almost 200 terabytes in size. From this we extracted and made available a compressed certificate of 68 gigabytes, that allows anyone to reconstruct the DRAT proof for checking.

## 1 Introduction

Propositional satisfiability (SAT, for short) is a formalism that allows for representation of all finite-domain constraint satisfaction problems. Consequently, all decision problems in the class NP, as well as all search problems in the class FNP [9, 19, 29, 35], can be polynomially reduced to SAT. Due to great progress with *SAT solvers*, many practically important problems are solved using such reductions. SAT is especially an important tool in hardware verification, for example model checking [8] and reactive systems checking [8]. In this paper we are, however, dealing with a different application of SAT, namely as a tool in computations of configurations in a part of Mathematics called *Extremal Combinatorics*, especially *Ramsey theory*. In this area, the researcher attempts to find various configurations that satisfy some combinatorial conditions, as well as values of various parameters associated with such configurations [49].

One important result of Ramsey theory, the van der Waerden Theorem [45], has been studied by the SAT community, started by [14]. That theorem says that for all natural numbers $k$ and $l$ there is a number $n$, so that whenever the integers $1, \dots, n$ are partitioned into $k$ sets, there is a set containing an

© Springer International Publishing Switzerland 2016
N. Creignou and D. Le Berre (Eds.): SAT 2016, LNCS 9710, pp. 228–245, 2016.
DOI: 10.1007/978-3-319-40970-2_15

arithmetic progression of length $l$. A good deal of effort has been spent on specific values of the corresponding number theoretic function, vdW$(k, l)$. Two results on specific values: vdW$(2, 6) = 1132$ and vdW$(3, 4) = 293$, were obtained by Kouril [31,32] using specialized FPGA-based SAT solvers. Other examples include the Schur Theorem [43] on sum-free subsets, its generalization known as Rado's Theorem [42], and a generalization of van der Waerden numbers [4]. In this paper we investigate two areas:

1. We show the "boolean Pythagorean triples partition theorem" (Theorem 1), or colouring of Pythagorean triples, an analogue of Schur's Theorem.
2. We develop methods to compute numbers in Ramsey theory by SAT solvers.

A triple $(a, b, c) \in \mathbb{N}^3$ is called *Pythagorean* if $a^2 + b^2 = c^2$. If for some $n > 2$ all partitions of the set $\{1, \ldots, n\}$ into two parts contain a Pythagorean triple in at least one part, then that property holds for all such partitions of $\{1, \ldots, m\}$ for $m \geq n$. A partition by Cooper and Overstreet [10] of the set $\{1, \ldots, 7664\}$ into two parts, with no part containing a Pythagorean triple, was previously the best result, thereby improving on earlier lower bounds [11,30,41].

**Theorem 1.** *The set $\{1, \ldots, 7824\}$ can be partitioned into two parts, such that no part contains a Pythagorean triple, while this is impossible for $\{1, \ldots, 7825\}$.*

Graham repeatedly offered a prize of \$100 for proving such a theorem, and the problem is explicitly stated in [10]. To emphasize, the situation of Theorem 1 is not as in previous applications of SAT to Ramsey theory, where SAT only "filled out the numerical details", but the existence of these numbers was not known (and as such is a good success of Automated Theorem Proving). It is natural to generalize our problem in a manner similar to the Schur Theorem:

*Conjecture 1.* For every $k \geq 1$ there exist Ptn$(k) \in \mathbb{N}$ (the "Pythagorean triple number"), such that $\{1, \ldots, \text{Ptn}(k) - 1\}$ can be partitioned into $k$ parts with no part containing a Pythagorean triple, while this is impossible for $\{1, \ldots, \text{Ptn}(k)\}$.

We prove Theorem 1 by considering two SAT problems. One showing that $\{1, \ldots, 7824\}$ can be partitioned into two parts such that no part contains a Pythagorean triple (i.e., the case $n = 7824$ is satisfiable). The other one showing that any partitioning of $\{1, \ldots, 7825\}$ into two parts contains a Pythagorean triple (i.e., the case $n = 7825$ is unsatisfiable). Now a Pythagorean triple-free partition for $n = 7824$ is checkable in a second, but the *absence* of such a partition for $n = 7825$ requires a more "durable proof" than just the statement that we run a SAT solver (in some non-trivial fashion!) which answered UNSAT — to become a mathematically accepted theorem, our assertion for $n = 7825$ carries a stronger burden of proof. Fortunately, the SAT community has spent a significant effort to develop techniques that allow to extract, out of a failed attempt to get a satisfying assignment, an actual *proof of the unsatisfiability.*

It is worth noting the similarities and differences to the endeavours of extending mathematical arguments into actual *formal proofs*, using tools like *Mizar* [1] and *Coq* [2]. Cases, where intuitions (or convictions) about completeness of mathematical arguments fail, are known [47]. So T. Hales in his project *flyspeck* [3] extracted and verified his own proof of the *Kepler Conjecture*. Now the core of the argument in such examples has been constructed by mathematicians. Very different from that, the proofs for unsatisfiability coming from SAT solvers are, from a human point of view, a giant heap of random information (no direct understanding is involved). But we don't need to search for the proof — the present generation of SAT solvers supports emission of unsatisfiability proofs and standards for such proofs exist [48], as well as checkers that the proof is valid. However the proof that we will encounter in our specific problem is of very large size. In fact, even *storing* it is a significant task, requiring significant compression. We will tackle these problems in this paper.

## 2    Preliminaries

**CNF Satisfiability.** For a Boolean variable $x$, there are two *literals*, the positive literal $x$ and the negative literal $\bar{x}$. A *clause* is a finite set of literals; so it may contain complementary literals, in which case the clause is tautological. The empty clause is denoted by $\bot$. If convenient, we write a clause as a disjunction of literals. Since a clause is a set, no literal occurs several times, and the order of literals in it does not matter. A (CNF) *formula* is a conjunction of clauses, and thus clauses may occur several times, and the order of clauses does matter; in many situations these distinctions can be ignored, for example in semantical situations, and then we consider in fact finite sets of clauses.

A *partial assignment* is a function $\tau$ that maps a finite set of literals to $\{0,1\}$, such that for $v \in \{0,1\}$ holds $\tau(x) = v$ if and only if $\tau(\bar{x}) = \neg v$. A clause $C$ is satisfied by $\tau$ if $\tau(l) = 1$ for some literal $l \in C$, while $\tau$ satisfies a formula $F$ if it satisfies every clause in $F$. If a formula $F$ contains $\bot$, then $F$ is unsatisfiable. A formula $F$ *logically implies* another formula $F'$, denoted by $F \models F'$, if every satisfying assignment for $F$ also satisfies $F'$. A transition $F \rightsquigarrow F'$ is *sat-preserving*, if either $F$ is unsatisfiable or both $F, F'$ are satisfiable, while the transition if *unsat-preserving* if either $F$ is satisfiable or both $F, F'$ are unsatisfiable. Stronger, $F, F'$ are *satisfiability-equivalent* if both formulas are satisfiable or both unsatisfiable, that is, iff the transition $F \rightsquigarrow F'$ is both sat- and unsat-preserving. We note that if $F \models F'$, then $F \rightsquigarrow F'$ is sat-preserving, and that $F \rightsquigarrow F'$ is sat-preserving iff $F' \rightsquigarrow F$ is unsat-preserving. Clause addition is always unsat-preserving, clause elimination is always sat-preserving.

**Resolution and Extended Resolution.** The resolution rule (see [18, Subsects. 1.15–1.16]) infers from two clauses $C_1 = (x \vee a_1 \vee \ldots \vee a_n)$ and $C_2 = (\bar{x} \vee b_1 \vee \ldots \vee b_m)$ the *resolvent* $C = (a_1 \vee \ldots \vee a_n \vee b_1 \vee \ldots \vee b_m)$, by resolving on variable $x$. $C$ is logically implied by any formula containing $C_1$ and $C_2$. For a given CNF formula $F$, the *extension rule* [44] allows one to iteratively add definitions of the form $x := a \wedge b$ by adding the *extended resolution clauses* $(x \vee \bar{a} \vee \bar{b}) \wedge (\bar{x} \vee a) \wedge (\bar{x} \vee b)$ to $F$, where $x$ is a new variable and $a$ and $b$ are literals in the current formula. The addition of these clauses is sat-equivalent.

**Unit Propagation.** For a CNF formula $F$, *unit propagation* simplifies $F$ based on unit clauses; that is, it repeats the following until fixpoint: if there is a unit clause $\{l\} \in F$, remove all clauses that contain the literal $l$ from the set $F$ and remove the literal $\bar{l}$ from the remaining clauses in $F$. This process is sat-equivalent. If unit propagation on formula $F$ produces complementary units $\{l\}$ and $\{\bar{l}\}$, we say that unit propagation *derives a conflict* and write $F \vdash_1 \bot$ (this relation also holds if $\bot$ is already in $F$).

Ordinary resolution proofs (or "refutations" – derivations of the empty clause) just add resolvents. This is too inefficient, and is extended via unit propagation as follows. For a clause $C$ let $\neg C$ denote the conjunction of unit clauses that falsify all literals in $C$. A clause $C$ is an *asymmetric tautology* with respect to a CNF formula $F$ if $F \wedge \neg C \vdash_1 \bot$. This is equivalent to the clause $C$ being derivable from $F$ via *input resolution* [20]: a sequence of resolution steps using for every resolution step at least one clause of $F$. So addition of resolvents is generalised by addition of asymmetric tautologies (where addition steps always refer to the current (enlarged) formula, the original axioms plus the added clauses). Asymmetric tautologies, also known as *reverse unit propagation* (RUP) clauses, are the most common learned clauses in *conflict-driven clause learning* (CDCL) SAT solvers (see [39, Subsect. 4.4]). This extension is irrelevant from the proof-complexity point of view, but for practical applications exploitation of the power of fast unit propagation algorithms is essential.

**RAT Clauses.** We are seeking to add sat-preserving clauses beyond logically implied clauses. The basic idea is as follows (proof left as instructive exercise):

**Lemma 1.** *Consider a formula $F$, a clause $C$ and a literal $x \in C$. If for all $D \in F$ such that $\bar{x} \in F$ it holds that $F \models C \cup (D \setminus \{\bar{x}\})$, then addition of $C$ to $F$ is sat-preserving.*

In order to render the condition $F \models C \cup (D \setminus \{\bar{x}\})$ polytime-decidable, we stipulate that the right-hand clause must be derivable by input resolution:

**Definition 1 [28].** *Consider a formula $F$, a clause $C$ and a literal $x \in C$ (the "pivot"). We say that $C$ has RAT ("Resolution asymmetric tautology") on $x$ w.r.t. $F$ if for all $D \in F$ with $\bar{x} \in D$ holds that $F \wedge \neg(C \cup (D \setminus \{\bar{x}\})) \vdash_1 \bot$.*

By Lemma 1, addition of RAT-clauses is sat-preserving. Every non-empty asymmetric tautology $C$ for $F$ has RAT on any $x \in C$ w.r.t. $F$. It is also easy to see that the three extended resolution clauses are RAT clauses (using the new variable for the pivot literals). All preprocessing, inprocessing, and solving techniques in state-of-the-art SAT solvers can be expressed in terms of addition and removal of RAT clauses [28].

## 3    Proofs of Unsatisfiability

A *proof of unsatisfiability* (also called a *refutation*) for a formula $F$ is a sequence of sat-preserving transitions which ends with some formula containing the empty

CNF formula     DRAT proof

```
p cnf 4 8
   1  2 -3 0            -1 0
  -1 -2  3 0         d -1 2 4 0
   2  3 -4 0             2 0
  -2 -3  4 0               0
  -1 -3 -4 0
   1  3  4 0
  -1  2  4 0
   1 -2 -4 0
```

**Fig. 1.** Left, a formula in DIMACS CNF format, the conventional input for SAT solvers which starts with `p cnf` to denote the format, followed by the number of variables and the number of clauses. Right, a DRAT refutation for that formula. Each line in the proof is either an addition step (no prefix) or a deletion step identified by the prefix "d". Spacing is used to improve readability. Each clause in the proof must be a RAT clause using the first literal as pivot, or the empty clause as an asymmetric tautology.

clause. There are currently two prevalent types of unsatisfiability proofs: *resolution proofs* and *clausal proofs*. Both do not display the sequence of transformed formulas, but only list the axioms (from $F$) and the additions and (possibly) deletions. Several formats have been designed for resolution proofs [5,17,50] (which only add clauses), but they all share the same disadvantages. Resolution proofs are often huge, and it is hard to express important techniques, such as conflict clause minimization, with resolution steps. Other techniques, such as bounded variable addition [38], cannot be polynomially-simulated by resolution at all. Clausal proof formats [23,46,48] are syntactically similar; they involve a sequence of clauses that are claimed to be sat-preserving, starting with the given formula. But now we might add clauses which are not logically implied, and we also might remove clauses (this is needed now in order to enable certain additions, which might depend on global conditions).

**Definition 2 [48].** *A DRAT proof ("Deletion Resolution Asymmetric Tautology") for a formula $F$ is a sequence of additions and deletions of clauses, starting with $F$, such that each addition is the addition of a RAT clause w.r.t. the current formula (the result of additions and deletions up to this point), or, in case of adding the empty clause, unit-clause propagation on the current formula yields a contradiction. A DRAT refutation is a DRAT proof containing $\perp$.*

DRAT refutations are correct proofs of unsatisfiability (based on Lemma 1 and the fact, that deletion of clauses is always sat-preserving; note that Definition 2 allows unrestricted deletions). Furthermore they are checkable in cubic time. Since the proof of Lemma 1 is basically the same as the proof for [33, Lemma 4.1], by adding unit propagation appropriately one can transfer [33, Corollary 7.2] and prove that the power of DRAT refutations is up to polytime transformations the same as the power of Extended Resolution.

*Example 1.* Figure 1 shows an example DRAT refutation. Consider the CNF formula $F = (a \vee b \vee \bar{c}) \wedge (\bar{a} \vee \bar{b} \vee c) \wedge (b \vee c \vee \bar{d}) \wedge (\bar{b} \vee \bar{c} \vee d) \wedge (a \vee c \vee d) \wedge (\bar{a} \vee \bar{c} \vee \bar{d}) \wedge (\bar{a} \vee b \vee d) \wedge (a \vee \bar{b} \vee \bar{d})$, shown in DIMACS format in Fig. 1 (left), where 1 represents $a$, 2 is $b$, 3 is $c$, 4 is $d$, and negative numbers represent negation. The first clause in the proof, $(\bar{a})$, is a RAT clause with respect to $F$ because all possible resolvents are asymmetric tautologies:

$$F \wedge (a) \wedge (\bar{b}) \wedge (c) \vdash_1 \bot \quad \text{using} \quad (a \vee b \vee \bar{c})$$
$$F \wedge (a) \wedge (\bar{c}) \wedge (\bar{d}) \vdash_1 \bot \quad \text{using} \quad (a \vee c \vee d)$$
$$F \wedge (a) \wedge (b) \wedge (d) \vdash_1 \bot \quad \text{using} \quad (a \vee \bar{b} \vee \bar{d})$$

## 4    Cube-and-Conquer Solving

Arguably the most effective method to solve many hard combinatorial problems via SAT technology is the *cube-and-conquer* paradigm [25], abbreviated by C&C, due to strong performance and easy parallelization, which has been demonstrated by the C&C solver `Treengeling` [6] in recent SAT Competitions. C&C consists of two phases. In the first phase, a look-ahead SAT solver [26] partitions the problem into many (potentially millions of) subproblems. These subproblems, expressed as "cubes" (conjunctions) of the decisions (the literals set to true), are solved using a CDCL solver, also known as the "conquer" solver. The intuition behind this combination of paradigms is that look-ahead heuristics focus on global decisions, while CDCL heuristics focus on local decisions. Global decisions are important to split the problem, while local decisions are effective when there exist a short refutation. So the idea behind C&C is to partition the problem until a short refutation arises. C&C can solve hard problems much faster than either pure look-ahead or pure CDCL. The problem with pure look-ahead solving is that global decisions become poor decisions when a short refutation is present, while pure CDCL tends to perform rather poor when there exist no short refutation. We will demonstrate that C&C outperforms pure CDCL and pure look-ahead in Sect. 6.2. Apart from improved performance on a single core, C&C allows for easy parallelization. The subproblems are solved independently, so they are distributed on a large cluster.

There are two C&C variants: solving one cube per solver and solving multiple cubes by an incremental solver. The first approach allows solving cubes in parallel, while the second approach allows for reusing heuristics and learned clauses while solving multiple cubes. The second approach works as follows: an incremental SAT solver receives the input formula and a sequence of cubes[1]. After solving the formula under the assumption that a cube is true, the solver does not terminate, but starts working on a next cube. The heuristics and the learned clause database are not reset when starting solving a new cube, but reused to potentially exploit similarities between cubes.

---

[1] In practice this is done using a single incremental CNF file. For details about the format, see http://www.siert.nl/icnf/.

In our computation we combined them, via a two-staged splitting, to exploit both parallelism and reusage. First the problem is split into $10^6$ cubes, and then for each cube, the corresponding subproblem is split again creating billions of sub-cubes. An incremental SAT solver solves all the sub-cubes generated from a single cube sequentially.

## 5   Solving the Boolean Pythagorean Triples Problem

Our framework for solving hard problems consists of five phases: encode, transform, split, solve, and validate. The focus of the encode phase is to make sure that representation of the problem as SAT instance is valid. The transform phase reformulates the problem to reduce the computation costs of the later phases. The split phase partitions the transformed formula into many, possibly millions of subproblems. The subproblems are tackled in the solve phase. The validation phase checks whether the proofs emitted in the prior phases are a valid refutation for the original formula. Figure 2 shows an illustration of the framework. The framework, including the specialized heuristics, have been developed by the first author, who also performed all implementations and experiments.

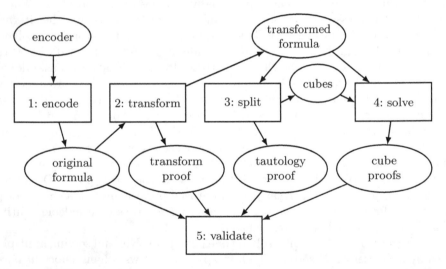

**Fig. 2.** Illustration of the framework to solve hard combinatorial problems. The phases are shown in the rectangle boxes, while the input and output files for these phases are shown in oval boxes.

### 5.1   Encode

The first phase of the framework focusses on making sure that the problem to be solved is correctly represented into SAT. In the second phase the representation will be optimized. The DRAT proof format can express all transformations.

Formula $F_n$ expresses whether the natural numbers up to $n$ can be partitioned into two parts with no part containing a triple $(a, b, c)$ such that $a^2 + b^2 = c^2$. One set will be called the positive part, while the other will be called the negative part. $F_n$ uses Boolean variables $x_i$ with $i \in \{1, \ldots, n\}$. The assignment $x_i$ to true/false, expresses that $i$ occurs in the positive/negative part, respectively. For each triple $(a, b, c)$ such that $a^2 + b^2 = c^2$, there is a constraint $\textsc{NotEqual}(a, b, c)$ in $F_n$, or in clausal form: $(x_a \vee x_b \vee x_c) \wedge (\bar{x}_a \vee \bar{x}_b \vee \bar{x}_c)$.

## 5.2   Transform

The goal of the transformation phase is to massage the initial encoding to execute the later phases more efficiently. A proof for the transformations is required to ensure that the changes are valid. Notice that a transformation that would be helpful for one later phase, might be harmful for another phase. Selecting transformations is therefore typically a balance between different trade-offs. For example, bounded variable elimination [16] is a preprocessing technique that tends to speed up the solving phase. However, this technique is generally harmful for the splitting phase as it obscures the look-ahead heuristics.

We applied two transformations. First, blocked clause elimination (BCE) [27]. BCE on $F_{7824}$ and $F_{7825}$ has the following effect: Remove $\textsc{NotEqual}(a, b, c)$ if $a$, $b$, or $c$ occurs only in this constraint, and apply this removal until fixpoint. Note that removing a constraint $\textsc{NotEqual}(a, b, c)$ because e.g. $a$ occurs once, reduces the occurrences of $b$ and $c$ by one, and as a result $b$ or $c$ may occur only once after the removal, allowing further elimination. We remark that a solution for the formula after the transformation may not satisfy the original formula, however this can be easily repaired [27]. The numerical effects of this reduction are as follows: $F_{7824}$ has 6492 (occurring) variables and 18930 clauses, $F_{7825}$ has 6494 variables and 18944 clauses, while after BCE-reduction we get 3740 variables and 14652 clauses resp. 3745 variables and 14672 clauses.

The second transformation is symmetry breaking [12]. The Pythagorean Triples encoding has one symmetry: the two parts are interchangeable. To break this, we can pick an arbitrary variable $x_i$ and assign it to true (or, equivalently, put in the positive part). In practice it is best to pick the variable $x_i$ that occurs most frequently in $F_n$. For the two formulas used during our experiments, the most occurring variable is $x_{2520}$ which was used for symmetry breaking. Symmetry breaking can be expressed in the DRAT format, but it is tricky. A recent paper [24] explains how to construct this part of the transformation proof.

Bounded variable elimination (a useful transformation in general) was not applied. Experiments showed that this transformation slightly increased the solving times. More importantly, applying bounded variable elimination transforms the problem into a non-3-SAT formula, thereby seriously harming the look-ahead heuristics, as the specialized 3-SAT heuristics can no longer be used.

## 5.3  Split

Partitioning is crucial to solve hard combinatorial problems. Effective partitioning is based on global heuristics [25] — in contrast to the "local" heuristics used in CDCL solvers. The result of partitioning is a binary branching tree of which the leaf nodes represent a subproblem of the original problem. The subproblem is constructed by adding the conjunction of decisions that lead to the leaf as unit clauses. Figure 3 shows such a partitioning as a binary tree with seven leaf nodes (left) and the corresponding list of seven cubes (right). The cubes are shown in the `inccnf` format that is used for incremental solvers to guide their search.

Splitting heuristics are crucial in solving a problem efficiently. In practice, the best heuristics are based on *look-aheads* [26,34]. In short, a look-ahead refers to assigning a variable to a truth value followed by unit propagation and measuring the changes to the formula during the propagation. It remains to find good measures. The simplest measure is to count the number of assigned variables; measures like that can be used for tie-breaking, but as has been realised in the field of heuristics [34], the expected future gains for unit-clause propagation, given by *new short clauses*, are more important than the current reductions. The default heuristic in C&C, which works well on most hard-combinatorial problems, weighs all new clauses using weights based on the length of the new clause (with an exponential decay of the weight in the length). However for our Pythagorean Triples encoding, using a refinement coming from random 3-SAT turned out to be more powerful. Here all newly created clauses are binary, i.e., ternary clauses that become binary during the look-ahead. The weight of a new binary clause depends on the occurrences of its two literals in the formula, estimating how likely they become falsified. This better performance is not very surprising as the formulas $F_n$ exhibit somewhat akin behavior to random 3-SAT formulas: (i) all clauses have length three; and (ii) the distribution of the occurrences of literals is similar. On the other hand, $F_n$ consists of clauses with either only positive literal or only negative literals — in contrast to random 3-SAT.

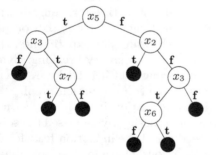

```
cube file in inccnf format

a   5  -3   0
a   5   3   7   0
a   5   3  -7   0
a  -5   2   0
a  -5  -2   3  -6   0
a  -5  -2   3   6   0
a  -5  -2  -3   0
```

**Fig. 3.** A binary branching tree (left) with the decision variables in the nodes and the polarity on the edges. The corresponding cube file (right) in the `inccnf` format. The prefix a denotes *assumptions*. Positive numbers express positive literals, while negative numbers express negative literals. Each cube (line) is terminated with a 0.

## 5.4   Details Regarding the Heuristics

The heuristics used for splitting extends the *recursive weight heuristics* [40], based on earlier work [13,15,36,37], by introducing minimal and maximal values $\alpha, \beta$, and choosing different parameters, optimized for the special case at hand. A look-ahead on literal $l$ measures the difference between a formula before and after assigning $l$ to true followed by simplification. Let $F$ (or $F_l$) denote the formula before (or after) the look-ahead on $l$, respectively. We assume that $F$ and $F_l$ are fully simplified using unit propagation. Thus $F_l \setminus F$ is the set of new clauses, and the task is to weigh them; we note that each clause in $F_l \setminus F$ is binary. Each literal is assigned a heuristic value $h(l)$ and the weight $w_{y \vee z}$ for $(y \vee z) \in F_l \setminus F$ is defined as $h(\bar{y}) \cdot h(\bar{z})$. The values of $h(l)$ are computed using multiple iterations $h_0(l), h_1(l), \ldots$, choosing the level with optimal performance, balancing the predictive power of the heuristics versus the cost to compute it. The idea of the heuristic values $h_i(l)$ is to approximate how strongly the literal $l$ is forced to true by the clauses containing $l$ (via unit propagation). First, for all literals $l$, $h_0(l)$ is initialized to 1: $h_0(x) = h_0(\bar{x}) = 1$. At each level $i \geq 0$, the average value $\mu_i$ is computed in order to scale the heuristics values $h_i(x)$:

$$\mu_i = \frac{1}{2n} \sum_{x \in \mathrm{var}(F)} \big( h_i(x) + h_i(\bar{x}) \big). \tag{1}$$

Finally, in each next iteration, the heuristic values $h_{i+1}(x)$ are computed in which literals $y$ get weight $h_i(\bar{y})/\mu_i$. The weight $\gamma$ expresses the relative importance of binary clauses. This weight could also be seen as the heuristic value of a falsified literal. Additionally, we have two other parameters, $\alpha$ expressing the minimal heuristic value and $\beta$ expressing maximum heuristic value.

$$h_{i+1}(x) = \max(\alpha, \min(\beta, \sum_{(x \vee y \vee z) \in F} \Big( \frac{h_i(\bar{y})}{\mu_i} \cdot \frac{h_i(\bar{z})}{\mu_i} \Big) + \gamma \sum_{(x \vee y) \in F} \frac{h_i(\bar{y})}{\mu_i})). \tag{2}$$

In each node of the branching tree we compute $h(l) := h_4(l)$ for all literals occurring in the formula. We use $\alpha = 8$, $\beta = 550$, and $\gamma = 25$. The "magic" constants differ significantly compared to the values used for random 3-SAT formulas where $\alpha = 0.1$, $\beta = 25$, and $\gamma = 3.3$ appear optimal [40]. The branching variable $x$ chosen is a variable with maximal $H(x) \cdot H(\bar{x})$, where $H(l) := \sum_{y \vee z \in F_l \setminus F} w_{y \vee z}$.

## 5.5   Solve

The solving phase is the most straightforward part of the framework. It takes the transformed formula and cube files as input and produces a proof of unsatisfiability of the transformed formula. Two different approaches can be distinguished in general: one for "easy" problems and one for "hard" problems. A problem is considered easy when it can be solved in reasonable time, say within a day on a single core. In that case, a single cube file can be used and the incremental SAT solver will emit a single proof file. The more interesting case is when problems are hard and two levels of splitting are required.

The boolean Pythagorean triples problem $F_{7825}$ is very hard and required two level splitting: the total runtime was approximately 4 CPU years (21,900 CPU hours for splitting and 13,200 CPU hours for solving). Any problem requiring that amount of resources has to be solved in parallel. The first level consists of partitioning the problem into $10^6$ subproblems, which required approximately 1000 s on a single core; for details see Sect. 6.2. Each subproblem is represented by a cube $\varphi_i$ with $i \in \{1, \ldots, 10^6\}$ expressing a conjunction of decisions. On the second level of splitting, each subproblem $F_{7825} \wedge \varphi_i$ is partitioned again using the same look-ahead heuristics. In contrast to the first level, the cubes generated on the second level are not used to create multiple subproblems. Instead, the second level cubes are provided to an incremental SAT solver together with a subproblem $F_{7825}$ and assumptions $\varphi_i$. The second level cubes are used to guide the CDCL solver. The advantage of guiding the CDCL solver is that learned clauses computed while solving one cube can be reused when solving another cube.

For each subproblem $F_{7825} \wedge \varphi_i$, the SAT solver produces a DRAT refutation. Most state-of-the-art SAT solvers currently support the emission of such proofs. One can check that the emitted proof of unsatisfiability is valid for $F_{7825} \wedge \varphi_i$. In this case, no changes to the proof logging of the solver are required. However, in order to create an unsatisfiability proof of $F_{7825}$ by concatenating the proofs of subproblems, all lemmas generated while solving $F_{7825} \wedge \varphi_i$ need to be extended with the clause $\neg\varphi_i$, and the SAT solver must not delete clauses from $F_{7825}$.

### 5.6 Validate

The last phase of the framework validates the results of the earlier phases. First, the encoding into SAT needs to be validated. This can be done by proving that the encoding tool is correct using a theorem prover. Alternatively, a small program can be implemented whose correctness can be checked manually. For example, our encoding tool consists of only 19 lines of C code. For details and validation files, check out http://www.cs.utexas.edu/~marijn/ptn/.

The second part consists of checking the three types of DRAT proofs produced in the earlier phases: the transformation, tautology, and the cube proofs. DRAT proofs can be merged easily by concatenating them. The required order for merging the proofs is: transformation proof, cube proofs, and tautology proof.

**Transformation Proof.** The transformation proof expresses how the initial formula, created by the encoder, is converted into a formula that is easier to solve. This part of the proof is typically small. The latest version of the `drat-trim` checker supports validating transformation proofs without requiring the other parts of the proof, based on a compositional argument [22].

**Cube Proofs.** The core of the validation is checking whether the negation of each cube, the clause $\neg\varphi_i$, is implied by the transformed formula. Since we partitioned the problem using $10^6$ cubes, there are $10^6$ of cube proofs. We generated and validated them all. However, their total size is too large to share: almost 200 terabyte in the DRAT format. We tried to compress the proof using a range of dedicated clause compression techniques [21] combined with state-of-the-art general purpose tools, such as `bzip2` or `7z`. After compression the total proof size was still 14 terabytes. So instead we provide the cube files for the subproblems as a certificate. Cube files can be compressed heavily, because they form a tree. Instead of storing all cubes as a list of literals, shown as in Fig. 3, it is possible to store only one literal per cube. Storing the literal in a binary format [21] followed by `bzip2` allowed us to store all the cube files using "only" 68 gigabytes of disk space. We added support for the `inccnf` format to `glucose` 3.0 in order to solve the cube files. This solver can also reproduce the DRAT proofs in about 13,000 CPU hours. Checking these proofs requires about 16,000 CPU hours, so reproducing the DRAT proofs almost doubles the validation effort. This is probably a smaller burden than downloading and decompressing many terabytes of data.

**Tautology Proof.** A cube partitioning is valid, i.e., covers the complete search space, if the disjunction of cubes is a tautology. This needs to be checked during the validation phase. Checking this can be done by negating the disjunction of cubes and feed the result to a CDCL solver which supports proof logging. If the solver can refute the formula, then the disjunction of cubes is a tautology. We refer to the proof emitted by the CDCL solver as the *tautology proof*. This tautology proof is part of the final validation effort.

# 6   Results

This section offers details of solving the boolean Pythagorean Triples problem[2]. All experiments were executed on the Stampede cluster[3]. Each node on this cluster consists of an Intel Xeon Phi 16-core CPU and 32 Gb memory. We used cube solver `march_cc` and conquer solver `glucose` 3.0 during our experiments.

## 6.1   Heuristics

In our first attempt to solve the Pythagorean triples problem, we partitioned the problem (top-level and subproblems) using the default decision heuristic in the cube solver `march_cc` for 3-SAT formulas. After some initial experiments, we estimated that the total runtime of solving (including splitting) $F_{7825}$ would be roughly 300,000 CPU hours on the Stampede cluster. To reduce the computation costs, we (manually) optimized the magic constants in `march_cc`, resulting in the heuristic presented in Sect. 5.4. The new heuristics reduced the total runtime to 35,000 CPU hours, so by almost an order of magnitude. Table 1 shows the results of various heuristics on five randomly selected subproblems. Here, we optimized `march_cc` in favor of the other heuristics to make the comparison more fair: we turned off look-ahead preselection, which is helpful for the new heuristics (and thus used in the computation), but harmful for the other heuristics.

---

[2] Files and tools can be downloaded at http://www.cs.utexas.edu/~marijn/ptn/.
[3] https://www.tacc.utexas.edu/systems/stampede.

**Table 1.** Solving times for C&C using different look-ahead heuristics and pure CDCL. The top left, bottom left, and right numbers expresses the cube, conquer, and their sum times, respectively. *Ptn 3-SAT* is 3-SAT heuristics optimized for Pythagorean triple problems; *rnd 3-SAT* is the 3-SAT heuristics optimized for random 3-SAT (default); #*bin* is the sum of new binary clauses; and #*var* is the number of assigned variables.

| cube # | Ptn 3-SAT | | rnd 3-SAT | | #bin | | #var | | pure CDCL |
|---|---|---|---|---|---|---|---|---|---|
| 104302 | 152.98 / 75.50 | **228.48** | 608.46 / 174.94 | 783.40 | 263.23 / 150.71 | 413.94 | 789.43 / 263.79 | 1053.22 | 1372.87 |
| 268551 | 74.03 / 33.83 | **107.86** | 92.09 / 48.82 | 140.91 | 98.93 / 55.83 | 154.76 | 487.45 / 220.27 | 707.72 | 150.06 |
| 934589 | 136.94 / 74.44 | **211.38** | 206.28 / 121.99 | 328.27 | 156.78 / 107.16 | 263.94 | 529.21 / 235.70 | 764.91 | 631.91 |
| 950025 | 143.69 / 74.09 | **217.78** | 152.49 / 99.67 | 252.16 | 203.18 / 138.09 | 341.27 | 550.47 / 226.99 | 777.46 | 330.61 |
| 980757 | 112.22 / 30.41 | **142.63** | 170.34 / 53.90 | 224.24 | 181.14 / 60.53 | 241.67 | 685.04 / 160.93 | 845.97 | 155.57 |

## 6.2   Cube and Conquer

The first step of the solving phase was partitioning the transformed formula into many subproblems using look-ahead heuristics. Our cluster account allowed for running on 800 cores in parallel. We decided to partition the problem into a multiple of 800 to perform easy parallel execution: exactly $10^6$. Partitioning the formula into $10^6$ subproblems ensured that the conquer time of solving most subproblems is less than two minutes, a runtime with the property that proof validating can be achieved in a time similar to the solving time.

A simple way of splitting a problem into $10^6$ subproblems is to build a balanced binary branching tree of depth 20. However, using a balanced binary branching tree results in poor performance on hard combinatorial problems [25]. A more effective partitioning heuristic picks the leaf nodes such that the number of assigned variables (including inferred variables) in those nodes are equal. Based on some initial experiments, we observed that the best heuristics for Pythagorean Triples formulas however is to count the number of binary clauses in each node. Recall that all clauses in the transformed formula are ternary. Selecting nodes in the decision tree that have about 3,000 binary clauses resulted in $10^6$ subproblems. Figure 4 (left) shows a histogram of the depth of the branching tree (or, equivalently, the size of the cube) of the selected nodes. Notice that the smallest cube has size 12 and the largest cubes have size 49.

Figure 4 (right) shows the time for the cube and conquer runtimes averaged per size of the cubes. The peak average of the cube runtime is around size 24, while the peak of the conquer runtime is around size 26. The cutoff heuristics of the cube solver for second level splitting were based on the number of unassigned variables, 3450 variables to be precise.

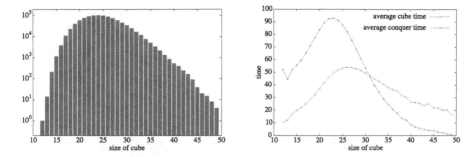

**Fig. 4.** Left, a histogram (logarithmic) of the cube size of the $10^6$ subproblems. Right, average runtimes per size for the split (cube) and solve (conquer) phases.

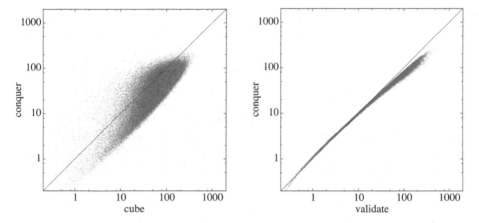

**Fig. 5.** Left, a scatter plot comparing the cube (split) and conquer (solve) time per subproblem. Right, a scatter plot comparing the validation and conquer time.

A comparison between the cube, conquer, and validation runtimes is shown in Fig. 5. The left scatter plot compares cube and conquer runtimes. It shows that within our experimental setup the cube computation is about twice as expensive compared to the conquer computation. The right scatter plot compares the validation and conquer runtimes. It shows that these times are very similar. Validation runtimes grow slightly faster compared to conquer runtimes. The average cube, conquer, and validation times for the $10^6$ subproblems are 78.87, 47.52, and 60.62 s, respectively.

Figure 6 compares the cube+conquer runtimes to solve the $10^6$ subproblems with the runtimes of pure CDCL (using `glucose` 3.0) and pure look-ahead (using `march_cc`). The plot shows that cube+conquer clearly outperforms pure CDCL. Notice that no heuristics of `glucose` 3.0 were changed during all experiments for both cube+conquer and pure CDCL. In particular, a variable decay of 0.8 was used throughout all experiments as this is the `glucose` default. However,

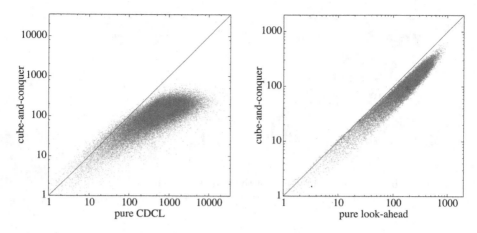

**Fig. 6.** Scatterplots comparing cube-and-conquer to pure CDCL (left) and pure look-ahead (right) solving methods on the Pythagorean Triples subproblems.

we observed that a higher variable decay (in between 0.95 and 0.99) would improve the performance of both cube+conquer and pure CDCL. We did not optimize glucose to keep it simple, and because the conquer part is already the cheapest phase of the framework (compared to split and validate); indeed frequently speed-ups of two orders or magnitude could be achieved on the harder instances. Pure look-ahead is also slower compared to cube+conquer, but the differences are smaller: on average cube+conquer is about twice as fast.

### 6.3    Extreme Solutions

Of the $10^6$ subproblems that were created during the splitting phase, only one subproblem is satisfiable for the extreme case, i.e., $n = 7824$. This suggests that the formula after symmetry breaking has a big *backbone*. A variable belongs to backbone of a formula if it is assigned to the same truth value in all solutions. We computed the backbone of $F_{7824}$, which consists of 2304 variables. The backbone reveals why it is impossible to avoid Pythagorean Triples indefinitely when partitioning the natural numbers into two parts: variables $x_{5180}$ and $x_{5865}$ are both positive in the backbone, forcing $x_{7825}$ to be negative due to $5180^2 + 5865^2 = 7825^2$. At the same time, variables $x_{625}$ and $x_{7800}$ are both negative in the backbone forcing $x_{7825}$ to be positive due to $625^2 + 7800^2 = 7825^2$.

A satisfying assignment does not necessarily assign all natural numbers up to 7824 that occur in Pythagorean Triples. For example, we found a satisfying assignment that assigns only 4925 out of the 6492 variables occurring in $F_{7824}$. So not only is $F_{7824}$ satisfiable, but it has a huge number of solutions.

## 7    Conclusions

We solved and verified the boolean Pythagorean Triples problem using C&C. The total solving time was about 35,000 h and the verification time about 16,000 h.

Since C&C allows for massive parallelization, resulting in almost linear-time speedups, the problem was solved altogether in about two days on the Stampede cluster. Apart from strong computational resources, dedicated look-ahead heuristics were required to achieve these results. In future research we want to further develop effective look-ahead heuristics that will work for such hard combinatorial problems out of the box. We expect that parallel computing combined with look-ahead splitting heuristics will make it feasible to solve many other hard combinatorial problems that are too hard for existing techniques. Moreover, we argue that solutions to such problems require certificates that can be validated by the community — similar to the certificate we provided for the boolean Pythagorean Triples problem. A fundamental question is whether Theorem 1 has a "mathematical" (human-readable) proof, or whether the gigantic (sophisticated) case-distinction, which is at the heart of our proof, is the best there is? It is conceivable that Conjecture 1 is true, but for each $k$ has only proofs like our proof, where the size of these proofs is growing so quickly, that Conjecture 1 is actually not provable in current systems of foundations of Mathematics (like ZFC).

**Acknowledgements.** The authors acknowledge the Texas Advanced Computing Center (TACC) at The University of Texas at Austin for providing grid resources that have contributed to the research results reported within this paper.

# References

1. Mizar proof checker. Accessed: November 2015
2. Coq proof manager. Accessed: November 2015
3. The site of *flyspeck* project, the formal verification of the proof of Kepler Conjecture. Accessed: November 2015
4. Ahmed, T., Kullmann, O., Snevily, H.: On the van der Waerden numbers w(2; 3, $t$). Discrete Appl. Math. **174**, 27–51 (2014)
5. Biere, A.: Picosat essentials. JSAT **4**(2–4), 75–97 (2008)
6. Biere, A.: Lingeling, Plingeling and Treengeling entering the SAT competition 2013. In: Proceedings of SAT Competition 2013, p. 51 (2013)
7. Biere, A., Heule, M.J.H., van Maaren, H., Walsh, T. (eds.): Handbook of Satisfiability. Frontiers in Artificial Intelligence and Applications, vol. 185. IOS Press, Amsterdam (2009)
8. Clarke, E.M., Grumberg, O., Peled, D.: Model Checking. MIT Press, Cambridge (1999)
9. Cook, S.A.: The complexity of theorem-proving procedures. In: Proceedings 3rd Annual ACM Symposium on Theory of Computing (STOC 1971), pp. 151–158 (1971)
10. Cooper, J., Overstreet, R.: Coloring so that no Pythagorean triple is monochromatic (2015). arXiv:1505.02222
11. Cooper, J., Poirel, C.: Note on the Pythagorean triple system (2008)
12. Crawford, J., Ginsberg, M., Luks, E., Roy, A.: Symmetry-breaking predicates for search problems. In: Proceedings of 5th International Conference on Knowledge Representation and Reasoning, KR 1996, pp. 148–159. Morgan Kaufmann (1996)

13. Dequen, G., Dubois, O.: *kcnfs*: an efficient solver for random *k*-SAT formulae. In: Giunchiglia, E., Tacchella, A. (eds.) SAT 2003. LNCS, vol. 2919, pp. 486–501. Springer, Heidelberg (2004)
14. Dransfield, M.R., Marek, V.W., Truszczyński, M.: Satisfiability and computing van der Waerden numbers. In: Giunchiglia, E., Tacchella, A. (eds.) SAT 2003. LNCS, vol. 2919, pp. 1–13. Springer, Heidelberg (2004)
15. Dubois, O., Dequen, G.: A backbone-search heuristic for efficient solving of hard 3-SAT formulae. In: International Joint Conferences on Artificial Intelligence (IJCAI), pp. 248–253 (2001)
16. Eén, N., Biere, A.: Effective preprocessing in SAT through variable and clause elimination. In: Bacchus, F., Walsh, T. (eds.) SAT 2005. LNCS, vol. 3569, pp. 61–75. Springer, Heidelberg (2005)
17. Eén, N., Sörensson, N.: An extensible SAT-solver. In: Giunchiglia, E., Tacchella, A. (eds.) SAT 2003. LNCS, vol. 2919, pp. 502–518. Springer, Heidelberg (2004)
18. Franco, J., Martin, J.: A history of satisfiability. In: Biere et al. [7], Chap. 1, pp. 3–74
19. Garey, M.R., Johnson, D.S.: Computers and Intractability/A Guide to the Theory of NP-Completeness. W.H. Freeman and Company, New York (1979)
20. Henschen, L.J., Wos, L.: Unit refutations and Horn sets. J. Assoc. Comput. Mach. **21**(4), 590–605 (1974)
21. Heule, M.J.H., Biere, A.: Clausal proof compression. In: 11th International Workshop on the Implementation of Logics (2015)
22. Heule, M.J.H., Biere, A.: Compositional propositional proofs. In: Davis, M., Fehnker, A., McIver, A., Voronkov, A. (eds.) LPAR-20 2015. LNCS, vol. 9450, pp. 444–459. Springer, Heidelberg (2015). doi:10.1007/978-3-662-48899-7_31
23. Heule, M.J.H., Hunt Jr., W.A., Wetzler, N.: Verifying refutations with extended resolution. In: Bonacina, M.P. (ed.) CADE 2013. LNCS, vol. 7898, pp. 345–359. Springer, Heidelberg (2013)
24. Heule, M.J.H., Hunt Jr., W.A., Wetzler, N.: Expressing symmetry breaking in DRAT proofs. In: Felty, A.P., Middeldorp, A. (eds.) CADE-25. LNCS, vol. 9195, pp. 591–606. Springer, Heidelberg (2015)
25. Heule, M.J.H., Kullmann, O., Wieringa, S., Biere, A.: Cube and conquer: guiding CDCL SAT solvers by lookaheads. In: Eder, K., Lourenço, J., Shehory, O. (eds.) HVC 2011. LNCS, vol. 7261, pp. 50–65. Springer, Heidelberg (2012)
26. Heule, M.J.H., van Maaren, H.: Look-ahead based SAT solvers. In: Biere et al. [7], Chap. 5, pp. 155–184
27. Järvisalo, M., Biere, A., Heule, M.J.H.: Blocked clause elimination. In: Esparza, J., Majumdar, R. (eds.) TACAS 2010. LNCS, vol. 6015, pp. 129–144. Springer, Heidelberg (2010)
28. Järvisalo, M., Heule, M.J.H., Biere, A.: Inprocessing rules. In: Gramlich, B., Miller, D., Sattler, U. (eds.) IJCAR 2012. LNCS, vol. 7364, pp. 355–370. Springer, Heidelberg (2012)
29. Karp, R.M.: Reducibility among combinatorial problems. In: Miller, R.E., Thatche, J.W. (eds.) Complexity of Computer Computations, pp. 85–103. Plenum Press, New York (1972)
30. Kay, W.: An overview of the constructive local lemma. Master's thesis, University of South Carolina (2009)
31. Kouril, M.: Computing the van der Waerden number $W(3,4) = 293$. INTEGERS: Electron. J. Comb. Number Theory **12**(A46), 1–13 (2012)
32. Kouril, M., Paul, J.L.: The van der Waerden number $W(2,6)$ is 1132. Exp. Math. **17**(1), 53–61 (2008)

33. Kullmann, O.: On a generalization of extended resolution. Discrete Appl. Math. **96–97**, 149–176 (1999)
34. Kullmann, O.: Fundaments of branching heuristics. In: Biere et al. [7], Chap. 7, pp. 205–244
35. Levin, L.: Universal search problems. Problemy Peredachi Informatsii **9**, 115–116 (1973)
36. Li, C.M.: A constraint-based approach to narrow search trees for satisfiability. Inf. Process. Lett. **71**(2), 75–80 (1999)
37. Li, C.M., Anbulagan: Heuristics based on unit propagation for satisfiability problems. In: Proceedings of 15th International Joint Conference on Artificial Intelligence (IJCAI 1997), pp. 366–371. Morgan Kaufmann Publishers (1997)
38. Manthey, N., Heule, M.J.H., Biere, A.: Automated reencoding of boolean formulas. In: Proceedings of Haifa Verification Conference 2012 (2012)
39. Marques-Silva, J.P., Lynce, I., Malik, S.: Conflict-driven clause learning SAT solvers. In: Biere et al. [7], Chap. 4, pp. 131–153
40. Mijnders, S., de Wilde, B., Heule, M.J.H.: Symbiosis of search and heuristics for random 3-SAT. In: Mitchell, D., Ternovska, E. (eds.) Third International Workshop on Logic and Search (LaSh 2010) (2010)
41. Myers, K.J.: Computational advances in Rado numbers. Ph.D. thesis, Rutgers University (2015)
42. Rado, R.: Some partition theorems. In: Colloquia Mathematica Societatis János Bolyai 4. Combinatorial Theory and Its Applications III, pp. 929–936. North-Holland, Amsterdam (1970)
43. Schur, I.: Über die Kongruenz $x^m + y^m = z^m$ (mod $p$). Jahresbericht der Deutschen Mathematikervereinigung **25**, 114–117 (1917)
44. Tseitin, G.S.: On the complexity of derivation in propositional calculus. In: Siekmann, J.H., Wrightson, G. (eds.) Automation of Reasoning 2, pp. 466–483. Springer, Heidelberg (1983)
45. van der Waerden, B.L.: Beweis einer Baudetschen Vermutung. Nieuw Archief voor Wiskunde **15**, 212–216 (1927)
46. Van Gelder, A.: Verifying RUP proofs of propositional unsatisfiability. In: ISAIM (2008)
47. Voevodski, V.: Lecture at ASC 2008, How I became interested in foundations of mathematics. Accessed: November 2015
48. Wetzler, N., Heule, M.J.H., Hunt Jr., W.A.: DRAT-trim: efficient checking and trimming using expressive clausal proofs. In: Sinz, C., Egly, U. (eds.) SAT 2014. LNCS, vol. 8561, pp. 422–429. Springer, Heidelberg (2014)
49. Zhang, H.: Combinatorial designs by SAT solvers. In: Biere et al. [7], Chap. 17, pp. 533–568
50. Zhang, L., Malik, S.: Validating SAT solvers using an independent resolution-based checker: practical implementations and other applications. In: DATE, pp. 10880–10885 (2003)

# Satisfiability Modulo Theory

# Deciding Bit-Vector Formulas with mcSAT

Aleksandar Zeljić[1], Christoph M. Wintersteiger[2]([✉]), and Philipp Rümmer[1]

[1] Uppsala University, Uppsala, Sweden
{aleksandar.zeljic,philipp.ruemmer}@it.uu.se
[2] Microsoft Research, Cambridge, UK
cwinter@microsoft.com

**Abstract.** The Model-Constructing Satisfiability Calculus (mcSAT) is a recently proposed generalization of propositional DPLL/CDCL for reasoning modulo theories. In contrast to most DPLL(T)-based SMT solvers, which carry out conflict-driven learning only on the propositional level, mcSAT calculi can also synthesise new theory literals during learning, resulting in a simple yet very flexible framework for designing efficient decision procedures. We present an mcSAT calculus for the theory of fixed-size bit-vectors, based on tailor-made conflict-driven learning that exploits both propositional and arithmetic properties of bit-vector operations. Our procedure avoids unnecessary bit-blasting and performs well on problems from domains like software verification, and on constraints over large bit-vectors.

## 1 Introduction

Fixed-length bit-vectors are one of the most commonly used datatypes in Satisfiability Modulo Theories (SMT), with applications in hardware and software verification, synthesis, scheduling, encoding of combinatorial problems, and many more. Bit-vector solvers are highly efficient, and typically based on some form of SAT encoding, commonly called *bit-blasting,* in combination with sophisticated methods for upfront simplification. Bit-blasting may be implemented with varying degree of laziness, ranging from eager approaches where the whole formula is translated to propositional logic in one step, to solvers that only translate conjunctions of bit-vector literals at a time (for an overview, see [22]). Despite the huge body of research, aspects of bit-vector solving are still considered challenging, including the combination of bit-vectors with other theories (e.g., arrays or uninterpreted functions), large bit-vector problems that are primarily of arithmetic character (in particular when non-linear), and problems involving very long bit-vectors. A common problem encountered in such cases is excessive memory consumption of solvers, especially for solvers that bit-blast eagerly.

A contrived yet instructive example with very long bit-vectors is given in Fig. 1, adapted from a benchmark of the SMT-LIB QF_BV 'pspace' subset [18]. The benchmark tests the overflow behavior of addition. Its model is simple, regardless of the size of bit-vectors $x$ and $y$ ($x$ should consist of only 1-bits, while $y$ should consist of only 0-bits). Finding a model for the formula should in principle be easy,

© Springer International Publishing Switzerland 2016
N. Creignou and D. Le Berre (Eds.): SAT 2016, LNCS 9710, pp. 249–266, 2016.
DOI: 10.1007/978-3-319-40970-2_16

```
(set-logic QF_BV)
(declare-fun x () (_ BitVec 29980))
(declare-fun y () (_ BitVec 29980))
(assert (and (bvuge x y) (bvule (bvadd x (_ bv1 29980)) y)))
```

**Fig. 1.** Simplified example from the 'pspace' subset [18] of SMT-LIB, QF_BV

but proves challenging for bit-blasting procedures. Other sources of very long bit-vectors are system memory in hardware verification or heap in software verification (e.g., [4]), chemical reaction networks, or gene regulatory networks (e.g., [35]).

Generally, memory consumption is a limiting factor in the application of bit-vector solvers. Increasing the size of bit-vectors that can efficiently be reasoned about would broaden the range of applications, while also simplifying system models for further analysis. With that in mind, we introduce a new model-constructing decision procedure for bit-vectors that is lazier than previous solvers. The procedure is defined as an instance of the *model-constructing satisfiability calculus* (mcSAT [30]), a framework generalizing DPLL and conflict-driven learning (CDCL) to non-Boolean domains. Like other SMT solvers for bit-vectors, our procedure is defined on top of well-understood SAT technology; unlike most existing solvers, we treat bit-vectors as first-class objects, which enables us to design tailor-made propagation and learning schemes for bit-vector constraints, as well as avoiding bit-blasting of bit-vector operations altogether.

The contributions of this paper are as follows: 1. a novel decision procedure for the theory of bit-vectors that avoids bit-blasting, 2. an extension of the mcSAT calculus to support partial model assignments, 3. a new mcSAT heuristic for generalizing explanations, and 4. an implementation of the procedure and preliminary experimental evaluation of its performance.

## 1.1 Motivating Examples

We start by illustrating our approach using two examples: a simple bit-vector constraint that illustrates the overall strategy followed by our decision procedure, and a simple family of bit-vector problems on which our procedure outperforms existing bit-blasting-based methods. Consider bit-vectors $x, y, z$ of length 4, and let $\oplus$ denote bit-wise exclusive-or and $\leq_u, <_u$ be unsigned comparison in

$$\phi \quad \equiv \quad x = y + z \ \wedge \ y <_u z \ \wedge \ (x \leq_u y + y \ \vee \ x \oplus z = 0001).$$

The goal is to find an assignment to $x, y, z$ such that formula evaluates to *true*. Figure 2 illustrates an application of our algorithm to $\phi$ (after clausification). Starting from an empty trail, we assert the unit clauses, denoted by implications with empty antecedents (lines 1 and 2 in Fig. 2a). At this point the procedure chooses between making a model assignment to one of the bit-vector variables, or Boolean decisions. Here, we choose to make a *decision* and assume $x \leq_u y + y$ (line 3). Decisions (and model assignments) are denoted with a horizontal line above them in the trail. The Boolean structure of the formula $\phi$ is now satisfied,

| Trail element | | Trail element | | Trail element | |
|---|---|---|---|---|---|
| 1 | $() \to x = y + z$ | 1 | $() \to x = y + z$ | 1 | $() \to x = y + z$ |
| 2 | $() \to y <_u z$ | 2 | $() \to y <_u z$ | 2 | $() \to y <_u z$ |
| 3 | $x \leq_u y + y$ | 3 | $x \leq_u y + y$ | 3 | $x \leq_u y + y$ |
| 4 | $y \mapsto 1111$ | 4 | $y <_u z \to \neg(y = 1111)$ | 4 | $y <_u z \to \neg(y = 1111)$ |
| 5 | $z \mapsto ?$ | 5 | $y \mapsto 1110$ | 5 | $(\ldots) \to \neg(y \geq_u 1000)$ |
|  |  | 6 | $z \mapsto 1111$ | 6 | $y \mapsto 0111$ |
|  |  | 7 | $x \mapsto 1101$ | 7 | $z \mapsto 1001$ |
|  |  |  |  | 8 | $x \mapsto 0000$ |

|  |  |  |
|---|---|---|
| (a) Infeasible trail | (b) Conflicted trail | (c) Satisfied trail |

**Fig. 2.** Critical trail states during the execution of our algorithm

so we search for satisfying model assignments to the bit-vector variables. Here, we decide on $y \mapsto 1111$ (line 4 in Fig. 2a). The literal $y <_u z$ now gives a lower bound for $z$. Our procedure immediately determines that the trail has become *infeasible,* since no value of $z$ will be consistent with $y = 1111$ and $y <_u z$.

We now need an *explanation* to restore the trail to a state where it is not infeasible anymore. In mcSAT, an explanation of a conflict is a valid clause with the property that the trail implies falsity of each of the clause literals. One possible explanation in our case is $\neg(y = 1111) \vee \neg(y <_u z)$. After resolving the explanation against the trail (in reverse order, similar to Boolean conflict resolution in SAT solvers), at the first point where at least one literal in the conflict clause no longer evaluates to *false*, the conflict clause becomes an implication and is put on the trail. In this example, as soon as we undo the assignment $y \mapsto 1111$, the literal $\neg(y = 1111)$ can be assumed (line 4 in Fig. 2b). The procedure makes the next legal assignment $y \mapsto 1110$ (line 5 in Fig. 2b). Bounds propagation using $y <_u z$ then implies the model assignment $z \mapsto 1111$ (line 6 in Fig. 2b). Values of $y$ and $z$ imply a unique value 1101 for $x$, however, the model assignment $x \mapsto 1101$ is not legal because it violates $x <_u y + y$ when $y = 1110$. By means of bounds propagation we have detected a conflict in $y = 1110 \wedge y <_u z \wedge x = z + y \wedge x <_u y + y$.

Our procedure tries to generalize conflicts, to avoid re-visiting conflicts of similar shape in the future. Generalization is done by weakening the literals of $y = 1110 \wedge y <_u z \wedge x = z + y \wedge x <_u y + y$, and checking if the conflict persists. First, $y = 1110$ is rewritten to $y \leq_u 1110 \wedge y \geq_u 1110$; it is then detected that $y \leq_u 1110$ is redundant, because bounds propagation derives unsatisfiability even without it. Now we weaken the literal $y \geq_u 1110$ by changing the constant, say to $y \geq_u 1000$, and verify using bounds propagation that the conflict persists (Example 3). Weakening $y \geq_u 1000$ further would lead to satisfiability. By negation we obtain a valid explanation $\neg(y \geq_u 1000) \vee \neg(y <_u z) \vee \neg(x = z + y) \vee \neg(x <_u y + y)$, which we use to backtrack the trail to a non-conflicted state (line 5 in Fig. 2c). From this point on straight-forward propagation yields a satisfying solution.

After presenting the basic ideas behind our procedure, we argue that it is well suited to problems that stem from model checking applications. Consider

**Algorithm 1.** Factorial

```
1  unsigned int factorial = 1u;
2  unsigned int n;
3  for (int i = n; i > 0u; i--) {
4    factorial = factorial * i;
5  }
6  assert ( n <= 1 || f % 2u == 0u)
```

the simple C program shown in Algorithm 1. The program computes the factorial of some value $n$ by multiplying the factors starting from $n$ and counting down. We add an assertion at the end, which checks whether `factorial` is even if the value of `n` is greater than one. We use bounded model checking and unwind the loop a fixed number of iterations, to generate formulas of increasing complexity. Figure 3 shows the performance of mcBV (our prototype) and state-of-the-art solvers on these benchmarks (Boolector [10] is the winner of the QF_BV track of the 2015 SMT competition; Z3 [29] is our baseline as mcBV uses the Z3 parser and preprocessor). On this class of benchmarks, mcBV performs significantly better than Z3 and comparably to Boolector.

## 1.2   Related Work

The most popular approach to solving bit-vector formulas is to translate them to propositional logic and further into conjunctive normal form via the Tseitin translation [33] (*bit-blasting*), such that an off-the-shelf SAT solver can be used to determine satisfiability. In contrast, our approach does not bit-blast, and we attempt to determine satisfiability directly on the word level. Our technique builds on the Model-Constructing Satisfiability Calculus recently developed by Jovanović and de Moura [23,30]. Our approach is similar in spirit to previous work by Bardin et al. [1], which avoids bit-blasting by encoding bit-vector problems into integer arithmetic, such that a (customized) CLP solver for finite

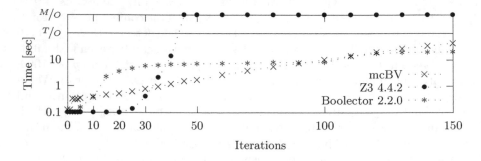

**Fig. 3.** Factorial example performance

domains can be used. A different angle is taken by Wille et al. [34] in their SWORD tool, which uses vector-level information to increase the performance of the SAT solver by abstracting some sub-formulas into 'modules' that are handled similar to custom propagators in a CLP solver.

On the one hand, various decisions problems involving bit-vectors have recently been shown to be of fairly high complexity [19,24,25] and on the other hand, some fragments are known to be decidable in polynomial time; for instance, Bruttomesso and Sharygina describe an efficient decision procedure for the 'core' theory of bit-vectors [11], based on earlier work by Cyrluk et al. [15], who defined this fragment via syntactic restriction to extraction and concatenation being the only bit-vector operators that are permitted. There is also a small body of work on the extension of decision procedures for bit-vectors that do not have a fixed size. For instance, Bjørner and Pichora [7] describe a unification-based calculus for (non-)fixed size bit-vectors, while Möller and Ruess [28] describe a procedure for (non-)fixed size bit-vectors that bit-blasts lazily ('splitting on demand').

Most SMT solvers implement lazy and/or eager bit-blasting procedures. These either directly, eagerly translate to a Boolean CNF and then run a SAT solver, or, especially when theory combination is required, they use a lazy bit-blaster that translates relevant parts of formulas on demand. This is the case, for instance in Boolector [10], MathSAT [13], CVC4 [2], STP [20], Yices [17], and Z3 [29]. Hadarean et al. [22] present a comparison and evaluation of eager, lazy, and combined bit-vector solvers in CVC4.

Griggio proposes an efficient procedure for the construction of Craig interpolants of bit-vector formulas ([27] via translation to QF_LIA, quantifier-free linear integer arithmetic [21]). Interpolants do have applications in mcBV, e.g., for conflict generalization, but we do not currently employ such methods.

Model checkers that do not use SMT solvers sometimes implement their own bit-blasting procedures and then use SAT solvers, BDD-based, or AIG-based solvers. This is often done in bounded model checking [5,6], but more recently also in IC3 [9], Impact [27], or $k$-induction [32]. Examples thereof include CBMC [14], EBMC [26], NuSMV2 [12]. In some cases model-checkers based on abstract interpretation principles use bit-vector solvers for counter-example generation when the proof fails; this is the case, for instance, for the separation-logic based memory analyzer SLAyer [3,4].

In recent times bit-vector constraints are also used for formal systems analysis procedures in areas other than verification, for instance in computational biology [35] where Dunn et al. [16] identify and analyze a bit-vector model for pluripotent stem-cells via an encoding of the model into bit-vector constraints. Similarly, bit-vector solvers are used within interactive and automated theorem provers to construct bit-vector proofs, for instance, in Isabelle/HOL [8].

## 2    Preliminaries: Bit-Vector Constraints

We consider a logic of quantifier-free fixed-width bit-vector constraints, defined by the following grammar, in which $\phi$ ranges over formulas and $t$ over bit-vector terms:

$$\phi \quad ::= \quad true \mid false \mid p \mid \neg\phi \mid \phi \wedge \phi \mid \phi \vee \phi \mid t \bullet t$$

$$t \quad ::= \quad 0_n \mid 1_n \mid \cdots \mid (0|1)^+ \mid x \mid \mathsf{extract}_p^n(t) \mid \, !t \mid t \circ t$$

Here, $p$ ranges over propositional variables; expressions $0_n, 1_n, \ldots$ are decimal bit-vector literals of width $n$; literals $(0|1)^+$ represent bit-vectors in binary; $x$ ranges over bit-vector variables of size $\alpha(x)$; predicates $\bullet \in \{=, \leq_s, \leq_u\}$ represent equality, signed inequality (2's complement format), and unsigned inequality, respectively; the operator $\mathsf{extract}_p^n$ represents extraction of $n$ bits starting from position $p$ (where the left-most bit has position 0); ! is bit-wise negation; binary operators $\circ \in \{+, \times, \div, \&, \mid, \oplus, \ll, \gg, \frown\}$ represent addition, multiplication, (unsigned) integer division, bit-wise and, bit-wise or, bit-wise exclusive or, left-shift, right-shift, and concatenation, respectively. We assume typing and semantics of bit-vector constraints are defined as usual.

An *atom* is a formula $\phi$ that does not contain $\neg, \wedge, \vee$. An atom is *flat* if it is of the form $x \leq_s y$, $x \leq_u y$, or $x = t$, and $t$ does not contain nested operators. A *literal* is an atom or its negation. A *clause* is a disjunction of literals. When checking satisfiability of a bit-vector constraint, we generally assume that the constraint is given in the form of a set of clauses containing only flat atoms.

## 3    mcSAT with Projections

We now introduce the framework used by our decision procedure for bit-vector constraints, based on a generalized version of mcSAT [30]. In contrast to previous formulations of mcSAT, we include the possibility to *partially assign* values to variables; this enables assignments that only affect some of the bits in a bit-vector, which helps us to define more flexible propagation operators. mcSAT with projections is first defined in general terms, and tailored to the setting of bit-vector constraints in the subsequent sections, resulting in mcBV.

We define our framework in the form of a transition system, following the tradition of DPLL(T) [31] and mcSAT [30]. The states of the system have the form $\langle M, C \rangle$, where $M$ is the *trail* (a finite sequence of *trail elements*), and $C$ is a set of clauses. The trail $M$ consists of: (1) *decided literals* $l$, (2) *propagated literals* $e \rightarrow l$, and (3) *model assignments* $\pi(x) \mapsto \alpha$. A literal $l$ (either decided or propagated) is considered to be true in the current state if it appears in $M$, which is denoted by $l \in M$. Model assignments $\pi(x) \mapsto \alpha$ denote a *partial assignment* of a value $\alpha$ to a variable $x$, where $\pi$ is a *projection function*.

We consider constraints formulated over a set of types $\{\mathcal{T}_1, \mathcal{T}_2, \ldots, \mathcal{T}_n\}$ with fixed domains $\{T_1, T_2, \ldots, T_n\}$, and a finite family $\{\pi_i\}_{i \in I}$ of surjective functions $\pi_i : T \mapsto T'$ (here called *projections*) between the domains. Types can for instance be bit-vector sorts of various lengths. A partial model assignment $\pi(x) \mapsto \alpha$ with projection $\pi : T \mapsto T'$ expresses that variable $x$ of type $\mathcal{T}$ is assigned some value $\beta \in T$ such that $\pi(\beta) = \alpha$, where $\alpha \in T'$. A trail can contain multiple partial assignments to the same variable $x$; we define the *partial domain* of a variable $x$ under a trail $M$ as

$$Domain(x, M) = \bigcap_{(\pi(x) \mapsto \alpha) \in M} \{\beta \in T \mid \pi(\beta) = \alpha\}.$$

We call a trail $M$ *assignment consistent* if the partial assignments to variables are mutually consistent, i.e., the partial domain $Domain(x, M)$ of each variable $x$ is non-empty. If the partial domain of a variable $x$ contains exactly one element, i.e., $Domain(x, M) = \{\beta\}$, then we say that all the partial assignments to $x$ in $M$ form a *full model assignment*; in the original mcSAT calculus, this is denoted by $x \mapsto \beta$. Assignment consistency is violated if $Domain(x, M) = \varnothing$. We generally require that projections $\{\pi_i\}_{i \in I}$ are chosen in such a way that assignment consistency and full assignments can be detected effectively for any trail $M$. In addition, projections are required to be *complete* in the sense that every full model assignment $x \mapsto \beta$ can be expressed as some finite combination of partial assignments $\{\pi_j(x) \mapsto \alpha_j\}_{j \in J}$. More formally, for every $\beta \in T$ there exists a finite set of partial model assignments $S$ such that $Domain(x, S) = \{\beta\}$. Inclusion of the *identity function* among projections enables expression of full model assignments directly.

Given a trail $M$, an interpretation $v[M] = \{x_1 \mapsto \beta_1, x_2 \mapsto \beta_2, \ldots, x_k \mapsto \beta_k\}$ is constructed by collecting all full model assignments $x_i \mapsto \beta_i$ implied by $M$. The value $v[M](t)$ of a term or formula $t$ is its value under the interpretation $v$, provided that all variables occurring in $t$ are interpreted by $v$; or *undef* otherwise. We define a *trail extension* $\bar{M}$ of a trail $M$ as any trail $\bar{M} = [M, M']$ such that $M'$ consists only of (partial) model assignments to variables already appearing in $M$, and furthermore each variable $x$ that appears in $M$ has a unique value assigned from its partial domain; $Domain(x, M) \neq \emptyset$ implies that $|Domain(x, \bar{M})| = 1$. This ensures that assignment consistency is maintained.

Evaluation of literals in respect to the trail $M$ is achieved using a pair of functions $value_B$ and $value_T$, defined as

$$value_B(l, M) = \begin{cases} true & l \in M \\ false & \neg l \in M \\ undef & \text{otherwise} \end{cases} \quad \text{and} \quad value_T(l, M) = v[M](l).$$

A trail $M$ is *consistent* if it is assignment consistent, and for all literals $l \in M$ it holds that $value_T(l, M) \neq false$. A trail $M$ is said to be *complete* if it is consistent and every literal $l$ on the trail $M$ can be evaluated in the theory, i.e. $value_T(l, M) = true$. A trail which has no complete extensions is called *infeasible*. Note that if a trail is inconsistent then it is also infeasible.

The value of a literal in a consistent state (consistent trail) is defined as

$$value(l, M) = \begin{cases} value_B(l, M) & value_B(l, M) \neq undef \\ value_T(l, M) & \text{otherwise} \end{cases},$$

which is extended to clauses in the obvious way.

*Evaluation strength.* We remark that there is some freedom in the way $value_T$ is defined: even if $v[M](l) = undef$ for some literal $l$ (because $l$ contains variables with undefined value), the trail $M$ might still uniquely determine the value of $l$. In general, our calculus can use any definition of $value_T^*$ that satisfies

(1) $v[M](l) \neq undef$ implies $value_T^*(l, M) = v[M](l)$, and
(2) $value_T^*(l, M) \neq undef$ implies that for every extension $\bar{M}$ of $M$ it holds that $value_T(l, \bar{M}) = value_T^*(l, M)$.

These properties leave room for a trade-off between the strength of reasoning and computational effort invested to discover such implications. For example, suppose that a bit-vector variable $x$ of length 3, under trail $M$ has the partial domain $Domain(x, M) = \{000, 001, 010\}$. For a literal $l = (x < 100)$, evaluation yields $value_T(l, M) = undef$. It is easy to see that $value_T(l, \bar{M}) = true$ in every trail extension $\bar{M}$ of $M$, however, so that a lazier mode of evaluation could determine $value_T^*(l, M) = true$. With a more liberal evaluation strategy, propagations and conflicts are detected earlier, though perhaps at higher cost.

### 3.1   A Calculus with Projections

The transitions of our calculus are the same as those of mcSAT [30], with the exception of the T-Decide rule, which we define in terms of partial assignments and partial domains (Fig. 4). As in mcSAT, it is assumed that a *finite basis* $B$ of literals is given, representing all literals that are taken into account in decisions, propagations, or when constructing conflict clauses and explanations. $B$ at least has to contain all atoms, and the negation of atoms occurring in the clause set $C$. The function *explain* is supposed to compute *explanations* of infeasible trails $M$ (which correspond to theory lemmas in DPLL(T) terminology). An explanation of $M$ is a clause $e$ such that 1. $e$ is valid; 2. all literals $l \in e$ evaluate to false on $M$ (i.e., $value(l, M) = false$); 3. all literals $l \in e$ occur in the basis $B$.

In order to state correctness of the calculus, we need one further assumption about the well-foundedness of partial assignments: for every sequence of partial assignments to a variable $x$, of the form $[\pi_1(x) \mapsto \alpha_1, \pi_2(x) \mapsto \alpha_2, \ldots]$, we assume that the sequence of partial prefix domains

$$Domain(x, [])$$
$$Domain(x, [\pi_1(x) \mapsto \alpha_1])$$
$$Domain(x, [\pi_1(x) \mapsto \alpha_1, \pi_2(x) \mapsto \alpha_2])$$
$$\ldots$$

eventually becomes constant. This ensures that partial assignment of a variable cannot be refined indefinitely. Correctness of mcSAT with projections is then be proven in largely the same manner as in mcSAT:

**Theorem 1 (Correctness [30]).** *Any derivation starting from the initial state $\langle [], C \rangle$ eventually terminates in state sat, if $C$ is satisfiable, or in state unsat, if $C$ is unsatisfiable.*

T-Decide

$$\langle M, C \rangle \longrightarrow \langle [M, \pi(x) \mapsto \alpha], C \rangle \quad \textbf{if} \quad \begin{array}{l} x \text{ is a (theory) variable in } C \\ Domain(x, [M, \pi(x) \mapsto \alpha]) \neq Domain(x, M) \\ [M, \pi(x) \mapsto \alpha] \text{ is consistent} \end{array}$$

**Fig. 4.** The modified T-Decide rule.

# 4    Searching for Models with mcBV

We now describe how the mcSAT calculus with projections is tailored to the theory of bit-vectors, leading to our procedure mcBV. The theory of bit-vectors already contains a natural choice for the projections, namely the extract functions, of which we use a finite subset as projections. To ensure *completeness* of this subset (in the sense that every full model assignment has a representation as a combination of partial model assignments), we include all one-bit projections $\pi_i^k = \textsf{extract}_i^1$, selecting the $i$-th bit of a bit-vector of length $k$.

In practice, our prototype implementation maintains a trail $M$ as part of its state, and attempts to extend the trail with literals and model assignments such that the trail stays consistent, every literal on the trail eventually becomes justified by a model assignment (i.e., $value_T(l, M) = true$ for every literal $l$ in $M$), and every clause in $C$ is eventually satisfied. A conflict is detected if either some clause in $C$ is found to be falsified by the chosen trail elements (which is due to literals or model assignments), or if infeasibility of the trail is detected.

Since the calculus is *model constructing,* there is a strong preference to justify all literals on the trail through model assignments, i.e., to make the trail complete, before making further Boolean decisions. Partial model assignments are instrumental for this strategy: they enable flexible implementation of propagation rules that extract as much information from trail literals as possible. For instance, if the trail contains the equation $x = \textsf{extract}_2^3(y)$ and a model assignment $x \mapsto 101$, propagation infers and puts a further partial assignment $\textsf{extract}_2^3(y) \mapsto 101$ on the trail, by means of the T-Decide rule. This partial assignment is subsequently used to derive further information about other variables. For this, we defined bit-precise propagation rules for all operators; our solver includes native implementations of those rules and does not have to resort to explicit bit-blasting. Similarly to Boolean Constraint Propagation (BCP), propagation on the level of bit-vectors is often able to detect inconsistency of trails (in particular variables with empty partial domain) very efficiently.

Once all possible bit-vector propagations have been carried out, but the trail $M$ is still not complete, the values of further variables have to be decided upon through T-Decide. In order to avoid wrong decisions and obvious conflicts, our implementation also maintains over-approximations of the set of feasible values of each variable, in the form of *bit-patterns* and *arithmetic intervals*. These sets are updated whenever new elements are pushed on the trail, and refined using BCP-equivalent propagation and interval constraint propagation (ICP). Besides indicating values of variables consistent with the trail, these sets offer cheap infeasibility detection (when they become empty), which is crucial for the T-Propagate and T-Conflict rules. Also, frequently one of these sets becomes singleton, in which case a unique model assignment for the variable is implied.

## 4.1   Efficient Representation of Partial Model Assignments

Our procedure efficiently maintains information about the partial domains of variables by tracking *bit-patterns*, which are strings over the 3-letter alphabet $\{0, 1, u\}$; the symbol $u$ represents undefined bits (*don't-cares*). We say that bit-vector $x$ *matches* bit-pattern $p$ (both of length $k$) iff $x$ is included in the set of vectors covered in the bit-pattern; formally we define

$$matches(x, p) = \bigwedge_{\substack{0 \leq i < k \\ p_i \neq u}} x_i = p_i,$$

where $x_i$ and $p_i$ denote $i$-th bit of $x$ and $p$. The atom $matches(x, p)$ is not a formula in the sense of our language of bit-vector constraints, but for the sake of presentation we treat it as such in this section.

For long bit-vectors, representation of partial domains using simple bit-patterns can be inefficient, since linear space is needed in the length of the bit-vector variable. To offset this, we use *run-length encoding (RLE)* to store bit-patterns. Besides memory compression, RLE speeds up propagation, as it is not necessary to process every individual bit of a bit-vector separately. The complexity then depends on the number of bit alternations, as shown in the following example demonstrating exclusive-or evaluation on both representations.

*Example 1.* Each digit in the output represents one bit operation, in standard bit-vectors (left) and run-length encoded bit-vectors (right):

$$
\begin{array}{rl}
x & 0000011111 \\
y & 1110000011 \\
\hline
x \oplus y & 1110011100
\end{array}
\qquad
\begin{array}{rl}
x & 0^3\, 0^2\, 1^3\, 1^2 \\
y & 1^3\, 0^2\, 0^3\, 1^2 \\
\hline
x \oplus y & 1^3\, 0^2\, 1^3\, 0^2
\end{array}
$$

## 4.2   Maintaining Partial Domain Over-Approximations

To capture arithmetic properties and enable efficient propagation, our implementation stores bounds $x \in [x_l, x_u]$ for each variable $x$. The bounds are updated when new elements occur on the trail, and bounds propagation is used to refine bounds. Note that bit-patterns and arithmetic bounds abstractions sometimes also refine each other. For example, lower and upper bounds are derived from a bit-pattern, by replacing all $u$ bits with 0 and 1 respectively. Conversely, if the lower and upper bound share a prefix, then they imply a bit-pattern with the same prefix and the remaining bits set to $u$.

# 5    Conflicts and Explanations

Explanations are the vehicle used by our calculus to generalize from conflicts. Given an infeasible trail $M$, an explanation $explain(M)$ is defined to be a valid clause $E = l_1 \vee \cdots \vee l_n$ over the finite basis $\boldsymbol{B}$, such that every literal $l_i$ evaluates to $false$ under the current trail, i.e., $value(l_i, M) = false$ for every $i \in \{1, \ldots, n\}$. Explanations encode contradictory assumptions made on the trail, and are needed in the T-Consume and T-Conflict rules to control conflict resolution and backtracking, as well as in T-Propagate to justify literals added to the trail as the result of theory propagation.

Since the trail $M$ is inconsistent at the beginning of conflict resolution, it is always possible to find explanations that are simply disjunctions of negated trail literals; to this end, every propagated literal $c \to l$ is identified with $l$, and every model assignment $\pi(x) \mapsto \alpha$ as the formula $\pi(x) = \alpha$.

## 5.1    Greedy Generalization

We present a greedy algorithm for creating explanations that abstract from concrete causes of conflict. To this end, we assume that we have already derived some correct (but not very general) explanation

$$e = \neg t_1 \vee \neg t_2 \vee \cdots \vee \neg t_n \vee \neg b_1 \vee \neg b_2 \vee \cdots \vee \neg b_m,$$

where $t_1, \ldots, t_n$ denote literals with $value_T(t_i, M) = true$, and $b_1, \ldots, b_m$ literals with $value_T(t_i, M) = undef$ but $value_B(b_i, M) = true$. The former kind of literal holds as a result of model assignments, whereas the latter literals occur on the trail either as decisions or as the result of propagation. The key observation is that the literals $t_1, \ldots, t_n$ allow over-approximations (replacements with logically weaker literals), as long as the validity of the overall explanation clause is maintained, in this way producing a more general explanation.

Our procedure requires the following components as input (apart from $e$):

- for each literal $t_i$ (for $i \in \{1, \ldots, n\}$), a finite lattice $(T_i, \Rightarrow)$ of conjunctions of literals $T_i \subseteq \{l_1 \wedge \cdots \wedge l_k \mid l_1, \ldots, l_k \in \boldsymbol{B}\}$ ordered by logical implication, with join $\sqcup_i$ and meet $\sqcap_i$, and the property that $t_i \in T_i$. The set $T_i$ provides constraints that are considered as relaxation of $t_i$.
- a heuristic satisfiability checker $hsat$ to determine the satisfiability of a conjunction of literals. The checker $hsat$ is required to (1) be be sound (i.e., $hsat(\phi) = false$ implies that $\phi$ is actually unsatisfiable), (2) to correctly report the validity of $e$,

$$hsat(t_1 \wedge \cdots \wedge t_n \wedge b_1 \wedge \cdots \wedge b_m) = false,$$

and (3) to be *monotonic* in the following sense: for all elements $l, l' \in T_i$ with $l \Rightarrow l'$ in one of the lattices, and for all conjunctions $\phi$, $\psi$ of literals, if $hsat(\phi \wedge l' \wedge \psi) = false$ then $hsat(\phi \wedge l \wedge \psi) = false$.

---

**Algorithm 2.** Explanation relaxation

---

**Input**: Raw explanation $\bigvee_{i=1}^{n} \neg t_i \vee \bigvee_{i=1}^{m} \neg b_i$;
        lattices $(T_i, \Rightarrow)_{i=1}^{n}$; satisfiability checker $hsat$.
**Output**: Refined explanation $\bigvee_{i=1}^{n} \neg t_i^a \vee \bigvee_{i=1}^{m} \neg b_i$.

1  $\phi_b \leftarrow b_1 \wedge \cdots \wedge b_m$;
2  **for** $i \leftarrow 1$ **to** $n$ **do**
3   |   $t_i^a \leftarrow t_i$;
4   |   $B_i \leftarrow \emptyset$;
5  **end**

6  $changed \leftarrow true$;
7  **while** $changed$ **do**
8   |   $changed \leftarrow false$;
9   |   **for** $i \leftarrow 1$ **to** $n$ **do**
10  |   |   **if** $\exists t \in T_i \setminus \{t_i^a\}$ *with* $t_i^a \Rightarrow t$ *and* $\forall l \in B_i.\, t \not\Rightarrow l$ **then**
11  |   |   |   **if** $hsat(\bigwedge_{j=1}^{i-1} t_j^a \wedge t \wedge \bigwedge_{j=i+1}^{n} t_j^a \wedge \phi_b)$ **then**
12  |   |   |   |   $B_i \leftarrow B_i \cup \{t\}$;
13  |   |   |   **else**
14  |   |   |   |   $t_i^a \leftarrow t$;
15  |   |   |   |   $changed \leftarrow true$;
16  |   |   |   **end**
17  |   |   **end**
18  |   **end**
19  **end**

20  **return** $\neg t_1^a \vee \cdots \vee \neg t_n^a \vee \neg b_1 \vee \cdots \vee \neg b_m$;

---

The pseudo-code of the procedure is shown in Algorithm 2, and consists mainly of a fixed-point loop in which the literals $t_1, \ldots, t_n$ are iteratively weakened, until no further changes are possible (lines 6–19). The algorithm keeps blocking sets $B_i$ of conjunctions of literals (for $i \in \{1, \ldots, n\}$) that have been considered as relaxation for $t_i$, but were found to be too weak to maintain a valid explanation.

**Lemma 1.** *Provided a correct explanation clause as input, Algorithm 2 terminates and produces a correct refined explanation.*

The next two sections describe two instances of our procedure: one targeting explanations that are primarily of arithmetic character, and one for explanations that mainly involve bit-wise operations.

## 5.2   Greedy Bit-wise Generalization

If the literals of an explanation clause are primarily bit-wise in nature, the relaxation considered in our method is to weaken the bit-patterns associated with the variables occurring in the conflict. For every literal $t_i$ of the form $matches(x, p)$,

| | | |
|---|---|---|
| $\neg matches(y,\ 0000) \lor \neg matches(x,\ 10) \lor \neg extract_2^2(y) = x$ | | Valid |
| $\neg matches(y,\ \mathbf{u}000) \lor \neg matches(x,\ 10) \lor \neg extract_2^2(y) = x$ | | Valid |
| $\neg matches(y,\ \mathbf{uuu}0) \lor \neg matches(x,\ 10) \lor \neg extract_2^2(y) = x$ | $y = 0010, x = 10$ |  |
| $\neg matches(y,\ \mathbf{uu}00) \lor \neg matches(x,\ 10) \lor \neg extract_2^2(y) = x$ | | Valid |
| $\neg matches(y,\ \mathbf{uu0u}) \lor \neg matches(x,\ 10) \lor \neg extract_2^2(y) = x$ | | Valid |
| $\neg matches(y,\ uu0u) \lor \neg matches(x,\ \mathbf{u}0) \lor \neg extract_2^2(y) = x$ | $y = 0000, x = 00$ |  |
| $\neg matches(y,\ uu0u) \lor \neg matches(x,\ 1\mathbf{u}) \lor \neg extract_2^2(y) = x$ | | Valid |

**Fig. 5.** Generalization based on bit-patterns

where $x$ is a bit-vector variable and $p$ a bit-pattern of width $k = \alpha(x)$ implied by the trail, we choose the lattice $(T_i, \Rightarrow)$ with

$$T_i = \{\textit{false}\} \cup \{matches(x,a) \mid a \in \{0,1,u\}^k\}.$$

This set contains a constraint that is equivalent to *true*, namely the literal $matches(x, u^k)$. Conflicts involving (partial) assignments allow us to start near the bottom of the lattice and we weaken the literal as much as possible in order to cover as many similar assignments as possible. Concretely, the weakening is performed by replacing occurrences of 1 or 0 in the bit-pattern by $u$.

Our prototype satisfiability checker *hsat* for this type of constraints is implemented by propagation (Sect. 4), which is able to show validity of raw explanation clauses, and similarly handles bit-pattern relaxations. Our implementation covers all operations, but it is imprecise for some arithmetic operations.

*Example 2.* Consider the trail

$$M = [\ldots, y \mapsto 0^4, x \mapsto 1^1 0^1, \mathsf{extract}_2^2(y) = x].$$

This trail is inconsistent because literal $\mathsf{extract}_2^2(y) = x$ evaluates to false in the theory. A simple explanation clause is $\neg matches(y, 0^4) \lor \neg matches(x, 1^1 0^1) \lor \neg(\mathsf{extract}_2^2(y) = x)$. The generalization procedure now tries to generalize this naive explanation by weakening the literals. Figure 5 shows steps of generalizing the conflict observed in trail $M$. One by one, bits in the pattern are set to $u$ and it is checked whether the new clause is valid. If it is valid, the new bit-pattern is altered further, otherwise we discard it and continue with the last successfully weakened pattern. Note that we are not restricted to changes to only one bit, or even only one literal at a time.

## 5.3 Greedy Arithmetic Generalization

If the literals of an explanation clause are primarily arithmetic, the relaxation considered in our method is to replace equations (that stem from model assignments on the trail) with inequalities or interval constraints: for every literal $t_i$

| | |
|---|---|
| $\neg(y \leq_u 14) \vee \neg(y \geq_u 14) \vee \neg(y <_u z) \vee \neg(x = z + y) \vee \neg(x <_u y + y)$ | Valid |
| $\neg(y \leq_u 15) \vee \neg(y \geq_u 14) \vee \neg(y <_u z) \vee \neg(x = z + y) \vee \neg(x <_u y + y)$ | Valid |
| $\neg(y \geq_u 0) \vee \neg(y <_u z) \vee \neg(x = z + y) \vee \neg(x <_u y + y)$ | $y = 1, \dots$ |
| $\neg(y \geq_u 7) \vee \neg(y <_u z) \vee \neg(x = z + y) \vee \neg(x <_u y + y)$ | $y = 7, \dots$ |
| $\neg(y \geq_u 10) \vee \neg(y <_u z) \vee \neg(x = z + y) \vee \neg(x <_u y + y)$ | Valid |
| $\neg(y \geq_u 8) \vee \neg(y <_u z) \vee \neg(x = z + y) \vee \neg(x <_u y + y)$ | Valid |

**Fig. 6.** Generalization based on arithmetic and interval bounds propagation

of the form $x = v_k$, where $x$ is a bit-vector variable and $v_k$ a literal bit-vector constant of width $k = \alpha(x)$, we choose the lattice $(T_i, \Rightarrow)$ with

$$T_i = \{false\} \cup \{x = a \mid a \in \{0,1\}^k\} \cup$$
$$\{a \leq_u x \wedge x \leq_u b \mid a, b \in \{0,1\}^k, a < b\}$$

This set contains a constraint that is equivalent to *true* $(0^k \leq_u x \wedge x \leq_u 1^k)$, and similarly constraints that only impose concrete lower or upper bounds on $x$, and equalities $(x = v_k) \in T_i$.

Our satisfiability checker *hsat* for this type of constraints implements interval constraint propagation, covering all bit-vector operations, although it tends to yield more precise results for arithmetic than for bit-wise operations. If ICP shows that an interval becomes empty, then the generalization succeeds because the conflict persists. Otherwise, generalization fails when ICP reaches a fix point or exceeds a fixed number of steps (to avoid problems with slow convergence).

*Example 3.* For readability purposes we switch to numerical notation in this example. Recall the basic explanation of the trail conflict shown in Fig. 2b in the motivating example (Sect. 1.1):

$$\neg(y = 14) \vee \neg(y <_u z) \vee \neg(x = z + y) \vee (x <_u y + y)$$

We rewrite the first negated equality as a disjunction of negated inequalities. Then the iterative generalization procedure starts. For each bound literal, the procedure first attempts to remove it (by weakening it to a *true*-equivalent in the lattice $T_i$). If unsuccessful, it navigates the lattice of literals using binary search over bounds. Figure 6 shows the steps in this particular example.

# 6  Experiments and Evaluation

To evaluate the performance of our mcBV implementation we conducted experiments on the SMT-LIB QF_BV benchmark set, using our implementation of mcBV in F#, on a Windows HPC cluster of Intel Xeon machines. The benchmark set contains 49971 files in SMT2 format, each of which contains a set of assertions and a single (check-sat) command. The timeout for all experiments is at 1200 s and the memory limit is 2 GB.

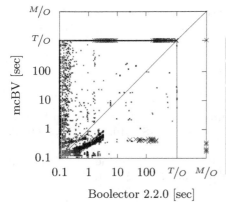

| Set | | Z3 4.4.2 | Boolector 2.2.0 | mcBV | # < |
|---|---|---|---|---|---|
| QF_BV | SAT | 16260 | 16793 | 6679 | 35 |
| | UNSAT | 30748 | 31534 | 17025 | 58 |
| brummayer-biere4 | SAT | 10 | 0 | 10 | 0 |
| | UNSAT | 0 | 0 | 0 | 0 |
| pspace | SAT | 0 | 21 | 21 | 21 |
| | UNSAT | 15 | 60 | 0 | 0 |
| sage | SAT | 8077 | 8077 | 6069 | 0 |
| | UNSAT | 18530 | 18530 | 16152 | 29 |
| Sage2 | SAT | 5104 | 5649 | 16 | 14 |
| | UNSAT | 9961 | 10612 | 176 | 29 |

**Fig. 7.** Runtime comparison on selected subsets of SMT QF_BV. Markers for 'sage' and 'sage2' are smaller to avoid clutter; #< shows the number of benchmarks that only mcBV solves or mcBV solves quicker than both Z3 and Boolector.

Currently we do not implement any advanced heuristics for clause learning, clause deletion, or restarts and thus mcBV does not outperform any other solver consistently. We present a runtime comparison of mcBV with the state-of-the-art SMT solvers Boolector and Z3 on a selected subset of the QF_BV benchmarks see Fig. 7. On the whole benchmark set, mcBV is not yet competitive with Boolector or Z3, but it is interesting to note that mcBV performs well on some of the benchmarks in the 'sage', 'sage2' and 'pspace' sets, as well as the entirety of the 'brummayerbiere4' set. Those sets contain a substantial number of benchmarks that mcBV could solve, but Z3 and Boolector cannot. The 'pspace' benchmarks are hard for all solvers as they contains very large bit-vectors (in the order of 20k bits) which will often result in the bit-blaster running out of memory; this is reflected in the small clusters at the bottom right in Fig. 7. The table in Fig. 7 gives the number of instances solved by each approach. While our prototype performs relatively well on selected subsets, it will need improvements and advanced heuristics to compete with the state-of-the-art on all of QF_BV.

## 7  Conclusion

We presented a new decision procedure of the theory of bit-vectors, which is based on an extension of the Model-Constructing Satisfiability Calculus (mcSAT). In contrast to state-of-the-art solvers, our procedure avoids unnecessary bit-blasting. Although our implementation is prototypical and lacks most of the more advanced heuristics used in solvers (e.g., variable selection/decision heuristics, lemma learning, restarts, deletion strategies), our approach shows promising performance, and is comparable with the best available solvers on a number of benchmarks. This constitutes a proof of concept for instantiation of the mcSAT framework for a new theory; we expect significantly improved performance as we further optimise our implementation. Additionally, we improve the flexibility of the mcSAT framework by introducing projection functions and partial assignments, which we believe to be crucial for the model-constructing approach for bit-vectors.

# References

1. Bardin, S., Herrmann, P., Perroud, F.: An alternative to SAT-based approaches for bit-vectors. In: Esparza, J., Majumdar, R. (eds.) TACAS 2010. LNCS, vol. 6015, pp. 84–98. Springer, Heidelberg (2010)
2. Barrett, C., Conway, C.L., Deters, M., Hadarean, L., Jovanović, D., King, T., Reynolds, A., Tinelli, C.: CVC4. In: Gopalakrishnan, G., Qadeer, S. (eds.) CAV 2011. LNCS, vol. 6806, pp. 171–177. Springer, Heidelberg (2011)
3. Berdine, J., Cook, B., Ishtiaq, S.: SLAYER: memory safety for systems-level code. In: Gopalakrishnan, G., Qadeer, S. (eds.) CAV 2011. LNCS, vol. 6806, pp. 178–183. Springer, Heidelberg (2011)
4. Berdine, J., Cox, A., Ishtiaq, S., Wintersteiger, C.M.: Diagnosing abstraction failure for separation logic–based analyses. In: Madhusudan, P., Seshia, S.A. (eds.) CAV 2012. LNCS, vol. 7358, pp. 155–173. Springer, Heidelberg (2012)
5. Biere, A., Cimatti, A., Clarke, E.M., Fujita, M., Zhu, Y.: Symbolic model checking using SAT procedures instead of BDDs. In: DAC (1999)
6. Biere, A., Cimatti, A., Clarke, E., Zhu, Y.: Symbolic model checking without BDDs. In: Cleaveland, W.R. (ed.) TACAS 1999. LNCS, vol. 1579, pp. 193–207. Springer, Heidelberg (1999)
7. Bjørner, N.S., Pichora, M.C.: Deciding fixed and non-fixed size bit-vectors. In: Steffen, B. (ed.) TACAS 1998. LNCS, vol. 1384, pp. 376–392. Springer, Heidelberg (1998)
8. Böhme, S., Fox, A.C.J., Sewell, T., Weber, T.: Reconstruction of Z3's bit-vector proofs in HOL4 and Isabelle/HOL. In: Jouannaud, J.-P., Shao, Z. (eds.) CPP 2011. LNCS, vol. 7086, pp. 183–198. Springer, Heidelberg (2011)
9. Bradley, A.R.: sat-based model checking without unrolling. In: Jhala, R., Schmidt, D. (eds.) VMCAI 2011. LNCS, vol. 6538, pp. 70–87. Springer, Heidelberg (2011)
10. Brummayer, R., Biere, A.: Boolector: an efficient SMT solver for bit-vectors and arrays. In: Kowalewski, S., Philippou, A. (eds.) TACAS 2009. LNCS, vol. 5505, pp. 174–177. Springer, Heidelberg (2009)
11. Bruttomesso, R., Sharygina, N.: A scalable decision procedure for fixed-width bit-vectors. In: ICCAD. ACM (2009)
12. Cimatti, A., Clarke, E., Giunchiglia, E., Giunchiglia, F., Pistore, M., Roveri, M., Sebastiani, R., Tacchella, A.: NuSMV 2: an opensource tool for symbolic model checking. In: Brinksma, E., Larsen, K.G. (eds.) CAV 2002. LNCS, vol. 2404, pp. 359–364. Springer, Heidelberg (2002)
13. Cimatti, A., Griggio, A., Schaafsma, B.J., Sebastiani, R.: The MathSAT5 SMT solver. In: Piterman, N., Smolka, S.A. (eds.) TACAS 2013 (ETAPS 2013). LNCS, vol. 7795, pp. 93–107. Springer, Heidelberg (2013)
14. Clarke, E., Kroning, D., Lerda, F.: A tool for checking ANSI-C programs. In: Jensen, K., Podelski, A. (eds.) TACAS 2004. LNCS, vol. 2988, pp. 168–176. Springer, Heidelberg (2004)
15. Cyrluk, D., Möller, M.O., Rueß, H.: An efficient decision procedure for the theory of fixed-sized bit-vectors. In: Grumberg, O. (ed.) CAV 1997. LNCS, vol. 1254, pp. 60–71. Springer, Heidelberg (1997)

16. Dunn, S.J., Martello, G., Yordanov, B., Emmott, S., Smith, A.: Defining an essential transcriptional factor program for naive pluripotency. Science **344**(6188), 1156–1160 (2014)
17. Dutertre, B.: System description: Yices 1.0.10. In: SMT-COMP 2007 (2007)
18. Froehlich, A., Kovasznai, G., Biere, A.: Efficiently solving bit-vector problems using model checkers. In: SMT Workshop (2013)
19. Fröhlich, A., Kovásznai, G., Biere, A.: More on the complexity of quantifier-free fixed-size bit-vector logics with binary encoding. In: Bulatov, A.A., Shur, A.M. (eds.) CSR 2013. LNCS, vol. 7913, pp. 378–390. Springer, Heidelberg (2013)
20. Ganesh, V., Dill, D.L.: A decision procedure for bit-vectors and arrays. In: Damm, W., Hermanns, H. (eds.) CAV 2007. LNCS, vol. 4590, pp. 519–531. Springer, Heidelberg (2007)
21. Griggio, A.: Effective word-level interpolation for software verification. In: FMCAD. FMCAD Inc. (2011)
22. Hadarean, L., Bansal, K., Jovanović, D., Barrett, C., Tinelli, C.: A tale of two solvers: eager and lazy approaches to bit-vectors. In: Biere, A., Bloem, R. (eds.) CAV 2014. LNCS, vol. 8559, pp. 680–695. Springer, Heidelberg (2014)
23. Jovanovic, D., de Moura, L.M.: Cutting to the chase - solving linear integer arithmetic. J. Autom. Reasoning **51**(1), 79–108 (2013)
24. Kovásznai, G., Fröhlich, A., Biere, A.: On the complexity of fixed-size bit-vector logics with binary encoded bit-width. In: SMT. EPiC Series, vol. 20. EasyChair (2013)
25. Kovásznai, G., Veith, H., Fröhlich, A., Biere, A.: On the complexity of symbolic verification and decision problems in bit-vector logic. In: Csuhaj-Varjú, E., Dietzfelbinger, M., Ésik, Z. (eds.) MFCS 2014, Part II. LNCS, vol. 8635, pp. 481–492. Springer, Heidelberg (2014)
26. Kroening, D.: Computing over-approximations with bounded model checking. In: BMC Workshop, vol. 144, January 2006
27. McMillan, K.L.: Interpolation and SAT-based model checking. In: Hunt Jr., W.A., Somenzi, F. (eds.) CAV 2003. LNCS, vol. 2725, pp. 1–13. Springer, Heidelberg (2003)
28. Möller, M.O., Rueß, H.: Solving bit-vector equations. In: Gopalakrishnan, G.C., Windley, P. (eds.) FMCAD 1998. LNCS, vol. 1522, pp. 36–48. Springer, Heidelberg (1998)
29. de Moura, L., Bjørner, N.S.: Z3: an efficient SMT solver. In: Ramakrishnan, C.R., Rehof, J. (eds.) TACAS 2008. LNCS, vol. 4963, pp. 337–340. Springer, Heidelberg (2008)
30. de Moura, L., Jovanović, D.: A model-constructing satisfiability calculus. In: Giacobazzi, R., Berdine, J., Mastroeni, I. (eds.) VMCAI 2013. LNCS, vol. 7737, pp. 1–12. Springer, Heidelberg (2013)
31. Nieuwenhuis, R., Oliveras, A., Tinelli, C.: Solving SAT and SAT modulo theories: from an abstract Davis-Putnam-Logemann-Loveland procedure to DPLL(T). J. ACM **53**(6), 937–977 (2006)
32. Sheeran, M., Singh, S., Stålmarck, G.: Checking safety properties using induction and a SAT-solver. In: Johnson, S.D., Hunt Jr., W.A. (eds.) FMCAD 2000. LNCS, vol. 1954, pp. 108–125. Springer, Heidelberg (2000)
33. Tseitin, G.: On the complexity of derivation in propositional calculus. Studies in Constructive Mathematics and Mathematical Logic, Part II, Seminars in Mathematics (1970), translated from Russian: Zapiski Nauchnykh Seminarov LOMI 8 (1968)

34. Wille, R., Fey, G., Große, D., Eggersglüß, S., Drechsler, R.: SWORD: a SAT like prover using word level information. In: International Conference on Very Large Scale Integration of System-on-Chip (VLSI-SoC 2007). IEEE (2007)
35. Yordanov, B., Wintersteiger, C.M., Hamadi, Y., Kugler, H.: SMT-based analysis of biological computation. In: Brat, G., Rungta, N., Venet, A. (eds.) NFM 2013. LNCS, vol. 7871, pp. 78–92. Springer, Heidelberg (2013)

# Solving Quantified Bit-Vector Formulas Using Binary Decision Diagrams

Martin Jonáš[(✉)] and Jan Strejček

Faculty of Informatics, Masaryk University, Brno, Czech Republic
{martin.jonas,strejcek}@mail.muni.cz

**Abstract.** We describe a new approach to deciding satisfiability of quantified bit-vector formulas using binary decision diagrams and approximations. The approach is motivated by the observation that the binary decision diagram for a quantified formula is typically significantly smaller than the diagram for the subformula within the quantifier scope. The suggested approach has been implemented and the experimental results show that it decides more benchmarks from the SMT-LIB repository than state-of-the-art SMT solvers for this theory, namely Z3 and CVC4.

## 1 Introduction

During the last decades, the area of *Satisfiability Modulo Theories* (SMT) [6] solving has undergone steep development in both theory and practice. Achieved advances of SMT solving opened new research directions in program analysis and verification, where SMT solvers are now seen as standard tools.

Common programming languages provide basic datatypes of fixed size. Program variables of these datatypes naturally correspond to variables of the bit-vector logic, which can easily express bit-wise operations or arithmetic overflows. In spite of this natural correspondence, most SMT-based program analysis techniques model program variables by variables in the theory of integers. This may look a bit strange considering the fact that the satisfiability problem for the theory of integers is undecidable whenever an arbitrary usage of addition and multiplication is allowed, while the same problem is decidable for the bit-vector theory. The reasons for using the integer logic instead of the bit-vector logic are twofold. First, the satisfiability problem is NEXPTIME-complete even for formulas of the *quantifier-free fragment of the bit-vector logic* (QF_BV) with binary encoding of bit-vector sizes [21]. In this paper, we consider formulas with quantifiers and without uninterpreted functions. The precise complexity of the problem for this logic is an open question: it is known to be NEXPTIME-hard [21] and trivially solvable in EXPSPACE. Second and from the practical point of view more important, the SMT solvers for the theory of integers are often more efficient than the solvers for the theory of fixed-size bit-vectors.

The research was supported by Czech Science Foundation, grant GBP202/12/G061.

© Springer International Publishing Switzerland 2016
N. Creignou and D. Le Berre (Eds.): SAT 2016, LNCS 9710, pp. 267–283, 2016.
DOI: 10.1007/978-3-319-40970-2_17

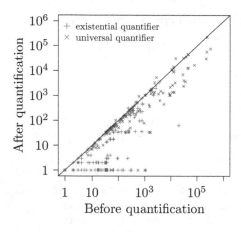
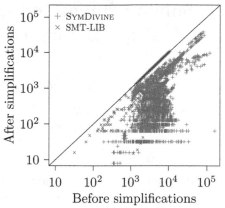

**Fig. 1.** Comparison of sizes (measured by the number of BDD nodes) of BDDs corresponding to all quantified subformulas in SMT-LIB benchmarks for BV logic, before and after quantification.

**Fig. 2.** Effect of simplifications on the number of bit variables in the formulas of the SMT-LIB and SYMDIVINE benchmarks. Formulas simplified to *true* or *false* are not represented.

While there are several SMT solvers for QF_BV formulas, only few of them can decide the *quantified bit-vector (BV) logic*. In particular, the logic is supported by CVC4 [3] and Z3 [16]. Relatively modest support of this logic from developers of SMT solvers is definitely not a consequence of a low demand from potential users. For example, in the program analysis community, BV formulas are suitable for description of various properties of program loops like loop invariants, ranking functions, or loop summaries [22], or to describe properties of symbolic representations of sets of program states, such as inclusion [7].

While current solvers for BV logic rely on *model-based quantifier instantiation* [25], we present a new algorithm based on *Binary Decision Diagrams* (BDDs) and approximations. BDDs have been previously used to implement satisfiability decision procedures for the propositional logic, however state-of-the-art CDCL-based solvers usually achieve much better performance. The main disadvantage of BDDs is low scalability: the size of a BDD corresponding to a propositional formula can be exponential in the number of propositional variables, and when a BDD becomes too large, some operations are very slow. Employment of BDDs in SMT solving makes more sense when formulas with quantifiers are considered: quantification usually reduces size of a BDD as it decreases the number of BDD variables. This can be documented by Fig. 1, which compares the BDD sizes for formulas before and after existential or universal quantification.

There already exist some BDD-based tools deciding validity of quantified boolean formulas with the performance similar to state-of-the-art solvers for this problem [2,23].

Our BDD-based algorithm for satisfiability of the BV logic consists of three main components:

- Formula simplifications, which reduce the number of variables in the formula and push quantifiers downwards in the syntax tree of the formula (which later helps to keep intermediate BDDs smaller as they are build in the bottom-up manner). Formula simplifications can reduce some formulas to *true* or *false* and thus immediately decide their satisfiability.
- Construction of BDD using a specific variable ordering. The ordering has a significant influence on the BDD size.
- Formula approximations, which reduce the width of bit-vector variables in the formula and thus lead to smaller BDDs. Unsatisfiability of a formula over-approximation implies unsatisfiability of the original formula and an analogous statement holds for satisfiability of an under-approximation.

We present a minor contribution in each component. The main contribution of the paper is the fact that the algorithm based on the three parts can compete with leading SMT solvers for the BV logic, which participated in the BV category of SMT-COMP 2015 [1], namely Z3 and CVC4.

In the next section, we recall the definition of BV logic and BDDs, and briefly explain the main idea of the model-based quantifier instantiation technique employed by CVC4 and Z3. The proposed algorithm including the three main components is presented in Sect. 3. Section 4 is devoted to the implementation of the algorithm and to experimental results showing separately the effect of formula simplification, variable ordering, and approximations. The section also provides an experimental comparison of our solver with Z3 and CVC4. The paper closes with conclusions and intended directions of future work.

## 2   Preliminaries

### 2.1   Quantified Bit-Vector Formulas

In what follows, we assume a knowledge of the multi-sorted first-order logic and the model theory [18,19]. Let $\mathbb{N}$ denote the set of positive integers.

The *bit-vector logic* is a multi-sorted first-order logic with the set of sort symbols $S = \{\mathtt{bitvec}_i \mid i \in \mathbb{N}\}$, where $\mathtt{bitvec}_i$ represents the sort of bit-vectors of length $i$, the set of function symbols

$$F = \{c^i_{[n]} \mid n, i \in \mathbb{N}\} \cup \bigcup_{n \in \mathbb{N}} \{0_{[n]}, 1_{[n]}, \ldots, (2^n - 1)_{[n]}\} \cup$$

$$\cup \bigcup_{n \in \mathbb{N}} \{\mathtt{not}_{[n]}, \mathtt{and}_{[n]}, \mathtt{or}_{[n]}, \mathtt{shl}_{[n]}, \mathtt{shr}_{[n]}, -_{[n]}, +_{[n]}, \times_{[n]}, /_{[n]}, \%_{[n]}\} \cup$$

$$\cup \{\mathtt{concat}_{[m,n]} \mid m, n \in \mathbb{N}\} \cup \{\mathtt{extract}_{[n,i,j]} \mid n, i, j \in \mathbb{N}, i \leq j < m\},$$

and the set of the predicate symbols $P = \{=_{[n]}, <_{[n]} \mid n \in \mathbb{N}\}$. Arities of function and predicate symbols are described in Table 1.

**Table 1.** Function and predicate symbols of the bit-vector logic.

| Symbol | Arity | Interpretation |
|---|---|---|
| $0_{[n]}, 1_{[n]}, \ldots$ | $\texttt{bitvec}_n$ | natural number constants |
| $c^1_{[n]}, c^2_{[n]}, \ldots$ | $\texttt{bitvec}_n$ | uninterpreted constants |
| $\texttt{not}_{[n]}$ | $\texttt{bitvec}_n \rightarrow \texttt{bitvec}_n$ | bit-wise negation |
| $\texttt{and}_{[n]}, \texttt{or}_{[n]}$ | $\texttt{bitvec}_n \times \texttt{bitvec}_n \rightarrow \texttt{bitvec}_n$ | bit-wise and, or |
| $\texttt{shl}_{[n]}, \texttt{shr}_{[n]}$ | $\texttt{bitvec}_n \times \texttt{bitvec}_n \rightarrow \texttt{bitvec}_n$ | bit-wise shift left, right |
| $-_{[n]}$ | $\texttt{bitvec}_n \rightarrow \texttt{bitvec}_n$ | two-complement negation |
| $+_{[n]}, \times_{[n]}$ | $\texttt{bitvec}_n \times \texttt{bitvec}_n \rightarrow \texttt{bitvec}_n$ | addition, multiplication |
| $/_{[n]}, \%_{[n]}$ | $\texttt{bitvec}_n \times \texttt{bitvec}_n \rightarrow \texttt{bitvec}_n$ | unsigned division, remainder |
| $\texttt{concat}_{[m,n]}$ | $\texttt{bitvec}_m \times \texttt{bitvec}_n \rightarrow \texttt{bitvec}_{m+n}$ | concatenation |
| $\texttt{extract}_{[m,i,j]}$ | $\texttt{bitvec}_m \rightarrow \texttt{bitvec}_{j-i+1}$ | extraction from $i$-th to $j$-th bit |
| $=_{[n]}, <_{[n]}$ | $\texttt{bitvec}_n \times \texttt{bitvec}_n$ | equality, unsigned less than |

The syntax of bit-vector formulas is defined in the standard way. Every formula can be transformed into the *negation normal form* (NNF), where negation is applied only to atomic subformulas and implication is not used at all.

A structure $M$ is said to be a *model* for formula $\varphi$, if the formula $\varphi$ is true in $M$ and if $M$ interprets all function and predicate symbols according to Table 1. Precise description of function and predicate symbols interpretation can be found in [4]. A closed formula is said to be *satisfiable* if it has a model.

We omit subscripts representing the sorts from the function and predicate symbols if the bit-width can be inferred from the context. If the sort of a variable or a constant is not specified, it is assumed to be $\texttt{bitvec}_{32}$. We also write $a, b, c, \ldots$ instead of uninterpreted constants $c^1, c^2, c^3, \ldots$. For example, $\forall x \, (x < a)$ denotes the formula $\forall x_{[32]} \, (x_{[32]} <_{[32]} c^1_{[32]})$. We write $\varphi[x_1, \ldots, x_n]$ for a formula $\varphi$, which may contain free variables $x_1, \ldots, x_n$. If $\varphi[x_1, \ldots, x_n]$ is a formula and $t_1, \ldots, t_n$ are terms of corresponding sorts, then $\varphi[t_1, \ldots, t_n]$ is the result of simultaneous substitution of free variables $x_1, \ldots, x_n$ in the formula $\varphi$ by terms $t_1, \ldots, t_n$, respectively.

## 2.2   Model-Based Quantifier Instantiation

Satisfiability of the quantifier-free fragment of the bit-vector logic is traditionally solved by eager or lazy reduction to a propositional formula (bit-blasting) and subsequent call of a SAT solver. In the following, we describe the *model-based quantifier instantiation* algorithm [25], which is used by existing solvers for the full bit-vector logic.

Given a closed formula with quantifiers, the first step is to convert the formula to the negation normal form and apply Skolemization to obtain equisatisfiable formula of the form

$$\varphi \ \wedge \ \forall x_1, x_2, \ldots, x_n \, (\psi[x_1, \ldots, x_n]),$$

where $\varphi$ and $\psi$ are quantifier-free formulas. Then the QF_BV solver is invoked to check the satisfiability of the formula $\varphi$. If $\varphi$ is unsatisfiable, then the entire formula is unsatisfiable. If $\varphi$ is satisfiable, the QF_BV solver returns its model $M$ and another call to the QF_BV solver is made to determine whether $M$ is also a model of $\forall x_1, x_2, \ldots, x_n\,(\psi)$. This is achieved by asking the solver whether the formula $\neg\widehat{\psi}$ is satisfiable, where $\widehat{\psi}$ is the formula $\psi$ with uninterpreted constants replaced by their corresponding values in $M$. If $\neg\widehat{\psi}$ is not satisfiable, then the structure $M$ is indeed a model of the formula $\forall x_1, x_2, \ldots, x_n\,(\psi)$, therefore the entire formula is satisfiable and $M$ is its model. If $\neg\widehat{\psi}$ is satisfiable, we get values $v_1, \ldots, v_n$ such that $\neg\widehat{\psi}[v_1, \ldots, v_n]$ holds. To rule out $M$ as a model, the instance $\psi[v_1, \ldots, v_n]$ of the quantified formula is added to the quantifier-free part, i.e. the formula $\varphi$ is modified to

$$\varphi' \equiv \varphi \wedge \psi[v_1, \ldots, v_n],$$

and the procedure is repeated.

**Example 1.** *Consider the formula $3 < a \;\wedge\; \forall x\,(\neg(a = 2 \times x))$. The subformula $3 < a$ is satisfiable and $a = 4$ is its model. However, it is not a model of the formula $\forall x\,(\neg(a = 2 \times x))$, since the QF_BV solver called on the formula $\neg(\neg(4 = 2 \times x))$ returns $x = 2$ as a model. The next step is to decide the satisfiability of the formula $3 < a \;\wedge\; \neg(a = 2 \times 2)$. This formula is satisfiable and $a = 5$ is its model. Moreover, it is also a model of $\forall x\,(\neg(a = 2 \times x))$ as $\neg(\neg(5 = 2 \times x))$ is unsatisfiable. Hence, the input formula is satisfiable and $a = 5$ is its model.*

This algorithm is trivially terminating, since there is only a finite number of distinct models $M$ of $\varphi$. However, in some cases exponentially many such models have to be ruled out before the solver is able to find a correct model or decide unsatisfiability of the whole formula. To overcome this issue, state-of-the-art SMT solvers do not use just instances of the form $\psi[v_1, \ldots, v_n]$ with concrete values, but employ heuristics such as E-matching [15,17] or symbolic quantifier instantiation [25] to choose instances with ground terms which can potentially rule out more spurious models and thus significantly reduce the number of iterations of the algorithm. In practice, suitable ground terms substituted for quantified variables are selected only from subterms of the input formula. This strategy brings some drawbacks. For example, the formula

$$a = 2^4 \times b + 2^4 \times c \;\wedge\; \forall x\,(\neg(a = 2^4 \times x))$$

is unsatisfiable as the subformula $\forall x\,(\neg(a = 2^4 \times x))$ is true precisely when the value of $a$ is not a multiple of $2^4$, while $a = 2^4 \times b + 2^4 \times c$ implies that $a$ is a multiple of $2^4$. The quantifier instantiation can prove the unsatisfiability easily by using the instance $\psi[b+c]$ of $\psi[x] \equiv \neg(a = 2^4 \times x)$. However, the current tools do not consider this instance as $b+c$ is not a subterm of the formula. As a result, current tools can not decide satisfiability of this formula within a reasonable time limits.

## 2.3    Binary Decision Diagrams

A *binary decision diagram (BDD)* is a data structure proposed by Bryant [12] to succinctly represent all satisfying assignments of a boolean formula.

A BDD is a rooted directed acyclic graph with inner nodes labeled by boolean variables of the formula and two leaf nodes 0 and 1. Every inner node has two outgoing edges, one labeled with *true* and the other with *false*. Every assignment of boolean variables determines a path from the root to a leaf: from every inner node we follow the edge labeled with the truth value assigned to the variable corresponding to the node. The BDD represents all assignments that determine paths to leaf 1. Fixing an order in which variables can occur on paths from the root yields an *Ordered Binary Decision Diagram (OBDD)* and merging identical subgraphs of an OBDD and deleting every node whose two children are identical yields a *Reduced Ordered Binary Decision Diagram (ROBDD)*. The main advantage of ROBDDs is that for the fixed variable order every set of assignments corresponds to a unique ROBDD [12]. In the following, BDD always stands for ROBDD.

A BDD for a boolean formula can be built from BDDs for atomic subformulas in a bottom-up manner. Application of negation corresponds to switching the leaf nodes 0 and 1. For binary operators, there is a function `Apply` that gets an operator and two BDDs corresponding to the operands and produces the desired BDD. Using this function, one can also build a BDD representing a quantified boolean formula: if $B$ is a BDD representing a formula $\varphi$, then the BDD for $\forall x\,(\varphi)$ is obtained by `Apply`$(\wedge, B[x \leftarrow \mathit{true}], B[x \leftarrow \mathit{false}])$ and the BDD for $\exists x\,(\varphi)$ by `Apply`$(\vee, B[x \leftarrow \mathit{true}], B[x \leftarrow \mathit{false}])$.

A BDD can also represent a set of all models of a BV formula. It is sufficient to decompose every bit-vector variable and every uninterpreted constant of bit-width $n$ into $n$ boolean variables and perform operations on individual bits. For example, all models of the formula $\forall x\,(\neg(a = 2^4 \times x))$ are represented by the BDD of Fig. 3a, where the 32 bits of the uninterpreted constant $a$ are denoted by boolean variables $a_0, a_1, \ldots, a_{31}$ in order from the least significant to the most significant bit. Boolean variables arising from bit-vector variables and uninterpreted constants are called *bit variables* henceforth. As usual, instead of labelling edges as *true* and *false*, edges are drawn as solid and dashed, respectively. Note that every unsatisfiable formula is represented by the BDD with the single node 0. By the BDD size we mean the number of its nodes.

# 3    Our Approach

This section first describes three main parts of our algorithm, namely formula simplifications, bit variable ordering for BDD construction, and approximations. Subsequently, the main algorithm is presented.

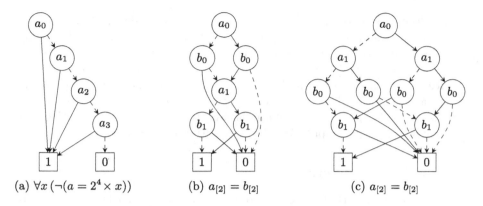

(a) $\forall x\,(\neg(a = 2^4 \times x))$    (b) $a_{[2]} = b_{[2]}$    (c) $a_{[2]} = b_{[2]}$

**Fig. 3.** Examples of BDDs representing bit-vector formulas.

## 3.1    Formula Simplifications

As in most of modern SMT solvers, the first step of deciding satisfiability is simplification of the input formula. Besides trivial simplifications (e.g. $\varphi \wedge \varphi$ reduces to $\varphi$), we apply the following simplification rules.

*Miniscoping.* Miniscoping [19] is a technique reducing the scope of universal quantifier over disjunctions whenever one disjunct has no free occurrences of the quantified variable, and over conjunctions by distributivity (existential quantifiers are handled analogously). The simplification rules are as follows:

$$\forall x\,(\varphi[x] \vee \psi) \rightsquigarrow \forall x\,(\varphi[x]) \vee \psi \qquad \forall x\,(\varphi[x] \wedge \psi[x]) \rightsquigarrow \forall x\,(\varphi[x]) \wedge \forall x\,(\psi[x])$$
$$\exists x\,(\varphi[x] \wedge \psi) \rightsquigarrow \exists x\,(\varphi[x]) \wedge \psi \qquad \exists x\,(\varphi[x] \vee \psi[x]) \rightsquigarrow \exists x\,(\varphi[x]) \vee \exists x\,(\psi[x])$$

*Destructive Equality Resolution.* Destructive equality resolution (DER) [25] eliminates a universally quantified variable $x$ in a formula $\forall x\,\overline{Qy}\,(\neg(x = t) \vee \varphi[x])$, where $t$ is a term that does not contain the variable $x$, and $\overline{Qy}$ is a sequence of variable quantifications. The formula is equivalent to $\forall x\,\overline{Qy}\,(x = t \rightarrow \varphi[x])$ and hence also to $\overline{Qy}\,(\varphi[t])$. The simplification rule is formulated as follows:

$$\forall x\,\overline{Qy}\,(\neg(x = t) \vee \varphi[x]) \quad \rightsquigarrow \quad \overline{Qy}\,(\varphi[t])$$

*Constructive Equality Resolution.* Constructive equality resolution (CER) is a dual version of DER. As far as we know, it was not considered before as solvers for quantified formulas typically work with formulas after Skolemization and thus without any existential quantifiers. CER can be formulated as the following simplification rule, where $t$ and $\overline{Qy}$ have the same meaning as above:

$$\exists x\,\overline{Qy}\,(x = t \wedge \varphi[x]) \quad \rightsquigarrow \quad \overline{Qy}\,(\varphi[t])$$

*Theory-Related Simplifications.* We also perform several simplifications related to the interpretation of the function and predicate symbols in the BV logic. Examples of such simplifications are reductions $a_{[n]} + (-a_{[n]}) \rightsquigarrow 0_{[n]}$, $a_{[n]} \times 0_{[n]} \rightsquigarrow 0_{[n]}$, $a_{[n]}$ and $0_{[n]} \rightsquigarrow 0_{[n]}$, or $\texttt{extract}_{[n,i,j]}(0_{[n]}) \rightsquigarrow 0_{[j-i+1]}$.

Note that all mentioned simplification rules have no effect on models of the formula and thus they have no direct effect on the resulting BDD. However, a simplified formula has simpler subformulas and thus the intermediate BDDs are often smaller and the computation of the resulting BDD is faster.

## 3.2    Bit Variable Ordering

When constructing a BDD, one has to specify an order of BDD variables. In our case, BDD variables precisely correspond to bit variables. The order of these variables has a significant effect on the BDD size and its construction time. In some cases, the size of a BDD for a formula is linear in the number of BDD variables with one variable ordering, but exponential with another ordering.

For example, consider the formula $\phi_1 \equiv a_{[n]} = b_{[n]}$ for arbitrary $n \in \mathbb{N}$ and let $a_0, a_1, \ldots, a_{n-1}$ be the bits of $a$ and $b_0, b_1, \ldots, b_{n-1}$ be the bits of $b$. We define two orderings:

$\leq_1$ All bit variables are ordered according to their significance (from the least to the most significant) and variables with the same significance are ordered by the order of the first occurrences of the corresponding bit-vector variables in the formula. For the considered formula $\phi_1$, we get:

$$a_0 \leq_1 b_0 \leq_1 a_1 \leq_1 b_1 \leq_1 \ldots \leq_1 a_{n-1} \leq_1 b_{n-1}$$

$\leq_2$ Bit variables are ordered by the order of the first occurrences of the corresponding bit-vector variables in the formula and bit variables corresponding to the same bit-vector variable are ordered according to their significance (from the least to the most significant). For the considered formula, we get:

$$a_0 \leq_2 a_1 \leq_2 \ldots \leq_2 a_{n-1} \leq_2 b_0 \leq_2 b_1 \leq_2 \ldots \leq_2 b_{n-1}$$

The BDD for $\phi_1$ using the ordering $\leq_1$ has $3n + 2$ nodes, while the BDD for the same formula and $\leq_2$ has $3 \cdot 2^n - 1$ nodes. Figure 3b and c show these BDDs for $n = 2$ and orderings $\leq_1$ and $\leq_2$, respectively.

These orderings can lead to opposite results with other formulas. For example, the size of the BDD for the formula

$$\phi_2 \equiv (c_{[2]}^1 = c_{[2]}^2 \ \texttt{shr} \ 1_{[2]}) \wedge (c_{[2]}^3 = c_{[2]}^4 \ \texttt{shr} \ 1_{[2]}) \wedge \ \ldots \ \wedge (c_{[2]}^{2n-1} = c_{[2]}^{2n} \ \texttt{shr} \ 1_{[2]})$$

using the ordering $\leq_1$ is $2^{n+2} - 1$, while it is only $4n + 2$ for $\leq_2$. In general, choosing the optimal variable ordering is an NP-complete problem [10]. In the following, we introduce an ordering $\leq_3$ combining advantages of $\leq_1$ and $\leq_2$.

Let $V$ be the set of bit-vector variables and uninterpreted constants appearing in an input formula $\varphi$. Elements $x, y \in V$ are *dependent*, written $x \sim y$, if they both appear in some atomic subformula of $\varphi$. Let $\simeq$ be the equivalence on $V$ defined as the transitive closure of $\sim$. Every $v \in V$ then defines an equivalence class $[v]_\simeq$ of transitively dependent elements.

$\leq_3$ Bit variables are first ordered according to $\leq_1$ within corresponding equivalence classes of $\simeq$ and the equivalence classes are then ordered by the first occurrences of BV variables in $\varphi$. In particular for $u \not\simeq v$, $u_i \leq_3 v_j$ if there is a BV variable in $[u]_\simeq$, which occurs in $\varphi$ before all BV variables of $[v]_\simeq$.

Note that for both formulas $\phi_1, \phi_2$ mentioned above, $\leq_3$ coincides with the better of the orderings $\leq_1$ and $\leq_2$.

In addition to the initial variable ordering, there are several techniques that dynamically reorder the BDD variables to reduce the BDD size. We use *sifting* [24] as usually the most successful one [20].

### 3.3 Approximations

For some BV formulas, e.g. formulas containing non-linear multiplication, the size of the BDD representation is exponential for every possible variable ordering [13]. Fortunately, satisfiability of these formulas can be often decided using their over-approximations or under-approximations. Given a formula $\varphi$, its *under-approximation* is any formula $\underline{\varphi}$ that logically entails $\varphi$, and its *over-approximation* is any formula $\overline{\varphi}$ logically entailed by $\varphi$. Clearly, every model of $\underline{\varphi}$ is also a model of $\varphi$ and if an under-approximation $\underline{\varphi}$ is satisfiable, so is the formula $\varphi$. Similarly, if an over-approximation $\overline{\varphi}$ is unsatisfiable, so is $\varphi$.

The model-based quantifier instantiation presented in Sect. 2.2 can be seen as a technique based on iterative over-approximation refinement: the formulas $\varphi$, $\varphi \wedge \psi[\bar{v}]$, ... are over-approximations of $\varphi \wedge \forall \bar{x} \, (\psi[\bar{x}])$. A different concepts of approximations can be found in SMT solvers for QF_BV formulas. For example, the SMT solver UCLID over-approximates a formula in the negation normal form by replacing some subformulas with fresh uninterpreted constants [14]. Further, SMT solvers UCLID and Boolector under-approximate a formula by restricting the value of $m$ most significant bits of a bit-vector variable while leaving the remaining bits unchanged [11,14]. The number of bit variables used to represent the bit-vector variable or uninterpreted constant is called its *effective bit-width*. This approach inspired both over- and under-approximation used in our algorithm.

Let $a_{[n]}$ be a variable or an uninterpreted constant of bit-width $n$ and $e \in \mathbb{N}$ be its desired effective bit-width. If $e \geq n$, we leave $a_{[n]}$ unchanged. Otherwise, we consider four different ways to reduce the effective bit-width of $a_{[n]}$ to $e$:

**zero-extension** uses the effective bit-width to represent the $e$ least significant bits and sets the $n - e$ most significant bits to 0.
**sign-extension** also uses the effective bit-width to represent the $e$ least significant bits and sets the $n - e$ most significant bits to the value of the $e$-th least significant bit.

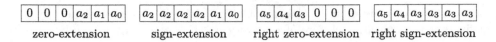

**Fig. 4.** Reductions of $a_{[6]} = a_5a_4a_3a_2a_1a_0$ to 3 effective bits.

**right zero-extension** uses the effective bit-width to represent the $e$ most significant bits and sets the $n - e$ least significant bits to 0.

**right sign-extension** also uses the effective bit-width to represent the $e$ most significant bits and sets the $n - e$ least significant bits to the value of the $e$-th most significant bit.

All considered extensions are illustrated in Fig. 4. The first two extensions are taken directly from [14], while the other two are original. One can easily see that each extension reduces the domain of $a_{[n]}$ to a different subdomain of size $2^e$. Another extensions are suggested in [11], e.g. *one-extension* defined analogously to the zero-extension. Our choice of considered extensions is motivated by exploration of values near corner cases as well as by reduction of BDD size. In particular, we do not consider one-extension because it produces only few zero bits which are desired as they tend to reduce the size of BDDs for multiplication.

In the following, the term *extension* always refers either to zero-extension, or to sign-extension. In an over-approximation, we apply a selected reduction to all universally quantified variables. Given a formula $\varphi$ and $e \in \mathbb{N}$, let $\overline{\varphi}_e$ denote the formula $\varphi$ where the effective bit-width of each universally quantified variable is reduced to $e$ by the chosen extension. Further, $\overline{\varphi}_{-e}$ denotes the formula obtained by application of the right counterpart of the chosen extension.

In an under-approximation, we apply the selected reduction to all existentially quantified variables and uninterpreted constants. Given a formula $\varphi$ and $e \in \mathbb{N}$, let $\underline{\varphi}_e$ and $\underline{\varphi}_{-e}$ denote the formula $\varphi$ where the effective bit-width of each existentially quantified variable and uninterpreted constant is reduced to $e$ by the chosen extension or its right counterpart, respectively.

The following theorem establishes that, for each formula $\varphi$ in the negation normal form, $\overline{\varphi}_e$, $\overline{\varphi}_{-e}$ are over-approximations (and analogously for under-approximations). The theorem can be easily proven by an induction on the structure of the formula $\varphi$.

**Theorem 1.** *For every formula $\varphi$ in the NNF and any $e \in \mathbb{N}$, it holds:*

1. *If $M$ is a model of $\varphi$, then $M$ is also a model of $\overline{\varphi}_e$ and $\overline{\varphi}_{-e}$.*
2. *If $M$ is a model of $\underline{\varphi}_e$ or $\underline{\varphi}_{-e}$, then $M$ is also a model of $\varphi$.*

**Corollary 1.** *For every formula $\varphi$ in the NNF and any $e \in \mathbb{N}$, it holds:*

1. *If the formula $\overline{\varphi}_e$ or $\overline{\varphi}_{-e}$ is unsatisfiable, so is the formula $\varphi$.*
2. *If the formula $\underline{\varphi}_e$ or $\underline{\varphi}_{-e}$ is satisfiable, so is the formula $\varphi$.*

## 3.4 The Algorithm

In this section, we present the complete algorithm deciding satisfiability of BV formulas. In the algorithm, we use a procedure ConvertToBDD which converts a formula to the corresponding BDD recursively on the formula structure. For a given input formula $\varphi$, the algorithm proceeds in the following steps:

1. Simplify the formula $\varphi$ using the rules discussed in Sect. 3.1 up to the fixpoint and convert it to the negation normal form. If the result is *true*, return SAT. If the result is *false*, return UNSAT.
2. Take the simplified formula in NNF $\varphi'$ and compute a chosen ordering $\leq$ as described in Sect. 3.2. This ordering will be used as the initial ordering in the procedure ConvertToBDD.
3. Call ConvertToBDD($\varphi'$) to compute the BDD corresponding to $\varphi'$. If the root node of the BDD has label 0, return UNSAT. Otherwise return SAT.
4. If the procedure ConvertToBDD called in the previous step has not finished within 0.1 s, additionally run in parallel:
   (a) *Under-approximations*: Sequentially compute ConvertToBDD($\underline{\varphi'}_i$) for $i = 1, -1, 2, -2, 4, -4, 6, -6, \ldots$ until reaching the greatest bit-width of a bitvector variable in $\varphi'$. If any of the resulting BDDs has a root node distinct from the leaf 0, return SAT.
   (b) *Over-approximations*: Sequentially compute ConvertToBDD($\overline{\varphi'}_i$) for $i = 1, -1, 2, -2, 4, -4, 6, -6, \ldots$ until reaching the greatest bit-width of a bitvector variable in $\varphi'$. If any of the produced BDDs has a root node labeled by 0, return UNSAT.

The algorithm is parametrized by the choice of an ordering and reductions for approximations. Regardless these parameters, the algorithm is sound and complete. The decision to start the solvers using approximations after 0.1 s is based on our experiments. In practice, the procedure ConvertToBDD may need exponential time and memory and thus the algorithm may not finish within reasonable limits.

## 4 Implementation and Experimental Results

We have implemented the presented algorithm in an experimental SMT solver called Q3B. The implementation is written in C++, relies on the BDD package BuDDy[1], and uses the API of Z3 to parse the input formula in the SMT-LIB 2.5 format [4] and to perform some formula simplifications. As the BuDDy package does not support allocation of multiple BDD instances, we run separate processes for the base solver and for computing over- and under-approximations. The execution of these three processes is controlled by a Python wrapper.

We have evaluated our solver on two sets of BV formulas. The first set consists of all 191 formulas in the category BV of the SMT-LIB benchmark repository [5]. The second set contains 5 461 formulas generated by the model

---

[1] http://sourceforge.net/projects/buddy.

checker SYMDIVINE [7] when run on verification tasks from SV-COMP [8]. These formulas correspond to checking equivalence of two symbolic states of the verified program. In total, SYMDIVINE generated 1 462 500 formulas. For tasks with more than 25 generated formulas, we randomly picked 25 formulas to keep the number of formulas reasonable.

All experiments were performed on a Debian machine with two six-core Intel Xeon E5-2620 2.00 GHz processors and 128 GB of RAM. Each benchmark run was limited to use 3 processor cores, 4 GB of RAM and 20 min of CPU time (if not stated otherwise). All measured times are CPU times. For reliable benchmarking we employed BENCHEXEC [9], a tool that allocates specified resources for a program execution and measures their use precisely.

All used benchmarks and detailed experimental results are available at http://www.fi.muni.cz/~xstrejc/sat2016.tar.gz. Q3B is available under the MIT License and hosted at GitHub: https://github.com/martinjonas/Q3B.

In the following, we demonstrate the effect of formula simplifications on the formulas and the effect of various algorithm parameters on its efficiency. At the end, we compare our solver with the best parameters against CVC4 and Z3.

*Formula Simplifications.* Considered formula simplifications reduced 108 of 191 SMT-LIB benchmarks and 300 of 5 461 SYMDIVINE benchmarks to *true* or *false*. Additionally, 1 276 SYMDIVINE benchmarks were reduced to a quantifier-free formulas, which is not the case for any SMT-LIB benchmark. Figure 2 shows the number of bit variables (i.e. the sum of bit-widths of all bit-vector variables and uninterpreted constants in the formula) of each formula before and after simplification.

*Variable Ordering.* To compare the effect of BDD variable orderings $\leq_1$, $\leq_2$, and $\leq_3$ defined in Sect. 3.2, we run our tool with each of these initial orderings on all considered benchmarks. Recall that sifting method is used for dynamic variable reordering. The solver has been executed without approximations (to ensure that approximations will not hide the effect of the initial ordering) and with CPU time limited to 3 min. The results are shown in Fig. 5.

When SMT-LIB are considered, the worst performing initial ordering is $\leq_2$. The results for $\leq_1$ and $\leq_3$ are almost identical as nearly all bit-vector variables in these benchmarks are mutually transitively dependent. For SYMDIVINE benchmarks, initial ordering $\leq_1$ performs the worst. The results for $\leq_2$ and $\leq_3$ are very similar, as SYMDIVINE formulas usually contain a large number of mutually independent groups of variables. The solver using $\leq_3$ decided 3 more formulas than the solver using $\leq_2$. To sum up, since now we always use the ordering $\leq_3$ as it provides better overall performance than $\leq_1$ and $\leq_2$.

Note that the solver runs usually faster when executed without sifting, as the dynamic reordering causes some computational overhead. However, with sifting it decides 2 more SMT-LIB benchmarks and 9 more SYMDIVINE benchmarks.

*Approximations.* To compare the effect of the considered effective bit-width reductions, we run the solver once with approximations based on (right) zero-

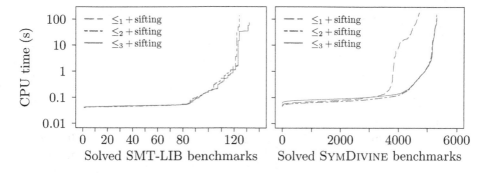

**Fig. 5.** Quantile plot of the number of solved benchmarks for each of three described initial variable orderings. (Color figure online)

extension, and again with approximations based on (right) sign-extension. Quantile plots in Fig. 6 show results for zero-extension and sign-extension on SMT-LIB benchmarks. The results are presented separately for satisfiable and unsatisfiable formulas. On satisfiable formulas, approximation using zero-extension performs better and can decide 3 more satisfiable formulas. On the contrary, on unsatisfiable formulas sign-extension performs better and can decide 5 formulas more. Corresponding plots for SYMDIVINE formulas are not presented, since the difference in CPU times was insignificant. Based on this observation and the fact that satisfiability can be decided by an under-approximation and unsatisfiability by an over-approximation, the default bit-width reduction method in our solver is zero-extension for under-approximations and sign-extension for over-approximations.

Further, to show the contribution of approximations, we compare the solver using the proposed algorithm as described in the Sect. 3.4 against the same algorithm without approximations. Figure 7 shows quantile plots corresponding to measured CPU times. With approximations, the solver was able to decide 54 more SMT-LIB formulas. The difference is less significant when SYMDIVINE formulas are considered, as they mostly do not contain difficult arithmetic; only 15 more of SYMDIVINE formulas were decided using approximations.

*Comparison.* Finally, we compare our solver (with the parameters selected by the previous experiments) to the current stable versions of leading SMT solvers for BV logic, namely to the version 4.4.1 of Z3 [16] and the version 1.4 of CVC4 [3]. We also tested the latest development version of CVC4 (2016-02-25), but it decided some SYMDIVINE benchmarks incorrectly. The solver Z3 was executed with the default settings, CVC4 was executed with settings supplied for the SMT-competition, where the benchmarks from the SMT-LIB benchmark repository are used.

Table 2 shows summary results of the solvers CVC4, Z3, and Q3B on the two benchmark sets. Additionally, Table 3 shows for each pair of solvers the number of formulas which were decided by one solver, but not by the other

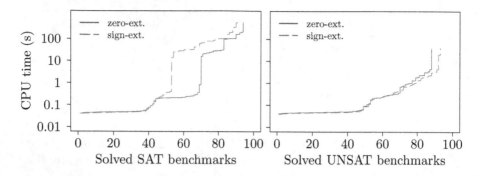

**Fig. 6.** Quantile plot of the number of solved SMT-LIB benchmarks using approximation via sign-extension and zero-extension compared by the CPU time. (Color figure online)

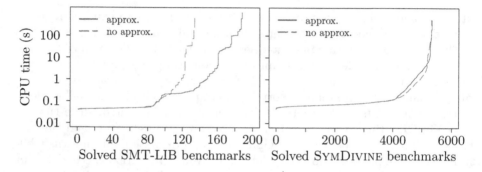

**Fig. 7.** Quantile plot of the number of solved benchmarks with and without approximations compared by the CPU time. (Color figure online)

one. Out of the 191 SMT-LIB benchmarks, CVC4 solves 84 benchmarks, Z3 decides 164 benchmarks, and our solver can decide 188 benchmarks. Out of the 5 461 SYMDIVINE benchmarks, CVC4 decides 4 969 benchmarks, Z3 solves 5 297 benchmarks, and our solver Q3B decides 5 339 benchmarks. To sum up, in the number of decided benchmarks Q3B outperforms both CVC4 and Z3. Moreover, only 1 of all considered formulas was solved by Z3 and not by Q3B, and no formula was solved by CVC4 and not by Q3B.

Further, quantile plots of Fig. 8 show numbers of input formulas each of the solvers was able to decide within different CPU time limits. Note that the $y$ axis has the logarithmic scale. On the easy instances, our experimental solver can not compete with highly optimized solvers as CVC4 and Z3. The initial delay of Q3B is caused by an overhead of a process creation within the Python wrapper. However, as the instances become harder, the difference in solving times decreases. In particular, Q3B solves more SMT-LIB benchmarks than CVC4 whenever the CPU time limit is longer than 0.05 s and more than Z3 for any CPU time limit over 0.39 s. For SYMDIVINE benchmarks, these thresholds

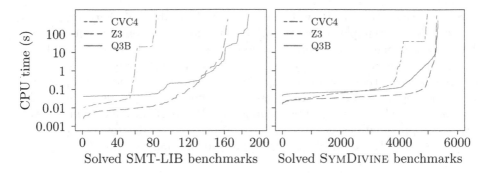

**Fig. 8.** Quantile plot of the number of benchmarks which CVC4, Q3B, and Z3 solved compared by the CPU time. (Color figure online)

**Table 2.** For each benchmark set and each solver, the table provides the numbers of formulas decided as satisfiable (*sat*), unsatisfiable (*unsat*), or undecided with the result unknown or because of an error (*unknown*), or a *timeout*.

|  | SMT-LIB | | | | SymDivine | | | |
|---|---|---|---|---|---|---|---|---|
|  | sat | unsat | unknown | timeout | sat | unsat | unknown | timeout |
| CVC4 | 29 | 55 | 32 | 75 | 1 124 | 3 845 | 2 | 490 |
| Z3 | 71 | 93 | 5 | 22 | 1 135 | 4 162 | 22 | 142 |
| Q3B | 94 | 94 | 0 | 3 | 1 137 | 4 202 | 0 | 122 |

**Table 3.** For each pair of solvers, the table shows the number of benchmarks, which were solved by the solver in the corresponding row, but not by the solver in the corresponding column.

|  | SMT-LIB | | | SymDivine | | |
|---|---|---|---|---|---|---|
|  | CVC4 | Z3 | Q3B | CVC4 | Z3 | Q3B |
| CVC4 | – | 0 | 0 | – | 21 | 0 |
| Z3 | 80 | – | 1 | 349 | – | 0 |
| Q3B | 104 | 25 | – | 370 | 42 | – |

are 0.08 s for CVC4 and 8.72 s for Z3. Note that Q3B uses 3 parallel processes and hence its wall times are usually three times shorter than presented CPU times, while wall times are the same as CPU times for Z3 and CVC4.

## 5  Conclusions

We presented a new SMT solving algorithm for quantified bit-vector formulas. While current SMT solvers for this logic typically rely on model-based quantifier instantiation and an SMT solver for quantifier-free bit-vector formulas, our algorithm is based on BDDs (with a specific initial variable ordering) and approximations. We have implemented the algorithm and experimental results indicate that our approach can compete with state-of-the-art SMT solvers CVC4 and Z3. In fact, it decides more formulas than the mentioned solvers.

We plan to further develop the algorithm and the tool. In particular, we plan to add a support for arrays and uninterpreted functions as these are useful for modelling some features of computer programs. We would also like to investigate possible approximations of bit-vector operations and predicates, or to develop some fine-grained methods for a targeted approximation refinement.

# References

1. The 10th International Satisfiability Modulo Theories Competition (SMT-COMP2015) (2015). http://smtcomp.sourceforge.net/2015/
2. Audemard, G., Sais, L.: SAT based BDD solver for quantified Boolean formulas. In: 16th IEEE International Conference on Tools with Artificial Intelligence, ICTAI 2004, pp. 82–89 (2004)
3. Barrett, C., Conway, C.L., Deters, M., Hadarean, L., Jovanović, D., King, T., Reynolds, A., Tinelli, C.: CVC4. In: Gopalakrishnan, G., Qadeer, S. (eds.) CAV 2011. LNCS, vol. 6806, pp. 171–177. Springer, Heidelberg (2011)
4. Barrett, C., Fontaine, P., Tinelli, C.: The SMT-LIB Standard: Version 2.5. Technical report, Department of Computer Science, The University of Iowa (2015). www.SMT-LIB.org
5. Barrett, C., Stump, A., Tinelli, C.: The Satisfiability Modulo Theories Library (SMT-LIB) (2010). www.SMT-LIB.org
6. Barrett, C.W., Sebastiani, R., Seshia, S.A., Tinelli, C.: Satisfiability modulo theories. In: Handbook of Satisfiability, pp. 825–885 (2009)
7. Bauch, P., Havel, V., Barnat, J.: LTL model checking of LLVM bitcode with symbolic data. In: Hliněný, P., Dvořák, Z., Jaroš, J., Kofroň, J., Kořenek, J., Matula, P., Pala, K. (eds.) MEMICS 2014. LNCS, vol. 8934, pp. 47–59. Springer, Heidelberg (2014)
8. Beyer, D.: Software verification and verifiable witnesses. In: Baier, C., Tinelli, C. (eds.) TACAS 2015. LNCS, vol. 9035, pp. 401–416. Springer, Heidelberg (2015)
9. Beyer, D., Löwe, S., Wendler, P.: Benchmarking and resource measurement. In: Proceedings Model Checking Software - 22nd International Symposium, SPIN 2015, pp. 160–178 (2015)
10. Bollig, B., Wegener, I.: Improving the variable ordering of OBDDs Is NP-complete. IEEE Trans. Comput. **45**(9), 993–1002 (1996)
11. Brummayer, R., Biere, A.: Effective bit-width and under-approximation. In: Moreno-Díaz, R., Pichler, F., Quesada-Arencibia, A. (eds.) EUROCAST 2009. LNCS, vol. 5717, pp. 304–311. Springer, Heidelberg (2009)
12. Bryant, R.E.: Graph-based algorithms for Boolean function manipulation. IEEE Trans. Comput. **35**(8), 677–691 (1986)
13. Bryant, R.E.: On the complexity of VLSI implementations and graph representations of Boolean functions with application to integer multiplication. IEEE Trans. Comput. **40**(2), 205–213 (1991)
14. Bryant, R.E., Kroening, D., Ouaknine, J., Seshia, S.A., Strichman, O., Brady, B.A.: Deciding bit-vector arithmetic with abstraction. In: Grumberg, O., Huth, M. (eds.) TACAS 2007. LNCS, vol. 4424, pp. 358–372. Springer, Heidelberg (2007)

15. de Moura, L., Bjørner, N.S.: Efficient E-matching for SMT solvers. In: Pfenning, F. (ed.) CADE 2007. LNCS (LNAI), vol. 4603, pp. 183–198. Springer, Heidelberg (2007)

16. de Moura, L., Bjørner, N.S.: Z3: an efficient SMT solver. In: Ramakrishnan, C.R., Rehof, J. (eds.) TACAS 2008. LNCS, vol. 4963, pp. 337–340. Springer, Heidelberg (2008)

17. Detlefs, D., Nelson, G., Saxe, J.B.: Simplify: a theorem prover for program checking. J. ACM **52**(3), 365–473 (2005)

18. Enderton, H.B.: A Mathematical Introduction to Logic. Harcourt/Academic Press, Burlington (2001)

19. Harrison, J.: Handbook of Practical Logic and Automated Reasoning, 1st edn. Cambridge University Press, New York (2009)

20. Knuth, D.E.: The Art of Computer Programming, Volume 4, Fascicle 1: Bitwise Tricks & Techniques; Binary Decision Diagrams, 12th edn. Addison-Wesley Professional, Boston (2009)

21. Kovásznai, G., Fröhlich, A., Biere, A.: Complexity of fixed-size bit-vector logics. Theor. Comput. Syst. **7913**, 1–54 (2015)

22. Kroening, D., Lewis, M., Weissenbacher, G.: Under-approximating loops in C programs for fast counterexample detection. In: Sharygina, N., Veith, H. (eds.) CAV 2013. LNCS, vol. 8044, pp. 381–396. Springer, Heidelberg (2013)

23. Olivo, O., Emerson, E.A.: A more efficient BDD-based QBF solver. In: Lee, J. (ed.) CP 2011. LNCS, vol. 6876, pp. 675–690. Springer, Heidelberg (2011)

24. Rudell, R.: Dynamic variable ordering for ordered binary decision diagrams. In: Proceedings of the 1993 IEEE/ACM International Conference on Computer-Aided Design, pp. 42–47 (1993)

25. Wintersteiger, C.M., Hamadi, Y., de Moura, L.: Efficiently solving quantified bit-vector formulas. Formal Methods Syst. Des. **42**(1), 3–23 (2013)

# Speeding up the Constraint-Based Method in Difference Logic

Lorenzo Candeago[1], Daniel Larraz[2], Albert Oliveras[2],
Enric Rodríguez-Carbonell[2(✉)], and Albert Rubio[2]

[1] SpazioDati, Trento, Italy
[2] Universitat Politècnica de Catalunya, Barcelona, Spain
erodri@cs.upc.edu

**Abstract.** Over the years the constraint-based method has been successfully applied to a wide range of problems in program analysis, from invariant generation to termination and non-termination proving. Quite often the semantics of the program under study as well as the properties to be generated belong to difference logic, i.e., the fragment of linear arithmetic where atoms are inequalities of the form $u - v \leq k$. However, so far constraint-based techniques have not exploited this fact: in general, Farkas' Lemma is used to produce the constraints over template unknowns, which leads to non-linear SMT problems. Based on classical results of graph theory, in this paper we propose new encodings for generating these constraints when program semantics and templates belong to difference logic. Thanks to this approach, instead of a heavyweight non-linear arithmetic solver, a much cheaper SMT solver for difference logic or linear integer arithmetic can be employed for solving the resulting constraints. We present encouraging experimental results that show the high impact of the proposed techniques on the performance of the VeryMax verification system.

## 1 Introduction

Since Colón's *et al.* seminal paper [1], the so-called *constraint-based method* has been applied with success to a wide range of problems in system verification, from invariant generation in Petri nets [2], hybrid systems [3] and programs with arrays [4,5], to termination [6,7] and non-termination proving [8]. In most of these applications, one is interested in generating *linear properties*, e.g., linear invariants or linear ranking functions. In these cases, Farkas' Lemma is employed for producing the constraints over the template unknowns. As a result, a *non-linear* SMT formula is obtained, for which a model has to be found. Despite the great advances in non-linear SMT [9–11], the applicability of the approach is still strongly conditioned by current solving technology.

A way to circumvent the bottleneck of using non-linear constraint solvers is to exploit the fragment of logics in which the program under study is described.

Partially supported by Spanish MINECO under grant TIN2015-69175-C4-3-R.

N. Creignou and D. Le Berre (Eds.): SAT 2016, LNCS 9710, pp. 284–301, 2016.
DOI: 10.1007/978-3-319-40970-2_18

Although this has not been explored so far in the constraint-based method, other more mature approaches for program analysis such as abstract interpretation [12] have profited from this sort of refinements since the early days of their inception. Indeed, there is a wide variety of non-relational and weakly-relational numerical abstract domains which cover different subsets of linear arithmetic, but whose complexity is lower than that of the full language [13]: *intervals* [14], *zones* [15] and *octagons* [16], to name a few. Also in the model checking community, it is common to focus on particular subclasses of linear inequalities as a means to improve efficiency. In particular, *potential constraints* have been employed in the verification of several kinds of timed and concurrent systems [17–19].

In this paper we restrict our attention to *difference logic* over the integers, in which atoms are inequalities of the form $u - v \leq k$, where $u$ and $v$ are integer variables, and $k \in \mathbb{Z}$. This fragment of linear arithmetic corresponds to the aforementioned zone abstract domain in abstract interpretation, and to the potential constraints in model checking. Our contributions in this work are:

- we propose an encoding for satisfiability and unsatisfiability of sets of inequalities in difference logic including templates, which results in formulas of difference logic. This is noteworthy since current approaches to equivalent problems in general full linear arithmetic lead to non-linear formulas.
- for the problem of, given a set of inequalities with free independent terms, choosing an invariant subset that proves an assertion, we present two encodings, one for full linear arithmetic and another specialized one for difference logic. While the former leads to non-linear formulas, again the latter falls into a more tractable fragment, in this case linear arithmetic.
- we present an experimental evaluation with the constraint-based verification system VeryMax [20]. We consider the problem of proving the absence of out-of-bounds array accesses in a benchmark suite of numerical programs, and our results show that the expressiveness of difference logic is sufficient to succeed in the majority of the cases, while a remarkable boost in performance is obtained thanks to the proposed techniques.

Two closely related works are [21,22], in which invariants of slightly more general classes than difference logic are generated following the constraint-based method, but with different strategies for producing the constraints over template unknowns. In [21], the authors discover *octagonal* invariants, i.e., of the form $\pm x_1 \pm x_2 \leq k$, as well as *max-plus* invariants $\max_{1 \leq i \leq n}(a_0, x_i + a_i) \leq \max_{1 \leq i \leq n}(b_0, x_i + b_i)$. While our approach cannot currently produce max-plus invariants, the standard technique of adding for each variable $x$ a copy standing for $-x$ [16,23] allows generating octagonal invariants too. However, the quantifier elimination method in [21] is incomplete and, as a result, invariants may be missed. Moreover, program assignments must be of the form $x := \pm x + K$ or $x := K$, where $K$ is a constant, and so unlike with our techniques the common case of assignments like $x := y$, where $x$, $y$ are different variables, is not allowed. As regards [22], in that paper *template domains* are considered, where templates are linear inequalities with free independent term but fixed dependent term. There the quantifier elimination procedure [24] is precise, but is not specialized for difference logic. Unfortunately, the available implementation cannot produce difference logic invariants and so cannot be used for an experimental comparison.

## 2  Background

**Programs, Invariants and Safety.** Let us fix a set of (integer) program *variables* $\mathcal{X} = \{x_1, \ldots, x_n\}$, and denote by $\mathcal{F}(\mathcal{X})$ the formulas consisting of conjunctions of linear inequalities[1] over the variables $\mathcal{X}$. Let $\mathcal{L}$ be the set of program *locations*, which contains a set $\mathcal{L}_0$ of *initial* locations. Program *transitions* $\mathcal{T}$ are tuples $(\ell_S, \tau, \ell_T)$, where $\ell_S$ and $\ell_T \in \mathcal{L}$ represent the *source* and *target* locations respectively, and $\tau \in \mathcal{F}(\mathcal{X} \cup \mathcal{X}')$ describes the transition relation. Here $\mathcal{X}' = \{x_1', \ldots, x_n'\}$ represent the values of the variables after the transition.[2] A transition is *initial* if its source location is initial. The set of initial transitions is denoted by $\mathcal{T}_0$. A *program* is a pair $\mathcal{P} = (\mathcal{L}, \mathcal{T})$, which can be viewed as a directed graph where the locations $\mathcal{L}$ are the nodes, and each transition $(\ell_S, \tau, \ell_T)$ from $\mathcal{T}$ leads to an edge in the graph from $\ell_S$ to $\ell_T$ labelled by $\tau$.

A *state* $s = (\ell, \boldsymbol{x})$ consists of a location $\ell \in \mathcal{L}$ and a *valuation* $\boldsymbol{x} : \mathcal{X} \to \mathbb{Z}$. A state is *initial* if its location is initial. We denote a *computation step* with transition $t = (\ell_S, \tau, \ell_T)$ by $(\ell_S, \boldsymbol{x}) \to_t (\ell_T, \boldsymbol{x}')$ when the valuations $\boldsymbol{x}, \boldsymbol{x}'$ satisfy the transition relation $\tau$ of $t$. We use $\to_{\mathcal{P}}$ if we do not care about the executed transition, and $\to_{\mathcal{P}}^*$ to denote the transitive-reflexive closure of $\to_{\mathcal{P}}$. We say that a state $s$ is *reachable* if there exists an initial state $s_0$ such that $s_0 \to_{\mathcal{P}}^* s$.

An *assertion* $(\ell, \varphi)$ is a pair of a location $\ell \in \mathcal{L}$ and a formula $\varphi$ with free variables $\mathcal{X}$. A program is *safe* with respect to the assertion $(\ell, \varphi)$ if for every reachable state $(\ell, \boldsymbol{x})$, we have that $\boldsymbol{x} \models \varphi$ holds.

A map $\mathcal{I} : \mathcal{L} \to \mathcal{F}(\mathcal{X})$ is an *invariant* if for every $\ell \in \mathcal{L}$, the program is safe with respect to $(\ell, \mathcal{I}(\ell))$. An important class of invariants are inductive invariants. A map $\mathcal{I}$ is an *inductive invariant* if the following two conditions hold:

**Initiation:**    For $(\ell_S, \tau, \ell_T) \in \mathcal{T}_0$:          $\tau \models \mathcal{I}(\ell_T)'$
**Consecution:**  For $(\ell_S, \tau, \ell_T) \in \mathcal{T} - \mathcal{T}_0$:  $\mathcal{I}(\ell_S) \wedge \tau \models \mathcal{I}(\ell_T)'$

If only the condition **Consecution** is fulfilled, the map $\mathcal{I}$ is called a *conditional inductive invariant*.

One of the key problems in program analysis is to determine whether a program is safe with respect to a given assertion $(\ell, \varphi)$. This is typically proved by computing an (inductive) invariant $\mathcal{I}$ such that the following condition holds:

**Safety:** $\mathcal{I}(\ell) \models \varphi$

In this case we say that the invariant $\mathcal{I}$ *proves* the assertion $(\ell, \varphi)$.

Finally, we say a transition $t = (\ell_S, \tau, \ell_T)$ is *disabled* if it can never be executed, i.e., if for any reachable state $(\ell_S, \boldsymbol{x})$, there does not exist any $\boldsymbol{x}'$ such that $(\boldsymbol{x}, \boldsymbol{x}')$ satisfies $\tau$. One can prove this by computing an invariant $\mathcal{I}$ such

---

[1] Note that equalities can be considered as conjunctions of inequalities.
[2] For $\varphi \in \mathcal{F}(\mathcal{X})$, the formula $\varphi' \in \mathcal{F}(\mathcal{X}')$ is the version of $\varphi$ using primed variables.

that $\mathcal{I}(\ell_S) \models \neg\tau$. Disabled transitions allow one to simplify the program under analysis, since they can be soundly removed from the program. In general, if $\mathcal{I}$ is an invariant map, then any transition $t = (\ell_S, \tau, \ell_T)$ can be soundly strengthened by replacing the transition relation $\tau$ by $\mathcal{I}(\ell_S) \wedge \tau$.

**Constraint-Based Invariant Generation.** Invariants can be generated using the *constraint-based* (also called *template-based*) method [1]. The idea is to consider *templates* for candidate invariant properties. These templates involve both the program variables as well as fresh template variables whose values have to be determined to ensure invariance. To this end, conditions **Initiation** and **Consecution** are enforced by means of *constraints*. Any solution to these constraints yields an invariant. If templates represent linear inequalities, Farkas' Lemma [25] is used to express the constraints in terms of the template variables:

**Theorem 1 (Farkas' Lemma).** *Let $S$ be a system of linear inequalities $Ax \leq b$ ($A \in \mathbb{R}^{m \times n}, b \in \mathbb{R}^m$) over real variables $x$. Then $S$ has no solution iff there is $\lambda \in \mathbb{R}^m$ (called the* multipliers*) such that $\lambda \geq 0$, $\lambda^T A = 0$ and $\lambda^T b \leq -1$.*

In general, an SMT formula over non-linear arithmetic is obtained. By assigning weights to the different conditions, invariant generation can be cast as an optimization problem in the Max-SMT framework [7,8,20].

*Example 1* Consider the program in Fig. 1, with the state variables $n$, $x_0$, $i$:

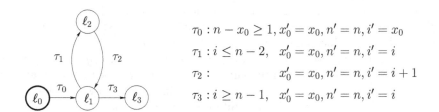

$$\tau_0 : n - x_0 \geq 1, x_0' = x_0, n' = n, i' = x_0$$
$$\tau_1 : i \leq n - 2, \quad x_0' = x_0, n' = n, i' = i$$
$$\tau_2 : \qquad\qquad x_0' = x_0, n' = n, i' = i + 1$$
$$\tau_3 : i \geq n - 1, \quad x_0' = x_0, n' = n, i' = i$$

**Fig. 1.** Program with a single initial location $\ell_0$.

Let us take the following 3 templates expressing general linear inequalities, one for each non-initial location:

$$T_j := c_{0j}\, x_0 + c_{1j}\, n + c_{2j}\, i \leq d_j \qquad \text{for all } j = 1 \ldots 3.$$

By imposing that these templates yield an invariant, we obtain the conditions (for simplicity, no assertion and thus no **Safety** condition is considered here):

| | | | |
|---|---|---|---|
| **Initiation:** | $\tau_0 \models T_1'$, | i.e., | $\tau_0 \wedge \neg T_1'$ unsatisfiable |
| **Consecution:** | $T_1 \wedge \tau_1 \models T_2'$, | i.e., | $T_1 \wedge \tau_1 \wedge \neg T_2'$ unsatisfiable |
| | $T_2 \wedge \tau_2 \models T_1'$, | i.e., | $T_2 \wedge \tau_2 \wedge \neg T_1'$ unsatisfiable |
| | $T_1 \wedge \tau_3 \models T_3'$, | i.e., | $T_1 \wedge \tau_3 \wedge \neg T_3'$ unsatisfiable . |

By fleshing out the transition relations, expanding the templates and simplifying, these four formulas are equivalent to

(1) $x_0 - n \leq -1 \wedge -(c_{01} + c_{21})x_0 - c_{11}n \leq -d_1 - 1$
(2) $c_{01}x_0 + c_{11}n + c_{21}i \leq d_1 \wedge i - n \leq -2 \wedge -c_{02}x_0 - c_{12}n - c_{22}i \leq -d_2 - 1$
(3) $c_{02}x_0 + c_{12}n + c_{22}i \leq d_2 \wedge \wedge -c_{01}x_0 - c_{11}n - c_{21}i \leq -d_1 - 1 + c_{21}$
(4) $c_{01}x_0 + c_{11}n + c_{21}i \leq d_1 \wedge n - i \leq 1 \wedge -c_{03}x_0 - c_{13}n - c_{23}i \leq -d_3 - 1$

respectively. Now Farkas' Lemma is applied to express unsatisfiability. Namely, for (1) we consider non-negative multipliers $\lambda_{11}, \lambda_{12}$ such that the linear combination that consists in multiplying the first inequality by $\lambda_{11}$ and the second inequality by $\lambda_{12}$ results in a trivially false inequality. For that, we need the coefficients of $x_0$ to cancel out, i.e., $\lambda_{11} - \lambda_{12}(c_{01} + c_{21}) = 0$, and the same for $n$, i.e., $-\lambda_{11} - \lambda_{12}c_{11} = 0$. With respect to the independent term, we force that it is smaller than or equal to $-1$, i.e., $-\lambda_{11} + \lambda_{12}(-d_1 - 1) \leq -1$, which will create a trivially false inequality. All in all, we get the non-linear formula

$$\exists \lambda_{11}\lambda_{12} \left( \lambda_{11}, \lambda_{12} \geq 0 \wedge \right.$$
$$\lambda_{11} - \lambda_{12}(c_{01} + c_{21}) = -\lambda_{11} - \lambda_{12}c_{11} = 0 \wedge \qquad (1)$$
$$\left. -\lambda_{11} + \lambda_{12}(-d_1 - 1) \leq -1 \right)$$

Similar constraints are obtained for (2)-(4). ☐

**Difference Logic and Graph Theory.** Given variables $u$ and $v$ and a numeric constant $k$, henceforth we will refer to an inequality of the form $u - v \leq k$ as a *difference inequality*. The fragment of (quantifier-free) first-order logic where atoms are difference inequalities is called *difference logic*.

Sets (conjunctions) of difference inequalities, also called *difference systems*, have long been studied in the literature (e.g., in [26], where they are referred to as *simple temporal problems*, STP's). For instance, they can be represented as graphs as follows. Given a difference system $S$ defined over variables $v_1, v_2, \ldots, v_n$, we consider the weighted graph $G$ with vertices $(v_1, v_2, \ldots, v_n)$ and an edge $v_i \xrightarrow{k} v_j$ for each inequality $v_i - v_j \leq k \in S$. This graph is called the *constraint graph* of $S$.

It is well-known that a constraint graph has interesting properties as regards to the solutions of the corresponding difference system [27]:

**Theorem 2** *Let $S$ be a difference system, and $G$ its constraint graph. Then $S$ has no solution iff $G$ has a negative cycle.*

This result is a particular case of Farkas' Lemma. It essentially ensures that, for difference systems, the multipliers of Farkas' Lemma are either 1 or 0 (the difference inequality belongs to the negative cycle or it does not, respectively).

One of the most important practical consequences of Theorem 2 is that any algorithm that is able to detect negative cycles in weighted graphs (such as, for instance, Bellman-Ford, or Floyd-Warshall [27]) can be used to determine the existence of solutions to a difference system.

Theorem 2 can be extended to allow also *bound* inequalities, i.e., inequalities of the form $v \leq k$ or $v \geq k$, where $v$ is a variable and $k$ is a numeric constant: Given a system $S$ that includes difference inequalities as well as bound inequalities, a fresh variable $v_0$ is introduced. Then a new system $S^*$ is defined, which is like $S$ but where each inequality of the form $v_i \leq k$ in $S$ is replaced by $v_i - v_0 \leq k$, and each $v_i \geq k$, or equivalently $-v_i \leq -k$, is replaced by $v_0 - v_i \leq -k$. It is not difficult to prove that $S$ has a solution iff $S^*$ has one.

## 3    Proving Safety of Difference Programs

In this paper we will focus on *difference programs*, that is, programs whose transition relations are conjunctions of difference inequalities.

Although this may seem rather restrictive, in fact more general programs can be cast into this form: for any program with difference as well as bound inequalities in the transition relations, there exists an equivalent difference program, as it is well-known in the literature [15]. The trick consists in introducing an artificial variable $x_0$, which intuitively is always zero, and then transform bound inequalities into difference inequalities by adding $x_0$ with the appropriate sign. Thus, e.g., $n \geq 1$ is transformed into $n - x_0 \geq 1$. Moreover, the equation $x'_0 = x_0$ has to be added to all transitions. For example, after this transformation the program in Fig. 2 leads to that in Fig. 1.

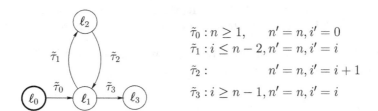

$$\tilde{\tau}_0 : n \geq 1, \qquad n' = n, i' = 0$$
$$\tilde{\tau}_1 : i \leq n - 2, n' = n, i' = i$$
$$\tilde{\tau}_2 : \qquad\qquad n' = n, i' = i + 1$$
$$\tilde{\tau}_3 : i \geq n - 1, n' = n, i' = i$$

**Fig. 2.** Program with difference and bound inequalities in the transition relations.

The problem we consider in this section is, given a location $\ell$ and a difference inequality $\varphi$, to prove that the program under consideration is safe with respect to the assertion $(\ell, \varphi)$. As the following theorem states, proving safety of a difference program is in general undecidable, and therefore we cannot hope for a sound and complete terminating algorithm that solves the problem:

**Theorem 3** *Given a difference program $\mathcal{P}$, a location $\ell \in \mathcal{L}$ and a difference inequality $\varphi$, the problem of deciding whether $\mathcal{P}$ is safe with respect to the assertion $(\ell, \varphi)$ is undecidable.*

## 3.1   Specialization of the Constraint-Based Method

Here we attempt to prove difference programs safe by finding invariants consisting of difference inequalities with a specialization of the constraint-based method.[3] Let us first illustrate the gist of our technique with an example.

*Example 2* Again let us consider the program in Fig. 1 and assign a template to each non-initial location: $T_j := c_{0j} x_0 + c_{1j} n + c_{2j} i \leq d_j$ for all $j = 1 \ldots 3$. This program is a difference program. Let us also consider the assignment $c_{0,j} = 0, c_{1,j} = -1, c_{2,j} = 1$ for all $j = 1 \ldots 3$, $d_1 = d_3 = -1$, $d_2 = -2$, which instantiates the templates as follows:

$$T_1 \equiv i - n \leq -1 \qquad T_2 \equiv i - n \leq -2 \qquad T_3 \equiv i - n \leq -1,$$

and check that they are invariant. Since the above inequalities belong to difference logic, we can use Theorem 2 to check that indeed the formulas $\tau_0 \wedge \neg T_1'$, $T_1 \wedge \tau_1 \wedge \neg T_2'$, $T_2 \wedge \tau_2 \wedge \neg T_1'$ and $T_1 \wedge \tau_3 \wedge \neg T_3'$ are unsatisfiable, as required by the **Initiation** and **Consecution** conditions. By the theorem, the unsatisfiability of each of these formulas is equivalent to the existence of a negative cycle in the corresponding graph. In Fig. 3 some of these graphs are shown for the particular solution considered here, and the respective negative cycles are highlighted. Solving the **Initiation** and **Consecution** constraints over the template coefficients can thus be seen as adding new weighted edges to the graphs of the transition relations so that, in the end, all graphs have a negative cycle. Note this must be done consistently for all **Initiation** and **Consecution** constraints, so that, e.g., the edge of $\neg T_1'$ is the same in $\tau_0 \wedge \neg T_1'$ and in $T_2 \wedge \tau_2 \wedge \neg T_1'$.

In what follows, we assume we have associated to each non-initial location $\ell$ a template invariant $T_\ell$ of the form

$$c_{0,\ell} x_0 + c_{1,\ell} x_1 + \ldots + c_{n,\ell} x_n \leq d_\ell$$

where the $c_{i,\ell}$ and the $d_\ell$ are template unknowns.[4] For obvious reasons we will refer to the $c_{i,\ell}$ as *left-hand side variables*, whereas the $d_\ell$ are called *right-hand side variables* (*LHS* and *RHS* variables, respectively). Here we focus on difference inequalities, and therefore the domain of LHS variables is $\{+1, 0, -1\}$, while the domain of RHS variables is $\mathbb{Z}$.

We propose to find appropriate values for the RHS and LHS variables following an *eager* approach: we encode all required constraints obtained from the **Initiation**, **Consecution** and **Safety** conditions into a single SMT formula, and then use an off-the-shelf SMT solver to solve the resulting problem. As will be seen next, in our particular case the atoms in the SMT formula will be either

---

[3] Here a simplified procedure for proving an assertion is described in order to highlight the key contribution of this work, that is, how to circumvent non-linearities.

[4] When the generated invariants consisting of a single inequality do not prove the assertion, as indicated in Section 2 the procedure can be iterated by strengthening the transitions, thereby allowing the synthesis of invariant conjunctions of inequalities.

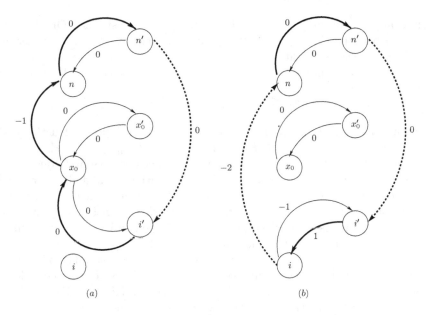

**Fig. 3.** Graphs for the formulas $\tau_0 \wedge \neg T_1'$ (a) and $T_2 \wedge \tau_2 \wedge \neg T_1'$ (b). The edges corresponding to the templates (or their negation) are dashed. The edges forming the negative cycles are highlighted with thicker lines.

Boolean variables or bound inequalities or difference inequalities. By virtue of the results reviewed in Sect. 2, the generated formula can be handled with an SMT solver of difference logic, for which efficient implementations are available.

The formula that expresses the constraints over template variables (LHS and RHS variables) is a conjunction of the following ingredients.

*Membership to Difference Logic.* First of all, we have to express that all templates are difference inequalities. To that end, for each LHS variable $c_i$ we introduce two auxiliary Boolean variables: $c_i^+$ and $c_i^-$. Intuitively, $c_i^+$ will be true iff $c_i$ is assigned to $+1$, and $c_i^-$ will be true iff $c_i$ is assigned to $-1$. If both $c_i^+$ and $c_i^-$ are false, then $c_i$ is 0. We need to enforce: (i) that the $c_i^+$ and $c_i^-$ cannot be true at the same time, (ii) that exactly one of the $c_i$ in each template is $+1$ (i.e., exactly one of the $c_i^+$ is true), and (iii) exactly one is $-1$ (i.e., exactly one of the $c_i^-$ is true). The constraints resulting from (ii) and (iii), which are of the form $\sum_{j=1}^m b_j = 1$, are encoded with a clause $\bigvee_{j=1}^m b_j$ that imposes $\sum_{j=1}^m b_j \geq 1$, together with additional clauses that express $\sum_{j=1}^m b_j \leq 1$ using one of the available encodings in the literature (e.g., quadratic, logarithmic [28] or ladder [29]).

*Unsatisfiability of Difference Systems.* When encoding the **Initiation, Consecution** and **Safety** conditions, essentially one has to impose the unsatisfiability of a set of difference inequalities, some of which may be templates. Namely, in **Initiation** and **Safety** one has a single template, but while in the former the template appears negatively, in the latter it appears positively. On the other hand, in **Consecution** two templates appear, one negatively and the other positively. Here we will elaborate on this latter case, being the others simpler.

Thus, let $\mathcal{S}$ be a difference system over program variables $\mathcal{X}$, $\mathcal{X}'$ such that

$$c_0 x_0 + \ldots + c_n x_n \leq d \wedge \mathcal{S} \wedge \neg(\tilde{c}_0 x_0' + \ldots + \tilde{c}_n x_n' \leq \tilde{d})$$

must be unsatisfiable. Our goal is to instantiate the templates so that this is the case. Note $\neg(\tilde{c}_0 x_0' + \ldots + \tilde{c}_n x_n' \leq \tilde{d})$ is equivalent to $-\tilde{c}_0 x_0' - \ldots - \tilde{c}_n x_n' \leq -\tilde{d} - 1$. To ensure unsatisfiability, i.e., that a negative cycle exists, we first construct $\mathcal{G}$, the constraint graph induced by $\mathcal{S}$. We then apply Floyd-Warshall algorithm in order to compute the distances $dist(y, z)$ for each pair of vertices $y$ and $z$ in $\mathcal{G}$.

If for some vertex $y$ we have $dist(y, y) < 0$, then $\mathcal{S}$ has a negative cycle and hence the unsatisfiability requirement is fulfilled independently from the templates. In this case, no clause needs to be added.

Otherwise $\mathcal{S}$ has no negative cycles, and the only possibility to construct one is to go through the edges induced by the templates. Let us consider an assignment such that $c_u = +1$, $c_v = -1$, $\tilde{c}_{\tilde{u}} = +1$ and $\tilde{c}_{\tilde{v}} = -1$ (i.e. $c_u^+$, $c_v^-$, $c_{\tilde{u}}^+$ and $c_{\tilde{v}}^-$ are true). In this case the instantiation of the positive template is $x_u - x_v \leq d$, and the instantiation of the negation of the other template is $x_{\tilde{v}}' - x_{\tilde{u}}' \leq -\tilde{d} - 1$. Hence, the former induces an edge from $x_u$ to $x_v$ with weight $d$, while the latter induces an edge from $x_{\tilde{v}}'$ to $x_{\tilde{u}}'$ with weight $-\tilde{d} - 1$.

To form a negative cycle, either (i) the cycle contains only the positive template, or (ii) contains only the negative template, or (iii) contains both. The first situation can be seen in Fig. 4(a), where it is needed that $dist(x_v, x_u) + d < 0$. The second situation is depicted in Fig. 4(b), where we need $dist(x_{\tilde{u}}', x_{\tilde{v}}') - \tilde{d} - 1 < 0$. Finally the third situation can be seen in Fig. 4(c), where we need $d + dist(x_v, x_{\tilde{v}}') - \tilde{d} - 1 + dist(x_{\tilde{u}}', x_u) < 0$. Hence, we add the following clause:

$$
\begin{aligned}
c_u^+ \wedge c_v^- \wedge \tilde{c}_{\tilde{u}}^+ \wedge \tilde{c}_{\tilde{v}}^- \implies d \quad &\leq -dist(x_v, x_u) - 1 \quad \vee \\
-\tilde{d} &\leq -dist(x_{\tilde{u}}', x_{\tilde{v}}') \quad \vee \\
d - \tilde{d} &\leq -dist(x_v, x_{\tilde{v}}') - dist(x_{\tilde{u}}', x_u)
\end{aligned}
\tag{2}
$$

Note that it might be the case that some of the paths represented in Fig. 4 do not actually exist. For example, if $x_u$ is unreachable from $x_v$, i.e., $dist(x_v, x_u) = \infty$, then there cannot be a negative cycle that only uses the positive template, independently of the value we give to its RHS variable. Hence the first inequality in the clause of Eq. 2 can be dropped.

This reasoning is applied to all vertices, namely, to all $u, v, \tilde{u}, \tilde{v}$ with $u \neq v$, $\tilde{u} \neq \tilde{v}$, and $u, v, \tilde{u}, \tilde{v} \in \{0, 1, \ldots, n\}$, adding in each case the respective clause.

*Satisfiability of Difference Systems.* The opposite problem to the previous one, that is, to enforce that a difference system is *satisfiable*, also arises in the constraint-based method. This is the case when, for example, one performs several rounds of invariant generation as described above, and requires that the

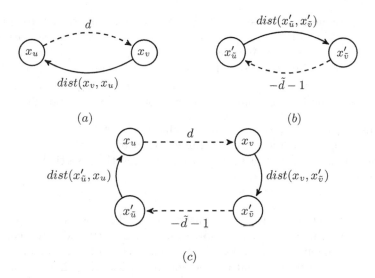

Fig. 4. The only three ways of creating a negative cycle.

newly generated invariants are not redundant with respect to the already computed ones: then there must exist a witness that certifies the non-redundancy.

Hence, let $\mathcal{S}$ be a difference system over program variables $\mathcal{X}$ such that

$$\mathcal{S} \wedge \neg(c_0\, x_0 + \ldots + c_n\, x_n \leq d) \equiv \mathcal{S} \wedge -c_0\, x_0 - \ldots - c_n\, x_n \leq -d - 1$$

must be satisfiable[5]. By Theorem 2, this amounts to proving that no negative cycle exists in the corresponding constraint graph. Again, we will start by constructing $\mathcal{G}$, the constraint graph induced by $\mathcal{S}$, and applying Floyd-Warshall.

If a negative cycle is already detected, the satisfiability requirement cannot be met. Otherwise $\mathcal{S}$ has no negative cycles, and the only possibility to achieve one is to go through the edge induced by the template. If $c_u = 1$ and $c_v = -1$, then the negation of the template is $x_v - x_u \leq -d - 1$, which induces an edge from $x_v$ to $x_u$ with weight $-d - 1$. This edge is part of a negative cycle iff $dist(x_u, x_v) - d - 1 < 0$. Since we want to avoid negative cycles, we should enforce that $d \leq dist(x_u, x_v) - 1$. Hence, we should add the clause:

$$c_u^+ \wedge c_v^- \implies d \leq dist(x_u, x_v) - 1\,.$$

Note that if $dist(x_u, x_v) = \infty$ the clause is trivially satisfied and can be dropped.

## 3.2 Experiments

To experimentally evaluate our techniques, we executed an implementation of the encoding presented in Sect. 3.1 on a benchmark suite obtained as follows[6]:

---

[5] This yields a non-linear formula if templates and $\mathcal{S}$ include general linear inequalities.

[6] Executables, benchmarks and detailed tables with the results of all the experiments in this paper can be found at www.cs.upc.edu/~erodri/sat16.tgz.

we first ran our verification system VeryMax [20] on numerical (possibly non-difference) programs from [30], checking whether all array accesses are within bounds. For each such check, VeryMax needs to process several safety subqueries, which consist of a small program with an assertion to be proved. Among them, we chose those where the program and the assertion can be expressed in difference logic. For these queries, VeryMax requires one of the next five possible outputs:

   I. An invariant at each location proving the assertion
  II. An invariant at each location disabling a transition
 III. A conditional invariant at each location proving the assertion
  IV. An invariant at each location
   V. None of the previous ones

Solving one such query using the constraint-based method generates an SMT formula with multiple **Initiation**, **Consecution**, **Safety** and other conditions (e.g. no redundant invariants are generated, conditional invariants are compatible with initial transitions) that can be encoded via Farkas' Lemma or via our novel difference logic encoding presented in the previous section. By making some of these conditions soft with the use of appropriate weights as in [31], we can order the five possible outputs from most desirable (I) to least desirable (V). For example, the optimal solution gives output (III) only if no solution exists that gives results (II) or (I).

The resulting Max-SMT formula can be processed with an off-the-shelf Max-SMT solver, such as Opti-Mathsat [32], Z3Opt [33] or Barcelogic [34]. Unfortunately, we had to discard Opti-Mathsat because it cannot deal with non-linearities. Between the remaining two, it was Barcelogic the one that showed a better performance, probably due to its novel method to deal with non-linearities [11]. Regarding the optimization part, Barcelogic implements a very simple branch-and-bound approach as explained in [35]. Due to its better performance, in what follows only experiments with Barcelogic will be reported.

Experiments were performed on an Intel i5 2.8 GHz CPU with 8 Gb of memory. For each of the 3270 generated queries and each encoding, we consider the best solution obtained within a time limit of 5 s[7]. In Table 1 we can see the output and the running time of four different encodings: **Farkas** (the standard encoding based on Farkas' Lemma), **FarkasDL** (the previous one additionally restricting the templates to be difference logic), **FarkasDL-$\lambda$** (the previous one additionally imposing that the $\lambda$ multipliers are 0 or 1), and **Diff Logic** (our novel encoding introduced in the previous section).

The experiments confirm our intuition that our specialized difference logic encoding outperforms Farkas both in runtime and in quality of solutions. Even if we try to improve Farkas with additional constraints that limit the search space, as in **FarkasDL** and **FarkasDL-$\lambda$**, the differences are still dramatic. We want to remark that in no query Farkas gave a better-quality result than **Diff Logic**.

More detailed results can be seen in Fig. 5, where in the scatter plots we display the timings (in seconds, logarithmic scale) over queries whose optimal

---

[7] This is the time limit used in VeryMax for this type of queries in previous works [20].

**Table 1.** Results on the 3270 generated queries with a time limit of 5 s.

| Method | (I)<br>Inv. prove | (II)<br>Disable tr | (III)<br>Cond. inv. prove | (IV)<br>Invariant | (V)<br>Nothing | Time |
|---|---|---|---|---|---|---|
| **Farkas** | 215 | 427 | 330 | 1024 | 1274 | 4 h 11 m 47 s |
| **FarkasDL** | 215 | 526 | 322 | 1042 | 1165 | 3 h 8 m 22 s |
| **FarkasDL-λ** | 217 | 594 | 324 | 1042 | 1039 | 3 h 1 m 52 s |
| **Diff Logic** | 786 | 1044 | 328 | 1112 | 0 | 56 m 20 s |

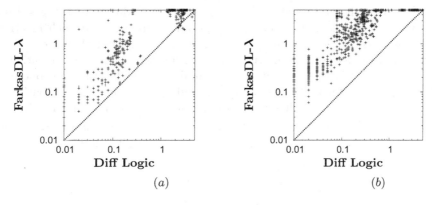

**Fig. 5.** Comparison of **Diff Logic** and **FarkasDL-λ** runtimes over queries whose optimal solution gives invariants proving the assertion $(a)$ or disabling a transition $(b)$.

solution finds invariants proving the assertion $(a)$ or disabling a transition $(b)$. One can see that even the best Farkas-based encoding is systematically slower than **Diff Logic**. We can also observe that in lots of queries Farkas times out, which means that the Max-SMT solver could not prove the solution to be optimal. One could think this is because proving optimality is equivalent to proving unsatisfiability, something at which Barcelogic non-linear techniques are particularly bad. However, a careful inspection of the results reveals the situation is worse, as in more than 80 % of the queries the found solution was not optimal.

## 4    Finding Invariant Subsets

Another important problem that we need to solve inside VeryMax is the *Invariant Subset Selection Problem*. Formally, we are given a program, an assertion $(\ell_{ass}, \varphi)$ and, for each location $\ell \in \mathcal{L}$, a set $Cand(\ell)$ of $m_\ell$ candidate invariants

$$c_1^{\ell,1}x_1 + \cdots + c_n^{\ell,1}x_n \leq d^{\ell,1}$$
$$\cdots$$
$$c_1^{\ell,m_\ell}x_1 + \cdots + c_n^{\ell,m_\ell}x_n \leq d^{\ell,m_\ell}$$

where the $c_i^{\ell,j}$ are fixed integer numbers and the $d^{\ell,j}$ are integer variables. The goal is to select, if it exists, a subset of $Cand(\ell)$ for each $\ell \in \mathcal{L}$, and find an assignment to the $d^{\ell,j}$'s such that (i) the chosen subsets are invariant and (ii) the invariants chosen at $\ell_{ass}$ imply $\varphi$.

As in Sects. 2 and 3 we will show that, in the general case, if we use Farkas' Lemma we obtain a non-linear formula, whereas non-linearities can be avoided when the program, the assertion and the candidate invariants are difference logic. In this case, the resulting formula belongs to linear arithmetic.

## 4.1  General Case

We can imagine the process of finding a solution as working in two stages. First of all, we have to select a subset of the candidate invariants at each location, together with their corresponding right-hand sides $d^{\ell,j}$. After that, we need to ensure that **Initiation**, **Consecution** and **Assertion** conditions are satisfied. To prove these conditions, we only need to find the right Farkas' multipliers.

More precisely, for each location $\ell \in \mathcal{L}$ and $1 \le j \le m_\ell$, let us consider a Boolean variable $chosen_{\ell,j}$ that indicates whether the $j$-th invariant in $Cand(\ell)$ is chosen. Additionally, to each coefficient $c_i^{\ell,j}$ we will associate a fresh integer variable $\widehat{c}_i^{\ell,j}$, and to each $d^{\ell,j}$ a fresh integer variable $\widehat{d}^{\ell,j}$. The following formulas

$$chosen_{\ell,j} \implies \bigwedge_{i=1}^{n} \widehat{c}_i^{\ell,j} = c_i^{\ell,j} \ \wedge \ \widehat{d}^{\ell,j} = d^{\ell,j} \qquad (3)$$

$$\neg chosen_{\ell,j} \implies \bigwedge_{i=1}^{n} \widehat{c}_i^{\ell,j} = 0 \ \wedge \ \widehat{d}^{\ell,j} = 0 \qquad (4)$$

constraint the shape of the invariants depending on whether they are chosen or not. In the following, we will consider that $\widehat{Cand}(\ell)$ consists of all elements of $Cand(\ell)$ where all $c$'s and $d$'s have been replaced by their respective $\widehat{c}$'s and $\widehat{d}$'s.

Let us explain how a **Consecution** condition will be enforced (for **Initiation** and **Assertion** an analogous idea applies). Let $(\ell_S, \tau, \ell_T)$ be the transition to which consecution refers. We want to enforce that $\widehat{Cand}(\ell_S) \wedge \tau \models \widehat{Cand}(\ell_T)'$, which amounts to checking, for each $\widehat{inv}' \in \widehat{Cand}(\ell_T)'$, that $\widehat{Cand}(\ell_S) \wedge \tau \models \widehat{inv}'$, or equivalently, that $\widehat{Cand}(\ell_S) \wedge \tau \wedge \neg \widehat{inv}'$ is unsatisfiable. The latter can be easily encoded into a non-linear formula by using Farkas' Lemma.

## 4.2  Difference Logic Case

Let us now assume that all candidate invariants, the formula in the assertion and the input program are expressed in difference logic. The idea of the encoding is similar. However, in Sect. 4.1 new inequalities were globally introduced standing for the original inequalities or the trivial inequality $0 \le 0$, depending on whether they had been chosen or not. Instead, here we exploit the fact that in Farkas' proofs of unsatisfiability of difference sets, multipliers are 0 or 1: for each unsatisfiability proof that must hold, new inequalities are locally introduced, standing for the *product* of the Farkas' multiplier with the original inequality.

As an example, let us explain how to encode a **Consecution** condition referring to a transition $(\ell_S, \tau, \ell_T)$. The $chosen_{\ell,j}$ variables will be as before, common to the overall encoding. However, for each $inv \in Cand(\ell_T)$, in order to enforce that $Cand(\ell_S) \wedge \tau \models inv'$, we will now introduce fresh $\widehat{c}$'s and $\widehat{d}$'s and add, for $1 \leq j \leq m_{\ell_S}$, the previous formula (4) and:

$$(\bigwedge_{i=1}^{n} \widehat{c}_i^{\,\ell_S,j} = 0 \ \wedge \ \widehat{d}^{\,\ell_S,j} = 0) \quad \vee \quad (\bigwedge_{i=1}^{n} \widehat{c}_i^{\,\ell_S,j} = c_i^{\ell_S,j} \ \wedge \ \widehat{d}^{\,\ell_S,j} = d^{\ell_S,j})$$

The intuition is that $\widehat{c}_1^{\,\ell_S,j} x_1 + \cdots + \widehat{c}_n^{\,\ell_S,j} x_n \leq \widehat{d}^{\,\ell_S,j}$ is the inequality resulting from multiplying $c_1^{\ell_S,j} x_1 + \cdots + c_n^{\ell_S,j} x_n \leq d^{\ell_S,j}$ by the corresponding multiplier in Farkas' proof of unsatisfiability of $Cand(\ell_S) \wedge \tau \wedge \neg inv'$. Similarly, let us assume that $inv$ is $c_1 x_1 + \cdots + c_n x_n \leq d$, with *chosen* being the variable that indicates whether we pick it or not. Then we will add the formula

$$(\bigwedge_{i=1}^{n} c_i^{\star} = 0 \ \wedge \ d^{\star} = 0) \quad \vee \quad (\bigwedge_{i=1}^{n} c_i^{\star} = -c_i \ \wedge \ d^{\star} = -1 - d),$$

which intuitively means that $c_1^{\star} x_1 + \cdots + c_n^{\star} x_n \leq d$ is the inequality resulting from multiplying $\neg(c_1 x_1 + \cdots + c_n x_n \leq d) \equiv -c_1 x_1 - \cdots - c_n x_n \leq -1 - d$ by the corresponding multiplier in the proof of unsatisfiability of $Cand(\ell_S) \wedge \tau \wedge \neg inv'$.

The encoding finishes by: (i) applying Farkas' Lemma to enforce unsatisfiability of $\widehat{Cand(\ell_S)} \wedge \tau \wedge c_1^{\star} x_1' + \cdots + c_n^{\star} x_n' \leq d'$ as in the general case, but now assuming that multipliers are 1, which gives a linear formula $F$; and (ii) adding the implication *chosen* $\Rightarrow F$ to the encoding. Detailed experiments comparing the general and the particular encoding for difference logic give similar results to Sect. 3.2, and we omit them here due to lack of space.

A final remark is that the previous encoding can be applied to solve the problem in Sect. 3 by exhaustively considering in $Cand(\ell)$ all differences of variables. This allows finding simultaneously more than one invariant inequality per location, in particular coinductive invariants. So the price to pay for a complete method is moving from a difference logic to a linear arithmetic formula.

## 5   Experiments

The goal here is to assess to which extent a constraint-based verifier like VeryMax can be *globally* improved by incorporating the novel encodings introduced in Sects. 3.1 and 4.2 (for handling safety subqueries and invariant subset selection problems, respectively). Note it is not uncommon that huge enhancements on the runtime of a theorem prover (SAT or SMT solver or first-order theorem prover) get diluted into insignificant improvements on the verifier that uses it.

We compared the original VeryMax safety prover, which uses Farkas as the encoding methodology to find linear invariants, with a new version VeryMaxDL where the problems described in Sects. 3 and 4 are solved using the novel encodings. A time limit of 900 s was given to each problem. Table 2 summarizes the experiments, where for each system we report the number of problems found to

**Table 2.** Results comparing VeryMax and VeryMaxDL.

| System | Yes | No | Only-yes | Time |
|---|---|---|---|---|
| VeryMax | 524 | 312 | 27 | 11 h 58 m 59 s |
| VeryMaxDL | 516 | 320 | 19 | 4 h 12 m 38 s |

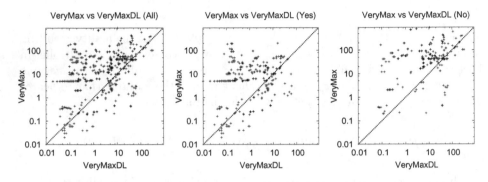

**Fig. 6.** Scatter plots comparing VeryMax and VeryMaxDL.

be safe (Yes), not found to be safe[8] (No), proved safe only by this version of the system (Only-yes) and the total runtime (including timeouts).

The results are extremely positive since the runtime is reduced to one third, and the loss of verification power due to generating only difference logic invariants, as opposed to linear invariants, is very limited. We analyzed all problems that VeryMax could prove safe whereas VeryMaxDL could not and they all need linear invariants outside difference logic, which means that they cannot be proved using the techniques on which VeryMaxDL is based.

Figure 6 contains scatter plots comparing VeryMax and VeryMaxDL on all problems, problems proved safe by VeryMaxDL, and problems not found to be safe by VeryMaxDL. At first glance, although the difference logic version is faster, we observe that the plots are not as clean as the ones of Sect. 3.2. This is not surprising: if the subproblems solved via Farkas or difference logic give different results (e.g. they disable different transitions), the overall behavior of the verification system changes and this has an impact on the overall runtime. The second observation is that VeryMaxDL is faster, independently of whether the problem can be found to be safe or not. This opens the way to run both versions in parallel, or even first run VeryMaxDL and if the program cannot be proved safe, run VeryMax in a second attempt.

## 6    Conclusions and Future Work

It is well acknowledged that the current bottleneck in the effectiveness of the constraint-based method compared to other approaches for verification is the

---

[8] Note that this does not mean that they are unsafe.

technology for solving non-linear constraints. In this paper we have introduced novel encodings that, if we restrict ourselves to programs and invariants in difference logic, allow one to replace non-linear solvers by cheaper ones. Experiments show that this yields a remarkable gain in terms of runtime at the expense of restricting the class of programs under consideration and a certain but acceptable loss of verification power.

As future work, we plan to extend the use of similar encodings in our verification system VeryMax. E.g., finding simple ranking functions in (non)-termination problems is a particularly interesting area where related ideas could be applied.

# References

1. Colón, M.A., Sankaranarayanan, S., Sipma, H.B.: Linear invariant generation using non-linear constraint solving. In: Hunt Jr., W.A., Somenzi, F. (eds.) CAV 2003. LNCS, vol. 2725, pp. 420–432. Springer, Heidelberg (2003)
2. Sankaranarayanan, S., Sipma, H.B., Manna, Z.: Petri net analysis using invariant generation. In: Dershowitz, N. (ed.) Verification: Theory and Practice. LNCS, vol. 2772, pp. 682–701. Springer, Heidelberg (2004)
3. Sankaranarayanan, S., Sipma, H.B., Manna, Z.: Constructing invariants for hybrid systems. Formal Methods Syst. Des. **32**(1), 25–55 (2008)
4. Beyer, D., Henzinger, T.A., Majumdar, R., Rybalchenko, A.: Invariant synthesis for combined theories. In: Cook, B., Podelski, A. (eds.) VMCAI 2007. LNCS, vol. 4349, pp. 378–394. Springer, Heidelberg (2007)
5. Larraz, D., Rodríguez-Carbonell, E., Rubio, A.: SMT-based array invariant generation. In: Giacobazzi, R., Berdine, J., Mastroeni, I. (eds.) VMCAI 2013. LNCS, vol. 7737, pp. 169–188. Springer, Heidelberg (2013)
6. Podelski, A., Rybalchenko, A.: A complete method for the synthesis of linear ranking functions. In: Steffen, B., Levi, G. (eds.) VMCAI 2004. LNCS, vol. 2937, pp. 239–251. Springer, Heidelberg (2004)
7. Larraz, D., Oliveras, A., Rodríguez-Carbonell, E., Rubio, A.: Proving termination of imperative programs using Max-SMT. In: Proceeding FMCAD 2013, pp. 218–225. IEEE (2013)
8. Larraz, D., Nimkar, K., Oliveras, A., Rodríguez-Carbonell, E., Rubio, A.: Proving non-termination using Max-SMT. In: Biere, A., Bloem, R. (eds.) CAV 2014. LNCS, vol. 8559, pp. 779–796. Springer, Heidelberg (2014)
9. Borralleras, C., Lucas, S., Oliveras, A., Rodríguez-Carbonell, E., Rubio, A.: SAT modulo linear arithmetic for solving polynomial constraints. Journal of Automated Reasoning **48**(1), 107–131 (2012)
10. Jovanović, D., de Moura, L.: Solving non-linear arithmetic. In: Gramlich, B., Miller, D., Sattler, U. (eds.) IJCAR 2012. LNCS, vol. 7364, pp. 339–354. Springer, Heidelberg (2012)
11. Larraz, D., Oliveras, A., Rodríguez-Carbonell, E., Rubio, A.: Minimal-model-guided approaches to solving polynomial constraints and extensions. In: Sinz, C., Egly, U. (eds.) SAT 2014. LNCS, vol. 8561, pp. 333–350. Springer, Heidelberg (2014)
12. Cousot, P., Cousot, R.: Abstract interpretation : a unified lattice model for the static analysis of programs by construction or approximation of fixpoints. In: Proceeding POPL 1977, pp. 238–252. ACM Press (1977)

13. Cousot, P., Halbwachs, N.: Automatic discovery of linear restraints among variables of a program. In: Proceeding POPL 1978, pp. 84–96. ACM Press (1978)
14. Cousot, P., Cousot, R.: Static determination of dynamic properties of programs. In: Proceeding 2nd International Symposium on Programming, pp. 106–130 (1976)
15. Miné, A.: A new numerical abstract domain based on difference-bound matrices. In: Danvy, O., Filinski, A. (eds.) PADO 2001. LNCS, vol. 2053, pp. 155–172. Springer, Heidelberg (2001)
16. Miné, A.: The octagon abstract domain. High. Order Symbolic Comput. **19**(1), 31–100 (2006)
17. Menasche, M., Berthomieu, B.: Time petri nets for analyzing and verifying time dependent communication protocols. In: Protocol Specification, Testing, and Verification, pp. 161–172 (1983)
18. Dill, D.L.: Timing assumptions and verification of finite-state concurrent systems. Automatic Verification Methods for Finite State Systems. LNCS, vol. 407, pp. 197–212. Springer, Heidelberg (1989)
19. Yovine, S.: Model checking timed automata. In: Lectures on Embedded Systems, European Educational Forum, School on Embedded Systems. vol. 1494. LNCS, pp. 114–152. Springer, Heidelberg (1996)
20. Brockschmidt, M., Larraz, D., Oliveras, A., Rodríguez-Carbonell, E., Rubio, A.: Compositional safety verification with Max-SMT. In: Proceeding FMCAD 2015, pp. 33–40. IEEE (2015)
21. Kapur, D., Zhang, Z., Horbach, M., Zhao, H., Lu, Q., Nguyen, T.V.: Geometric quantifier elimination heuristics for automatically generating octagonal and max-plus invariants. In: Bonacina, M.P., Stickel, M.E. (eds.) Automated Reasoning and Mathematics. LNCS, vol. 7788, pp. 189–228. Springer, Heidelberg (2013)
22. Monniaux, D.: Automatic modular abstractions for linear constraints. In: Proceeding POPL 2009, pp. 140–151. ACM (2009)
23. Lahiri, S.K., Musuvathi, M.: An efficient decision procedure for UTVPI constraints. In: Gramlich, B. (ed.) FroCos 2005. LNCS (LNAI), vol. 3717, pp. 168–183. Springer, Heidelberg (2005)
24. Monniaux, D.: A quantifier elimination algorithm for linear real arithmetic. In: Cervesato, I., Veith, H., Voronkov, A. (eds.) LPAR 2008. LNCS (LNAI), vol. 5330, pp. 243–257. Springer, Heidelberg (2008)
25. Schrijver, A.: Theory of Linear and Integer Programming. Wiley, Amsterdam (1998)
26. Dechter, R., Meiri, I., Pearl, J.: Temporal constraint networks. Artif. Intell. **49**(1–3), 61–95 (1991)
27. Cormen, T.H., Stein, C., Rivest, R.L., Leiserson, C.E.: Introduction to Algorithms, 2nd edn. McGraw-Hill Higher Education, New York (2001)
28. Frisch, A.M., Peugniez, T.J., Doggett, A.J., Nightingale, P.W.: Solving non-boolean satisfiability problems with stochastic local search: a comparison of encodings. J. Autom. Reasoning **35**(1–3), 143–179 (2005)
29. Argelich, J., Manyà, F.: Solving over-constrained problems with SAT technology. In: Bacchus, F., Walsh, T. (eds.) SAT 2005. LNCS, vol. 3569, pp. 1–15. Springer, Heidelberg (2005)
30. Press, W.H., Teukolsky, S.A., Vetterling, W.T., Flannery, B.P.: Numerical Recipes: The Art of Scientific Computing. Cambridge Univ. Press, NewYork (1989)
31. Marques-Silva, J., Argelich, J., Graça, A., Lynce, I.: Boolean lexicographic optimization: algorithms & applications. Ann. Math. Artif. Intell. **62**(3–4), 317–343 (2011)

32. Sebastiani, R., Trentin, P.: OptiMathSAT: a tool for optimization modulo theories. In: Kroening, D., Păsăreanu, C.S. (eds.) CAV 2015. LNCS, vol. 9206, pp. 447–454. Springer, Heidelberg (2015)
33. Bjørner, N., Phan, A.-D., Fleckenstein, L.: $\nu z$ - an optimizing SMT solver. In: Baier, C., Tinelli, C. (eds.) TACAS 2015. LNCS, vol. 9035, pp. 194–199. Springer, Heidelberg (2015)
34. Bofill, M., Nieuwenhuis, R., Oliveras, A., Rodríguez-Carbonell, E., Rubio, A.: The Barcelogic SMT solver. In: Gupta, A., Malik, S. (eds.) CAV 2008. LNCS, vol. 5123, pp. 294–298. Springer, Heidelberg (2008)
35. Nieuwenhuis, R., Oliveras, A.: On SAT modulo theories and optimization problems. In: Biere, A., Gomes, C.P. (eds.) SAT 2006. LNCS, vol. 4121, pp. 156–169. Springer, Heidelberg (2006)

# Synthesis of Domain Specific CNF Encoders for Bit-Vector Solvers

Jeevana Priya Inala$^{(\boxtimes)}$, Rohit Singh, and Armando Solar-Lezama

Massachusetts Institute of Technology, Cambridge, MA, USA
{jinala,rohitsingh,asolar}@csail.mit.edu

The theory of bit-vectors in SMT solvers is very important for many applications due to its ability to faithfully model the behavior of machine instructions. A crucial step in solving bit-vector formulas is the translation from high-level bit-vector terms down to low-level boolean formulas that can be efficiently mapped to CNF clauses and fed into a SAT solver. In this paper, we demonstrate how a combination of program synthesis and machine learning technology can be used to automatically generate code to perform this translation in a way that is tailored to particular problem domains. Using this technique, the paper shows that we can improve upon the basic encoding strategy used by CVC4 (a state of the art SMT solver) and automatically generate variants of the solver tailored to different domains of problems represented in the bit-vector benchmark suite from SMT-COMP 2015.

## 1 Introduction

SMT solvers are at the heart of a number of software engineering tools ranging from automatic test generators [42,44,49] to deterministic replay tools [16], just to name two applications among many others [11,25,53]. Of particular importance to these applications is the theory of bit-vectors, which is widely used [17,26,43,47] because of its ability to faithfully represent the full range of machine arithmetic.

One of the most important steps in a bit-vector solver is the mapping of high-level bit-vector terms down to low-level CNF clauses that can be fed to a SAT solver—a process often referred to as *bit-blasting*. One approach to bit-blasting is to use the known efficient encodings for simpler boolean terms (such as AND or XOR) and compose them to generate CNF clauses for complex terms [50]. This approach can have a huge impact on the performance of the solver [39,40], but generally, it relies on having optimal encodings for the simpler terms, and even then it does not guarantee any kind of optimality of the overall encoding.

In this paper, we propose OPTCNF, a new approach to automatically generate the code that converts high-level bit-vector terms into low-level CNF clauses. In addition to the obvious benefits of having the code automatically generated instead of having to write it by hand, OPTCNF has three novel aspects that together significantly improve the quality of the overall encoding: (a) OPTCNF uses synthesis technology to automatically generate efficient encodings from high-level formulas to CNF (b) OPTCNF relies on auto-tuning

© Springer International Publishing Switzerland 2016
N. Creignou and D. Le Berre (Eds.): SAT 2016, LNCS 9710, pp. 302–320, 2016.
DOI: 10.1007/978-3-319-40970-2_19

to choose encodings that produce the best results for problems from a given domain. (c) OPTCNF identifies commonly occurring clusters of terms in a given domain and focuses on finding optimal encodings for such clusters.

The synthesis of encodings balances optimality among three criteria: number of clauses, number of variables and propagation completeness. The propagation completeness requirement has been proposed as an important criterion in order for the encoded constraints to solve efficiently in the SAT solver [10]. Modern SAT solvers rely heavily on unit propagation to infer the values of variables without having to search for them. Propagation completeness means that if a given partial assignment implies that another unassigned variable should have a particular value, then the solver should be able to discover this value through unit propagation alone. Prior work has demonstrated the synthesis of propagation complete encodings for terms involving a small number of variables [12]. OPTCNF, however, is more flexible and is able to produce propagation complete encodings even for relatively large bit-vector terms by taking advantage of high-level hypothesis about the structure of the encoding (See Sect. 2).

In practice, however, propagation completeness does not *always* improve the performance of an encoding. For certain classes of problems, for example, the additional unit propagations caused by a propagation complete encoding can actually slow the solver down. Similarly, there is often a trade-off between the number of auxiliary variables and the number of clauses used by an encoding; for some problems having more variables but fewer clauses can be better, but for other problems, having fewer variables at the expense of more clauses can be better. In order to cope with this variability, OPTCNF uses auto-tuning to make choices about which encodings are best for problems from a particular domain. Prior work has demonstrated the value of tuning solver parameters in order to achieve optimal performance for problems from particular domains [32], but ours is the first work we know of where auto-tuning is used to make high-level decisions about how to encode particular terms (see Sect. 5).

Finally, OPTCNF is able to better leverage its ability to synthesize optimal encodings by focusing on larger clusters of terms, as opposed to focusing on individual bit-vector operations independently. Given a corpus of sample problems from a domain, OPTCNF is able to identify common recurring patterns in the formulas from those problems and then generate specialized encodings for those patterns.

Figure 1 shows how these ideas come together as OPTCNF. The input to OPTCNF is a collection of formulas represented as DAGs (Directed Acyclic Graphs) extracted from a set of benchmarks from a given problem domain. OPTCNF samples these DAGs to extract representative clusters of terms—what the figure refers to as patterns. OPTCNF then leverages SKETCH synthesis system [46] to synthesize "optimal" encodings for those patterns and generates C++ code for the encodings that can be linked with a modified version of CVC4 solver [6]. The auto-generated code contains a set of switches to turn different encodings on or off. Finally, the auto-tuner searches for the optimal configuration of those switches in order to produce the best performing domain-specific version of CVC4.

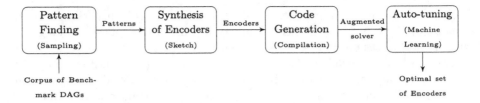

**Fig. 1.** OPTCNF: System overview

Our evaluation shows that the resulting domain-specific encodings are able to significantly improve the performance of CVC4 when run in eager bit-blasting mode. Using OPTCNF, we generated a separate solver for 7 different domains represented in the quantifier-free bit-vector benchmarks from the SMT-COMP 15 benchmark suite [7]; using these specialized solvers on their respective domains, we were able to solve 83 problems from the test set (see Sect. 6) that CVC4 could not solve.

## 2  Synthesis of Encoders

Previous work [12] has attacked the problem of generating optimal propagation complete encodings for a given term by starting with an initial encoding and then exhaustively checking for violations of propagation completeness and incrementally adding more clauses to fix these violations. The resulting propagation complete encoding is then minimized to produce an equivalent but smaller encoding. Our approach to generating encodings is quite different because it relies on program synthesis technology, allowing us to symbolically search for an encoding based on a formal specification. An important advantage of our approach is flexibility. In particular, it allows us to generate *encoders* that generate encodings at solver run-time from terms that have parameters that will only be known at run-time (for example, the bit-width for a bit-vector operation).

OPTCNF frames the task of generating these encoders as a Syntax Guided Synthesis problem (SyGuS) [2]. A SyGuS problem is a combination of a template or grammar that represents the space of the candidate solutions and a specification that constrains the solution. The goal of a SyGuS solver is to find a candidate in the template that satisfies the specification. The two components, template and specification, are very crucial in determining the scalability of the problem. Here, we first describe the templates that represent the space of CNF encoders for booleans and bit-vector terms. Then, we formalize the correctness and the optimality specification that constraints the template. Finally, we describe an efficient but equivalent specification that makes the SyGus synthesis problem more scalable.

### 2.1  CNF Encoders and Templates

The encoders generated by OPTCNF work in two passes. Given a formula to be encoded into SAT, OPTCNF first identifies terms for which it has learned to generate CNF constraints and replaces them by special placeholder operators $N_i$. Then, the pass that would normally have generated low-level constraints from the bit-vector terms is extended to recognize these placeholder operators and generate the specialized constraints for them.

The pass that identifies the known terms, and the scaffolding that iterates through the different operators in a DAG representation of the formula and identifies the placeholder nodes are all produced using relatively straightforward code-generation techniques. The synthesis problem focuses on the code that executes when one of these placeholder nodes is found. This is the encoder code that generates the CNF encoding for a previously identified term $T$.

$t \equiv and(x, or(y, z))$     $t \equiv bvAND_N(x, bvOR_N(y, z))$     $t \equiv bvEQ_N(x, y)$

```
clause({x,~t})
clause({~x,~y,t})
clause({~x,~z,t})
clause({y,z,~t})
```

```
for i from 1 to N:
    clause({x[i],~t[i]})
    clause({~x[i],~y[i],t[i]})
    clause({~x[i],~z[i],t[i]})
    clause({y[i],z[i],~t[i]})
```

```
t1 = true
for i from 1 to N:
    t2 = i == N ? t : newVar
    clause({x[i],y[i],~t1,t2})
    clause({x[i],~y[i],~t2})
    clause({~x[i],y[i],~t2})
    clause({~x[i],~y[i],~t1,t2})
    clause({t1,~t2})
    t1 = t2
```

**Fig. 2.** Encoders for three different kinds of terms

The term $T$ for which OPTCNF is generating an encoding is known at synthesis time, so OPTCNF can choose a template or a set of templates for this code depending on the properties of $T$. Figure 2 illustrates the three different kind of terms and the encodings that represent the terms. If $T$ is not parametric—for example if it is just a collection of boolean operators—then the encoder just needs to generate a fixed set of clauses corresponding to the constraint represented by $T$, and the template will reflect that. On the other hand, many terms will be parameterized by bit-widths, so the encoder will have to produce clauses in one or more loops.

For bit-vector terms, which are parametric on the bit-width of their different operators, we differentiate between two different kinds – bit-parallel and non bit-parallel. Bit-parallel terms are those that are composed entirely of operations, such as bitwise AND, OR or XOR, where there is no dependency from one column of the bit-vector to another. For these kinds of terms, generating the encoding for a single column and then enumerating them over all columns will still preserve optimality. Hence, it is sufficient to just synthesize the encoding for the boolean term that represents operations in a single column. This is, however, not the case for all bit-vector terms.

Terms involving bitwise PLUS, for example, cannot be dealt in the same way because there are dependencies that flow from one column to another. These operations can still be represented as a loop of encodings, but there will be auxiliary variables that are threaded from one iteration of the loop to another. Figure 3 shows one such template for a bit-vector formula involving two bit-vector inputs of size N (taken as a parameter) and outputs another bit-vector of size N. For each column in the bit-vectors, the template calls encode_column which is another template for explicit encodings, but this template can be instantiated with variables specific to loop iteration. This template has one auxiliary variable per column. Every column has an incoming auxiliary variable (a constant for the first column) which carries information from the previous columns and an outgoing auxiliary variable that carries information forward. This same template represents multiple formulas depending on how the encode_column template is instantiated. For example, this same template is used to generate encodings for both bitwise PLUS and bitwise MINUS operations.

```
Lit [N] encode(Lit [N] mval, Lit [N] fval ) {
    Lit [N] out = newVar(N)     /* creates an array of out  literals  */
    Lit [N] aux = newVar(N)     /* creates an array of auxiliary  literals  */
    /* Specialize the  first  column */
    encode_column(mval[1],  fval [1],  out [1],  const?, aux[1])
    for i in 2 to N:
        encode_column(mval[i], fval [ i ],  out[i]  aux[i−1], aux[ i ])
    return out
}
```

**Fig. 3.** Template for a bitwise operation on two bit-vectors (with one auxiliary variable per column)

The templates in OPTCNF are all written in the SKETCH language, which allows us to leverage the SKETCH synthesis engine for the synthesis problem. A template in SKETCH is a piece of code with integer and boolean holes to represent the set of candidate solutions that the synthesizer should consider. The standard template for an encoding is a list of clauses with holes representing the number of clauses, and the length and the literals present in each clause. We significantly reduce the size of the search space by enforcing an order among the literals in each clause and among clauses themselves and thus, eliminating symmetries. This canonical representation captures any general CNF encoding, but it does not impose any structure on the clauses. We found that this model is scalable enough for boolean formulas that expand into a small number of CNF clauses (about 20 to 30). But, in order to deal with bigger formulas like bit-vector operations, we need to represent the search space using loops to capture the structure.

OPTCNF has a library of templates for different kinds of input types, output types and number of auxiliary variables per column. When running SKETCH on a term, OPTCNF runs different instances of SKETCH with a different template and chooses the one that provides the best encoding (based on heuristics like number of clauses and number of auxiliary variables). Due to the scalability limits of SKETCH, OPTCNF can currently only synthesize encodings for non bit-parallel terms that have at most two input bit-vectors, at most two auxiliary variables per column and no nested loops in the template.

## 2.2   Problem Formulation

In addition to the template, the other important component of a SyGus problem is the specification. Unlike the templates, which are very different for parameterized and non-parameterized terms, the specifications for both are actually very similar; the only difference is that for parameterized terms, the parameters must be threaded through to all the relevant predicates. Therefore, the rest of the section will omit these bit-width parameters in the interest of clarity.

A term $T(in)$ can be represented by a predicate $P(in, out)$ defined as $P(in, out) \Leftrightarrow out = T(in)$. For notational convenience, we will just write $P(x)$, where $x$ is understood to be a vector containing both the input and the output variables. The goal is to generate an alternative representation of the predicate in terms of CNF clauses $C(x)$.

**Definition 1 (Correctness Specification).** *A set of CNF clauses "represents" a boolean predicate $P$ iff $P(x) \Leftrightarrow C(x)$.*

In addition to the correctness specification, however, we want to ensure propagation completeness which needs to be defined in terms of the behavior of the encoding under *partial* assignments. A partial assignment $\sigma$ maps every variable to one of $\{true, false, \top\}$ where $\top$ indicates that the value has not been assigned by the solver and could be *true* or *false*. A partial assignment can be understood as the set of all complete assignments that are consistent with the partial assignment. Therefore, it is standard to define a partial order among partial assignments as:

$$\sigma \sqsupseteq \sigma' \iff \forall i.\ \sigma(x_i) \neq \top \Rightarrow \sigma'(x_i) = \sigma(x_i)$$

We generalize the predicate to be a function from partial assignments to the set $\{true, false, \top\}$, and define $P(\sigma) = \top$ for any partial assignment where some variable $x_i$ is set to $\top$.

**Definition 2.** *We define the following predicates on partial assignments:*

$$
\begin{aligned}
complete(\sigma) &\equiv \forall i.\ \sigma(x_i) \neq \top \\
satisfiable(\sigma, P) &\equiv \exists \sigma' \sqsubseteq \sigma.\ P(\sigma') = true \\
unsatisfiable(\sigma, P) &\equiv \forall \sigma' \sqsubseteq \sigma.\ P(\sigma') \neq true \\
forces(\sigma, P, x_i, b) &\equiv (\sigma' = extend(\sigma, x_i, \neg b)) \Rightarrow unsatisfiable(\sigma', P) \\
maypropagate(\sigma, P) &\equiv \exists i, b.\ forces(\sigma, P, x_i, b)
\end{aligned}
$$

*where $extend(\sigma, x_i, b)$ is defined as extending an assignment with $\sigma(x_i) = \top$ to one where variable $x_i$ has value $b$. The predicate $maypropagate(\sigma, P)$ indicates that the partial assignment $\sigma$ forces the value of some currently unassigned variable.*

**Lemma 1.** *The forces() predicate has the following property.*

$$
\begin{aligned}
forces(\sigma, P, \hat{x}_i, \hat{b}) &\land \sigma' = extend(\sigma, \hat{x}_i, \hat{b}) \\
&\Rightarrow \forall_{(x_i, b) \neq (\hat{x}_i, \hat{b})}\ forces(\sigma, P, x_i, b) \Rightarrow forces(\sigma', P, x_i, b)
\end{aligned}
$$

*This means that if $P$ and a partial assignment $\sigma$ force $\hat{x}_i$ to take a particular value $\hat{b}$, then any other variable that was also forced by $\sigma$ and $P$ will also be forced after extending the assignment with $\sigma(\hat{x}_i) = \hat{b}$.*

**Lemma 2.** *Another important property of forces() is the following.*

$$satisfiable(\sigma, P) \;\wedge\; forces(\sigma, P, \hat{x}_i, \hat{b}) \;\wedge\; \sigma' = extend(\sigma, \hat{x}_i, \hat{b})$$
$$\Rightarrow satisfiable(\sigma', P)$$

*This means that if $P$ and a partial assignment $\sigma$ force $\hat{x}_i$ to take a particular value, then after extending the assignment with $\sigma(\hat{x}_i) = \hat{b}$, the new assignment is still satisfiable.*

A clause $c$ can be applied to a partial assignment as well, resulting in a value $c(\sigma) \in \{true, false, \mu, \top\}$. A clause is unit ($\mu$) if one of the literals in the clause has an unknown value and all others are $false$. A CNF encoding is a collection of clauses $C$. $C(\sigma)$ can either be $true$ if $c(\sigma) = true$ for all the clauses, $false$ if $c(\sigma) = false$ for at least one of the clauses, $\mu$ if $\sigma$ makes at least one clause unit (and $\sigma$ does not falsify any others), or $\top$ if none of the above. Thus, the result of applying $C$ to a partial assignment helps identify the case when at least one of the clauses is a unit clause, and it is, therefore, possible to propagate further assignments. This is useful in describing unit propagation.

**Definition 3 (UP).** *The function UP captures the unit propagation in SAT solvers. We say that $C$ propagates $\sigma$ to $UP(C, \sigma)$ under unit propagation according to the following rules:*

1. *if $C(\sigma) \neq \mu$, then $UP(C, \sigma) = \sigma$.*
2. *else, $C(\sigma)$ has a unit clause. If the unit clause forces $\sigma(x_i) = b$, then $UP(C, \sigma) = UP(C, \sigma')$ where $\sigma' = extend(\sigma, x_i, b)$.*

The definitions above give rise to an important lemma.

**Lemma 3.** *A set of CNF clauses $C$ "represents" a boolean predicate $P$ iff it satisfies the following two conditions:*

$$\begin{aligned} 1.\; & satisfiable(\sigma, P) \Rightarrow C(UP(C,\ \sigma)) \neq false \\ 2.\; & unsatisfiable(\sigma, P) \Rightarrow C(UP(C,\ \sigma)) \neq true \end{aligned} \qquad (2.1)$$

*i.e. if an assignment can be extended to a satisfiable assignment for $P$, then unit propagation should not lead to a contradiction. And similarly, if an assignment (possibly partial) already contradicts $P$, then unit propagation should not lead to a satisfiable assignment for the CNF clauses.*

With the definitions above, we can now state the requirement for propagation completeness.

**Definition 4 (Propagation Completeness).** *C is a set of propagation complete CNF clauses representing P if C "represents" P and*

$$\forall \sigma. \ satisfiable(\sigma, P)$$
$$\Rightarrow \forall \ x_i, b_i. \ (forces(\sigma, P, x_i, b_i) \Rightarrow UP(C, \sigma) \sqsubseteq extend(\sigma, x_i, b_i)) \quad (2.2)$$

*In other words, if a partial assignment can be completed into a satisfying assignment, and if there are unassigned variables $x_i$ that if set to $\neg b_i$ would make the partial assignment unsatisfiable, then unit propagation must set all such $x_i$ to $b_i$.*

### 2.3 Synthesis-Friendly Propagation Completeness

The above definition captures the notion of propagation complete encodings, but it is unsuitable as a specification for synthesis because the recursive definition of *UP* essentially defines a small SAT solver, making it too complex for a state of the art synthesizer. Instead, OPTCNF relies on an equivalent but simpler specification that does not require implementing a SAT solver. The idea is that instead of thinking in terms of full unit propagation, we now verify propagation only one step at a time. Specifically, the claim is that the following three rules guarantee propagation completeness.

$$1. \forall \sigma. \ satisfiable(\sigma, P) \ \Rightarrow \ C(\sigma) \neq false$$
$$2. \forall \sigma. \ maypropagate(\sigma, P) \ \Rightarrow C(\sigma) = \mu \quad (2.3)$$
$$3. \forall \sigma. \ unsatisfiable(\sigma, P) \ \Rightarrow \ C(\sigma) = false \ \lor \ C(\sigma) = \mu$$

**Theorem 1.** *Formula* (2.3) $\Longleftrightarrow$ *Correctness $\land$ Formula* (2.2)

*Proof:* Formula (2.3) $\Rightarrow$ Correctness
This follows directly from Formula (2.3), because when $\sigma$ is complete, $satisfiable(\sigma, P)$ implies $P(\sigma) = true$ and similarly, $unsatisfiable(\sigma, P)$ implies $P(\sigma) = false$.

*Proof:* Formula (2.3) $\Rightarrow$ Formula (2.2)
This can be proved by induction on the number of times $\sigma$ can be extended before it fails $maypropagate(\sigma, P)$. For the base case, $\neg maypropagate(\sigma, P)$, (2.2) is vacuously satisfied because $forces()$ fails for all variables. For the inductive case, $maypropagate(\sigma, P)$, $C(\sigma) = \mu$ (by 2.3-2). Let $\sigma' = extend(\sigma, \hat{x}_i, \hat{b})$ which is obtained by propagating the unit clause in $C(\sigma)$. Note that $UP(C, \sigma) = UP(C, \sigma')$ by Definition 3. Applying Lemma 2 tells us that $satisfiable(\sigma', P)$, so applying the inductive hypothesis together with Lemma 1, we can prove the inductive case.

*Proof:* Correctness $\land$ Formula (2.2) $\Rightarrow$ Formula (2.3)
First, we use the fact that correctness is equivalent to Formula (2.1). If $satisfiable(\sigma, P)$, then $C(UP(C, \sigma)) \neq false$ and this implies $C(\sigma) \neq false$.
If $\sigma$ can be propagated, then $\exists x_i, b. \ forces(\sigma, P, x_i, b)$. And hence, $UP(C, \sigma) \neq \sigma$ and this implies $C(\sigma) = \mu$.
If $unsatisfiable(\sigma, P)$, then let $\sigma' \sqsupseteq \sigma$ be the maximal satisfying subset of $\sigma$ i.e. $\sigma'$ is satisfiable and $\forall \sigma' \sqsupseteq \sigma'' \sqsupseteq \sigma. \ \sigma''$ is unsatisfiable. Then, $C(\sigma') = \mu$ and since $\sigma'$ is maximal subset, $C(\sigma) = false \ \lor \ C(\sigma) = \mu$.

## 2.4   Introducing Auxiliary Variables

In some cases, the encoding $C$ will involve auxiliary variables $t_i$ in addition to the variables $x_i$, in such cases, we write $C((x,t))$. In that case, the correctness specification must be generalized to

$$\forall x. \quad P(x) \iff \exists t. \; C((x,t))$$

Similarly, the conditions in Formula (2.3) generalize to the conditions below.

1. $\forall \sigma. \; satisfiable(\sigma, P) \; \Rightarrow \; \exists \sigma_t. \; C((\sigma, \; \sigma_t)) \neq false$
2. $\forall \sigma, \sigma_t. \; maypropagate(\sigma, P) \; \wedge \; C((\sigma, \; \sigma_t)) \neq false \; \Rightarrow \; C((\sigma, \; \sigma_t)) = \mu$ (2.4)
3. $\forall \sigma, \sigma_t. \; unsatisfiable(\sigma, P) \; \Rightarrow \; C((\sigma, \; \sigma_t)) = false \; \vee \; C((\sigma, \; \sigma_t)) = \mu$

The proof for this has a similar structure to the previous proof. Basically, once we establish the first rule above, auxiliary variables can be treated just as the other variables in $P$. It should be noted that this specification is more complex than Formula (2.3) because of the existential quantifier in the R.H.S of rule 1. The CEGIS algorithm employed by solvers like SKETCH is designed to deal with the outer universal quantifiers, but cannot handle inner existential quantifiers. Hence, this existential quantifier should be translated into an explicit loop over all auxiliary assignments, which makes the synthesis problem hard. In practice, we found that this overhead is not significant when the number of auxiliaries used in the encodings is low.

## 2.5   Clause Minimization

Another important optimality criterion for the encodings is the clause minimization. If there are two propagation complete encodings having different number of clauses representing the same predicate, then the encoding with the lower number of clauses is preferred. OPTCNF relies on binary search to find an encoding with an optimal number of clauses. This requires solving a logarithmic number of synthesis problems to generate a single encoding, which has proven to be reasonably efficient in practice.

## 2.6   Guarantees of the Synthesized Solution

When the formula is a boolean term or a bit-parallel term, SKETCH performs full verification and hence, the output is guaranteed to be correct and propagation complete. When the input formula is a non bit-parallel bit-vector term, OPTCNF does bounded verification on the size of the bit-width parameters. The correctness specification is easier to verify than the propagation completeness requirement, so OPTCNF allows the user to separately specify the checking bounds for both specifications. In our experiments, we check correctness for all inputs up to 6-bits and propagation completeness for up to 3-bits. Beyond these bounds, OPTCNF relies on verifying the output (sat/unsat) of the solver on all the benchmarks used in our experiments to provide confidence on the correctness of the synthesized encodings. We did not encounter a single instance where OPTCNF resulted in an incorrect output.

# 3   Pattern Finding

In this phase, we identify commonly occurring patterns in the formulas arising from a given domain. For this, we build on prior work on representative sampling from DAG-based representations of formulas [45]. The original sampling work on which we build takes as input a size $k$ and produces a representative sample of all sub-terms of size $k$ that appear in the corpus. When $k = 1$, for example, the process will return a sample of all the operations that appear in the corpus; the frequency with which a given operation appears in the sample will be approximately the same as the frequency with which it appears in the corpus. When sampling with higher values of $k$, the sampling process takes into account the fact that some operations are commutative, but not others.

Given a corpus, OPTCNF collects representative samples for values of $k \leq 5$ for bit-parallel formulas and $k \leq 3$ for non bit-parallel formulas. The upper bounds are determined by the capabilities of our encoding synthesis algorithm, which is unable to generate encodings for larger terms.

# 4   Encoder Code Generation

OPTCNF uses CVC4 as the target solver and generates the code for implementing the synthesized encoders in two phases: (1) Pattern matching in the decreasing order of the pattern size and (2) Extending the existing encoding phase in CVC4. OPTCNF generates code for a straight-forward pattern matching phase while handling symmetries by enumerating all equivalent permutations of patterns with commutative operations. The generated code for augmenting CVC4 implements the synthesized encoder for each matched pattern and provides a command-line interface for switching them on or off individually.

However, there is scope for optimizing this code by implementing: (1) fast pattern matching that reuses common terms in the matched patterns (2) caching and reusing newly generated literals in the encoding phase (3) reduction in number of function calls in the generated code and (4) simplifying the encodings for patterns with constant inputs. Even without these optimizations, we are able to show significant improvement in CVC4's performance on certain domains (Sect. 6).

# 5   Auto-Tuning Encoders

For each domain, we use OpenTuner [3] to auto-tune the set of encoders (one for each pattern) obtained from the synthesis phase according to a performance metric based on the number of benchmark problems solved and the time taken to solve them. The evaluation function (**fopt**) to be optimized takes as input a set of encoders to be used and returns a real number. The number is the sum of all the times taken by the benchmarks to solve; for any benchmarks that time out, their time is counted as the timeout bound times two. The auto-tuner tries to minimize this value by trying out various subsets of encoders provided to it as input while learning a model of the dependence of **fopt** on the selection of encoders.

# 6    Evaluation

We extend CVC4 solver (ranked 2 in the bit-vector category of SMT-COMP 2015 [8]) with synthesized encoders for each domain and evaluate the impact on its performance. Each generated solver is evaluated on the non-incremental quantifier free bit-vector (QF_BV) benchmark suite from SMT-COMP 2015. This benchmark suite consists of 26320 benchmarks that are grouped into 36 sub-categories. In most cases, these sub-categories represent a particular domain of problems–for example, the log − slicing category represents benchmarks that verify bit-vector translation from operations like addition and multiplication to a set of base operations. Some other sub-categories like asp are themselves a collection of benchmarks from multiple different sources. We treat these sub-categories as domains irrespective of whether they really represent a single domain.

## 6.1    Experimental Setup

OPTCNF generates a domain specific solver in four stages:

1. Randomly sampling 10 % of the benchmarks from the domain and running CVC4 to collect all the formulas just before they are encoded to SAT.
2. Pattern finding (Sect. 3) on these formulas and filtering the terms based on capabilities of the synthesis phase of OPTCNF.
3. Translation of each term to multiple SyGus problems one for each possible template that is suitable for the type and the size of the term. For problems involving non bit-parallel terms, OPTCNF uses SKETCH with 4 cores to parallelize the clause minimization algorithm (Sect. 2.5). All other problems use a single core. Each problem is also given a timeout of 3 h.
4. Augmenting CVC4 code with the generated encoders (Sect. 4) and auto-tuning to find a subset of encoders that improve the performance (Sect. 5).

Different parts of OPTCNF system were run on different machines. Pattern finding and synthesis of encoders were run on a machine with forty 2.4 GHz Intel Xeon processors and 96 GB RAM. For auto-tuning, we used a private cluster running OpenStack with parallelism of 150 on 75 virtual machines each with 4 cores and 8 GB RAM of processing power. Finally, the performance experiment evaluating the solvers on QF_BV benchmarks was run on the StarExec [48] cluster infrastructure with a timeout of 900 s and a memory limit of 200 GB (similar to the resources used for the SMT competition).

## 6.2    Domains and Benchmarks

We generate a total of 7 domain-specific solvers and a general solver which is obtained by using the entire QF_BV benchmark suite for pattern finding and synthesis. For the general solver, we enable all the generated encoders and do not auto-tune them. The 7 domains are chosen from the 36 categories in QF_BV. We chose these categories based on the criteria that the number of benchmarks in the domain is at least 20 and the average run-time is significant enough to see an improvement. The solvers for these domains are referred by their category name.

## 6.3  Experiments

We focus on the following questions: (1) Can OPTCNF generate domain-specific solvers in reasonable amount of time? (2) How does the performance of the domain-specific optimal solvers generated by OPTCNF compare to CVC4? (3) How domain-specific are the encoders generated by OPTCNF?

**Time Taken to Generate Optimal Encoders:** Table 1 shows the number of generated (gen) and selected (sel) encoders (selected after auto-tuning, differentiated by the type of patterns), and, the total time taken to synthesize these encoders (both cpu time and clock time). In addition to this, Pattern Finding was run for an hour per domain and Auto-tuning was run for 7.5 h per domain. In total, OPTCNF was able to generate domain-specific encoders in $10 - 22$ h per domain which is a reasonable amount of time as compared to a software engineer implementing and debugging encoders in a solver.

**Table 1.** Encoder statistics and SKETCH running times

| Domain | # boolean | | # bit parallel | | # non bit parallel | | Total patterns | | Synthesis time | |
|---|---|---|---|---|---|---|---|---|---|---|
| | gen | sel | gen | sel | gen | sel | gen | sel | (cpu hrs) | (clock hrs) |
| general | 336 | 336 | 334 | 334 | 12 | 12 | 682 | 682 | 497 | 17 |
| asp | 29 | 22 | 0 | 0 | 4 | 3 | 33 | 25 | 8 | 2 |
| brummayerbiere2 | 66 | 0 | 12 | 7 | 2 | 2 | 80 | 9 | 16 | 2 |
| brummayerbiere3 | 35 | 0 | 13 | 3 | 5 | 3 | 53 | 6 | 15 | 3 |
| bruttomesso | 21 | 4 | 1 | 0 | 1 | 0 | 23 | 4 | 5 | 2 |
| float | 272 | 17 | 294 | 18 | 3 | 0 | 569 | 35 | 360 | 13 |
| log-slicing | 19 | 0 | 86 | 60 | 5 | 5 | 110 | 65 | 49 | 4 |
| mcm | 13 | 3 | 2 | 1 | 4 | 1 | 19 | 5 | 7 | 2 |

**Impact of Domain-Specific Solvers:** With the exception of the general solver, all the other solvers are auto-tuned to select a subset of the generated encodings that improves the performance. For all domains except asp and bruttomesso, the training set for auto-tuning contains 50 % benchmarks chosen randomly from the domain. For these domains, we perform 2-fold cross-validation i.e. we swap training/test sets and run auto-tuning again. For asp and bruttomesso, the training set contains only 20 % benchmarks due to resource constraints for auto-tuning resulting from them having a large number of benchmarks. For these two domains, we run auto-tuning again for approximating cross-validation with another disjoint training set that contains 20 % benchmarks form the domain.

**Table 2.** Performance comparison: Domain-specific, general and CVC4 solvers on 7 categories of QF_BV benchmark suite (first training set)

| Benchmark category | CVC4 | | general | | Domain-Specific | | Boolector | |
|---|---|---|---|---|---|---|---|---|
| | solved | time (s) | solved | time (s) | solved | time (s) | solved | time (s) |
| asp (365) | 240 | 32652.8 | 238 | 33291.8 | **288** | **34971.5** | 308 | 29821.6 |
| brummayerbiere2 (33) | 28 | 1202.8 | 24 | 1653.2 | **29** | **1691.0** | 33 | 1371.2 |
| brummayerbiere3 (40) | 23 | 1165.2 | 23 | 2239.4 | **24** | **1272.1** | 32 | 1760.7 |
| bruttomesso (676) | 623 | 32880.8 | 604 | 35808.6 | **623** | **32840.2** | 774 | 8461.1 |
| float (62) | 59 | 4015.9 | 55 | 3599.6 | **60** | **4395.5** | 58 | 6152.9 |
| log-slicing (79) | 33 | 12636.1 | 57 | 17290.6 | **62** | **21115.4** | 53 | 9534.8 |
| mcm (61) | 40 | 3933.9 | 38 | 3355.0 | **43** | **4193.0** | 39 | 8333.1 |
| | 1046 | 88487.5 | 1039 | 97238.2 | **1129** | **100479.8** | 1297 | 65435.4 |

We compare the performance of the domain-specific solvers (auto-tuned on the first training set) with the general solver and CVC4 in Table 2. Only the benchmarks from the first test set are considered for evaluation in the table. The best-performing solver for every domain is marked as bold. The auto-tuned solver solves 83 benchmarks more than CVC4 in total. For all domains, the domain-specific solvers outperform CVC4. The domain-specific solvers auto-tuned on the second training set for each domain also outperform CVC4 and solve 73 more benchmarks on their corresponding test sets (the details are omitted due to lack of space).

Table 2 also presents the performance of the Boolector solver (the best bit-vector solver in SMT-COMP'15) on the same test set benchmarks for reference. CVC4 is already better than Boolector on two domains (mcm, float) and OPTCNF improves it slightly further. On one domain (log − slicing), CVC4 is notably worse than Boolector, but OPTCNF makes it outperform Boolector. In addition, OPTCNF significantly bridges the gap between CVC4 and Boolector on the mcm domain.

The run-times for benchmarks from domains where we did not perform auto-tuning can be found in the full technical report [37]. The general solver performs better on some domains but not the others, and, slightly worse than CVC4 overall. In all cases where we performed auto-tuning, the domain-specific solvers beat the general solver (Table 2). Two scatter plots showing the performance of CVC4 versus general and the domain-specific solvers on these 7 domains can be found in Fig. 4. It is evident from the graphs that the domain-specific solvers reduce the number of negative points (in the upper left triangle) thereby improving the performance when compared to CVC4 overall.

**Domain Specificity:** We ran each domain-specific solver (obtained from the first training set) on all the other domains and the results are summarized in Table 3. The best performing result for each domain is marked as bold and the results that are worse than CVC4 are underlined. 5 out of 7 of the domains are very domain-specific; the solvers that are tuned specially for them perform

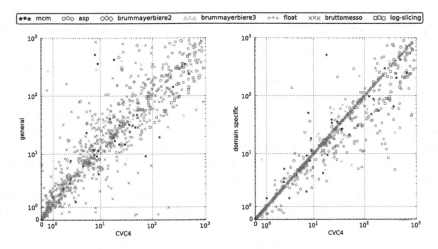

**Fig. 4.** Scatter plots showing run-times (log scale) for different solvers on the 7 domains

**Table 3.** Cross-domain performance

| solver → <br> domain ↓ | asp | | brummayerbiere2 | brummayerbiere3 | bruttomesso | float | log-slicing | mcm |
|---|---|---|---|---|---|---|---|---|
| | solved | time (s) | solved time (s) | solved time (s) | solved time (s) | solved time (s) | solved time (s) | solved time (s) |
| asp | **288** | **34971.5** | 227 28173.5 | 253 34061.6 | 240 33118.9 | 236 29491.3 | 227 28230.5 | 255 35159.5 |
| brummayerbiere2 | 28 | 786.9 | **29** **1691.0** | 29 2363.0 | 29 2174.8 | 29 1804.2 | 29 1705.3 | 28 1706.6 |
| brummayerbiere3 | 22 | 1206.7 | 22 1149.3 | 24 1272.1 | 23 1169.2 | 23 1410.1 | 22 911.3 | **25** **1945.6** |
| bruttomesso | 606 | 37216.1 | 609 38744.1 | 623 32809.8 | 623 32840.1 | 623 32867.5 | 607 37164.7 | 623 32683.5 |
| float | 57 | 1650.8 | 57 2179.3 | 60 4853.1 | 59 3599.5 | **60** **4395.5** | 57 1832.4 | 59 4100.9 |
| log-slicing | 58 | 20816.6 | 59 20125.7 | 35 12955.7 | 35 14640.7 | 32 11796.1 | **62** **21115.4** | 36 14021.6 |
| mcm | 38 | 4301.6 | 40 3413.1 | 39 3411.2 | 41 3940.7 | 39 3759.5 | 39 5313.0 | **43** **4193.0** |

significantly better than all the other solvers. In some cases, using one solver on another domain makes it worse than CVC4. However, mcm domain has a solver optimal for other two domains performing almost identical to their respective solvers.

## 7 Related Work

A recent paper [12] on automatically generating propagation complete encodings is the closest to this work. Encodings generated through OPTCNF are propagation complete and OPTCNF also minimizes the number of clauses across the template being used for the encoder similar to [12]. But, OPTCNF is different in two important ways: (1) Instead of encodings, OPTCNF generates *encoders* which produce encodings at run-time (enabled by program synthesis) (2) The generated encoders are specialized for a particular domain (enabled by pattern finding and auto-tuning).

Different notions of propagation strength of encodings have been considered in both Knowledge Compilation [18] (e.g. unit-refutation completeness [20] and its generalizations [27–29]) and Constraint Programming [5,13] communities.

Propagation complete encodings (PCEs) have been established [10] to be "well-posed" for a SAT solver's deduction mechanism, which provides a tractable reasoning on the constraints. [10] reduces the problem of generating PCEs to iteratively solving QBF formulas whereas OPTCNF relies on CEGIS based program synthesis [46] to generate encoders producing PCEs at run-time. There has also been some recent work on using SAT solvers for enumeration of prime implicants in the Knowledge Compilation community [27,28]. In Constraint Programming, Generalized Arc-Consistency (GAC) [5] is connected to propagation completeness and has been adopted in SAT [24] but is usually only enforced on input/output variables and not on auxiliary variables which provides a weaker notion of propagation strength as compared to PCEs. [9] shows that certain global constraints can require exponential sized formulas for PCEs. In our work, we do not encounter this issue since we consider only small patterns as constraints.

Reducing the size of the CNF encodings derived from SAT formulas has been shown to be an effective way of optimizing SAT solvers [15,22,23,30,41,52]. There has been a lot of work on optimal encodings for specific kinds of constraints like cardinality constraints [1], sequence constraints [13], verification of microprocessors [52]. There is also some work on logic minimization techniques like Beaver [38]. But, to our knowledge, we are the first ones to generate domain specific encodings that are propagation complete and minimal for multiple challenging domains using program synthesis technology.

OPTCNF can be extended to other SMT solvers besides CVC4 such as Z3 [19], Beaver [38], Boolector [14] and Yices [21]. In Beaver and Boolector, intermediate data structures like And-Inverter graphs (AIGs) are employed and are later on transformed to CNF efficiently. Consequently, they have numerous optimizations on the AIG representation before translating it to CNF. Applying OPTCNF directly to such solvers can override these optimizations and hence, requires more work. These solvers can also use lazy bit-blasting strategy as opposed to eager bit-blasting that we use in our experiments. OPTCNF can be extended to solvers employing lazy bit-blasting by using the generated encodings at the time of bit-blasting.

Finally, algorithm configuration [4,33,34], an active area of research in artificial intelligence, has been used in generation of encodings for Planning Domain Models [51] and improving CSP solving by searching for optimal solver choices and the different encodings for the CSP constraints [31]. It has also been shown to be successful for tuning parameters for SAT solvers [36]. Unlike OpenTuner [3], where the optimization function is a black-box, algorithm configuration can use the structure of certain types of functions and employ additional heuristics [35,36] to optimize them.

# 8   Conclusion

In this paper, we presented a technique to generate propagation complete CNF encoders for bit-vector terms. We combined it with machine learning based techniques namely pattern finding and auto-tuning to generate domain-specific solvers. Our evaluation showed that this technique can significantly improve CVC4, a state of the art SMT solver, on the domains represented in the bit-vector benchmark suite from SMT-COMP 2015.

**Acknowledgments.** This research was partially supported by NSF award #1139056 (ExCAPE) and by DARPA MUSE award #FA8750-14-2-0270.

# References

1. Abío, I., Nieuwenhuis, R., Oliveras, A., Rodríguez-Carbonell, E.: A parametric approach for smaller and better encodings of cardinality constraints. In: Schulte, C. (ed.) CP 2013. LNCS, vol. 8124, pp. 80–96. Springer, Heidelberg (2013)
2. Alur, R., Bodik, R., Juniwal, G., Martin, M.M., Raghothaman, M., Seshia, S.A., Singh, R., Solar-Lezama, A., Torlak, E., Udupa, A.: Syntax-guided synthesis. Dependable Softw. Syst. Eng. **40**, 1–25 (2015)
3. Ansel, J., Kamil, S., Veeramachaneni, K., Ragan-Kelley, J., Bosboom, J., O'Reilly, U., Amarasinghe, S.P.: OpenTuner: an extensible framework for program auto-tuning. In: Amaral, J.N., Torrellas, J., (eds.) International Conference on Parallel Architectures and Compilation, PACT 2014, Edmonton, AB, Canada, 24–27 August 2014, pp. 303–316. ACM (2014)
4. Ansótegui, C., Sellmann, M., Tierney, K.: A gender-based genetic algorithm for the automatic configuration of algorithms. In: Gent, I.P. (ed.) CP 2009. LNCS, vol. 5732, pp. 142–157. Springer, Heidelberg (2009)
5. Bacchus, F.: GAC via unit propagation. In: Bessière, C. (ed.) CP 2007. LNCS, vol. 4741, pp. 133–147. Springer, Heidelberg (2007)
6. Barrett, C., Conway, C.L., Deters, M., Hadarean, L., Jovanović, D., King, T., Reynolds, A., Tinelli, C.: CVC4. In: Gopalakrishnan, G., Qadeer, S. (eds.) CAV 2011. LNCS, vol. 6806, pp. 171–177. Springer, Heidelberg (2011)
7. Barrett, C., Deters, M., Moura, L., Oliveras, A., Stump, A.: 6 years of SMT-COMP. J. Autom. Reasoning **50**(3), 243–277 (2012)
8. Barrett, C.W., de Moura, L., Stump, A.: SMT-COMP: satisfiability modulo theories competition. In: Etessami, K., Rajamani, S.K. (eds.) CAV 2005. LNCS, vol. 3576, pp. 20–23. Springer, Heidelberg (2005)
9. Bessiere, C., Katsirelos, G., Narodytska, N., Walsh, T.: Circuit complexity and decompositions of global constraints. In: Boutilier, C. (ed.) IJCAI 2009, Proceedings of the 21st International Joint Conference on Artificial Intelligence, Pasadena, 11–17 July 2009, pp. 412–418 (2009)
10. Bordeaux, L., Marques-Silva, J.: Knowledge compilation with empowerment. In: Bieliková, M., Friedrich, G., Gottlob, G., Katzenbeisser, S., Turán, G. (eds.) SOF-SEM 2012. LNCS, vol. 7147, pp. 612–624. Springer, Heidelberg (2012)
11. Bounimova, E., Godefroid, P., Molnar, D.: Billions and billions of constraints: whitebox fuzz testing in production. In: Proceedings of the 2013 International Conference on Software Engineering, ICSE 2013, Piscataway, pp. 122–131. IEEE Press (2013)
12. Brain, M., Hadarean, L., Kroening, D., Martins, R.: Automatic generation of propagation complete SAT encodings. In: Jobstmann, B., Leino, K.R.M. (eds.) VMCAI 2016. LNCS, vol. 9583, pp. 536–556. Springer, Heidelberg (2016). doi:10.1007/978-3-662-49122-5_26

13. Brand, S., Narodytska, N., Quimper, C.-G., Stuckey, P.J., Walsh, T.: Encodings of the sequence constraint. In: Bessière, C. (ed.) CP 2007. LNCS, vol. 4741, pp. 210–224. Springer, Heidelberg (2007)
14. Brummayer, R., Biere, A.: Boolector: an efficient SMT solver for bit-vectors and arrays. In: Kowalewski, S., Philippou, A. (eds.) TACAS 2009. LNCS, vol. 5505, pp. 174–177. Springer, Heidelberg (2009)
15. Chambers, B., Manolios, P., Vroon, D.: Faster SAT solving with better CNF generation. In: Proceedings of the Conference on Design, Automation and Test in Europe, DATE 2009, pp. 1590–1595. European Design and Automation Association, Belgium (2009)
16. Cheung, A., Solar-Lezama, A., Madden, S.: Partial replay of long-running applications. In: Proceedings of the 19th ACM SIGSOFT Symposium and the 13th European Conference on Foundations of software engineering, ESEC/FSE 2011, pp. 135–145. ACM, New York (2011)
17. Cook, B., Kroening, D., Rümmer, P., Wintersteiger, C.M.: Ranking function synthesis for bit-vector relations. In: Esparza, J., Majumdar, R. (eds.) TACAS 2010. LNCS, vol. 6015, pp. 236–250. Springer, Heidelberg (2010)
18. Darwiche, A., Marquis, P.: A knowledge compilation map. J. Artif. Intell. Res. (JAIR) **17**, 229–264 (2002)
19. de Moura, L., Bjørner, N.S.: Z3: an efficient SMT solver. In: Ramakrishnan, C.R., Rehof, J. (eds.) TACAS 2008. LNCS, vol. 4963, pp. 337–340. Springer, Heidelberg (2008)
20. del Val, A.: Tractable databases: how to make propositional unit resolution complete through compilation. In: Doyle, J., Sandewall, E., Torasso, P., (eds.) Proceedings of the 4th International Conference on Principles of Knowledge Representation and Reasoning (KR 1994), Bonn, Germany, 24–27 May 1994, pp. 551–561. Morgan Kaufmann (1994)
21. Dutertre, B.: Yices 2.2. In: Biere, A., Bloem, R. (eds.) CAV 2014. LNCS, vol. 8559, pp. 737–744. Springer, Heidelberg (2014)
22. Eén, N., Biere, A.: Effective preprocessing in SAT through variable and clause elimination. In: Bacchus, F., Walsh, T. (eds.) SAT 2005. LNCS, vol. 3569, pp. 61–75. Springer, Heidelberg (2005)
23. Eén, N., Mishchenko, A., Sörensson, N.: Applying logic synthesis for speeding up SAT. In: Marques-Silva, J., Sakallah, K.A. (eds.) SAT 2007. LNCS, vol. 4501, pp. 272–286. Springer, Heidelberg (2007)
24. Gent, I.P.: Arc consistency in SAT. In: van Harmelen, F. (ed.) Proceedings of the 15th European Conference on Artificial Intelligence, ECAI 2002, Lyon, July 2002 pp. 121–125. IOS Press (2002)
25. Godefroid, P.: Test generation using symbolic execution. In: D'Souza, D., Kavitha, T., Radhakrishnan, J. (eds.) IARCS Annual Conference on Foundations of Software Technology and Theoretical Computer Science, FSTTCS 2012, 15–17 December 2012, Hyderabad, vol. 18. LIPIcs, pp. 24–33. Schloss Dagstuhl - Leibniz-Zentrum fuer Informatik (2012)
26. Gulwani, S., Srivastava, S., Venkatesan, R.: Constraint-based invariant inference over predicate abstraction. In: Jones, N.D., Müller-Olm, M. (eds.) VMCAI 2009. LNCS, vol. 5403, pp. 120–135. Springer, Heidelberg (2009)
27. Gwynne, M., Kullmann, O.: Generalising and unifying SLUR and unit-refutation completeness. In: van Emde Boas, P., Groen, F.C.A., Italiano, G.F., Nawrocki, J., Sack, H. (eds.) SOFSEM 2013. LNCS, vol. 7741, pp. 220–232. Springer, Heidelberg (2013)

28. Gwynne, M., Kullmann, O.: Towards a theory of good SAT representations. CoRR, abs/1302.4421 (2013)
29. Gwynne, M., Kullmann, O.: Generalising unit-refutation completeness and SLUR via nested input resolution. J. Autom. Reasoning **52**(1), 31–65 (2014)
30. Heule, M., Järvisalo, M., Biere, A.: Clause elimination procedures for CNF formulas. In: Fermüller, C.G., Voronkov, A. (eds.) LPAR-17. LNCS, vol. 6397, pp. 357–371. Springer, Heidelberg (2010)
31. Hurley, B., Kotthoff, L., Malitsky, Y., O'Sullivan, B.: Proteus: a hierarchical portfolio of solvers and transformations. In: Simonis, H. (ed.) CPAIOR 2014. LNCS, vol. 8451, pp. 301–317. Springer, Heidelberg (2014)
32. Hutter, F., Babic, D., Hoos, H.H., Hu, A.J.: Boosting verification by automatic tuning of decision procedures. In: Proceedings of the Formal Methods in Computer Aided Design, FMCAD 2007, pp. 27–34. IEEE Computer Society, Washington, DC (2007)
33. Hutter, F., Hoos, H.H., Leyton-Brown, K.: Sequential model-based optimization for general algorithm configuration. In: Coello, C.A.C. (ed.) LION 2011. LNCS, vol. 6683, pp. 507–523. Springer, Heidelberg (2011)
34. Hutter, F., Hoos, H.H., Leyton-Brown, K., Stützle, T.: ParamILS: an automatic algorithm configuration framework. J. Artif. Int. Res. **36**(1), 267–306 (2009)
35. Hutter, F., Hoos, H.H., Stützle, T.: Automatic algorithm configuration based on local search. In: Proceedings of the Twenty-Second AAAI Conference on Artificial Intelligence, 22–26 July 2007, Vancouver, pp. 1152–1157. AAAI Press (2007)
36. Hutter, F., Lindauer, M.T., Balint, A., Bayless, S., Hoos, H.H., Leyton-Brown, K.: The configurable SAT solver challenge (CSSC). CoRR, abs/1505.01221 (2015)
37. Inala, J.P., Singh, R., Solar-Lezama, A.: Technical report: Synthesis of Domain Specific CNF Encoders for Bit-Vector Solvers (2016). http://jinala.github.io/assets/papers/sat2016tr.pdf. (Accessed on 24 April 2016)
38. Jha, S., Limaye, R., Seshia, S.A.: Beaver: engineering an efficient SMT solver for bit-vector arithmetic. In: Bouajjani, A., Maler, O. (eds.) CAV 2009. LNCS, vol. 5643, pp. 668–674. Springer, Heidelberg (2009)
39. Manthey, N., Heule, M.J.H., Biere, A.: Automated reencoding of Boolean formulas. In: Biere, A., Nahir, A., Vos, T. (eds.) HVC. LNCS, vol. 7857, pp. 102–117. Springer, Heidelberg (2013)
40. Martins, R., Manquinho, V.M., Lynce, I.: Exploiting cardinality encodings in parallel maximum satisfiability. In: IEEE 23rd International Conference on Tools with Artificial Intelligence, ICTAI 2011, Boca Raton, 7–9 November 2011, pp. 313–320. IEEE Computer Society (2011)
41. Moskewicz, M.W., Madigan, C.F., Zhao, Y., Zhang, L., Malik, S.: Chaff: engineering an efficient SAT solver. In: Proceedings of the 38th Annual Design Automation Conference, DAC 2001, pp. 530–535. ACM, New York (2001)
42. Nguyen, C., Yoshida, H., Prasad, M.R., Ghosh, I., Sen, K.: Generating succinct test cases using don't care analysis. In: Proceedings of the Eighth IEEE International Conference on Software Testing, Verification and Validation, pp. 1–10. IEEE (2015)
43. Pnueli, A., Rosner, R.: On the synthesis of a reactive module. In: Proceedings of the 16th ACM SIGPLAN-SIGACT Symposium on Principles of Programming Languages, POPL 1989, pp. 179–190. ACM, New York (1989)
44. Sen, K., Kalasapur, S., Brutch, T., Gibbs, S.: Jalangi: A selective record-replay and dynamic analysis framework for Javascript. In: Proceedings of the 2013 9th Joint Meeting on Foundations of Software Engineering, ESEC/FSE 2013, pp. 488–498. ACM, New York (2013)

45. Singh, R., Solar-Lezama, A.: Automatic generation of formula simplifiers based on conditional rewrite rules arXiv:1602.07285 (2016)
46. Solar-Lezama, A.: Program Synthesis By Sketching. PhD thesis, EECS Dept., UC Berkeley (2008)
47. Srivastava, S., Gulwani, S., Foster, J.S.: From program verification to program synthesis. In: Proceedings of the 37th Annual ACM SIGPLAN-SIGACT Symposium on Principles of Programming Languages, POPL 2010, pp. 313–326. ACM, New York (2010)
48. Stump, A., Sutcliffe, G., Tinelli, C.: Introducing StarExec: a cross-community infrastructure for logic solving. In: Klebanov, V., Beckert, B., Biere, A., Sutcliffe, G. (eds.) COMPARE, CEUR Workshop Proceedings, vol. 873, p. 2 (2012). CEUR-WS.org
49. Tanno, H., Zhang, X., Hoshino, T., Sen, K.: TesMa and CATG: automated test generation tools for models of enterprise applications. In: Proceedings of the 37th International Conference on Software Engineering, ICSE 2015, vol. 2, pp. 717–720. IEEE Press, Piscataway (2015)
50. Tseitin, G.S.: On the complexity of derivation in propositional calculus. In: Siekmann, J.H., Wrightson, G. (eds.) Automation of Reasoning, pp. 466–483. Springer, Heidelberg (1983)
51. Vallati, M., Hutter, F., Chrpa, L., McCluskey, T.L.: On the effective configuration of planning domain models. In: Yang, Q., Wooldridge, M. (eds.) Proceedings of the Twenty-Fourth International Joint Conference on Artificial Intelligence, IJCAI 2015, Buenos Aires, 25–31 July 2015, pp. 1704–1711. AAAI Press (2015)
52. Velev, M.N.: Efficient translation of boolean formulas to cnf in formal verification of microprocessors. In: Proceedings of the 2004 Asia and South Pacific Design Automation Conference, ASP-DAC 2004, pp. 310–315. IEEE Press, Piscataway (2004)
53. Wang, X., Zeldovich, N., Kaashoek, M.F., Solar-Lezama, A.: A differential approach to undefined behavior detection. Commun. ACM **59**(3), 99–106 (2016)

# Beyond SAT

# Finding Finite Models in Multi-sorted First-Order Logic

Giles Reger[1]([✉]), Martin Suda[1], and Andrei Voronkov[1,2,3]

[1] University of Manchester, Manchester, UK
giles.reger@manchester.ac.uk
[2] Chalmers University of Technology, Gothenburg, Sweden
[3] EasyChair, Manchester, UK

**Abstract.** This work extends the existing MACE-style finite model finding approach to multi-sorted first-order logic. This existing approach iteratively assumes increasing domain sizes and encodes the related ground problem as a SAT problem. When moving to the multi-sorted setting each sort may have a different domain size, leading to an explosion in the search space. This paper focusses on methods to tame that search space. The key approach adds additional information to the SAT encoding to suggest which domains should be grown. Evaluation of an implementation of techniques in the Vampire theorem prover shows that they dramatically reduce the search space and that this is an effective approach to find finite models in multi-sorted first-order logic.

## 1 Introduction

There have been a number of approaches looking at finding finite models for First-Order Logic (FOL), however there has not been much work on finding such models for Multi-Sorted FOL where symbols are given *sorts*. We consider a model finding method, pioneered by MACE [12], that encodes the search as a SAT problem. We show how this method can be modified to deal directly with multi-sorted input, rather than translating the problem to the unsorted setting, which is the most common current method.

There are two main motivations for this work. Firstly, many problems are more naturally expressed in multi-sorted FOL than in unsorted FOL (although their theoretical expressive power is equivalent). Therefore, it is useful to be able to reason in this setting and translations from multi-sorted FOL to unsorted FOL often make this reasoning harder. Secondly, MACE-style model finders can use sort information to make the SAT encoding smaller. However, as we discuss below, finding finite models of multi-sorted formulas also presents significant challenges.

This work was supported by EPSRC grant EP/K032674/1. Martin Suda and Andrei Voronkov were partially supported by ERC Starting Grant 2014 SYMCAR 639270. Andrei Voronkov was also partially supported by the Wallenberg Academy Fellowship 2014 - TheProSE.

N. Creignou and D. Le Berre (Eds.): SAT 2016, LNCS 9710, pp. 323–341, 2016.
DOI: 10.1007/978-3-319-40970-2_20

The MACE-style approach, later extended in the Paradox [5] work, involves selecting a domain size for the finite model, grounding the first-order problem with this domain and translating the resulting formulas into a SAT problem, which, if satisfied, gives a finite model of the selected size. Search for a finite model then involves considering iteratively larger domain sizes. In the multi-sorted setting it is necessary to consider the size of each sort separately. This can be demonstrated by the following example, which is an extension of the much used *Monkey Village* example [2,4].

*Example 1 (Organised Monkey Village).* Imagine a village of monkeys where each monkey owns at least two bananas. As the monkeys are well-organised, each tree contains exactly three monkeys. Monkeys are also very friendly, so they pair up to make sure they will always have a partner. We can represent this problem as follows:

$(\forall M : monkey)(\mathsf{owns}(M, \mathsf{b}_1(M)) \wedge \mathsf{owns}(M, \mathsf{b}_2(M)) \wedge \mathsf{b}_1(M) \neq \mathsf{b}_2(M))$
$(\forall M_1, M_2 : monkey)(\forall B : banana)(\mathsf{owns}(M_1, B) \wedge \mathsf{owns}(M_2, B) \rightarrow M_1 = M_2)$
$(\forall T : tree)(\exists M_1, M_2, M_3 : monkey)((\bigwedge_{i=1}^{3} \mathsf{sits}(M_i) = T) \wedge \mathsf{distinct}(M_1, M_2, M_3))$
$(\forall M_1, M_2, M_3, M_4 : monkey)(\forall T : tree)((\bigwedge_{i=1}^{4} \mathsf{sits}(M_i) = T) \Rightarrow \neg\mathsf{distinct}(M_1, M_2, M_3, M_4))$
$(\forall M : monkey)(\mathsf{partner}(M) \neq M \wedge \mathsf{partner}(\mathsf{partner}(M)) = M)$

where the predicates owns associates monkeys with bananas, the functions $\mathsf{b}_1$ and $\mathsf{b}_2$ witness the existence of each monkey's minimum two bananas, the function sits maps monkeys to the tree that they sit in, the function partner associates a monkey with its partner, and the (meta-)predicate distinct is true if all of its arguments are distinct.

This problem requires the domain of *monkey* to be exactly three times larger than the domain of *tree*, the domain of *banana* to be at least twice as large as the domain of *monkey*, and the domain of *monkey* to be even. The main model finding effort then becomes searching for an assignment of domain sizes that satisfies this problem. Here, the smallest such assignment is $|tree| = 2$, $|monkey| = 6$, and $|banana| = 12$. In the general case it is necessary to try all combinations of domain sizes. In the worst case this will mean trying a number of assignments exponential in the number of sorts.

The techniques introduced in this paper tackle this issue by introducing a number of ways to constrain this search space. The main contributions can be summarised as:

1. We show how the MACE-style approach (described in Sect. 3) can be extended to the multi-sorted setting via a novel extension of the SAT encoding (Sect. 5). This encoding uses information from the SAT solver to guide the search through the space of domain size assignments.
2. We use *monotonic sorts* introduced in [4] in a new way for the multi-sorted case to reduce the search space further (Sect. 5.5).
3. We utilise the saturation-based approach for first-order logic to detect further constraints on the search space introduced by injective and surjective functions (Sect. 6).

4. We present an alternative to (1), a complementary search strategy, utilising a different SAT encoding, that only needs to expand (and never shrink) the sizes of sort domains (Sect. 8).

These ideas have been realised within the VAMPIRE theorem prover [11] and evaluated on problems taken from the TPTP and SMT-LIB benchmark suites. Our experimental evaluation (Sect. 9) shows that these techniques can be used to (i) improve finite model finding in the unsorted setting, (ii) effectively and efficiently find finite models in the multi-sorted setting, and (iii) detect cases where no models exist.

## 2  Preliminaries

**Multi-sorted First-Order Logic.** We consider a multi-sorted first-order logic with equality. A term is either a variable, a constant, or a function symbol applied to terms. A literal is either a propositional symbol, a predicate applied to terms, an equality of two terms, or a negation of either. Function and predicate symbols are *sorted* i.e. their arguments (and the return value in the case of functions) have a unique *sort* drawn from a finite set of sorts $S$. We only consider well-sorted literals. There is an equality symbol per sort and equalities can only be between terms of the same sort. Formulas may use the standard notions of quantification and boolean connectives, but in this work we assume all formulas are *clausified* using standard techniques. A *clause* is a disjunction of literals where all variables are universally quantified (existentially quantified variables can be replaced by skolem functions during clausification).

**SAT Solvers.** The technique we present later will make use of a black-box SAT solver and we assume the reader is familiar with their general properties. We assume that a SAT solver supports *solving under assumptions* [7,8]. This means the SAT solver can be asked to search for a model of a set of clauses $N$ additionally satisfying a conjunction of assumption literals $A$ and is able, in case the answer is UNSAT, to provide a subset $A_0 \subseteq A$ of those assumptions which were sufficient for the unsatisfiability proof.

## 3  MACE-Style Finite Model Finding in an Unsorted Setting

We describe the finite model finding procedure in a single sorted setting. This is a variation of the approach taken by Paradox [5]. The general idea is to create, for each integer $n \geq 1$, a SAT problem that is satisfiable if the problem has a finite model of size $n$. To find a finite model we therefore iterate the approach for domain sizes $n = 1, 2, 3, \ldots$.

## 3.1 $DC$-Models

Let $S$ be a set of clauses. Let us fix an integer $n \geq 1$. Let $DC = \{c_1, \ldots, c_n\}$ be a set of distinct constants not occurring in $S$, we will call the elements of $DC$ *domain constants*. We extend the language by adding the domain constants and say that an interpretation is a $DC$-*interpretation*, if (i) the domain of this interpretation is $DC$ and (ii) every domain constant $c_i$ is interpreted in it by itself. Every model of $S$ that is also a $DC$-interpretation will be called a $DC$-*model* of $S$. It is not hard to argue that, if $S$ has a model of size $n$, then it also has a $DC$-model. We say that $S$ is $n$-satisfiable if it has a model of size $n$.

Let $C$ be a clause. A $DC$-*instance* of $C$ is a ground clause obtained by replacing every variable in $C$ by a constant in $DC$. For example, if $p(x) \vee x = y$ is a clause and $n \geq 2$, then $p(c_1) \vee c_1 = c_2$ and $p(c_1) \vee c_1 = c_1$ are $DC$-instances, while $p(c_1) \vee c_2 = c_3$ is not a $DC$-instance. A clause with $k$ different variables has exactly $n^k$ $DC$-instances.

**Theorem 1.** *Let $I$ be a $DC$-interpretation and $C$ a clause. Then $C$ is true in $I$ if and only if all $DC$-instances of $C$ are true in $I$.*

Let us denote by $S^*$ the set of all $DC$-instances of the clauses in $S$. Consider an example. Let $S$ consist of three clauses

$$p(b), \qquad f(a) \neq b, \qquad f(f(x)) = x.$$

The smallest model of $S$ has a domain of size two. Take $n = 2$, then $DC = \{c_1, c_2\}$. By the above theorem, $S$ has a model of size two if an only if $S^*$ has a DC-model. The set $S^*$ consists of four ground clauses:

$$p(b), \qquad f(a) \neq b, \qquad f(f(c_1)) = c_1, \qquad f(f(c_2)) = c_2.$$

Note that $DC$-models are somehow similar to Herbrand models used in logic programming and resolution theorem proving, except that they are built using (domain) constants instead of all ground terms and $DC$-instances instead of ground instances.

Theorem 1 is not directly applicable to encode the existence of models of size $n$ as a SAT problem, because $DC$-instances can contain complex terms. We will now introduce a special kind of ground atom which contains no complex subexpressions. We call a *principal term* any term of the form $f(d_1, \ldots, d_m)$, where $m \geq 0$, $f$ is a function symbol, which is not a domain constant, and $d_1, \ldots, d_m$ are domain constants. In our example there are four principal terms: $a, b, f(c_1), f(c_2)$. A ground atom is called *principal* if it either has the form $p(d_1, \ldots, d_m)$ where $m \geq 0$, $p$ is a predicate symbol different from equality and $d_1, \ldots, d_m$ are domain constants or has the form $t = d$, where $t$ is a principal term and $d$ a domain constant. We call a *principal literal* a principal atom or its negation.

**Theorem 2.** *Let $I_1, I_2$ be $DC$-interpretations. If they satisfy the same principal atoms, then $I_1$ coincides with $I_2$.*

Theorem 1 reduces $n$-satisfiability of $S$ to the existence of a $DC$-interpretation of the set $S^*$ of ground clauses. Theorem 2 shows that $DC$-interpretations can be identified by the set of principal atoms true in them. What we will do next is to introduce a propositional variable for every principal atom and reduce the existence of a $DC$-model of $S^*$ to satisfiability of a set of clauses using only principal literals.

## 3.2   The SAT Encoding

The main step in the reduction is to transform every non-ground clause $C$ into an equivalent clause $C'$ such that $DC$-instances of $C'$ consist (almost) only of principal literals. We will explain what "almost" means below. This transformation is known as *flattening*.

**Flattening.** A literal is called *flat* if it has one of the following forms:

1. $p(x_1, \ldots, x_m)$ or $\neg p(x_1, \ldots, x_m)$, where $m \geq 0$ and $p$ is a predicate symbol;
2. $f(x_1, \ldots, x_m) = y$ or $f(x_1, \ldots, x_m) \neq y$, where $m \geq 0$ and $f$ is a function symbol, which is not a domain constant.
3. an equality between variables $x = y$.

Every $DC$-instance of a flat literal is either a principal literal (for the first two cases), or an equality $c_i = c_j$ between domain constants.

To flatten clauses in $S$, we first get rid of all inequalities between variables, replacing every clause of the form $x \neq y \vee C[x]$ by the equivalent clause $C[y]$. Then we repeatedly replace every clause $C[t]$, where $t$ is not a variable and $t$ occurs as an argument to a predicate or a function symbol, by the equivalent clause $t \neq x \vee C[x]$, where $x$ is a fresh variable.

Our example clauses can be flattened as follows:

$$p(y) \vee b \neq y, \qquad f(y_1) = y_2 \vee a \neq y_1 \vee b \neq y_2, \qquad f(y) = x \vee f(x) \neq y$$

**$DC$-Instances.** We can now produce the $DC$-instances of each flattened clause $C[x_1, \ldots, x_k]$. For our running example (with $n = 2$) this produces the following ten $DC$-instances:

| | | |
|---|---|---|
| $p(c_1) \vee b \neq c_1$ | $f(c_1) = c_1 \vee a \neq c_1 \vee b \neq c_1$ | $f(c_1) = c_1 \vee f(c_1) \neq c_1$ |
| $p(c_2) \vee b \neq c_2$ | $f(c_1) = c_2 \vee a \neq c_1 \vee b \neq c_2$ | $f(c_1) = c_2 \vee f(c_2) \neq c_1$ |
| | $f(c_2) = c_2 \vee a \neq c_2 \vee b \neq c_2$ | $f(c_2) = c_2 \vee f(c_2) \neq c_2$ |
| | $f(c_2) = c_1 \vee a \neq c_2 \vee b \neq c_1$ | $f(c_2) = c_1 \vee f(c_1) \neq c_2$ |

Note that all literals are principal. If we treat principal atoms as propositional variables, the two leftmost clauses can be satisfied by making $b \neq c_1$ and $b \neq c_2$ both true, but this violates the assumption that $b$ should equal one of the domain constants. Additionally, the two rightmost topmost clauses can be satisfied by making $f(c_1) = c_1$ and $f(c_1) = c_2$ true but this violates the assumption that $f$ is a function. We would like to prevent both situations. To do this we introduce additional definitions.

**Functionality Definitions.** For each principal term $p$ and distinct domain constants $d_1, d_2$ we produce the following clause

$$p \neq d_1 \vee p \neq d_2,$$

These clauses are satisfied by every $DC$-interpretation and guarantee that all function symbols are interpreted as (partial) functions.

For our running example we introduce four new definitions:

$$a \neq c_1 \vee a \neq c_2, \quad b \neq c_1 \vee b \neq c_2, \quad f(c_1) \neq c_1 \vee f(c_1) \neq c_2, \quad f(c_2) \neq c_1 \vee f(c_2) \neq c_2$$

**Totality Definitions.** For each principal term $p$ we produce the following clause

$$p = c_1 \vee \ldots \vee p = c_n$$

These clauses are satisfied by every $DC$-interpretation of size $n$ and guarantee, together with functionality axioms, that all function symbols are interpreted as total functions.

For our running example we introduce four new definitions:

$$a = c_1 \vee a = c_2, \quad b = c_1 \vee b = c_2, \quad f(c_1) = c_1 \vee f(c_1) = c_2, \quad f(c_2) = c_1 \vee f(c_2) = c_2$$

The resulting SAT clauses have a model, meaning that the original clauses have a finite model with a domain of size 2, which can be extracted from the SAT encoding.

**Equalities Between Variables.** Flattening can result in equalities between variables, that is, clauses of the form $C \vee x = y$. $DC$-instances of such clauses can have, in addition to principal literals, equalities between domain constants $d_1 = d_2$, which are not principal literals. Since we only want to deal with principal literals, we will get rid of such equalities in an obvious way: delete clauses containing tautologies $d = d$ and delete from clauses literals $d_1 = d_2$, where $d_1$ are distinct $d_2$ domain constants.

The following theorem underpins the SAT-based finite model building method:

**Theorem 3.** *Let $S$ be a set of flat clauses and $S'$ be the set of clauses obtained from $S^*$ by removing equalities between domain constants as described above and adding all functionality and totality definitions. Then (i) all literals in $S'$ are principal and (ii) $S$ is $n$-satisfiable if and only if $S'$ is propositionally satisfiable.*

**Incrementality.** In [5] the authors describe a method for incremental finite model finding which advocates keeping (parts of) the contents of the SAT solver when increasing $n$. However, in previous experiments we discovered that the technique of *variable and clause elimination* [6] is useful at reducing the size of the SAT problem. As this is not compatible with incremental solving, our general approach is non-incremental.

## 3.3   Reducing the Number of Variables

The number of instances produced is exponential in the number of variables in a flattened clause. We describe two approaches that aim to reduce this number.

**Definition Introduction.** This reduces the size of clauses produced by flattening. Complex ground subterms are removed from clauses by introducing definitions. For example, a clause $p(f(a, b), g(f(a, b)))$ becomes $p(e_1, e_2)$ and we introduce the definition clauses $e_1 = f(a, b)$ and $e_2 = g(e_1)$, where $e_1, e_2$ are new constants. One can also introduce definitions for non-ground subterms.

**Clause Splitting.** Clauses with $k$ variables are split into subclauses having less than $k$ variables each. New predicate symbols applied to the shared variables are then added to join the subclauses. For example, the clause $p(x, y) \vee q(y, z)$ with three variables is replaced by the two clauses $p(x, y) \vee s(y)$ and $\neg s(y) \vee q(y, z)$ where $s$ is a new predicate symbol. These new clauses have two variables each. For large domain sizes splitting can drastically reduce the size of the resulting propositional problem. This was first used for finite model finding by Gandalf [17] and later in Eground [14] for EPR problems.

## 3.4   Symmetry Breaking

The SAT problem produced above can contain many symmetries. For example, every permutation of $DC$ applied to a $DC$-model will give a $DC$-model, and there are $n!$ such permutations. We can (partially) break these symmetries as follows. Firstly, if the input contains constants $a_1, \dots, a_l$ we can add the clauses

$$a_i \neq c_m \vee a_1 = c_{m-1} \vee \dots \vee a_{i-1} = c_{m-1}$$

for $1 < i \leq l$ and $1 < m \leq n$, where we have arbitrarily ordered the constants and captured the constraint that if the $i$-th constant is equal to a domain element then some earlier constant must be equal to the next smallest domain element. Secondly, we can tell the SAT solver about this order on constants by adding the clauses

$$a_i = c_1 \vee \dots \vee a_i = c_i$$

for $i \leq \min(m, n)$, which captures the constraint that the $i$-th constant must be equal to one of the first $i$-th domain elements. If $1 < m < n$ then we can also use principal terms other than constants in the second case, but not in the first.

## 3.5   Determining Unsatisfiability

If it is possible to detect the *maximum* domain size then it is possible to show there is no model for a formula if all domain sizes up to, and including, this maximum size have been explored. There are two straightforward ways to detect maximum domain sizes. Firstly, we can look for axioms such as $(\forall x)(x = a \vee x = b)$ and $(\forall x)(\forall y)(\forall z)(x = y \vee x = z \vee z = y)$. Both indicate that the problem has a maximum domain size of 2. Secondly, we can look for so-called EPR problems that only use constant function symbols, in this case, the domain size is bounded by the number of constants.

# 4    Previous Work in the Multi-sorted Setting

We review previous work related to finite model finding for multi-sorted FOL.

**Translating Sorts Away.** One approach to dealing with multi-sorted FOL is to translate the sorts away. We discuss two well-known translations, see [2] for further discussions of such translations.

*Sort Predicates.* One can *guard* the use of sorted variables by a *sort predicate* that indicates whether a variable is of that sort. This predicate can be set to false in a model for all constants not of the appropriate sort. For example, the last formula in the Organised Monkey Village problem can be rewritten using the sort predicate isMonkey.

$$(\forall M)(\mathsf{isMonkey}(M) \to \mathsf{partner}(M) \neq M \land \mathsf{partner}(\mathsf{partner}(M)) = M)$$

One also needs to add additional axioms that say that sorts are non-empty and that functions return the expected sort. For the *monkey* sort we need to add

$$(\exists M)(\mathsf{isMonkey}(M)) \qquad (\forall M)(\mathsf{isMonkey}(\mathsf{partner}(M))).$$

*Sort Functions or Tags.* One can *tag* all values of a sort using a *sort function* for that sort. The idea is that in a model the function can map all constants (of any sort) to a constant of the given sort. For example, the last formula from the Organised Monkey Village problem can be rewritten using $\mathsf{f_m}$ as a sort function for *monkey*:

$$(\forall M)(\mathsf{f_m}(\mathsf{partner}(\mathsf{f_m}(M))) \neq \mathsf{f_m}(M) \land \mathsf{f_m}(\mathsf{partner}(\mathsf{f_m}(\mathsf{partner}(\mathsf{f_m}(M))))) = \mathsf{f_m}(M))$$

The authors of [2] suggest conditions that allow certain sort predicates and functions to be omitted. However, their arguments relate to resolution proofs and do not apply here.

**Sorting it Out with Monotonicity.** In [4] Claessen et al. introduce a monotonicity analysis and show how it can help translate multi-sorted formulas to unsorted ones by only applying the above translations to non-monotonic sorts. A sort $\tau$ is monotonic for a multi-sorted FOL formula $\phi$ if for any model of $\phi$ one can add an element to the domain of $\tau$ to produce another model of $\phi$. For example, in the Organised Monkey Village example the *banana* sort is monotonic as we can add more bananas once we have enough. However, *monkey* and *tree* are not monotonic as increasing either requires more trees, monkeys and bananas.

In [4] they observe that if there is no positive equality between elements of a sort then a new domain constant can be added and made to behave like an existing domain constant and there is no way to detect this i.e. positive equalities are required to bound a sort. They refine this notion further by noting that a positive equality can be guarded by a predicate, if that predicate can be forced to be true for all new domain elements. They introduce a calculus and associated SAT encoding capturing these ideas that can be used to detect monotonic sorts, which we use in our work.

**Using a Theory of Sort Cardinalities.** In the single-sorted setting there is a family of techniques called SEM-style after the SEM model finder [18] based on constraint satisfaction methods. There exists a technique in this direction for the multi-sorted setting implemented in the CVC4 SMT solver [13]. The idea behind this approach is to introduce a theory of sort cardinality constraints and to incorporate this theory into the standard SMT solver structure. Briefly, this approach introduces cardinality constraints (upper bounds) for sorts and searches for a set of constraints that is consistent with the axioms. To check a cardinality constraint $k$ for sort $s$, a congruence relation is built for $s$-terms and an attempt made to merge congruence classes so that there are at most $k$. Cardinality constraints are then increased if found to be inconsistent. Quantified formulas are then instantiated with representative constants from the equivalence classes.

# 5 A Framework for the Multi-sorted Setting

In this section we introduce our framework able to build models of multi-sorted formulas directly, in contrast to translating the sorts away. The key challenge is dealing with a large and growing search space of domain sizes.

## 5.1 Using Sorts in the SAT Encoding

The SAT encoding in Sect. 3 can be updated to become sort-aware. First, instead of the domain size $n$ we use finite domain sizes $n_s$ for every sort $s$. Second, instead of considering $DC = \{c_1, \ldots, c_n\}$, we consider domains for each sort $DC_s = \{c_1, \ldots, c_{n_s}\}$. We can now define $n$ as a function (called *domain size assignment*) mapping each sort $s$ to $n_s$ and likewise, define $DC$ as the function mapping each sort $s$ to $DC_s$. After that we can speak about $DC$-models and $n$-satisfiability in the multi-sorted case.

All the definitions for the one-sorted case are modified to respect sorts. This means, in particular, that in a $DC$-instance of a clause a variable of a sort $s$ can only be replaced by a domain constant in $DC_s$. For example, for the Organised Monkey Village problem (see page 324) we could consider the domain size assignment $n$ such that $n_{tree} = 1$, $n_{monkey} = 2$ and $n_{banana} = 2$. The first formula in this description can be split into three clauses, the first of which would be flattened as $\mathsf{owns}(M, x) \lor \mathsf{b}_1(M) \neq x$, which would have the two $DC$-instances $\mathsf{owns}(c_1, c_1) \lor \mathsf{b}_1(c_1) \neq c_1$ and $\mathsf{owns}(c_1, c_2) \lor \mathsf{b}_1(c_1) \neq c_2$. We can use $c_1$ for both monkeys and bananas here as monkeys and bananas are never compared. For this reason we can also break symmetries on a per-sort basis.

Once we have updated the SAT encoding, finite model finding can then proceed as before where we construct the SAT problem for the current domain size assignment, check for satisfiability, and then either return a model or repeat the process with an updated domain size assignment. The problem then becomes how to generate the next domain size assignment to try.

## 5.2   A Search Strategy

We will view the search space of domain size assignments as an infinite directed graph whose nodes are domain size assignments and the children of an assignment are all the nodes that have exactly one domain size that is one larger. Thus, the number of children of every node is the number of sorts. A child of a node $n$ having a larger domain size than $n$ for a sort $s$ is called the $s$-*child* of $n$. The $s$-*descendant* relation is the transitive closure of the $s$-child relation.

A *search strategy* will explore this graph node by node in such a way that a node is always visited before its children. For each node $n$ that we visit, we can either check $n$-satisfiability or *ignore* this node. To decide whether a node can be ignored, we will maintain a set of *constraints*. Abstractly, a constraint is a predicate on domain size assignments and nodes that do not satisfy the current set of constraints will be ignored. Concretely, we will use a language of (boolean combinations of) arithmetical comparison literals such as $|s| < b$, $|s| \leq b$, ... to represent the constraints. Here $b$ stands for a concrete integer and $|s|$ is a symbolic placeholder variable for the "intended size" of the domain of sort $s$. The semantics of the this representation is the obvious one.

We will work with a queue $Q$ of nodes and a set $\mathcal{C}$ of *constraints*. Initially, $Q$ consists of a single node assigning 1 to all sorts and $\mathcal{C}$ is empty. We then repeat the following steps:

1. If $Q$ is empty, return "unsatisfiable".
2. Remove the node $q$ from the front of $Q$. Do nothing if $q$ was visited before at this step. Otherwise, continue with the following steps.
3. If $q$ satisfies all constraints in $\mathcal{C}$, perform finite model finding for $q$, terminating if a model is found. In variations of this algorithm considered later, we can add some constraints to $\mathcal{C}$ at this step: these constraints will be obtained by analyzing the proof of $q$-unsatisfiability.
4. Add to $Q$ all children of $q$.

We will now introduce an important notion helping us to prevent exploring large parts of the search space. A constraint is said to have the $s$-*beam* property at a node $n$, if all $s$-descendants of $n$ violate this constraint. For example, the constraint $|s| < 3$ has the $s$-beam property at any node $q$ having $q_s = 2$. We can generalize this notion to more than one sort.

With this notion we can improve step 4 of the algorithm as follows:

4.′ If there is a constraint in $\mathcal{C}$ having an $s$-beam property at the $s$-child $n$ of $q$, if $n$ violates this constraint, add to $Q$ all children of $q$ apart from $n$.

For example, if we have the constraint $|s| < 3$ and $q_s = 2$, this constraint will prevent us from considering the $s$-child $n$ of $q$ having $n_s = 3$.

As a small refinement, we introduce a heuristic for deciding which node in the queue to consider next, rather than processing them in the first-in-first-out order. The idea is to estimate how difficult a domain size assignment is to check and to prioritise exploration of the easier parts of the search space. Under this variation,

$Q$ is a priority queue ordered by some size measure of the corresponding SAT encoding (in the experiment, we measured size in the number of clauses). This setup is complete, as long as this size grows strictly from a parent to its child (which is trivially satisfied for number of clauses).

## 5.3   Encoding the Search Problem

We now show how an extension of the SAT encoding can be used to produce constraints and therefore indicate areas of the search space that should be avoided. This is done by marking certain clauses of the encoding with certain special variables and using the mechanism for solving under assumptions to detect which of these clauses were actually used in the unsatisfiability proof. We will design the names of these special marking variables in such a way that the detected set of used assumptions will immediately correspond to a (disjunctive) constraint.

Let us assume we are encoding for the domain size assignment $n$. For each sort $s$ we introduce two new propositional variables "$|s| > n_s$" and "$|s| < n_s$", which can be understood as stating that the intended size of the domain of $s$ should be larger, respectively smaller, than current $n_s$. The marking of clauses is now done as follows.

The totality definition for each principal term $p$ becomes

$$p = c_1 \vee \ldots \vee p = c_{n_s} \vee \text{``}|s| > n_s\text{''}$$

i.e. either the principal term equals one of the domain constants or the domain is currently too small. $DC$-instances can be similarly updated. Let $C$ be a $DC$-instance and let $\mathsf{sorts}(C)$ be the set of sorts of variables occurring in $C$. We replace $C$ by

$$C \vee \bigvee_{s \in \mathsf{sorts}(C)} \text{``}|s| < n_s\text{''}$$

i.e. either the $DC$-instance holds or the domain is too large.

We then attempt to solve the updated SAT problem under the assumptions

$$A = \bigwedge_{s \in S}(\neg\text{``}|s| > n_s\text{''}) \wedge (\neg\text{``}|s| < n_s\text{''})$$

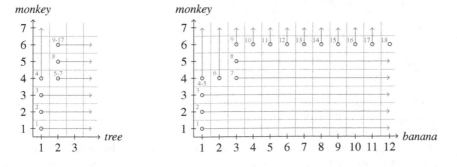

**Fig. 1.** Finding a finite model for the Organised Monkey Village problem.

i.e. we assume that we are using the correct domain sizes. These added assumptions ensure that the logical meaning of the updated encoding is exactly the same as before. However, in the unsatisfiable case the solver now returns a subset $A_0 \subseteq A$ of the assumptions that were sufficient to establish unsatisfiability. Equivalently, $\neg A_0$ is a conflict clause over the marking variables implied by the encoded problem. This clause can now be understood as the newly derived constraint. We just need to interpret the marking variables in their "unquoted" form, i.e., as arithmetic comparison literals.

The argument why this interpretation is correct is best done with the set $A_0$, which contains the marking variables negated. It consists of two main observations:

1. If $A_0$ contains $\neg$"$|s| > n_s$", the unsatisfiability relies on a totality clause for the sort $s$. Because a totality clause gets logically stronger when the domain size is decreased, essentially the same unsatisfiability proof could be repeated for the domain size $|s|$ smaller or equal to the current $n_s$ (given the other conditions from $A_0$).
2. If $A_0$ contains $\neg$"$|s| < n_s$", the unsatisfiability relies on a $DC$-instance with a variable of sort $s$. Because we only add more instances of a clause if a domain size is increased, the same unsatisfiability proof would also work for the domain size $|s|$ greater or equal to the current $n_s$.

Thus at least one of the (atomic) constraints represented by the literals in $A_0$ must be violated by a domain size assignment, if we want to have a chance of finding a model.

### 5.4   An Example

Let us consider the Organised Monkey Village example (page 2). Running the initial search strategy on this problem requires checking 2,661 different domain sort assignments. Using the encoding described above means that only 18 assignments are tried. The search carried out by our approach is illustrated in Fig. 1. We give two projections of the 3-dimensional search space and the arrows show the parts of the search space ruled out by $s$-beam constraints. On each step the constraints rule out all but one neighbour, meaning that we take a direct path to the solution through the search space.

### 5.5   Using Monotonicity

In our framework we can use the notion of monotonicity (see Sect. 4) in two ways.

– *Collapsing Monotonic Sorts.* All monotonic sorts can be collapsed into a single sort as this sort can grow to the size of the largest monotonic sort. It is never safe to collapse a monotonic sort into a non-monotonic one as the monotonic sort may depend on the non-monotonic one. For example, whilst *banana* is a monotonic sort it must always be twice as large as the non-monotonic sort *monkey*.

– *Refining the Search Encoding.* If a sort $s$ is monotonic then if no model exists for domain size $n_s$ then no model can exist where $n_s$ is smaller. We reflect this in our encoding by not marking $DC$-instances with marking variables for monotonic sorts. This leads to a derivation of potentially stronger constraints.

## 6   Detecting Constraints Between Sorts

In this section we discuss how properties of functions between sorts can be used to further constrain the domain size search space. Consider the following set of formulas

$$\text{distinct}(a_1, a_2, a_3, a_4, a_5) \qquad (\forall x : s_1)(f(x) \neq b)$$
$$(\forall x, y : s_1)(f(x) = f(y) \rightarrow x = y) \qquad (\forall x, y : s_2)(g(x) = g(y) \rightarrow x = y)$$

where $a_i$ are constants of sort $s_1$, $b$ is a constant of sort $s_2$, $f : s_1 \rightarrow s_2$, and $g : s_2 \rightarrow s_3$. The previous approach would try increasing the size of each sort by 1, discovering that one sort must grow at each step. However, we can see from the bottom two formulas that $f$ and $g$ are injective and therefore that $|s_1| \leq |s_2| \leq |s_3|$, furthermore, the second formula tells us that $f$ is non-surjective and therefore that $|s_1| < |s_2|$. Using these constraints we can immediately discount 9 of the 15 domain size assignments considered without them.

To find constraints between the sizes of sorts $s_1$ and $s_2$ we look for four cases:

1. If a function $f : s_1 \rightarrow s_2$ is injective, then $|s_1| \leq |s_2|$.
2. If a function $f : s_1 \rightarrow s_2$ is injective and non-surjective, then $|s_1| < |s_2|$.
3. If a function $f : s_1 \rightarrow s_2$ is surjective, then $|s_1| \geq |s_2|$.
4. If a function $f : s_1 \rightarrow s_2$ is surjective and non-injective, then $|s_1| > |s_2|$.

These constraints can be added to the constraints used in the search described in Sect. 5.

Our method for detecting bounds was inspired by Infinox [3], a method for showing no finite model can exist for unsorted FOL formulas if there is a strict bound within a sort. We detect bounds by attempting to prove properties of functions *between* sorts. For a unary function $f : s_1 \rightarrow s_2$ occurring in the problem we can simply make a claim such as

$$(\forall x : s_1)(\forall y : s_2)(f(x) = f(y) \rightarrow x = y) \land (\exists y : s_2)(\forall x : s_1)(f(x) \neq y)$$

for each case (this is case (2) above), and then ask whether this claim follows from the axioms of the input problem. For non-unary functions it is necessary to existentially quantify over one of the arguments, details of how to do this can be found in [3].

To check each claim $C$ we could use standard techniques to check $A \models C$ where $A$ are the input axioms. Any black box solver could be used for this. However, doing this on a per-claim basis is inefficient and we implement an optimisation of Vampire's saturation loop to establish multiple claims in a single proof attempt. Recall that the saturation loop will search for consequences of its input. Therefore, we saturate $A \cup \{C_i \rightarrow l_i\}$ where $l_i$ is a fresh propositional symbol labelling claim $C_i$. If the unit $l_i$ is derived then we can conclude that the claim $C_i$ is a consequence of $A$. This approach was inspired by the consequence elimination mode of Vampire [9] (see this work for technical details).

# 7   Getting More Sorts

Previously we have seen how sort information can be used to reduce the size of the SAT encoding by only growing the domain sizes of sorts that need to be grown. In this section we recall a technique first described in [4] for inferring new sorts and explain how these new sorts can be useful.

**Inferring Subsorts.** Consider the Organised Monkey Village example. The *monkey* sort can be split into three separate subsorts as there are three parts (assigning bananas to monkeys, assigning monkeys to trees and assigning monkeys to their partners) where the signatures do not overlap. Abstractly, we can use different monkeys in these different places as they do not interact – later we will see why this is useful. To infer such subsorts we can use the standard union-find method on positions in the signature.

**Using Inferred Subsorts.** Claessen et al. [4] describe two uses for inferred subsorts:

–*Removing Instances.* If a subsort $\tau$ is monotonic and all function symbols with the return sort $\tau$ are constants, then we can bound the subsort by the number of constants. It is easy to argue that any ground clauses (instances, totality or functionality) for a domain constant larger than the bound of the subsort can be omitted as they will necessarily be equivalent to an existing clause. This helps reduce the size of the SAT encoding.

–*Symmetry breaking.* For the same reasons that symmetry breaking can occur per sort, symmetry breaking can now occur per inferred subsort. This is safe due to the above observation that values for different subsorts will never be compared.

**Making Subsorts Proper Sorts.** Proper sorts and inferred subsorts are treated differently as we only grow the sizes of proper sorts. If an inferred subsort is not bounded as described above then it is forced to grow to the same size as its parent sort. To understand why this can be problematic consider the FOL formula

$$\text{distinct}(a_1, \ldots, a_{50}) \wedge (\forall x)(f(f(f(f(f(f(f(f(f(f(x)))))))))) \neq x)$$

which has an overall finite model size of 50. Establishing this finite model requires a SAT problem consisting of 1,187,577 clauses. However, there are two subsorts: that of the constants $a_1$ to $a_{50}$ and that of $f$. The second subsort is monotonic and does not need to grow beyond size 3. If this had been declared as a separate sort then the required SAT encoding would only consist of 125,236 clauses.

It is only safe to treat an inferred subsort as a proper sort if we can translate any resulting model into one where elements of the inferred subsort belong to the original sort. This is possible when (i) the inferred subsort is monotonic, and

(ii) the size of the inferred subsort is not larger than the size of its parent sort. To ensure (ii) we add constraints to the search strategy in the same way as for sort bounds detected previously.

# 8  An Alternative Growing Search

The previous search strategy considers each domain size assignment separately (we therefore refer to it as a *pointwise* encoding). However, we can modify the encoding so that it captures the current assignment *and all smaller ones* at the same time. Thus we no longer talk of a domain size but rather of a domain size upper *bound*, as the parameter of the encoding. These bounds never need to shrink and thus grow monotonically for each sort. We call this encoding a *contour* encoding as we can think of it drawing a contour around the explored part and growing this outwards.

This alternative encoding works as follows. For each sort $s$ with domain size bound $n_s$ we introduce $n_s$ propositional variables $bound_s(1)$ to $bound_s(n_s)$. Then instead of single totality constraint for each principal term $p$ we introduce all totality constraints for domain sizes up to $n_s$ guarded by the appropriate bound i.e.

$$p = c_1 \vee bound_s(1), \quad \ldots, \quad p = c_1 \vee \ldots \vee p = c_{n_s} \vee bound_s(n_s)$$

We guard $DC$-instances of clauses with negations of these guards in the following way. For each sort $s$ let $s_{max}$ be the index of the largest domain constant in this instance used to replace a variable of sort $s$. Then if $s_{max}$ is defined, i.e. there is at least one such variable, and $s_{max} > 1$ we guard the instance with a literal $\neg bound_s(s_{max} - 1)$. For example, given a function symbol $f : s_1 \rightarrow s_2$, a constant $b : s_2$, and a flattened clause $f(x) \neq y \vee b \neq y$, its $DC$-instance $f(c_3) \neq c_1 \vee b \neq c_1$ would be guarded as $f(c_3) \neq c_1 \vee b \neq c_1 \vee \neg bound_{s_1}(2)$.

In this encoding the SAT solver can satisfy the clauses for a domain size smaller than $n_s$ i.e. if it can satisfy a stricter totality constraint then it can effectively ignore some of the instances. As a further variation, if a sort is monotonic then we do not need to consider the possibility that a sort is smaller than its current bound. Therefore, we only need the largest totality constraint and do not need constraints on instances.

In a similar way as before, we solve the problem under the assumptions that the sort sizes are big enough i.e.

$$A = \bigwedge_{s \in S} \neg bound_s(n_s).$$

If this is shown unsatisfiable the subset of assumptions $A_0$ will suggest the sorts that could be grown; growing a sort not mentioned in $A_0$ would allow the same proof of unsatisfiability to be produced. If $A_0$ is empty, the SAT-solver has shown that the given first-order formula is unsatisfiable. Otherwise, we can either arbitrarily select a sort to grow out of the ones mentioned in $A_0$. This approach is significantly different from the previous approach as now we only consider one

**Table 1.** Experimental results for unsorted problems.

|  | Vampire | | | | | |
|---|---|---|---|---|---|---|
|  | CVC4 | Paradox | iProver | Ignore | Use | Expand |
| FOF+CNF: sat | 1181 | 1444 | 1348 | 1421 | 1463 | **1503** |
| FOF+CNF: unsat | - | - | 1337 | 1400 | 1604 | **1628** |

next domain size assignment. However, the SAT problems may be considerably harder to solve as the SAT solver is now considering a much larger set of models. In essence, each new SAT problem contains all the previous ones as sub-problems.

Finally, if the SAT problem is satisfiable then the actual size of a sort $s$ is given by its smallest totality constraint that is "enabled"; more precisely, by the smallest $i$ such that $bound_s(i)$ is false in the computed model.

## 9   Experimental Evaluation

In this section we evaluate the different techniques for finite model finding in multi-sorted FOL described in this paper and compare our approach to other tools.

**Experimental Setup.** We considered two sets of problems. From the TPTP [16] library (version 6.3.0) we took unsorted problems in the FOF or CNF format. From the SMT-LIB library [1] we took problems from the UF (Uninterpreted Functions) logic. Experiments were run on the StarExec cluster [15], whose nodes are equipped with Intel Xeon 2.4 GHz processors and 128 GB of memory. For each experiment we will report the number of problems solved with the time limit of 60 s.

On satisfiable problems we compare our implementation with version 3.0 of Paradox [5] and version 1.4 of CVC4 [13]; Paradox does not establish unsatisfiability and CVC4 runs more than a finite model finding approach, making a comparison on unsatisfiable problems difficult. On the TPTP problems, we also compare to version 2.0 of iProver [10]. The techniques described in this paper were implemented in Vampire.

**Adding Sorts to Unsorted Problems.** Our first experiment considers the effect of sort inference on unsorted problems. We consider three settings: (i) inferred subsorts are ignored, (ii) inferred subsorts are used to reduce the problem size and break symmetries only, and (iii) inferred subsorts are expanded to proper sorts where possible. Table 1 presents the results. This shows that sort information can be used to solve more problems. For satisfiable problems the best Vampire strategy solves more problems than CVC4, Paradox or iProver. For both satisfiable and unsatisfiable problems, expanding subsorts into proper sorts and treating the problems as multi-sorted problems helps solve the most problems. We note that 4 problems found unsatisfiable using this approach could not be solved by any other technique in Vampire, this is significant as Vampire is one of the best theorem provers available for such problems.

**Table 2.** Experimental results for translations from multi-sorted to unsorted.

| | Sort Predicates | | | | Sort Functions | | | |
|---|---|---|---|---|---|---|---|---|
| | Plain | Monotonicity | Subsorts | Both | Plain | Monotonicity | Subsorts | Both |
| UF: sat | 813 | 810 | 872 | **874** | 710 | 771 | 834 | 873 |
| UF: unsat | 101 | 112 | 221 | **232** | 67 | 67 | 171 | 171 |

**Table 3.** Experimental results for multi-sorted problems.

| | CVC4 | Pointwise | | | | Contour | | | Without |
|---|---|---|---|---|---|---|---|---|---|
| | | Default | Expand | Collapse | Bounds | Default | Collapse | Bounds | Constraints |
| UF: sat | 764 (8) | 795 | 789 | **901 (12)** | 810 | 886 (3) | 899 (1) | 886 (1) | 154 |
| UF: unsat | - | 212 | 215 | 241 | 218 | 270 | 261 | 267 | 66 |

**Removing Sorts from Sorted Problems.** Next we consider the translation techniques described in Sect. 4 applied to multi-sorted problems. Table 2 shows the results of running variations of these translations on the multi-sorted UF problems described above. Plain applies the translation to the whole problem, ignoring subsorts in the result. With Monotonicity only non-monotonic sorts are translated and with Subsorts the resulting problem is solved using inferred subsorts. Both adds both variations.

These results show that, for these problems, sort predicates are more useful and that the techniques of monotonicity detection and subsort inference are useful in improving the translation and reasoning with it.

**Finding Models of Multi-sorted Problems.** Finally, we consider our framework for reasoning with multi-sorted problems directly. Table 3 gives the results for CVC4 and eight variations of the techniques presented in this paper (we use only CVC4 since Paradox does not work on sorted problems). At the top level these are split into the Pointwise and Contour encodings and a version where no constraints were added. Then Expand refers to subsort expansion (Sect. 7), Collapse refers to collapsing monotonic sorts together (Sect. 5.5), and Bounds refers to sort bound extraction (Sect. 6). These results can also be compared to Table 2 as the problems are the same.

The three main conclusions from this information are (i) overall the approach taken in this paper is able to solve more problems than the approach taken by CVC4, (ii) collapsing monotonic sorts is very useful, and (iii) including the search problem as part of the SAT encoding is vital. Bracketed numbers show unique problems solved by an approach. This shows that although CVC4 solves fewer problems it does solve some uniquely. The contour encoding was generally more successful, however in UF there are 15 and 19 problems that are only solvable using the pointwise and contour encodings respectively. As a further point, we note that the heuristic introduced on page 9 is useful, without it the default pointwise approach solved 61 fewer problems. Finally, comparing with the results in Table 2, we see that finding models for multi-sorted problems directly performs better than translating the problem to an unsorted one.

# 10    Conclusions and Further Work

We have introduced a new framework for MACE-style finite model finding for multi-sorted first-order logic. This involved two complementary SAT encodings that capture the search for a satisfying domain size assignment and techniques aimed at decreasing the size of this search space. We have demonstrated experimentally that these techniques are effective at improving finite model finding in the unsorted setting and finding finite models for multi-sorted first-order formulas. Further work will consider possible extensions to uninterpreted sorts and infinite, but finitely representable, models.

# References

1. Barrett, C., Stump, A., Tinelli, C.: The Satisfiability Modulo Theories Library (SMT-LIB) (2010). http://www.SMT-LIB.org
2. Blanchette, J.C., Böhme, S., Popescu, A., Smallbone, N.: Encoding monomorphic and polymorphic types. In: Piterman, N., Smolka, S.A. (eds.) TACAS 2013 (ETAPS 2013). LNCS, vol. 7795, pp. 493–507. Springer, Heidelberg (2013)
3. Claessen, K., Lillieström, A.: Automated inference of finite unsatisfiability. J. Autom. Reasoning **47**(2), 111–132 (2011)
4. Claessen, K., Lillieström, A., Smallbone, N.: Sort it out with monotonicity. In: Bjørner, N., Sofronie-Stokkermans, V. (eds.) CADE 2011. LNCS, vol. 6803, pp. 207–221. Springer, Heidelberg (2011)
5. Claessen, K., Sörensson, N.: New techniques that improve MACE-style model finding. In: CADE-19 Workshop: Model Computation - Principles, Algorithms and Applications (2003)
6. Eén, N., Biere, A.: Effective preprocessing in SAT through variable and clause elimination. In: Bacchus, F., Walsh, T. (eds.) SAT 2005. LNCS, vol. 3569, pp. 61–75. Springer, Heidelberg (2005)
7. Eén, N., Sörensson, N.: An extensible SAT-solver. In: Giunchiglia, E., Tacchella, A. (eds.) SAT 2003. LNCS, vol. 2919, pp. 502–518. Springer, Heidelberg (2004)
8. Eén, N., Sörensson, N.: Temporal induction by incremental SAT solving. Electr. Notes Theor. Comput. Sci. **89**(4), 543–560 (2003)
9. Hoder, K., Kovács, L., Voronkov, A.: Case studies on invariant generation using a saturation theorem prover. In: Batyrshin, I., Sidorov, G. (eds.) MICAI 2011, Part I. LNCS, vol. 7094, pp. 1–15. Springer, Heidelberg (2011)
10. Korovin, K.: iProver – an instantiation-based theorem prover for first-order logic (System Description). In: Armando, A., Baumgartner, P., Dowek, G. (eds.) IJCAR 2008. LNCS (LNAI), vol. 5195, pp. 292–298. Springer, Heidelberg (2008)
11. Kovács, L., Voronkov, A.: First-order theorem proving and VAMPIRE. In: Sharygina, N., Veith, H. (eds.) CAV 2013. LNCS, vol. 8044, pp. 1–35. Springer, Heidelberg (2013)
12. Mccune, W.: A Davis-Putnam Program and its Application to Finite First-Order Model Search: Quasigroup Existence Problems. Technical report, Argonne National Laboratory (1994)
13. Reynolds, A., Tinelli, C., Goel, A., Krstić, S.: Finite model finding in SMT. In: Sharygina, N., Veith, H. (eds.) CAV 2013. LNCS, vol. 8044, pp. 640–655. Springer, Heidelberg (2013)

14. Schulz, S.: A comparison of different techniques for grounding near-propositional CNF formulae. In: Proceedings of the Fifteenth International Florida Artificial Intelligence Research Society Conference, May 14–16, 2002, Pensacola Beach, Florida, USA, pp. 72–76 (2002)
15. Stump, A., Sutcliffe, G., Tinelli, C.: StarExec, a cross community logic solving service (2012). https://www.starexec.org
16. Sutcliffe, G.: The TPTP problem library and associated infrastructure. J. Autom. Reasoning **43**(4), 337–362 (2009)
17. Tammet, T.: Reasoning. Gandalf. J. Autom **18**(2), 199–204 (1997)
18. Zhang, J., Zhang, H.: SEM: a system for enumerating models. In: Proceedings of the Fourteenth International Joint Conference on Artificial Intelligence, IJCAI 95, Montréal Québec, Canada, August 20–25 1995, vol. 2s, pp. 298–303 (1995)

# MCS Extraction with Sublinear Oracle Queries

Carlos Mencía[2(✉)], Alexey Ignatiev[1,3], Alessandro Previti[1],
and Joao Marques-Silva[1]

[1] LaSIGE, Faculty of Science, University of Lisbon, Lisbon, Portugal
{aignatiev,jpms}@ciencias.ulisboa.pt, apreviti.research@gmail.com
[2] University of Oviedo, Gijón, Spain
cmencia@gmail.com
[3] ISDCT SB RAS, Irkutsk, Russia

**Abstract.** Given an inconsistent set of constraints, an often studied problem is to compute an irreducible subset of the constraints which, if relaxed, enable the remaining constraints to be consistent. In the case of unsatisfiable propositional formulas in conjunctive normal form, such irreducible sets of constraints are referred to as Minimal Correction Subsets (MCSes). MCSes find a growing number of applications, including the approximation of maximum satisfiability and as an intermediate step in the enumeration of minimal unsatisfiability. A number of efficient algorithms have been proposed in recent years, which exploit a wide range of insights into the MCS extraction problem. One open question is to find the best worst-case number of calls to a SAT oracle, when the calls to the oracle are kept simple, and given reasonable definitions of simple SAT oracle calls. This paper develops novel algorithms for computing MCSes which, in specific settings, are guaranteed to require asymptotically fewer than linear calls to a SAT oracle, where the oracle calls can be viewed as simple. The experimental results, obtained on existing problem instances, demonstrate that the new algorithms contribute to improving the state of the art.

## 1 Introduction

The analysis of over-constrained systems finds a wide range of practical applications [8,17,20]. Given an inconsistent set of constraints, one is often interested in finding minimal explanations of inconsistency, or in finding maximally consistent sets of constraints (and so minimal sets of constraints to discard). For propositional formulas, a minimal set of clauses to discard to achieve consistency of the remaining clauses is referred to as a minimal correction subset (MCS). In general, selecting minimal sets of constraints is of prime importance, since this enables irrelevant constraints not to be considered. Many application areas of inconsistency analysis, including minimal explanations of inconsistency and maximally consistent subsets of constraints, can be viewed as instantiations of model-based diagnosis [36]. These include, among others, product configuration, axiom pinpointing in description logics, fault localization in software, and design debugging. Other applications of MCSes include the computation of minimal

N. Creignou and D. Le Berre (Eds.): SAT 2016, LNCS 9710, pp. 342–360, 2016.
DOI: 10.1007/978-3-319-40970-2_21

and maximal models [6,23], enumeration of minimal unsatisfiability, approximation of maximum satisfiability and, as a result, also in relational inference [26]. In the concrete case of propositional formulas, the enumeration of MCSes has been used for approximating maximum satisfiability (MaxSAT), enumeration of minimal unsatisfiability and also solving function problems in the second level of the function polynomial hierarchy, including minimum unsatisfiability [19]. Although the subject of over-constrained system analysis has been studied in different settings, in some cases for more than a century[1], recent years have seen a multitude of algorithms and algorithmic optimizations being proposed, both for finding minimal explanations of inconsistency [5,17,18,22] and for finding minimal corrections of inconsistency (which enable the consistency of the remaining constraints) [1,3,14,16,25,27,32,33,39].

This paper studies the problem of computing MCSes of unsatisfiable propositional formulas in conjunctive normal form (CNF). Recent work proposed algorithms that analyze variables instead of clauses [16,27,32]. The main goal of most novel algorithms is to reduce the number of calls to a SAT solver (i.e. the oracle queries) in such a way that the SAT solver calls remain *simple*, i.e. changes to the original formula are mostly negligible. Accordingly, most recent algorithms require a worst-case number of SAT oracle queries that is linear in the number of variables. Given that the number of clauses can be exponentially larger than the number of variables, these algorithms are in the worst-case asymptotically as efficient as using maximum satisfiability [35], while ensuring *simpler* oracle queries, in the sense that no new variables are used and encodings of cardinality constraints are not required [24].

The main contribution of this paper is to develop three novel algorithms for MCS extraction, aiming at reducing the number of oracle queries, both in theory and in practice. Two of the proposed algorithms are shown to require asymptotically fewer than linear SAT oracle queries in the number of (soft) clauses. Nevertheless, for problem domains where the number of (soft) clauses is linear in the number of (interesting) variables, the algorithms are shown to require asymptotically fewer oracle queries than recent algorithms [16,27,32]. Concrete examples of problem domains where the number of (soft) clauses is linear on the number of relevant variables include computing minimal and maximal models [6,23]. Besides providing theoretical guarantees in terms of the worst-case number of SAT oracle queries, the new algorithms are shown to be effective in practice. On the one hand, the new algorithms are shown to be more robust, being less dependent on practical optimizations commonly used in practice. In addition, the new algorithms contribute to improving the performance of portfolios of MCS extraction algorithms.

It should also be emphasized that the proposed new algorithms find application in settings other than MCS extraction. As shown elsewhere [21,28,29], MCS extraction is an instantiation of the problem of computing a minimal set subject to a monotone predicate (MSMP), with a specific predicate form. The

---

[1] This is the case for example with the analysis of infeasible systems of linear inequalities [10].

algorithms proposed in this paper can be applied to any computational problem that can be reduced to the same MSMP predicate form. A list of problems with the same MSMP predicate form as MCS extraction can be found in [28], and include finding a minimal distinguishing subset (MDS), computing a minimal (and a maximal) model, and computing a maximum autarky.

The paper is organized as follows. Section 2 introduces the notation and definitions used throughout. Section 3 provides a brief overview of recent work on MCS extraction. Section 4 builds on recent algorithms and develops an alternative MCS extraction algorithm that requires a worst-case linear number of queries to the SAT oracle on the number of clauses. The insights provided by this new algorithm enable the development of two novel algorithms in Sect. 5 which require worst-case sublinear number of queries to a SAT oracle, on the number of clauses. This section also shows that a number of important classes of problems are shown to require sublinear number of queries to a SAT oracle, on the number of variables. Section 6 evaluates the proposed algorithms on standard problem instances [1,2,16,27,32]. The results demonstrate that the new algorithms proposed in this paper enable further improvements to the state of the art in MCS extraction. Section 7 concludes the paper.

## 2    Preliminaries

This section introduces the notation and definitions used throughout the paper. Standard propositional logic definitions apply (e.g. [7]). CNF formulas are defined over a set of propositional variables. A CNF formula $\mathcal{E}$ is a finite conjunction of clauses, also interpreted as a finite set of clauses. A clause is a disjunction of literals, also interpreted as a set of literals. A literal is a variable or its complement. $m \triangleq |\mathcal{E}|$ represents the number of clauses. The set of variables associated with a CNF formula $\mathcal{E}$ is denoted by $X \triangleq \text{var}(\mathcal{E})$, with $n \triangleq |X|$. A CNF formula $\mathcal{E}$ is partitioned into a set of soft clauses $\mathcal{F}$, which can be relaxed (i.e. may not be satisfied), and a(n optional) set of hard clauses $\mathcal{H}$, which cannot be relaxed (i.e. must be satisfied). Thus, a CNF formula $\mathcal{E}$ is represented by a pair $\langle \mathcal{H}, \mathcal{F} \rangle$. Moreover, MUSes, MCSes and MSSes are defined over $\mathcal{F}$ taking into account the hard clauses in $\mathcal{H}$ as follows:

**Definition 1 (Minimal Unsatisfiable Subset (MUS)).** *Let $\mathcal{F}$ denote the set of soft clauses and $\mathcal{H}$ denote the set of hard clauses, such that $\mathcal{H} \cup \mathcal{F}$ is unsatisfiable. $\mathcal{M} \subseteq \mathcal{F}$ is a* Minimal Unsatisfiable Subset *(MUS) iff $\mathcal{H} \cup \mathcal{M}$ is unsatisfiable and $\forall_{\mathcal{M}' \subsetneq \mathcal{M}}, \mathcal{H} \cup \mathcal{M}'$ is satisfiable.*

**Definition 2 (Minimal Correction Subset (MCS)).** *Let $\mathcal{F}$ denote the set of soft clauses and $\mathcal{H}$ denote the set of hard clauses, such that $\mathcal{H} \cup \mathcal{F}$ is unsatisfiable. $\mathcal{C} \subseteq \mathcal{F}$ is a* Minimal Correction Subset *(MCS) iff $\mathcal{H} \cup \mathcal{F} \setminus \mathcal{C}$ is satisfiable and $\forall_{\mathcal{C}' \subsetneq \mathcal{C}}, \mathcal{H} \cup \mathcal{F} \setminus \mathcal{C}'$ is unsatisfiable.*

**Definition 3 (Maximal Satisfiable Subset (MSS)).** *Let $\mathcal{F}$ denote the set of soft clauses and $\mathcal{H}$ denote the set of hard clauses, such that $\mathcal{H} \cup \mathcal{F}$ is unsatisfiable. $\mathcal{S} \subseteq \mathcal{F}$ is a* Maximal Satisfiable Subset *(MSS) iff $\mathcal{H} \cup \mathcal{S}$ is satisfiable and $\forall_{\mathcal{S}' \supsetneq \mathcal{S}}, \mathcal{H} \cup \mathcal{S}'$ is unsatisfiable.*

For query complexity analyses, the size of either the largest or the smallest MCS is denoted by $k$. Recent years have seen extensive work on extracting MUSes and MCSes. Recent work on MUS extraction is summarized for example in [2,5,29]. Recent work on MCS extraction is overviewed in the next section. Among the many applications of MCS extraction, a representative example is the computation of minimal and maximal models [6,23]. Throughout the paper special emphasis will be given to problems for which the number of soft clauses is linear on the number of variables, with the computation of minimal and maximal models representing concrete examples.

Unless stated otherwise, in the remainder of the paper clauses are assumed to be half-reified [15], and the algorithms are described in terms of the reification variables. Concretely, the input to an MCS extraction algorithm is a pair of clauses $\langle \mathcal{H}, \mathcal{F} \rangle$, where $\mathcal{H}$ denotes the *hard* clauses, and $\mathcal{F}$ denotes the soft clauses, and such that $\mathcal{H} \nvDash \bot$ and $\mathcal{H} \wedge \mathcal{F} \vDash \bot$. An initial assignment is selected from which two sets $\mathcal{S} \supseteq \mathcal{H}$ and $\mathcal{U} \subseteq \mathcal{F}$ are obtained, that denote respectively the satisfied and falsified clauses of $\langle \mathcal{H}, \mathcal{F} \rangle$. The clauses in $\mathcal{U}$ are half-reified using a set $\mathcal{R}$ of $\mathcal{O}(m)$ fresh variables [2,31]. Each (soft) clause $c_i \in \mathcal{U}$ is replaced by a (hard) clause $(\neg r_i \vee c_i)$, which is added to $\mathcal{S}$, and the new variable $r_i$ is added to $\mathcal{R}$. The set $\mathcal{U}$ of falsified soft clauses is replaced by a new set of unit (soft) clauses $(r_i)$, one for each $r_i \in \mathcal{R}$. Given $\mathcal{U}$, each variable occurs with a single polarity (otherwise one of the clauses would be satisfied) [27]. The $D$ clause is the disjunction of all literals in the clauses of $\mathcal{U}$; either the original literals or the reification literals can be considered. The $D$ clause check step is to run a SAT oracle on $\mathcal{S} \wedge D$ (or on a subset of $D$), being used in recent MCS extraction algorithms [16,27,32].

Given a formula $\mathcal{E}$, a backbone literal $l$ of $\mathcal{E}$ is such that $\mathcal{E} \vDash l$. Given an MCS $\mathcal{C}$ of a pair $\langle \mathcal{H}, \mathcal{F} \rangle$, any literal in $\mathcal{C}$ is the complement of a backbone literal of $\mathcal{H} \wedge (\mathcal{F} \setminus \mathcal{C})$ [27].

A SAT oracle is modeled as a function call that, in this paper, returns an outcome, which is either true (or SAT), for a satisfiable formula, or false (or UNSAT), for an unsatisfiable formula. For the cases the outcome is true, the SAT oracle also returns a witness of satisfiability $\mu$ (i.e. a satisfying truth assignment). Given a pair $\langle \mathcal{H}, \mathcal{F} \rangle$, a SAT solver call is thus represented as $(\text{st}, \mu) \leftarrow \text{SAT}(\mathcal{H} \wedge \mathcal{F})$, where st is either *true* (or SAT) or *false* (or UNSAT).

## 3   Related Work

The extraction of MCSes can be traced at least to the work of Reiter [36], in the form of minimal diagnoses[2]. A wealth of work has been developed since

---

[2] Nevertheless, irreducible inconsistent sets of linear inequalities have been studied for more than a century [10].

then. We emphasize the work more related with MCS extraction. Besides the work on algorithms for MCS and MUS extraction, there is a well-known hitting set duality relationship between MCSes and MUSes, which has been studied in different settings [3, 8, 25, 36].

A simple approach for MCS extraction is to use MaxSAT [25], with optimality guarantees in terms of the worst-case number of SAT oracle queries [9, 35]. Recent experimental results [27] indicate that the use of MaxSAT can perform poorly in practice, on representative classes of problem instances. As a result, most recent work focused on *simple* SAT oracle queries, where the modifications to the original formula are minimal. (For example, the use of cardinality constraints and the auxiliary variables required for CNF encoding cardinality constraints yields CNF formulas fairly different from the original CNF formula, which can be significantly harder.).

One of the most widely used MCS extraction algorithms is *linear search* [3, 27, 33]. Starting from a set $S$ of satisfiable clauses and a set $U$ of falsified clauses, the linear search algorithm iteratively tests the satisfiability of $S \cup \{c\}$, for each clause $c \in U$. If the outcome is true, the clause is added to $S$; otherwise it is discarded, i.e. it is included in the MCS. The number of SAT oracle calls grows with the number of clauses, i.e. $\Theta(m)$. Recent work [21, 28, 29] has shown that linear search essentially corresponds to the well-known deletion-based MUS extraction algorithm [4, 11].

Inspired by the work on QuickXplain [22], FastDiag [14] represents an alternative for MCS extraction, which often requires a number of oracle calls smaller than linear (on the number of clauses), but which is linear in the worst-case. An alternative, similar to dichotomic search for MUS extraction [18], is the generation of corrective explanations [34], which is worse than linear in the worst-case.

The modification of the actual oracle, by introducing a preference on the literals was considered in different settings [1, 37, 38]. These approaches require a *single* SAT oracle call, with the main drawback being that the oracle is modified to solve SAT problems with preferences. Existing experimental evidence [32] (but also the results in this paper) indicate that the use of SAT with preferences does not scale for the more challenging MCS extraction problem instances.

Recent years have seen a number of alternative algorithms being proposed. The CLD algorithm [27] exploits the fact that falsified clauses do not have complemented literals, and so analyzes a single clause (i.e. the $D$ clause) at each iteration. The $D$ clause represents the disjunction of the literals in the falsified clauses of $U$. An important observation is that the worst-case number of oracle calls for the CLD algorithm grows with the number of variables [32]. Moreover, it has been shown that the CLD can be applied to the more general setting of constraint programming [31], by using half reification [15]. The connection of backbone literals with MCSes was first investigated in [27] and further refined in [32]. These insights are used extensively throughout this paper. A variant of CLD was also recently proposed [2]. Similar to earlier work [31], half reification is also used.

---

**Algorithm 1.** Literal-Based eXtractor (LBX) [32]

---

1  **Function** LBX$(\mathcal{H}, \mathcal{F})$
2    $(\mathcal{S}, \mathcal{U}) \leftarrow \texttt{InitialAssignment}(\mathcal{H}, \mathcal{F})$
3    $(\mathcal{L}, \mathcal{B}) \leftarrow (\texttt{Literals}(\mathcal{U}), \emptyset)$
4    **while** $\mathcal{L} \neq \emptyset$ **do**
5      **if** $\texttt{CheckDClause}(\mathcal{S}, \mathcal{L})$ **then break**;
6      $l \leftarrow \texttt{RemoveLiteral}(\mathcal{L})$
7      $(st, \mu) = \texttt{SAT}(\mathcal{S} \cup \mathcal{B} \cup \{l\})$
8      **if** $st$ **then**
9        $(\mathcal{S}, \mathcal{U}) \leftarrow \texttt{UpdateSATClauses}(\mu, \mathcal{S}, \mathcal{U})$
10       $\mathcal{L} \leftarrow \mathcal{L} \cap \texttt{Literals}(\mathcal{U})$
11     **else**
12       $\mathcal{B} \leftarrow \mathcal{B} \cup \{\neg l\}$
13   **return** $\mathcal{F} \setminus \mathcal{S}$                    // $\mathcal{F} \setminus \mathcal{S}$ is MCS of $\mathcal{F}$

---

---

**Algorithm 2.** Set up phase

---

1  **Function** Setup$(\mathcal{H}, \mathcal{F})$
2    $(\mathcal{S}, \mathcal{U}) \leftarrow \texttt{InitialAssignment}(\mathcal{H}, \mathcal{F})$
3    $(\mathcal{S}, \mathcal{U}, \mathcal{R}) \leftarrow \texttt{ReifyClauses}(\mathcal{U})$
4    **return** $(\mathcal{S}, \mathcal{U}, \mathcal{R})$

---

The CMP algorithm [16] revisits the well-known insertion-based MUS extraction algorithm [12] in the context of MCS extraction[3]. Although apparently the CMP algorithm requires a close to quadratic number of oracle calls [16], it is also the case that the CMP algorithm checks the satisfiability of a (partial) $D$ clause at every major iteration. The main consequence, as shown elsewhere [32], is that this technique reduces the worst-case number of oracle calls to linear, on the number of clauses.

A recent line of work has investigated algorithms that analyze falsified literals (which are bound by the number of variables) instead of clauses [32]. Although not claimed in earlier work, but an immediate consequence of this earlier work [32], is the observation that the number of oracle calls can be exponentially smaller than the size of the formula, because the number of clauses can be exponentially larger than the number of variables. The current state of the art algorithms for MCS extraction [16,32] both require a number of oracle calls that grows linearly with the number of variables. Concretely, LBX [32] requires $\Theta(n)$ SAT oracle calls and CLD requires $\mathcal{O}(\min\{n - r, m - k\})$, where $k$ is the size of the smallest MCS and $r$ is the smallest size of backbone literals in an MCS [32]. LBX, the currently best performing MCS extraction algorithm, is summarized in Algorithm 1. Inspired by CLD [27], and similarly to CMP [16], LBX performs

---

[3] By exploiting the reduction of MCS extraction to MSMP [29], a wealth of different MUS extraction algorithms can be used for MCS extraction, including the insertion-based algorithm.

a $D$ clause check (i.e. the satisfiability of $S$ conjoined with a global $D$ clause) at line 5. If the function call returns true, then $S \land D$ is unsatisfiable, and each literal in $D$ is the complement of a backbone literal of $S$. It is important to note that the linear complexity of CMP is achieved only when a similar $D$ clause check is executed at each iteration of the algorithm. (The experimental results Sect. 6 demonstrate the practical importance of the clause $D$ check.) The extraction of preferred MCSes (and MUSes as well) has been investigated in [30,39], extending earlier work on computing preferred explanations [22]. Finally, the algorithms proposed in this paper can be related with recent work on computing maximum autarkies [24].

## 4   MCS Extraction with Linear Oracle Calls

The algorithms described in this and the next section start with a set up phase shown in Algorithm 2. The set up phase picks an initial assignment that satisfies the clauses in $\mathcal{H}$ and some clauses in $\mathcal{F}$. The falsified clauses in $\mathcal{F}$, represented as set $\mathcal{U}$, are half-reified and added to $S$. The set of reification variables is $\mathcal{R}$. Set $\mathcal{U}$ is rewritten to include the reification variables of the clauses originally in $\mathcal{U}$.

This section develops a new algorithm for MCS extraction. The algorithm is referred to as *Unit Clauses D* (UCD). Similarly to CLD [27], UCD uses $D$ clauses. However, in contrast with CLD, all of the $D$ clauses are *unit*. UCD is also tightly related with FastDiag [14] in that UCD iteratively splits a working set of clauses. Similar to earlier algorithms, an initial SAT call (see Algorithm 2) will split $\langle \mathcal{H}, \mathcal{F} \rangle$ into $S \supseteq \mathcal{H}$ and $\mathcal{U} \subseteq \mathcal{F}$, respectively the satisfied and falsified clauses of $\langle \mathcal{H}, \mathcal{F} \rangle$. Let $X_U$ be the set of falsified literals in the clauses of $\mathcal{U}$. Let $D$ be the set of unit $D$ clauses. A stack $\mathbb{T}$ of sets $\mathcal{D}$ is maintained, initially with a single set $\mathcal{D}$ containing one unit $D$ clause for each falsified literal in $\mathcal{U}$. Moreover, $S$ denotes the set of clauses that must be satisfied, starting from an initial assignment that includes the hard clauses $\mathcal{H}$. Algorithm 3 summarizes the organization of UCD. As can be observed, UCD shares a number of ideas with FastDiag [14] (and so with QuickXplain [22]), with the main differences summarized by the pseudo-code invariants. One invariant is that any set of $D$ clauses in the stack is inconsistent with $S$. Each set is split into two approximately equal sets of $D$ clauses. Each such set is tested for consistency with $S$. If the SAT formula is satisfiable, the set $S$ is updated. The satisfied unit soft clauses (each with a single reified variable) become unit hard clauses. As a result, at least one call to the SAT solver is *guaranteed* to return false (i.e. the formula is unsatisfiable). Thus, at each iteration, if $|\mathcal{D}| > 1$, then at least one set of $D$ clauses is added to the stack. The size of the sets of $D$ clauses is strictly decreasing. Whenever a set with a single $D$ clause is found, the complements of the associated literals are added to the set of backbone literals.

**Proposition 1.** *(Correctness of UCD) Algorithm 3 computes an MCS $C$ of $\langle \mathcal{H}, \mathcal{F} \rangle$.*

---

**Algorithm 3.** Unit Clauses $D$ (UCD)

---

1  **Function** UCD($\mathcal{H}$, $\mathcal{F}$)
2      $(\mathcal{S}, \mathcal{U}, \mathcal{R}) \leftarrow$ Setup($\mathcal{H}, \mathcal{F}$)
3      $(\mathcal{B}, \mathcal{D}) \leftarrow (\emptyset, \text{CreateUnitClausesD}(\mathcal{R}))$
4      Push($\mathbb{T}, \mathcal{D}$)
5      **while** $\mathbb{T} \neq \emptyset$ **do**
6          **if** CheckDClause($\mathcal{S}, \mathcal{R}$) **then break**;
7          $\mathcal{D} \leftarrow$ Pop($\mathbb{T}$)
8          $\mathcal{D} \leftarrow$ UpdateDClauses($\mathcal{D}, \mathcal{R}, \mathcal{B}$)
            // **Invariant:** $\neg\text{SAT}(\mathcal{S} \wedge \mathcal{D})$
9          **if** $|\mathcal{D}| \leq 1$ **then**
10             $(\mathcal{B}, \mathcal{R}) \leftarrow$ UpdateBackbones($\mathcal{B}, \mathcal{R}, \mathcal{D}$)
11             **continue**
12         $(\mathcal{D}_1, \mathcal{D}_2) \leftarrow$ HalfSplit($\mathcal{D}$)
13         $(\text{st}_1, \mu) \leftarrow \text{SAT}(\mathcal{S} \cup \bigcup_{D_i \subset \mathcal{D}_1}\{D_i\})$
14         **if** $\text{st}_1$ **then** $(\mathcal{S}, \mathcal{R}) \leftarrow$ UpdateSATClauses($\mu, \mathcal{S}, \mathcal{R}$)
15         $(\text{st}_2, \mu) \leftarrow \text{SAT}(\mathcal{S} \cup \bigcup_{D_i \in \mathcal{D}_2}\{D_i\})$
16         **if** $\text{st}_2$ **then** $(\mathcal{S}, \mathcal{R}) \leftarrow$ UpdateSATClauses($\mu, \mathcal{S}, \mathcal{R}$)
            // **Invariant:** $\neg\text{st}_1 \vee \neg\text{st}_2$
17         **if** $\neg\text{st}_1$ **then** Push($\mathbb{T}, \mathcal{D}_1$)
18         **if** $\neg\text{st}_2$ **then** Push($\mathbb{T}, \mathcal{D}_2$)
19      **return** $\mathcal{F} \setminus \mathcal{S}$                              // $\mathcal{F} \setminus \mathcal{S}$ is MCS of $\mathcal{F}$

---

*Proof.* (Sketch) By inspection the following invariants hold. First, the set of clauses $\mathcal{S}$ is satisfiable. Second, any set of unit $D$ soft clauses consistent with $\mathcal{S}$ is added to $\mathcal{S}$, and consistency of $\mathcal{S}$ is preserved. If the set of unit clauses $D$ is unsatisfiable given $\mathcal{S}$, then the set is split and the resulting sets of unit $D$ clauses are pushed onto the stack. The size of the sets of $D$ clauses in $\mathbb{T}$ is decreasing, and the stack is not updated when the size of the $D$ clauses is less than or equal to 0. Thus the algorithm is terminating. Any clause $c \in \mathcal{F}$ not added to $\mathcal{S}$ is such that $\mathcal{S} \cup \{c\}$ is unsatisfiable. Finally, sets of $D$ clauses of size not greater than 1 are either discarded (if there are no literals) or added as the (complement) of backbone literals, with $\mathcal{S} \wedge D \vDash \bot$. Thus, the UCD algorithm computes an MCS of $\langle \mathcal{H}, \mathcal{F} \rangle$.                                                                    □

**Proposition 2.** *(Query Complexity of UCD) The worst case number of SAT oracle calls for the UCD algorithm is $\mathcal{O}(k + k \log(\frac{m}{k}))$, where $k$ is the size of the largest MCS.*

*Proof.* (Sketch) The query complexity analysis ignores the additional SAT oracle query made by the $D$ clause check set (see line 6). This additional call at most doubles the total number of SAT oracle queries. The analysis develops an upper bound on the number of oracle queries. The algorithm searches a tree with $m$ possible leaves, with a total number of nodes bounded by $2m$. Since each node can make two oracle queries, except the leaves, then an upper bound on the number of oracle queries is $2m - 1$. Groups of literals that are not in the MCS, and so the outcome of the SAT query is true, are not pushed onto the stack. Groups of literals that contain one or more literals in the MCS, and so the outcome of the SAT query is false, are pushed onto the stack. Therefore, the number of oracle queries grows with the number of literals in the (largest) MCS, $k$ with $k \geq 1$. To maximize the number of oracle queries the literals in the MCS must be as far apart as possible. Thus, we consider groups of $\frac{m-k}{k} + 1 = \frac{m}{k}$ literals, where the last one is a literal in the MCS. We now consider how each literal in the MCS is identified. The literal in the MCS needs to be selected after a number of queries that will select that literal out of $\frac{m}{k}$ literals. This corresponds to a tree with an upper bound of $2(\frac{m}{k}) - 1$ nodes. Of these, the SAT query will return false in $\log\left(\frac{m}{k}\right) + 1$ nodes and true in $\log\left(\frac{m}{k}\right)$ nodes, because the subtree for the $\frac{m}{k}$ elements starts with an unsatisfiable call. Thus, the number of saved oracle queries becomes $2(\frac{m}{k}) - 1 - 2\log\left(\frac{m}{k}\right) - 1$. Since there are $k$ literals in the MCS, then the total number of oracle queries becomes $2m - 2k(\frac{m}{k}) + 2k\log\left(\frac{m}{k}\right) + 2k$, which can be simplified to $2k + 2k\log\left(\frac{m}{k}\right)$. Thus, an upper bound on the number of oracle queries is $\mathcal{O}(k + k\log\left(\frac{m}{k}\right))$.                           □

For problem domains where the number of clauses are of the order of the number of variables (e.g. computing minimal and maximal models), the query complexity becomes asymptotically competitive with LBX [32] and CLD [27].

**Corollary 1.** *The query complexity of UCD is* $\mathcal{O}(k + k\log\left(\frac{n}{k}\right))$ *when* $m = \Theta(n)$.

# 5   MCS Extraction with Sublinear Oracle Calls

This section develops two novel algorithms for MCS extraction which, for specific classes of problems, will require asymptotically less than linear oracle calls. The two algorithms exploit the insights provided by the UCD algorithm (see Sect. 4). The algorithms also build on recent work on computing maximum autarkies [24].

## 5.1   Uniform Splitting with Binary Search

Instead of creating unit clauses $D$, the *Uniform Spliting with Binary Search* (UBS) algorithm, divides the reified variables into (approximately) $\sqrt{m}$ clauses $D$, each containing (approximately) $\sqrt{m}$ literals. The SAT oracle is run on $\mathcal{S}$ together with all the $D$ clauses. If the formula is satisfiable, then at least $\sqrt{m}$ reification variables are satisfied, indicating that the corresponding clauses in $\mathcal{U}$ are consistent with $\mathcal{S}$. As a result, the sets of satisfied and falsified clauses, respectively $\mathcal{S}$ and $\mathcal{U}$, are updated. Moreover, the reification variables are dropped from $\mathcal{R}$. Otherwise, if the formula is unsatisfiable, a binary search step is used to find a culprit, i.e. a $D$ clause which, together with $\mathcal{S}$ is unsatisfiable (or alternatively, $\mathcal{S}$ entails its complement). Algorithm 4 shows the main steps of the proposed approach. The binary search step is shown in Algorithm 5.

The operation of the binary search step used for finding the culprit is essential for correctness, and exploits the ideas used previously for the UCD algorithm. As shown in Algorithm 5, the loop invariant for the binary search procedure is

---

**Algorithm 4.** Uniform Splitting Binary Search (UBS)

---

1  **Function** UBS($\mathcal{H}$, $\mathcal{F}$)
2      $(\mathcal{S}, \mathcal{U}, \mathcal{R}) \leftarrow$ Setup($\mathcal{H}, \mathcal{F}$)
3      $\mathcal{B} \leftarrow \emptyset$
4      **while** $\mathcal{R} \neq \emptyset$ **do**
5          **if** CheckDClause($\mathcal{S}, \mathcal{R}$) **then break**;
6          $\mathcal{D} \leftarrow$ CreateClausesD($\mathcal{R}$)          // Uniform split reif. variables
7          $(\text{st}, \mu) \leftarrow \text{SAT}(\mathcal{S} \cup \bigcup_{D_i \in \mathcal{D}} \{D_i\})$
8          **if** st **then**
9              $(\mathcal{S}, \mathcal{R}) \leftarrow$ UpdateSATClauses($\mu, \mathcal{S}, \mathcal{R}$)
10         **else**
11             $(\mathcal{S}, \mathcal{B}, \mathcal{R}) \leftarrow$ FindCulprit($\mathcal{S}, \mathcal{B}, \mathcal{R}, \mathcal{D}$)
12     **return** $\mathcal{F} \setminus \mathcal{S}$          // $\mathcal{F} \setminus \mathcal{S}$ is MCS of $\mathcal{F}$

---

**Algorithm 5.** Binary Search Step

---

1  **Function** FindCulprit($\mathcal{S}, \mathcal{B}, \mathcal{R}, \mathcal{D}$)
2      **while** $|\mathcal{D}| > 1$ **do**
           // **Invariant:** $\neg\text{SAT}(\mathcal{S} \cup \bigcup_{D_i \in \mathcal{D}} D_i)$
3          $(\mathcal{D}_1, \mathcal{D}_2) \leftarrow$ HalfSplit($\mathcal{D}$)
4          $(\text{st}_1, \mu) \leftarrow \text{SAT}(\mathcal{S} \cup \bigcup_{D_i \in \mathcal{D}_1} \{D_i\})$
5          **if** $\text{st}_1$ **then** $(\mathcal{S}, \mathcal{R}) \leftarrow$ UpdateSATClauses($\mu, \mathcal{S}, \mathcal{R}$)
6          $(\text{st}_2, \mu) \leftarrow \text{SAT}(\mathcal{S} \cup \bigcup_{D_i \in \mathcal{D}_2} \{D_i\})$
7          **if** $\text{st}_2$ **then** $(\mathcal{S}, \mathcal{R}) \leftarrow$ UpdateSATClauses($\mu, \mathcal{S}, \mathcal{R}$)
           // **Invariant:** $\neg\text{st}_1 \vee \neg\text{st}_2$
8          $\mathcal{D} \leftarrow (\neg\text{st}_1) ? \mathcal{D}_1 : \mathcal{D}_2$
9      $(\mathcal{B}, \mathcal{R}) \leftarrow$ UpdateBackbones($\mathcal{B}, \mathcal{R}, \mathcal{D}$)
10     **return** $(\mathcal{S}, \mathcal{B}, \mathcal{R})$

---

such that the set of $D$ clauses being analyzed is inconsistent with the current set $\mathcal{S}$.

**Lemma 1.** *The loop invariants of Algorithm 5 hold.*

*Proof.* (Sketch) Observe that any SAT oracle query with a satisfiable outcome causes the set $\mathcal{S}$ to be updated. Thus, one of the SAT oracle queries must return an unsatisfiable outcome in each loop iteration.          □

Lemma 1 is used for arguing the correctness of the two algorithms investigated in this section.

**Proposition 3.** *(Correctness of UBS) Given $\langle \mathcal{H}, \mathcal{F} \rangle$, UBS (see Algorithm 4) computes an MCS $\mathcal{C}$ of $\langle \mathcal{H}, \mathcal{F} \rangle$.*

*Proof.* (Sketch) By inspection of Algorithm 4, any clause added to $\mathcal{S}$ is such that the resulting set of clauses is consistent. Given a SAT oracle query with an unsatisfiable outcome, by Lemma 1 the binary search step in Algorithm 5 will find a single $D$ clause which is inconsistent with (a possibly updated set) $\mathcal{S}$. Thus $\mathcal{S} \vDash \neg D$. This means that each literal in $D$ is the complement of a backbone literal of $\mathcal{S}$, and so clauses composed of the complements of backbone literals are not consistent with $\mathcal{S}$. □

**Proposition 4.** *(Query Complexity of UBS) The number of SAT oracle calls for the UBS algorithm is $\mathcal{O}(\sqrt{m}\log m)$.*

*Proof.* (Sketch) The query complexity analysis ignores the additional SAT oracle query made by the $D$ clause check (see line 5). This additional call at most doubles the total number of SAT oracle queries. By inspection, the algorithm removes $\mathcal{O}(\sqrt{m})$ literals in each iteration of the main loop. In the worst case, the algorithm runs the binary search step in each iteration of the main loop. The worst-case number of loop iterations for the binary search step is $\mathcal{O}(\log m)$. Thus, the overall number of calls to the SAT oracle is $\mathcal{O}(\sqrt{m}\log m)$. □

**Corollary 2.** *The query complexity of UBS is $\mathcal{O}(\sqrt{n}\log n)$ when $m = \Theta(n)$.*

*Remark 1.* For computing minimal or maximal models, the query complexity of UBS is sublinear on the number of variables.

### 5.2 Literal-Oriented Geometric Progression

For the UBS algorithm, there is no bias towards SAT or UNSAT outcomes. However, UNSAT outcomes have far greater cost, due to the need to find a culprit. As a result, it would be preferable to give preference to obtaining more SAT outcomes. Moreover, for the cases where the number of backbone literals in $\mathcal{U}$ is large, the uniform splitting of UBS may represent a drawback. A modification to the UBS algorithm is to start with larger $D$ clauses, trying to mimic the behavior of the original CLD algorithm. The number of $D$ clauses is increased by a geometric progression while instances are satisfied. Similarly, the number of literals in each $D$ clause is reduced by a geometric progression. When an unsatisfiable instance is identified, the algorithm runs a binary search to identify a culprit. Thus, in contrast with UBS, the size and number of $D$ clauses is not kept constant given the number of literals in $\mathcal{U}$. Algorithm 6 illustrates the proposed approach.

**Proposition 5.** *(Correctness of LOPZ) Given $\langle \mathcal{H}, \mathcal{F} \rangle$, LOPZ (see Algorithm 6) computes an MCS $\mathcal{C}$ of $\langle \mathcal{H}, \mathcal{F} \rangle$.*

*Proof.* (Sketch) By inspection of Algorithm 6, any clause added to $\mathcal{S}$ is such that the resulting set of clauses is consistent. Given a SAT oracle query with an unsatisfiable outcome, by Lemma 1 the binary search procedure in Algorithm 5 will find a single $D$ clause which is inconsistent with (a possibly updated set) $\mathcal{S}$. Thus $\mathcal{S} \vDash \neg D$. This means that each literal in $D$ is the complement of a backbone of $\mathcal{S}$, and so clauses composed of the complements of backbone literals are not consistent with $\mathcal{S}$. □

---

**Algorithm 6.** Literal-Oriented Geometric Progression (LOPZ)

---

1 **Function** UBS($\mathcal{H}$, $\mathcal{F}$)
2     $(\mathcal{S},\mathcal{U},\mathcal{R}) \leftarrow$ Setup($\mathcal{H},\mathcal{F}$)
3     $(i,\mathcal{B}) \leftarrow (0,\emptyset)$
4     **while** $\mathcal{R} \neq \emptyset$ **do**
5         **if** CheckDClause($\mathcal{S},\mathcal{R}$) **then break**
6         $\mathcal{D} \leftarrow$ CreateClausesD($\mathcal{R}, \min(1, |\mathcal{R}|/2^i)$)
7         $(\text{st},\mu) \leftarrow$ SAT($\mathcal{S} \cup \bigcup_{D_i \in \mathcal{D}}\{D_i\}$)
8         **if** st **then**
9             $(\mathcal{S},\mathcal{R}) \leftarrow$ UpdateSATClauses($\mu,\mathcal{S},\mathcal{R}$)
10            $i \leftarrow i+1$
11        **else**
12            $(\mathcal{S},\mathcal{B},\mathcal{R}) \leftarrow$ FindCulprit($\mathcal{S},\mathcal{B},\mathcal{R},\mathcal{D}$)
13            $i \leftarrow 0$
14    **return** $\mathcal{F} \setminus \mathcal{S}$                    // $\mathcal{F} \setminus \mathcal{S}$ is MCS of $\mathcal{F}$

---

**Proposition 6.** *(Query Complexity of LOPZ) The number of SAT oracle calls for the LOPZ algorithm is $\mathcal{O}(\sqrt{m}\log m)$.*

*Proof.* (Sketch) The query complexity analysis ignores the additional SAT oracle query made by the $D$ clause check (see line 5). This additional call at most doubles the total number of SAT oracle queries. We first count the number of removed literals for a sequence of satisfiable oracle calls. If the current number of literals is $m_i$ and the $i^{\text{th}}$ iteration takes $j_i$ steps, i.e. the number of satisfiable instances until the first unsatisfiable instance, then the number of removed literals is at least, $\Delta_i = 2^{j_i} - 1 + \frac{m_i}{2^{j_i}+1}$. Thus, we can define the variation in the number of elements with a recurrence:

$$m_i = \begin{cases} m & i = 0 \\ m_{i-1} - 2^{j_{i-1}} + 1 - \frac{m_{i-1}}{2^{j_{i-1}+1}} & i \geq 1 \end{cases}$$

where $m$ is the initial number of distinct literals in $\mathcal{R}$. Observe that the value of each $j_i$ cannot exceed $\log m$. The LOPZ algorithm can be viewed as a sequence of phases, where a phase ends when the index $i$ is reset to 0. For each phase, an upper bound on the number of queries to the SAT oracle is $\mathcal{O}(\log m)$. This results from the number of SAT oracle calls returning true being $\mathcal{O}(\log m)$ and, to find a culprit, binary search also makes $\mathcal{O}(\log m)$ oracle calls. Now, the number of phases is maximized when the number of elements removed in each phase is minimized. Thus, we want to find the value of $j_i$ that minimizes $\Delta_i$ given some initial number of elements $m_i$. To find the minima of the function above, we can differentiate with respect to $j_i$ and equate to 0. Thus, the minimum number of removed elements is achieved for $j_i = \frac{1}{2}\log\frac{m_i}{2} = \log\sqrt{\frac{m_i}{2}}$. This means that the minimum number of literals removed in each phase is $\sqrt{\frac{m_i}{2}} - 1 + \sqrt{\frac{m_i}{2}} = 2\sqrt{\frac{m_i}{2}} - 1$, and so the number of phases is $\mathcal{O}(\sqrt{m})$. Thus, an upper bound on the number of queries to the SAT oracle is $\mathcal{O}(\sqrt{m}\log m)$. □

Although the query complexity of LOPZ and UBS are the same, we expect LOPZ to perform better in practice, since preference is given to SAT outcomes.

**Corollary 3.** *The query complexity of LOPZ is $\mathcal{O}(\sqrt{n}\log n)$ when $m = \Theta(n)$.*

*Remark 2.* For computing minimal or maximal models, the query complexity of LOPZ is sublinear on the number of variables.

*Remark 3.* As pointed out earlier (see Sect. 1), the algorithms described in this and the previous section, namely UCD, UBS and LOPZ can be used in settings other than MCS extraction. Concretely, the algorithms can be formulated in the general setting of solving the MSMP problem [29], and used with any computational problem with the same predicate form as MCS extraction [21,28]. Concretely, the UCD, UBS and LOPZ algorithms can be formulated in terms of computing a minimal set subject to a monotone predicate for predicates of form $\mathscr{L}$.

# 6    Experimental Results

This section conducts an experimental evaluation of the algorithms proposed in this paper with state-of-the-art MCS extractors. Representative sets of problem instances were considered [32]. The experiments were performed in Ubuntu Linux running on an Intel Xeon E5-2630 2.60 GHz processor with 64GByte of memory. The time limit was set to 1800 s and the memory limit to 10GByte. All the described algorithms were implemented in a prototype tool named MCSXL[4] (***MCS eXtractor with subLinear number of oracle calls***). The MCSXL tool is written in C++ on top of the MiniSat SAT solver [13]. Besides, the experiments also used the state-of-the-art MCS extractors: CLD [27], CMP (with all its optimizations) [16], RS [1], as well as LBX [32]. In all cases, the original binaries provided by the authors were run according to their instructions. Other well-known MCS extraction algorithms including basic linear search [3,33], MaxSAT [25], SAT with preferences [37,38] (resp. nOPTSAT and SAT&PREF), and FastDiag [14] have been shown to be outperformed by recent approaches. A recently proposed MCS extraction algorithm [2] is not publicly available. In addition, the experimental results in [2] show gains with respect to CLD [27] and RS [1], but these gains are not competitive with other recent approaches [16,32].

## 6.1    Performance Comparison

The benchmark suite considered in this paper is referred to as *IJCAI15* and comprises three different sets of unsatisfiable instances considered previously in [32]. These include *ijcai13-bench* containing 866 plain and partial MaxSAT instances taken from [27], *mus-bench* with 295 plain instances from the 2011 MUS competition proposed in [16], as well as 179 plain and partial instances from the

---

[4] Available at http://logos.ucd.ie/web/doku.php?id=mcsxl.

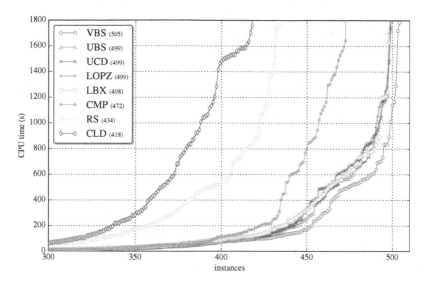

**Fig. 1.** Results for the IJCAI15 [32] problem instances.

**Table 1.** Average number of SAT oracle calls for the IJCAI15 [32] problem instances.

|   | CLD | CMP | LBX | UCD | UBS | LOPZ | LBX$_{w/\circ D}$ | UCD$_{w/\circ D}$ | UBS$_{w/\circ D}$ | LOPZ$_{w/\circ D}$ |
|---|---|---|---|---|---|---|---|---|---|---|
| **S** | 91.9 | 5.9 | 16.5 | 11.5 | 9.3 | 8.4 | 131.9 | 30.5 | 9.1 | 7.4 |
| **U** | 3651.3 | 24.2 | 5.5 | 7.4 | 2.5 | 1.8 | 799.8 | 1641.9 | 71.5 | 2.8 |
| **A** | 3743.2 | 30.0 | 21.9 | 18.8 | 11.8 | 10.2 | 931.6 | 1672.4 | 80.6 | 10.2 |

*maxsat-bench* considered in [32] and taken from the 2014 MaxSAT evaluation. All the three benchmark sets were filtered by removing *easy* instances. An instance was declared to be easy if it was solved by each of the considered algorithms within 5 s. As a result, after filtering easy instances the total number of instances in the IJCAI15 benchmark suite is 556.

The cactus plot reporting the performance of the considered algorithms measured for the IJCAI15 problem instances is shown in Fig. 1. The algorithms proposed in this paper perform comparably to LBX, solving one more instance than LBX. The plot also includes a VBS (virtual best solver) based on picking the best result out of LBX, UCD, UBS and LOPZ. It is important to emphasize that the good performance of some algorithms, more concretely of LBX and CMP, but also of UCD, is in part explained by the regular clause $D$ checks. The cactus plot shown in Fig. 2 illustrates the effect of removing the clause $D$ check step in LBX, UCD, UBS and LOPZ. As can be observed, the performance of LBX and UCD is strongly dependent on the use of the clause $D$ check step. However, this is not the case with LOPZ and UBS. Our interpretation of these results is that both LOPZ and UBS are more robust than either LBX and UCD, since both are less dependent upon the clause $D$ check step. Another measure of robustness of the MCS extraction algorithms is the number of SAT oracle calls. This informa-

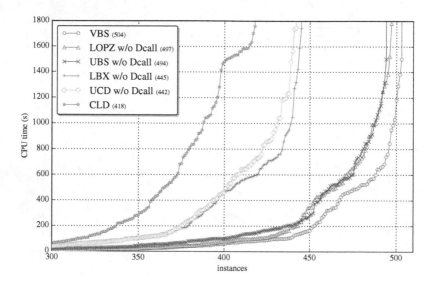

**Fig. 2.** Results for the IJCAI15 [32] problem instances, without the $D$ clause check.

**Table 2.** Contribution to the VBS by LBX, UCD, UBS, and LOPZ.

|          | LBX$^a$ | UCD$^a$ | UBS$^a$ | LOPZ$^a$ | Total |
|----------|---------|---------|---------|----------|-------|
| Figure 1 | 82      | 108     | 120     | 194      | 504   |
| Figure 2 | 66      | 81      | 124     | 232      | 503   |

$^a$ Observe that the algorithms make use of D calls for Fig. 1 but do not use D calls for Fig. 2.

tion is summarized in Table 1 (**S, U, A** represent the number of SAT, UNSAT and total calls). The reduction in the number of SAT oracle calls observed in practice confirms the theory, with LOPZ and UBS achieving the smallest number of SAT oracle calls. The large number of SAT oracle calls for CLD results for the identification of disjoint unsatisfiable cores as a preprocessing step [27], being used in the default configuration. Furthermore, it can be observed that the number of SAT oracle calls increases significantly (i.e. between one and two orders of magnitude) when the clause $D$ check step is not used in LBX and UCD. For UBS there is a modest increase, whereas for LOPZ the average number of SAT oracle calls remains unchanged (with a reduction in the number of SAT outcomes and an increase in the number of UNSAT outcomes).

## 6.2   Analysis of the VBS

The VBS in both Figs. 1 and 2 shows gains when compared to each separate algorithm. This confirms that the new algorithms have an observable contribution to improving the state of the art in MCS extraction. Table 2 summarizes the

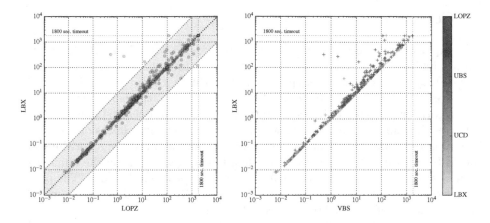

**Fig. 3.** Scatter plots comparing LBX with LOPZ and the VBS.

**Table 3.** Comparison of the percentage of pairwise wins between LBX, UCD, UBS, and LOPZ.

|            | LBX  | UCD  | UBS  | LOPZ |
|------------|------|------|------|------|
| LBX wins   |      | 46.2 | 44.4 | 35.1 |
| UCD wins   | 53.6 |      | 48.2 | 36.5 |
| UBS wins   | 55.0 | 51.4 |      | 34.5 |
| LOPZ wins  | 64.3 | 63.1 | 65.1 |      |

contribution to the VBS. Regarding the VBS in Fig. 1, and as can be observed, the percentage of instances for which the selected solver is either LOPZ, UBS, UCD or LBX is, respectively, 38.5 %, 33.8 %, 21.4 % and 16.3 %. For Fig. 2, without the use of the clause $D$ check step, the numbers are even more conclusive. The data shown provides additional evidence of the contribution of LOPZ, UBS and UCD to the state of the art in MCS extraction.

Figure 3 shows a scatter plot comparing LOPZ with LBX. As can be observed, there is an observable performance gain of LOPZ when compared to LBX. More importantly, LOPZ adds to the robustness of MCS extraction, with fewer outliers than LBX. Figure 3 also includes a scatter plot highlighting the run times of the instances for which the selected solver is not LBX. As can be observed, the new algorithms LOPZ, UBS and UCD contribute for the VBS for the larger run times. Again, the conclusion is that the new algorithms contribute to increase the robustness of MCS extraction. Table 3 complements the scatter plots of Fig. 3 and shows an analysis of the percentage of pairwise performance wins between the different algorithms. As can be concluded, LOPZ performs better than LBX for every around 2 out of 3 problem instances.

# 7  Conclusions

The paper extends recent work on MCS extraction [1,16,27,32]. Recently proposed highly efficient algorithms have been shown to be linear on the number of variables [32]. This paper develops a new algorithm with worst-case linear number of oracle calls, on the number of clauses. In addition, the paper develops two new algorithms which require asymptotically fewer than linear number of oracle calls, on the number of clauses. However, for representative classes of problems, including the computation of minimal and maximal models, the algorithms are shown to require asymptotically fewer than linear oracle calls on the number of variables. This improves earlier results for specific problem domains. The paper also argues that the proposed algorithms can be easily adapted to solving specific predicate forms of the MSMP problem [21,28,29].

The experimental results show that the new algorithms perform comparably to the state of the art in MCS extraction. More importantly, the new algorithms are shown to enable observable performance gains when portfolios of algorithms are considered. The experimental results also confirm that, in practice, far fewer oracle calls may not guarantee large performance gains, even when the calls are expected to be simple. Earlier work [27] had reached similar conclusions when using MaxSAT for MCS extraction. Finding the best possible balance between fewer oracle calls and best SAT solver performance remains an open research subject.

**Acknowledgement.** This work was partly funded by grant TIN2013-46511-C2-2-P.

# References

1. Bacchus, F., Davies, J., Tsimpoukelli, M., Katsirelos, G.: Relaxation search: a simple way of managing optional clauses. In: AAAI, pp. 835–841 (2014)
2. Bacchus, F., Katsirelos, G.: Using minimal correction sets to more efficiently compute minimal unsatisfiable sets. In: Kroening, D., Păsăreanu, C.S. (eds.) CAV 2015. LNCS, vol. 9207, pp. 70–86. Springer, Heidelberg (2015)
3. Bailey, J., Stuckey, P.J.: Discovery of minimal unsatisfiable subsets of constraints using hitting set dualization. In: Hermenegildo, M.V., Cabeza, D. (eds.) PADL 2004. LNCS, vol. 3350, pp. 174–186. Springer, Heidelberg (2005)
4. Bakker, R.R., Dikker, F., Tempelman, F., Wognum, P.M.: Diagnosing and solving over-determined constraint satisfaction problems. In: IJCAI, pp. 276–281 (1993)
5. Belov, A., Lynce, I., Marques-Silva, J.: Towards efficient MUS extraction. AI Commun. **25**(2), 97–116 (2012)
6. Ben-Eliyahu, R., Dechter, R.: On computing minimal models. Ann. Math. Artif. Intell. **18**(1), 3–27 (1996)
7. Biere, A., Heule, M., van Maaren, H., Walsh, T. (eds.): Handbook of Satisfiability. IOS Press, Amsterdam (2009)
8. Birnbaum, E., Lozinskii, E.L.: Consistent subsets of inconsistent systems: structure and behaviour. J. Exp. Theor. Artif. Intell. **15**(1), 25–46 (2003)
9. Cayrol, C., Lagasquie-Schiex, M.-C., Schiex, T.: Nonmonotonic reasoning: from complexity to algorithms. Ann. Math. Artif. Intell. **22**(3–4), 207–236 (1998)

10. Chinneck, J.W.: Feasibility Infesasibility in Optimization: Algorithms and Computational Methods. International Series in Operations Research & Management Science, vol. 118. Springer, Heidelberg (2008)

11. Chinneck, J.W., Dravnieks, E.W.: Locating minimal infeasible constraint sets in linear programs. INFORMS J. Comput. **3**(2), 157–168 (1991)

12. de Siqueira J.L.N., Puget, J.: Explanation-based generalisation of failures. In: ECAI, pp. 339–344 (1988)

13. Eén, N., Sörensson, N.: An extensible SAT-solver. In: Giunchiglia, E., Tacchella, A. (eds.) SAT 2003. LNCS, vol. 2919, pp. 502–518. Springer, Heidelberg (2004). https://github.com/niklasso/minisat

14. Felfernig, A., Schubert, M., Zehentner, C.: An efficient diagnosis algorithm for inconsistent constraint sets. AI EDAM **26**(1), 53–62 (2012)

15. Feydy, T., Somogyi, Z., Stuckey, P.J.: Half reification and flattening. In: Lee, J. (ed.) CP 2011. LNCS, vol. 6876, pp. 286–301. Springer, Heidelberg (2011)

16. Grégoire, É., Lagniez, J., Mazure, B.: An experimentally efficient method for (MSS, coMSS) partitioning. In: AAAI, pp. 2666–2673 (2014)

17. Grégoire, É., Mazure, B., Piette, C.: On approaches to explaining infeasibility of sets of boolean clauses. In: ICTAI, pp. 74–83 (2008)

18. Hemery, F., Lecoutre, C., Sais, L., Boussemart, F.: Extracting MUCs from constraint networks. In: ECAI, pp. 113–117 (2006)

19. Ignatiev, A., Previti, A., Liffiton, M.H., Marques-Silva, J.: Smallest MUS extraction with minimal hitting set dualization. In: CP, pp. 173–182 (2015)

20. Jampel, M.: A brief overview of over-constrained systems. In: Jampel, M., Maher, M.J., Freuder, E.C. (eds.) CP-WS 1995. LNCS, vol. 1106, pp. 1–22. Springer, Heidelberg (1996)

21. Janota, M., Marques-Silva, J.: On the query complexity of selecting minimal sets for monotone predicates. Artif. Intell. **233**, 73–83 (2016)

22. Junker, U.: QuickXplain: Preferred explanations and relaxations for over-constrained problems. In: AAAI, pp. 167–172 (2004)

23. Kavvadias, D.J., Sideri, M., Stavropoulos, E.C.: Generating all maximal models of a boolean expression. Inf. Process. Lett. **74**(3–4), 157–162 (2000)

24. Kullmann, O., Marques-Silva, J.: Computing maximal autarkies with few and simple oracle queries. In: Heule, M. (ed.) SAT 2015. LNCS, vol. 9340, pp. 138–155. Springer, Heidelberg (2015). doi:10.1007/978-3-319-24318-4_11

25. Liffiton, M.H., Sakallah, K.A.: Algorithms for computing minimal unsatisfiable subsets of constraints. J. Autom. Reasoning **40**(1), 1–33 (2008)

26. Mangal, R., Zhang, X., Kamath, A., Nori, A.V., Naik, M.: Scaling relational inference using proofs and refutations. In: AAAI (2016)

27. Marques-Silva, J., Heras, F., Janota, M., Previti, A., Belov, A.: On computing minimal correction subsets. In: IJCAI, pp. 615–622 (2013)

28. Marques-Silva, J., Janota, M.: Computing minimal sets on propositional formulae I: problems & reductions. CoRR, abs/1402.3011 (2014)

29. Marques-Silva, J., Janota, M., Belov, A.: Minimal sets over monotone predicates in boolean formulae. In: Sharygina, N., Veith, H. (eds.) CAV 2013. LNCS, vol. 8044, pp. 592–607. Springer, Heidelberg (2013)

30. Marques-Silva, J., Previti, A.: On computing preferred MUSes and MCSes. In: Sinz, C., Egly, U. (eds.) SAT 2014. LNCS, vol. 8561, pp. 58–74. Springer, Heidelberg (2014)

31. Mencía, C., Marques-Silva, J.: Efficient relaxations of over-constrained CSPs. In: ICTAI, pp. 725–732 (2014)

32. Mencía, C., Previti, A., Marques-Silva, J.: Literal-based MCS extraction. In: IJCAI, pp. 1973–1979 (2015)
33. Nöhrer, A., Biere, A., Egyed, A.: Managing SAT inconsistencies with HUMUS. In: VaMoS, pp. 83–91 (2012)
34. O'Callaghan, B., O'Sullivan, B., Freuder, E.C.: Generating corrective explanations for interactive constraint satisfaction. In: van Beek, P. (ed.) CP 2005. LNCS, vol. 3709, pp. 445–459. Springer, Heidelberg (2005)
35. Papadimitriou, C.H.: Computational Complexity. Addison Wesley, Boston (1994)
36. Reiter, R.: A theory of diagnosis from first principles. Artif. Intell. **32**(1), 57–95 (1987)
37. Rosa, E.D., Giunchiglia, E.: Combining approaches for solving satisfiability problems with qualitative preferences. AI Commun. **26**(4), 395–408 (2013)
38. Rosa, E.D., Giunchiglia, E., Maratea, M.: Solving satisfiability problems with preferences. Constraints **15**(4), 485–515 (2010)
39. Walter, R., Felfernig, A., Kuchlin, W.: Inverse QUICKXPLAIN vs. MAXSAT - a comparison in theory and practice. In: Configuration Workshop, volume CEUR 1453, pp. 97–104 (2015)

# Predicate Elimination for Preprocessing in First-Order Theorem Proving

Zurab Khasidashvili[1] and Konstantin Korovin[2(✉)]

[1] Intel Israel Design Center, 31015 Haifa, Israel
zurabk@iil.intel.com
[2] The University of Manchester, Manchester, UK
korovin@cs.man.ac.uk

**Abstract.** Preprocessing plays a major role in efficient propositional reasoning but has been less studied in first-order theorem proving. In this paper we propose a predicate elimination procedure which can be used as a preprocessing step in first-order theorem proving and is also applicable for simplifying quantified formulas in a general framework of satisfiability modulo theories (SMT). We describe how this procedure is implemented in a first-order theorem prover iProver and show that many problems in the TPTP library can be simplified using this procedure. We also evaluated our preprocessing on the HWMCC'15 hardware verification benchmarks and show that more than 50 % of predicates can be eliminated without increasing the problem sizes.

## 1 Introduction

Preprocessing techniques for Boolean satisfiability apply rewriting and other simplification rules to a propositional formula in conjunctive normal form (CNF) with the aim of making SAT solving easier [3,6,10,14,27,37]. Since the development of the SatElite pre-processing technique [14], the usefulness of preprocessing has become evident and it is now an integrated part of SAT solving. We refer to [20] for an overview of recent advances in CNF pre-processing and in-processing techniques where simplification steps are also applied in interleaving the SAT solving.

Variable elimination [6,12,14,37] is an important simplification technique for CNF and QBF formulas. In this paper we aim at generalizing this approach to first-order logic such that it becomes a predicate elimination technique. Predicate elimination is a special case of second-order quantifier elimination which has been investigated starting from the work of Ackermann [1] and more recently in, e.g., [15,16]. Second-order quantifier elimination in first-order logic has a number of applications ranging from invariant generation and interpolation in program analysis [19,25] to correspondence theory in modal logics [33] and uniform interpolation in description logics [22].

In this paper we propose to use predicate elimination as a preprocessing technique for first-order reasoning and show that it is also applicable for simplifying quantified formulas in a general framework of satisfiability modulo theories

© Springer International Publishing Switzerland 2016
N. Creignou and D. Le Berre (Eds.): SAT 2016, LNCS 9710, pp. 361–372, 2016.
DOI: 10.1007/978-3-319-40970-2_22

(SMT). The main goal of our algorithm is to eliminate as many predicates as possible without increasing the complexity of the clause set. The second goal is to simplify the original set of clauses as much as possible using newly generated clauses. Let us note that in contrast to variable elimination in propositional logic, effective predicate elimination is not always possible for first-order formulas with quantifiers. In order obtain a terminating procedure we propose to restrict predicate elimination to a specific case of non-self-referential predicates as defined in the paper. We show that using non-self-referential predicate elimination one can eliminate predicates from 77 % of problems in the TPTP library [38] and considerably reduce the number of predicates and clauses in many cases. In the case of HWMCC'15 benchmarks [9] we show that in total more than 50 % of predicates can be eliminated without increasing the problem size.

This paper is organized as follows: In the next section we set the terminology and the framework. In Sect. 3 we introduce predicate elimination and prove its soundness and correctness as a simplification technique for first-order formulas in clausal normal form. In Sect. 4 we introduce an algorithm for predicate elimination, called NSR-Pred-Elim. In Sect. 5 we present experimental results which show that a number of previously unsolved problems in the TPTP library for first-order logic problems can be solved after applying predicate elimination.

## 2   Preliminaries

In this paper we are mainly focused on theorem proving in first-order logic but our considerations are also applicable in a more general setting of satisfiability of quantified first-order formulas modulo theories (SMT) [5, 28, 29, 31], as briefly introduced below. Let $\Sigma$ be a first-order signature consisting of predicate and function symbols. Let $\Sigma$ be split into a theory part $\Sigma_T$, containing interpreted theory symbols and $\Sigma_F$ containing uninterpreted symbols. We assume that the equality symbol $\simeq$ is in $\Sigma_T$ and is interpreted as equality. A theory $T$ is defined by a non-empty class of first-order interpretations in the signature $\Sigma_T$, closed under isomorphisms. We say that a first-order formula $F$ is satisfiable modulo $T$ if there exists an interpretation $I$ satisfying $F$ such that the reduction of $I$ to the signature $\Sigma_T$ (forgetting symbols in $\Sigma_F$) is in $T$. In the rest of the paper when we say satisfiability we assume satisfiability modulo some arbitrary but fixed theory $T$. If $\Sigma_T$ is empty then we have usual first-order logic without equality.

Satisfiability of quantified first-order formulas (modulo theories) can be reduced to satisfiability of sets of universally quantified first-order clauses using Skolemization and CNF transformation [18, 30]. In the rest of the paper we will consider the satisfiability of sets of first-order clauses and assume that all variables in clauses are implicitly universally quantified. Given an interpretation $I$, a *variable assignment* $\sigma$ is a mapping from variables $X$ into $I$. We denote by $\sigma_{\bar{x}}$ a restriction of $\sigma$ onto variables $\bar{x}$. A substitution is a mapping from variables $X$ into the set of terms over $X$. We will use $\sigma, \rho, \gamma$ to denote variable assignments and also substitutions when this does not cause confusion.

# 3 Predicate Elimination

Let $S$ be a set of first-order clauses over the signature $\Sigma$. Consider a predicate $P$ of arity $n$ in $\Sigma_{\mathcal{F}}$ which we aim to eliminate from $S$. A literal which contains $P$ will be called a $P$-literal. Our predicate elimination procedure will be based on two rules: flattening, where we abstract all terms from $P$-literals, and flat resolution, where we resolve the flattened predicates.

A $P$-literal is flat if it is of the form $P(x_1, \ldots, x_n)$ or $\neg P(x_1, \ldots, x_n)$, where $x_1, \ldots, x_n$ are pairwise distinct variables. A predicate $P$ is flat in a set of clauses $S$ if all occurrences of $P$-literals in $S$ are flat. We define the *flattening* rule on clauses containing $P$ as follows:

$$\frac{C \vee (\neg)P(t_1, \ldots, t_n)}{C \vee x_1 \not\approx t_1 \vee \ldots \vee x_n \not\approx t_n \vee (\neg)P(x_1, \ldots, x_n)} \text{ (Flattening)}$$

where $x_1, \ldots, x_n$ are fresh variables that do not occur in $C \vee P(t_1, \ldots, t_n)$ and literal $(\neg)P(t_1, \ldots, t_n)$ is not flat. It is clear that flattening is equivalence preserving and therefore we can remove the premise after adding the conclusion of the rule.

We eliminate predicates after flattening using *flat resolution* defined as follows:

$$\frac{C \vee P(x_1, \ldots, x_n) \qquad D \vee \neg P(x_1, \ldots, x_n)}{C \vee D} \text{ (FRes)}$$

We assume that before applying flat resolution, variables in clauses are renamed such that resolved literals are of the form $P(x_1, \ldots, x_n)$ and $\neg P(x_1, \ldots, x_n)$ respectively and all common variables in $C$ and $D$ are in $x_1, \ldots, x_n$.

We call a predicate $P$ *self-referential in a clause* $C$ if the number of occurrences of $P$ in $C$ is greater than 1. A predicate $P$ is *self-referential in a set of clauses* $S$ if $P$ is self-referential in at least one clause in $S$. A predicate which is not self-referential in a set of clauses $S$ will be called *non-self-referential in* $S$.

In order to obtain a terminating elimination procedure we restrict ourselves to eliminating non-self-referential predicates. Let $S$ be a set of clauses and let $P \in \Sigma_{\mathcal{F}}$ be a flat and non-self-referential predicate in $S$. Partition $S$ into three disjoint sets $S_P, S_{\neg P}, S_{\bar{P}}$ – the set of clauses containing $P$ positively, negatively and not containing $P$, respectively. Denote by $S_P \bowtie S_{\neg P}$ the set of all resolvents obtained by pairwise flat resolutions between clauses in $S_P$ and $S_{\neg P}$ upon $P$.

Then, the *non-self-referential predicate elimination transformation* is defined as the following rule on sets of clauses:

$$\frac{S_{\bar{P}} \cup S_P \cup S_{\neg P}}{S_{\bar{P}} \cup (S_P \bowtie S_{\neg P})} \text{ (NSR-Pred-Elim)}$$

Let us note that since $P$ is non-self-referential in $S_{\bar{P}} \cup S_P \cup S_{\neg P}$, $P$ does not occur in the conclusion set of the NSR-Pred-Elim.

The following theorem essentially follows from results in [5, 16]. Here we give a simple proof and would like to emphasise that predicate elimination is a satisfiability preserving transformation in a general setting of satisfiability modulo theories.

**Theorem 1.** *The non-self-referential predicate elimination transformation is satisfiability preserving.*

*Proof.* Since flat resolution is a sound rule, satisfiability of the set of clauses in the premise implies satisfiability of the set of clauses in the conclusion. Let us show that the reverse direction also holds. Let us note that $P$ does not occur in $S_{\tilde{P}} \cup (S_P \bowtie S_{\neg P})$. Assume that $S_{\tilde{P}} \cup (S_P \bowtie S_{\neg P})$ is satisfiable and let $I$ be a model of $S_{\tilde{P}} \cup (S_P \bowtie S_{\neg P})$ in the signature $\Sigma \setminus \{P\}$. We construct an interpretation $I_P$ over the signature $\Sigma$ which will be an expansion of $I$ (i.e., coincide with $I$ on all predicates other than $P$) and satisfy all clauses in $S_{\tilde{P}} \cup S_P \cup S_{\neg P}$.

We define the predicate $P$ in $I_P$ by defining the truth value of $P(\bar{x})\sigma$ for each variable assignment $\sigma$ in $I$. Consider a variable assignment $\sigma$ in $I$. We define $P(\bar{x})\sigma$ to be true in $I_P$ whenever there is a clause $D \vee P(\bar{x}) \in S_P$ such that $D\sigma$ is false in $I$. Likewise, define $P(\bar{x})\sigma$ to be false in $I_P$ if there is a clause $C \vee \neg P(\bar{x}) \in S_{\neg P}$ such that $C\sigma$ is false in $I$. In all other cases we define the truth value of $P(\bar{x})\sigma$ arbitrarily. Let us show that this definition is well-defined, i.e., we never assign true to $P(\bar{x})\sigma$ and false to $P(\bar{x})\rho$ under assignments $\sigma$ and $\rho$ such that $\sigma_{\bar{x}} = \rho_{\bar{x}}$. Assume otherwise. Let $\sigma$ and $\rho$ be such that $\sigma_{\bar{x}} = \rho_{\bar{x}}$ and for some clauses $D \vee P(\bar{x}) \in S_P$ and $C \vee \neg P(\bar{x}) \in S_{\neg P}$, both $D\sigma$ and $C\rho$ are false in $I$. Then $D \vee C$ is in $S_P \bowtie S_{\neg P}$ and since all common variables in $D$ and $C$ are in $\bar{x}$ we have $(D\sigma) \vee (C\rho) = (D \vee C)\gamma$ for an assignment $\gamma$ that coincides with $\sigma$ on variables of $D$ and with $\rho$ on variables of $C$. We have $I \models (D \vee C)\gamma$, hence $I \models (D\sigma) \vee (C\rho)$ which contradicts to the assumption that $D\sigma$ and $C\rho$ are false in $I$. By construction, $I_p$ satisfies all clauses in $S_P \cup S_{\neg P}$ and since $I_P$ is an expansion of $I$ we also have $I_P$ satisfies all clauses in $S_{\tilde{P}}$.

Predicate elimination allows us to eliminate a non-self-referential predicate from a set of clauses by first applying flattening and then NSR-Pred-Elim. After application of NSR-Pred-Elim we can apply the equality substitution rule to eliminate negative equalities which were introduced during flattening. Equality substitution is defined as follows:

$$\frac{C \vee x \not\approx t}{C[t/x]} \text{ (Eq-Subst)}$$

where $x$ does not occur in $t$. Equality substitution is an equivalence preserving rule so we can remove the premise when we add the conclusion to a set of clauses.

Let us note that in the case of first-order logic without theories (including equality) we do not need to apply flattening and instead we can use most general unifiers when applying resolution in the elimination. The proof of Theorem 1 carries over by considering Herbrand interpretations rather than arbitrary interpretations. Already in the presence of equality flattening is essential.

*Example 1.* Consider the following set of unit clauses:

$$P(a), \neg P(b), a \simeq b.$$

This set is unsatisfiable, but we can not eliminate $P$ based on unification. When flattening is applied to $P$-literals we obtain:

$$P(x_1) \vee x_1 \not\simeq a, \neg P(x_1) \vee x_1 \not\simeq b, a \simeq b.$$

After applying flat resolution and equality substitution we obtain an unsatisfiable set $a \simeq b, a \not\simeq b$ where $P$ is eliminated.

## 4  Predicate Elimination for Preprocessing

Let us note that if we start with a finite set of clauses then after eliminating a non-self-referential predicate we obtain a clause set that is at most quadratic in the number of clauses compared to the original set. As we will see in Sect. 5, in practice this set is usually much smaller after eliminating redundant clauses. Let us also note that after eliminating a non-self-referential predicate some previously non-self-referential predicates can become self-referential and conversely some self-referential predicates can become non-self-referential due to removal of clauses during elimination and simplification steps. Thus, the order of elimination can affect which predicates can be eliminated in the process.

*Example 2.* Consider the following set of clauses $S$, where predicates $P, Q, R$ are uninterpreted and function symbols $f, g$ can be either interpreted or uninterpreted.

$$
\begin{array}{ll}
1 & P(f(u), u) \vee f(f(u)) \simeq g(u) \\
2 & \neg P(v, g(u)) \vee Q(f(g(u))) \vee \neg Q(v) \\
3 & Q(v) \vee R(v) \\
4 & \neg Q(f(v)) \vee \neg R(f(v)) \vee R(f(f(v)))
\end{array}
$$

In this set of clauses predicate $P$ is non-self-referential while $Q$ and $R$ are self-referential. In order to eliminate $P$ we first flatten $P$-literals in clauses 1 and 2 and rename variables, obtaining:

$$
\begin{array}{ll}
1_{flat} & P(x_1, x_2) \vee x_1 \not\simeq f(u_1) \vee x_2 \not\simeq u_1 \vee f(f(u_1)) \simeq g(u_1) \\
2_{flat} & \neg P(x_1, x_2) \vee x_1 \not\simeq v_2 \vee x_2 \not\simeq g(u_2) \vee Q(f(g(u_2))) \vee \neg Q(v_2)
\end{array}
$$

After applying flat resolution and repeatedly applying equality substitution we obtain:

$$5.\ f(f(g(u_2))) \simeq g(g(u_2)) \vee Q(f(g(u_2))) \vee \neg Q(f(g(u_2)))$$

Clause 5 is a tautology and therefore we can eliminate $P$ by simply removing clauses 1 and 2. After eliminating $P$, $Q$ becomes non-self-referential and therefore can also be eliminated. After eliminating $Q$ we obtain the empty set of clauses, hence the original set of clauses is satisfiable.

Our main goal is to use predicate elimination for preprocessing to simplify sets of first-order clauses. For this we consider the non-self-referential predicate elimination (NSR-Pred-Elim), presented below. Clause simplifications play a central role in the NSR-Pred-Elim algorithm. Let us first briefly describe simplifications that were used in our implementation.

*Tautology elimination.* Clauses of the form $P(\bar{t}) \vee \neg P(\bar{t}) \vee C$ are *tautologies* and can be eliminated. In the presence of equality we also eliminate equational tautologies of the form $t \simeq t \vee C$. We refer to [26] for a more general notion of equational tautologies.

*Subsumption.* A clause $C$ *subsumes* a clause $D$ if $C\sigma \subseteq D$ for some substitution $\sigma$, considering clauses as literal multi-sets. Subsumed clause $D$ can be removed in the presence of clause $C$.

*Subsumption resolution.* Subsumption resolution of two clauses can be seen as an application of resolution to these clauses followed by subsumption of one of its premises by the conclusion [4]. In this case, the subsumed premise is replaced by the conclusion.

*Global subsumption.* A set of clauses $S$ *globally subsumes* a clause $D$ if there is a clause $C$ such that $S \vDash_{gr} C$ and $C\sigma \subsetneq D$ for a substitution $\sigma$, where $\vDash_{gr}$ is a propositional approximation of $\vDash$. In this case $C$ is called a *witness for global subsumption* and we can replace subsumed clause $D$ by $C$. A global subsumption witness $C$ can be obtained from $S$ and $D$ by adjoining negations of subclauses of $D$ to $S$ and applying propositional reasoning. We refer to [24] for details.

Let us note that we are not restricted to using only these simplifications, the method allows one to use any other collection of sound simplifications.

**The NSR-Pred-Elim algorithm.** The input of the NSR-Pred-Elim (Algorithm 1) is a set of first-order clauses and a predicate elimination queue. The main goal is to eliminate as many predicates as possible from the elimination queue without increasing the complexity of the clause set. The second goal is to simplify the original set of clauses as much as possible using newly generated clauses. There are different possible complexity measures, we use one of the most restrictive and require that after each predicate elimination the number of literals should not increase and in addition that the *normalised variable complexity* of the set of clauses should also not increase. We define normalised variable complexity of a set of clauses as a sum of squares of the number of variables in each clause. The idea behind the last restriction is to give a greater weight to clauses with higher number of variables since reasoning with such clauses is generally more expensive.

During a run of the algorithm we maintain a set *global-clauses* which is equi-satisfiable to the input set and a set *local-clauses* which is generated during elimination of a predicate. Clause simplifications play a central role in the NSR-Pred-Elim algorithm. Each newly generated clause is first *self-simplified* by simplifications such as equality substitution and tautology elimination (line 14).

Then the *forward-simplify* procedure (line 15) is applied, where the clause is simplified by local clauses and global clauses using simplifications such as subsumption, subsumption resolution and global subsumption. If the clause is eliminated by, e.g., tautology elimination or subsumption then we proceed to the next generated clause, otherwise we apply the *backward-simplify* procedure which uses (a simplification of) this clause to simplify both local and global clauses using simplifications such as subsumption and subsumption resolution (lines 17, 18). After processing all clauses in the resolvent set $S_P \bowtie S_{\neg P}$ we check whether to keep the result of the elimination or to discard it based on the complexity of the generated clause set (line 26). Even when we discard the result of the predicate elimination we still benefit from the simplifications of the global set by the clauses generated in the process. For this, in the case when we simplify a clause in the global set we keep *simp-witnesses* of this simplification (line 32): the set of clauses which are used to simplify the clause and imply the simplified clause. A simplification witness usually consists of a clause derived by flat resolution and equality substitution during the elimination or by simplifications such as subsumption resolution or global subsumption.

The algorithm maintains a map (*pred-map*) which maps predicates to sets of clauses where the predicate occurs positively and negatively and a set of suspended predicates (*suspended-pred-set*) for which the elimination was suspended either due to self-referential occurrences in clauses (line 32) or to complexity of the clause set after elimination (line 27). After updating the set of global clauses we also update *pred-map* and *suspended-pred-set*. During this update some predicates from *suspended-pred-set* can be moved back to the elimination queue due to clauses eliminated from the global set, which contain predicates from *suspended-pred-set*.

## 5  Implementation and Evaluation

We implemented the NSR-Pred-Elim algorithm in iProver[1] – a theorem prover for first-order logic [23,24]. iProver is a general purpose theorem prover for first-order logic which incorporates SAT solvers at its core, currently MiniSAT [13] and optionally PicoSAT [7]. One of the main challenges in efficient implementation of NSR-Pred-Elim are efficient local and global simplifications. In contrast to propositional subsumption, clause-to-clause first-order subsumption is an NP-complete problem. In order to deal with subsumption and subsumption resolution efficiently we employed the compressed feature vector index (CFVI) which is an extension of the feature vector index proposed in [34]. This allowed us to successfully apply NSR-Pred-Elim to problems containing hundreds of thousands of clauses. One of the important parameters of the algorithm is a criterion deciding when to keep the result of a predicate elimination.

For our experiments we used machines with Intel Xeon L5410 2.33 GHz CPU, 4 cores, 12Gb. Each problem was run on a single core with time limit

---

[1] iProver is available at http://www.cs.man.ac.uk/~korovink/iprover.

---

**Algorithm 1.** NSR-Pred-Elim

---

1: input: $S$ – input clause set
2: input: $elim\text{-}queue$ – predicate elimination priority queue
3: output: a simplified set of clauses equi-satisfiable with $S$
4: $global\text{-}clauses \leftarrow S$
5: $eliminated\text{-}pred\text{-}set \leftarrow \emptyset$
6: $suspended\text{-}pred\text{-}set \leftarrow \emptyset$
7: $pred\text{-}map.create(global\text{-}clauses)$
8: **while** $elim\text{-}queue \neq \emptyset$ **do**
9:      $P \leftarrow pop(elim\text{-}queue)$
10:     **if** $non\text{-}self\text{-}referential(P)$ **then**
11:         $(S_P, S_{\neg P}) \leftarrow pred\text{-}map.find(P)$
12:         $local\text{-}clauses \leftarrow \emptyset$
13:         **for** $C \in S_P \bowtie S_{\neg P}$ **do**
14:             $self\text{-}simplify(C)$
15:             $C \leftarrow forward\text{-}simplify(C, local\text{-}clauses \cup global\text{-}clauses)$
16:             **if** $\neg is\text{-}eliminated(C)$ **then**
17:                 $local\text{-}clauses \leftarrow backward\text{-}simplify(C, local\text{-}clauses)$
18:                 $global\text{-}clauses \leftarrow backward\text{-}simplify(C, global\text{-}clauses)$
19:                 $local\text{-}clauses \leftarrow local\text{-}clauses \cup \{C\}$
20:             **end if**
21:         **end for**
22:         **if** $keep\text{-}elim(local\text{-}clauses)$ **then**
23:             $global\text{-}clauses \leftarrow (global\text{-}clauses \setminus (S_P \cup S_{\neg P})) \cup local\text{-}clauses$
24:             $eliminated\text{-}pred\text{-}set \leftarrow eliminated\text{-}pred\text{-}set \cup \{P\}$
25:         **else**
26:             $global\text{-}clauses \leftarrow global\text{-}clauses \cup simp\text{-}witnesses(global\text{-}clauses)$
27:             $suspended\text{-}pred\text{-}set \leftarrow suspended\text{-}pred\text{-}set \cup \{P\}$
28:         **end if**
29:         $pred\text{-}map.update(global\text{-}clauses)$
30:         $suspended\text{-}pred\text{-}set.update(global\text{-}clauses)$
31:     **else**
32:         $suspended\text{-}pred\text{-}set \leftarrow suspended\text{-}pred\text{-}set \cup \{P\}$
33:     **end if**
34: **end while**
35: **return** $global\text{-}clauses$

---

300 s and memory limit 3.5Gb. We evaluated predicate elimination over first-order problems in FOF and CNF formats over the TPTP-v6.1.0 library. iProver accepts first-order problems in CNF form, for problems in general first-order form we used Vampire's [26] clausifier to transform them into CNF. TPTP-v6.1.0 contains 15897 problems in FOF and CNF formats, 13708 (86 %) of problems contain predicates other than equality. The left-hand table in Fig. 1 shows that our predicate elimination procedure is able to eliminate predicates in 10617 problems in TPTP, which is 77 % of all first-order problems in TPTP containing non-equality predicates. In particular, from this it follows that at least 77 % of

problems contain non-self-referential predicates which can be eliminated without increasing the size of the problem.

| eliminated predicates | problems |
|---|---|
| 1 − 10 | 4759 |
| 11 − 100 | 3682 |
| 101 − 1000 | 1473 |
| 1001 − 10000 | 607 |
| 10001 − 100000 | 82 |
| 100001 − 1000000 | 14 |
| total | 10617 |

| reduction in clauses | problems |
|---|---|
| 0.01% − 20% | 8042 |
| 20% − 40% | 930 |
| 40% − 60% | 515 |
| 60% − 80% | 399 |
| 80% − 100% | 188 |
| total | 10074 |

**Fig. 1.** The number of TPTP problems with the number of eliminated predicates in the specified range (left) and the specified reduction in the number of clauses (right).

The right-hand table in Fig. 1 shows the percentage of reduction in the number of clauses due to the predicate elimination. Let us note that even when there is no reduction in the number of clauses the set of clauses may change after the predicate elimination process. This is the case when the number of removed clauses due to simplifications is the same as the number of generated clauses due to predicate elimination and the addition of simplification witnesses.

Table 1 compares the performance of iProver with predicate elimination and without. Ratings of problems range from 0 − easy problems to 1 − problems that cannot be solved by any solver so far (including previous versions of iProver). We

**Table 1.** iProver performance on TPTP problems with and without predicate elimination

| Ratings | Total | pred elim Solved | % | w/o pred elim Solved | % |
|---|---|---|---|---|---|
| 0.0 - 0.1 | 3495 | 3123 | 89 | 3110 | 89 |
| 0.1 - 0.2 | 1744 | 1512 | 87 | 1498 | 86 |
| 0.2 - 0.3 | 1302 | 931 | 71 | 916 | 70 |
| 0.3 - 0.4 | 1092 | 602 | 55 | 587 | 54 |
| 0.4 - 0.5 | 999 | 512 | 51 | 506 | 51 |
| 0.5 - 0.6 | 830 | 321 | 39 | 294 | 35 |
| 0.6 - 0.7 | 913 | 268 | 29 | 241 | 26 |
| 0.7 - 0.8 | 607 | 141 | 23 | 129 | 21 |
| 0.8 - 0.9 | 1115 | 109 | 10 | 110 | 10 |
| 0.9 - 1.0 | 993 | 55 | 6 | 44 | 4 |
| 1.0 - 1.0 | 1892 | 20 | 1 | 18 | 1 |
| total | 14982 | 7594 | 51 | 7453 | 50 |

can see from this table that iProver with predicate elimination considerably out-performs iProver without predicate elimination on problems of all ratings except for one case where it loses one problem. iProver can also be used as a preprocessing tool and combined with other theorem provers. We evaluated the effect of the NSR-Pred-Elim preprocessing on top-of-the-range first-order theorem provers Vampire [26] and E [35], which also have their own advanced preprocessors. In both cases the NSR-Pred-Elim preprocessing considerably increased the number of solved problems over TPTP: 130 in the case of E and 32 in the case of Vampire.

Our second set of experiments focuses on the model checking problems in the Hardware Model Checking Competition, HWMCC'15 [9]. In [21], the authors proposed two new model checking algorithms based on an encoding of the model-checking problem into the effectively propositional fragment of first-order logic (EPR). The first one is a bounded model checking [8] algorithm called UCM-BMC1, and the second one is a k-induction algorithm [36] called UCM-k-ind. These algorithms employ an unsat core and model-based (UCM) abstraction refinement scheme inspired by [2,11,17], based on approximations of the transition relation by first-order predicates. These algorithms are implemented in iProver and the implementation uses the iProver theorem prover to solve the BMC and induction formulas of UCM-BMC1 and UCM-k-ind.

In these experiments, we apply predicate elimination to the BMC and induction formulas of UCM-BMC1 and UCM-k-ind as part of preprocessing and compare them with UCM-BMC1 and UCM-k-ind where predicate elimination is not performed in iProver. With predicate elimination iProver solves more problems and reaches deeper bounds: on the 547 available problems, the total number of bounds reached with predicate elimination is 24244, without predicate elimination it is 21820. Furthermore, predicate elimination eliminated 27 million predicates from the total of 54 million, and eliminated 57 million clauses from the total of 160 million.

# References

1. Ackermann, W.: Untersuchungen über das Eliminationsproblem der matheraatis-chen Logik. Math. Ann. **110**, 390–413 (1935)
2. Amla, N., McMillan, K.L.: Combining abstraction refinement and SAT-based model checking. In: Grumberg, O., Huth, M. (eds.) TACAS 2007. LNCS, vol. 4424, pp. 405–419. Springer, Heidelberg (2007)
3. Bacchus, F., Winter, J.: Effective preprocessing with hyper-resolution and equality reduction. In: Giunchiglia, E., Tacchella, A. (eds.) SAT 2003. LNCS, vol. 2919, pp. 341–355. Springer, Heidelberg (2004)
4. Bachmair, L., Ganzinger, H.: Resolution theorem proving. In: Robinson and Voronkov [32], pp. 19–99
5. Bachmair, L., Ganzinger, H., Waldmann, U.: Refutational theorem proving for hierarchic first-order theories. Appl. Algebra Eng. Commun. Comput. **5**, 193–212 (1994)
6. Biere, A.: Resolve and expand. In: H. Hoos, H., Mitchell, D.G. (eds.) SAT 2004. LNCS, vol. 3542, pp. 59–70. Springer, Heidelberg (2005)

7. Biere, A.: Picosat essentials. JSAT **4**(2–4), 75–97 (2008)
8. Biere, A., Cimatti, A., Clarke, E., Zhu, Y.: Symbolic model checking without BDDs. In: Cleaveland, W.R. (ed.) TACAS 1999. LNCS, vol. 1579, pp. 193–207. Springer, Heidelberg (1999)
9. Biere, A., Heljanko, K.: Hardware model checking competition report (2015). http://fmv.jku.at/hwmcc15/Biere-HWMCC15-talk.pdf
10. Brafman, R.I.: A simplifier for propositional formulas with many binary clauses. IEEE Trans. Syst. Man Cybern. Part B **34**(1), 52–59 (2004)
11. Clarke, E.M., Grumberg, O., Jha, S., Lu, Y., Veith, H.: Counterexample-guided abstraction refinement for symbolic model checking. J. ACM **50**(5), 752–794 (2003)
12. Davis, M., Putnam, H.: A computing procedure for quantification theory. J. ACM **7**(3), 201–215 (1960)
13. Eén, N., Sörensson, N.: An extensible SAT-solver. In: Giunchiglia, E., Tacchella, A. (eds.) SAT 2003. LNCS, vol. 2919, pp. 502–518. Springer, Heidelberg (2004)
14. Eén, N., Biere, A.: Effective preprocessing in SAT through variable and clause elimination. In: Bacchus, F., Walsh, T. (eds.) SAT 2005. LNCS, vol. 3569, pp. 61–75. Springer, Heidelberg (2005)
15. Gabbay, D.M., Schmidt, R.A., Szalas, A.: Second-Order Quantifier Elimination: Foundations, Computational Aspects and Applications, Studies in Logic: Mathematical Logic andFoundations, vol. 12. College Publications (2008)
16. Gabbay, D.M., Ohlbach, H.J.: Quantifier elimination in second-order predicate logic. In: Proceedings of the 3rd International Conference on Principles of Knowledge Representation and Reasoning (KR 1992), pp. 425–435 (1992)
17. Gupta, A., Ganai, M.K., Yang, Z., Ashar, P.: Iterative abstraction using sat-based BMC with proof analysis. In: International Conference on Computer-Aided Design, ICCAD, pp. 416–423 (2003)
18. Hoder, K., Khasidashvili, Z., Korovin, K., Voronkov, A.: Preprocessing techniques for first-order clausification. In: Cabodi, G., Singh, S. (eds.) Formal Methods in Computer-Aided Design, FMCAD, pp. 44–51. IEEE (2012)
19. Hoder, K., Kovács, L., Voronkov, A.: Interpolation and symbol elimination in vampire. In: Giesl, J., Hähnle, R. (eds.) IJCAR 2010. LNCS, vol. 6173, pp. 188–195. Springer, Heidelberg (2010)
20. Järvisalo, M., Heule, M.J.H., Biere, A.: Inprocessing rules. In: Gramlich, B., Miller, D., Sattler, U. (eds.) IJCAR 2012. LNCS, vol. 7364, pp. 355–370. Springer, Heidelberg (2012)
21. Khasidashvili, Z., Korovin, K., Tsarkov, D.: EPR-based k-induction with counterexample guided abstraction refinement. In: Gottlob, G., Sutcliffe, G., Voronkov, A. (eds.) GCAI 2015. Global Conference on Artificial Intelligence. EPiC Series in Computing, vol. 36, pp. 137–150. EasyChair (2015)
22. Koopmann, P., Schmidt, R.A.: Uniform interpolation and forgetting for $\mathcal{ALC}$ ontologies with aboxes. In: Bonet, B., Koenig, S. (eds.) Proceedings of the AAAI-2015, pp. 175–181. AAAI Press (2015)
23. Korovin, K.: iProver – an instantiation-based theorem prover for first-order logic (system description). In: Armando, A., Baumgartner, P., Dowek, G. (eds.) IJCAR 2008. LNCS (LNAI), vol. 5195, pp. 292–298. Springer, Heidelberg (2008)
24. Korovin, K.: Inst-Gen – a modular approach to instantiation-based automated reasoning. In: Voronkov, A., Weidenbach, C. (eds.) Programming Logics. LNCS, vol. 7797, pp. 239–270. Springer, Heidelberg (2013)
25. Kovács, L., Voronkov, A.: Interpolation and symbol elimination. In: Schmidt, R.A. (ed.) CADE-22. LNCS, vol. 5663, pp. 199–213. Springer, Heidelberg (2009)

26. Kovács, L., Voronkov, A.: First-Order theorem proving and VAMPIRE. In: Sharygina, N., Veith, H. (eds.) CAV 2013. LNCS, vol. 8044, pp. 1–35. Springer, Heidelberg (2013)
27. Lynce, I., Silva, J.P.M.: Probing-based preprocessing techniques for propositional satisfiability. In: 15th IEEE International Conference on Tools with Artificial Intelligence ICTAI, p. 105. IEEE Computer Society (2003)
28. de Moura, L., Bjørner, N.S.: Efficient E-matching for SMT solvers. In: Pfenning, F. (ed.) CADE 2007. LNCS (LNAI), vol. 4603, pp. 183–198. Springer, Heidelberg (2007)
29. Nieuwenhuis, R., Oliveras, A., Tinelli, C.: Solving SAT and SAT modulo theories: from an abstract davis-putnam-logemann-loveland procedure to DPLL(T). J. ACM **53**(6), 937–977 (2006)
30. Nonnengart, A., Weidenbach, C.: Computing small clause normal forms. In: Robinson and Voronkov [32], pp. 335–367
31. Reynolds, A., Tinelli, C., de Moura, L.M.: Finding conflicting instances of quantified formulas in SMT. In: Formal Methods in Computer-Aided Design, FMCAD, pp. 195–202. IEEE (2014)
32. Robinson, J.A., Voronkov, A. (eds.): Handbook of Automated Reasoning (in volume 2s). Elsevier and MIT Press, Cambridge (2001)
33. Schmidt, R.A.: The Ackermann approach for modal logic, correspondence theory and second-order reduction. J. Appl. Logic **10**(1), 52–74 (2012)
34. Schulz, S.: Simple and efficient clause subsumption with feature vector indexing. In: Bonacina, M.P., Stickel, M.E. (eds.) Automated Reasoning and Mathematics. LNCS, vol. 7788, pp. 45–67. Springer, Heidelberg (2013)
35. Schulz, S.: System description: E 1.8. In: McMillan, K., Middeldorp, A., Voronkov, A. (eds.) LPAR-19 2013. LNCS, vol. 8312, pp. 735–743. Springer, Heidelberg (2013)
36. Sheeran, M., Singh, S., Stålmarck, G.: Checking safety properties using induction and a SAT-Solver. In: Johnson, S.D., Hunt Jr., W.A. (eds.) FMCAD 2000. LNCS, vol. 1954, pp. 108–125. Springer, Heidelberg (2000)
37. Subbarayan, S., Pradhan, D.K.: NiVER: non-increasing variable elimination resolution for preprocessing SAT instances. In: H. Hoos, H., Mitchell, D.G. (eds.) SAT 2004. LNCS, vol. 3542, pp. 276–291. Springer, Heidelberg (2005)
38. Sutcliffe, G.: The TPTP World – Infrastructure for automated reasoning. In: Clarke, E.M., Voronkov, A. (eds.) LPAR-16 2010. LNCS, vol. 6355, pp. 1–12. Springer, Heidelberg (2010)

# Quantified Boolean Formula

# Incremental Determinization

Markus N. Rabe$^{(\boxtimes)}$ and Sanjit A. Seshia$^{(\boxtimes)}$

University of California, Berkeley, USA
{rabe,sseshia}@berkeley.edu

**Abstract.** We present a novel approach to solve quantified boolean formulas with one quantifier alternation (2QBF). The algorithm incrementally adds new constraints to the formula until the constraints describe a unique Skolem function - or until the absence of a Skolem function is detected. Backtracking is required if the absence of Skolem functions depends on the newly introduced constraints. We present the algorithm in analogy to search algorithms for SAT and explain how propagation, decisions, and conflicts are lifted from *values* to *Skolem functions*. The algorithm improves over the state of the art in terms of the number of solved instances, solving time, and the size of the certificates.

## 1 Introduction

Solvers for quantified boolean formulas (QBFs) have been considered as an algorithmic backend in a variety of application areas, such as planning in uncertain environments [3,32,38], chess [2,3,44], program verification [5,14], model checking of Markov chains [42], circuit analysis [17,18,35], and synthesis [12,16,46]. However, the performance of the currently available solvers can be unsatisfactory. For example, competitive solvers such as DepQBF [34], RAReQS [26], and Qesto [27] cannot solve the quantified boolean formula $\forall X.\, \exists Y.\, X = Y$ in a reasonable timeframe, where $X$ and $Y$ are 32-bit words and $=$ states their bitwise equivalence. Even though preprocessors like Bloqqer [11] help to solve this formula, the example suggests that there is a fundamental problem with the solving principle of state-of-the-art QBF solvers.

The formula describes a trivial problem. We can see that for every assignment to $X$ there is exactly one assignment to $Y$ that satisfies the constraint. That is, the formula describes the Skolem function that is the solution to the problem. This reasoning, however, requires us to detect functional dependencies in formulas that are typically given in conjunctive normal form.

In this paper we present an algorithm to determine the truth of formulas with one quantifier alternation (2QBF) that detects existing functional dependencies among variables and incrementally builds new Skolem functions whenever the problem does not imply a unique Skolem function. We employ the view that the propositional part $\varphi$ of a 2QBF $\forall x_1, \ldots, x_n \in \mathbb{B}.\, \exists y_1, \ldots, y_m \in \mathbb{B}.\, \varphi$ is a binary *relation* $R_\varphi$ over assignments $\mathbf{x}$ and $\mathbf{y}$ to the variables $x_1, \ldots, x_n$ and $y_1, \ldots, y_m$: $R_\varphi = \{(\mathbf{x}, \mathbf{y}) \mid \varphi(\mathbf{x}, \mathbf{y})\}$. We call $R_\varphi$ the *Skolem relation*. The

© Springer International Publishing Switzerland 2016
N. Creignou and D. Le Berre (Eds.): SAT 2016, LNCS 9710, pp. 375–392, 2016.
DOI: 10.1007/978-3-319-40970-2_23

solution to a true 2QBF is a *Skolem function* $f$ that assigns values to the existentially quantified variables depending on the universally quantified variables such that the constraints are satisfied for all pairs of assignments $(\mathbf{x}, f(\mathbf{x}))$. Also a Skolem function can be seen as a relation over assignments and it is a subset of the Skolem relation $R_\varphi$. The difference between the Skolem relation $R_\varphi$ and a Skolem function $f$ is that $R_\varphi$ may still provide multiple possible assignments $\mathbf{y}$ for some assignment $\mathbf{x}$, while $f$ has to provide exactly one $\mathbf{y}$ for every $\mathbf{x}$. The presented algorithm adds constraints to $\varphi$ to eliminate the remaining nondeterminism - we *determinize* the Skolem relation to obtain a Skolem function.

The algorithm is a generalization of the DPLL algorithm [15] with conflict-driven clause learning (CDCL) [45]. We lift the concepts of propagation, decisions, and conflicts from *values* for variables to *Skolem functions* for variables. We thereby break the search for Skolem functions down to single variables, which allows us to determinize the relation incrementally, giving rise to the name of the algorithm - *incremental determinization*.

After presenting an overview of the algorithm in Sect. 3, we present a propagation procedure in Sect. 4, which identifies variables that already have unique Skolem functions and whether there is a conflicted variable. In Sect. 5 we discuss how to introduce additional constraints to fix a Skolem function for a variable in case propagation cannot derive a unique Skolem function. Section 6 covers how to compute a conflict clause after a conflicted variable is detected. Termination, correctness, and the generation of certificates is covered in Sect. 7. In Sect. 8 we describe the implementation and give an experimental evaluation of the approach. We sketch out relations to other algorithms and preprocessing techniques for QBF in Sect. 9 and conclude with Sect. 10.

## 2   Quantified Boolean Formulas

We assume that the reader is familiar with the natural semantics of propositional boolean formulas and summarize the basic notation for quantified boolean formulas in the following. Quantified boolean formulas over a finite set of variables $x \in X$ with domain $\mathbb{B} = \{0, 1\}$ are generated by the following grammar:

$$\varphi := 0 \mid 1 \mid x \mid \neg\varphi \mid (\varphi) \mid \varphi \vee \varphi \mid \varphi \wedge \varphi \mid \exists x.\,\varphi \mid \forall x.\,\varphi,$$

We abbreviate multiple quantifications $Qx_1.Qx_2.\ldots Qx_n.\varphi$ to the quantification over a set of variables $QX.\varphi$, where $x_i \in X$ and $Q \in \{\forall, \exists\}$.

An *assignment* $\mathbf{x}$ to a set of variables $X$ is a function $\mathbf{x} : X \to \mathbb{B}$ that maps each variable $x \in X$ to either 1 or 0. Given a propositional formula $\varphi$ over variables $X$ and an assignment $\mathbf{x}'$ for $X' \subseteq X$, we define $\varphi(\mathbf{x}')$ to be the formula obtained by replacing the variables $X'$ by their truth value in $\mathbf{x}'$. By $\varphi(\mathbf{x}', \mathbf{x}'')$ we denote the replacement by multiple assignments for disjoint sets $X', X'' \subseteq X$.

The dependency set of an existentially quantified variable $y$, denoted by $dep(y)$, is the set of universally quantified variables $x$ such that $\exists y.\,\varphi$ is a subformula of $\forall x.\varphi'$. A *Skolem function* $f_y$ maps assignments to $dep(y)$ to assignments to $y$. We define the truth of a QBF $\varphi$ as the existence of Skolem functions

$f_Y = \{f_{y_1}, \ldots, f_{y_n}\}$ for the existentially quantified variables $Y = \{y_1, \ldots, y_n\}$, such that $\varphi(\mathbf{x}, f_Y(\mathbf{x}))$ holds for every $\mathbf{x}$, where $f_Y(\mathbf{x})$ is the assignment to $Y$ that the Skolem functions $f_Y$ provide for $\mathbf{x}$.

A quantifier $Q\,x.\,\varphi$ for $Q \in \{\exists, \forall\}$ *binds* the variable $x$ in its subformula $\varphi$. A *closed* QBF is a formula in which all variables are bound. A formula is in prenex normal form, if the formula is closed and starts with a sequence of quantifiers followed by a propositional subformula. A formula $\varphi$ is in the $k$QBF fragment for $k \in \mathbb{N}^+$ if it is closed, in prenex normal form, and has exactly $k - 1$ alternations between $\exists$ and $\forall$ quantifiers.

A *literal* $l$ is either a variable $x \in X$, or its negation $\neg x$. Given a set of literals $\{l_1, \ldots, l_n\}$, their disjunction $(l_1 \vee \ldots \vee l_n)$ is called a *clause* and their conjunction $(l_1 \wedge \ldots \wedge l_n)$ is called a *cube*. A propositional formula is in conjunctive normal form (CNF), if it is a conjunction of clauses. A prenex QBF is in prenex conjunctive normal form (PCNF) if its propositional subformula is in CNF. W.l.o.g. we assume for all PCNF formulas that none of the clauses contains two opposite literals, which would trivially satisfy the clause, and that all clauses contain at least one literal from an existentially quantified variable. To simplify the notation, we treat the propositional formulas $\psi$ as sets of clauses $\psi = \{C_1, \ldots, C_n\}$, clauses $C$ as sets of literals $C = \{l_1, \ldots, l_m\}$, and use set operations like intersection and union for their manipulation. Every QBF $\varphi$ can be transformed into an equivalent PCNF with size $O(|\varphi|)$ [47].

We assume that the reader is familiar with unit propagation and define $UP(\varphi)$ as the partial assignment to the variables in a propositional $\varphi$ resulting from applying the unit propagation rule until a fixpoint is reached. We define $UP(\varphi) = \perp$ if unit propagation results in a conflicting assignment for a variable.

## 3   Algorithm

Let $\forall X.\exists Y.\varphi$ be a 2QBF in PCNF, where $\varphi$ is the propositional part. The algorithm INCREMENTALDETERMINIZATION determines whether the formula is true. The key principle of the algorithm is to maintain a set of variables $D \subseteq Y$ for which the set of clauses $\mathcal{D} = \{C \in \varphi \mid C \subseteq D \cup X\}$ that only have variables in $D$ and $X$ defines a Skolem function for each variable in $D$: We say that $\varphi$ is $D$-*consistent* if for each assignment $\mathbf{x}$ to $X$, $UP(\mathcal{D}(\mathbf{x}))$ is not $\perp$ and assigns a value to all variables in $D$. (In particular, $\forall X \exists ! D.\,\mathcal{D}$.) It is clear that a $Y$-consistent 2QBF is true and for each true 2QBF $\forall X.\exists Y.\varphi$ there exists a set of clauses $\psi$, such that $\forall X.\exists Y.\varphi \wedge \psi$ is $Y$-consistent.

Given a $D$-consistent formula $\forall X.\exists Y.\varphi$, we say a variable $v \in Y$ has a *unique Skolem function*, if $\forall X.\exists Y.\varphi$ is also $(D \cup \{v\})$-consistent. For determining $(D \cup \{v\})$-consistency we have to extend the clauses $\mathcal{D}$ by the clauses $\mathcal{U}_v$ in which $v$ is the only variable not in $D$ and not in $X$. Clauses in $\mathcal{U}_v$ can be read as implications where the consequence is a literal of $v$, because we know that all other variables are already determined for all assignments $\mathbf{x}$. We say that a clause $C \in \mathcal{U}_v$ has the *unique consequence* $v$.

The algorithm checks for unique Skolem functions in two steps which require the following definitions: Variable $v$ is *deterministic*, if $UP(\mathcal{D}(\mathbf{x}) \wedge \mathcal{U}_v(\mathbf{x}))$ is $\perp$

or gives a unique assignment to $v$ for all assignments $\mathbf{x}$ to $X$, and $v$ is *conflicted*, if $UP(\mathcal{D}(\mathbf{x}) \wedge \mathcal{U}_v(\mathbf{x})) = \bot$ for some assignment $\mathbf{x}$ to $X$. Deterministic variables that are not conflicted have a unique Skolem function.

```
1: procedure INCREMENTALDETERMINIZATION(∀X.∃Y.φ)
2:     dlvl ← 0;  D ← ∅
3:     while true do
4:         D, φ, conflict, x ← PROPAGATE(∀X.∃Y.φ, D, dlvl)
5:         if conflict then
6:             c ← ANALYZECONFLICT(∀X.∃Y. φ, x, D, dlvl)
7:             if c only contains variables in X then
8:                 return false
9:             dlvl ← (maximal decision level in c) − 1
10:            φ, D ← BACKTRACK(φ, D, dlvl)
11:            φ ← φ ∧ c
12:        else
13:            if D = Y then
14:                return true
15:            v ← PICKVAR(Y \ D)
16:            dlvl ← dlvl + 1
17:            φ ← φ ∧ DECISION(v, φ, D)
```

The elements of the algorithm are as follows: The procedure PROPAGATE extends $D$ with variables with unique Skolem functions until the procedure returns an updated set $D$, or until a conflicted variable is detected upon which propagation reports an assignment to $X$ for which the conflict occurs (see Sect. 4). In case of a conflict, ANALYZECONFLICT computes a conflict clause (Sect. 6).

Variables that are detected to have a unique Skolem function get labeled with the current decision level ($dlvl$) during propagation, which is used during backtracking (lines 10 to 12). The procedure BACKTRACK($\varphi$, $dlvl$) resets the set $D$ and the formula $\varphi$ to a certain decision level, but keeps the learnt clauses. In case no conflict is detected during propagation, the procedures PICKVAR and DECISION fix a Skolem function for an additional variable (see Sect. 5).

## 4    Propagation and Conflicts

During *propagation* search algorithms for SAT consider which assignments to the yet unassigned variables are entailed by the current partial assignment. In this section we generalize propagation of values to a notion of propagation of Skolem functions. To develop some intuition on the determinicity check let us consider the following example:

$$\forall x_1.\forall x_2.\ \exists y_1.\exists y_2.\ \underbrace{(x_1 \vee \neg y_1) \wedge (x_2 \vee \neg y_1) \wedge (\neg x_1 \vee \neg x_2 \vee y_1)}_{f_{y_1}:\ (x_1,x_2) \mapsto x_1 \wedge x_2} \tag{1}$$

$$\wedge \underbrace{(\neg x_1 \vee \neg y_2) \wedge (\neg y_1 \vee \neg y_2) \wedge (x_1 \vee y_1 \vee y_2)}_{f_{y_2}:\ (x_1,y_1) \mapsto \neg(x_1 \vee y_1)} \tag{2}$$

The first three clauses (clause group (1)) of the formula can easily be identified as a definition of a Skolem function for $y_1$: For each assignment to $x_1$ and $x_2$, one of the first three clauses entails a unique value for $y_1$. It helps to consider each of these clauses as an implication, that is $(\neg x_1 \rightarrow \neg y_1) \wedge (\neg x_2 \rightarrow \neg y_1) \wedge (x_1 \wedge x_2 \rightarrow y_1)$. Variable $y_1$ thus satisfies the determinicity condition for the empty set $D$. It is also easy to see that the antecedents of the implications described by clause group 1 do not overlap; variable $v$ is not conflicted. We can conclude that there is no solution to the formula above in which $y_1$ has a Skolem function different from $f_{y_1} : (x_1, x_2) \mapsto x_1 \wedge x_2$.

After we identified that $y_1$ has a unique Skolem function, we see that also the Skolem function for variable $y_2$ is unique. The second group of clauses (clause group (2)) allows no Skolem function other than $f_{y_2} : (x_1, y_1) \mapsto \neg(x_1 \vee y_1)$. The use of an existentially quantified variable in the definition of this Skolem function is a short form for the Skolem function $f_{y_2} : (x_1, x_2) \mapsto \neg(x_1 \vee f_{y_1}(x_1, x_2))$.[1] Note that variable $y_2$ does not have a unique Skolem function relative to an empty set $D$. Identifying Skolem functions for some variables can thus help to identify variables for further variables.

## 4.1   Checking for Determinicity

We consider $D$-consistent 2QBF in PCNF $\forall X. \exists Y. \varphi$, where $\varphi$ is the propositional part and $D \subset Y$. For determining determinicity of a variable $v \in Y \setminus D$, only the clauses $\mathcal{D}$ and $\mathcal{U}_v$ play a role. As explained in Sect. 3 we can see each clause $C$ in $\mathcal{U}_v$ as an implication $\neg(C \setminus \{v, \neg v\}) \implies v$ from variables in $X$ and variables with a fixed Skolem function to a literal of $v$. If, and only if, for every pair of assignments $(\mathbf{x}, \mathbf{d})$ (to variables $X$ and $D$) satisfying $\mathcal{D}$ one of the antecedents described by $\mathcal{U}_v$ applies, variable $v$ is deterministic: $\mathcal{D} \implies \bigvee_{C \in \mathcal{U}_v} \neg(C \setminus \{v, \neg v\})$. To enable the use of SAT solvers we avoid the validity in the formulation and negate the formula.

**Lemma 1.** *Variable $v$ is deterministic w.r.t. $D$ iff the following is unsatisfiable:*

$$\mathcal{D} \wedge \bigwedge_{C \in \mathcal{U}_v} C \setminus \{v, \neg v\}$$

## 4.2   Local Under-Approximation of Determinicity

Checking determinicity with the formula above can be costly, as the check involves the potentially large set of clauses $\mathcal{D}$. We thus suggest to drop $\mathcal{D}$ and obtain the under-approximation: $\bigwedge_{C \in U_v} C \setminus \{v, \neg v\}$ . That is, the local under-approximation does not take into account that certain assignments to $D$ may violate the definitions of the Skolem functions. We call a variable that satisfies the local determinicity check *locally deterministic*.

---

[1] The algorithm only constructs Skolem functions that depend on universally quantified variables. The use of existentially quantified variables is just an abbreviation, so there is no risk of circular dependencies.

Let us revisit the example in the beginning of Sect. 4. Assuming that none of the variables have been identified yet to have a unique Skolem function ($D = \emptyset$), we check local determinicity for variable $y_1$ with the following propositional formula: $x_1 \wedge x_2 \wedge (\neg x_1 \vee \neg x_2)$. As the formula is unsatisfiable, $y_1$ is deterministic. If we checked $y_2$ first, there would be only a single clause with unique consequence $y_2$ and thus the local determinicity check consists of the single clause $\neg x_1$, which is trivially satisfiable. Only after identifying a Skolem function for $y_1$, i.e. with $y_1 \in D$, also the other two clauses in which $y_2$ occurs have a unique consequence. In this case we formulate the query $\neg x_1 \wedge \neg y_1 \wedge (x_1 \vee y_1)$, determine its unsatisfiability, and conclude that $y_2$ satisfies condition (1) as well.

### 4.3 Pure Literals

If an existentially quantified variable occurs in only one polarity in a formula in PCNF we call it a *pure literal* and we can assign it this polarity while preserving the truth of the formula. When we additionally consider a partial assignment, we can easily generalize this to the following: when all literals of one polarity are in clauses that are satisfied by literals of other variables, we can assign the variable the opposite literal. We encode this condition as a constraint:

**Lemma 2.** *Given a PCNF* $\forall X.\exists Y.\varphi$ *and a literals* $l$ *of* $v \in Y$, *we have:*

$$\forall X.\exists Y.\varphi \iff \forall X.\exists Y.\varphi \wedge \left( \left( \bigwedge_{C \in \varphi \text{ with } l \in C} C \setminus \{l\} \right) \implies \bar{l} \right)$$

We call $\left( \bigwedge_{C \in \varphi \text{ with } l \in C} C \setminus \{l\} \right) \implies \bar{l}$ the *pure literal constraint*. We could add the constraint for all variables, but this would increase the formula size significantly. Instead we add the constraint only if during propagation $v$ is not locally deterministic and a literal $l$ of $v$ only occurs in clauses in $\mathcal{U}_v$. Adding the pure literal constraint then guarantees that $v$ is deterministic.

Consider a variation of the previous example, where we only flip the negation of the second occurrence of $y_2$:

$$\forall x_1.\forall x_2. \ \exists y_1.\exists y_2. \ \underbrace{(x_1 \vee \neg y_1) \wedge (x_2 \vee \neg y_1) \wedge (\neg x_1 \vee \neg x_2 \vee y_1)}_{f_{y_1}: \ (x_1, x_2) \mapsto x_1 \wedge x_2} \tag{3}$$

$$\wedge \underbrace{(\neg x_1 \vee \neg y_2)}_{f_{y_2}: \ x_1 \mapsto \neg x_1} \wedge (\neg y_1 \vee y_2) \wedge (x_1 \vee y_1 \vee y_2) \tag{4}$$

Even when we test $y_2$ for determinicity before we establish that $y_1$ is deterministic, we can now fix a Skolem function for $y_2$: The only negative occurrence of variable $y_2$ is in the clause $\neg x_1 \vee \neg y_2$, which happens to have $\neg y_2$ as a unique consequence. We can thus fix $y_2$ to be positive in all remaining cases and set $f_{y_2}: \ x_1 \mapsto \neg x_1$ by adding the clause $x_1 \vee y_2$ to the formula.

## 4.4   Checking for Conflicts

In algorithms for propositional SAT, we call variables conflicted, if unit propagation resulted in conflicting assignments to the variable. In the incremental
determinization algorithm a variable $v$ is conflicted in a $D$-consistent 2QBF,
if there is an assignment to the universally quantified variables that propagates conflicting assignments to $v$. We consider the example from Subsect. 4.3 to
develop some intuition on conflicted variables. Assume we first identified variable $y_2$ to have the unique Skolem function $f_{y_2} : x_1 \mapsto \neg x_1$. Then all clauses in
clause group 3 and the latter two clauses of clause group 4 have a literal of $y_1$
as a unique consequence. The determinicity check determines that the formula
$x_1 \wedge x_2 \wedge (\neg x_1 \vee \neg x_2) \wedge \neg y_2 \wedge (x_1 \vee y_2)$ is unsatisfiable and concludes that $y_1$
is deterministic. To see that the variable is conflicted, i.e. there is no Skolem
function satisfying all constraints, consider the assignment $x_1 \wedge x_2$, which sets
$y_2$ to false according to $f_{y_2}$. In this case we cannot give variable $y_1$ a value that
satisfies all constraints: The last clause of clause group 3 requires $y_1$ to be set
to true while the second clause of clause group 4 requires $y_1$ to be set false. We
found a conflict!

As for the determinicity check, we can easily construct a formula for the
*global conflict check* that represents the assignments to $X$ that prove a variable
$v \in Y \setminus D$ to be conflicted. Let us consider a $D$-consistent 2QBF in PCNF
$\forall X.\exists Y.\varphi$, where $\varphi$ is the propositional part and $D \subset Y$. The clauses $\mathcal{D}$ represent
the known Skolem functions and are guaranteed to provide unique values to $D$
for every assignment $\mathbf{x}$ to $X$. In particular, $UP(\mathbf{x})$ cannot result in $\bot$ and so it
suffices to check for conflicting assignments to variable $v$. Variable $v$ can only be
propagated if one of the antecedents of the clauses $\mathcal{U}_v$ with unique consequence $v$
is true. We thus know that variable $v$ is conflicted if, and only if, there is a pair of
assignments $(\mathbf{x}, UP(\mathbf{x}))$ that satisfies the antecedents of two clauses $C, C' \in \mathcal{U}_v$
with $v \in C$ and $\neg v \in C'$.

**Lemma 3.** *Variable $v$ is conflicted if, and only if, the following is satisfiable:*

$$\mathcal{D} \wedge \left( \bigvee_{C \in \mathcal{U}_v \text{ with } v \in C} \neg(C \setminus \{v\}) \right) \wedge \left( \bigvee_{C \in \mathcal{U}_v \text{ with } \neg v \in C} \neg(C \setminus \{\neg v\}) \right)$$

## 4.5   Local Over-Approximation for Conflict Detection

The global conflict check is a relatively expensive step, as it involves the potentially large set of clauses $\mathcal{D}$. Similar to the local determinicity check, we drop
$\mathcal{D}$ from the conflict check to first check for local conflicts. If the local conflict
check returns an assignment to $X$ and $D$, we cannot be sure that the assignment
satisfies $\mathcal{D}$, so we then resort to the global conflict check.

## 4.6   The Propagation Procedure

The procedure PROPAGATE extends a given set $D$ of variables that have a unique
Skolem function and it checks whether there is a conflicted variable. The returned

set $D$ may be an under-approximation of the set of variables with unique Skolem functions, just as propagation for SAT computes an under-approximation of the set of variables having a unique value. This may lead to unnecessary decisions, but avoids the costly global determinicity check. Also, the procedure does not check all variables for conflicts. Instead it only makes sure that deterministic variables are not conflicted, so no conflicted variable gets added to the set $D$. In this way all variables will still be checked for conflicts eventually (unless the algorithm terminates with false).

Given a set $D$ of variables, a $D$-consistent 2QBF, and a decision level, PROP-AGATE returns a 4-tuple indicating the updated set of variables $D$, whether there is a conflict, the formula (which may be modified by pure literal detection), and an assignment **x** to the universally quantified variables:

```
1:  procedure PROPAGATE(∀X.∃Y. φ, D, dlvl)
2:      U ← variables occurring as unique consequence in φ
3:      while U ∩ (Y \ D) ≠ ∅ do
4:          v ← pick a variable in U ∩ (Y \ D)
5:          U ← U \ v
6:          if v is locally deterministic then
7:              if local conflict for v then
8:                  if global conflict for v for assignment x then
9:                      return (D, true, φ, x)
10:             v.dlvl ← dlvl
11:             D ← D ∪ {v}
12:             check for new clauses with unique consequences; update U
13:         else
14:             if v occurs only in Uv or ¬v occurs only in Uv then
15:                 φ ← φ ∧ pure literal constraint
16:                 U ← U ∪ {v}
17:     return (D, false, φ, N/A)
```

With the set $U$ we remember which variables we still have to check for determinicity. Whenever a variable is detected to have a unique Skolem function, we check for clauses that now have a unique consequence and update $U$ (line 12). It is possible (and desirable) to start with a smaller set $U$ than shown above: only variables $v$ for which we added a new clause with unique consequence $v$ since the last propagation phase can possibly become deterministic. For the sake of simplicity we omitted the additional bookkeeping in this exposition.

## 5    Decisions

Decisions are made when the propagation procedure comes to a stop and no conflict was detected. The procedure PICKVAR picks a variable $v \in Y \setminus D$, which we call the *decision variable*. The procedure DECISION then adds clauses, the *decision clauses*, that make variable $v$ locally deterministic. Note the decision variable may be conflicted, though not yet detected as such, at the time of the decision, as the propagation procedure does not guarantee that all variables are conflict free. In the next propagation phase, after the decision variable is detected to be deterministic, it may thus be detected to be conflicted.

In this section we propose a simple way to take decisions that avoids introducing additional conflicts—between decision clauses and clauses in $\mathcal{U}_v$—for the decision variable. We simply fix the Skolem function that assigns 1 to $v$ whenever the clauses $\mathcal{U}_v$ do not require otherwise. That is, we consider the clauses with unique consequence that may require $v$ to be set to 0, i.e. $C \in \mathcal{U}_v$ with $\neg v \in C$, and define the result of DECISION as the constraint that sets $v$ to 1 when all their antecedents are false:

$$\left( \bigwedge_{C \in \mathcal{U}_v \, with \, \neg v \in C} C \setminus \{\neg v\} \right) \implies v$$

We again consider the example of Subsect. 4.3 with an empty set $D$. In case we were to take a decision over variable $y_2$ instead of considering the pure literal rule, we would fix the cases described by the only clause in $\mathcal{U}_v$, which is $\neg x_1 \lor \neg y_2$. Then we fix $y_2$ to be true in all remaining cases, i.e. by adding the clause $x_1 \lor y_2$.

Given a $D$-consistent 2QBF $\forall X.\exists Y.\varphi$ it is clear that a variable $v \in Y \setminus D$ is deterministic in $\forall X.\exists Y.\varphi \land$ DECISION$(v, \varphi, D)$. It is also easy to see that the procedure does not introduce additional conflicts for $v$. Also, for all assignments $\mathbf{x}$ to $X$ and $\mathbf{d}$ to $D$ we have:

$$UP\big(\mathcal{D} \land \mathcal{U}_v \land \text{DECISION}(v, \varphi, D)(\mathbf{x}, \mathbf{d})\big) = \bot \implies UP\big(\mathcal{D} \land \mathcal{U}_v(\mathbf{x}, \mathbf{d})\big) = \bot$$

This property guarantees that the conflict analysis, which we cover in Sect. 6, always results in a *new* clause and thereby provides us with an argument for the termination of the algorithm. The procedure DECISION also marks the added clauses as *decision clauses*. During conflict analysis and during backtracking we have to distinguish decision clauses from learnt clauses.

# 6    Conflict Analysis

*Conflict analysis* for incremental determinization stays remarkably similar to CDCL. Once a (global) conflict is detected, we compute a conflict clause along an implication graph. If the conflict clause contains only universally quantified variables, we proved the formula to be false. Otherwise, we have to backtrack to the largest decision level that contributed to the conflict, add the conflict clause to the formula, and continue with propagation.

Let us consider a $D$-consistent 2QBF $\forall X.\exists Y.\ \varphi$, and let $\delta \subseteq \varphi$ be the set of decision clauses for the decision variables $E \subseteq Y$. Assume we detected a conflict for a variable $v \in Y \setminus D$ for which the global conflict check returned the assignment $\mathbf{x}$ to $X$. The procedure ANALYZECONFLICT first computes $UP(\mathcal{D}(\mathbf{x}))$ to obtain an assignment $\mathbf{e}$ for the variables $D \cap E$. Then the algorithm computes a conflict clause as for CDCL [45] along the implication graph of:

$$UP\Big(\big((\mathcal{D} \land \mathcal{U}_v) \setminus \delta\big)(\mathbf{x}, \mathbf{e})\Big)$$

which is guaranteed to return $\bot$. That is, we omit the decision clauses and instead treat decision variables as a decisions in the sense of CDCL with the values obtained by $UP(\mathcal{D}(\mathbf{x}))$.

**Lemma 4.** *Let $\forall X.\exists Y.\ \varphi$ be a $D$-consistent 2QBF and let $\delta \subseteq \varphi$ be the set of decision clauses in $\varphi$. The algorithm* ANALYZECONFLICT($\forall X.\exists Y.\ \varphi, D, dlvl$) *returns a clause $C$ such that $\varphi \setminus \delta \Leftrightarrow (\varphi \setminus \delta) \wedge C$ and $C \notin \varphi \setminus \delta$.*

*Proof.* Let $v \in Y \setminus D$ be a conflicted variable, provoked by an assignment $\mathbf{x}$ to $X$. After every decision the first variable that is checked for determinicity and conflictedness, is the decision variable. So we have $E \subseteq (D \cup \{v\})$ and if $v \notin E$ then we even have $E \subseteq D$. In either case $UP(\mathcal{D}(\mathbf{x}))$ returns an assignment to all decision variables that are not conflicted. Since $v$ is conflicting with the decision clauses, it is also conflicting when we replace the decision clauses by the values of the decision variables: $UP\Big(\big((\mathcal{D} \wedge \mathcal{U}_v) \setminus \delta\big)(\mathbf{x}, \mathbf{e})\Big) = \bot$.

Any conflict clause $C$ derived by CDCL can be derived by a sequence of resolution steps [40] and we thus know $(\mathcal{D} \wedge \mathcal{U}_v) \setminus \delta \Leftrightarrow ((\mathcal{D} \wedge \mathcal{U}_v) \setminus \delta) \wedge C$ by the soundness of the resolution rule [43]. Since $C$ is over the same variables as $\mathcal{D}$ this extends to $\varphi \Leftrightarrow \varphi \wedge C$. Also, a conflict clause computed as for CDCL is guaranteed to be not contained in the formula it is derived from, so it is not in $(\mathcal{D} \wedge \mathcal{U}_v) \setminus \delta$. Since $C$ only contains variables from $X$ and $D$ and since $\mathcal{D}$ includes all clauses over $X$ and $D$ in $\varphi$, $C$ cannot be in $\varphi \setminus (\mathcal{D} \cup \mathcal{U}_v)$ either.                    $\square$

Consider the following example:

$$\forall x_1.\forall x_2.\ \exists y_1.\exists y_2.\quad (x_2 \vee \neg y_1) \wedge (\neg x_1 \vee \neg x_2 \vee y_1) \tag{5}$$
$$\wedge\ (x_1 \vee \neg y_2) \wedge (\neg y_1 \vee \neg y_2) \wedge (x_1 \vee y_1 \vee y_2) \tag{6}$$

This time, neither $y_1$ nor $y_2$ can be propagated. Let us pick $y_2$ as the decision variable. According to the decisions discussed in Sect. 5 we add the clause $\neg x_1 \vee y_2$ to complement the only other clause in which $y_2$ occurs as its unique consequence, i.e., $x_1 \vee \neg y_2$. The variable $y_2$ is thus assigned the Skolem function $f_{y_2} : x_1 \mapsto x_1$. Now variable $y_1$ is conflicted: The conflict test could return the assignment for the universal variables represented by the conjunction $x_1 \wedge x_2$, which determines $y_2$ to be true according to $f_{y_2}$. The second clause of clause group 5 and the second clause of clause group 6 then require conflicting assignments to $y_1$. Consider the implication graph:

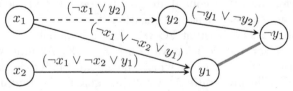

The presence of both nodes $y_1$ and $\neg y_1$ represents a conflict, indicated by the red edge. The labels on the edges indicate the clauses with unique consequence that correspond to the edges. The clause $\neg x_1 \vee \neg y_2$ added in the decision corresponds to the dashed edge from $x_1$ to $y_2$. ANALYZECONFLICT considers the implication graph for the formula without the dashed edge and instead assumes the value 1 for the decision variable $y_2$ as an additional decision. The only conflict clause that could be computed in this case is $\neg x_1 \vee \neg x_2 \vee \neg y_2$.

# 7    Correctness, Termination, Certificates

**Theorem 1.** *Incremental determinization is correct and terminates.*

*Proof Sketch:* We first show that the algorithm maintains $D$-consistency and terminates as there are only finitely many clauses that could be learnt. If the algorithm terminates with true, the computed formula is $Y$-consistent and contains the original formula. This represents a Skolem function satisfying all original constraints. Lemmas 4 and 2 imply soundness for the case that the algorithm terminates with false.

**Certificates.** Once terminated, the correctness of the result can be checked by an independent, simpler algorithm. This is important in practice since highly optimized logic solvers can easily contain programming errors. In some applications the proof object, that is the Skolem function, may also represent an interesting object like an implementation or a strategy.

For the presented method, certification is straightforward: The final set of clauses is $Y$-consistent and we can thus obtain an assignment to the existentially quantified variables $Y$ for every assignment to the universally quantified variables through propagation - the clauses represents the Skolem function. To check the correctness of the Skolem function via existing QBF proof checkers such as QBFCert [36,41], we translate the function into a circuit in the AIGER format [1]. The circuit reads the values of the universally quantified variables and its outputs provide the values for the existentially quantified variables.

We can exploit the order in which the algorithm added variables to the set $D$. This order provides us a unique direction in which the clauses can propagate. Let $U_v$ be the set of clauses with unique consequence at the time variable $v$ was added last to the set $D$. For every variable $v$ we then define the circuit's output for variable $v$ as

$$\bigvee_{C \in \mathcal{U}_v \text{ with } v \in C} \neg (C \setminus \{v\}) \, .$$

The order among the existentially quantified variables then guarantees the absence of circular dependencies.

In case the algorithm determines that the given QBF is false, the implementation provides the assignment to the universally quantified variables from the last global conflict check.

# 8    Implementation and Experimental Evaluation

We implemented the algorithm in a tool we named the <u>Ca</u>l incremental <u>determi</u>nizer (CADET)[2], using PicoSAT [10] to solve the propositional problems. The

---

[2] CADET is available via https://eecs.berkeley.edu/~rabe/cadet.html.

implementation emphasizes simplicity over speed and consists of about 4000 lines of code (not counting the SAT solver) in the programming language C. For global conflict detection, we make use of the incremental solving features of PicoSAT. We maintain an instance of PicoSAT containing the clauses $\mathcal{D}$ that have only literals of universally quantified variables and deterministic existentially quantified variables. Each global conflict check can then be performed by adding only few clauses. Whenever we detect new variables to be deterministic, we push more clauses into the SAT solver, and accordingly pop clauses during backtracking. We implemented two further optimizations:

*Restarts:* After a certain threshold of conflicts, we backtrack all decisions and continue with propagation. The restart threshold is then increased by a constant factor >1. Thus there will eventually be a large enough interval such that the termination argument holds.

*Constant propagation:* When the algorithm starts and whenever we identify variables as deterministic, we check for unit clauses. Unit clauses imply a unique value for a variable, which we propagate among the not yet determined variables.

**Experimental Evaluation.** We performed experiments on several sets of benchmarks of 2QBF instances to compare CADET to state-of-the-art QBF solvers. The experiments were conducted on machines with a quadcore 3.6 GHz Intel Xeon processor, with a 10 min timeout and 4 GB memory limit. Table 1 summarizes the number of instances solved for different benchmark families.

**Table 1.** Number of instances solved within 10 min and 4 GB memory for various benchmarks. For columns labeled with +b we applied the preprocessor Bloqqer before running the solver. Numbers in bold font indicate the best result for a benchmark.

| Family | Total | CADET - | CADET +b | RAReQS - | RAReQS +b | quantor - | quantor +b | Qesto - | Qesto +b | CAQE - | CAQE +b | DepQBF - | DepQBF +b | GhostQ - |
|---|---|---|---|---|---|---|---|---|---|---|---|---|---|---|
| Terminator | 590 | **583** | 513 | 0 | 382 | 12 | 138 | 1 | 491 | 32 | 288 | 547 | 572 | 480 |
| Hardware Fixpoint | 131 | **110** | 92 | 8 | 67 | 16 | 62 | 8 | 67 | 7 | 78 | 22 | 68 | 8 |
| Ranking Functions | 365 | **365** | **365** | 0 | 363 | 13 | 360 | 0 | 363 | 0 | 363 | 168 | 360 | 6 |
| Reduction Finding | 48 | 0 | 4 | 21 | **38** | 2 | 5 | 22 | 36 | 16 | 26 | 14 | 16 | 2 |
| Circuit Underst. | 78 | 20 | 44 | 2 | **50** | 2 | 6 | 2 | 34 | 2 | 48 | 36 | 17 | 36 |
| Partial Equivalence | 300 | 217 | 201 | 2 | 246 | 80 | 102 | 20 | 235 | 56 | **261** | 86 | 159 | 11 |
| Reactive Synthesis | 153 | **153** | **153** | 133 | **153** | **153** | **153** | 129 | **153** | 106 | **153** | 137 | **153** | 149 |
| Random | 254 | 242 | 243 | 246 | 238 | 90 | 91 | 226 | 220 | 249 | **250** | 231 | 224 | 183 |

The first group of benchmarks was taken from QBFLIB [19]. These benchmarks consist of interesting software and hardware verification problems (Terminator [5], Hardware Fixpoint [48], Ranking Functions [14]) and demonstrate the strength of incremental determinization compared to existing algorithms. CADET solved more instances than any other approach while not being dependent on preprocessing.

The second group of benchmarks is from recent papers on QBF applications: Reduction Finding [29] (also used in QBFGallery 2014 [25]), Circuit Understanding [18], Partial Equivalence [17], Reactive Synthesis [12]. We included the second group of benchmarks to also present cases in which incremental determinization is not superior to the existing approaches. The last benchmark, Random, consists of all 2QBFs from the randomly generated instances listed on QBFLIB [19].

Figure 1 shows that CADET is the strongest solver overall. The log-scale allows us to observe that there is a substantial gap between the solving times of CADET and the other approaches. This is also reflected in the overall solving times, where CADET leads with 144009 s before Qesto+b, DepQBF+b, and RAReQS+b taking 224220, 243575, and 259125 s.

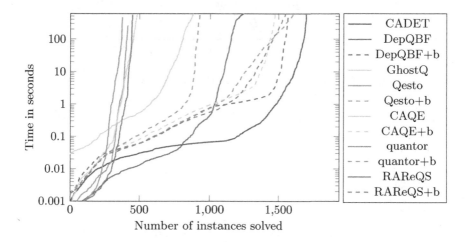

**Fig. 1.** Log-scale cactus plot comparing the performance over all instances.

The scatter plot in Fig. 2 compares the relative runtimes of RAReQS (with Bloqqer) and CADET. The scatter plot suggests that incremental determinization and abstraction refinement perform quite differently and that their relative performance highly depends on the benchmark. The comparison to the other solvers shows similar features.

The certificates computed by CADET are typically much smaller than those by DepQBF. In Fig. 3 we compare the certificate sizes of DepQBF and CADET. For CADET we present the size of the certificates before and after simplification with the ABC model checker. For DepQBF we used the `--simplify` during the generation with `qrpcert`. DepQBF's certificate cannot be simplified further using ABC, as DepQBF encodes some information in the structure of the certificates. The plot reveals an enormous difference in the ability to certify and the quality of the produced certificates.

## 9   Related Work

Early approaches to solving QBFs focused on *expanding* the quantifiers using data structures like CNF [9,33], BDDs [4,37] or AIGs [39]. *Skolemization* helps

Runtime of RAReQS+Bloqqer in s

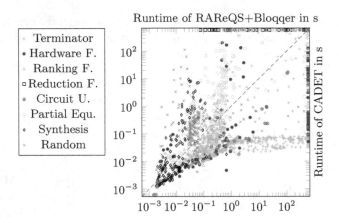

**Fig. 2.** Relative performance of CADET and RAReQS+Bloqqer.

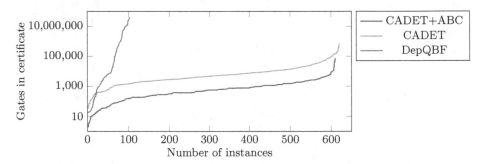

**Fig. 3.** Log-scale cactus plot showing the distribution of sizes of the certificates.

to reduce the expansion effort [7]. *QDPLL* is a generalization of DPLL to QBF where we propagate and decide on values of variables, as for propositional SAT [6, 13, 20, 34]. QDPLL solvers often explore a large portion of the exponentially many assignments to the universally quantified variables as the example from the introduction shows. The incremental determinization algorithm encapsulates the reasoning about the assignments to the universal variables in the global conflict check, which is a propositional formula and can thus be offloaded to SAT solvers.

The *CEGAR* approach for QBF [26–28, 41] was most competitive in recent evaluations [19, 25]. The basic idea is to maintain one SAT solver for each quantifier level. In each iteration these algorithms exclude all assignments to universal variables that can be matched with one assignment to the existential variables. Skolem functions that require many different outputs, like the identity example $\forall X.\, \exists Y.\, X = Y$ from the introduction, thus need an exponential number of iterations. Incremental determinization does not have to explore single assignments to $Y$ in cases, like the identity example, where we can derive parts of the Skolem function directly.

*Non-CNF solvers.* Recently there has been a surge of interest in QBF solvers that are not based on the PCNF representation of formulas. Exploiting the formula structure [21,30,31] or even the word-level structure [48] has been argued to be superior to pure CNF-level reasoning. The experimental results in this paper suggest that this question is not yet settled. For example, the circuit-based solver GhostQ is significantly less effective than CADET on most benchmarks. After the initial circuit-extraction current circuit-based solvers suffer from the same principal problem as solvers using QDPLL or CEGAR.

*Proof systems for QBF.* Resolution based proof calculi, like *resolve and expand* [9] and IR [8], also add clauses to the QBF. The generation of conflict clauses in this work relies on the instantiation of the universally quantified variables, which resembles the proof rules in IR. In contrast, incremental determinization adds constraints that possibly change the truth of the formula. This comes at the cost of backtracking, in case the added constraints (during decisions) were too strong.

*Certification.* Previous certifying QBF solvers produce proof traces from which certificates (Skolem functions) can be reconstructed [36,41]. Also preprocessing techniques such as QBCE [11] can be certified [22,23] and combined with certificates for solvers [24]. In this work, the final set of clauses is the certificate and can easily be translated into a circuit representing the Skolem function.

# 10    Conclusions

Quantified boolean formulas are a natural choice to encode problems in verification, synthesis, and artificial intelligence. We give a completely new algorithm to solve problems in 2QBF that is based on incrementally adding constraints until the Skolem relation collapses to a Skolem function. The algorithm employs a propagation step that directly constructs Skolem functions out of the constraints of the formula. We thereby exploit the formula structure despite working on the level of the CNF. The example from the introduction, $\forall X. \exists Y. X = Y$, is thus solved instantly. Working on the level of Skolem functions avoids situations in which other approaches have to explore an exponential number of cases and also helps to certify the results.

Most parts of the algorithm are kept simplistic on purpose. Likely there are stronger and more efficient ways to propagate, take decisions, and to analyze conflicts in this setting. The purpose of this work is to introduce a framework for the direct manipulation of Skolem functions in a search-based algorithm. Still, the experimental evaluation suggests that the algorithm already improves over existing approaches.

**Acknowledgements.** We thank our anonymous reviewers for their comments and Marcell Vazquez-Chanlatte, Leander Tentrup, Ashutosh Trivedi, and Christoph Wintersteiger for fruitful discussions on this work. Adrià Gascón and Leander Tentrup kindly provided 2QBF benchmarks. This work was supported in part by NSF grants CCF-1139138 and CCF-1116993, by SRC contract 2460.001, NSF STARSS grant 1528108, and by gifts from Toyota and Microsoft.

# References

1. AIGER toolset. http://fmv.jku.at/aiger/
2. Alur, P., Madhusudan, W.N.: Symbolic computational techniques for solving games. Softw. Tools Technol. Transf. **7**(2), 118–128 (2005)
3. Ansotegui, C., Gomes, C.P., Selman, B.: The Achilles' heel of QBF. In: Proceedings of AAAI, vol. 2, pp. 275–281 (2005)
4. Audemard, G., Saïs, L.: A symbolic search based approach for quantified boolean formulas. In: Bacchus, F., Walsh, T. (eds.) SAT 2005. LNCS, vol. 3569, pp. 16–30. Springer, Heidelberg (2005)
5. Basler, G., Kroening, D., Weissenbacher, G.: SAT-based summarization for boolean programs. In: Bošnački, D., Edelkamp, S. (eds.) SPIN 2007. LNCS, vol. 4595, pp. 131–148. Springer, Heidelberg (2007)
6. Bayless, S., Hu, A.J.: Single-solver algorithms for 2QBF. In: Cimatti, A., Sebastiani, R. (eds.) SAT 2012. LNCS, vol. 7317, pp. 487–488. Springer, Heidelberg (2012)
7. Benedetti, M.: sKizzo: A suite to evaluate and certify QBFs. In: Nieuwenhuis, R. (ed.) CADE 2005. LNCS (LNAI), vol. 3632, pp. 369–376. Springer, Heidelberg (2005)
8. Beyersdorff, O., Chew, L., Janota, M.: On unification of QBF resolution-based calculi. In: Csuhaj-Varjú, E., Dietzfelbinger, M., Ésik, Z. (eds.) MFCS 2014, Part II. LNCS, vol. 8635, pp. 81–93. Springer, Heidelberg (2014)
9. Biere, A.: Resolve and expand. In: H. Hoos, H., Mitchell, D.G. (eds.) SAT 2004. LNCS, vol. 3542, pp. 59–70. Springer, Heidelberg (2005)
10. Biere, A.: PicoSAT essentials. JSAT **4**(2–4), 75–97 (2008)
11. Biere, A., Lonsing, F., Seidl, M.: Blocked clause elimination for QBF. In: Bjørner, N., Sofronie-Stokkermans, V. (eds.) CADE 2011. LNCS, vol. 6803, pp. 101–115. Springer, Heidelberg (2011)
12. Bloem, R., Egly, U., Klampfl, P., Könighofer, R., Lonsing, F.: SAT-based methods for circuit synthesis. In: Proceedings of FMCAD, pp. 31–34 (2014)
13. Cadoli, M., Schaerf, M., Giovanardi, A., Giovanardi, M.: An algorithm to evaluate quantified boolean formulae and its experimental evaluation. J. Automated Reason. **28**(2), 101–142 (2002)
14. Cook, B., Kroening, D., Rümmer, P., Wintersteiger, C.M.: Ranking function synthesis for bit-vector relations. In: Esparza, J., Majumdar, R. (eds.) TACAS 2010. LNCS, vol. 6015, pp. 236–250. Springer, Heidelberg (2010)
15. Davis, M., Logemann, G., Loveland, D.: A machine program for theorem-proving. Commun. ACM **5**(7), 394–397 (1962)
16. Faymonville, P., Finkbeiner, B., Rabe, M.N., Tentrup, L.: 3 encodings of reactive synthesis. In: Proceedings of QUANTIFY, pp. 20–22 (2015)
17. Finkbeiner, B., Tentrup, L.: Fast DQBF refutation. In: Sinz, C., Egly, U. (eds.) SAT 2014. LNCS, vol. 8561, pp. 243–251. Springer, Heidelberg (2014)
18. Gascón, A., Subramanyan, P., Dutertre, B., Tiwari, A., Jovanovic, D., Malik, S.: Template-based circuit understanding. In: Proceedings of FMCAD, pp. 83–90. IEEE (2014)
19. Giunchiglia, E., Narizzano, M., Pulina, L., Tacchella, A.: Quantified Boolean Formulas satisfiability library (QBFLIB) (2005). www.qbflib.org
20. Giunchiglia, E., Narizzano, M., Tacchella, Q.A.: A system for deciding Quantified Boolean Formulas satisfiability. In: Proceedings of IJCAR, pp. 364–369 (2001)

21. Goultiaeva, A., Bacchus, F.: Recovering and utilizing partial duality in QBF. In: Järvisalo, M., Van Gelder, A. (eds.) SAT 2013. LNCS, vol. 7962, pp. 83–99. Springer, Heidelberg (2013)
22. Heule, M., Seidl, M., Biere, A.: Efficient extraction of skolem functions from QRAT proofs. In: Proceedings of FMCAD, pp. 107–114 (2014)
23. Heule, M.J.H., Seidl, M., Biere, A.: A unified proof system for QBF preprocessing. In: Demri, S., Kapur, D., Weidenbach, C. (eds.) IJCAR 2014. LNCS, vol. 8562, pp. 91–106. Springer, Heidelberg (2014)
24. Janota, M., Grigore, R., Marques-Silva, J.: On QBF proofs and preprocessing. In: McMillan, K., Middeldorp, A., Voronkov, A. (eds.) LPAR-19 2013. LNCS, vol. 8312, pp. 473–489. Springer, Heidelberg (2013)
25. Janota, M., Jordan, C., Klieber, W., Lonsing, F., Seidl, M., Van Gelder, A.: The QBFGallery 2014: The QBF competition at the FLoC olympic games. J. Satisfiability Boolean Modeling Comput. **9**, 187–206 (2016)
26. Janota, M., Klieber, W., Marques-Silva, J., Clarke, E.: Solving QBF with counterexample guided refinement. In: Cimatti, A., Sebastiani, R. (eds.) SAT 2012. LNCS, vol. 7317, pp. 114–128. Springer, Heidelberg (2012)
27. Janota, M., Marques-Silva, J.: Solving QBF by clause selection. In: Proceedings of IJCAI, pp. 325–331. AAAI Press (2015)
28. Janota, M., Marques-Silva, J.: Abstraction-based algorithm for 2QBF. In: Sakallah, K.A., Simon, L. (eds.) SAT 2011. LNCS, vol. 6695, pp. 230–244. Springer, Heidelberg (2011)
29. Jordan, C., Kaiser, L.: Experiments with reduction finding. In: Järvisalo, M., Van Gelder, A. (eds.) SAT 2013. LNCS, vol. 7962, pp. 192–207. Springer, Heidelberg (2013)
30. Jordan, C., Klieber, W., Seidl, M.: Non-CNF QBF solving with QCIR. In: Proceedings of BNP (Workshop) (2016)
31. Klieber, W., Sapra, S., Gao, S., Clarke, E.: A non-prenex, non-clausal QBF solver with game-state learning. In: Strichman, O., Szeider, S. (eds.) SAT 2010. LNCS, vol. 6175, pp. 128–142. Springer, Heidelberg (2010)
32. Kronegger, M., Pfandler, A., Pichler, R.: Conformant planning as a benchmark for QBF-solvers. In: Report on the International Workshop on QBF, pp. 1–5 (2013)
33. Lonsing, F., Biere, A.: Nenofex: Expanding NNF for QBF solving. In: Kleine Büning, H., Zhao, X. (eds.) SAT 2008. LNCS, vol. 4996, pp. 196–210. Springer, Heidelberg (2008)
34. Lonsing, F., Biere, A.: DepQBF: A dependency-aware QBF solver. JSAT **7**(2–3), 71–76 (2010)
35. Miller, C., Scholl, C., Becker, B.: Proving QBF-hardness in bounded model checking for incomplete designs. In: Proceedings of MTV, pp. 23–28 (2013)
36. Niemetz, A., Preiner, M., Lonsing, F., Seidl, M., Biere, A.: Resolution-based certificate extraction for QBF. In: Cimatti, A., Sebastiani, R. (eds.) SAT 2012. LNCS, vol. 7317, pp. 430–435. Springer, Heidelberg (2012)
37. Olivo, O., Allen Emerson, E.: A more efficient BDD-based QBF solver. In: Proceedings of CP, pp. 675–690 (2011)
38. Otwell, C., Remshagen, A., Truemper, K.: An effective QBF solver for planning problems. In: MSV/AMCS, pp. 311–316. Citeseer (2004)
39. Pigorsch, F., Scholl, C.: An AIG-based QBF-solver using SAT for preprocessing. In: Proceedings of DAC, pp. 170–175. IEEE (2010)
40. Pipatsrisawat, K., Darwiche, A.: On the power of clause-learning SAT solvers as resolution engines. Artif. Intell. **175**(2), 512–525 (2011)

41. Rabe, M.N., Tentrup, L.: CAQE: A certifying QBF solver. In: Proceedings of FMCAD, pp. 136–143 (2015)
42. Rabe, M.N., Wintersteiger, C.M., Kugler, H., Yordanov, B., Hamadi, Y.: Symbolic approximation of the bounded reachability probability in large Markov Chains. In: Norman, G., Sanders, W. (eds.) QEST 2014. LNCS, vol. 8657, pp. 388–403. Springer, Heidelberg (2014)
43. Robinson, J.A.: A machine-oriented logic based on the resolution principle. J. ACM **12**(1), 23–41 (1965)
44. Sabharwal, A., Ansótegui, C., Gomes, C.P., Hart, J.W., Selman, B.: QBF modeling: Exploiting player symmetry for simplicity and efficiency. In: Biere, A., Gomes, C.P. (eds.) SAT 2006. LNCS, vol. 4121, pp. 382–395. Springer, Heidelberg (2006)
45. Marques Silva, J.P., Sakallah, K.A.: GRASP - A new search algorithm for satisfiability. In: Proceedings of CAD, pp. 220–227. IEEE (1997)
46. Solar-Lezama, A.: Program synthesis by sketching. Ph.D. thesis, University of California, Berkeley (2008)
47. Tseitin, G.S.: On the complexity of derivation in propositional calculus. Stud. Constructive Math. Math. Logic **2**(115–125), 10–13 (1968)
48. Wintersteiger, C.M., Hamadi, Y., De Moura, L.: Efficiently solving quantified bit-vector formulas. In: Proceedings of FMSD, vol. 42, no. 1, pp. 3–23, (2013). Observation of strains. Infect. Dis. Ther. **3**(1), 35–43 (2011)

# Non-prenex QBF Solving Using Abstraction

Leander Tentrup[✉]

Reactive Systems Group, Saarland University, Saarbrücken, Germany
`tentrup@react.uni-saarland.de`

**Abstract.** In a recent work, we introduced an abstraction based algorithm for solving quantified Boolean formulas (QBF) in prenex negation normal form (PNNF) where quantifiers are only allowed in the formula's prefix and negation appears only in front of variables. In this paper, we present a modified algorithm that lifts the restriction on prenex quantifiers. Instead of a linear quantifier prefix, the algorithm handles tree-shaped quantifier hierarchies where different branches can be solved independently. In our implementation, we exploit this property by solving independent branches in parallel. We report on an evaluation of our implementation on a recent case study regarding the synthesis of finite-state controllers from $\omega$-regular specifications.

## 1 Introduction

In recent work [18], we introduced an algorithm for solving quantified Boolean formulas (QBF) in prenex negation normal form (PNNF). For each maximal consecutive block of quantifiers of the same type, we build an *abstraction*, i.e., a propositional formula that combines valuations of inner and outer quantifier blocks into valuations of special literals, called *interface literals*. The algorithm employs a counterexample guided abstraction refinement (CEGAR) loop that does recursion over the quantifier blocks. In every block, we use a SAT solver as an oracle to generate new abstraction entries and to provide us with witnesses for unsatisfiable queries. This algorithm, however, is limited to prenex QBF where quantifier are only allowed in the formula's prefix. This can be problematic for non-prenex formulas since the task of prenexing a QBF is non-deterministic and different prenexing strategies lead to different solving times [3]. On the other hand, miniscoping can be used to translate prenex formulas into non-prenex form. We have observed [17] that this is very effective for splitting instances into independent parts on some benchmark families.

In this paper, we extend our previous algorithm to handle non-prenex QBFs in negation normal form. Instead of a linear quantifier prefix, the algorithm is optimized to handle tree-shaped quantifier hierarchies. These optimizations include identifying parts of the formula that belongs only to one quantifier block,

This work was partially supported by the German Research Foundation (DFG) as part of the Transregional Collaborative Research Center "Automatic Verification and Analysis of Complex Systems" (SFB/TR 14 AVACS).

© Springer International Publishing Switzerland 2016
N. Creignou and D. Le Berre (Eds.): SAT 2016, LNCS 9710, pp. 393–401, 2016.
DOI: 10.1007/978-3-319-40970-2_24

hence, eliminating the need for interface literals. Further, for a branching node, i.e., a quantifier block which has multiple children, it is possible to solve the children independently. Our implementation exploits this independence by solving the different branches in parallel.

## 2   Quantified Boolean Formulas

A quantified Boolean formula (QBF) is a propositional formula over a finite set of variables $\mathcal{X}$ with domain $\mathbb{B} = \{0, 1\}$ extended with quantification. The syntax is given by the grammar

$$\varphi := x \mid \neg\varphi \mid \varphi \vee \varphi \mid \varphi \wedge \varphi \mid \exists x.\, \varphi \mid \forall x.\, \varphi,$$

where $x \in \mathcal{X}$. For readability, we lift the quantification over variables to the quantification over sets of variables and denote a maximal consecutive block of quantifiers of the same type $\forall x_1.\forall x_2.\cdots\forall x_n.\,\varphi$ by $\forall X.\,\varphi$ and $\exists x_1.\exists x_2.\cdots\exists x_n.\,\varphi$ by $\exists X.\,\varphi$, accordingly, where $X = \{x_1, \ldots, x_n\}$.

Given a subset of variables $X \subseteq \mathcal{X}$, an *assignment* of $X$ is a function $\alpha : X \to \mathbb{B}$ that maps each variable $x \in X$ to either true (1) or false (0). We identify $\alpha$ as the conjunctive formula $\bigwedge_{x \in X \mid \alpha(x)=1} x \wedge \bigwedge_{x \in X \mid \alpha(x)=0} \neg x$. When the domain of $\alpha$ is not clear from context, we write $\alpha_X$. A partial assignment $\beta : X \to \mathbb{B} \cup \{\bot\}$ may additional set variables $x \in X$ to an undefined value $\bot$. We say that $\beta$ is *compatible* with $\alpha$, $\beta \sqsubseteq \alpha$ for short, if they have the same domains $(\mathrm{dom}(\alpha) = \mathrm{dom}(\beta))$ and $\alpha(x) = \beta(x)$ for all $x \in X$ where $\beta(x) \neq \bot$. For two assignments $\alpha$ and $\alpha'$ with disjoint domains $X = \mathrm{dom}(\alpha)$ and $X' = \mathrm{dom}(\alpha')$ we define the combination $\alpha \sqcup \alpha' : X \cup X' \to \mathbb{B}$ as $\alpha \sqcup \alpha'(x) = \begin{cases} \alpha(x) & \text{if } x \in X, \\ \alpha'(x) & \text{otherwise.} \end{cases}$
We define the *complement* $\overline{\alpha}$ to be $\overline{\alpha}(x) = \neg\alpha(x)$ for all $x \in \mathrm{dom}(\alpha)$. The *set of assignments* of $X$ is denoted by $\mathcal{A}(X)$.

A quantifier $Q\,x.\,\varphi$ for $Q \in \{\exists, \forall\}$ *binds* the variable $x$ in the *scope* $\varphi$. Variables that are not bound by a quantifier are called *free*. The set of free variables of formula $\varphi$ is defined as $\mathit{free}(\varphi)$. We assume the natural semantics of the satisfaction relation $\alpha_X \vDash \varphi$ for QBF $\varphi$ and assignments $\alpha_X$ where $X$ are the free variables of $\varphi$. *QBF satisfiability* is the problem to determine, for a given QBF $\varphi$, the existence of an assignment $\alpha$ for the free variables of $\varphi$, such that the relation $\vDash$ holds.

A *closed* QBF is a formula without free variables. Closed QBFs are either true or false. A formula is in prenex form, if the formula consists of a quantifier prefix followed by a propositional formula.

A *literal* $l$ is either a variable $x \in X$, or its negation $\neg x$. Given a set of literals $\{l_1, \ldots, l_n\}$, the disjunctive combination $(l_1 \vee \ldots \vee l_n)$ is called a *clause*. Given a literal $l$, the polarity of $l$, $\mathit{sign}(l)$ for short, is 1 if $l$ is positive and 0 otherwise. The variable corresponding to $l$ is defined as $\mathit{var}(l) = x$ where $x = l$ if $\mathit{sign}(l) = 1$ and $x = \neg l$ otherwise.

A QBF is in negation normal form (NNF) if negation is only applied to variables. Every QBF can be transformed into NNF by at most doubling the size of the formula and without introducing new variables. For formulas in NNF, we treat literals as atoms.

## 3   Algorithm

We introduce additional notation to facilitate working with arbitrary Boolean formulas. Let $\mathcal{B}$ be the set of quantified Boolean formulas and let $sf(\varphi) \subset \mathcal{B}$ ($dsf(\varphi) \subset \mathcal{B}$) be the set of (direct) subformulas of $\varphi$ (note that $\varphi \in sf(\varphi)$ but $\varphi \notin dsf(\varphi)$). Further, $dqsf(\varphi) \subset \mathcal{B}$ denotes the direct quantified sub-formulas of $\varphi$, i.e., a quantifier $Q\,X'.\,\psi$ is in $dqsf(\varphi)$ if $Q\,X'.\,\psi$ is in the scope of $\varphi$ and there is no other quantifier $Q\,X''.\,\psi'$ such that $Q\,X''.\,\psi'$ is in the scope of $\varphi$ and $Q\,X'.\,\psi$ is in the scope of $\psi'$. For a subformula $\psi$, $type(\psi) \in \{lit, \vee, \wedge, Q\}$ returns the Boolean connector if $\psi$ is not a literal nor a quantifier. For example, given $\psi = \exists x.\,(\forall y.\,\exists z.\,(x \vee y \vee \neg x)) \vee (\forall y.\,(y \wedge x))$, it holds that $type(\psi) = Q$, $dsf(\psi) = \{\forall y.\,\exists z.\,(x \vee y \vee \neg x)) \vee (\forall y.\,(y \wedge x)\}$, and $dqsf(\psi) = \{\forall y.\,\exists z.\,(x \vee y \vee \neg x), \forall y.\,(y \wedge x)\}$.

For this section, we assume w.l.o.g. that all quantifier blocks in the QBF are strictly alternating, even for quantifiers not in the prefix. That means that for every quantified formula $Q\,X.\,\psi$, the quantifier type of all $\psi' \in dqsf(Q\,X.\,\psi)$ is $\overline{Q}$. We use a generic solving function $\text{SAT}(\theta, \alpha)$ for propositional formula $\theta$ under assumptions $\alpha$, that returns whether $\theta \wedge \alpha$ is satisfiable and either a satisfying assignment $\alpha_V$ for variables $V \subseteq free(\theta)$ or a partial assignment $\beta_{failed} \sqsubseteq \alpha$ such that $\theta \wedge \beta_{failed}$ is unsatisfiable. Further, we define $\text{SAT}_Q$ to be SAT if $Q = \exists$ and UNSAT otherwise ($\text{UNSAT}_Q$ analogously).

The non-prenex algorithm works on the principle of communicating the sat-isfaction of subformulas between quantifier blocks in the QBF. This communi-cation is realized by two special types of literals which we call *interface literals*. Only the valuation of those literals are communicated between the quantifier levels. For a given quantifier $Q\,X$ and a subformula $\psi$, the $T$ literal $t_\psi$ represents the assignments made by the outer quantifiers while the $B$ literal $b_\psi$ represents the assignments from the current quantifier $Q\,X$ including assumptions on the satisfaction of subformulas by inner quantifiers. Thus, $t_\psi$ is true if $\psi$ is satisfied by the outer quantifiers and a valuation that sets $b_\psi$ to true indicates that $\psi$ is satisfied by the quantifier $Q\,X$. Before going in more detail on the abstraction, we introduce the basic algorithm first.

The algorithm ABSTRACTION-QBF is depicted in Algorithm 1. The algorithm uses a *dual abstraction* for optimization of abstraction entries [18]. Given a quan-tifier $Q\,X$, the sets $T_X$ and $B_X$ contain the $T$ and $B$ literals corresponding to this quantifier. When translating a $B$ literal to a $T$ literal, we use the the same index, e.g., in line 12, the $B$ literals $b_{\psi''}$ are translated to $T$ literals $t_{\psi''}$ of the inner quantifier. We initialize $\theta_X$ with the abstraction described below, which is a propositional formula over variables in $X$, as well as $T$ and $B$ literals. For the innermost quantifier, it holds that $B_X = \emptyset$. Further, the *dual abstraction* $\overline{\theta}_X$ is defined as the abstraction for $\overline{Q}\,X$.

In every iteration of the while loop, a $B$ literal assignment $\alpha_{B_X}$ is generated according to the outer $T$ assignment $\alpha_{T_X}$ (line 3). For every direct quantified

---

**Algorithm 1.** Non-prenex Abstraction Based Algorithm

---

1: **procedure** ABSTRACTION-QBF($Q\,X.\,\psi, \alpha_{T_X}$)
2:     **while** true **do**
3:         $result, \alpha_X \sqcup \alpha_{B_X}, \beta_{failed} \leftarrow \text{SAT}(\theta_X, \alpha_{T_X})$                    ▷ $\beta_{failed} \sqsubseteq \alpha_{T_X}$
4:         **if** $result = \text{UNSAT}$ **then**
5:             **return** $\text{UNSAT}_Q, \beta_{failed}$
6:         **else if** $\psi$ is propositional **then**
7:             $result, \_, \beta_{failed} \leftarrow \text{SAT}(\overline{\theta}_X, \alpha_X \sqcup \overline{\alpha_{T_X}})$                    ▷ $result = \text{UNSAT}$
8:             **return** $\text{SAT}_Q, \overline{\beta_{failed}}$                    ▷ $\overline{\beta_{failed}} \sqsubseteq \alpha_{T_X}$
9:         $sub\text{-}result \leftarrow \text{SAT}_Q$
10:        Let $\beta_{sub}$ be the empty assignment
11:        **for** $\psi' = \overline{Q}\,Y.\,\psi^*$ in $dqsf(Q\,X.\,\psi)$ **where** $\alpha_{B_X}(b_{\psi'}) = 0$ **do**
12:            Define $\alpha_{T_Y}$ s.t. $\alpha_{T_Y}(t_{\psi''}) = \neg\alpha_{B_X}(b_{\psi''})$ for all $t_{\psi''} \in T_Y$
13:            $result, \beta_{T_Y} \leftarrow \text{ABSTRACTION-QBF}(\psi', \alpha_{T_Y})$                    ▷ $\beta_{T_Y} \sqsubseteq \alpha_{T_Y}$
14:            **if** $result = \text{UNSAT}_Q$ **then**
15:                $\theta_X \leftarrow \theta_X \wedge \left(\bigvee_{b_{\psi''} \in B_X | \beta_{T_Y}(t_{\psi''})=1} b_{\psi''}\right)$
16:                $sub\text{-}result \leftarrow \text{UNSAT}_Q$
17:            **else**
18:                $\beta_{sub} \leftarrow \beta_{sub} \sqcup \beta_{T_Y}$
19:        **if** $sub\text{-}result = \text{SAT}_Q$ **then**
20:            $\overline{\theta}_X \leftarrow \overline{\theta}_X \wedge \left(\bigvee_{b_{\psi''} \in B_X | \beta_{sub}(t_{\psi''})=1} b_{\psi''}\right)$
21:            $result, \_, \beta_{failed} \leftarrow \text{SAT}(\overline{\theta}_X, \alpha_X \sqcup \overline{\beta_{sub}})$                    ▷ $result = \text{UNSAT}$
22:            **return** $\text{SAT}_Q, \overline{\beta_{failed}}$                    ▷ $\overline{\beta_{failed}} \sqsubseteq \alpha_{T_X}$

---

subformula $\psi' = \overline{Q}\,Y.\,\psi^*$ (line 11) which is assumed to be satisifed ($\alpha_{B_X}(b_{\psi'}) = 0$), we translate $\alpha_{B_X}$ into a $T$ assignment for the inner quantifier (line 12) and proceed recursively (line 13). If the recursive call is UNSAT (w.r.t the current quantifier), we refine the abstraction to exclude the counterexample $\beta_{T_Y}$, i.e., in the following iterations one of the $b_{\psi''}$ such that $\beta_{T_Y}(t_{\psi''}) = 1$ must be set to true. Due to the negation during translation, those $b_{\psi''}$ were set to false in $\alpha_{B_X}$. In case the recursive call is SAT, we update $\beta_{sub}$ to include the optimized $T$ assignment returned from the inner quantifier. When all recursive calls returned SAT, we update the dual abstraction $\overline{\theta}_X$ and use it to optimize the witness $\beta_{sub}$ for the outer scope. If the query in line 3 fails, $\text{UNSAT}_Q$ together with an assignment $\beta_{failed} \sqsubseteq \alpha_{T_X}$ witnessing the unsatisfiability is returned.

To ensure correctness, we need requirements on the abstraction being used. Given a quantifier $\exists X$, we say that a subformula $\psi$ is good if it is not yet falsified ($type(\psi) = \wedge$), respectively satisfied ($type(\psi) = \vee$). For every quantifier $Q\,X$ and $B$ literal $b_\psi$ it must hold that if $b_\psi$ is set to true then $\psi$ is good for quantifier $X$. This gives us proper refinement semantics (line 15). The same property holds for $t$ literals $t_\psi$. For every direct quantified subformula $\psi' = Q\,Y.\,\varphi^*$ in the current scope, the $B$ literal $b_{\psi'}$ can be only set to true if the subformula $\psi'$ is not assumed to be true. Intuitively, this means that the result of subformula $\psi'$ is not used to satisfy the abstraction $\theta_X$. Further, for a quantifier alternation, it must hold that the set of outer $B$ literals $B_X$ matches the union of all inner $T$ literals $T_Y$

to enable the translation in line 12. Combining these properties gives us that a good subformula of quantifier $Q\,X$ is a bad subformula of quantifier $Q\,Y$ and vice versa. Termination then follows from progress due to refinements (line 15) and correctness can be showed by induction over the quantifiers.

*Abstraction.* We now give a formal definition of the abstraction. Given a QBF $\varphi$ in NNF and a quantifier $\exists X.\,\varphi'$, we build the following propositional formula in conjunctive normal form representing the structural abstraction $\theta_X = out(\varphi) \wedge \bigwedge_{\psi \in sf(\varphi) \wedge type(\psi) \neq lit} enc(\psi)$ for this quantifier, where $out$ encodes the entry point of the formula and $enc$ defines a CNF formula that encodes the truth of subformula $\psi$ with respect to the valuations of the current, inner and outer quantifiers represented by $B$ and $T$ literals, respectively:

$$
enc(\psi) = \begin{cases} \bigwedge\limits_{\psi' \in\, dsf(\psi)} (\neg b_\psi \vee enc_\psi(\psi')) & \text{if } type(\psi) = \wedge \\[2ex] \neg b_\psi \vee \bigvee\limits_{\psi' \in\, dsf(\psi)} enc_\psi(\psi') & \text{if } type(\psi) = \vee \\[2ex] (b_\psi \vee out(\psi')) & \text{if } \psi = Q\,X.\,\psi' \end{cases}
$$

$$
enc_\psi(\psi') = \begin{cases} \psi' & \text{if } type(\psi') = lit \wedge var(\psi') \in X \\ t_\psi & \text{if } type(\psi') = lit \wedge var(\psi') \text{ bound by outer scope} \\ \neg b_\psi & \text{if } type(\psi') = lit \wedge var(\psi') \text{ bound by inner scope} \\ b_{\psi'} & \text{if } type(\psi') \neq lit \wedge \psi' \text{ only influenced by current or outer scope} \\ \neg b_{\psi'} & \text{if } type(\psi') = Q \\ \bot & \text{otherwise} \end{cases}
$$

$$
enc_\vee(\psi) = \bigvee\limits_{\substack{\psi' \in dsf(\psi) \\ type(\psi')=lit}} enc_\psi(\psi') \vee \bigvee\limits_{\substack{\psi' \in\, dsf(\psi) \\ type(\psi')=\wedge}} b_{\psi'} \vee \bigvee\limits_{\substack{\psi' \in\, dsf(\psi) \\ type(\psi')=\vee}} enc_\vee(\psi') \vee \bigvee\limits_{\substack{\psi' \in\, dsf(\psi) \\ type(\psi')=Q}} \neg b_{\psi'}
$$

$$
out(\psi) = \begin{cases} b_\psi & \text{if } type(\psi) = \wedge \\ enc_\vee(\psi) & \text{if } type(\psi) = \vee \\ \neg b_\psi & \text{if } type(\psi) = Q \end{cases}
$$

An undefined result $\bot$ form $enc_\psi(\psi')$ means that the subformula $\psi'$ is ignored in the encoding $enc(\psi)$. The abstraction of a scope $\forall X$ is defined as the existential abstraction for $\neg\varphi$. The dual abstraction $\overline{\theta}_X$ is defined as the abstraction for $\neg\, Q\,X$. Note that not every $B$ literal that is used in the abstraction may be exposed as an interface literal.

For disjunctive formulas, $enc(\psi)$ enforces that $b_\psi$ can be only set to true if a direct subformula that is (1) a (possibly negated) variable of the current or outer scope, (2) a subformula that is not influenced by an inner variable, or (3) a quantified subformula, is set to true. The encoding $enc(\psi)$ of a conjunctive formula enforces likewise that if $b_\psi$ is true, the encodings $enc_\psi(\psi')$ of all such direct subformulas $\psi' \in dsf(\psi)$ are true. The abstraction has the required refinement semantics: If we want to ensure that one of the subformulas in a set

$R = \{\psi_1, \ldots, \psi_k\}$ is guaranteed to be true at the current scope, we add the clause $(b_{\psi_1} \vee \cdots \vee b_{\psi_k})$.

*Optimizations.* The optimizations from the prenex algorithm [18] can be applied to this algorithm as well. Additionally, we preprocess the formula using the well-known miniscoping rules in order to decompose quantifier blocks.

In the algorithm, satisfying results of direct quantified subformulas are discarded if one of them is UNSAT. Instead, we found that modifying the decision heuristic of the underlying SAT solver to regenerate this subassignment reduced the number of iterations overall.

## 4   Case Study: Reactive Synthesis

For our case study, we consider the *reactive synthesis* problem, i.e., the problem of synthesizing a finite-state controller from an $\omega$-regular specification. Formally, we have a specification $\varphi$ that defines a language $\mathcal{L}(\varphi) \subseteq (2^{I \cup O})^\omega$ over the atomic propositions that are partitioned into a finite set of inputs $I$ to the controller and a finite set of outputs $O$ of the controller. An implementation of a controller is a $2^O$-*labeled* $2^I$-*transition system* $\mathcal{S} = \langle S, s_0, \delta, l \rangle$ where $S$ is a finite set of states, $s_0 \in S$ is the designated initial state, $\delta \colon S \times 2^I \to S$ is the transition function, and $l \colon S \to 2^O$ is the state-labeling. The run of $\mathcal{S}$ on a sequence $\pi \in (2^I)^\omega$ is $run(\mathcal{S}, \pi) = s_0 \pi_0 s_1 \pi_1 \cdots \in (S \cdot 2^I)^\omega$ where $s_{i+1} = \delta(s_i, \pi_i)$ for every $i \geq 0$. The corresponding trace, denoted by $trace(\mathcal{S}, \pi)$, is $(l(s_0) \cup \pi_0)(l(s_1) \cup \pi_1) \cdots \in (2^{I \cup O})^\omega$. A transition system $\mathcal{S}$ satisfies the specification $\varphi$ if $trace(\mathcal{S}, \pi) \in \mathcal{L}(\varphi)$ for all input sequences $\pi \in (2^I)^\omega$. By bounding the number of states that the implementation of the controller may use, one can derive a QBF encoding [4] from this problem using the *bounded synthesis* approach [5]. The synthesis instances used in this case study where taken from the Acacia benchmark set [2].

The exact encoding is out of scope for this paper, so we are only giving a high level overview. The QBF query has a quantifier prefix of the form $\exists\forall\exists$. The variables in the top level existential correspond to a global constraint that cannot be split syntactically. However, the constraints regarding the inner quantifiers $\forall\exists$ are local to the state of the implementation, so one gets a QBF with a top level existentially quantifier and $n$ independent $\forall\exists$ quantifiers below by using miniscoping rules, where $n$ is the number of states in the implementation. This is merely a new observation and not particularly special for this kind of benchmark as we have made similar observations regarding competitive benchmark suites for CNF [17].

We implemented Algorithm 1 and its optimizations in a prototype tool called PQUABS (Parallel Quantified Abstraction Solver)[1] that takes QBFs in the standard format QCIR [16]. We use PicoSAT [1] as the underlying SAT solver and the POSIX pthreads library for thread creation and synchronization. For every quantifier $Q\,X$ that branches more than once, we create a thread for each child

---

[1] Available at https://www.react.uni-saarland.de/tools/quabs/.

**Table 1.** Cumulated solving time of PQuAbS with respect to number of used threads. There are 443 instances in total.

|  | 1 thread | 2 threads | 3 threads | 4 threads | Prenex |
|---|---|---|---|---|---|
| # solved instances | 397 | 403 | 407 | 409 | 325 |
| cumulated solving time | 100 % | 64.51 % | 54.15 % | 49.94 % | - |

quantifier. The loop in line 11 is then implemented by passing $\alpha_{T_Y}$ to the sub-quantifier and waking the corresponding thread. Before line 19, there is a barrier where we wait for all children to finish. For our experiments, we used a machine with a 3.6 GHz quad-core Intel Xeon processor and 32 GB of memory. The time-out was set to 10 min.

Table 1 shows the overall results of our experiments. It depicts the number of solved instances and the cumulated solving times with respect to the number of threads used. For comparison, we also included the number of solved instances from the single threaded version of PQuAbS without miniscoping, i.e., linear prenex solving. One cannot expect linear speedup due to the non-parallelizable parts, like preprocessing and solving of the top-level existential quantifier, as well as the fact that the solving time of the children $\forall\exists$ quantifiers are not uniform.

Nevertheless, already using 2 threads, the speedup compared to single thread solving is more than 1.5 and using 4 threads reduces the solving time by a factor of 2 on average. Table 2 gives detailed results for select instances from the scatter plot of Fig. 1. These examples are the two "outliers" *load-full-6* and *ltl2dba-05*, the hardest commonly solved instance *ltl2dba-23*, and two instances with close to optimal speedup (*ltl2dpa-12* and *ltl2dpa-11*).

**Table 2.** Detailed solving results for example instances.

| Instance | Branching | 1 thread | 2 threads | 3 threads | 4 threads |
|---|---|---|---|---|---|
| ltl2dba-23 | 10 | 598.20 s | 393.68 s | 335.59 s | 312.70 s |
|  |  | 100 % | 65.81 % | 56.10 % | 52.27 % |
| ltl2dpa-12 | 15 | 521.35 s | 302.13 s | 233.98 s | 202.27 s |
|  |  | 100 % | 57.95 % | 44.88 % | 38.80 % |
| ltl2dba-05 | 4 | 476.12 s | 359.40 s | 331.87 s | 322.59 s |
|  |  | 100 % | 75.49 % | 69.70 % | 67.75 % |
| load-full-6 | 3 | 386.94 s | 332.15 s | 314.37 s | 321.75 s |
|  |  | 100 % | 85.84 % | 81.25 % | 83.15 % |
| ltl2dpa-11 | 18 | 252.61 s | 143.13 s | 107.54 s | 92.43 s |
|  |  | 100 % | 56.67 % | 42.57 % | 36.59 % |

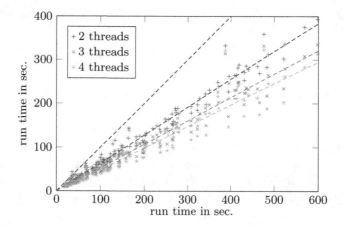

**Fig. 1.** Scatter plot of solving times with multiple threads against single thread baseline. Here, we consider only instances with more than 1 s of solving time.

## 5    Related Work

There are other solving techniques that use structural information, but they are conceptional very different, including DPLL like [3,6,10] and expansion [7,13, 15]. Further, some of them can be applied to non-prenex setting as well [3,10]. Employing SAT solver to solve propositional queries with quantifier alternations has been used before [7,8,17,19]. We extend our own work on abstraction based QBF solving [18] that itself originated from techniques that communicate the satisfaction of clauses through a recursive refinement algorithm [8,17] that were limited to conjunctive normal form. MPIDepQBF [9] is the most recent parallel solver for QBF. Their approach differs from our as they start instances of a sequential solver without synchronization. Mota et al. [14] proposed methods to split a QBF at the top level and solve the resulting QBF instances in parallel by a sequential CNF algorithm. In contrast, our approach can handle branches at every node in the quantifier hierarchy and our solving step is tightly integrated into the algorithm. Other parallel solving approaches [11,12] are conceptionally very different to our solution.

## 6    Conclusion and Future Work

We presented a QBF solving algorithm for QBFs in negation normal form, which extends our previous algorithm [18] to non-prenex formulas together with new optimizations and parallelization. Our evaluation shows that the parallelization is beneficial for our case study and other experiments suggests this method is more broadly applicable. Adapting the certification from [18] for non-prenex formulas is left for future work. Further, it would be interesting to try non-syntactic unprenexing methods to improve the parallelization, e.g., expansion of variables that combine otherwise independent subformulas.

# References

1. Biere, A.: PicoSAT essentials. JSAT **4**(2–4), 75–97 (2008)
2. Bohy, A., Bruyère, V., Filiot, E., Jin, N., Raskin, J.-F.: Acacia+, a tool for LTL synthesis. In: Madhusudan, P., Seshia, S.A. (eds.) CAV 2012. LNCS, vol. 7358, pp. 652–657. Springer, Heidelberg (2012)
3. Egly, U., Seidl, M., Woltran, S.: A solver for QBFs in negation normal form. Constraints **14**(1), 38–79 (2009)
4. Faymonville, P., Finkbeiner, B., Rabe, M.N., Tentrup, L.: Encodings of reactive synthesis. In: Proceedings of QUANTIFY (2015)
5. Finkbeiner, B., Schewe, S.: Bounded synthesis. STTT **15**(5–6), 519–539 (2013)
6. Goultiaeva, A., Iverson, V., Bacchus, F.: Beyond CNF: a circuit-based QBF solver. In: Kullmann, O. (ed.) SAT 2009. LNCS, vol. 5584, pp. 412–426. Springer, Heidelberg (2009)
7. Janota, M., Klieber, W., Marques-Silva, J., Clarke, E.: Solving QBF with counterexample guided refinement. In: Cimatti, A., Sebastiani, R. (eds.) SAT 2012. LNCS, vol. 7317, pp. 114–128. Springer, Heidelberg (2012)
8. Janota, M., Marques-Silva, J.: Solving QBF by clause selection. In: Proceedings of IJCAI, pp. 325–331. AAAI Press (2015)
9. Jordan, C., Kaiser, L., Lonsing, F., Seidl, M.: MPIDepQBF: towards parallel QBF solving without knowledge sharing. In: Sinz, C., Egly, U. (eds.) SAT 2014. LNCS, vol. 8561, pp. 430–437. Springer, Heidelberg (2014)
10. Klieber, W., Sapra, S., Gao, S., Clarke, E.: A non-prenex, non-clausal QBF solver with game-state learning. In: Strichman, O., Szeider, S. (eds.) SAT 2010. LNCS, vol. 6175, pp. 128–142. Springer, Heidelberg (2010)
11. Lewis, M.D.T., Schubert, T., Becker, B.: QmiraXT - a multithreaded QBF solver. In: Methoden und Beschreibungssprachen zur Modellierung und Verifikation von Schaltungen und Systemen (MBMV), Berlin, Germany, 2–4 March 2009, pp. 7–16. Universitätsbibliothek Berlin, Germany (2009)
12. Lewis, M.D.T., Schubert, T., Becker, B., Marin, P., Narizzano, M., Giunchiglia, E.: Parallel QBF solving with advanced knowledge sharing. Fundam. Inf. **107**(2–3), 139–166 (2011)
13. Lonsing, F., Biere, A.: Nenofex: expanding NNF for QBF solving. In: Kleine Büning, H., Zhao, X. (eds.) SAT 2008. LNCS, vol. 4996, pp. 196–210. Springer, Heidelberg (2008)
14. Mota, B.D., Nicolas, P., Stéphan, I.: A new parallel architecture for QBF tools. In: Proceedings of HPCS, pp. 324–330. IEEE (2010)
15. Pigorsch, F., Scholl, C.: Exploiting structure in an AIG based QBF solver. In: Proceedings of DATE, pp. 1596–1601. IEEE (2009)
16. QBF Gallery 2014: QCIR-G14: a non-prenex non-CNF format for quantified Boolean formulas. http://qbf.satisfiability.org/gallery/qcir-gallery14.pdf
17. Rabe, M.N., Tentrup, L.: CAQE: a certifying QBF solver. In: Proceedings of FMCAD, pp. 136–143. IEEE (2015)
18. Tentrup, L.: Solving QBF by abstraction. CoRR abs/1604.06752 (2016). https://arxiv.org/abs/1604.06752
19. Tu, K.-H., Hsu, T.-C., Jiang, J.-H.R.: QELL: QBF reasoning with extended clause learning and levelized SAT solving. In: Heule, M., Weaver, S. (eds.) SAT 2015. LNCS, vol. 9340, pp. 343–359. Springer, Heidelberg (2015)

# On Q-Resolution and CDCL QBF Solving

Mikoláš Janota$^{(\boxtimes)}$

Microsoft Research, Cambridge, UK
mikolas.janota@gmail.com

**Abstract.** The proof system Q-resolution and its variations provide the underlying proof systems for the DPLL-based QBF solvers. While (long-distance) Q-resolution models a conflict driven clause learning (CDCL) QBF solver, the inverse relation is not well understood. This paper shows that CDCL solving not only does not simulate Q-resolution, but also that this is deeply embedded in the workings of the solver. This contrasts with SAT solving, where CDCL solvers have been shown to simulate resolution.

## 1 Introduction

*Conflict driven clause learning (CDCL)* has been established as an efficient and practical method for SAT solving [26]. The relation between CDCL and propositional resolution has been extensively studied. It has been shown that, under various assumptions, modern SAT solvers simulate propositional resolution [2,4,5,27].

CDCL, with certain modifications, enables also solving quantified Boolean formulas (QBF) [11,22,30]. Analogously, propositional resolution also has its quantified counterpart *Q-resolution* [18] and a popular extension *long distance Q-resolution* [3,30]. As of today, the relation between CDCL solving and the underlying proof systems leaves a number of open questions.

The objective of this paper is to explore the relation between CDCL solving and (long-distance) Q-resolution. In particular, the paper gives a negative answer to the question whether a CDCL solver can simulate any Q-resolution proof. By this is meant that there is an infinite enumerable family of formulas $\Phi_1,\dots,\Phi_n,\dots$, for which there exist Q-resolution refutations polynomial in $n$ but any CDCL solver requires computation time super-polynomial (or even exponential) in $n$. In fact, it is shown that not even tree-like Q-resolution can be simulated by CDCL.

QBF solving brings in the complication that propagation can also be performed on the universal variables, which is done with the aid of *solution driven cube learning (SDCL)*. When combined, CDCL and SDCL influence one another because order of propagation determines which clauses/cubes are learned. While the presented result focuses on CDCL solving, we show that it is still relevant in a restricted form for solving with the combined learning and some hints for future work are explored.

© Springer International Publishing Switzerland 2016
N. Creignou and D. Le Berre (Eds.): SAT 2016, LNCS 9710, pp. 402–418, 2016.
DOI: 10.1007/978-3-319-40970-2_25

The paper is organized as follows. The paper reviews notation and basic concepts in Sect. 2. Section 3 reviews the family of formulas $CR_n$, which are studied in the remainder of the paper. This section also builds on a previous result, which shows that level-ordered refutations of the said formula must be exponential [16]. Here we strengthen this result by showing that also any derivation of unit clauses must already be exponential (Sect. 3.1). This will let us study runs of the solver only until the first unit clause is derived. Section 4 studies properties of unit propagation on the studied family of formulas $CR_n$. Section 5 uses these properties to show exponential behavior on $CR_n$. Section 6 reviews a polynomial tree-like, ordered Q-resolution refutation of $CR_n$. Finally, Sect. 7 discusses results of the paper and Sect. 8 concludes the paper.

Two distinctive proof techniques are used in the paper. The construction and motivation for the formula $CR_n$ relies on the concept of two-player games perspective for QBF. This is somewhat similar to *Ehrenfeucht-Fraïssé games* [10] used in other domains (e.g. [21]). Second is to force the QBF solver to derive level-order proofs, which must be necessarily exponential on $CR_n$. Further, we show that not only they are exponential for the whole formula but as early as a unit clause is derived. This greatly simplifies the lower-bound proof because propagation in a QBF solver is simpler with no unit clauses in its database.

# 2   Preliminaries

A *literal* is a Boolean variable or its negation. The literal complementary to a literal $l$ is denoted as $\bar{l}$, i.e. $\bar{x} = \neg x$, $\overline{\neg x} = x$. A *clause* is a disjunction of zero or more non-complementary literals. A formula in *conjunctive normal form* (CNF) is a conjunction of clauses. Whenever convenient, a clause is treated as a set of literals and a CNF formula as a set of sets of literals. For a literal $l = x$ or $l = \neg x$, we write $\mathsf{var}(l)$ for $x$. For a clause $C$, we write $\mathsf{var}(C)$ to denote $\{\mathsf{var}(l) \mid l \in C\}$ and for a CNF $\psi$, $\mathsf{var}(\psi)$ denotes $\{l \mid l \in \mathsf{var}(\psi), C \in \psi\}$

A complementary concept to clause, is *cube*, which is a conjunction of zero or more non-complementary literals.

## 2.1   Quantified Boolean Formulas

*Quantified Boolean Formulas* (QBFs) [17] extend propositional logic by enabling quantification over Boolean variables. Any propositional formula $\phi$ is also a QBF with all variables *free*. If $\Phi$ is a QBF with a free variable $x$, the formulas $\exists x.\,\Phi$ and $\forall x.\,\Phi$ are QBFs with $x$ *bound*, i.e. not free. Note that we disallow expressions such as $\exists x.\exists x.\,x$. Whenever possible, we write $\exists x_1 \ldots x_k$ instead of $\exists x_1 \ldots \exists x_k$; analogously for $\forall$. For a QBF $\Phi = \forall x.\,\Psi$ we say that $x$ is *universal* in $\Phi$ and is existential in $\exists x.\,\Psi$. Analogously, a literal $l$ is universal (resp. existential) if $\mathsf{var}(l)$ is universal (resp. existential).

The application of an assignment $\tau$ is defined for a QBF $\Phi$ if all variables of $\mathsf{dom}(\tau)$ are free in $\Phi$, and, it is defined as $(\mathcal{Q}x.\,\Phi)\tau = \Phi\tau$ for $\mathcal{Q} \in \{\exists, \forall\}$. QBFs can be seen as compact representations of propositional formulas. In particular, the formula $\forall x.\ \Psi$ is satisfied by the same truth assignments as $\Psi[x\to0]\wedge\Psi[x\to1]$ and $\exists x.\ \Psi$ by $\Psi[x\to0]\vee\Psi[x\to1]$. Since $\forall x\forall y.\ \Phi$ and $\forall y\forall x.\ \Phi$ are semantically equivalent, we allow writing $\forall X$ for a set of variables $X$; analogously for $\exists$. A QBF with no free variables is *false* (resp. *true*), iff it is semantically equivalent to the constant 0 (resp. 1).

A QBF is *closed* if it does not contain any free variables. A QBF is in *prenex form* if it is of the form $\mathcal{Q}_1 X_1 \ldots \mathcal{Q}_k X_k.\ \phi$, where $\mathcal{Q}_i \in \{\exists,\forall\}$, $\mathcal{Q}_i \neq \mathcal{Q}_{i+1}$, and $\phi$ is propositional. The propositional part $\phi$ is called the *matrix* and the rest the *prefix*. If a variable $x$ is in the set $X_i$, we say that $x$ is at *level* $i$ and write $\mathsf{lv}(x) = i$; we write $\mathsf{lv}(l)$ for $\mathsf{lv}(\mathsf{var}(l))$.

We write QCNF for the class of QBFs in prenex form where the matrix is in CNF. Unless specified otherwise, QBFs are assumed to be closed and with CNF matrix.

## 2.2    Q-Resolution

*Q-resolution (Q-Res)*, by Kleine Büning et al. [18], is a resolution-like calculus that operates on QBFs in prenex form where the matrix is a CNF. The rules are given in Fig. 1.

$$\frac{}{C}\ (\text{Axiom}) \qquad\qquad \frac{C_1 \cup \{x\} \qquad C_2 \cup \{\neg x\}}{C_1 \cup C_2}\ (\text{Res})$$

$C$ is a clause in the matrix. Variable $x$ is existential. If $z \in C_1$, then $\neg z \notin C_2$.

$$\frac{C \cup \{u\}}{C}\ (\forall\text{-Red}) \qquad\qquad \begin{array}{l}\text{Variable } u \text{ is universal. If } x \in C \text{ is}\\ \text{existential, then } \mathsf{lv}(x) < \mathsf{lv}(u).\end{array}$$

**Fig. 1.** The rules of Q-Res [18]

*Long-distance resolution (LD-Q-Res)* appears originally in the work of Zhang and Malik [30] and was formalized into a calculus by Balabanov and Jiang [3]. It merges complementary literals of a universal variable $u$ into the special literal $u^*$. These special literals prohibit certain resolution steps. In particular, different literals of a universal variable $u$ may be merged only if $\mathsf{lv}(x) < \mathsf{lv}(u)$, where $x$ is the resolution variable. The rules are given in Fig. 2. In practice, solvers do not maintain a literal of the form $u^*$ but rather two complementary literals, e.g. $e \vee u \vee \neg u \vee z$. Such clauses arise naturally by learning and give propagation due to universal reduction (see below). An alternative formulation via no-goods is suggested by Klieber [19,20].

For a clause $C$, a universal literal $l \in C$ is *blocked* by an existential literal $k \in C$ iff $\mathsf{lv}(l) < \mathsf{lv}(k)$. $\forall$-*reduction* is the operation of removing from a clause $C$ all universal literals that are *not* blocked by some literal.

$$\frac{}{C} \text{ (Axiom)} \qquad \frac{D \cup \{u\}}{D} \text{ (}\forall\text{-Red)} \qquad \frac{D \cup \{u^*\}}{D} \text{ (}\forall\text{-Red}^*)$$

$C$ is a clause in the matrix. Literal $u$ is universal and $\mathsf{lv}(u) \geq \mathsf{lv}(l)$ for all $l \in D$.

$$\frac{C_1 \cup U_1 \cup \{x\} \qquad C_2 \cup U_2 \cup \{\neg x\}}{C_1 \cup C_2 \cup U} \text{ (Res)}$$

Variable $x$ is existential. If for $l_1 \in C_1, l_2 \in C_2$, $\mathsf{var}(l_1) = \mathsf{var}(l_2) = z$ then $l_1 = l_2 \neq z^*$. $U_1, U_2$ contain only universal literals with $\mathsf{var}(U_1) = \mathsf{var}(U_2)$. For each $u \in \mathsf{var}(U_1)$ we require $\mathsf{lv}(x) < \mathsf{lv}(u)$. If for $w_1 \in U_1, w_2 \in U_2$, $\mathsf{var}(w_1) = \mathsf{var}(w_2) = u$ then $w_1 = \neg w_2$, $w_1 = u^*$ or $w_2 = u^*$. $U$ is defined as $\{u^* \mid u \in \mathsf{var}(U_1)\}$.

**Fig. 2.** The rules of LD-Q-Res [3]

For a QCNF $\mathcal{P} . \phi$, a *Q-resolution proof* of a clause $C$ is a sequence of clauses $C_1, \ldots, C_n$ where $C_n = C$ and any $C_i$ in the sequence is part of the given matrix $\phi$; or was obtained from one of the preceding clauses by $\forall$-reduction; or it is a Q-resolvent of some pair of preceding clauses. A Q-resolution proof is called a *refutation* iff $C$ is the empty clause, denoted $\bot$.

In this article Q-resolution and plain resolution proofs are treated as connected directed acyclic graphs (DAG). Any graph representing a resolution or Q-resolution proof has one and only one node with in-degree 0, which we call the *root* (the final clause in the proof). All the nodes with out-degree 0 are labeled with clauses from the original formula and we call them *leafs* of the proof or *axiom* clauses. Any non-axiom clause has two outgoing edges pointing to its respective antecedents.

A Q-resolution proof is called *tree-like* if the corresponding graph forms a tree (rooted in the final clause). It is known that at the propositional level, tree resolution does *not* p-simulate DAG resolution [7].

A Q-resolution proof is called *ordered*, if there exists a sequence of variables $S$ such that for any path from a leaf to the root the sequence of variables being eliminated form a sub-sequence of $S$. Sometimes ordered propositional resolution is referred to as Davis-Putnam resolution as it corresponds to the Davis-Putnam algorithm. It is known that at the propositional level, ordered resolution does *not* p-simulate unrestricted resolution [1,13].

**Definition 1 (level-ordered proof).** Let $\pi$ be a Q-resolution proof of a QCNF formula $\Phi$. We say that $\pi$ is *level-ordered* iff the following holds. Consider an arbitrary path from the root to some leaf and some resolution steps on that path on literals $x_1$ and $x_2$ such that the resolution on $x_1$ is closer to the root. Then, $\mathsf{lv}(x_1) \leq \mathsf{lv}(x_2)$.

*Remark 1.* Janota and Silva define *level-order refutations* [16], which the above definition generalizes for an arbitrary proof.

## 2.3   CDCL and SDCL Solving

Basic understanding of CDCL+SDCL QCNF solving is assumed; for further details see [12,22]. We assume that a solver's state is uniquely determined by a sequence of *decisions* $\mathcal{D}$, a *trail* $\mathcal{T}$, and a *database of clauses and cubes*. The input formula is always in the database of clauses.

For the purpose of this paper we only assume CDCL, i.e. universal variables are handled by traditional chronological backtracking and there are no cubes in the database. SDCL is also briefly discussed for completeness.

The trail $\mathcal{T}$ and decisions $\mathcal{D}$ are modeled as sequences of literals, where $\mathcal{D}$ is a subsequence of $\mathcal{T}$. Any literal $l$ in $\mathcal{T}$ but not in $\mathcal{D}$ is said to be *propagated*, and is obtained by unit propagation from $\mathcal{D}$. For a literal $l$ we write $\mathcal{T}(l)$ to denote the value of $l$ in $\mathcal{T}$, i.e. $\mathcal{T}(l) = 1$ if $l \in \mathcal{T}$ and $\mathcal{T}(l) = 0$ if $\bar{l} \in \mathcal{T}$.

**CDCL.** A *conflict* is reached whenever unit propagation gives an existential literal two opposing values. This corresponds to violating a clause in the database. Upon a conflict, *clause learning* is invoked. Learning performs Q-resolution steps in the reverse chronological order of propagations. We assume that any ∀-reduction is carried out as soon as possible.

The learning process stops when the derived clause $C$ fulfills the *unique implication point* property, which for QBF has the following conditions [30]. There exists a literal $l \in C$ such that all the following properties are fulfilled:

1. $l$ is existential and it has the highest decision level in $C$.
2. $l$ is at a decision level where the decision is an existential literal.
3. All universal literals with quantification level smaller than $l$ are decided at a decision level smaller than $l$.

CDCL QBF solving is simulated by long-distance Q-resolution. More precisely, if a solver decides given formula as false, a long-distance Q-resolution refutation can be constructed for it in time polynomial to the running time of the solver.

Certain propagation can yield long-distance resolution steps, which can be avoided by modifying the UIP scheme [11], this however potentially leads to exponential blowup [28]. For the purpose of this paper, this difference is not important as we later see that long-distance steps do not occur in the considered formula.

**SDCL.** Solution driven cube learning (SDCL) is symmetrical to CDCL with the difference that the initial cube database is empty and new cubes are also created when the current assignment gives a model to the initial formula. Hence, in SDCL initially the solver has an empty database of cubes. Once it reaches an assignment $\mu$ that satisfies the original matrix, it uses it to generate a cube. This is done by conjoining the literals that form the assignment $\mu$. The cube is then existentially reduced and inserted into the database. Such cube is from now on used for propagation. If a cube is fully satisfied by the current assignment, this

| $a_1$ | $\bullet \ \bullet \ \bullet$ | $a_1$ | $\bullet \ \bullet \ \bullet$ | $a_N$ | $\bullet \ \bullet \ \bullet$ | $a_N$ |
|-------|------|-------|------|-------|------|-------|
| $b_1$ | $\bullet \ \bullet \ \bullet$ | $b_N$ | $\bullet \ \bullet \ \bullet$ | $b_1$ | $\bullet \ \bullet \ \bullet$ | $b_N$ |

**Fig. 3.** Completion principle

constitutes a losing situation for the universal player. Hence, unit cubes are used for propagation and fully satisfied cubes to jumpstart learning just as conflicting clauses do. See [22] for further details. We note that symmetrical approaches to SDCL+CDCL solving exist [14, 29].

**Model for the Paper.** For the results in the paper we are assuming a CDCL solver with *no* SDCL. This means that the solver makes decisions along the quantifier prefix until it either reaches a conflict and then it applies clause learning, or, it satisfies the matrix and then it backtracks to the last universal variable and flips its polarity (if such exists). We do not make any assumptions about restarts, i.e., restarts may be issued arbitrarily.

## 3 Formula

The formula used to show the main result is taken from Janota and Silva [16].[1] Its construction derives from a principle named *completion principle*. Two sets are considered: $A = \{a_1, \ldots, a_n\}$, $B = \{b_1, \ldots, b_n\}$ and their Cartesian-product $A \times B$. Let us visualize the cross-product as in Fig. 3. The following game is played. In the first round, the $\exists$-player deletes one and only one cell from each column. In the second round, the $\forall$-player chooses one of the two rows. The $\forall$-player wins if the chosen row contains either the complete set $A$ or the set $B$.

There is the following winning strategy for the $\forall$-player. If the $\exists$-player wants to make sure that the bottom row (the $b_i$ row) does *not* contain the complete set $B$, it must delete at least one element from each of the $n$ copies of $B$. Hence, for the $j$-th copy of $B$ there is an element $a_j$ that was not deleted and thus forming the complete set $A$. Hence, the winning strategy for the $\forall$-player is to look at the bottom row and see if it contains a complete copy of $B$. If it does, the $\forall$-player selects the bottom row and otherwise he selects the top row.

Let us construct a formula based on this principle. For each column $(i, j)$ introduce a variable $x_{ij}$ that determines which cell is deleted by the $\exists$-player in the first round. For the $\forall$-player introduce a single universal variable $z$, which determines the selected row. Further, add clauses that make sure that whenever one of the sets $A$ or $B$ is complete, the formula becomes false.

In the remainder of the paper we denote the set of variables $x_{ij}$, $i, j \in 1..n$ by $\mathcal{X}$, and $a_i, b_i$, $i \in 1..n$ by $\mathcal{L}$ (the letter $\mathcal{L}$ is chosen to evoke "last").

---

[1] See http://sat.inesc-id.pt/~mikolas/cdcl16 for a formula generator.

The formula $\mathtt{CR}_n$ is defined by the prefix $\exists \mathcal{X} \forall z \exists \mathcal{L}$ and the following matrix.

$$x_{ij} \vee z \vee a_i, \ i, j \in 1..n \tag{1}$$

$$\neg x_{ij} \vee \neg z \vee b_j, \ i, j \in 1..n \tag{2}$$

$$\bigvee_{i \in 1..n} \neg a_i \tag{3}$$

$$\bigvee_{i \in 1..n} \neg b_i \tag{4}$$

The first two types of clauses (1) and (2) represent the effects of the moves. The last two clauses (3) and (4) disable setting all $a_i$ and $b_i$ to true, respectively. Hence, the whole formula $\mathtt{CR}_n$ is false because once $z$ is set according to the strategy outlined above, the variables $\mathcal{L}$ must be set such that variables from one of the sets $a_i$ and $b_i$ will be all true. Consequently, one of the clauses (3) and (4) must be falsified.

## 3.1   Lower Bounds for Level-Ordered Q-Resolution

An exponential lower-bound for level-ordered resolution for $\mathtt{CR}_n$ is known [16].

**Proposition 1 [16, Proposition 5].** *Any level-ordered Q-resolution refutation of $CR_n$ is exponential in $n$.*

For the purpose of this paper a stronger result is needed. Here we show that any level-ordered proof of any *unit clause* is already exponential.

**Lemma 1.** *Let $\pi$ be a level-ordered Q-resolution proof from $CR_n$, where $\pi$ is a proof of a unit clause $\{l\}$ with $\mathsf{var}(l) \in \mathcal{L}$. Let $P$ be a path from the root of $\pi$ to some clause $C$ where $P$ does not contain any resolutions on $\mathcal{L}$. Let $\pi_C \subseteq \pi$ be the proof of $C$.*

*Then, $\pi_C$ contains one of the clauses (3), (4).*

*Proof.* We consider the following cases:

1. $\pi_C$ contains at least one resolution step on $\mathcal{L}$. Then one of (3), (4) must appear in $\pi_C$ since they are the only clauses containing negative occurrences of the $\mathcal{L}$ variables.
2. $\pi_C$ does not contain any resolution step on $\mathcal{L}$. Then, if there are no clauses (3), (4), than all axiom clauses are of type (1) or (2). This would be a contradiction with the requirement that $\pi$ proves a unit clause since the variable $z$ would always remain blocked as there are no more $\mathcal{L}$ resolutions on the path from the root of $\pi$ to $C$.    □

**Lemma 2.** *Let $\pi$ be a level-ordered Q-resolution proof from $CR_n$, where $\pi$ is a proof of a unit clause $\{l\}$ with $\mathsf{var}(l) \in \mathcal{L}$. Let $P$ be a path from the root of $\pi$ to some clause $C$ where $P$ does not contain any resolutions on $\mathcal{L}$ and $P$ is maximal in that respect.*

*Then, $C$ contains at least $n - 1$ different $\mathcal{X}$ variables.*

*Proof.* Observe that all $\mathcal{L}$ variables are treated symmetrically in the formula so without loss of generality, consider $l \in \{a_k, \neg a_k\}$ for some $k \in 1..n$.

Let $\pi_C \subseteq \pi$ be the proof of $C$. From Lemma 1, one of the clauses (3), (4) must appear in $\pi_C$.

Let us assume that (3) is in $\pi_C$. Since $\pi$ is level-ordered and it derives the clause $\{l\}$, all $a_i$ with $i \in 1..n$, $i \neq k$ must be resolved away in $\pi_C$. This means that $\pi_C$ contains the clause $x_{ij} \vee z \vee a_i$ for each $i \in 1..n$, $i \neq k$. Since $\pi_C$ has no resolutions on $\mathcal{X}$, the corresponding $\mathcal{X}$ variables also appear in $C$.

If (4) is in $\pi_C$, the reasoning is analogous to the above.                           $\square$

**Proposition 2.** *Let $\pi$ be a level-ordered Q-resolution proof from $CR_n$, where $\pi$ is a proof of a unit clause $\{l\}$ with $\mathsf{var}(l) \in \mathcal{L}$. Then $\pi$ is exponential in $n$.*

*Proof.* Pick an assignment $\tau$ to all the $\mathcal{X}$ variables. Construct a path $P$ from the root of $\pi$ that contains $\forall$-reductions and $\mathcal{X}$ resolutions only and is maximal in that respect such that it respects the assignment $\tau$.

Due to Lemma 2, $P$ ends in a clause $C$ containing at least $n-1$ $\mathcal{X}$ variables. There are $2^{n^2}$ assignments to $\mathcal{X}$ variables and $C$ covers at most $2^{n^2-(n-1)}$ of those. Hence, there are at least $2^{n^2}/2^{n^2-(n-1)} = 2^{n-1}$ clauses in $\pi$.                           $\square$

**Proposition 3.** *Let $\pi$ be a level-ordered Q-resolution proof from $CR_n$, where $\pi$ is a proof of a unit clause $\{l\}$ with $\mathsf{var}(l) \in \mathcal{X}$. Then $\pi$ is exponential in $n$.*

*Proof.* Since $x_{ij}$ and $\neg x_{ij}$ are treated symmetrically in $CR_n$, any level-ordered sub-exponential proof of $x_{ij}$ could be rewritten to a level-ordered sub-exponential proof of $\neg x_{ij}$. Resolving the two would give a level-ordered sub-exponential refutation of $CR_n$, which would be a contradiction with Proposition 1.                           $\square$

**Corollary 1.** *Let $C$ be a unit or empty clause with a level-ordered Q-resolution proof $\pi$ from $CR_n$. Then $\pi$ is exponential with respect to $n$.*

# 4  Properties of Propagation on $CR_n$

To be able to reason about the clauses that are learned during solving of $CR_n$, several properties of unit propagation are needed. A crucial property of the input formula is that there are no clauses enabling propagation "across quantification levels". Namely, while decisions are being made on the $\mathcal{X}$ variables, no propagation happens in the $\mathcal{L}$ variables. These get value only once $z$ gets a value. For this purpose we introduce the concept of mixed clauses.

**Definition 2 (mixed clause).** We say that a clause is *mixed* if it contains both $\mathcal{X}$ variables and $\mathcal{L}$ variables, i.e., if $\mathsf{var}(C) \cap \mathcal{X} \neq \emptyset$ and $\mathsf{var}(C) \cap \mathcal{L} \neq \emptyset$.

The following lemma shows that Q-resolution does not enable us to derive mixed clauses without the variable $z$.

**Lemma 3.** *Let $\pi$ be an arbitrary Q-resolution proof from $CR_n$. For any mixed clause $C \in \pi$ it holds that $z \in \mathsf{var}(C)$.*

*Proof (By induction on the derivation depth).* The hypothesis holds for all the axiom clauses of $CR_n$.

Let $C$ be a new mixed clause derived by resolution from some clauses $D_1$ and $D_2$, for which the hypothesis holds. Since $C$ is mixed, at least one of $D_1$, $D_2$ must be mixed. Therefore $C$ also contains $z$.

Let $C$ be derived by $\forall$-reduction from an existing clause $D$. Since $\mathcal{L}$ variables block $\forall$-reduction of $z$, the clauses $C$ and $D$ do not contain $\mathcal{L}$ variables and therefore are not mixed.     □

*Remark 2.* The above-lemma can be easily generalized. Indeed, if the input formula only contains mixed clauses that have a universal variable in the middle, that variable cannot be reduced unless all the variables at the higher quantification level are resolved away.

The following lemma is crucial for our result. As long as there are no unit clauses, there cannot be propagation across quantification levels since all the mixed clauses need $z$ to have a value to give propagation.

**Lemma 4.** *Let $\mathcal{T}$ be the trail for a CDCL solver in a state before any unit clauses are learned. If $\mathsf{var}(\mathcal{T}) \subseteq \mathcal{X}$ then there is no propagation on $\mathcal{L}$ variables.*

*Proof.* Since $CR_n$ does not contain any unit clauses and no unit clauses have been learned so far, any propagation on $\mathcal{L}$ must come from a clause in the database that has at least two literals. Since $\mathsf{var}(\mathcal{T}) \subseteq \mathcal{X}$, such clause would have to contain an $\mathcal{X}$ variable. However, due to Lemma 3, any mixed clauses contain also the $z$ variable, which is unassigned and therefore cannot give a unit $\mathcal{L}$ clause. There is no propagation on $z$ since we're assuming only CDCL (not SDCL).     □

## 5   Exponential Lower Bound for CDCL QBF Learning

This section shows that a run of a CDCL solver on the formula $CR_n$ is exponential. This is done by showing that the proofs of all learned clauses are level-ordered. Due to Corollary 1, it is sufficient to show that the proofs of learned clauses are level-ordered until a unit clause is learned. Indeed, even if the solver derives some non-level-ordered proofs after the first unit clause has been learned, the proof of the unit clause is already exponential. Therefore, the solver must perform an exponential number of steps to derive the unit clause. Note also that the input formula has no unit clauses as long as $n \geq 2$.

Note that the reasoning below needs to account for all the clauses derived during learning, not just the learned clauses. We start by a couple of technical lemmas characterizing the learning process.

**Lemma 5.** *Let $\mathcal{T}$ be the trail for a CDCL solver in a state before any unit clauses are learned. Let $\mathcal{T}$ be such that it leads to a conflict and let $C$ be some clause that is derived during the learning of the pertaining learned clause. If $l \in C$ with $\mathsf{var}(l) = z$, then $\mathcal{T}(l) = 0$.*

*Proof.* By construction, $C$ is derived by resolution steps on clauses that participated in the propagation leading to the conflict.

While $z$ is not assigned, propagation is only on clauses that contain $\mathcal{X}$ variables only because any clauses that contain also some $\mathcal{L}$ variables also contain $z$ due to Lemma 3.

Once $z$ is assigned, all clauses that contain a literal $l$ with $var(l) = z$ and $\mathcal{T}(l) = 1$ cannot be used during propagation (they are effectively deleted from the propagation process). □

The above lemma immediately gives us the consequence that there are no long-distance resolution steps in learned clauses.

**Corollary 2.** *For the formula $CR_n$, a CDCL solver never learns clauses derived by long-distance Q-resolution steps.*

**Lemma 6.** *Let $\mathcal{T}$ be a trail for a CDCL solver in a state before any unit clauses are learned. Let $\mathcal{T}$ be such that it leads to a conflict and let $C$ be some clause that is derived during the learning of the pertaining learned clause.*

*Let $C$ be such that it does not contain any* propagated *$\mathcal{L}$ literals, but it contains some $\mathcal{L}$ literals. Then $C$ is a UIP.*

*Proof.* Due to the precondition, all $\mathcal{L}$ are decisions in $C$ therefore there must be one literal $k$ with $\mathsf{var}(k) \in \mathcal{L}$ with the highest decision level as all $\mathcal{L}$ variables are decided after $\mathcal{X}$ variables due to Lemma 4. From Lemma 5, if $C$ contains the variable $z$, then the corresponding literal is set to 0. This fulfills the conditions for $C$ be a UIP. □

**Proposition 4.** *Let $\mathcal{T}$ be the trail for a CDCL solver in a state before any unit clauses are learned. Let $\mathcal{T}$ be such that it leads to a conflict and let $C$ be some clause that is derived during the learning of the pertaining learned clause.*

*If $C$ contains any $\mathcal{L}$ variables, then it is derived by resolution steps only on the $\mathcal{L}$ variables.*

*Proof.* The hypothesis is trivially true for all the axiom clauses.

If $\mathsf{var}(\mathcal{T}) \subseteq \mathcal{X}$ then the derivation of the learned clause only contains $\mathcal{X}$ variables due to Lemma 4.

If $\mathsf{var}(\mathcal{T}) \cap (\mathcal{L} \cup \{z\}) \neq \emptyset$ then consider the following cases:

1. $C$ contains some $\mathcal{L}$ literal, forced to 0 by propagation. Then, resolution is performed on one such literal as propagation on $\mathcal{L}$ takes place later chronologically than propagation on $\mathcal{X}$ variables due to Lemma 4.
2. $C$ does not contain any $\mathcal{L}$ literal, then the hypothesis is trivially satisfied.
3. $C$ does not contain any *propagated* $\mathcal{L}$ literal, then $C$ is a UIP due to Lemma 6.

□

Finally we need to show that as long as clauses containing $\mathcal{L}$ variables are only derived by resolution steps on $\mathcal{L}$, the corresponding proofs are level-ordered. For such we utilize the following two observations.

**Observation 1.** *Let $C$ be derived by a resolution step over an $\mathcal{X}$ variable from clauses with level-ordered proofs. Then the proof of $C$ is also level-ordered.*

**Observation 2.** *A proof that contains resolution steps only on $\mathcal{L}$ variables is level-ordered.*

**Proposition 5.** *Let $\mathcal{T}$ be the trail that leads to a conflict for a CDCL solver in a state before any unit clauses are learned. The proof of the corresponding learned clauses is level-ordered.*

*Proof.* Prove from Proposition 4 by induction on derivation depth.

The hypothesis is trivially true for axiom clauses and is trivially preserved by ∀-reduction. Split on the following two cases.

1. If a clause $C$ is derived by resolution on a $\mathcal{L}$ variable, then both antecedents must contain at least one $\mathcal{L}$ variable, From Proposition 4, the antecedents are derived only by $\mathcal{L}$ resolutions only. Therefore, the proof is level-ordered (Observation 2).
2. If a clause $C$ is derived by resolution on an $\mathcal{X}$ variable, then the derivation of $C$ is level-ordered because the antecedents are level-ordered (IH) and due to Observation 1.                                                                                   □

*Remark 3.* Proving that learned clauses are themselves level-ordered is not itself inductive. The solver could learn a clause containing an $\mathcal{L}$ variable while using $\mathcal{X}$ resolutions and then use this learned clause in a level-ordered manner.

**Theorem 1.** *Solving $\mathit{CR}_n$ by a CDCL QBF solver requires time exponential in $n$.*

*Proof.* Due to Proposition 5, proof of any learned clause is level-ordered while no unit clauses are learned. Consider the first unit clause learned by the solver. Due to Corollary 1, the Q-resolution proof of the clause is exponential in $n$. Therefore, the solver must have carried out an exponential number of steps to learn this clause.

Observe that restarting the solver does not affect in any way the above results. Indeed, the whole proof hinges on the fact that learned clauses are always level-ordered independently of the content of the current trail—as long as the trail observes the condition that a variable is decided only if all the variables that precede it in the prefix are valued. More broadly speaking, restarts in QBF do not change the fact that the solver has to assign variables according to quantification levels; unlike in SAT, where restarts can come up with an arbitrary order.

Note that the proof relies on the UIP learning scheme (Lemma 6). we conjecture, however, that this precondition can be weakened. Other schemas, like QPUP [24, 28] could be considered.

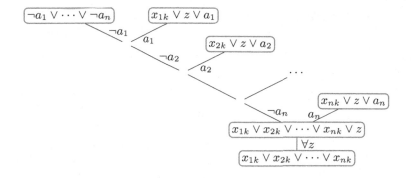

**Fig. 4.** Derivation of an "all $x$" clause for $k \in 1..n$ in $n$ resolution steps.

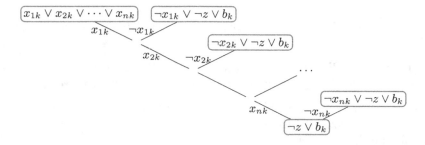

**Fig. 5.** Derivation of an $\neg z \vee b_k$ clause for $k \in 1..n$ in $n$ steps, using Fig. 4.

## 6    Short Tree-Like Q-Resolution Refutation of $CR_n$

This section shows that $CR_n$ has a polynomial tree-like Q-resolution refutation. We do so in a constructive manner and the proof is conceptually divided into three parts.

First, derive clauses $\bigvee_{i \in 1..n} x_{ik}$ for $k \in 1..n$ (Fig. 4). Each of these clauses lets us derive a clause $\neg z \vee b_k$ for $k \in 1..n$ (Fig. 5). Finally, using the clause $\neg b_1 \vee \cdots \vee \neg b_n$, the empty clause is derived (Fig. 6).

The first phase requires $n^2$ resolution steps and $n$ ∀-reduction steps. The second phase requires $n^2$ resolution steps. Finally, the last phase requires $n$ resolution steps and one ∀-reduction. Since the size of the input formula has $2n^2 + 2$ clauses, the proof size is linear in the formula's size.

Observe that each clause appears exactly once in the proof—the proof forms a tree. Also note that the proof is *not* level-ordered because it starts with resolutions on $a_i$ variables, continues with $\mathcal{X}$ resolutions, and finishes with resolutions on $b_i$ variables. However, the proof is *ordered* with the following order.

$$a_1, \ldots, a_n, x_{(1,1)}, \ldots, x_{(n,n)}, b_1, \ldots, b_n$$

**Theorem 2.** *$CR_n$ has a polynomial refutation in ordered tree-like Q-resolution.*

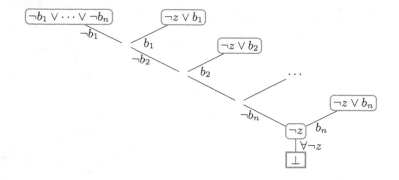

**Fig. 6.** Derivation of $\bot$, using Fig. 5.

**Corollary 3.** *Level-ordered resolution does not simulate ordered or tree-like Q-resolution.*

**Corollary 4.** *CDCL QBF solving does not simulate Q-resolution even under the restriction that the Q-resolution proofs are tree-like and ordered.*

*Remark 4.* Mahajan and Shukla in fact show that level-ordered Q-resolution and tree-like Q-resolution are incomparable [25]. In fact, the short resolution proof above also appears in their paper.

# 7    Discussion

QBF solving presents us with some subtleties and complications due to the two types of propagation and learning. Here we discuss how these relate to the presented results.

## 7.1    SDCL

The presented result is concerned only with CDCL and so we may ask whether SDCL can speed up the solving of $\mathrm{CR}_n$. A pivotal point in our proof is Lemma 4, which shows that there is no propagation from $\mathcal{X}$ variables to $\mathcal{L}$ variables, i.e., propagation across levels. The lemma relies on the fact that $z$ will not be given a value by propagation from $\mathcal{X}$ variables. This can happen if SDCL is employed. More precisely, if the solver learns a cube containing only a subset of $\mathcal{X}$ variables and $z$, the variable $z$ may be given a value *before* all $\mathcal{X}$ variable are assigned, which may subsequently give propagation on $\mathcal{L}$ variables. Such propagation could potentially lead to learned clauses with non-level-ordered proofs.

However, this can only happen when the universal player made a *wrong* choice for the value of $z$ in the past. Since there is a winning strategy for the universal player, there always is a value for $z$ that does not lead to a solution and consequently no cube learning takes place if the universal player follows the strategy.

**Table 1.** Number of backtracks for DepQBF on $\mathrm{CR}_n$ with a 1-hr. timeout. *Unsolved* instances are marked with $>$.

| $n$ | CDCL | CDCL + SDCL | CDCL + SDCL − pure lits |
|---|---|---|---|
| 4 | 101 | 101 | 101 |
| 5 | 1081 | 1081 | 751 |
| 6 | 19611 | 19611 | 3531 |
| 7 | 370811 | 370811 | 36411 |
| 8 | > 9995451 | > 10000981 | 5464551 |
| 9 | > 10612011 | > 10619361 | > 931211 |
| 10 | > 10303551 | > 10313901 | > 8608251 |

**Observation 3.** *For a false QBF $\Phi$, if universal variables are given values according to a winning strategy for the universal player, a SDCL+CDCL QBF solver behaves identically to a CDCL QBF solver. Consequently, the Corollary 4 also holds for a SDCL+CDCL QBF solver under such restriction.*

## 7.2 Pure Literals

Another relevant technique is *pure literals* [8,22]. Those enable assigning values to variables out of the quantification order. This can again influence what kind of clauses and cubes are learned. However, there is also a potential adversarial effect of pure literals. If the universal player makes better choices, it learns fewer cubes—which could have otherwise potentially speed up the proof.

While at this point we do not have a definite answer to the above questions, experimental evaluation might provide some hints. I have run the solver DepQBF [23] on $\mathrm{CR}_n$ and recorded the number of backtracking steps—these are presented for various configurations in Table 1. All configurations have the switches `--traditional-qcdcl --long-dist-res --dep-man=simple` to disable advanced features of DepQBF but also allow long-distance Q-resolution. The leftmost configuration only performs CDCL, the middle CDCL+SDCL, and the last one also combines CDCL and SDCL but switches off the pure-literal technique.

Instances that were not solved within 1 hour, are marked with $> B$ where $B$ is the number of backtracking steps performed up to that point.

The configuration CDCL and CDCL+SDCL behave identically and can only solve $\mathrm{CR}_n$ for $n \in 1..7$. Interestingly enough, turning off pure literals leads to a significant improvement and also $n = 8$ is solved.

Further inspection reveals that the CDCL+SDCL configuration never learned any cubes. Because, as indicated above, it never makes a wrong decision for the universal variable $z$. Somewhat paradoxically, this is disadvantageous.

## 7.3  Other Work on Separation

In his thesis, Lonsing presents a formula that shows that Q-resolution is more powerful than standard CDCL solving—see Example 3.3.6 in [22]. The formula is of the form $\exists X \forall U \exists Z. \phi_1 \wedge \phi_2$, where $\phi_1$ depends only on the X variables and it is a hard unsatisfiable propositional formula, e.g. pigeonhole principle [15]. The formula $\phi_2$ depends on $U$ and $Z$ and is easy to refute. While a Q-resolution proof can simply use $\phi_2$ to construct a refutation, a CDCL solver will try to construct an assignment satisfying $\phi_1$, which is impossible and will take an exponential amount of time. As discussed in Lonsing's work, this can be overcome by analyzing variable dependencies such as in the DepQBF solver [23]. A solver could also refute this formula by deciding $U$ out-of-order and soundly conclude that it is false. Hence, in some sense, the hardness of this formula really lies in the propositional formula $\phi_1$ rather than the QBF structure. This contrasts with the $CR_n$ formulas where pulling the universal variable to the front makes the formula true. Nevertheless, this notion should be better understood and the recent work of Chen could possibly provide pointers in this direction [9].

## 8  Summary and Future Work

The paper compares the strength of QBF conflict driven clause learning (CDCL) to Q-resolution. In contrast to its propositional counterparts, CDCL QBF solving appears to be quite weak compared to general Q-resolution. Indeed, even if we impose the limit that the resolution should be tree-like and ordered, CDCL cannot simulate the refutation. Our result strengthens an existing separation between CDCL solving and Q-resolution, which can be overcome by variable dependencies (see Sect. 7.3). We further conjecture that the presented separation stems from the QBF hardness of the problem rather than propositional hardness.

The crux of our proof is that the investigated formula does not permit propagation across levels, which consequently leads to level-ordered derivations of the learned clauses. This observation suggests a number of interesting questions for future research. Can solution driven cube learning (SDCL) speed-up the proof? The experimental evaluation suggest that it might. However, only if the pure literal technique is turned *off*. This observation also has a practical consequence. Pure literals may lead to fewer learned cubes and consequently a decrease in the quality of the *clausal* proof. Can such adversarial effect be avoided?

## References

1. Alekhnovich, M., Johannsen, J., Pitassi, T., Urquhart, A.: An exponential separation between regular and general resolution. Theory Comput. **3**(5), 81–102 (2007)
2. Atserias, A., Fichte, J.K., Thurley, M.: Clause-learning algorithms with many restarts and bounded-width resolution. J. Artif. Intell. Res. (JAIR) **40**, 353–373 (2011)
3. Balabanov, V., Jiang, J.H.R.: Unified QBF certification and its applications. Formal Methods Syst. Des. **41**(1), 45–65 (2012)

4. Beame, P., Kautz, H.A., Sabharwal, A.: Towards understanding and harnessing the potential of clause learning. J. Artif. Intell. Res. (JAIR) **22**, 319–351 (2004)
5. Beame, P., Sabharwal, A.: Non-restarting SAT solvers with simple preprocessing can efficiently simulate resolution. In: Brodley, C.E., Stone, P. (eds.) Proceedings of the Twenty-Eighth AAAI Conference on Artificial Intelligence, pp. 2608–2615. AAAI Press (2014)
6. Biere, A., Heule, M., van Maaren, H., Walsh, T. (eds.): Handbook of Satisfiability. Frontiers in Artificial Intelligence and Applications, vol. 185. IOS Press, Amsterdam (2009)
7. Bonet, M.L., Esteban, J.L., Galesi, N., Johannsen, J.: Exponential separations between restricted resolution and cutting planes proof systems. In: 39th Annual Symposium on Foundations of Computer Science, FOCS, pp. 638–647. IEEE Computer Society (1998)
8. Cadoli, M., Schaerf, M., Giovanardi, A., Giovanardi, M.: An algorithm to evaluate quantified Boolean formulae and its experimental evaluation. J. Autom. Reasoning **28**(2), 101–142 (2002)
9. Chen, H.: Proof complexity modulo the polynomial hierarchy: understanding alternation as a source of hardness. In: 43rd International Colloquium on Automata, Languages, and Programming (ICALP) (2016)
10. Ehrenfeucht, A.: An application of games to the completeness problem for formalized theories. Fund. Math. **49**, 129–141, 13 (1961)
11. Giunchiglia, E., Narizzano, M., Tacchella, A.: Clause/term resolution and learning in the evaluation of quantified Boolean formulas. J. Artif. Intell. Res. **26**(1), 371–416 (2006)
12. Giunchiglia, E., Marin, P., Narizzano, M.: Reasoning with quantified boolean formulas. In: Biere et al. [6], pp. 761–780
13. Goerdt, A.: Davis-Putnam resolution versus unrestricted resolution. Ann. Math. Artif. Intell. **6**(1–3), 169–184 (1992)
14. Goultiaeva, A., Seidl, M., Biere, A.: Bridging the gap between dual propagation and CNF-based QBF solving. In: DATE, pp. 811–814 (2013)
15. Haken, A.: The intractability of resolution. Theoret. Comput. Sci. **39**, 297–308 (1985)
16. Janota, M., Marques-Silva, J.: Expansion-based QBF solving versus Q-resolution. Theoret. Comput. Sci. **577**, 25–42 (2015)
17. Kleine Büning, H., Bubeck, U.: Theory of quantified boolean formulas. In: Biere et al. [6], pp. 735–760
18. Büning Kleine, K., Karpinski, M., Flögel, A.: Resolution for quantified Boolean formulas. Inf. Comput. **117**(1), 12–18 (1995)
19. Klieber, W.: Formal Verification Using Quantified Boolean Formulas (QBF). Ph.D. thesis, Carnegie Mellon University (2014). http://www.cs.cmu.edu/~wklieber/thesis.pdf
20. Klieber, W., Sapra, S., Gao, S., Clarke, E.: A non-prenex, non-clausal QBF solver with game-state learning. In: Strichman, O., Szeider, S. (eds.) SAT 2010. LNCS, vol. 6175, pp. 128–142. Springer, Heidelberg (2010)
21. Kolaitis, P.G.: The expressive power of stratified programs. Inf. Comput. **90**(1), 50–66 (1991)
22. Lonsing, F.: Dependency Schemes and Search-Based QBF Solving: Theory and Practice. Ph.D. thesis, Johannes Kepler Universität (2012). http://www.kr.tuwien.ac.at/staff/lonsing/diss/
23. Lonsing, F., Biere, A.: DepQBF: a dependency-aware QBF solver. JSAT **7**(2–3), 71–76 (2010)

24. Lonsing, F., Egly, U., Van Gelder, A.: Efficient clause learning for quantified boolean formulas via QBF pseudo unit propagation. In: Järvisalo, M., Van Gelder, A. (eds.) SAT 2013. LNCS, vol. 7962, pp. 100–115. Springer, Heidelberg (2013)
25. Mahajan, M., Shukla, A.: Level-ordered Q-resolution and tree-like Q-resolution are incomparable. Inf. Process. Lett. **116**(3), 256–258 (2016)
26. Marques Silva, J.P., Sakallah, K.A.: GRASP: a search algorithm for propositional satisfiability. IEEE Trans. Comput. **48**(5), 506–521 (1999)
27. Pipatsrisawat, K., Darwiche, A.: On the power of clause-learning SAT solvers as resolution engines. Artif. Intell. **175**(2), 512–525 (2011)
28. Van Gelder, A.: Contributions to the theory of practical quantified boolean formula solving. In: Milano, M. (ed.) CP 2012. LNCS, vol. 7514, pp. 647–663. Springer, Heidelberg (2012)
29. Zhang, L.: Solving QBF by combining conjunctive and disjunctive normal forms. In: AAAI. AAAI Press (2006)
30. Zhang, L., Malik, S.: Conflict driven learning in a quantified Boolean satisfiability solver. In: ICCAD, pp. 442–449 (2002)

# On Stronger Calculi for QBFs

Uwe Egly[(✉)]

Institut für Informationssysteme 184/3, Technische Universität Wien,
Favoritenstrasse 9–11, 1040 Vienna, Austria
uwe@kr.tuwien.ac.at

**Abstract.** Quantified Boolean formulas (QBFs) generalize propositional formulas by admitting quantifications over propositional variables. QBFs can be viewed as (restricted) formulas of first-order predicate logic and easy translations of QBFs into first-order formulas exist. We analyze different translations and show that first-order resolution combined with such translations can polynomially simulate well-known deduction concepts for QBFs. Furthermore, we extend QBF calculi by the possibility to instantiate a universal variable by an existential variable of smaller level. Combining such an enhanced calculus with the propositional extension rule results in a calculus with a universal quantifier rule which essentially introduces propositional formulas for universal variables. In this way, one can mimic a very general quantifier rule known from sequent systems.

## 1 Introduction

Quantified Boolean formulas (QBFs) generalize propositional formulas by admitting quantifications over propositional variables. QBFs can be viewed in two different ways, namely (i) as a generalization of propositional logic and (ii) as a restriction of first-order predicate logic (where we interpret over a two element domain). A number of calculi are available for QBFs: the ones based on variants of resolution for QBFs [2,3,12,14], the ones based on instantiating universal variables with truth constants combined with propositional resolution and an additional instantiation rule [4], and different sequent systems [7,9,11,15].

In all these calculi (except the latter ones from [7,9,15]), the possibility to instantiate a given formula is limited. In purely resolution-based calculi, formulas (or more precisely universal variables) are never instantiated. In instantiation-based calculi, instantiation is restricted to truth constants. In contrast, sequent systems possess flexible quantifier rules, and (existential) variables as well as (propositional) formulas can be used for instantiation with tremendous speed-ups in proof complexity. This motivates why we are interested in strengthening instantiation techniques for instantiation-based calculi.

This work was supported by the Austrian Science Fund (FWF) under grant S11409-N23. Partial results have been announced at the QBF Workshop 2014 (http://www.easychair.org/smart-program/VSL2014/QBF-program.html). We thank the reviewers for valuable comments. An extended version with proofs is available [10].

© Springer International Publishing Switzerland 2016
N. Creignou and D. Le Berre (Eds.): SAT 2016, LNCS 9710, pp. 419–434, 2016.
DOI: 10.1007/978-3-319-40970-2_26

We allow to replace (some) universal variables $x$ not only by truth constants but by existential variables left of $x$ in the quantifier prefix. This approach mimics the effect of quantifier rules introducing atoms in sequent calculi from [9]. We add a propositional extension principle (known from extended resolution [20]), which enables the introduction of propositional formulas for universal variables via extension variables (or names for the formula). Contrary to [9], where we proposed propositional extensions of the form $\exists q(q \leftrightarrow F)$ which can be eliminated if the cut rule is available in the sequent calculus, such an elimination is not possible here for which reason we have to use (classical) extensions.

*Contributions.*

1. We consider different translations from QBFs to first-order logic [18] and provide a proof-theoretical analysis of the translation in combination with first-order resolution ($R_1$). We exponentially separate two variants of the translation in Theorem 4.
2. We show that such combinations can polynomially simulate Q-resolution with resolution over existential and universal variables (QU-res [12], Theorem 1), Q-resolution (Q-res [14], Corollary 1) and the instantiation-based calculus IR-calc [4] (Theorem 2, Corollary 2). The latter simulation provides a soundness proof for IR-calc independent from strategy extraction.
3. We show in Theorem 3 that neither Q-res nor QU-res, the long-distance Q-resolution variants LDQ-res, LDQU-res, LDQU$^+$-res [2,3,21], different instantiation-based calculi [4] nor Q(D)-res [19] can polynomially simulate $R_1$ with one of the considered translations.
4. We generalize IR-calc by the possibility to instantiate universal variables not only with truth constants but also with existential variables (similar to the corresponding quantifier rule in [9]). We show in Proposition 12 that this generalized calculus is actually stronger than the original one.
5. We combine generalized IR-calc by a propositional extension rule [6,20] essentially enabling the introduction of Boolean functions (instead of atoms and truth constants) for universal variables.

*Structure.* In Sect. 2 we introduce necessary definitions and notations. In Sect. 3 different translations from QBFs to (restrictions of) first-order logic [18] are reconsidered. In Sect. 4 different calculi based on (variants of) the resolution calculus are described. Here, we introduce our calculi generalized from IR-calc. In Sect. 5 we present our results on polynomial simulations between considered calculi and in Sect. 6 we provide exponential separations. In the last section we conclude and discuss future research possibilities.

## 2  Preliminaries

We assume familiarity with the syntax and semantics of propositional logic, QBFs and first-order logic (see, e.g., [16] for an introduction). We recapitulate some notions and notations which are important for the rest of the paper.

We consider a propositional language based on a set $\mathcal{PV}$ of Boolean variables and truth constants $\top$ (true) and $\bot$ (false), both of which are not in $\mathcal{PV}$. A variable or a truth constant is called *atomic* and connectives are from $\{\neg, \wedge, \vee, \rightarrow, \leftrightarrow, \oplus\}$. A *literal* is a variable or its negation. A *clause* is a disjunction of literals, but sometimes we consider it as a set of literals. *Tautological clauses* contain a variable and its negation and the *empty clause* is denoted by $\square$. Propositional formulas are denoted by capital Latin letters like $A, B, C$ possibly annotated with subscripts, superscripts or primes.

We extend the propositional language by Boolean quantifiers. Universal ($\forall^b$) and existential ($\exists^b$) quantification is allowed within a QBF. The superscript $b$ is used to distinguish Boolean quantifiers from first-order quantifiers introduced later. QBFs are denoted by Greek letters. Observe that we allow non-prenex formulas, i.e., quantifiers may occur deeply in a QBF. An example for a non-prenex QBF is $\forall^b p \, (p \rightarrow \forall^b q \exists^b r \, (q \wedge r \wedge s))$, where $p$, $q$, $r$ and $s$ are variables. Moreover, free variables (like $s$) are allowed, i.e., there might be occurrences of variables in the formula for which we have no quantification. Formulas without free variables are called *closed*; otherwise they are called *open*. The *universal (existential)* closure of $\varphi$ is $\forall^b x_1 \ldots \forall^b x_n \varphi$ ($\exists^b x_1 \ldots \exists^b x_n \varphi$), for which we often write $\forall^b \boldsymbol{X} \varphi$ ($\exists^b \boldsymbol{X} \varphi$) if $\boldsymbol{X} = \{x_1, \ldots, x_n\}$ is the set of all free variables in $\varphi$. A formula in *prenex conjunctive normal form* (PCNF) has the form $Q_1^b p_1 \ldots Q_n^b p_n \, M$, where $Q_1^b p_1 \ldots Q_n^b p_n$ is the *quantifier prefix*, $Q \in \{\forall, \exists\}$ and $M$ is the (propositional) *matrix* which is in CNF. Often we write a QBF as $Q_1^b X_1 \ldots Q_k^b X_k \, M$ ($Q_i \neq Q_{i+1}$ for all $i = 1, \ldots, k-1$ and the elements of $\{X_1, \ldots, X_k\}$ are pairwise disjoint). We define the *level of a literal* $\ell$, $lv(\ell)$, as the index $i$ such that the variable of $\ell$ occurs in $X_i$. The *logical complexity* of a formula $\Phi$, $lc(\Phi)$, is the number of occurrences of connectives and quantifiers.

We use a first-order language consisting of (objects) *variables, function symbols* (FSs), *predicate symbols* (PSs), together with the truth constants and connectives mentioned above. Quantifiers $\forall$ and $\exists$ bind object variables. *Terms* and *formulas* are defined according to the usual formation rules. We identify 0-ary PSs with propositional atoms, and 0-ary FSs with *constants*. Clauses, tautological clauses and the empty clause are defined as in the propositional case.

Let $V$ be the set of first-order variables and $T$ be the set of terms. A *substitution* is a mapping $\sigma$ of type $V \rightarrow T$ such that $\sigma(v) \neq v$ only for finitely many variables $v \in V$. We represent $\sigma$ by a finite set of the form $\{v_1 \backslash t_1, \ldots, v_n \backslash t_n\}$. The *domain* of $\sigma$, $dom(\sigma)$, is the set $\{v \mid v \in V, \sigma(v) \neq v\}$. The *range* of $\sigma$, $rg(\sigma)$, is the set $\{\sigma(v) \mid v \in dom(\sigma)\}$. We call $\sigma$ a *variable substitution* if $rg(\sigma) \subseteq V$. The *empty* substitution $\epsilon$ is denoted by $\{\}$. We often write substitutions post-fix, e.g., we use $x\sigma$ instead of $\sigma(x)$. Algebraically, substitutions define a monoid with $\epsilon$ being the neutral element under the usual composition of substitutions.

Substitutions are extended to terms and formulas in the usual way, e.g., $f(t_1, \ldots, t_n)\sigma = f(t_1\sigma, \ldots, t_n\sigma)$, $(\neg)p(t_1, \ldots, t_n)\sigma = (\neg)p(t_1\sigma, \ldots, t_n\sigma)$, and $(F \circ G)\sigma = F\sigma \circ G\sigma$, where $f$ is an $n$-place FS, $p$ is an $n$-place PS, $t_1, \ldots, t_n$ are terms, $F$ and $G$ are (quantifier-free) formulas and $\circ$ is a binary connective. For substitutions $\sigma$ and $\tau$, $\sigma$ is *more general than* $\tau$ if there is a substitution $\mu$ such

$$[\![\perp]\!]_p^f = p(0) \qquad\qquad [\![\top]\!]_p^f = p(1) \qquad\qquad [\![x]\!]_p^f = p(x)$$

$$[\![\neg\Phi]\!]_p^f = \neg[\![\Phi]\!]_p^f \qquad [\![\Phi_1 \circ \Phi_2]\!]_p^f = [\![\Phi_1]\!]_p^f \circ [\![\Phi_2]\!]_p^f \qquad [\![Q^b x \Phi]\!]_p^f = Qx[\![\Phi]\!]_p^f$$

**Fig. 1.** The translation of QBFs to first-order formulas. The connective ∘ is a binary connective present in both languages and $Q \in \{\forall, \exists\}$. The symbols $p$ and $f$ do not occur in the source QBF; $p$ is a unary predicate symbol and $f$ is used to construct constant and function symbols by indices.

that $\sigma\mu = \tau$. A substitution $\sigma$ is called a *permutation* if $\sigma$ is one-one and a variable substitution. A permutation $\sigma$ is called a *renaming* (substitution) of an expression $E$ (i.e., $E$ is a term or a quantifier-free formula) if $var(E) \cap rg(\sigma) = \{\}$, where $var(E)$ is the set of all variables occurring in $E$. For an expression $G$, $G\sigma$ is a *variant* of $G$ provided $\sigma$ is a renaming substitution.

Let $E = \{E_1, \dots, E_n\}$ be a non-empty set of expressions. A substitution $\sigma$ is called a *unifier of $E$* if $|\{E_1\sigma, \dots, E_n\sigma\}| = 1$. Unifier $\sigma$ is called *most general unifier* (mgu), if for every unifier $\tau$ of $E$, $\sigma$ is more general than $\tau$.

Let $P_1$ and $P_2$ be two proof systems. $P_1$ *polynomially simulates* (p-simulates) $P_2$ if there is a polynomial $p$ such that, for every natural number $n$ and every formula $\Phi$, the following holds. If there is a proof of $\Phi$ in $P_2$ of size $n$, then there is a proof of $\Phi$ (or a suitable translation of it) in $P_1$ whose size is less than $p(n)$.

## 3    Different Translations of QBFs to First-Order Logic

We introduce different translations of (closed) QBFs to (closed) formulas in (restrictions of) first-order logic. We start with the basic translation from [18] in Fig. 1. Obviously, the QBF $\Phi$ and the first-order formula $[\![\Phi]\!]_p^f$ enjoy a very similar structure. Especially the variable dependencies expressed by the quantifier prefix are exactly the same and the connectives are preserved.

**Proposition 1.** *Let $\Phi$ be a (closed) QBF and let $[\![\Phi]\!]_p^f$ be its (closed) first-order translation. Then $\Phi \cong [\![\Phi]\!]_p^f$, i.e., $\Phi$ and $[\![\Phi]\!]_p^f$ are isomorphic.*

The proof in the appendix of [10] is by induction on the logical complexity of $\Phi$.

The basic translation from Fig. 1 can be extended to $Sk[\![\Phi]\!]_p^f$ generating a skolemized version of $[\![\Phi]\!]_p^f$. We restrict our attention here to QBFs in PCNF.

**Definition 1.** *Let $\Phi$ be a closed QBF in PCNF with matrix $M$ and let $[\![\Phi]\!]_p^f$ be its closed first-order translation. For any existential variable $a$ in the quantifier prefix of $\Phi$, let $dep(a)$ be the sequence of universal variables left of $a$ (in exactly the same order in which they occur in the prefix). Let $f_a$ be the Skolem function symbol associated to $a$. We call $[\![M]\!]_p^f \sigma$ the skolemized form of $[\![M]\!]_p^f$ and denote it by $Sk[\![M]\!]_p^f$, where the substitution $\sigma$ is as follows.*

$$\sigma = \{a \backslash f_a(dep(a)) \mid \text{for all existential variables } a \text{ in } \Phi\}$$

$$\frac{}{C}\ \text{Axiom} \qquad \frac{x \vee C_1 \quad \neg x \vee C_2}{C_1 \vee C_2}\ \text{Res} \qquad \frac{C \vee \ell \vee \ell}{C \vee \ell}\ \text{Fac} \qquad \frac{D \vee m}{D}\ \forall \text{R}$$

$C$ is a non-tautological clause from the matrix. If $y \in C_1$ then $\neg y \notin C_2$. Variable $x$ is existential (Q-res) and existential or universal (QU-res), $\ell$ is a literal and $m$ is a universal literal. If $e \in D$ is existential, then $lv(e) < lv(m)$ holds.

**Fig. 2.** The rules of Q-res and QU-res [12,14]

Traditionally, $Sk[\![ M ]\!]_p^f$ is denoted as a quantifier-free formula with the assumption that all free variables are (implicitly) universally quantified.

The number of universal variables a Skolem function depends on can be optimized, e.g., by using miniscoping or dependency schemes [18]. As we will see later on, most of our results do not depend on such optimizations.

**Proposition 2.** *Let $\Phi$ be a closed QBF in PCNF with matrix $M$ and let $[\![ \Phi ]\!]_p^f$ be its closed first-order translation. Let $Sk[\![ M ]\!]_p^f$ be the skolemized form of $[\![ M ]\!]_p^f$. Then $M \cong Sk[\![ M ]\!]_p^f$.*

Due to Propositions 1 and 2, we can relate each literal of each clause from $M$ to *its* isomorphic counterpart in $[\![ M ]\!]_p^f \sigma$.

Since we interpret over a two-element domain, proper Skolem function symbols (i.e., the arity is greater than 0) can be eliminated by introducing new predicate symbols. The resulting formula belongs to EPR (Effectively PRopositional logic or more traditionally it belongs to the Bernays-Schoenfinkel class).

**Definition 2.** *Let $\Phi$ be a closed QBF in PCNF with matrix $M$ and let $[\![ \Phi ]\!]_p^f$ be its closed first-order translation. Let $Sk[\![ M ]\!]_p^f$ the skolemized form of $[\![ M ]\!]_p^f$. Replace any occurrence of a predicate of the form $p(f_b(X))$ by $p_b(X)$ where $f_b$ is a proper function symbol and $X$ is a non-empty list of universal variables. The formula resulting after all possible replacements is the EPR formula $EPR[\![ M ]\!]_p^f$.*

We will see later that the first-order and the EPR translation have different proof-theoretical properties because some resolutions are blocked by different predicate symbols. Proposition 3 is Lemma 1 in [18] (stated without a proof).

**Proposition 3.** *Let $\Phi$ be a closed QBF. Then*

$$\Phi \text{ is satisfiable} \quad \text{iff} \quad [\![ \Phi ]\!]_p^f \wedge p(1) \wedge \neg p(0) \text{ is satisfiable.}$$

A proof can be found in the appendix of [10].

## 4   Different Calculi Based on Resolution

We introduce different calculi used in this paper. We start with two resolution calculi, Q-res and QU-res, for QBFs in Fig. 2. Observe that the consequence

$$\frac{}{\{e^{[\sigma]} \mid e \in C, e \text{ is existential}\}} \text{Axiom}$$

$C$ is a non-tautological clause from the matrix $M$, $\sigma = \{u \backslash 0 \mid u \in C \text{ universal}\}$
where $u \backslash 0$ is a shorthand for $x \backslash 0$ if $u = x$ and $x \backslash 1$ if $u = \neg x$.

$$\frac{x^\tau \vee C_1 \quad \neg x^\tau \vee C_2}{C_1 \vee C_2} \text{Res} \qquad \frac{C \vee \ell^\tau \vee \ell^\tau}{C \vee \ell^\tau} \text{Fac} \qquad \frac{C}{\text{inst}(\tau, C)} \text{Inst}$$

$\tau$ is an assignment to universal variables and $rg(\tau) \subseteq \{0,1\}$.

**Fig. 3.** The rules of IR-calc(P,M) taken from [4]

of each rule is non-tautological. We continue with the calculus IR-calc$(P, M)$ in Fig. 3, where we use the same presentation as in [4]. $P$ is the quantifier prefix and $M$ is the quantifier-free matrix in CNF. In instantiation-based calculi, inference rules do not work on usual clauses but on *annotated clauses* based on *extended assignments*. An extended assignment is a partial mapping from the Boolean variables to $\{0, 1\}$. An annotated clause consists of *annotated literals* of the form $\ell^{[\tau]}$, where $\tau$ is an extended assignment to *universal* variables and $[\tau] = \{u \backslash c \mid (u \backslash c) \in \tau, lv(u) < lv(\ell)\}$ with $c \in \{0, 1\}$. Composition of extended assignments is defined using *completion*. The expression $\mu \veebar \tau$ is called the completion of $\mu$ by $\tau$. Then $\sigma$, the completion of $\mu$ by $\tau$, is defined as follows.

$$\sigma(x) = \begin{cases} \mu(x) & \text{if } x \in dom(\mu); \\ \tau(x) & \text{if } x \notin dom(\mu) \text{ and } x \in dom(\tau). \end{cases} \tag{1}$$

The function inst$(\tau, C)$ produces instantiations of clauses; it computes $\{\ell^{[\mu \veebar \tau]} \mid \ell^\mu \in C\}$ for an extended assignment $\tau$ and an annotated clause $C$. Later, we will clarify the relation between annotations and substitutions in first-order logic.

We extend IR-calc$(\cdot, \cdot)$ by the possibility to instantiate universal variables by existential ones. Technically the instantiation is performed by a global substitution $\sigma_v$. If a universal variable $x$ is replaced by some existential variable $e$, i.e., $(x \backslash e) \in \sigma_v$, then $lv(e) < lv(x)$ must hold. We name the calculus equipped with the substitution $\sigma_v$ IR-calc$(P, M, \sigma_v)$ and depict the rules in Fig. 4.

It is immediately apparent that this calculus is sound and complete. We get completeness, when we use the empty substitution as $\sigma_v$ because then, IR-calc$(\cdot, \cdot, \cdot)$ reduces to IR-calc$(\cdot, \cdot)$ which is sound and complete [4]. Soundness follows from the validity of QBFs of the form

$$\mathcal{Q}_1 \exists e \mathcal{Q}_2 \forall x \mathcal{Q}_3 \, \varphi(e, x) \;\to\; \mathcal{Q}_1 \exists e \mathcal{Q}_2 \mathcal{Q}_3 \, \varphi(e, e).$$

If the right formula has an IR-calc$(\cdot, \cdot)$ refutation, then it is false and therefore the left formula has to be false.

We further enhance IR-calc$(\cdot, \cdot, \cdot)$ by the possibility to use propositional extensions [6,20]. This extension operation is a generalization of the well-known

---

$$\frac{}{\{e^{[\sigma]} \mid e \in C\sigma_v, e \text{ is existential}\}} \text{ Axiom}$$

1. $C$ is a non-tautological clause from the matrix $M$.
2. $\sigma_v = \{x \backslash e \mid x \text{ is universal}, e \text{ is existential}, lv(e) < lv(x)\}$.
3. $C\sigma_v = \{e \mid e \in C \text{ existential}\} \cup \{x\sigma_v \mid x \in C \text{ universal}, x \in dom(\sigma_v)\} \cup \{x \mid x \in C \text{ universal}, x \notin dom(\sigma_v)\}$.
4. $\sigma$, Res, Fac and Inst are the same as in IR-calc$(\cdot, \cdot)$, but $C\sigma_v$ is used instead of $C$ in the definition of $\sigma$.

**Fig. 4.** The rules of IR-calc$(P, M, \sigma_v)$

---

$$\frac{}{\{e^{[\sigma]} \mid e \in C\sigma_v, e \text{ is existential}\}} \text{ Axiom}$$

1. $C$ is a non-tautological clause from the matrix $M$ or from $\Delta$.
2. $\sigma_v$, $C\sigma_v$, $\sigma$, Res, Fac and Inst are the same as in IR-calc$(\cdot, \cdot, \cdot)$.
3. If $C \in \Delta$ then $\sigma = \epsilon$ and $C = C\sigma_v$ by construction.

**Fig. 5.** The rules of IR-calc$(P, M, \Delta, \sigma_v)$

structure-preserving translation to (conjunctive) normal form in propositional logic. For presentational reasons, we require to have all extensions at the very beginning of the deduction in order to allow extension variables as replacements for universal variables. Figure 5 shows the inference rules of this calculus IR-calc$(P, M, \Delta, \sigma_v)$, where $\Delta$ is a sequence $\delta_1, \ldots, \delta_d$ of (clausal representations of) extensions of the form $\delta_i \colon q_i \leftrightarrow F$ with $F$ being of the form $\neg p$ or of the form $p \circ r$ ($\circ \in \{\wedge, \vee, \rightarrow, \leftrightarrow, \oplus\}$) and $q_i$ is a variable neither occurring in $M$ nor in $F$ nor in $\delta_1, \ldots, \delta_{i-1}$. The variables $q_i, p, r$ are existential. The quantification $\exists q_i$ extends the quantifier prefix $P$ such that $lv(v) \leq lv(q_i)$ for all variables $v$ occurring in $F$ and $lv(q_i)$ is minimal. Due to the requirements on the extension variables $q_i$ and the placement of $\exists q_i$, the resulting calculus is sound. Completeness is not an issue here, because we can use an empty $\Delta$.

*Remark 1.* The usual formalization of clauses and resolvents as sets of literals can be simulated in our formalizations by the factoring rule Fac. We assume in the following that Fac is applied as soon as possible.

We finally introduce first-order resolution. Let $C$ be a clause and let $K$ and $L$ be two distinct literals in $C$ both of which are either negated or unnegated. If there is an mgu $\sigma$ of $K$ and $L$, then the clause $D = C\sigma = \{N\sigma \mid N \in C\}$ is called a *factor* of $C$. The clause $C$ is called the *premise* of the factoring operation.

Let $C$ and $D$ be two clauses and let $D'$ be a variant of $D$ which has no variable in common with $C$. A clause $E$ is a *resolvent* of the parent clauses $C$ and $D$ if the following conditions hold:

1. $K \in C$ and $L' \in D'$ are literals of opposite sign whose atoms are unifiable by an mgu $\sigma$.
2. $E = (C\sigma \setminus \{K\sigma\}) \cup (D'\sigma \setminus \{L'\sigma\})$.

Let $C$ be a set of clauses. A sequence $C_1, \ldots, C_n$ is called $\mathsf{R}_1$ *deduction* (first-order resolution deduction) of a clause $C$ from $\mathcal{C}$ if $C_n = C$ and for all $i = 1, \ldots, n$, one of the following conditions hold.

1. $C_i$ is an input clause from $\mathcal{C}$.
2. $C_i$ is a factor of a $C_j$ for $j < i$.
3. $C_i$ is a resolvent of $C_j$ and $C_k$ for $j, k < i$.

An $\mathsf{R}_1$ *refutation of* $\mathcal{C}$ is an $\mathsf{R}_1$ deduction of the empty clause $\square$ from $\mathcal{C}$. The *size* of a deduction is given by $\sum_{i=1}^{n} size(C_i)$, where $size(C_i)$ is the number of character occurrences in $C_i$. An $\mathsf{R}_1$ deduction has *tree form* if every occurrence of a clause is used at most once as a premise in a factoring operation or as a parent clause in a resolution operation.

Next we introduce the *subsumption rule* taken from Definition 2.3.4 in [8]. Contrary to the usual use of subsumption in automated deduction as a deletion rule, here we *add* clauses which are (factors of) instantiations of clauses.

**Definition 3.** *If $C$ and $D$ are clauses, then $C$ subsumes $D$ or $D$ is subsumed by $C$, if there is a substitution $\sigma$ such that $C\sigma \subseteq D$. A set $S'$ of clauses is obtained from a set $S$ by subsumption if $S' = S \cup \{D\}$ where $D$ is subsumed by a clause of $S$.*

Resolution can be extended by the subsumption rule (Definition 3.2.3 in [8]).

**Definition 4.** *By a derivation of a set of clauses $S_2$ from a set of clauses $S_1$ by $\mathsf{R}_1$ plus subsumption, we mean a sequence $C_1, \ldots, C_n$ of clause such that the following conditions are fulfilled.*

1. $S_2 \subseteq S_1 \cup \{C_1, \ldots, C_n\}$.
2. *For all $k = 1, \ldots, n$ there is a clause $C \in S_1 \cup \{C_1, \ldots, C_{k-1}\}$ subsuming the clause $C_k$ or there exist clauses $C, D \in S_1 \cup \{C_1, \ldots, C_{k-1}\}$ such that $C_k$ is subsumed by a resolvent of $C$ and $D$.*

Factors are not needed in item 2, because the factor of $C$ can be generated by subsumption. We need a simplified version of Proposition 3.2.1 from [8].

**Proposition 4.** $\mathsf{R}_1$ *polynomially simulates* $\mathsf{R}_1$ *plus subsumption.*

The subsumption rule is not necessary but makes proofs of polynomial simulation results much more convenient. It allows instantiated deductions for which eventually the lifting theorem provides a deduction "on the most general level".

# 5   Polynomial Simulations of Calculi

In this section we show that $R_1$ together with a suitable translation $\mathcal{T}$ and the clauses $p(1)$ and $\neg p(0)$ (denoted by $R_1 + \mathcal{T}$) polynomially simulates QU-res, Q-res and IR-calc$(\cdot, \cdot)$.

**Theorem 1.** $R_1 + Sk[\![ \cdot ]\!]_p^f$ *polynomially simulates* QU-res.

The proof is by induction on the number of clauses in the QU-res deduction. It can be found in the appendix of [10]. It shows that first-order literals obtained from universal literals in the QBF and eliminated by $\forall R$ are eliminated by resolutions with $p(1)$ and $\neg p(0)$ *without* instantiating the first-order resolvent.

**Corollary 1.** *The following results are immediate consequences of Theorem 1.*

1. $R_1 + EPR[\![ \cdot ]\!]_p^f$ *polynomially simulates* QU-res.
2. $R_1 + Sk[\![ \cdot ]\!]_p^f$ *as well as* $R_1 + EPR[\![ \cdot ]\!]_p^f$ *polynomially simulates* Q-res.

We present a soundness proof of IR-calc$(\cdot, \cdot)$ independent from strategy extraction by a polynomial simulation of IR-calc$(\cdot, \cdot)$ by $R_1 + Sk[\![ \cdot ]\!]_p^f$.

**Definition 5.** *Let* $\tau = \{x_1 \backslash s_1, \ldots, x_k \backslash s_k\}$ *and* $\mu = \{y_1 \backslash t_1, \ldots, y_l \backslash t_l\}$ *be two substitutions. The composition of* $\tau$ *and* $\mu$, $\tau\mu$, *is obtained from*

$$\{x_1 \backslash s_1\mu, \ldots, x_k \backslash s_k\mu, y_1 \backslash t_1, \ldots, y_l \backslash t_l\}$$

*by deleting all* $y_i \backslash t_i$ *for which* $y_i \in \{x_1, \ldots, x_k\}$ *holds.*

**Lemma 1.** *Let* $\tau$ *and* $\mu$ *be two substitutions as defined in Definition 5, where* $x_1, \ldots, x_k, y_1, \ldots, y_l$ *are universal variables and* $\{s_1, \ldots, s_k, t_1, \ldots, t_l\} \subseteq \{0, 1\}$. *Then* $\tau \veebar \mu$ *is the composition* $\tau\mu$.

*Proof.* Let $\sigma$ be the completion of $\tau$ by $\mu$ defined in (1). Since $dom(\tau)$ as well as $dom(\mu)$ is a subset of the set of universal variables and $rg(\tau)$ as well as $rg(\mu)$ is a subset of $\{0, 1\}$, $rg(\tau) \cap dom(\mu) = \{\}$ and therefore $s_i\mu = s_i$ for all $i = 1, \ldots, k$. Hence, the completion $\sigma$ of the two substitutions $\tau$ and $\mu$ is exactly their composition $\tau\mu$.  □

In the following, we deal with annotated clauses $C$ of the form $\{l_1^{[\sigma_1]}, \ldots, l_k^{[\sigma_k]}\}$ where any $l_i$ is an existential literal and any $[\sigma_i]$ is the restriction of assignment $\sigma_i$ to exactly those universal variables $x \in dom(\sigma_i)$ for which $lv(x) < lv(l_i)$ holds. We denote the sequence of *all* universal variables $x$ with $lv(x) < lv(l_i)$ by $dep(l_i) = \overline{X}_{l_i}$ where we assume the same order as in the quantifier prefix. A first-order clause $D$ corresponding to $C$ is constructed as follows

$$\{(\neg)p(f_e(\overline{X}_e))\sigma \mid (\neg)e^{[\sigma]} \in C \quad \text{and} \quad p(f_e(\overline{X}_e)) \cong e\},$$

where $p(f_e(\overline{X}_e))$ is the isomorphic counterpart of $e$ (cf. the remark after Proposition 2). Using $\overline{X}_e$ together with $\sigma$ mimics the effect of $[\sigma]$; the difference is the explicit notation of *all* universal variables $\overline{X}_e$ left of $e$ and not only the variables in $\overline{X}_e \cap dom(\sigma)$.

**Theorem 2.** $R_1 + Sk[\![ \cdot ]\!]_p^f$ *polynomially simulates* IR-calc$(\cdot, \cdot)$.

In the proof, we construct by induction on the number of derived clauses in the IR-calc deduction stepwisely a deduction in $R_1$ plus subsumption. We consider the sequence of first-order clauses obtained from the original clauses as a skeleton for the final proof. Since not all clauses in the skeleton follow by a single application of an inference rule, we have to provide a short deduction of the clauses.

*Proof.* We utilize Proposition 4 and allow subsumption in the simulation. The proof is by strong mathematical induction on the number of derived clauses in the IR-calc deduction. Let $P(n)$ denote the statement "Given a IR-calc deduction $C_1, \ldots, C_n$ from a QBF $Q.M$ and a sequence of first-order clauses $D_1, \ldots, D_n$, the clause $D_n$ has a short deduction in $R_1$ plus subsumption from $p(1), \neg p(0), Sk[\![ M ]\!]_p^f, D_1, \ldots, D_{n-1}$".
Base: $n = 1$. $C_1$ is a consequence of the axiom rule using clause $C$ from the matrix $M$. Let $\sigma$ be the assignment induced by $C$. Then we have a clause $D \in Sk[\![ M ]\!]_p^f$ from which we can derive $D_1 \sigma$ by resolution steps using $p(1)$ and $\neg p(0)$. The number of these steps is equal to the number of universal variables in $C$.
IH: Suppose $P(1), \ldots, P(n)$ hold for some $n \geq 1$.
Step: We have to show $P(n+1)$. Consider $C_1, \ldots, C_{n+1}$ and $D_1, \ldots, D_{n+1}$.
CASE 1: $C_{n+1}$ is derived by the axiom rule. Then proceed like in the base case.
CASE 2: $C_{n+1}$ is a consequence of the rule Inst with premise $C_i$ (for some $i$ with $1 \leq i \leq n$) and assignment $\tau$. By IH and Remark 1, we have a short $R_1$ plus subsumption deduction of $D_i = \{(\neg)p(f_e(\overline{X}_e))\sigma \mid (\neg)e^{[\sigma]} \in C_i\}$. $C_{n+1}$ is of the form $\{(\neg)e^{[\sigma \vee \tau]} \mid (\neg)e^{[\sigma]} \in C_i\}$. By Lemma 1, $x(\sigma \vee \tau) = x\sigma\tau$ for any universal variable $x$ with $lv(x) < lv(e)$. Therefore $D_{n+1}$ is of the form $\{(\neg)p(f_e(\overline{X}_e))\sigma\tau \mid (\neg)e^{[\sigma \vee \tau]} \in C_{n+1}\}$. Now $D_{n+1} = D_i\tau$ and $D_{n+1}$ can be derived by subsumption.
CASE 3: $C_{n+1}$ is a consequence of the rule Fac with premise $C_i : \widetilde{C}_i \vee \ell^\tau \vee \ell^\tau$ (for some $i$ with $1 \leq i \leq n$). By IH, we have a short $R_1$ plus subsumption deduction of $D_i : \widetilde{C}_i \vee L \vee L$, where $L$ is of the form $(\neg)p(f_e(\overline{X}_e)\tau$. We generate a factor $D_{n+1}$ of $D_i$ simply by omitting one of the duplicates.
CASE 4: $C_{n+1}$ is a consequence of the resolution rule with parent clauses $C_i, C_j$ (for some $i, j$ with $1 \leq i, j \leq n$). By IH, we have two clauses

$$D_i = \{p(f_e(\overline{X}_e))\sigma\} \cup D_i' \qquad \text{and} \qquad D_j = \{\neg p(f_e(\overline{X}_e))\sigma\} \cup D_j'$$

We use $\lambda$ of the form $\{x \backslash y\}$ as a renaming of the variables in $D_j$ such that $D_j\lambda$ does not share any variable with $D_i$. The resolvent is $D_i' \cup D_j'\lambda\mu$ where $\mu$ is the mgu of the form $\{y \backslash x \mid x \in \overline{X}_e, x \notin dom(\sigma)\}$. We add $D_i' \cup D_j'\lambda\mu\lambda'$ by subsumption, where $\lambda'$ maps all remaining variables $y$ to their $x$ counterpart. □

**Corollary 2.** $R_1 + EPR[\![ \cdot ]\!]_p^f$ *polynomially simulates* IR-calc$(\cdot, \cdot)$.

When we inspect the translation of (axiom) clauses, we observe that a universal variable $x$ is translated to an atom of the form $p(x)$. With the subsumption rule we can instantiate the clause by a substitution of the form $\{x \backslash t\}$ for a term $t$.

This observation was the trigger to introduce the stronger calculus $\mathsf{IR\text{-}calc}(\cdot,\cdot,\cdot)$, where universal variables cannot be replaced only by 0 or 1 but also by any existential variable $e$ with $lv(e) < lv(x)$.

## 6    Exponential Separation of Resolution Calculi

We constructed in [9] a family $(\Phi_n)_{n\geq 1}$ of short closed QBFs in PCNF for which any Q-res refutation of $\Phi_n$ is superpolynomial. We recapitulate the construction here. The formula $\Phi_n$ is

$$\exists^b X_n \forall^b Y_n \exists^b Z_n \left( \mathsf{TPHP}_n^{Y_n, Z_n} \wedge \mathsf{CPHP}_n^{X_n} \right). \tag{2}$$

$\mathsf{CPHP}_n^{X_n}$ is the pigeon hole formula for $n$ holes and $n+1$ pigeons in *conjunctive* normal form and denoted over the variables $X_n = \{x_{1,1}, \ldots, x_{n+1,n}\}$. Variable $x_{i,j}$ is intended to denote that pigeon $i$ is sitting in hole $j$. $\mathsf{CPHP}_n^{X_n}$ is

$$\left( \bigwedge_{i=1}^{n+1} \left( \bigvee_{j=1}^{n} x_{i,j} \right) \right) \wedge \left( \bigwedge_{j=1}^{n} \bigwedge_{1 \leq i_1 < i_2 \leq n+1} (\neg x_{i_1,j} \vee \neg x_{i_2,j}) \right).$$

The number of clauses in $\mathsf{CPHP}_n^{X_n}$ is $l_n = (n+1) + n^2(n+1)/2$ and $size(\mathsf{CPHP}_n^{X_n})$ is $O(n^3)$. The formula $\mathsf{TPHP}_n^{Y_n, Z_n}$ is obtained from the pigeon hole formula in *disjunctive* normal form, $\mathsf{DPHP}_n^{Y_n}$, by a structure-preserving polarity-sensitive translation to clause form [17]. The formula $\mathsf{DPHP}_n^{Y_n}$ is simply the negation of $\mathsf{CPHP}_n^{Y_n}$ where negation has been pushed in front of atoms and double-negation elimination has been applied.

We use new variables of the form $z_{i_1,i_2,j}$ for disjuncts in $\mathsf{DPHP}_n^{Y_n}$. For the first $n+1$ disjuncts of the form $\bigwedge_{j=1}^{n} \neg y_{i,j}$ with $1 \leq i \leq n+1$, we use variables $z_{1,0,0}, \ldots, z_{n+1,0,0}$. For the second part, for any $1 \leq j \leq n$ and the $n(n+1)/2$ disjuncts, we use

$$z_{1,2,j}, \ldots, z_{1,n+1,j}, z_{2,3,j}, \ldots, z_{2,n+1,j}, \ldots, z_{n,n+1,j}. \tag{3}$$

The set of these variables for $\mathsf{DPHP}_n$ is denoted by $Z_n$. Due to this construction, we can speak about the conjunction corresponding to the variable $z_{i_1,i_2,j}$.

We construct the conjunctive normal form $\mathsf{TPHP}_n^{Y_n, Z_n}$ of $\mathsf{DPHP}_n^{Y_n, Z_n}$ as follows. First, we take the clause $D_n^{Z_n} = \bigvee_{z \in Z_n} \neg z$ over all variables in $Z_n$. The formula $P_n^{Y_n, Z_n}$ for the first $(n+1)$ disjuncts of $\mathsf{DPHP}_n^{Y_n}$ is of the form

$$\bigwedge_{i=1}^{n+1} \bigwedge_{j=1}^{n} (z_{i,0,0} \vee \neg y_{i,j}) .$$

For the remaining $n^2(n+1)/2$ disjuncts of $\mathsf{DPHP}_n^{Y_n}$, we have the formula $Q_n^{Y_n, Z_n}$

$$\bigwedge_{j=1}^{n} \bigwedge_{1 \leq i_1 < i_2 \leq n+1} \left( (z_{i_1,i_2,j} \vee y_{i_1,j}) \wedge (z_{i_1,i_2,j} \vee y_{i_2,j}) \right) .$$

Then $\mathsf{TPHP}_n^{Y_n,Z_n}$ is $D_n^{Z_n} \wedge P_n^{Y_n,Z_n} \wedge Q_n^{Y_n,Z_n}$ and $size(\mathsf{TPHP}_n^{Y_n,Z_n})$ is $O(n^3)$. It is easy to check that $\mathsf{DPHP}_n^{Y_n} \leftrightarrow \exists^b Z_n \, \mathsf{TPHP}_n^{Y_n,Z_n}$ is valid.

Let us modify the quantifier prefix of $\Phi_n$. By quantifier shifting rules we get, in an "antiprenexing" step, the equivalent formula $(\forall^b Y_n \exists^b Z_n \mathsf{TPHP}_n^{Y_n,Z_n}) \wedge (\exists^b X_n \mathsf{CPHP}_n^{X_n})$. Prenexing yields the equivalent QBF $\Omega_n$

$$\forall^b Y_n \exists^b Z_n \exists^b X_n \left( \mathsf{TPHP}_n^{Y_n,Z_n} \wedge \mathsf{CPHP}_n^{X_n} \right) \tag{4}$$

which has only one quantifier alternation instead of two. In [9] we showed that $\Phi_n$ and $\Omega_n$ have short cut-free tree proofs in a sequent system $\mathsf{Gqve}^*$, where weak quantifiers introduce atoms. The following extends Proposition 3 in [9].

**Proposition 5.** *Any* Q-res *refutation of* $\Phi_n$ *from (2) or* $\Omega_n$ *from (4) has superpolynomial size.*

The proof is based on the fact that (i) the two conjuncts belong to languages with different alphabets and (ii) that the alphabets cannot be made identical by instantiation of quantifiers in Q-res. Therefore we have to refute either $\mathsf{TPHP}_n^{Y_n,Z_n}$ or $\mathsf{CPHP}_n^{X_n}$ under the given quantifier prefix. Since $\forall^b Y_n \exists^b Z_n \mathsf{TPHP}_n^{Y_n,Z_n}$ is true, there is no Q-res refutation and we have to turn to $\exists^b X_n \mathsf{CPHP}_n^{X_n}$. But then, we essentially have to refute $\mathsf{CPHP}_n^{X_n}$ with propositional resolution and consequently, by Haken's famous result [13], any Q-res refutation of $\mathsf{CPHP}_n^{X_n}$ is superpolynomial in $n$.

Since QU-res, LDQ-res, LDQU-res, LDQU$^+$-res, and Q(D)-resolution (Q(D)-res) [19] are based on the same quantifier-handling mechanism as Q-res, the following corollary is obvious.

**Corollary 3.** *Any refutation of* $\Phi_n$ *from (2) or* $\Omega_n$ *from (4) in the* QU-res, LDQ-res, LDQU-res, LDQU$^+$-res, *or* Q(D)-res *calculus has superpolynomial size.*

For IR-calc$(\cdot,\cdot)$ the situation is not better. Since universal literals are only replaced by 0, no unification of the two alphabets can happen. Moreover, even the factorization rule of IRM-calc [4] does not help.

**Proposition 6.** *Any refutation of* $\Phi_n$ *from (2) or* $\Omega_n$ *from (4) in* IR-calc$(\cdot,\cdot)$ *or* IRM-calc *has size superpolynomial in* $n$.

The quantifier prefix is unfortunate if one expects $\Omega_n$ being false. Actually, the initial universal quantifier block prevents any non-empty $\sigma_v$ and consequently, any IR-calc$(\cdot,\cdot,\cdot)$ refutation of $\Omega_n$ reduces to an IR-calc$(\cdot,\cdot)$ refutation of $\Omega_n$.

**Proposition 7.** *Any refutation of* $\Omega_n$ *from (4) in* IR-calc$(\cdot,\cdot,\cdot)$ *has size superpolynomial in* $n$.

In the following we show that $Sk[\![\Omega_n]\!]_p^f$ has a short tree refutation in $R_1$. We use $f_{x_{i,j}}$ to denote the Skolem function symbol corresponding to $x_{i,j} \in X_n$ and $f_{z_{i,j,k}}$ to denote the Skolem function symbols corresponding to $z_{i,j,k} \in Z_n$.

All the Skolem function symbols have arity $|Y_n| = n(n+1)$. Let $\overline{F}$ denote the formula $F$ under the first-order translation. We have

$$\overline{\mathsf{CPHP}_n^{X_n}} : \left( \bigwedge_{i=1}^{n+1} \left( \bigvee_{j=1}^{n} p(f_{x_{i,j}}(Y_n)) \right) \right) \wedge$$

$$\left( \bigwedge_{j=1}^{n} \bigwedge_{1 \le i_1 < i_2 \le n+1} (\neg p(f_{x_{i_1,j}}(Y_n)) \vee \neg p(f_{x_{i_2,j}}(Y_n))) \right).$$

$$\overline{D_n^{Z_n}} : \bigvee_{z \in Z_n} \neg p(f_z(Y_n))$$

$$\overline{P_n^{Y_n,Z_n}} : \bigwedge_{i=1}^{n+1} \bigwedge_{j=1}^{n} (p(f_{z_{i,0,0}}(Y_n)) \vee \neg p(y_{i,j}))$$

$$\overline{Q_n^{Y_n,Z_n}} : \bigwedge_{j=1}^{n} \bigwedge_{1 \le i_1 < i_2 \le n+1} ((p(f_{z_{i_1,i_2,j}}(Y_n)) \vee p(y_{i_1,j})) \wedge$$

$$(p(f_{z_{i_1,i_2,j}}(Y_n)) \vee p(y_{i_2,j}))).$$

The refutation of $Sk[\![\, \Omega_n \,]\!]_p^f$ is constructed as follows.

1. We use $\overline{P_n^{Y_n,Z_n}}$ together with the first $n+1$ clauses from $\overline{\mathsf{CPHP}_n^{X_n}}$ to derive $p(f_{z_{i,0,0}}(Y_n))\mu_i$ (for all $i = 1,\dots,n+1$). The deduction consists of $O(n^2)$ clauses and applies resolution and factoring. The substitution $\mu_i$ is $\bigcup_{j=1}^{n} \{y_{i,j} \backslash f_{x_{i,j}}(Y_n)\sigma_{i,j}\}$, where $\sigma_{i,j}$ is a variable renaming from the variant generation in resolution.

2. We use $\overline{Q_n^{Y_n,Z_n}}$ together with the binary clauses from $\overline{\mathsf{CPHP}_n^{X_n}}$ to derive $p(f_{z_{i_1,i_2,j}}(Y_n))\nu_{i_1,i_2,j}$ (for all $j = 1,\dots,n$ and $i_1,i_2$ with $1 \le i_1 < i_2 \le n+1$). The deduction consists of $O(n^3)$ clauses and applies resolution and factoring. Then $\nu_{i_1,i_2,j}$ is $\{y_{i_1,j}\backslash f_{x_{i_1,j}}(Y_n)\sigma_{i_1,i_2,j}, y_{i_2,j}\backslash f_{x_{i_2,j}}(Y_n)\sigma_{i_1,i_2,j}\}$. Again $\sigma_{i_1,i_2,j}$ is a variable renaming like above.

3. We use $\overline{D_n^{Z_n}}$ together with the derived instance of $p(f_{z_{k,l,m}}(Y_n))$ to derive $\square$ by resolution. Since any variable $y_{i,j}$ is assigned to a variant of $f_{x_{i,j}}(Y_n)$ for all $i = 1,\dots,n+1$ and all $j = 1,\dots,n$, all resolution steps are possible. The deduction consists of $O(n^3)$ clauses.

The formulas $Sk[\![\, \Phi_n \,]\!]_p^f$ and $EPR[\![\, \Phi_n \,]\!]_p^f$ can be refuted in a similar way in $\mathsf{R}_1$. Short refutations of $\Phi_n$ can be obtained similarly in $\mathsf{IR}\text{-calc}(\cdot,\cdot,\cdot)$ by instantiating $y_{i,j}$ by $x_{i,j}$ (for all $i = 1,\dots,n+1$ and all $j = 1,\dots,n$).

**Proposition 8.** *Let $(\Phi_n)_{n \ge 1}$ and $(\Omega_n)_{n \ge 1}$ be the families of closed QBFs from above. Then $Sk[\![\, \Phi_n \,]\!]_p^f$, $Sk[\![\, \Omega_n \,]\!]_p^f$ and $EPR[\![\, \Phi_n \,]\!]_p^f$ have short tree refutations in $\mathsf{R}_1$ consisting of $O(n^3)$ clauses. Moreover the size of the refutation is $O(n^8)$. The same is true for $\Phi_n$ and $\mathsf{IR}\text{-calc}(\cdot,\cdot,\cdot)$.*

**Theorem 3.** QU-res, LDQ-res, LDQU-res, $\mathsf{LDQU}^+$-res, Q(D)-res, $\mathsf{IR}\text{-calc}(\cdot,\cdot)$, $\mathsf{IR}\text{-calc}(\cdot,\cdot,\cdot)$ and $\mathsf{IRM}\text{-calc}$ *cannot polynomially simulate tree* $\mathsf{R}_1 + Sk[\![\, \cdot \,]\!]_p^f$.

We use $(\Omega_n)_{n\geq 1}$ to exponentially separate $\mathsf{R}_1$ combined with the two translations, i.e., we compare $Sk[\![\,\cdot\,]\!]_p^f$ with $EPR[\![\,\cdot\,]\!]_p^f$.

**Proposition 9.** *Let $(\Omega_n)_{n\geq 1}$ be the family of closed QBFs from (4). Then any refutation of $EPR[\![\,\Omega_n\,]\!]_p^f \wedge p(1) \wedge \neg p(0)$ in $\mathsf{R}_1$ has size superpolynomial in $n$.*

*Proof (Sketch).* Similar arguments as in Proposition 5 apply, because the EPR translations of $\mathsf{TPHP}_n^{Y_n,Z_n}$ and $\mathsf{CPHP}_n^{X_n}$ are denoted in different languages and literals from the former cannot be resolved with literals from the latter. Again, the pigeonhole formula has to be refuted. Consequently, the (essentially propositional) resolution proof has size superpolynomial in $n$. □

**Theorem 4.** $\mathsf{R}_1 + EPR[\![\,\cdot\,]\!]_p^f$ *cannot polynomially simulate tree* $\mathsf{R}_1 + Sk[\![\,\cdot\,]\!]_p^f$.

Let us reconsider the family $(\Psi_t)_{t\geq 1}$ of QBFs from [14]. Formula $\Psi_t$ has the prefix $P_t \colon \exists d_0 d_1 e_1 \forall x_1 \exists d_2 e_2 \forall x_2 \exists d_3 e_3 \dots \forall x_{t-1} \exists d_t e_t \forall x_t \exists f_1 \dots f_t$ and the matrix $M_t$ consisting of the following clauses:

$$
\begin{array}{ll}
C_0 \;\; : \overline{d}_0 & C_1 \qquad : d_0 \vee \overline{d}_1 \vee \overline{e}_1 \\
C_{2j} : d_j \vee \overline{x}_j \vee \overline{d}_{j+1} \vee \overline{e}_{j+1} \quad & C_{2j+1} : e_j \vee x_j \vee \overline{d}_{j+1} \vee \overline{e}_{j+1} \quad j = 1,\dots,t-1 \\
C_{2t} : d_t \vee \overline{x}_t \vee \overline{f}_1 \vee \cdots \vee \overline{f}_t \quad & C_{2t+1} : e_t \vee x_t \vee \overline{f}_1 \vee \cdots \vee \overline{f}_t \\
B_{2j} : \overline{x}_{j+1} \vee f_{j+1} & B_{2j+1} : x_{j+1} \vee f_{j+1} \qquad\qquad j = 0,\dots,t-1
\end{array}
$$

By Theorem 3.2 in [14] and Theorem 6 in [5], any Q-res refutation and any IR-calc$(\cdot,\cdot)$ refutation of $\Psi_t$ is exponential in $t$. The formula $\Psi_t$ has a polynomial size Q-resolution refutation if universal pivot variables are allowed [12].

Let us extract Herbrand functions from such a short QU-res refutation of $\Psi_t$ with the method of [2] resulting in $\overline{d}_i \wedge e_i$ for $x_i$. We explain in the following how we can produce short IR-calc$(P_t, M_t, \Delta, \sigma_v)$ refutations using such functions.

Let $\Delta \colon \delta_1, \dots, \delta_t$ where $\delta_i$ is $q_i \vee d_i \vee \overline{e}_i, \overline{q}_i \vee \overline{d}_i, \overline{q}_i \vee e_i$, i.e., $\delta_i$ is the clausal representation of $q_i \leftrightarrow \overline{d}_i \wedge e_i$. The quantifier $\exists q_i$ is in the same quantifier block as $d_i$ and $e_i$ and thus $lv(q_i) < lv(x_i)$. Consequently, $\sigma_v$ can replace $x_i$ by $q_i$.

**Proposition 10.** *Let $\Delta \colon \delta_1, \dots, \delta_t$ where $\delta_i$ is $q_i \vee d_i \vee \overline{e}_i, \overline{q}_i \vee \overline{d}_i, \overline{q}_i \vee e_i$, i.e., $\delta_i$ is the clausal representation of $q_i \leftrightarrow \overline{d}_i \wedge e_i$. Let $\sigma_{v,t} = \{x_i \backslash q_i \mid 1 \leq i \leq t\}$. There is a tree refutation of $\Psi_t$ in IR-calc$(P_t, M_t, \Delta, \sigma_{v,t})$ of size polynomial in $t$.*

*Proof (sketch).* Derive $\overline{d}_1 \vee \overline{e}_1, \dots, \overline{d}_t \vee \overline{e}_t$. The first clause is derived by a resolution step between $C_0$ and $C_1$. Then we derive $\overline{d}_{j+1} \vee \overline{e}_{j+1}$ from $\overline{d}_j \vee \overline{e}_j$, $C_{2j}\sigma_{v,t}$, $C_{2j+1}\sigma_{v,t}$, and the clauses obtained from $q_j \leftrightarrow \overline{d}_j \vee e_j$ as follows. Resolve $d_j \vee \overline{q}_j \vee \overline{d}_{j+1} \vee \overline{e}_{j+1}$ with $q_j \vee d_j \vee \overline{e}_j$ and derive $d_j \vee \overline{e}_j \vee \overline{d}_{j+1} \vee \overline{e}_{j+1}$ by factoring. Then continue with $\overline{d}_j \vee \overline{e}_j$ and obtain $R \colon \overline{e}_j \vee \overline{d}_{j+1} \vee \overline{e}_{j+1}$ by resolution and factoring. Use $e_j \vee q_j \vee \overline{d}_{j+1} \vee \overline{e}_{j+1}$, resolve it with $\overline{q}_j \vee e_j$ and factor the resolvent resulting in $e_j \vee \overline{d}_{j+1} \vee \overline{e}_{j+1}$. Resolve $R$ with the latter clause, factor the resolvent and obtain $\overline{d}_{j+1} \vee \overline{e}_{j+1}$.

Each of the 15 clauses has at most 5 literals. For $j + 1 = t$, we have a similar deduction but with at most $2t + 3$ literals per clause. We obtain $\overline{f}_1 \vee \cdots \vee \overline{f}_t$ which can be resolved by $f_i$ obtained from $\overline{q}_i \vee f_i$ and $q_i \vee f_i$. Finally, it is easy to check that the refutation has tree structure and is of size polynomial in $t$. □

The Herbrand functions obtained from Q-res or QU-res refutations by the method in [2] are often (too) complex. It is easy to check that atomic Herbrand functions $e_i$ for $x_i$ are sufficient and therefore a short tree IR-calc$(\cdot, \cdot, \cdot)$ refutation of $\Psi_t$ is possible. The proof of the next proposition can be found in the appendix of [10].

**Proposition 11.** *Let* $\sigma_{v,t} = \{x_i \backslash e_i \mid 1 \leq i \leq t\}$. *Then there is a tree refutation of* $\Psi_t$ *in* IR-calc$(P_t, M_t, \sigma_{v,t})$ *of size polynomial in* $t$.

**Proposition 12.** IR-calc$(\cdot, \cdot)$ *cannot polynomially simulate* IR-calc$(\cdot, \cdot, \cdot)$.

According to Proposition 11, there are not only short tree refutations of $\Psi_t$, but also the search space is limited if a simple heuristic restricting the number of possible variable replacements $\sigma_{v,t}$ is employed during proof search. The heuristic requires that for each $(x \backslash e) \in \sigma_{v,t}$, there is at least one clause $C\sigma_{v,t}$, which contain duplicate literals.

# 7 Conclusion

We studied various calculi for QBFs with respect to their relative strength. We provided polynomial simulations using first-order translations in order to clarify the possibility to employ (non-trivial) instantiations in refutations. By a simulation of Q-res and QU-res by $R_1$, we have seen that the former ones avoid instantiations. The simulation of simple instantiation-based calculi by $R_1$ revealed that instantiation of universal variables is possible by resolutions with $p(1)$ and $\neg p(0)$ together with the usual propagation of substitutions, and clarified the purpose of the employed framework of assignments and annotated clauses. We showed that enabling instantiations with existential variables and formulas increase the strength of instantiation-based calculi.

*Open problems and future research directions:* In all our comparisons, we did not optimize the quantifier prefix by (advanced) dependency schemes. It is well known that less dependencies between variables can considerably shorten proofs, for which reason one would like to integrate these techniques into calculi. We have left open some proof-theoretical comparisons like sequent systems for prenex formulas with propositional cuts and IR-calc$(\cdot, \cdot, \cdot, \cdot)$, or IRM-calc [4] with our new calculi or $R_1$. The problem here is that $R_1$ is probably not strong enough because inference rules for Skolem function manipulation [1,8] are not available but seem to be necessary for a polynomial simulation. The ultimate goal is to make instantiation-based calculi ready for proof search. A first step has been accomplished by showing (in the simulation) that unrestricted instantiations in IR-calc$(\cdot, \cdot)$ can be restricted to minimal ones by simply using unification and mgus like in the first-order case. Achieving the goal for strong cacluli is not an easy exercise because some techniques like extensions are hard to control.

# References

1. Baaz, M., Egly, U., Leitsch, A.: Normal form transformations. In: Robinson, J.A., Voronkov, A. (eds.) Handbook of Automated Reasoning, pp. 273–333. Elsevier and MIT Press, Cambridge (2001)

2. Balabanov, V., Jiang, J.-H.R.: Unified QBF certification and its applications. Formal Methods Syst. Des. **41**(1), 45–65 (2012)
3. Balabanov, V., Widl, M., Jiang, J.-H.R.: QBF resolution systems and their proof complexities. In: Sinz, C., Egly, U. (eds.) SAT 2014. LNCS, vol. 8561, pp. 154–169. Springer, Heidelberg (2014)
4. Beyersdorff, O., Chew, L., Janota, M.: On unification of QBF resolution-based calculi. In: Csuhaj-Varjú, E., Dietzfelbinger, M., Ésik, Z. (eds.) MFCS 2014, Part II. LNCS, vol. 8635, pp. 81–93. Springer, Heidelberg (2014)
5. Beyersdorff, O., Chew, L., Janota, M.: Proof complexity of resolution-based QBF calculi. In: Mayr, E.W., Ollinger, N. (eds.) 32nd International Symposium on Theoretical Aspects of Computer Science, STACS. LIPIcs, Garching, Germany, 4–7 March 2015, vol. 30, pp. 76–89. Schloss Dagstuhl - Leibniz-Zentrum fuer Informatik (2015)
6. Beyersdorff, O., Chew, L., Janota, M.: Extension variables in QBF resolution. In: AAAI 2016 Workshop Beyond NP (2016)
7. Cook, S.A., Morioka, T.: Quantified propositional calculus and a second-order theory for $NC^1$. Arch. Math. Log. **44**(6), 711–749 (2005)
8. Eder, E.: Relative complexities of first order calculi. Artificial intelligence = Künstliche Intelligenz. Vieweg (1992)
9. Egly, U.: On sequent systems and resolution for QBFs. In: Cimatti, A., Sebastiani, R. (eds.) SAT 2012. LNCS, vol. 7317, pp. 100–113. Springer, Heidelberg (2012)
10. Egly, U.: On stronger calculi for QBFs. CoRR, abs/1604.06483 (2016)
11. Egly, U., Seidl, M., Woltran, S.: A solver for QBFs in negation normal form. Constraints **14**(1), 38–79 (2009)
12. Van Gelder, A.: Contributions to the theory of practical quantified boolean formula solving. In: Milano, M. (ed.) CP 2012. LNCS, vol. 7514, pp. 647–663. Springer, Heidelberg (2012)
13. Haken, A.: The intractability of resolution. Theor. Comput. Sci. **39**, 297–308 (1985)
14. Kleine Büning, H., Karpinski, M., Flögel, A.: Resolution for quantified Boolean formulas. Inf. Comput. **117**(1), 12–18 (1995)
15. Krajíček, J.: Bounded Arithmetic, Propositional Logic, and Complexity Theory. Encyclopedia of Mathematics and its Application, vol. 60. Cambridge University Press, Cambridge (1995)
16. Leitsch, A.: The Resolution Calculus. Texts in Theoretical Computer Science. Springer, Heidelberg (1997)
17. Plaisted, D.A., Greenbaum, S.: A structure-preserving clause form translation. J. Symb. Comput. **2**(3), 293–304 (1986)
18. Seidl, M., Lonsing, F., Biere, A.: qbf2epr: A tool for generating EPR formulas from QBF. In: Fontaine, P., Schmidt, R.A., Schulz, S. (eds.) PAAR@IJCAR. EPiC Series, vol. 21, pp. 139–148. EasyChair (2012)
19. Slivovsky, F., Szeider, S.: Variable dependencies and Q-resolution. In: Sinz, C., Egly, U. (eds.) SAT 2014. LNCS, vol. 8561, pp. 269–284. Springer, Heidelberg (2014)
20. Tseitin, G.S.: On the complexity of derivation in propositional calculus. In: Slisenko, A.O. (ed.) Studies in Constructive Mathematics and Mathematical Logic, Part II, vol. 8, pp. 234–259. Seminars in Mathematics, V.A. Steklov Mathematical Institute, Leningrad (1968)
21. Zhang, L., Malik, S.: Conflict driven learning in a quantified boolean satisfiability solver. In: Pileggi, L.T., Kuehlmann, A. (eds.) Proceedings of the 2002 IEEE/ACM International Conference on Computer-aided Design, ICCAD 2002, San Jose, California, USA, 10–14 November 2002, pp. 442–449. ACM/IEEE Computer Society (2002)

# Q-Resolution with Generalized Axioms

Florian Lonsing[1]($\boxtimes$), Uwe Egly[1], and Martina Seidl[2]

[1] Knowledge-Based Systems Group, Vienna University of Technology,
Vienna, Austria
florian.lonsing@tuwien.ac.at
[2] Institute for Formal Models and Verification, JKU, Linz, Austria

**Abstract.** Q-resolution is a proof system for quantified Boolean formulas (QBFs) in prenex conjunctive normal form (PCNF) which underlies search-based QBF solvers with clause and cube learning (QCDCL). With the aim to derive and learn stronger clauses and cubes earlier in the search, we generalize the axioms of the Q-resolution calculus resulting in an exponentially more powerful proof system. The generalized axioms introduce an interface of Q-resolution to any other QBF proof system allowing for the direct combination of orthogonal solving techniques. We implemented a variant of the Q-resolution calculus with generalized axioms in the QBF solver DepQBF. As two case studies, we apply integrated SAT solving and resource-bounded QBF preprocessing during the search to heuristically detect potential axiom applications. Experiments with application benchmarks indicate a substantial performance improvement.

## 1 Introduction

In the same way as SAT, the decision problem of propositional logic, is the archetypical problem complete for the complexity class NP, QSAT, the decision problem of *quantified Boolean formulas (QBF)*, is the archetypical problem complete for the complexity class PSPACE. The fact that many important practical reasoning, verification, and synthesis problems fall into the latter complexity class (cf. [3] for an overview) strongly motivates the quest for efficient QBF solvers.

As the languages of propositional logic and QBF only marginally differ from a syntactical point of view, namely the quantifiers, it is a natural approach to take inspiration from SAT solving and lift powerful SAT techniques to QSAT. Motivated by the success of *conflict-driven clause learning* (CDCL) in SAT solving [31], a generalized version of CDCL called *conflict/solution-driven clause/cube learning* (often abbreviated by QCDCL) is applied in QSAT solving [11]. Given a propositional formula in conjunctive normal form (CNF), a CDCL-based SAT solver enriches the original CNF with clauses—already found and justified conflicts—which force the solver into a different area of the search space until either a model, i.e., a satisfying variable assignment, is found or until the CNF is proven to be

Supported by the Austrian Science Fund (FWF) under grants S11408-N23 and S11409-N23.

© Springer International Publishing Switzerland 2016
N. Creignou and D. Le Berre (Eds.): SAT 2016, LNCS 9710, pp. 435–452, 2016.
DOI: 10.1007/978-3-319-40970-2_27

unsatisfiable. If a QBF in prenex conjunctive normal form (PCNF) is unsatisfiable then QCDCL works similar, apart from technical details. In the case of satisfiability, however, it is not sufficient to find one assignment satisfying the formula. To respect the semantics of universal quantification, QBF models have to be described either by assignment trees or by Skolem functions. Hence, a QBF solver may not abort the search if a satisfying assignment is found. Dual to clause learning, a cube (a conjunction of literals) is learned and the search is resumed. QCDCL is implemented in several state-of-the-art QBF solvers [12,19,21,34].

Apart from QCDCL, orthogonal approaches to QBF solving have been developed. QBF competitions like the QBF Galleries 2013 [25] and 2014 [15] revealed the power of *expansion-based approaches* [1,6,18], which are based on a different proof system than search-based solving with QCDCL. We refer to related work [4,5] for an overview of QBF proof systems. QCDCL relies on Q-resolution [19]. Traditionally, Q-resolution calculi[1] offer two kinds of axioms with limited deductive power: (i) the *clause axiom* stating that any clause in the CNF part of a QBF can be immediately derived and (ii) the *cube axiom* allowing to derive cubes which are propositional implicants of the CNF. In previous work [22], we generalized the cube axiom such that quantified blocked clause elimination (QBCE) [8], a clause elimination procedure for preprocessing, could be tightly integrated in QCDCL for learning smaller cubes earlier in the search.

To overcome the restrictions of the traditional axioms of Q-resolution, we extend previous work [22] on the cube axiom and present more powerful clause axioms. We generalize the traditional clause and cube axioms such that their application relies on checking the satisfiability of the PCNF under the current assignment in QCDCL. This way, the axioms can be applied earlier in the search. Further, they provide a framework to combine Q-resolution with any other (complete or incomplete) QBF proof system. We implemented the generalized axioms in the QCDCL solver DepQBF. As a case study, we integrated bounded expansion and SAT-based abstraction [9] in QCDCL as incomplete QBF solving techniques to detect potential axiom applications. Experimental results indicate a substantial performance increase, particularly on application benchmarks.

This paper is structured as follows. In Sects. 2 and 3, we introduce preliminaries and recapitulate search-based QBF solving with QCDCL and traditional Q-resolution. Then we generalize the axioms of Q-resolution in Sect. 4 allowing for the integration of other proof systems. In Sect. 5 we integrate SAT-based abstraction into QCDCL. Implementation and evaluation are discussed in Sect. 6. We conclude with a summary and an outlook to future work in Sect. 7.

## 2   Preliminaries

We introduce the concepts and terminology used in the rest of the paper. A *literal* is a variable $x$ or its negation $\bar{x}$. By $\bar{l}$ we denote the negation of literal $l$ and $\mathsf{var}(l) := x$ if $l = x$ or $l = \bar{x}$. A disjunction, resp. conjunction, of literals

---

[1] Note that there are different variants of Q-resolution, e.g., long-distance resolution [34], QU-resolution [33], etc. [2,5].

is called *clause*, resp. *cube*. A propositional formula in *conjunctive normal form* (CNF) is a conjunction of clauses. If convenient, we interpret a CNF as a set of clauses, and clauses and cubes as sets of literals. A QBF in *prenex conjunctive normal form* (PCNF) has the form $\Pi.\psi$ with prefix $\Pi := Q_1 X_1 \ldots Q_n X_n$ and matrix $\psi$, where $\psi$ is a propositional CNF over the variables defined in $\Pi$. The variable sets $X_i$ are pairwise disjoint and for $Q_i \in \{\forall, \exists\}$, $Q_i \neq Q_{i+1}$. We define $\mathsf{var}(\Pi) := X_1 \cup \ldots \cup X_n$. The quantifier $\mathsf{Q}(\Pi, l)$ of a literal $l$ is $Q_i$ if $\mathsf{var}(l) \in X_i$. If $\mathsf{Q}(\Pi, l) = Q_i$ and $\mathsf{Q}(\Pi, k) = Q_j$, then $l \leq_\Pi k$ iff $i \leq j$. For a clause or cube $C$, $\mathsf{var}(C) := \{\mathsf{var}(l) \mid l \in C\}$ and for CNF $\psi$, $\mathsf{var}(\psi) := \{\mathsf{var}(l) \mid l \in C, C \in \psi\}$.

An *assignment* $A$ is a mapping from the variables $\mathsf{var}(\Pi)$ of a QBF $\Pi.\psi$ to truth values *true* and *false*. We represent $A$ as a set of literals $A = \{l_1, \ldots, l_n\}$ with $\{\mathsf{var}(l_i) \mid l_i \in A\} \subseteq \mathsf{var}(\Pi)$ such that if a variable $x$ is assigned *true* then $l_i \in A$ and $l_i = x$, and if $x$ is assigned *false* then $l_i \in A$ and $l_i = \bar{x}$. Further, for any $l_i, l_j \in A$ with $i \neq j$, $\mathsf{var}(l_i) \neq \mathsf{var}(l_j)$. An assignment $A$ is *partial* if it does not map every variable in $\mathsf{var}(\Pi)$ to a truth value, i.e., $\{\mathsf{var}(l_i) \mid l_i \in A\} \subset \mathsf{var}(\Pi)$. A QBF $\phi$ *under assignment* $A$, written as $\phi[A]$, is the QBF obtained from $\phi$ in which for all $l \in A$, all clauses containing $l$ are removed, all occurrences of $\bar{l}$ are deleted, and $\mathsf{var}(l)$ is removed from the prefix. If the matrix of $\phi[A]$ is empty, then the matrix is satisfied by $A$ and $A$ is a *satisfying assignment* (written as $\phi[A] = \mathsf{T}$). If the matrix of $\phi[A]$ contains the empty clause, then the matrix is falsified by $A$ and $A$ is a *falsifying assignment* (written as $\phi[A] = \mathsf{F}$). A QBF $\Pi.\psi$ with $Q_1 = \exists$ (resp. $Q_1 = \forall$) is satisfiable iff $\Pi.\psi[\{x\}]$ or (resp. and) $\Pi.\psi[\{\bar{x}\}]$ is satisfiable where $x \in X_1$. Two QBFs $\phi$ and $\phi'$ are *satisfiability-equivalent*, written as $\phi \equiv_{sat} \phi'$, iff $\phi$ is satisfiable whenever $\phi'$ is satisfiable. Two propositional CNFs $\psi$ and $\psi'$ are *logically equivalent*, written as $\psi \equiv \psi'$, iff they have the same set of propositional models, i.e., satisfying assignments. Two simplification rules preserving satisfiability equivalence are *unit* and *pure literal detection*. If a QBF $\phi$ contains a unit clause $C = (l)$, where $\mathsf{Q}(\Pi, l) = \exists$, then $\phi \equiv_{sat} \phi[\{l\}]$. If a literal is pure in QBF $\phi$, i.e., $\phi$ contains $l$ but not $\bar{l}$, then $\phi \equiv_{sat} \phi[\{l\}]$ if $\mathsf{Q}(\Pi, l) = \exists$ and $\phi \equiv_{sat} \phi[\{\bar{l}\}]$ otherwise.

## 3    QCDCL-Based QBF Solving

Figure 1 shows an abstract workflow of traditional search-based QBF solving with QCDCL [12,19,21,34]. Given a PCNF $\phi$, assignments $A$ are successively generated (box in top left corner of Fig. 1). In general, variables must be assigned in the ordering of the quantifier prefix. Variables may either be assigned tentatively as *decisions* or by a QBF-specific variant of *Boolean constraint propagation (QBCP)*. QBCP consists of unit and pure literal detection. Assignments of variables carried out in QBCP do not have to follow the prefix ordering. We formalize the assignments generated during a run of QCDCL as follows.

**Definition 1 (QCDCL Assignment).** *Given a QBF $\phi = \Pi.\psi$. Let assignment $A = A' \cup A''$ where $A'$ are variables assigned as decisions and $A''$ are variables assigned by unit/pure literal detection. $A$ is a QCDCL assignment if (1) for a maximal $l \in A'$ with $\forall l' \in A' : l' \leq_\Pi l$ it holds that*

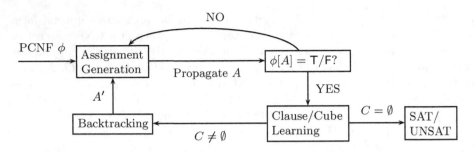

**Fig. 1.** Abstract workflow of QCDCL with traditional Q-resolution axioms.

$\forall x \in \text{var}(\Pi) <_\Pi l : x \in \text{var}(A)$ *and (2) all* $l \in A''$ *are unit/pure in* $\phi[A']$ *after applying QBCP until completion.*

QCDCL generates only QCDCL assignments by Definition 1. Assignment generation by decisions and QBCP continues until the current assignment $A$ is either falsifying or satisfying by checking whether $\phi[A] = \text{F}$ or $\phi[A] = \text{T}$ (box in top right corner of Fig. 1). In these cases, a new *learned clause* or *learned cube* is derived in a *learning* phase, which is based on the *Q-resolution calculus*.

**Definition 2 (Q-Resolution Calculus).** *Let* $\phi = \Pi.\psi$ *be a PCNF. The rules of the Q-resolution calculus (QRES) are as follows.*

$$\frac{C_1 \cup \{p\} \qquad C_2 \cup \{\bar{p}\}}{C_1 \cup C_2} \quad \begin{array}{l} \textit{if for all } x \in \Pi : \{x, \bar{x}\} \not\subseteq (C_1 \cup C_2), \\ \bar{p} \notin C_1, \ p \notin C_2, \ \textit{and either} \\ \textit{(1) } C_1, C_2 \textit{ are clauses and } \mathsf{Q}(\Pi, p) = \exists \textit{ or} \\ \textit{(2) } C_1, C_2 \textit{ are cubes and } \mathsf{Q}(\Pi, p) = \forall \end{array} \quad (res)$$

$$\frac{C \cup \{l\}}{C} \quad \begin{array}{l} \textit{if for all } x \in \Pi : \{x, \bar{x}\} \not\subseteq (C \cup \{l\}) \textit{ and either} \\ \textit{(1) } C \textit{ is a clause, } \mathsf{Q}(\Pi, l) = \forall, \\ \qquad l' <_\Pi l \textit{ for all } l' \in C \textit{ with } \mathsf{Q}(\Pi, l') = \exists \textit{ or} \\ \textit{(2) } C \textit{ is a cube, } \mathsf{Q}(\Pi, l) = \exists, \\ \qquad l' <_\Pi l \textit{ for all } l' \in C \textit{ with } \mathsf{Q}(\Pi, l') = \forall \end{array} \quad (red)$$

$$\frac{}{C} \quad \textit{if for all } x \in \Pi : \{x, \bar{x}\} \not\subseteq C, \ C \textit{ is a clause and } C \in \psi \qquad (cl\text{-}init)$$

$$\frac{}{C} \quad \begin{array}{l} A \textit{ is a QCDCL assignment,} \\ \phi[A] = \mathsf{T}, \\ \textit{and } C = (\bigwedge_{l \in A} l) \textit{ is a cube} \end{array} \qquad (cu\text{-}init)$$

QRES is a proof system which underlies QCDCL. Rule *cl-init* is an axiom to derive clauses which are already part of the given PCNF $\phi$. In practice, the clause $C$ selected by axiom *cl-init* is falsified under the current QCDCL assignment. Axiom *cu-init* allows to derive cubes based on a QCDCL assignment $A$ which satisfies all the clauses of the matrix $\psi$ of $\phi = \Pi.\psi$ (i.e., $\phi[A] = \top$). A cube $C$ derived by axiom *cu-init* is an *implicant* of $\psi$, i.e., the implication $C \Rightarrow \psi$ is valid.

The *resolution* and *reduction* rules *res* and *red*, respectively, are applied either to clauses or cubes. Rule *red* is called *universal (existential) reduction* when applied to clauses (cubes). We write $UR(C)$ $(ER(C))$ to denote the clause (cube) resulting from universal (existential) reduction of clause (cube) $C$. The PCNF $UR(\phi)$ is obtained by universal reduction of all clauses in the PCNF $\phi$.

Q-resolution of clauses [19] generalizes propositional resolution, which consists of rules *cl-init* and *res*, by the reduction rule *red*. Q-resolution of cubes was introduced for *cube learning* [12,21,34], the dual variant of clause learning.

QRES is sound and refutationally complete for PCNFs [12,19,21,34]. The empty clause (cube) is derivable from a PCNF $\phi$ in QRES if and only if $\phi$ is unsatisfiable (satisfiable). A derivation of the empty clause (cube) from $\phi$ is a *clause (cube) resolution proof* of $\phi$.

In QCDCL, the rules of QRES are applied to derive new learned clauses or cubes. A learned clause (cube) $C$ is added conjunctively (disjunctively) to the PCNF $\phi = \Pi.\psi$ to obtain $\Pi.(\psi \wedge C)$ $(\Pi.(\psi \vee C))$. After $C$ has been added, certain assignments in the current assignment $A$ are retracted during *backtracking*, resulting in assignment $A'$ ($C \neq \emptyset$ in Fig. 1). Assignment generation based on $A'$ continues, where learned clauses and cubes participate in QBCP. Typically, only *asserting* learned clauses and cubes are generated in QCDCL. A clause (cube) $C$ is asserting if $UR(C)$ $(ER(C))$ is unit under $A'$ after backtracking. QCDCL terminates if and only if the empty clause or cube is learned ($C = \emptyset$ in Fig. 1).

*Example 1 ([22]).* Given a PCNF $\phi$ with prefix $\exists z, z' \forall u \exists y$ and matrix $\psi$:

$$\psi := (u \vee \bar{y}) \wedge (\bar{u} \vee y) \wedge$$
$$(z \vee u \vee \bar{y}) \wedge (z' \vee \bar{u} \vee y) \wedge$$
$$(\bar{z} \vee \bar{u} \vee \bar{y}) \wedge (\bar{z}' \vee u \vee y)$$

$$\frac{\dfrac{(\bar{z} \wedge \bar{z}' \wedge \bar{u} \wedge \bar{y})}{(\bar{z} \wedge \bar{z}' \wedge \bar{u})} \quad \dfrac{(\bar{z} \wedge \bar{z}' \wedge u \wedge y)}{(\bar{z} \wedge \bar{z}' \wedge u)}}{\dfrac{\bar{z} \wedge \bar{z}'}{\emptyset}}$$

Let $A_1 := \{\bar{z}, \bar{z}', \bar{u}, \bar{y}\}$ and $A_2 := \{\bar{z}, \bar{z}', u, y\}$ be satisfying QCDCL assignments to be used for applications of axiom *cu-init*. A derivation of the empty cube by rules *cu-init*, *red*, *res*, and *red* (from top to bottom) is shown on the right.     ◇

## 4   Generalizing the Axioms of QRES

The axioms *cl-init* and *cu-init* of QRES have limited deductive power. Any clause derived by *cl-init* already appears in the matrix $\psi$ of the PCNF $\phi = \Pi.\psi$. Any cube derived by *cu-init* is an implicant of $\psi$.

To overcome these limitations, we equip QRES with two additional axioms— one to derive clauses and one to derive cubes—which generalize *cl-init* and *cu-init*. *Generalized model generation (GMG)* [22] was presented as a new axiom to derive learned cubes. The combination of QRES with GMG is stronger than QRES with *cu-init* in terms of the sizes of cube resolution proofs it is able

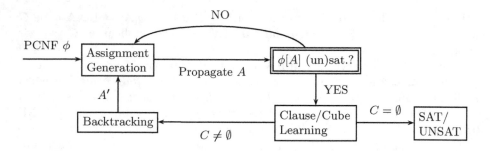

**Fig. 2.** Abstract workflow of QCDCL with generalized Q-resolution axioms.

to produce. In the following, we formulate a generalized clause axiom which we combine with QRES in addition to GMG. Thereby, we obtain a variant of QRES which is stronger than traditional QRES also in terms of sizes of clause resolution proofs.

Figure 2 shows an abstract workflow of search-based QBF solving with QCDCL relying on QRES *with* generalized axioms. This workflow is the same as in Fig. 1 except for applications of axioms (box in top right corner). The generalized axioms are applied if the PCNF $\phi[A]$ under a QCDCL assignment $A$ is *(un)satisfiable*. This is in contrast to the more restricted conditions $\phi[A] = \mathsf{T}$ or $\phi[A] = \mathsf{F}$ in Fig. 1. We show that the generalized axioms allow to combine *any* sound (but maybe incomplete) QBF solving technique with QCDCL based on QRES. First, we define *QCDCL clauses* and recapitulate *QCDCL cubes* [22].

**Definition 3 (QCDCL Clause/Cube).** *Given a QBF* $\phi = \Pi.\psi$. *The QCDCL clause* $C$ *of QCDCL assignment* $A$ *is defined by* $C = (\bigvee_{l \in A} \bar{l})$. *The QCDCL cube* $C$ *of QCDCL assignment* $A$ *is defined by* $C = (\bigwedge_{l \in A} l)$.

By Definition 1, a QCDCL clause or cube cannot contain complementary literals $x$ and $\bar{x}$ of some variable $x$. According to QCDCL assignments, we split QCDCL clauses and cubes into decision literals and literals assigned by unit and pure literal detection. Let $C$ be a QCDCL clause or QCDCL cube. Then $C = C' \cup C''$ where $C'$ is the maximal subset of $C$ such that $X_1 \cup \ldots \cup X_{i-1} \subset \mathsf{var}(C')$ and $C' \cap X_i \neq \emptyset$. The literals in $C'$ are the first $|C'|$ consecutive variables of $\Pi$ which are assigned, i.e., $C'$ contains all the variables in $C$ assigned as decisions.[2] The literals in $C''$ are assigned due to pure and unit literal detection and may occur anywhere in $\Pi$ starting from $X_{i+1}$. Further we define $\mathsf{dec}(C) = C'$ and $\mathsf{der}(C) = C''$. We review *generalized model generation* [22] as an axiom to derive cubes.

**Definition 4 (Generalized Model Generation [22]).** *Given a PCNF* $\phi$ *and a QCDCL assignment* $A$ *according to Definition 1. If* $\phi[A]$ *is satisfiable, then the QCDCL cube* $C = (\bigwedge_{l \in A} l)$ *is obtained by generalized model generation.*

---

[2] $C'$ can also contain literals assigned by pure/unit literal detection, but as they are left to the maximal decision variable in the prefix, we treat them like decision variables.

**Theorem 1 ([22]).** *Given PCNF $\phi = \Pi.\psi$ and a QCDCL cube $C$ obtained from $\phi$ by generalized model generation. Then it holds that $\Pi.\psi \equiv_{sat} \Pi.(\psi \vee C)$.*

**Corollary 1 ([22]).** *By Theorem 1, a cube $C$ obtained from PCNF $\Pi.\psi$ by generalized model generation can be used as a learned cube in QCDCL.*

Dual to generalized model generation, we define *generalized conflict generation* to derive clauses which can be added to a PCNF in a satisfiability-preserving way.

**Definition 5 (Generalized Conflict Generation).** *Given a PCNF $\phi$ and a QCDCL assignment $A$ according to Definition 1. If $\phi[A]$ is unsatisfiable, then the QCDCL clause $C = (\bigvee_{l \in A} \bar{l})$ is obtained by* generalized conflict generation.

**Theorem 2.** *Given PCNF $\phi = \Pi.\psi$ and a QCDCL clause $C$ obtained from $\phi$ by generalized conflict generation using QCDCL assignment $A$. Then it holds that $\Pi.\psi \equiv_{sat} \Pi.(\psi \wedge C)$.*

*Proof (Sketch).* We argue that if $\Pi.\psi$ is satisfiable, so is $\Pi.(\psi \wedge C)$. The case for unsatisfiability is trivial. Let $C = C' \cup C''$ with $C' = \mathsf{dec}(C)$ and $C'' = \mathsf{der}(C)$. Further, let $A = A' \cup A''$ such that $\mathsf{var}(A') = \mathsf{var}(C')$ and $\mathsf{var}(A'') = \mathsf{var}(C'')$. Now assume that $\Pi.\psi$ is satisfiable, but $\Pi.(\psi \wedge C)$ is not. In order to falsify $C$, its subclause $C'$ has to be falsified, i.e., the first $|C'|$ variables of $\Pi$ have to be set according to $A'$. Then, due to pure and unit, also $C''$ is falsified, and therefore, each assignment falsifying $C$ has to contain $A$. But $\Pi.\psi[A]$ is unsatisfiable. Since $\Pi.\psi$ is satisfiable, there have to be other decisions than the decisions of $A$ to show its satisfiability, but these also satisfy $\Pi.(\psi \wedge C)$. $\qquad\qquad\square$

**Corollary 2.** *By Theorem 2, a clause $C$ obtained from PCNF $\Pi.\psi$ by generalized conflict generation can be used as a learned clause in QCDCL.*

Based on Corollaries 1 and 2, we formulate axioms to derive learned clauses (cubes) from QCDCL assignments $A$ under which the PCNF $\phi$ is (un)satisfiable.

**Definition 6 (Generalized Axioms).** *Let $\phi = \Pi.\psi$ be a PCNF. The generalized clause* and *cube axioms are as follows.*

$$\frac{}{C} \quad \begin{array}{l} A \text{ is a QCDCL assignment,} \\ \phi[A] \text{ is unsatisfiable,} \\ \text{and } C = (\bigvee_{l \in A} \bar{l}) \text{ is a QCDCL clause} \end{array} \qquad (\textit{gen-cl-init})$$

$$\frac{}{C} \quad \begin{array}{l} A \text{ is a QCDCL assignment,} \\ \phi[A] \text{ is satisfiable,} \\ \text{and } C = (\bigwedge_{l \in A} l) \text{ is a QCDCL cube} \end{array} \qquad (\textit{gen-cu-init})$$

The generalized axioms *gen-cl-init* and *gen-cu-init* are added to QRES in addition to the traditional axioms *cl-init* and *cu-init* from Definition 2.

*Example 2.* Consider the PCNF from Example 1. Let $A := \{\bar{z}, \bar{z}'\}$ be a QCDCL assignment where $z$ and $z'$ are assigned as decisions. The PCNF $\phi[A] = \forall u \exists y.(u \vee \bar{y}) \wedge (\bar{u} \vee y)$ is satisfiable. We apply axiom *gen-cu-init* to derive the cube $C := (\bar{z} \wedge \bar{z}')$ and finally the empty cube $ER(C) = \emptyset$ (proof shown on the right).

$$\frac{\bar{z} \wedge \bar{z}'}{\emptyset}$$

$\diamond$

In contrast to axioms *cl-init* and *cu-init* (the latter corresponds to *model generation* [12]), the generalized axioms allow to derive clauses that are not part of the given PCNF $\phi$ and cubes that are not implicants of the matrix of $\phi$.

Given the empty assignment $A = \{\}$ and a PCNF $\phi$, the empty clause or cube can be derived using $A$ by axioms *gen-cl-init* or *gen-cu-init* right away if $\phi[A]$ is unsatisfiable or satisfiable, respectively. However, checking the satisfiability of the PCNF $\phi[A]$ as required in the side conditions of the generalized axioms is PSPACE-complete. Therefore, *in practice* it is necessary to consider non-empty QCDCL assignments $A$ and apply either complete approaches in a bounded way, like the successful expansion-based approaches [1,6,13,18], or incomplete polynomial-time procedures, e.g., as used in preprocessing [13], to check the satisfiability of $\phi[A]$. *Sign abstraction* [21] can be regarded as a first approach towards more powerful cube learning as formalized by axiom *gen-cu-init*.

Axioms *gen-cl-init* and *gen-cu-init* provide a formal framework for combining Q-resolution in QRES with *any* QBF decision procedure $\mathcal{D}$ by using $\mathcal{D}$ to check $\phi[A]$. This framework also applies to related combinations of search-based QBF solving with variable elimination [27]. Regarding proof complexity, decision procedures like expansion and Q-resolution are incomparable as the lengths of proofs they are able to produce for certain PCNFs differ by an exponential factor [2,5,16]. Due to this property, the combination of incomparable procedures in QRES via the generalized axioms allows to benefit from their individual strengths. For example, the use of expansion to check the satisfiability of $\phi[A]$ in axioms *gen-cl-init* and *gen-cu-init* results in a variant of QRES which is exponentially stronger than traditional QRES. For satisfiable PCNFs, QBCE, originally a preprocessing technique to eliminate redundant clauses in a PCNF, was shown to be effective to solve $\phi[A]$ for applications of axiom *gen-cu-init* [22], resulting in an exponentially stronger cube proof system.

If a decision procedure $\mathcal{D}$ is applied as a black box to check $\phi[A]$, then QRES extended by *gen-cl-init* and *gen-cu-init* is not a proof system as defined by Cook and Reckhow [10] because the final proof $P$ of $\phi$ cannot be checked in polynomial time. However, $\mathcal{D}$ can be augmented to return a proof $P'$ of $\phi[A]$ for every application of *gen-cl-init* and *gen-cu-init*. Such proof $P'$ may be formulated, e.g., in the QRAT proof system [14]. Finally, the proof $P$ of $\phi$ contains subproofs $P'$, all of which can be checked in polynomial time, like $P$ itself (the size of $P$ may blow up exponentially in the worst case depending on the decision procedures that are used to produce the subproofs $P'$).

The QCDCL framework (Fig. 2) readily supports applications of the generalized axioms *gen-cl-init* and *gen-cu-init*. A clause (resp. cube) $C$ derived by these axioms is first reduced by universal (resp. existential) reduction to obtain a reduced clause (cube) $C' \subseteq C$. Then $C'$ is used to derive an asserting learned clause (cube) *in the same way* as in clause learning by traditional QRES (Definition 2).

# 5  An Abstraction-Based Clause Axiom

Axioms *gen-cl-init* and *gen-cu-init* by Definition 6 are based on QCDCL assignments, where decision variables have to be assigned in prefix ordering. To overcome the order restriction, we introduce a clause axiom which allows to derive clauses based on an abstraction of a PCNF and *arbitrary* assignments.

**Definition 7 (Existential Abstraction).** *Let $\phi = \Pi.\psi$ be a PCNF with prefix $\Pi := Q_1 X_1 Q_2 X_2 \ldots Q_n X_n$ and matrix $\psi$. The existential abstraction $Abs_\exists(\phi) := \Pi'.\psi$ of $\phi$ has prefix $\Pi' := \exists(X_1 \cup X_2 \cup \ldots \cup X_n)$.*

**Lemma 1.** *Let $\phi = \Pi.\psi$ be a PCNF, $Abs_\exists(\phi)$ its existential abstraction, and $A$ a partial assignment of the variables in $Abs_\exists(\phi)$. If $Abs_\exists(\phi)[A]$ is unsatisfiable then $\psi \equiv \psi \wedge (\bigvee_{l \in A} \bar{l})$.*

*Proof.* Obviously, every model $M$ of $\psi \wedge (\bigvee_{l \in A} \bar{l})$ is also a model of $\psi$. To show the other direction, let $M$ be a model of $\psi$, but $(\psi \wedge (\bigvee_{l \in A} \bar{l}))[M] = \mathsf{F}$. Then $A \subseteq M$. Since $Abs_\exists(\phi)[A]$ is unsatisfiable, also $\psi[A]$ is unsatisfiable. Then $M$ cannot be a model of $\psi$. □

**Theorem 3 (cf. [29,30]).** *For a PCNF $\phi = \Pi.\psi$, $Abs_\exists(\phi)$ its existential abstraction, and a partial assignment $A$ of the variables in $Abs_\exists(\phi)$ such that $Abs_\exists(\phi)[A]$ is unsatisfiable, it holds that $\Pi.\psi \equiv_{sat} \Pi.(\psi \wedge (\bigvee_{l \in A} \bar{l}))$.*

*Proof.* By Lemma 1, $\psi$ and $\psi \wedge (\bigvee_{l \in A} \bar{l})$ have the same sets of propositional models. As argued in the context of SAT-based QBF solving [29] and QBF preprocessing [30], model-preserving manipulations of the matrix of a PCNF result in a satisfiability-equivalent PCNF.[3] □

**Definition 8 (Abstraction-Based Conflict Generation).** *Given a PCNF $\phi$, its existential abstraction $Abs_\exists(\phi)$ and an assignment $A$ (not necessarily being a QCDCL assignment). If $Abs_\exists(\phi)[A]$ is unsatisfiable, then the clause $C = (\bigvee_{l \in A} \bar{l})$ is obtained by* abstraction-based conflict generation.

We formulate a new axiom to derive clauses by abstraction-based conflict generation, which can be used as ordinary learned clauses in QCDCL (Theorem 3).

**Definition 9 (Abstraction-Based Clause Axiom).** *For a PCNF $\phi = \Pi.\psi$ and $Abs_\exists(\phi)$ by Definition 7, the abstraction-based clause axiom is as follows:*

$$\frac{}{C} \quad \begin{array}{l} A \text{ is an assignment,} \\ Abs_\exists(\phi)[A] \text{ is unsatisfiable,} \\ and\ C = (\bigvee_{l \in A} \bar{l})\ is\ a\ clause \end{array} \qquad (abs\text{-}cl\text{-}init)$$

---

[3] In fact, a stronger result is proved in [30]: model-preserving manipulations of the matrix of a PCNF result in a PCNF having the same set of *tree-like QBF models*.

Axiom *abs-cl-init* can be added to QRES in addition to all the other axioms. In the side condition of axiom *abs-cl-init*, the propositional CNF $Abs_\exists(\phi)[A]$ has to be solved, which naturally can be carried out by integrating a SAT solver in QCDCL. SAT solving has been applied in the context of QCDCL to derive learned clauses [29] and to overcome the ordering of the prefix of a PCNF. Further, many QBF solvers rely on SAT solving [17,18,26,32]. Integrating axiom *abs-cl-init* in QRES by Definition 2 results in a variant of QRES which is exponentially stronger than traditional QRES, as illustrated by the following example.

*Example 3.* Consider the following family $(\phi_t)_{t \geq 1}$ of PCNFs defined by Kleine Büning et al. [19]. A formula $\phi_t$ in $(\phi_t)_{t \geq 1}$ has the quantifier prefix

$$\exists d_0 d_1 e_1 \forall x_1 \exists d_2 e_2 \forall x_2 \exists d_3 e_3 \ldots \forall x_{t-1} \exists d_t e_t \forall x_t \exists f_1 \ldots f_t$$

and a matrix consisting of the following clauses:

$$
\begin{array}{llll}
C_0 & := \overline{d}_0 & C_1 & := d_0 \vee \overline{d}_1 \vee \overline{e}_1 \\
C_{2j} & := d_j \vee \overline{x}_j \vee \overline{d}_{j+1} \vee \overline{e}_{j+1} & C_{2j+1} := e_j \vee x_j \vee \overline{d}_{j+1} \vee \overline{e}_{j+1} & \text{for } 1 \leq j < t \\
C_{2t} & := d_t \vee \overline{x}_t \vee \overline{f}_1 \vee \ldots \vee \overline{f}_t & C_{2t+1} := e_t \vee x_t \vee \overline{f}_1 \vee \ldots \vee \overline{f}_t & \\
B_{2j-1} := x_j \vee f_j & B_{2j} & := \overline{x}_j \vee f_j & \text{for } 1 \leq j \leq t
\end{array}
$$

The size of every clause resolution proof of $\phi_t$ in traditional QRES (Definition 2) is exponential in $t$ [5,19]. We show that QRES with axiom *abs-cl-init* allows to generate proofs of $\phi_t$ which are polynomial in $t$. To this end, we apply *abs-cl-init* to derive unit clauses $(f_j)$ for all existential variables $f_j$ in $\phi_t$ using assignments $A := \{\overline{f}_j\}$, respectively. Since $Abs_\exists(\phi_t)[A]$ contains complementary unit clauses $(x_j)$ and $(\overline{x}_j)$ resulting from the clauses $B_{2j-1}$ and $B_{2j}$ in $\phi_t$, the unsatisfiability of $Abs_\exists(\phi_t)[A]$ can be determined in polynomial time without invoking a SAT solver. The derived unit clauses $(f_j)$ are resolved with clauses $C_{2t}$ and $C_{2t+1}$ to produce further unit clauses $(d_t)$ and $(e_t)$ after universal reduction. This process continues with $C_{2j}$ and $C_{2j+1}$ until the empty clause is derived using $C_0$ and $C_1$.                                                                          ◇

Abstraction-based failed literal detection [23], where certain universal quantifiers of a PCNF are treated as existential ones, implicitly relies on *QU-resolution*. QU-resolution allows universal variables as pivots in rule *res* and can generate the same proofs of $(\phi_t)_{t \geq 1}$ as in Example 3 [33]. Applying axiom *abs-cl-init* for clause learning in QCDCL harnesses the power of SAT solving. Furthermore, the combination of QRES (Definition 2) and *abs-cl-init* polynomially simulates[4] QU-resolution, which has not been applied systematically to learn clauses in QCDCL. Like with the axioms *gen-cl-init* and *gen-cu-init*, clauses derived by axiom *abs-cl-init* can readily be used to derive asserting learned clauses in QCDCL.

---

[4] We refer to an appendix of this paper with additional results [24].

# 6   Case Study and Experiments

DepQBF[5] is a QCDCL-based QBF solver implementing the Q-resolution calculus as in Definition 2. Since version 5.0, DepQBF additionally applies the generalized cube axiom *gen-cu-init* based on dynamic blocked clause elimination (QBCE) [22]. The case where QBCE reduces the PCNF $\phi[A]$ under the current assignment $A$ to the empty formula constitutes a successful application of axiom *gen-cu-init*. DepQBF comes with a sophisticated analysis of variable dependencies in a PCNF [28] to relax their linear prefix ordering. However, we disabled dependency analysis to focus the evaluation on axiom applications. In the following, we evaluate the impact of (combinations of) the generalized axioms *gen-cl-init* and *gen-cu-init* and the abstraction-based clause axiom *abs-cl-init* in practice.

## 6.1   Axiom Applications in Practice

In DepQBF, we attempt to apply the generalized axioms after QBCP has saturated in QCDCL, i.e., before assigning a variable as decision. We integrated the preprocessor Bloqqer [8] to detect applications of *gen-cl-init* and *gen-cu-init*. Bloqqer implements techniques such as equivalence reasoning, variable elimination, (variants of) QBCE, and expansion of universal variables. Since these techniques are applied in bounded fashion, Bloqqer can be regarded as an incomplete QBF solver. If the PCNF $\phi[A]$ is satisfiable (unsatisfiable) and Bloqqer solves it, then a QCDCL cube (clause) is generated by axiom *gen-cu-init* (*gen-cl-init*), which is used to derive a learned cube (clause). Otherwise, QCDCL proceeds as usual with assigning a decision variable. Bloqqer is explicitly provided with the entire PCNF $\phi[A]$ before each call. To limit the resulting run time overhead in practice, Bloqqer is called in intervals of $2^n$ decisions, where $n := 11$ in our experiments. Further, Bloqqer is never called on PCNFs with more than 500,000 original clauses, and it is disabled at run time if the average time spent to complete a call exceeds 0.125 s.

To detect applications of the abstraction-based clause axiom *abs-cl-init*, we use the SAT solver PicoSAT [7] to check the satisfiability of the existential abstraction $Abs_\exists(\phi)[A]$ of the PCNF $\phi = \Pi.\psi$ under the current QCDCL assignment $A$. The matrix $\psi$ is imported to PicoSAT once before the entire solving process starts. For each check of $Abs_\exists(\phi)[A]$, the QCDCL assignment $A$ is passed to PicoSAT via assumptions, and PicoSAT is called incrementally. If $Abs_\exists(\phi)[A]$ is unsatisfiable, then we try to minimize the size of $A$ by extracting the set $A' \subseteq A$ of *failed assumptions*. Failed assumptions are those assumptions that were relevant for the SAT solver to determine the unsatisfiability of $Abs_\exists(\phi)[A]$. Note that in general $A'$ is not a QCDCL assignment. It holds that $Abs_\exists(\phi)[A']$ is unsatisfiable and hence we derive the clause $C = (\bigvee_{l \in A'} \bar{l})$ by axiom *abs-cl-init*.

In addition to Bloqqer and dynamic QBCE (which is part of DepQBF 5.0 [22]) used to detect applications of the generalized cube axiom *gen-cu-init*, we implemented a *trivial truth* [9] test based on the following abstraction.

---
[5] DepQBF is free software: http://lonsing.github.io/depqbf/.

**Definition 10 (Universal Literal Abstraction, cf. *Trivial Truth* [9]).**
*Let $\phi = \Pi.\psi$ be a PCNF. The universal literal abstraction $Abs_\forall(\phi) := \Pi'.\psi'$ of $\phi$ is obtained by removing all universal literals from all the clauses in $\psi$ and by removing all universal variables and universal quantifiers from $\Pi$.*

**Lemma 2 ([9]).** *For a PCNF $\phi = \Pi.\psi$, $Abs_\forall(\phi)$, and a QCDCL assignment $A$ of variables in $Abs_\forall(\phi)$: if $Abs_\forall(\phi)[A]$ is satisfiable, then $\phi[A]$ is satisfiable.*

By Lemma 2, we can check the side condition of axiom *gen-cu-init* whether $\phi[A]$ is satisfiable under a QCDCL assignment $A$ by checking whether $Abs_\forall(\phi)[A]$ is satisfiable. To this end, we use a second instance of PicoSAT. Note that while Definition 10 corresponds to trivial truth, the existential abstraction (Definition 7) corresponds to *trivial falsity* [9]. Hence by axiom applications, we apply trivial truth and falsity, which originate from purely search-based QBF solving without learning, to derive clauses and cubes in QCDCL.

Like Bloqqer, we call the two instances of PicoSAT to detect applications of *abs-cl-init* and *gen-cu-init* in QCDCL before assigning a decision variable. PicoSAT is called in intervals of $2^m$ decisions, where $m := 10$. PicoSAT is never called on PCNFs with more than 500,000 original clauses, and it is disabled at run time if the average time spent to complete a call exceeds five seconds.

## 6.2  Experimental Results

The integration of Bloqqer and SAT solving to detect axiom applications results in several variants of DepQBF. We use the letter code "DQ-{nQ|B|A|T}" to label the variants, where "DQ" represents DepQBF 5.0 with dynamic QBCE used for axiom *gen-cu-init* [22]. Variant "nQ" indicates that dynamic QBCE is disabled. Letters, "B", "A", and "T" represent the additional application of Bloqqer for axioms *gen-cl-init* and *gen-cu-init*, SAT solving to check the existential abstraction for axiom *abs-cl-init*, and SAT solving to carry out the trivial truth test for *gen-cu-init*, respectively.

For the empirical evaluation, we used the original benchmark sets from the QBF Gallery 2014 [15][6] preprocessing track (243 instances), QBFLIB track (276 instances), and applications track (735 instances). We compare the variants of DepQBF to RAReQS [18] and GhostQ [20], which showed top performance in the QBF Gallery 2014, and to the recent solvers CAQE [26][7], QESTO [17], and QELL [32]. We tested QELL with (QELL-c) and without (QELL-nc) exploiting circuit information and show only the results of the better variant of the two in terms of solved instances. All experiments reported in the following were run on an AMD Opteron 6238 at 2.6 GHz under 64-bit Ubuntu Linux 12.04 with time and memory limits of 1800 s and 7 GB, respectively.

Tables 1, 2 and 3 illustrate solver performance by solved instances and total wall clock time. For DepQBF, the variant where only dynamic QBCE is applied

---

[6] http://qbf.satisfiability.org/gallery/.

[7] The authors [26] provided us with an updated version which we used in our tests.

**Table 1.** Preprocessing track. Solved instances ($\#T$), solved unsatisfiable ($\#U$) and satisfiable ones ($\#S$), and total wall clock time in seconds including time outs.

| Solver | $\#T$ | $\#U$ | $\#S$ | Time |
|---|---|---|---|---|
| RAReQS | 107 | 44 | 63 | 255K |
| DQ-nQAT | 105 | 46 | 59 | 266K |
| QESTO | 104 | 46 | 58 | 267K |
| DQ-nQ | 101 | 44 | 57 | 271K |
| DQ-AT | 99 | 45 | 54 | 273K |
| DQ-BAT | 98 | 43 | 55 | 276K |
| DQ | 95 | 43 | 52 | 278K |
| DQ-A | 95 | 44 | 51 | 280K |
| DQ-T | 94 | 41 | 53 | 278K |
| DQ-B | 94 | 42 | 52 | 284K |
| QELL-c | 87 | 34 | 53 | 290K |
| CAQE | 74 | 24 | 50 | 319K |
| GhostQ | 61 | 18 | 43 | 338K |

**Table 2.** QBFLIB track. Same column headers as Table 1.

| Solver | $\#T$ | $\#U$ | $\#S$ | Time |
|---|---|---|---|---|
| GhostQ | 139 | 62 | 77 | 265K |
| DQ-AT | 110 | 58 | 52 | 314K |
| DQ-BAT | 109 | 56 | 53 | 314K |
| DQ-T | 108 | 56 | 52 | 318K |
| QELL-c | 106 | 48 | 58 | 320K |
| DQ-A | 106 | 58 | 48 | 321K |
| DQ | 105 | 57 | 48 | 326K |
| DQ-B | 104 | 56 | 48 | 326K |
| DQ-nQAT | 88 | 49 | 39 | 352K |
| DQ-nQ | 82 | 44 | 38 | 362K |
| RAReQS | 80 | 47 | 33 | 361K |
| QESTO | 73 | 46 | 27 | 378K |
| CAQE | 53 | 32 | 21 | 406K |

(DQ) is the baseline of the comparison. In the QBFLIB (Table 2) and applications track (Table 3), DepQBF with Bloqqer and SAT solving for axioms *gen-cl-init*, *gen-cu-init*, and *abs-cl-init* solves substantially more instances than DQ.

Disabling dynamic QBCE used for axiom *gen-cu-init* (variants with "nQ" in the tables) results in a considerable performance decrease, except in the preprocessing track (Table 1). There, dynamic QBCE is harmful to the performance. We attribute this phenomenon to massive preprocessing, after which QBCE does not pay off. However, SAT solving for axioms *gen-cl-init* and *gen-cu-init* is crucial as the variant DQ-nQAT outperforms DQ-nQ without SAT solving.

In general, combinations of dynamic QBCE, Bloqqer, and SAT solving (for solving the existential abstraction and for testing trivial truth) outperform variants where only one of these techniques is applied. Examples are DQ-AT, DQ-A and DQ-T in Table 2 and DQ-BAT, DQ-B, DQ-A, and DQ-T in Table 3. The results in the applications track are most pronounced, where six out of eight variants of DepQBF outperform the other solvers (Fig. 3 shows a related cactus plot of the run times). In the following we focus on the applications track.

Consider the best performing variant DQ-BAT in Table 3. Table 4 shows statistics on the number of attempted and successful applications of axioms *gen-cl-init*, *gen-cu-init* and *abs-cl-init* by Bloqqer and SAT solving. On the 466 instances solved by DQ-BAT, Bloqqer was called on $\phi[A]$ at least once on 185 instances and successfully solved $\phi[A]$ at least once on 184 instances, thus allowing applications of axiom *gen-cl-init* or *gen-cu-init*. Bloqqer was disabled at run time on 143 instances due to the predefined limits. SAT solving for the trivial truth test for *gen-cu-init* (respectively, to solve the existential abstraction for *abs-cl-init*) was applied at least once on 364 (445) instances, was successful at least once on 177 (226) instances, and was disabled at run time on 21 (70) instances. While Bloqqer is applied less frequently than SAT solving by a factor

**Table 3.** Applications track. Same column headers as Table 1.

| Solver | #T | #U | #S | Time |
|--------|-----|-----|-----|------|
| DQ-BAT | 466 | 236 | 230 | 553K |
| DQ-AT | 461 | 234 | 227 | 555K |
| DQ-A | 459 | 237 | 222 | 561K |
| DQ-B | 449 | 222 | 227 | 563K |
| DQ-T | 441 | 220 | 221 | 571K |
| DQ | 441 | 224 | 217 | 575K |
| QELL-nc | 434 | 302 | 132 | 563K |
| RAReQS | 414 | 272 | 142 | 611K |
| CAQE | 370 | 192 | 178 | 708K |
| GhostQ | 347 | 166 | 181 | 752K |
| QESTO | 331 | 188 | 143 | 767K |
| DQ-nQBAT | 293 | 140 | 153 | 848K |
| DQ-nQ | 279 | 127 | 152 | 880K |

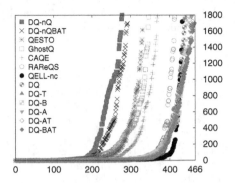

**Fig. 3.** Sorted run times (y-axis) of instances (x-axis) related to Table 3. (Color figure online)

of two, applications of Bloqqer have much higher success rates (97 %) than SAT solving (8 % and 22 %).

In the following, we analyze applications of the abstraction-based clause axiom in more detail. The extraction of failed assumptions in SAT solving for *abs-cl-init* allows to reduce the size of the clauses learned by abstraction-based conflict generation. On 145 instances solved by DQ-BAT (Table 3), axiom *abs-cl-init* was applied more than once. Per instance, on average (median) 3,336K (70.7K) assumptions were passed to the SAT solver when solving $Abs_\exists(\phi)[A]$, 28.8K (2.3K) failed assumptions were extracted, and the clauses finally learned had 20.7K (1.5K) literals. The difference in the number of failed assumptions and the size of learned clauses is due to additional, heuristic minimization of the set of failed assumptions which we apply. Given that $Abs_\exists(\phi)[A]$ is unsatisfiable, it may be possible to remove assignments from $A$, thus resulting in a smaller assignment $A'$, while preserving unsatisfiability of $Abs_\exists(\phi)[A']$. Additionally, universal reduction by rule *red* may remove literals from the clause learned by generalized conflict generation. Figure 4 shows related average statistics.

The abstraction-based clause axiom *abs-cl-init* is particularly effective on instances from the domain of conformant planning. With variant DQ-BAT (Table 3), 81 unsatisfiable instances from conformant planning were solved by a *single* application of axiom *abs-cl-init* where the empty clause was derived immediately. On 13 of these 81 instances, solving $Abs_\exists(\phi)$ was hard for the SAT solver, which took more than 900 s. In contrast to DQ-BAT, DQ does not use axiom *abs-cl-init* and failed to solve 15 of the 81 instances.

Additionally, we evaluated the variants of DepQBF and the other solvers on the benchmarks of the applications and QBFLIB tracks *with* preprocessing

**Table 4.** Related to variant DQ-BAT in Table 3: statistics on applications of Bloqqer (B), SAT solving for *abs-cl-init* (A), and SAT solving to test trivial truth for *gen-cu-init* (T) with respect to total solved instances ($\#T$) and solved satisfiable ($\#S$) and unsatisfiable ones ($\#U$).

|            | $\#T$   | $\#S$   | $\#U$   |
|------------|---------|---------|---------|
| B tried:   | 18,559  | 12,052  | 6,507   |
| B success: | 18,150  | 11,946  | 6,204   |
| B sat:     | 10,917  | 10,405  | 512     |
| B unsat:   | 7,233   | 1,541   | 5,692   |
| T tried:   | 241,180 | 88,623  | 152,557 |
| T success: | 20,494  | 19,276  | 1,218   |
| A tried:   | 301,652 | 122,929 | 178,723 |
| A success: | 67,129  | 34,306  | 32,823  |

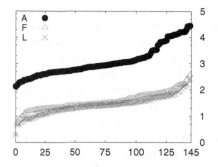

**Fig. 4.** Average SAT solver assumptions per successful application of *abs-cl-init* ("A") on 145 selected instances solved by DQ-BAT (Table 3), failed assumptions ("F"), and literals in the clauses learned by *abs-cl-init* ("L"), $log_{10}$ scale on y-axis.

by Bloqqer before solving.[8] In the QBFLIB track, RAReQS and DQ-T solved the largest number of instances (134 in total each instead of 80 and 108 in Table 2). However, here it is important to remark that already the plain variant DepQBF solved 132 instances if Bloqqer is applied before solving. With partial preprocessing by Bloqqer (using only QBCE and universal expansion), on the applications track QELL-nc and DQ-AT each solved 483 instances, i.e., 49 and 22 more instances than without preprocessing (Table 3). Note that the best variant DQ-BAT of DepQBF in Table 3 solved 480 instances. Partial preprocessing increases the number of instances solved by the variants of DepQBF. In contrast to that, with full preprocessing the performance of the variants of DepQBF on the applications track considerably decreases. If Bloqqer is applied to the full extent (enabling all techniques), then RAReQS, QELL-nc, and QESTO solve 547, 501, and 463 instances, respectively. The variant DQ-AT of DepQBF, however, which solved 483 instances with partial preprocessing, solves only 434 instances. The phenomenon that preprocessing is not always beneficial was also observed in the QBF Galleries [15,25]. When applied without restrictions, Bloqqer rewrites a formula and thus destroys or blurs structural information. For some approaches structural information is essential to fully exploit their individual strengths.

## 7  Conclusion

The Q-resolution calculus QRES is a proof system which underlies clause and cube learning in QCDCL-based QBF solvers. In QCDCL, the traditional axioms of QRES either select clauses which already appear in the input PCNF $\phi$ or construct cubes which are implicants of the matrix of $\phi$.

---

[8] We refer to an appendix of this paper with additional tables [24].

To overcome the limited deductive power of the traditional axioms, we presented two generalized axioms to derive clauses and cubes based on checking the satisfiability of $\phi$ under an assignment $A$ generated in QCDCL. We also formulated a new axiom to derive clauses which relies on an existential abstraction of $\phi$ and on SAT solving. This abstraction-based axiom leverages QU-resolution and allows to overcome the prefix order restriction in QCDCL to some extent. The new axioms can be integrated in QRES and used for clause and cube learning in the QCDCL framework. They are compatible with any variant of Q-resolution, like long-distance resolution [35], QU-resolution [33], and combinations thereof [2].

For axiom applications in practice, *any* complete or incomplete QBF decision procedure can be applied to check the satisfiability of $\phi$ under assignment $A$. In this respect, the generalized axioms act as an interface to combining Q-resolution with other QBF decision procedures in QRES. The combination of orthogonal techniques like expansion via the generalized axioms results in variants of QRES which are stronger than traditional QRES with respect to proof complexity. A proof $P$ produced by such variants of QRES can be checked in time which is polynomial in the size of $P$ if subproofs of all clauses and cubes derived by the generalized axioms are provided by the QBF decision procedures.

In order to demonstrate the effectiveness of the newly introduced axioms, we made case studies using the QCDCL solver DepQBF. We applied the preprocessor Bloqqer and SAT solving as incomplete QBF decision procedures in DepQBF to detect axiom applications. Overall, our experiments showed a considerable performance improvement of QCDCL, particularly on application instances.

As future work, it would be interesting to integrate techniques like expansion-based QBF solving more tightly in QCDCL than what we achieved with Bloqqer in our case study. A tighter integration would allow to reduce the run time overhead we observed in practice. Further research directions include axiom applications based on different QBF solving techniques in parallel QCDCL, and potential relaxations of the prefix order in assignments used for axiom applications.

# References

1. Ayari, A., Basin, D.A.: QUBOS: Deciding quantified boolean logic using propositional satisfiability solvers. In: Aagaard, M.D., O'Leary, J.W. (eds.) FMCAD 2002. LNCS, vol. 2517, pp. 187–201. Springer, Heidelberg (2002)
2. Balabanov, V., Widl, M., Jiang, J.-H.R.: QBF resolution systems and their proof complexities. In: Sinz, C., Egly, U. (eds.) SAT 2014. LNCS, vol. 8561, pp. 154–169. Springer, Heidelberg (2014)
3. Benedetti, M., Mangassarian, H.: QBF-based formal verification: Experience and perspectives. JSAT **5**(1–4), 133–191 (2008)
4. Beyersdorff, O., Chew, L., Janota, M.: On unification of QBF resolution-based calculi. In: Csuhaj-Varjú, E., Dietzfelbinger, M., Ésik, Z. (eds.) MFCS 2014, Part II. LNCS, vol. 8635, pp. 81–93. Springer, Heidelberg (2014)
5. Beyersdorff, O., Chew, L., Janota, M.: Proof complexity of resolution-based QBF calculi. In: STACS. Leibniz International Proceedings in Informatics (LIPIcs), vol. 30, pp. 76–89. Schloss Dagstuhl-Leibniz-Zentrum fuer Informatik (2015)

6. Biere, A.: Resolve and expand. In: H. Hoos, H., Mitchell, D.G. (eds.) SAT 2004. LNCS, vol. 3542, pp. 59–70. Springer, Heidelberg (2005)

7. Biere, A.: PicoSAT essentials. JSAT **4**(2–4), 75–97 (2008)

8. Biere, A., Lonsing, F., Seidl, M.: Blocked clause elimination for QBF. In: Bjørner, N., Sofronie-Stokkermans, V. (eds.) CADE 2011. LNCS, vol. 6803, pp. 101–115. Springer, Heidelberg (2011)

9. Cadoli, M., Giovanardi, A., Schaerf, M.: An algorithm to evaluate quantified boolean formulae. In: AAAI, pp. 262–267. AAAI Press/The MIT Press (1998)

10. Cook, S.A., Reckhow, R.A.: The relative efficiency of propositional proof systems. J. Symbolic Logic **44**(1), 36–50 (1979)

11. Giunchiglia, E., Marin, P., Narizzano, M.: Reasoning with quantified boolean formulas. In: Biere, A., Heule, M., Van Maaren, H., Walsh, T. (eds.) Handbook of Satisfiability. FAIA, vol. 185, pp. 761–780. IOS Press, Amsterdam (2009)

12. Giunchiglia, E., Narizzano, M., Tacchella, A.: Clause/term resolution and learning in the evaluation of quantified boolean formulas. JAIR **26**, 371–416 (2006)

13. Heule, M., Järvisalo, M., Lonsing, F., Seidl, M., Biere, A.: Clause elimination for SAT and QSAT. JAIR **53**, 127–168 (2015)

14. Heule, M.J.H., Seidl, M., Biere, A.: A unified proof system for QBF preprocessing. In: Demri, S., Kapur, D., Weidenbach, C. (eds.) IJCAR 2014. LNCS, vol. 8562, pp. 91–106. Springer, Heidelberg (2014)

15. Janota, M., Jordan, C., Klieber, W., Lonsing, F., Seidl, M., Van Gelder, A.: The QBF Gallery 2014: The QBF competition at the FLoC olympic games. JSAT **9**, 187–206 (2015)

16. Janota, M., Marques-Silva, J.: Expansion-based QBF solving versus Q-resolution. Theor. Comput. Sci. **577**, 25–42 (2015)

17. Janota, M., Marques-Silva, J.: Solving QBF by clause selection. In: IJCAI, pp. 325–331. AAAI Press (2015)

18. Janota, M., Klieber, W., Marques-Silva, J., Clarke, E.: Solving QBF with counterexample guided refinement. Artif. Intell. **234**, 1–25 (2016)

19. Kleine Büning, H., Karpinski, M., Flögel, A.: Resolution for quantified boolean formulas. Inf. Comput. **117**(1), 12–18 (1995)

20. Klieber, W., Sapra, S., Gao, S., Clarke, E.: A non-prenex, non-clausal QBF solver with game-state learning. In: Strichman, O., Szeider, S. (eds.) SAT 2010. LNCS, vol. 6175, pp. 128–142. Springer, Heidelberg (2010)

21. Letz, R.: Lemma and model caching in decision procedures for quantified boolean formulas. In: Egly, U., Fermüller, C. (eds.) TABLEAUX 2002. LNCS (LNAI), vol. 2381, pp. 160–175. Springer, Heidelberg (2002)

22. Lonsing, F., Bacchus, F., Biere, A., Egly, U., Seidl, M.: Enhancing search-based QBF solving by dynamic blocked clause elimination. In: Davis, M., Fehnker, A., McIver, A., Voronkov, A. (eds.) LPAR-20 2015. LNCS, vol. 9450, pp. 418–433. Springer, Heidelberg (2015)

23. Lonsing, F., Biere, A.: Failed literal detection for QBF. In: Sakallah, K.A., Simon, L. (eds.) SAT 2011. LNCS, vol. 6695, pp. 259–272. Springer, Heidelberg (2011)

24. Lonsing, F., Egly, U., Seidl, M.: Q-resolution with generalized axioms. CoRR abs/1604.05994, SAT 2016 proceedings version with appendix (2016). http://arxiv.org/abs/1604.05994

25. Lonsing, F., Seidl, M., Van Gelder, A.: The QBF Gallery: Behind the scenes. Artif. Intell. **237**, 92–114 (2016)

26. Rabe, M.N., Tentrup, L.: CAQE: A certifying QBF solver. In: FMCAD, pp. 136–143. IEEE (2015)

27. Reimer, S., Pigorsch, F., Scholl, C., Becker, B.: Enhanced Integration of QBF Solving Techniques. In: Methoden und Beschreibungssprachen zur Modellierung und Verifikation von Schaltungen und Systemen (MBMV), pp. 133–143. Verlag Dr. Kovac (2012)
28. Samer, M., Szeider, S.: Backdoor sets of quantified boolean formulas. JAR **42**(1), 77–97 (2009)
29. Samulowitz, H., Bacchus, F.: Using SAT in QBF. In: van Beek, P. (ed.) CP 2005. LNCS, vol. 3709, pp. 578–592. Springer, Heidelberg (2005)
30. Samulowitz, H., Davies, J., Bacchus, F.: Preprocessing QBF. In: Benhamou, F. (ed.) CP 2006. LNCS, vol. 4204, pp. 514–529. Springer, Heidelberg (2006)
31. Silva, J.P.M., Lynce, I., Malik, S.: Conflict-driven clause learning SAT solvers. In: Biere, A., Heule, M., Van Maaren, H., Walsh, T. (eds.) Handbook of Satisfiability. FAIA, vol. 185, pp. 131–153. IOS Press, Amsterdam (2009)
32. Tu, K.-H., Hsu, T.-C., Jiang, J.-H.R.: QELL: QBF reasoning with extended clause learning and levelized SAT solving. In: Heule, M., Weaver, S. (eds.) SAT 2015. LNCS, vol. 9340, pp. 343–359. Springer, Heidelberg (2015)
33. Van Gelder, A.: Contributions to the theory of practical quantified boolean formula solving. In: Milano, M. (ed.) CP 2012. LNCS, vol. 7514, pp. 647–663. Springer, Heidelberg (2012)
34. Zhang, L., Malik, S.: Conflict driven learning in a quantified boolean satisfiability solver. In: ICCAD, pp. 442–449. ACM/IEEE Computer Society (2002)
35. Zhang, L., Malik, S.: Towards a symmetric treatment of satisfaction and conflicts in quantified boolean formula evaluation. In: Van Hentenryck, P. (ed.) CP 2002. LNCS, vol. 2470, pp. 200–215. Springer, Heidelberg (2002)

# 2QBF: Challenges and Solutions

Valeriy Balabanov[1]([✉]), Jie-Hong Roland Jiang[2], Christoph Scholl[3],
Alan Mishchenko[4], and Robert K. Brayton[4]

[1] Calypto Systems Division, Mentor Graphics, Fremont, USA
balabasik@gmail.com
[2] National Taiwan University, Taipei, Taiwan
jhjiang@ntu.edu.tw
[3] University of Freiburg, Freiburg, Germany
scholl@informatik.uni-freiburg.de
[4] UC Berkeley, Berkeley, USA
{alanmi,brayton}@berkeley.edu

**Abstract.** 2QBF is a special form $\forall x \exists y.\phi$ of the quantified Boolean formula (QBF) restricted to only two quantification layers, where $\phi$ is a quantifier-free formula. Despite its restricted form, it provides a framework for a wide range of applications, such as artificial intelligence, graph theory, synthesis, etc. In this work, we overview two main 2QBF challenges in terms of solving and certification. We contribute several improvements to existing solving approaches and study how the corresponding approaches affect certification. We further conduct an extensive experimental comparison on both competition and application benchmarks to demonstrate strengths of the proposed methodology.

## 1 Introduction

Satisfiability (SAT) solving recently attracted much attention due to its numerous applications in computer science [14]. Some problems (e.g., in the domains of artificial intelligence and games), however, are beyond the reach of SAT solving alone but are naturally expressible in terms of quantified Boolean formulas (QBFs) [18]. 2QBF is a restriction of general QBF problems to just two quantification levels, i.e., to the form $\forall x \exists y.\phi$ or $\exists y \forall x.\phi$, where $\phi$ is a quantifier-free propositional formula. Despite this restriction many applications can be naturally expressed in 2QBF language [15,16,18]. The main goal of this study is to evaluate and improve scalability of the existing 2QBF methodology to enable QBF as a competitive framework for both existing and new applications.

Recent QBF evaluation showed that there are two main robust algorithms for 2QBF solving: search-based (i.e., QDPLL [13]) and expansion-based (i.e., CEGAR [11,12]). Both have their own strengths and weaknesses, depending on the problem domain, and both can be generally used with either CNF and/or circuit problem representation. Conjunctive normal form (CNF) is a commonly accepted format for propositional satisfiability problems. It is as well extended to QCNF and used to represent QBF problems [3]. It is not an uncommon case,

© Springer International Publishing Switzerland 2016
N. Creignou and D. Le Berre (Eds.): SAT 2016, LNCS 9710, pp. 453–469, 2016.
DOI: 10.1007/978-3-319-40970-2_28

however, that originally QBF is specified on a circuit, rather on a CNF. E.g., in [15] FPGA synthesis benchmarks are formulated on And-Inverter Graphs (AIGs), which is an efficient way to represent general Boolean networks. It is known that any Boolean circuit can be transformed into an equisatisfiable CNF formula, by the various *CNFization* procedures, e.g., by Tseitin transform [20]. The same procedure extends to the QBF context. In this work we shall provide a detailed comparison of different solving techniques over different representations, introduce several algorithmic and implementational improvements, and outline the important observations that must be taken in account by 2QBF users and developers (Sects. 4 and 5).

It is often not enough to only determine the answer to the QBF problem, but also to provide a certificate, that either could be used to prove the validity of the answer, or to be used for other application-specific purposes. The problem of finding certificates with QDPLL solvers was addressed in [4]. In the later sections we are going to show that certificates for 2QBFs have a very restricted form compared to general QBFs, and will show how they could be found using both QDPLL or CEGAR (Sect. 6).

In Sect. 7 we will evaluate all the proposed techniques on the existing competition benchmarks as well as on the application benchmarks, and show that we contribute a significant improvement over the existing 2QBF solvers. Conclusions will be drawn in Sect. 8.

## 2   Preliminaries

In this work we shall use commonly accepted notations from Boolean algebra and logic. A *Boolean variable* is interpreted over the binary domain $\{0, 1\}$. A *literal* is either a variable or its negation. A *clause* (resp. *cube*) is a disjunction (resp. conjunction) of literals (sometimes we might use set operations on clause/cube literals as well for convenience). A Boolean formula in conjunctive normal form (CNF) is a conjunction of clauses. A Boolean formula in disjunctive normal form (DNF) is a disjunction of cubes. Both CNF and DNF might be a subject to set operations for convenience. As a notational convention, we may also sometimes (where the context allows) omit the conjunction symbol ($\wedge$) and represent negation ($\neg$) by an overline.

For a Boolean formula $\phi(x_1, \ldots, x_i, \ldots, x_n)$, we say its positive (resp. negative) cofactor with respect to variable $x_i$, denoted $\phi|_{x_i}$ (resp. $\phi|_{\bar{x}_i}$), is the formula $\phi(x_1, \ldots, x_{i-1}, 1, x_{i+1}, \ldots, x_n)$ (resp. $\phi(x_1, \ldots, x_{i-1}, 0, x_{i+1}, \ldots, x_n)$). The cofactor definition also extends to a cube of literals $\alpha = l_1 \wedge \cdots \wedge l_m$, with the following recursive definition $\phi|_\alpha = (\phi|_{\alpha \setminus l_m})|_{l_m}$, where $\phi|_{l_m}$ is a positive (resp. negative) cofactor with respect to variable $var(l_m)$ if $l_m$ is a positive (resp. negative) literal. If the context allows we may as well drop the vertical line and simply write $\phi_\alpha$. We say that formula $\phi$ is satisfiable if there is an assignment to its variables that evaluates $\phi$ to true. We say $\phi$ is unsatisfiable otherwise. We denote $ON(\phi)$ the *onset* of $\phi$, i.e., the set of assignments under which $\phi$ evaluates to true. We define an *unsatisfiable core* of an unsatisfiable CNF formula $\phi$ as an arbitrary unsatisfiable subset of clauses of $\phi$.

A *quantified Boolean formula* (QBF) $\Phi$ over universal variables
$\boldsymbol{x} = \{x_{11}, \ldots, x_{ki_k}\}$ and existential variables $\boldsymbol{y} = \{y_{11}, \ldots, y_{kj_k}\}$ in *prenex form*
is of the form $\Phi = \forall x_{11} \ldots x_{1i_1} \exists y_{11} \ldots y_{1j_1} \ldots \forall x_{k1} \ldots x_{ki_k} \exists y_{k1} \ldots y_{kj_k}.\phi$ where
the quantification part is called the *prefix*, denoted $\Phi_{\mathrm{pfx}}$, and $\phi$, a quantifier-free
formula in terms of variables $\boldsymbol{x}$ and $\boldsymbol{y}$, is called the *matrix*, denoted $\Phi_{\mathrm{mtx}}$. Further
$\Phi$ is said to be a $k$-QBF, or a QBF with $k$ quantification levels. 2QBF (resp.
3QBF) is a special case of QBF with $k = 2$ (resp. $k = 3$).

We say that 2QBF $\Phi = \forall \boldsymbol{x} \exists \boldsymbol{y}.\phi$ is false if there is an assignment $\alpha_{\boldsymbol{x}}$ to
$\boldsymbol{x}$ variables (also referred to as the winning strategy for the universal player
or a constant Herbrand-functions countermodel) such that $\Phi_{\mathrm{mtx}} |_{\alpha_{\boldsymbol{x}}}$ (which is
a function of $\boldsymbol{y}$ variables) is unsatisfiable. We say $\Phi$ is true otherwise (i.e., a
winning strategy for the universal player does not exist). Alternatively, 2QBF
$\Phi = \forall \boldsymbol{x} \exists \boldsymbol{y}.\phi$ is true if and only if there exists a set of the so-called Skolem
functions $S_{\boldsymbol{y}}(\boldsymbol{x})$ that renders $\phi$ to a tautology (i.e., $\phi(\boldsymbol{x}, S_{\boldsymbol{y}}(\boldsymbol{x})) \equiv 1$). For more
general 3QBF $\Phi = \exists \boldsymbol{w} \forall \boldsymbol{x} \exists \boldsymbol{y}.\phi(\boldsymbol{w}, \boldsymbol{x}, \boldsymbol{y})$, it is true if and only if there exists a set
of constant functions $S_{\boldsymbol{w}}$, and a set of functions $S_{\boldsymbol{y}}(\boldsymbol{x})$ (i.e., depending on $\boldsymbol{x}$),
such that $\phi(S_{\boldsymbol{w}}, \boldsymbol{x}, S_{\boldsymbol{y}})$ is a tautology. Here $S_{\boldsymbol{w}}$ and $S_{\boldsymbol{y}}$ form the Skolem-functions
model, certifying the validity of $\Phi$. There are other forms of certificates (e.g., Q-
resolution). For more details on general QBF solving and certification please
refer to [4,5,8,13].

# 3  Overview of Prior Work

Among other possibilities, there are two main approaches to QBF solving: search
based and expansion based. The former is referred to the QDPLL style search
algorithm, and the latter to the counterexample guided abstraction refinement
(CEGAR). Recent 2QBF evaluation showed that CEGAR solvers are generally
more robust than QDPLL solvers. In this work, we focus on the CEGAR-based
approach, and will provide an intuition later how it is more beneficial than
QDPLL.

Figure 1 outlines the generic CEGAR-based algorithm `Cegar2QBF` for solving
an arbitrary 2QBF formula in the prenex form, introduced in [11]. In Line 1,
two SAT solving managers, i.e., *synthesis manager* `synMan` (to guess a candidate
winning move of the universal player) and *verification manager* `verMan` (to verify
if the guessed move of the universal player is indeed winning), are initialized.
Initially `synMan` contains variables $\boldsymbol{x}$ and `verMan` contains variables $\boldsymbol{x}$ and $\boldsymbol{y}$. In
Line 3, we search for a candidate winning move $\alpha_{\boldsymbol{x}}$. If all candidates have been
blocked, the 2QBF is determined to be true in Line 4. In Line 5, a counterexample
to the candidate winning move is searched, which renders $\phi$ true (i.e., it disproves
the candidate winning move $\alpha_{\boldsymbol{x}}$). Note that if CNF is used as an underlying
data structure for `verMan`, then $\alpha_{\boldsymbol{x}}$ can be easily passed to the SAT solver via
its assumptions interface (which is commonly available in modern SAT solvers,
e.g., in MINISAT [7]). If no counterexample can be found, the QBF is determined
to be false in Line 6. Otherwise, cofactor $\phi|_{\alpha_{\boldsymbol{y}}}$ is performed in Line 7, and block
all the known wrong candidates $\boldsymbol{x}'$, such that $\phi|_{\alpha_{\boldsymbol{y}}}(\boldsymbol{x}') = 1$, in Line 8.

---

Algorithm **Cegar2QBF**
**input**: a QBF $\Phi = \forall x \exists y.\phi$
**output**: True or False
**begin**
01   synMan$[x]$ := 1; verMan$[x, y]$ := $\phi$;
02   **while** True
03     $\alpha_x$ := SatSolve(synMan);
04     **if** $\alpha_x = \emptyset$ **then return** True;
05     $\alpha_y$ := SatSolve(verMan, $\alpha_x$);
06     **if** $\alpha_y = \emptyset$ **then return** False;
07     negCof := $\neg\phi|_{\alpha_y}$;
08     synMan := synMan $\wedge$ negCof;
**end**

---

**Fig. 1.** CEGAR algorithm for generic 2QBF solving.

The above CEGAR algorithm differs from QDPLL-based algorithms [9,13] in that in QDPLL negCof is simply substituted with $\neg\alpha_x'$, where $\alpha_x'$ is obtained from $\alpha_x$ by a single minimal hitting set generalization, to block the failed candidate within synMan. The strength of Cegar2QBF is in that besides $\alpha_x'$ it potentially blocks several other hitting set assignments.

Algorithm Cegar2QBF may be applied to an arbitrary 2QBF in prenex form regardless of the representation of its matrix. Some QBF solvers use And-Inverter graphs (AIGs) as an underlying matrix structure [15,17]. On the other hand, as CNF has been proven to be an efficient data structure for SAT solving (e.g., due to an efficient representation of learnt information in the form of learnt clauses [7]), CNF is also the most commonly accepted QBF matrix representation format [3]. We therefore distinguish between *CNF* and *circuit* 2QBF solvers, depending on the matrix input format that they accept. Both CNF and circuit representations have their advantages and disadvantages, which will be discussed in later sections.

Below we mention the idea of QESTO [12] for efficient implementation of CEGAR-based 2QBF solving on CNF matrix. Note that verMan is initialized to $\phi$ in Line 1 of Fig. 1 and is never changed, while synMan is constantly changed by conjunction with negCof in Line 8. If the matrix is already represented in CNF, then the main complication of the algorithm is the CNFization of negCof prior to conjunction with synMan. The approach of [11] suggested to use Tseitin transform [20] to perform a syntactic negation, at the cost of introducing fresh variables. This approach, however, suffers from variable blow up within synMan after a large number of iterations. In QESTO, the variable blow-up problem is overcome by efficient representation of a larger number of cofactors within synMan as follows. Consider a matrix $\phi = C_1 \wedge C_2 \wedge \cdots \wedge C_n$, where each $C_i$ is split into existential literals $C_{ei}$ and universal literals $C_{ui}$. Notice that regardless of the specifics of the assignment $\alpha_y$, `negCof` always takes the form $\neg C_{uj_1} \vee \neg C_{uj_2} \vee \cdots \vee \neg C_{uj_k}$. By defining $d_i \equiv \neg C_{ui}$ for each $i \in [1..n]$, one can conveniently represent `negCof` as a clause $(d_{j_1} \vee d_{j_2} \vee \cdots \vee d_{j_k})$. Consequently, at most $n$ fresh variables need to be introduced, independent of the number of iterations of the while-loop in Fig. 1. Under this scenario, the algorithm in Fig. 1 can be modified to initialize the synthesis manager by

$$\texttt{synMan}[x, d] := \bigwedge_{i \in [1..n]} (d_i \equiv \neg C_{ui}).$$

QESTO was experimentally shown to be superior to others [12]. In the next section we will examine the advantages and disadvantages of CNF CEGAR-based 2QBF solving, and introduce a heuristics, inspired by QESTO, to improve existing circuit CEGAR-based 2QBF algorithms.

# 4   Heuristics for CNF and Circuit 2QBF Solving

In this section we introduce several implementation improvements to the QESTO algorithm and also show how the QESTO CNF ideas can be lifted to the circuit 2QBF solving.

## 4.1   Improvements for CNF-based 2QBF Solving

The following three observations could be used to enhance the performance of the QESTO implementation of the `Cegar2QBF` algorithm:

1. In case $C_{uj_1} = x_i$ (i.e., $C_{j_1}$ is a universally unit clause), there is no need to introduce a fresh variable for this clause, but rather just using $x_i$ itself in computing `negCof`. Note that this condition may occur quite often if CNF was obtained through the Tseitin transform. For example AND-gate $c = a \wedge b$, where $c$ is existential and both $a$ and $b$ are universal, will be transformed into clauses $(\bar{c} \vee a)(\bar{c} \vee b)(c \vee \bar{a} \vee \bar{b})$, therefore producing two universally unit clauses.
2. In case $C_{uj_1} = C_{uj_2}$, there is no need to introduce two definition variables $d_1$ and $d_2$, but just one. Similar to the previous case, this may happen after Tseitin transform, e.g., if the universally quantified primary input had several fanouts in the circuit.
3. In case $C_{uj_1} \subseteq C_{uj_2}$ and `negCof` $= d_{uj_1} \vee d_{uj_2} \vee \cdots$, we can simply drop $C_{uj_1}$ from `negCof` because of the logical implication $d_{uj_1} \to d_{uj_2}$. This often may be seen after existential variable elimination, which is a common preprocessing technique in both QBF and SAT solving. For example given QBF

$$\forall ab \exists cde \,.\, (\bar{c} \vee a)_1 (c \vee \bar{a})_2 (d \vee \bar{b} \vee \bar{c})_3 (\bar{d} \vee b)_4 (\bar{d} \vee c)_5 (a \vee e)_6 (b \vee \bar{e})_7,$$

after elimination of variable $c$ we get QBF

$$\forall ab \exists de \,.\, (d \vee \bar{b} \vee \bar{a})_3 (\bar{d} \vee b)_4 (\bar{d} \vee a)_5 (\bar{a} \vee e)_6 (\bar{b} \vee \bar{e})_7,$$

where clause $C_{u3} = (\bar{b} \vee \bar{a})$ now subsumes both $C_{u6} = \bar{a}$ and $C_{u7} = \bar{b}$ (note that $C_{u3}$ does not trigger any of the first two heuristics).

The first two enhancements are simpler, and intend to decrease the number of definitions added to the synMan at the initialization step. As a consequence, underlying SAT solver has to deal with less variables, and hopefully find the solution faster. On the other hand the third enhancement does not change the number of definitions, but rather simplifies the blocking constraint negCof by observing that some literals could be effectively dropped from the underlying blocking clause. As learned clause size is an important criteria in SAT solving, we speculate that this could also potentially speed up the solving process. All the three refinements are to be evaluated in Sect. 7.

## 4.2   Improvements for AIG-based 2QBF Solving

The main idea behind QESTO could be formulated in a different manner: the cofactors computed in Line 7 of algorithm Cegar2QBF on Fig. 1 have a lot in common. If certain universal subclause $C_{ui}$ is present in several cofactors then the Tseitin definition for this clause could be reused multiple times when performing the CNFization of the complemented cofactors. This idea could be efficiently lifted to the circuit solvers. Following the notion of *structural hashing* of nodes in And-Inverter graphs (AIGs), we propose algorithm AigShare2Qbf as sketched in Fig. 2 for circuit-based 2QBF solving.

The core CEGAR procedure of algorithm AigShare2Qbf is the same as that of Cegar2QBF. Furthermore Cegar2QBF may use AIGs as the underlying data structure of its verification manager and the negated cofactors as well. The only difference is the *cofactor hashing* heuristics in lines 8 and 9 of Fig. 2.

---

Algorithm **AigShare2Qbf**
**input**: a QBF $\Phi = \forall X \exists Y.\phi$
**output**: True or False
**begin**
01    synMan$[X] := 1$; verMan$[X, Y] := \phi$; aigMan $:= \emptyset$;
02    **while** True
03        $\alpha_X := $ SatSolve(synMan);
04        **if** $\alpha_X = \emptyset$ **then return** True;
05        $\alpha_Y := $ SatSolve(verMan, $\alpha_X$);
06        **if** $\alpha_Y = \emptyset$ **then return** False;
07        negCof $:= $ AIG($\neg\phi|_{\alpha_Y}$);
08        newAnd $:= $ NotHashed(negCof, aigMan);
09        Hash(aigMan, newAnd);
10        synMan $:= $ synMan $\wedge$ CNF(newAnd);
**end**

---

**Fig. 2.** CEGAR algorithm for AIG-based 2QBF solving with cofactor sharing heuristics.

More specifically, in Line 8 we extract a subset `newAnd` of AND gates from `negCof` that have not been hashed previously. Then in Line 9 we hash them and add them to the AIG manager `aigMan`. In Line 10 we add CNFized `newAnd` gates to `synMan`. For the AND gates in Line 8 which have been hashed previously, `synMan` already contains clauses for the corresponding definitions. As to be seen from experiments, cofactor sharing heuristic gives a significant improvement on the benchmarks where a large number of iterations are needed for algorithm `Cegar2QBF` to converge.

## 5    CNF Versus Circuit Solvers

In this section we compare the CNF and circuit 2QBF solvers. We outline the strengths of one over the other, and address the question in which applications which one should be used.

Given a non-CNF formula $\phi$ (e.g., represented as an AIG) one could *CNFize* it (i.e., transform it to an equisatisfiable CNF formula, for example by using the Tseitin transform [20]) to get $\phi_{CNF}$, and then run the QESTO algorithm introduced in Sect. 3. However, this approach could be inefficient as the following example suggests.

*Example 1.* Consider the following simple true 2QBF formula $\Phi = \forall abc \; \exists d \, . \, \phi$, where $\phi$ is given as a *nested XOR tree* circuit $\phi = a \oplus b \oplus c \oplus d$. Assume that the synthesis manager in algorithm `Cegar2QBF` of Fig. 1 comes up with a candidate $\alpha_x = \overline{a}\overline{b}c$ (i.e., assigns $a = b = c = 0$), and the verification manager returns a counterexample $\alpha_y = d$ (i.e., assigns $d = 1$). Note that in this case $\text{negCof}_1 = a \oplus b \oplus c$. After the second iteration with, e.g., $\alpha_x = \overline{a}\overline{b}c$ and $\alpha_y = \overline{d}$, we have $\text{negCof}_2 = \neg(a \oplus b \oplus c)$, thus blocking all the universal candidate assignments, and conclude that $\Phi$ is true. The same number of iterations would be required for an arbitrary large nested XOR tree.

In contrast, consider the same 2QBF, but CNFized by Tseitin transform prior to solving, introducing definitions $x = a \oplus b$ and $y = x \oplus c$, and resulting into QBF

$$\Phi_{CNF} = \forall abc \; \exists d \; \exists xy \, . \, \phi_{CNF}, \text{ where}$$
$$\phi_{CNF} = (x \vee \overline{a} \vee b)(x \vee a \vee \overline{b})(\overline{x} \vee a \vee b)(\overline{x} \vee \overline{a} \vee \overline{b})$$
$$\wedge (y \vee \overline{x} \vee c)(y \vee x \vee \overline{c})(\overline{y} \vee x \vee c)(\overline{y} \vee \overline{x} \vee \overline{c})$$
$$\wedge (y \vee d)(\overline{y} \vee \overline{d}).$$

Suppose that the synthesis manager in algorithm `Cegar2QBF` guesses the same candidate $\alpha_x = \overline{a}\overline{b}c$, and the verification manager replies with a counterexample $\alpha_y = d\overline{x}y$. In this case we compute $\text{negCof}_1 = (a \oplus b) \vee c$. After the second iteration with $\alpha_x = \overline{a}\overline{b}c$ and $\alpha_y = \overline{d}\overline{x}y$, we have $\text{negCof}_2 = (a \oplus b) \vee \neg c$, thus resulting in formula $a \oplus b$ in the synthesis manager. Without any conclusion, we have to proceed with further iterations. In fact, for a general nested XOR tree the computation may require an exponential number of iterations to terminate.

The rationale behind Example 1 is as follows. Whenever the counterexample (Line 5 in Fig. 1) is computed under the CNF representation, it fixes a value assignment to the auxiliary variables (i.e., the intermediate variables introduced during CNFization). The negated cofactor (Line 7 in Fig. 1) in this case will only block $X$ assignments that respect the auxiliary variable assignment. This phenomenon is summarized in Proposition 1.

**Proposition 1.** *Given a 2QBF $\Phi = \forall x \exists y.\phi$, with $\phi$ represented as a circuit, let $\alpha_y$ be an assignment to variables $y$ and $\alpha_{x_1}$ and $\alpha_{x_2}$ be two assignments to variables $x$ and let $g = G(x, y)$ be such an intermediate gate in the circuit of $\phi$, such that*

$$\phi|_{\alpha_y \alpha_{x_1}} \wedge \phi|_{\alpha_y \alpha_{x_2}} \wedge G|_{\alpha_y \alpha_{x_1}} \oplus G|_{\alpha_y \alpha_{x_2}}.$$

*(That is, the output of $\phi$ evaluates to the same value under the two input assignments, but there is an internal gate disagreement for the two assignments.) Then $\alpha_{x_1}$ and $\alpha_{x_2}$ will be blocked within the same iteration computing counterexample $\alpha_y$ in Line 5 of Fig. 1, if algorithm Cegar2QBF is applied to $\Phi$. On the other hand, $\alpha_{x_1}$ and $\alpha_{x_2}$ will not be blocked within the same iteration, if Cegar2QBF is applied to a CNFized version (by Tseitin transform) of $\Phi$.*

Example 1 and Proposition 1 show that flattening the circuit structure into CNF affects the 2QBF solving process more than it does for propositional SAT. In theory one could avoid cofactoring on Tseitin variables, and modify the algorithm Cegar2QBF accordingly to eliminate the problem described in Proposition 1. In practice, however, CNF based QBF solvers can not easily distinguish auxiliary variables from the original primary inputs quantified in the same quantification layer. It is therefore advised to use CNF for underlying SAT queries (e.g., for efficient clause learning), while cofactoring on circuit level instead of CNF. As will be confirmed experimentally later, the gained reduction in number of iterations needed for completion of algorithm Cegar2QBF even overcomes the benefits of efficient cofactor representation in QESTO algorithm applied after circuit CNFization.

Please note that in circuit 2QBF solvers the SAT queries in Line 3 of algorithm Cegar2QBF (Fig. 1) are made to a CNF-based SAT solver as well. This choice is caused by a specific "incremental" nature of the underlying SAT calls: Please recall that synthesis manager synMan is updated by iterative conjunction with negCof in Line 8 of Fig. 1, i.e., in each iteration clauses from the CNFized negCof are added to manager synMan. Therefore it will be highly benefitial to use the learned information from the previous solving iterations in the later ones.

Despite the ineffectiveness of CNF 2QBF solvers compare to their circuit counterparts, we speculate that there is a better encoding of a circuit 2QBF problem into QCNF, as is described below. Given circuit 2QBF formula $\Phi = \forall x \exists y.\phi(x, y)$, instead of computing its CNFized version $\Phi_{CNF} = \forall x \exists y \exists t . \phi_{CNF}(x, y, t)$ we obtain it's negation as $\Phi' = \neg\Phi = \exists x \forall y.\phi'(x, y)$, with $\phi' = \neg\phi$. Now the translation to an equisatisfiable CNF can be done as following: $\Phi'_{CNF} = \exists x \forall y \exists t . \phi'_{CNF}(x, y, t)$. By construction, the truth of $\Phi$ could be determined as an inverse of $\Phi'_{CNF}$. Further any model (resp. countermodel) for $\Phi'_{CNF}$ could be mapped to corresponding model (resp. countermodel) for $\Phi$.

Note that $\Phi'_{CNF}$ above is a 3QBF formula, rather than 2QBF. Despite the increase in the number of quantifier alternations (which in general correlates with the formula complexity), $\Phi'_{CNF}$ may be much easier to solve compare to $\Phi_{CNF}$. The intuition here is that innermost quantification level in $\Phi'_{CNF}$ contains only Tseitin variables. Values for these variables shall be uniquely determined upon the assignments to variables $x$ and $y$.

Generally CEGAR based algorithm for 3QBF is slightly more complex than for 2QBF, but let us briefly outline how the above procedure applies to the earlier Example 1. First compute $\Phi'_{CNF}$ in the same way as it was done for $\Phi_{CNF}$:

$$\Phi'_{CNF} = \exists abc\ \forall d\ \exists xy\ .\ \phi'_{CNF},\ \text{where}$$
$$\phi'_{CNF} = (x \vee \overline{a} \vee b)(x \vee a \vee \overline{b})(\overline{x} \vee a \vee b)(\overline{x} \vee \overline{a} \vee \overline{b})$$
$$\wedge (y \vee \overline{x} \vee c)(y \vee x \vee \overline{c})(\overline{y} \vee x \vee c)(\overline{y} \vee \overline{x} \vee \overline{c})$$
$$\wedge (\overline{y} \vee d)(y \vee \overline{d}).$$

It is worthy to mention that the negation on circuit level is a very cheap operation, and as one can see $\Phi'_{CNF}$ differs from $\Phi_{CNF}$ only by the last two clauses. Let us now examine candidate solution $\alpha_{x} = \overline{a}b\overline{c}$, and the (potential) counterexample to this solution $\alpha_{y} = d$. First observe that values of $x$ and $y$ are uniquely determined to be $x = y = 0$. Second, under the completed assignment $\phi'$ evaluates to false. So at this point no further computations are required to conclude that $\alpha_{y} = d$ is a counterexample to candidate $\alpha_{x} = \overline{a}b\overline{c}$. To block this candidate we basically add formula $\texttt{negCof}_1 = \neg\phi'_{CNF}|_d$ to the outermost existential solver, and rename variables $x, y$ to $x_1, y_1$. After the second iteration with $\alpha_{x} = \overline{a}b c$ and $\alpha_{y} = \overline{d}$ we shall again conclude invalidity of $\alpha_{x}$ and block it by formula $\texttt{negCof}_2 = \neg\phi'_{CNF}|_{\overline{d}}$, where variables $x, y$ are renamed to $x_2, y_2$. Intuitively $\texttt{negCof}_1(a, b, x_1, y_1)$ and $\texttt{negCof}_2(a, b, x_2, y_2)$ are the CNFized versions of circuit cofactors $\phi|_d$ and $\phi|_{\overline{d}}$. One could verify that $\texttt{negCof}_1(a, b, x_1, y_1) \wedge \texttt{negCof}_2(a, b, x_2, y_2)$ is unsatisfiable. Generally speaking, solving 3QBF $\Phi'_{CNF}$ should not take more CEGAR iterations than circuit 2QBF $\Phi$. On the other hand each iteration could be more inefficient due to the naive cofactoring and variable renaming. In circuit solving this can be done more efficiently using e.g., cofactor sharing heuristics introduced in the previous section. This phenomenon is to be examined in the experimental section.

# 6   Certificate Generation for 2QBF

As mentioned previously, recent evaluation on QBF solvers suggested the superiority of CEGAR-based QBF solvers. In contrast to search-based approaches (e.g., DEPQBF [13]), however, there exists no methodology to certify their answer with semantic winning strategies in a closed form (e.g., Skolem-functions for true QBFs, which are essential for many QBF applications). CEGAR-based QBF solver RAREQS [10], can produce partial winning moves for both existential and universal players at each turn of an abstraction-refinement game [10]. One straightforward use of this ability is that RAREQS returns the winning

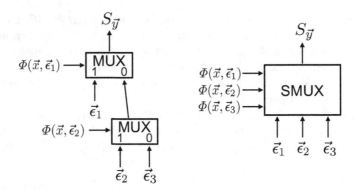

**Fig. 3.** Multiplexer construction [on the left], SMUX cell [on the right].

assignment to the outermost existential (resp. universal) variables for true (resp. false) QBFs upon completion. In this section we describe how to construct Skolem/Herbrand functions for 2QBFs, based on the partial winning-move information deduced from CEGAR-based QBF solvers. We present several heuristics to enhance the construction procedure as well as the produced certificate quality.

### 6.1   Construction Procedure

Consider a true 2QBF formula $\Phi$. Assume that the CEGAR 2QBF solver needs three, say, refinement iterations to prove its validity. It means that three candidate solutions for the universal player were found, leading to existential counterexamples $\epsilon_1$, $\epsilon_2$, and $\epsilon_3$. Let $\Phi_{cof}$ be the formula refined by the three corresponding cofactors as shown below.

$$\Phi = \forall \boldsymbol{x} \exists \boldsymbol{y}. \phi(\boldsymbol{x}, \boldsymbol{y}) \qquad \Phi_{cof} = \forall \boldsymbol{x}. \{ \phi(\boldsymbol{x}, \epsilon_1) \vee \phi(\boldsymbol{x}, \epsilon_2) \vee \phi(\boldsymbol{x}, \epsilon_3) \}$$

Now the search for a candidate solution fails, i.e., the universal player does not find a candidate solution which falsifies all cofactors generated so far. Consequently $\Phi_{cof}$ is true, which is determined by an unsatisfiable SAT call $\neg \Phi_{cof}$ (which is propositional as it has only existentially quantified variables $\boldsymbol{x}$).

Effectively, validity of $\Phi_{cof}$ says that an arbitrary assignment to $\boldsymbol{x}$ is included in $ON(\phi(\boldsymbol{x}, \epsilon_1))$, $ON(\phi(\boldsymbol{x}, \epsilon_2))$, or $ON(\phi(\boldsymbol{x}, \epsilon_3))$. This information in fact is sufficient to get Skolem functions $S_{\boldsymbol{y}}(\boldsymbol{x})$ for any assignment $\boldsymbol{\alpha}$ to $\boldsymbol{x}$, by the following steps:

1. For all $\boldsymbol{\alpha} \in ON(\phi(\boldsymbol{x}, \epsilon_1))$, define $S_{\boldsymbol{y}}(\boldsymbol{\alpha}) = \epsilon_1$.
2. For all $\boldsymbol{\alpha} \in ON(\phi(\boldsymbol{x}, \epsilon_2)) \setminus ON(\phi(\boldsymbol{x}, \epsilon_1))$, define $S_{\boldsymbol{y}}(\boldsymbol{\alpha}) = \epsilon_2$.
3. For all $\boldsymbol{\alpha} \in ON(\phi(\boldsymbol{x}, \epsilon_3)) \setminus (ON(\phi(\boldsymbol{x}, \epsilon_1)) \cup ON(\phi(\boldsymbol{x}, \epsilon_2)))$, define $S_{\boldsymbol{y}}(\boldsymbol{\alpha}) = \epsilon_3$.

The above computation of the Skolem functions is visualized by a multiplexer construction as shown on the left of Fig. 1. We abbreviate the multiplexer construction by a cell "SMUX" which means that we have a series of multiplexers defining a prioritization in case that the sets $ON(\Phi(x, \epsilon_i))$ overlap. SMUX cell is shown on the right of Fig. 1. By the following proposition we ensure the soundness of returned Skolem functions Figs. 3 and 5 .

**Proposition 2.** *The functions $S_y(x)$ constructed by the above procedure form a valid model for $\Phi$.*

*Proof.* Since $\Phi_{cof}$ is true, for every assignment $\alpha$ to $x$, some $\phi(\alpha, \epsilon_i)$), $i \in [1..3]$, must be true. Our construction ensures that $S_y(\alpha) = \epsilon_i$, i.e., that $\phi(\alpha, S_y(\alpha))$ evaluates to true. By definition, the functions $S_y$ form a valid set of Skolem functions.                                                                                            □

The proposed procedure can be easily extended to true 2QBFs with an arbitrary number of refinement steps, just by replacing three cofactors from the above example with the cofactors returned by the solver. Further, the method can be extended for true 3QBFs as follows. Suppose we are given a true 3QBF $\Phi = \exists w \forall x \exists y. \phi(w, x, y)$, and a winning move (assignment) $\beta$ for $w$ variables (which is returned upon completion of RAREQS as a by-product of solving process). Because the 2QBF $\Phi = \forall x \exists y. \phi(\beta, x, y)$ must be true, Skolem functions $S_y$ can be extracted using the previous method for 2QBFs. The complete Skolem model now consists of $\{S_w = \beta, S_y\}$. Although to achieve efficient implementation of this extension is slightly more sophisticated than the 2QBF case, the essential idea is as described above.

## 6.2   Certificate Optimization

Below we propose three optimization techniques in order to minimize the certificates returned by the proposed Skolem-function construction procedure.

1. Observe that any reordering of the multiplexers in a SMUX cell still maintains a valid set of Skolem functions. In our implementation, we allow an option to use cofactors either in a forward or backward order with respect to the derivation sequence of the corresponding winning moves. The backward order turns out to be empirically superior.
2. It was observed that cofactors often share identical clauses. Therefore we implement a hashing procedure that detects and substitutes repeating clauses.
3. Although each next counterexample returned by CEGAR QBF solving approach covers at least one new (so-far unblocked) universal winning move candidate, in practice it happens that older cofactors are fully covered by newer ones. The problem of identifying redundant cofactors can be done using the so-called group minimal unsatisfiability subset extraction (group MUS, or GMUS). The GMUS framework partitions the clauses of a CNF formula into groups (cofactors in our case), and returns the minimal subset of them, which is still unsatisfiable (which is clearly a requirement for our extracted Skolem functions to be sound). More information on GMUS extraction can be found at [19].

# 7    Experimental Results

We performed experimental evaluation of various 2QBF solving and certification techniques both on the competition as well as application benchmarks. The
experiments are divided into the solver performance testing and the certificate
quality testing. To test solvers' performance, we used two sets of benchmarks:
2QBF track of QBFEVAL'10 [2] QBF competition formulas and FPGA tech-
nolgy mapping benchmarks from [15]. For certificate generation, due to the few
number of true 2QBFs in QBFEVAL'10, we used the application track benchmarks of QBF Gallery 2014 [1].[1]

## 7.1    2QBF Solving

**2QBF Track of QBFEVAL'10 [2] .**  We used CEGAR-based QBF solvers
RAREQS [11] and QESTO [12] for comparison. In this section we refer to QESTO
as to the tool from [12], rather than to the algorithm itself. Please note that
according to [12], tool QESTO dominates other 2QBF solvers, including both
QDPLL-based and circuit-based solvers. Since the source code for QESTO is not
publicly available, we have reimplemented its simplified 2QBF version (further
referred to as MINI2QBF) on top of the MINISAT SAT solver [7]) to test the
enhancements proposed in Sect. 4.

The benchmark suit contains 200 QBFs in prenex CNF from the 2QBF
track of QBFEVAL'10. Preprocessor BLOQQER [6] was used to preprocess the
formulas, and 135 formulas solved directly by BLOQQER were excluded from
further experiments. All CNF-based solvers, namely RAREQS, QESTO, and
MINI2QBF, ran on the preprocessed benchmarks. To compare against the circuit 2QBF solvers, we implemented a tool MINICNF2BLIF to extract circuits
from unpreprocessed CNF formulas, and then ran the existing circuit 2QBF
solver (command "&qbf") embedded into the synthesis tool ABC [15] under
two settings: without cofactor sharing (referred to as ABC-) and with cofactor
sharing (referred to as ABC+). A third version, called PREABC+ is similar to
ABC+, with the only difference that the QBF preprocessor BLOQQER is used
for preprocessing the QBFs where MINICNF2BLIF did not find any circuit structure (after preprocessing the resulting CNF formulas are then just translated
to product-of-sums circuits and then solved by ABC+). All the solvers were
limited by the 4GB memory and 1200 second time limit.

Table 1 summarizes the results. A cactus plot of time performance is shown in
Fig. 4. The extraction time of MINICNF2BLIF was negligible compared to solving,
and we omitted reporting it. From Table 1 and Fig. 4, we see that the circuit-
based 2QBF solver PREABC+ outperforms existing CNF based solvers. On the
17 benchmarks where no circuit structure was found, PREABC+ was on average
3 times slower compared to QESTO. The remaining 48 benchmarks were found
highly structural. On these ABC+ was on an average of 28 times faster than

---

[1] All the tools, benchmarks, and experimental results can be found at
   https://www.dropbox.com/s/xyfb2i9xl1pvrv3/sat16_tools_2qbf.zip?dl=0.

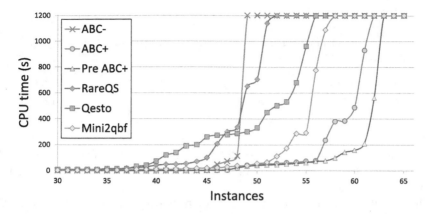

**Fig. 4.** Cactus plot of solvers performance on QBFEVAL'10 2QBF benchmark set.

**Table 1.** Statistics for 2QBF track from QBFEVAL'10.

|          | RAREQS | QESTO | MINI2QBF | ABC- | ABC+ | PREABC+ |
|----------|--------|-------|----------|------|------|---------|
| Solved   | 50     | 55    | 57       | 48   | 61   | 62      |
| Time, s  | 20658  | 17842 | 12725    | 20677| 7748 | 4124    |
| Iterations | NA   | 7.26M | NA       | 46.0K| 368K | 153K    |

QESTO. ABC+ in comparison to ABC- was on average 30 % faster; however if we only consider problems solved within more than 100 iterations (or alternatively solved in about more than 1 second) cofactor sharing gives about an order of magnitude speed up. This phenomenon is well explained by the fact that within first few iterations cofactor sharing occurs rarely, while on the larger scale AIG nodes from the new cofactors are found to be previously hashed practically all the time. On the other hand we can also see that our reimplementation of QESTO algorithm performs quite well too. If we switch off the last heuristic from Sect. 4, MINI2QBF solves 2 instances less and is of similar performance to QESTO (the first two heuristics from Sect. 4 are hardcoded so we cannot switch them on or off).

**FPGA Mapping Benchmarks From [15] .** We picked 100 (50 SAT and 50 UNSAT) 2QBF benchmarks from an FPGA mapping application [15]. For the comparison with CNF solvers we encoded these benchmarks into both 2QBF and negated 3QBF forms (introduced in Sect. 4). Original AIG circuits have 50 primary inputs and 250 AIG nodes on average. After CNFization by ABC's internal engine resulting QCNFs on average have 93 variables and 240 clauses.

For the circuit solving we used ABC to solve original problems, and RAREQS and QESTO for the corresponding CNF problems. Further we also used RAREQS for the negated 3QBF problems. The BLOQQER preprocessor was found to degrade the performance of CNF QBF solvers significantly on this benchmark set, therefore we do not include it in this set of experiments.

**Table 2.** Statistics for FPGA mapping benchmark set.

|            | RAREQS | QESTO | BLOQQER+RAREQS+3QBF | ABC  |
|------------|--------|-------|---------------------|------|
| Solved     | 22     | 44    | 100                 | 100  |
| Time, s    | 96.9K  | 75.0K | 63.5                | 64.3 |
| Iterations | NA     | 6.46M | NA                  | 1241 |

Solving statistics are shown in Table 2. As one can observe from Table 2 CNF 2QBF solvers required several orders of magnitude more iterations, and significantly larger solving time.

The cumulated number of iterations shown in Table 2 confirms that in 2QBF solving CNF representation is good for carrying out SAT queries, but cofactoring should be done on the circuit and not on the CNF level. On the other hand an alternative, equally efficient to circuit 2QBF solving, way is to use complemented 3QBF encoding presented in Sect. 4.

### 7.2 Certificate Derivation

We patched RAREQS solver to emit countermoves associated with the CEGAR QBF solving process. Proposed in Sect. 6 algorithm was implemented into a tool CEGARSKOLEM. In order to test the unsat core optimization from Sect. 6 we used HAIFA-HLMUC group MUS extractor [19]. Experimental setup consisted of QBFs from "QBFLIB" and "Applications" tracks taken from QBF Gallery 2014 [1]. We used DEPQBF QBF solver [13] and RESQU [4] to compare CEGARSKOLEM against existing Q-resolution based model computation in search-based QBF-solvers framework. An additional limitation of 1 Gb was imposed on Q-resolution proofs produced by DEPQBF and moves information emitted by RAREQS.

We selected 29 and 137 true 3QBFs either solved by DEPQBF or RAREQS, for experiments from "QBFLIB" and "Applications" tracks, respectively. The number of true 2QBF formulas was insufficient so we decided to focus only on 3QBFs instead. Table 3 shows the solving and certification time statistics (rightmost "#cert" and "#mincert" columns stand for CEGARSKOLEM w/o and w/ optimization heuristics, respectively). Note that as certification requires additional effort, both DepQBF and RareQs were not able to solve some of the instances that they could solve w/o certification. As we could see in general RAREQS solved more instances, but DEPQBF has much smaller runtime-per-instance. On contrary, even with optimizations, CEGARSKOLEM constructed Skolem functions much quicker than RESQU, which is explained by large overhead in the size of Q-refutations produced by DEPQBF, in comparison to (relatively small) number of existential counterexamples emitted by RAREQS.

Figure 2 compares certificates quality in terms of numbers of AIG nodes. X-axis in figures corresponds to certificates produced by RESQU, and Y-axis to those by CEGARSKOLEM and CEGARSKOLEMMIN. We omit the details of

**Table 3.** Solving and certification statistics.

| | DEPQBF+RESQU | | | | RAREQS+CEGARSKOLEM | | | | | |
|---|---|---|---|---|---|---|---|---|---|---|
| | #solved | time, s | #cert | time, s | #solved | time, s | #cert | time, s | #mincert | time, s |
| QBFLIB [29] | 15 | 424.1 | 14 | 943.2 | 19 | 2710.6 | 19 | 49.5 | 19 | 67.5 |
| Appl [137] | 94 | 457.9 | 86 | 5162.8 | 102 | 4351.6 | 97 | 533.8 | 97 | 589.0 |

the impact of various optimizations in CEGARSKOLEMMIN, but as one can see from Table 3 the computational overhead they introduce is small anyway. The certificate sizes, on the other hand, are reduced much in some cases. Another observation to make is that certificates for DEPQBF and RAREQS are quite scattered across the figures. This means that for some benchmarks there exist simple Skolem-functions found by RESQU but not found by CEGAR-SKOLEM and vice-versa.

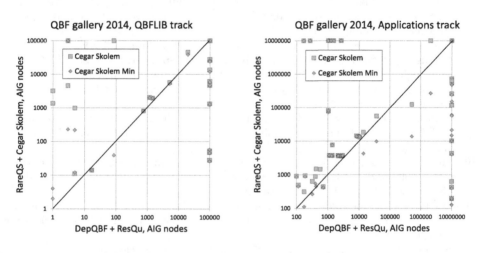

**Fig. 5.** Comparison of Skolem functions AIG sizes.

## 8  Conclusions and Future Work

In this work we overviewed different methods of solving and certifying 2QBF formulas. Based on our experience we introduced several improvements both to the CNF and circuit based solving techniques. We as well performed an extensive comparison to the state of the art algorithms. Experiments showed that the proposed solving improvements outperform existing 2QBF solvers, as well as the introduced approach for semantic certificate generation. In conclusion we summarize that circuit CEGAR-based 2QBF solvers generally scale better compare to the CNF based solvers. On the other hand, CNF preprocessing may be one of the levers, infeasible to the circuit solvers, that could lift the off-the-shelf CNF QBF solvers to a competitive level, given that circuit problem is properly encoded into CNF (e.g., with the negated 3QBF encoding).

**Acknowledgments.** This work was partly supported by NSF/NSA grant "Enhanced equivalence checking in cryptoanalytic applications" at University of California, Berkeley.

# References

1. QBF Gallery 2014. http://qbf.satisfiability.org/gallery/
2. QBF solver evaluation portal. http://www.qbflib.org/qbfeval/
3. QDIMACS: Standard QBF input format. http://www.qbflib.org/qdimacs
4. Balabanov, V., Jiang, J.-H.R.: Unified QBF certification and its applications. Formal Meth. Syst. Des. **41**, 45–65 (2012)
5. Benedetti, M.: sKizzo: a suite to evaluate and certify QBFs. In: Nieuwenhuis, R. (ed.) CADE 2005. LNCS (LNAI), vol. 3632, pp. 369–376. Springer, Heidelberg (2005)
6. Biere, A., Lonsing, F., Seidl, M.: Blocked clause elimination for QBF. In: Bjørner, N., Sofronie-Stokkermans, V. (eds.) CADE 2011. LNCS, vol. 6803, pp. 101–115. Springer, Heidelberg (2011)
7. Eén, N., Sörensson, N.: An extensible SAT-solver. In: Giunchiglia, E., Tacchella, A. (eds.) SAT 2003. LNCS, vol. 2919, pp. 502–518. Springer, Heidelberg (2004)
8. Giunchiglia, E., Narizzano, M., Tacchella, A.: QuBE++: an efficient QBF solver. In: Hu, A.J., Martin, A.K. (eds.) FMCAD 2004. LNCS, vol. 3312, pp. 201–213. Springer, Heidelberg (2004)
9. Giunchiglia, E., Narizzano, M., Tacchella, A.: Clause/term resolution and learning in the evaluation of quantified Boolean formulas. J. Artif. Intell. Res. (JAIR) **26**, 371–416 (2006)
10. Janota, M., Klieber, W., Marques-Silva, J., Clarke, E.: Solving QBF with counterexample guided refinement. In: Cimatti, A., Sebastiani, R. (eds.) SAT 2012. LNCS, vol. 7317, pp. 114–128. Springer, Heidelberg (2012)
11. Janota, M., Marques-Silva, J.: Abstraction-based algorithm for 2QBF. In: Sakallah, K.A., Simon, L. (eds.) SAT 2011. LNCS, vol. 6695, pp. 230–244. Springer, Heidelberg (2011)
12. Janota, M., Marques-Silva, J.: Solving QBF by clause selection. In: Proceedings International Joint Conference on Artificial Intelligence (IJCAI), pp. 325–331 (2015)
13. Lonsing, F., Biere, A.: DepQBF: a dependency-aware QBF solver (system description). J. Satisfiability Boolean Model. Comput. **7**, 71–76 (2010)
14. Marques-Silva, J.P., Sakallah, K.A.: Boolean satisfiability in electronic design automation. In: Proceednigs Design Automation Conference (DAC), pp. 675–680 (2000)
15. Mishchenko, A., Brayton, R.K., Feng, W., Greene, J.W.: Technology mapping into general programmable cells. In: Proceedings International Symposium on Field-Programmable Gate Arrays (FPGA), pp. 70–73 (2015)
16. Mneimneh, M., Sakallah, K.A.: Computing vertex eccentricity in exponentially large graphs: QBF formulation and solution. In: Giunchiglia, E., Tacchella, A. (eds.) SAT 2003. LNCS, vol. 2919, pp. 411–425. Springer, Heidelberg (2004)
17. Pigorsch, F., Scholl, C.: Exploiting structure in an AIG based QBF solver. In: Proceedings Design, Automation and Test in Europe (DATE), pp. 1596–1601 (2009)
18. Remshagen, A., Truemper, K.: An effective algorithm for the futile questioning problem. J. Autom. Reasoning **34**(1), 31–47 (2005)

19. Ryvchin, V., Strichman, O.: Faster extraction of high-level minimal unsatisfiable cores. In: Sakallah, K.A., Simon, L. (eds.) SAT 2011. LNCS, vol. 6695, pp. 174–187. Springer, Heidelberg (2011)
20. Tseitin, G.: On the complexity of derivation in propositional calculus. Studies in Constructive Mathematics and Mathematical Logic (1970)

# Dependency QBF

# Dependency Schemes for DQBF

Ralf Wimmer[1,2](✉), Christoph Scholl[1], Karina Wimmer[1], and Bernd Becker[1]

[1] Albert-Ludwigs-Universität Freiburg im Breisgau, Freiburg im Breisgau, Germany
{wimmer,scholl,wimmerka,becker}@informatik.uni-freiburg.de
[2] Dependable Systems and Software, Saarland University, Saarbrücken, Germany

**Abstract.** Dependency schemes allow to identify variable independencies in QBFs or DQBFs. For QBF, several dependency schemes have been proposed, which differ in the number of independencies they are able to identify. In this paper, we analyze the spectrum of dependency schemes that were proposed for QBF. It turns out that only some of them are sound for DQBF. For the sound ones, we provide a correctness proof, for the others counter examples. Experiments show that a significant number of dependencies can either be added to or removed from a formula without changing its truth value, but with significantly increasing the flexibility for modifying the representation.

## 1 Introduction

During the last two decades an enormous progress in the solution of quantifier-free Boolean formulas (SAT) has been observed. Nowadays, SAT solving is successfully used in many application areas, e.g., in formal verification of hard- and software systems [1,5,10], automatic test pattern generation [11,12], or planning [22]. Motivated by the success of SAT solvers, more general formalisms like quantified Boolean formulas (QBFs) have been studied. However, for applications like the verification of partial circuits [17,24], the synthesis of safe controllers [6], and the analysis of games with incomplete information [21], even QBF is not expressive enough to provide a compact and natural formulation. The reason is that the dependencies of existential variables on universal ones are restricted in QBF: Each existential variable implicitly depends on all universal variables in whose scope it is, i.e., in a model for a QBF in prenex normal form the value of an existential variable can be chosen depending on the values of the universal variables to the left. Consequently, the dependency sets of the existential variables (i.e., the sets of universal variables they may depend on) are linearly ordered w.r.t. set inclusion in QBF. In so-called *dependency quantified Boolean formulas (DQBFs)* this restriction is removed and for each existential variable a dependency set is explicitly specified. The more general DQBF formulations can be tremendously more compact than equivalent QBF formulations; on the other hand the decision problem is NEXPTIME-complete for DQBF [21] instead of

---

This work was partly supported by the German Research Council (DFG) as part of the project "Solving Dependency Quantified Boolean Formulas".

© Springer International Publishing Switzerland 2016
N. Creignou and D. Le Berre (Eds.): SAT 2016, LNCS 9710, pp. 473–489, 2016.
DOI: 10.1007/978-3-319-40970-2_29

PSPACE-complete for QBF. Driven by the needs of the applications mentioned above, research on DQBF solving has started during the last few years, leading to first solvers like ɪDQ and HQS [14,15,18,28].

Although the dependency sets of existential variables are fixed by the syntax of DQBFs, it might be the case that those dependency sets can be reduced or extended without changing the DQBF's truth value. Manipulating dependency sets can be beneficial in different ways: In search-based DQBF solvers [14] extending the DPLL algorithm, manipulation of the dependencies may be used in a similar way as in QBF solvers [20], e.g., for detecting unit literals and conflicts earlier (due to possible universal reductions), for enabling decisions earlier, etc. In DQBF solvers relying on universal expansions like HQS [18], minimizing the number of dependencies to be considered leads to fewer copies of existential variables and thus to faster solving times with lower memory consumption. Manipulating the number of dependencies might lead to similar advantages for ɪDQ [15]. ɪDQ uses instantiation-based solving, i.e., it reduces deciding a DQBF to deciding a series of SAT problems which correspond to partial universal expansions. HQS [18] processes a DQBF by several methods until the resulting formula is a QBF, which then can be solved by an arbitrary QBF solver. Therefore also adding dependencies can be beneficial for HQS in order to make the formula more QBF-like.

An existential variable $y$, which contains a universal variable $x$ in its syntactic dependency set, is called independent of $x$ iff removing $x$ from $y$'s dependency set preserves the truth value of the DQBF. Using the same proof idea as described in [23] for QBF, it can be shown that deciding whether an existential variable is independent of a universal one has the same complexity as deciding the DQBF itself. Therefore one resorts to sufficient criteria to show independencies. We mainly look into generalizations of so-called dependency schemes, which were devised for QBF [16,23,25,26] and can be computed efficiently. A dependency scheme $\mathrm{ds}(\psi)$ for a DQBF $\psi$ gives pairs of universal variables $x$ and existential variables $y$ such that '$y$ potentially depends on $x$'. If $(x,y) \notin \mathrm{ds}(\psi)$, then $y$ is definitely independent of $x$.

The contributions of this paper are as follows:

- For DQBF, the paper provides generalizations of dependency schemes known from QBF and provides for the first time a comprehensive characterization of the dependency schemes which are sound for proving independencies in DQBFs. For the dependency schemes which are not sound for DQBF counterexamples are given.
- The paper proves for all DQBF dependency schemes that both adding and removing dependencies has a unique fixed point.
- Dependency schemes and an orthogonal method based on detection of functional definitions are seamlessly integrated in order to profit from each other.
- We present first experimental results showing an enormous amount of flexibility w.r.t. adding and removing dependencies in numerous benchmark instances.

*Related work.* Several dependency schemes have been introduced for QBF. Based on earlier ideas in [4, 9, 23] defined the so-called standard dependency scheme and the more precise triangle dependency scheme for QBF. In [16, 25, 26] these ideas have been refined further, leading to the even more precise resolution path dependency scheme. In [19, 20] applications of dependency schemes to expansion-based and search-based QBF solvers have been intensively discussed. The correctness of using dependency schemes in search-based QBF solvers like DepQBF is studied in [27], showing that using quadrangle and triangle dependencies in that context is unsound. Instead (sound) reflexive variants of them are proposed. In [28] dependency schemes for DQBF has been considered first. [28] contains a generalization of the simple standard dependency scheme to DQBF, but neither a comprehensive characterization of the dependency schemes which are sound for DQBF nor any deeper analysis.

*Structure of this paper.* In the next section, we introduce the necessary foundations on DQBFs, Sect. 3 contains the main part of the paper on dependency schemes for DQBF. Section 4 shows first experimental results evaluating the flexibility provided by the different methods. Finally, Sect. 5 concludes the paper with a summary and directions for future research.

## 2   Foundations

Let $\varphi, \kappa$ be quantifier-free Boolean formulas over the set $V$ of variables and $v \in V$. We denote by $\varphi[\kappa/v]$ the Boolean formula which results from $\varphi$ by replacing all occurrences of $v$ (simultaneously) by $\kappa$. For a set $V' \subseteq V$, we denote by $\mathcal{A}(V')$ the set of Boolean assignments for $V'$, i.e., $\mathcal{A}(V') = \{\nu \mid \nu : V' \to \{0,1\}\}$. For each formula $\varphi$ over $V$, a variable assignment $\nu$ to the variables in $V$ induces a truth value 0 or 1 of $\varphi$, which we call $\nu(\varphi)$.

**Definition 1 (Syntax of DQBF).** *Let $V = \{x_1, \ldots, x_n, y_1, \ldots, y_m\}$ be a set of Boolean variables. A* dependency quantified Boolean formula *(DQBF) $\psi$ over $V$ has the form $\psi := \forall x_1 \ldots \forall x_n \exists y_1(D_{y_1}) \ldots \exists y_m(D_{y_m}) : \varphi$, where $D_{y_i} \subseteq \{x_1, \ldots, x_n\}$ for $i = 1, \ldots, m$ is the* dependency set *of $y_i$, and $\varphi$ is a quantifier-free Boolean formula over $V$, called the* matrix *of $\psi$.*

$V_\psi^\forall = \{x_1, \ldots, x_n\}$ is the set of universal and $V_\psi^\exists = \{y_1, \ldots, y_m\}$ the set of existential variables of $\psi$. We often write $\psi = Q : \varphi$ with the quantifier prefix $Q$ and the matrix $\varphi$. $Q \setminus \{v\}$ denotes the prefix that results from removing a variable $v \in V$ from $Q$ together with its quantifier. If $v$ is existential, then its dependency set is removed as well; if $v$ is universal, then all occurrences of $v$ in the dependency sets of existential variables are removed. Similarly we use $Q \cup \{\exists y(D_y)\}$ to add existential variables to the prefix. In this paper, we always assume that a DQBF $\psi = Q : \varphi$ as in Definition 1 with $\varphi$ in conjunctive normal form (CNF) is given. A formula is in CNF if it is a conjunction of (non-tautological) *clauses*; a clause is a disjunction of *literals*, and a literal is either a variable $v$ or its negation $\neg v$. As usual, we identify a formula in CNF with its set of clauses and a clause with its set of literals. For a formula $\varphi$ (resp. clause $C$, literal $l$), $\mathrm{var}(\varphi)$ (resp. $\mathrm{var}(C)$, $\mathrm{var}(l)$) means the set of variables occurring in $\varphi$ (resp. $C$, $l$); $\mathrm{lit}(\varphi)$ ($\mathrm{lit}(C)$) denotes the set of literals occurring in $\varphi$ ($C$).

A QBF (in prenex normal form) is a DQBF such that $D_y \subseteq D_{y'}$ or $D_{y'} \subseteq D_y$ holds for any two existential variables $y, y' \in V_\psi^\exists$. Then the variables in $V$ can be ordered resulting in a linear quantifier prefix, such that for each $y \in V_\psi^\exists$, $D_y$ equals the set of universal variables which are to the left of $y$.

The semantics of a DQBF is typically defined by so-called Skolem functions.

**Definition 2 (Semantics of DQBF).** *Let $\psi$ be a DQBF as above. It is satisfiable iff there are functions $s_y : \mathcal{A}(D_y) \to \{0,1\}$ for $y \in V_\psi^\exists$ such that replacing each $y \in V_\psi^\exists$ by (a Boolean expression for) $s_y$ turns $\phi$ into a tautology. The functions $(s_y)_{y \in V_\psi^\exists}$ are called* Skolem functions *for $\psi$.*

The elimination of universal variables in solvers like HQS [18] is done by *universal expansion* [2,7,8,17]:

**Definition 3 (Universal expansion).** *For a DQBF $\psi = \forall x_1 \ldots \forall x_n \exists y_1(D_{y_1}) \ldots \exists y_m(D_{y_m}) : \varphi$ with $Z_{x_i} = \{y_j \in V_\psi^\exists \mid x_i \in D_{y_j}\}$, the* universal expansion *w.r.t. variable $x_i \in V_\psi^\forall$, is defined by*

$$(Q \backslash \{x_i\}) \cup \{\exists y_j'(D_{y_j} \backslash \{x_i\}) \mid y_j \in Z_{x_i}\} : \varphi[1/x_i] \wedge \varphi[0/x_i][y_j'/y_j] \text{ for all } y_j \in Z_{x_i}].$$

$\psi$ and its universal expansion have the same truth value, they are called 'equi-satisfiable'. It can be seen from Definition 3 that the universal expansion w.r.t. a universal variable requires to double all existential variables which depend on it. This shows that reducing the dependency sets can be beneficial (as long as an equisatisfiable DQBF results). Increasing dependency sets may be beneficial as well, if adding dependencies makes the DQBF more QBF-like (see Sect. 3.2).

Manipulating the dependency sets is done using so-called dependency schemes in QBF. In the following section, we investigate which of the QBF dependency schemes can be generalized to DQBF and which cannot.

# 3   Dependency Schemes

Let $\psi = \forall x_1 \ldots \forall x_n \exists y_1(D_{y_1}) \ldots \exists y_m(D_{y_m}) : \varphi$ be a DQBF. For $x \in V_\psi^\forall$ and $y \in V_\psi^\exists$, we denote by $\psi \ominus (x, y)$ the formula which results from $\psi$ by removing the dependency of $y$ on $x$, i.e., by replacing $D_y$ with $D_y \backslash \{x\}$. Accordingly, $\psi \oplus (x, y)$ results from $\psi$ by replacing $D_y$ with $D_y \cup \{x\}$.

**Definition 4.** *An existential variable $y \in V_\psi^\exists$ is* independent *of a universal variable $x \in V_\psi^\forall$ in $\psi$, iff $\psi \oplus (x, y)$ and $\psi \ominus (x, y)$ have the same truth value. In this case, the pair $(x, y)$ is called a* pseudo-dependency.

Dependency schemes provide sufficient criteria to show variable independencies.

**Definition 5 (Dependency scheme).** *A dependency scheme for a DQBF* $\psi$ *is a relation* $\mathrm{ds}(\psi) \subseteq V_\psi^\forall \times V_\psi^\exists$ *such that* $(x, y) \notin \mathrm{ds}(\psi)$ *implies that* $y$ *is independent of* $x$ *in* $\psi$.

Dependency schemes for QBF have been considered in several papers like [16, 19, 20, 23, 25–27]. They encompass the standard, strict standard, (reflexive) triangle, (reflexive) quadrangle and the resolution path dependency schemes. In the following, we generalize them to DQBF.

Most dependency schemes are based on 'connections' between clauses:

**Definition 6 (Connected).** *Let* $\psi = Q : \varphi$ *be a DQBF. A Z-path for* $Z \subseteq V$ *between two clauses* $C, C' \in \varphi$ *is a sequence* $C_1, \ldots, C_n$ *of clauses with* $C = C_1$, $C' = C_n$ *such that* $\mathrm{var}(C_i) \cap \mathrm{var}(C_{i+1}) \cap Z \neq \emptyset$ *for all* $1 \leq i < n$. *A sequence* $v_1, \ldots, v_{n-1}$ *of variables with* $v_i \in \mathrm{var}(C_i) \cap \mathrm{var}(C_{i+1}) \cap Z$ *for all* $1 \leq i < n$ *is called a* connecting sequence *of the Z-path. Two clauses* $C, C' \in \varphi$ *are connected w.r.t. Z (written* $C \overset{Z}{\leftrightarrow} C'$*) if there is a Z-path between* $C$ *and* $C'$.

Using the notion of connected clauses, Samer et al. defined the standard (sdep) and triangle (tdep) dependency schemes [23]. Van Gelder [16] generalized standard dependencies to strict standard dependencies (ssdep) and triangle dependencies to quadrangle dependencies (qdep), using connected clauses as well.

For the definition of stronger dependency schemes (i.e., of dependency schemes leading to *smaller* relations $\mathrm{ds}(\psi)$, and thus detecting more independencies), in [16, 25, 26] a more restricted notion of connected clauses has been introduced which leads to so-called resolution path dependencies:

**Definition 7 (Resolution path connected).** *Let* $\psi = Q : \varphi$ *be a DQBF. A resolution Z-path for* $Z \subseteq V$ *between two clauses* $C, C' \in \varphi$ *is a sequence* $C_1, \ldots, C_n$ *of clauses with* $C = C_1$, $C' = C_n$ *such that for all* $1 \leq i < n$ *there is a literal* $l_i$ *with* $\mathrm{var}(l_i) \in Z$, $l_i \in C_i$, $\neg l_i \in C_{i+1}$, *and for all* $1 \leq i < n-1$ *we have* $\mathrm{var}(l_i) \neq \mathrm{var}(l_{i+1})$.[1] *The sequence* $\mathrm{var}(l_1), \ldots, \mathrm{var}(l_{n-1})$ *is called a* connecting sequence *of the resolution Z-path. Two clauses* $C, C' \in \varphi$ *are* resolution path connected *w.r.t. Z (written* $C \overset{Z}{\underset{\mathrm{rp}}{\leftrightarrow}} C'$*) if there is a resolution Z-path between* $C$ *and* $C'$.

Figure 1 gives an overview of the dependency schemes that have been proposed for QBF. The figure also shows the relation between the different dependency schemes. An arrow from a dependency scheme $\mathrm{ds}_1$ to a scheme $\mathrm{ds}_2$ means that $\mathrm{ds}_1(\psi) \subseteq \mathrm{ds}_2(\psi)$ holds for all DQBFs $\psi$ and that there is at least one DQBF for which the subset relation is strict. $\mathrm{ds}_1$ is more precise than $\mathrm{ds}_2$ in the sense that every independence identified by $\mathrm{ds}_2$ is also identified by $\mathrm{ds}_1$, but not necessarily vice versa. Arrows that can be derived by transitivity have been omitted. Because the resolution path dependency scheme [16, 25, 26] is the quadrangle dependency scheme

---

[1] In [16] there is an additional constraint 'the resolvent of $C_i$ and $C_{i+1}$ w.r.t. $l_i$ is non-tautologous'. This constraint has to be removed according to [25, 26]. If it is not removed, resolution path dependencies are not sound.

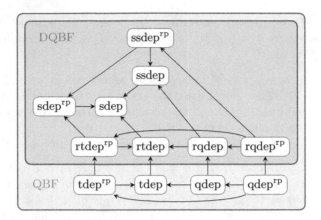

**Fig. 1.** The different dependency schemes for QBF and DQBF. Only those in the blue box are sound for DQBF, while all in the green box are sound for QBF. (Color figure online)

using 'resolution path connectivity' (Definition 7) instead of simple connectivity (Definition 6), we renamed it into *quadrangle resolution path dependency scheme* (qdep$^{\mathrm{rp}}$). For each dependency scheme in $\{\text{tdep}, \text{tdep}^{\mathrm{rp}}, \text{qdep}, \text{qdep}^{\mathrm{rp}}\}$ there is a so-called 'reflexive' counterpart (called rtdep, rtdep$^{\mathrm{rp}}$, rqdep, rqdep$^{\mathrm{rp}}$, resp.) which is weaker than the original scheme. Reflexivity has been introduced in [27]. Exact definitions for the different dependency schemes for (D)QBF follow in Definition 8.

After giving the definition, we will have a closer look into the different dependency schemes. The main result of the paper will be the fact that for DQBF only those dependency schemes in the blue box are sound and the others are not.

**Definition 8 (Dependency schemes).** *Let $\psi$ be a DQBF as in Def. 1. Furthermore, let $x^* \in V_\psi^\forall$ and $y^* \in V_\psi^\exists$ such that $x^* \in D_{y^*}$. We set $Z_{x^*} := \{z \in V_\psi^\exists \mid x^* \in D_z\}$.*

*For the following dependency schemes, $(x^*, y^*) \in \mathrm{ds}(\psi)$ holds iff there are clauses $C_1, C_2, C_3, C_4 \in \varphi$ with $x^* \in C_1$, $\neg x^* \in C_2$, $y^* \in C_3$, and $\neg y^* \in C_4$ such that the following requirements hold:*

*1.* standard dependency scheme sdep [23]:

$$C_1 \xleftrightarrow{Z_{x^*}} C_3 \vee C_2 \xleftrightarrow{Z_{x^*}} C_3 \vee C_1 \xleftrightarrow{Z_{x^*}} C_4 \vee C_2 \xleftrightarrow{Z_{x^*}} C_4$$

*2.* strict standard dependency scheme ssdep [16]:

$$(C_1 \xleftrightarrow{Z_{x^*}} C_3 \vee C_1 \xleftrightarrow{Z_{x^*}} C_4) \wedge (C_2 \xleftrightarrow{Z_{x^*}} C_3 \vee C_2 \xleftrightarrow{Z_{x^*}} C_4)$$

*3.* reflexive triangle dependency scheme rtdep [27]:

$$(C_1 \xleftrightarrow{Z_{x^*}} C_3 \vee C_2 \xleftrightarrow{Z_{x^*}} C_3) \wedge (C_1 \xleftrightarrow{Z_{x^*}} C_4 \vee C_2 \xleftrightarrow{Z_{x^*}} C_4)$$

4. triangle dependency scheme tdep [23]:

$$(C_1 \xleftarrow{Z_{x^*}\setminus\{y^*\}} C_3 \vee C_2 \xleftarrow{Z_{x^*}\setminus\{y^*\}} C_3) \wedge (C_1 \xleftarrow{Z_{x^*}\setminus\{y^*\}} C_4 \vee C_2 \xleftarrow{Z_{x^*}\setminus\{y^*\}} C_4)$$

5. reflexive quadrangle dependency scheme rqdep [27]:

$$(C_1 \xleftrightarrow{Z_{x^*}} C_3 \wedge C_2 \xleftrightarrow{Z_{x^*}} C_4) \vee (C_1 \xleftrightarrow{Z_{x^*}} C_4 \wedge C_2 \xleftrightarrow{Z_{x^*}} C_3)$$

6. quadrangle dependency scheme qdep [16]:

$$(C_1 \xleftrightarrow{Z_{x^*}\setminus\{y^*\}} C_3 \wedge C_2 \xleftrightarrow{Z_{x^*}\setminus\{y^*\}} C_4) \vee (C_1 \xleftrightarrow{Z_{x^*}\setminus\{y^*\}} C_4 \wedge C_2 \xleftrightarrow{Z_{x^*}\setminus\{y^*\}} C_3)$$

The references given for each of these dependency schemes refer to the original definition for QBF. Only the standard dependency scheme has been considered for DQBF so far [28]. Its correctness is proven in [29].

For each of these dependency schemes sdep, ssdep, rtdep, tdep, rqdep, qdep, a stronger variant is obtained by replacing connectedness $C_i \xleftrightarrow{Z_{x^*}} C_j$ ($C_i \xleftrightarrow{Z_{x^*}\setminus\{y^*\}} C_j$) with resolution path connectedness $C_i \xleftrightarrow[\mathrm{rp}]{Z_{x^*}} C_j$ ($C_i \xleftrightarrow[\mathrm{rp}]{Z_{x^*}\setminus\{y^*\}} C_j$, resp.). This yields the standard resolution path dependency scheme, the strict standard resolution path dependency scheme, etc. [16, 25–27]. They are denoted with $\mathrm{sdep}^{\mathrm{rp}}$, $\mathrm{ssdep}^{\mathrm{rp}}$, $\mathrm{rtdep}^{\mathrm{rp}}$, $\mathrm{tdep}^{\mathrm{rp}}$, $\mathrm{rqdep}^{\mathrm{rp}}$, and $\mathrm{qdep}^{\mathrm{rp}}$, respectively. We use the following sets of dependency schemes:

$$\Delta^{\mathrm{dqbf}} = \{\mathrm{sdep}, \mathrm{sdep}^{\mathrm{rp}}, \mathrm{ssdep}, \mathrm{ssdep}^{\mathrm{rp}}, \mathrm{rtdep}, \mathrm{rtdep}^{\mathrm{rp}}, \mathrm{rqdep}, \mathrm{rqdep}^{\mathrm{rp}}\},$$

$$\Delta^{\mathrm{qbf}} = \Delta^{\mathrm{dqbf}} \cup \{\mathrm{tdep}, \mathrm{tdep}^{\mathrm{rp}}, \mathrm{qdep}, \mathrm{qdep}^{\mathrm{rp}}\},$$

The quadrangle resolution path dependency scheme has been introduced as 'resolution path dependency scheme' in [16, 25, 26]. For a clear categorization, however, we prefer to call it the 'quadrangle resolution path dependency scheme' $\mathrm{qdep}^{\mathrm{rp}}$.

The relations between the dependency schemes shown in Fig. 1 are the immediate consequences of the different dependency schemes' definition. As each resolution path connection according to Definition 7 is also a simple connection according to Definition 6, but not vice versa, each variant using resolution paths is stronger than its counterpart that uses simple paths.

For all of the defined dependency schemes, it is known that the following result holds for QBF:

**Theorem 1** ([16, 23, 25–27]). *Let $\psi = Q : \varphi$ be a **QBF**. Let $x^* \in V_\psi^\forall$ and $y^* \in V_\psi^\exists$ such that $x^* \in D_{y^*}$. If, for some dependency scheme $\mathrm{ds} \in \Delta^{\mathrm{qbf}}$, $(x^*, y^*) \notin \mathrm{ds}(\psi)$ and $\psi \ominus (x^*, y^*)$ is a **QBF** as well, then $y^*$ is independent of $x^*$.*

In the following, we will prove (1) that the reflexive resolution path dependency scheme (and all weaker schemes) are sound for DQBF as well, and (2) that the triangle dependency scheme (and all stronger schemes) are unsound for DQBF.

We start by showing that the triangle dependency scheme is unsound for arbitrary DQBFs:

**Theorem 2.** *There is a DQBF $\psi = Q : \varphi$ with $x^* \in V_\psi^\forall$, $y^* \in V_\psi^\exists$ such that $x^* \in D_{y^*}$, with the following property: $y^*$ is not independent of $x^*$, but $(x^*, y^*) \notin$ tdep$(\psi)$, $(x^*, y^*) \notin$ tdep$^{\mathrm{rp}}(\psi)$, $(x^*, y^*) \notin$ qdep$(\psi)$, and $(x^*, y^*) \notin$ qdep$^{\mathrm{rp}}(\psi)$.*

*Proof.* Let

$$D_1 = (x_1 \vee x_2 \vee \neg y_1), \quad D_2 = (x_2 \vee y_1 \vee y_2)$$
$$D_3 = (\neg x_2 \vee y_1 \vee \neg y_2), \quad D_4 = (\neg x_1 \vee \neg x_2 \vee \neg y_1), \text{ and}$$
$$\varphi = D_1 \wedge D_2 \wedge D_3 \wedge D_4.$$

Consider the DQBF $\psi = \forall x_1 \forall x_2 \exists y_1(x_1, x_2) \exists y_2(x_1) : \varphi$. First we show for $\psi$ that $y_1$ is *not* independent of $x_1$, i.e., we have to prove that $\psi$ is satisfiable, but $\psi' = \psi \ominus (x_1, y_1) = \forall x_1 \forall x_2 \exists y_1(x_2) \exists y_2(x_1) : \varphi$ is unsatisfiable. We prove this by considering the full universal expansions of $\psi$ and $\psi'$ w.r.t. variables $x_1$ and $x_2$:

- $\psi$ is equisatisfiable to

$$[(\neg y_1^{00}) \wedge (y_1^{00} \vee y_2^0)] \wedge [(y_1^{01} \vee \neg y_2^0)] \wedge [(y_1^{10} \vee y_2^1)] \wedge [(y_1^{11} \vee \neg y_2^1) \wedge (\neg y_1^{11})]$$

A satisfying assignment is given by $y_2^0 = y_1^{01} = y_1^{10} = 1$, $y_1^{11} = y_2^1 = y_1^{00} = 0$.
- $\psi'$ is equisatisfiable to

$$[(\neg y_1^0) \wedge (y_1^0 \vee y_2^0)] \wedge [(y_1^1 \vee \neg y_2^0)] \wedge [(y_1^0 \vee y_2^1)] \wedge [(y_1^1 \vee \neg y_2^1) \wedge (\neg y_1^1)]$$

Due to the unit clause $(\neg y_1^0)$, $y_1^0$ has to be 0. Therefore $y_2^0$ has to be 1 and thus $y_1^1$ has to be 1. This contradicts the unit clause $(\neg y_1^1)$.

On the other hand, it is easy to see that $(x_1, y_1) \notin$ tdep$(\psi)$. For $(x_1, y_1) \in$ tdep$(\psi)$ we would need clauses $C_1, C_2, C_3, C_4 \in \varphi$ with $x_1 \in$ var$(C_1)$, $x_1 \in$ var$(C_2)$, $y_1 \in C_3$, $\neg y_1 \in C_4$, such that $C_1 \xleftrightarrow{\{y_2\}} C_3$ and $C_2 \xleftrightarrow{\{y_2\}} C_4$. The only clauses in $\varphi$ containing $\neg x_1$ or $x_1$ are $D_1$ and $D_4$. Since $\neg y_1$ is the only existential literal in $D_1$ and $D_4$, the only $\{y_2\}$-path starting with $D_1$ $(D_4)$ is the sequence $D_1$ $(D_4)$ of length 1. Therefore a suitable clause $C_3$ cannot be found in $\varphi$.

$(x_2, y_1) \notin$ tdep$(\psi)$ implies $(x_2, y_1) \notin$ tdep$^{\mathrm{rp}}(\psi)$, $(x_2, y_1) \notin$ qdep$(\psi)$, and $(x_2, y_1) \notin$ qdep$^{\mathrm{rp}}(\psi)$. $\qquad \square$

The next step is to prove that rqdep$^{\mathrm{rp}}$ is sound for DQBF.

**Theorem 3.** *Let $x^* \in V_\psi^\forall$ and $y^* \in V_\psi^\exists$ such that $x^* \in D_{y^*}$. If $(x^*, y^*) \notin$ rqdep$^{\mathrm{rp}}(\psi)$, then $y^*$ is independent of $x^*$.*

For the proof, we assume that $y^*$ is *not* independent of $x^*$ and show that then $(x^*, y^*) \in$ rqdep$^{\mathrm{rp}}(\psi)$. The main idea of the proof consists of using universal expansion on the DQBF until known results for QBFs become applicable. Then these results imply $(x^*, y^*) \in$ rqdep$^{\mathrm{rp}}(\psi)$. Before we come to the proof, we have to consider the following technical lemma on DQBFs resulting from universal expansion:

**Lemma 1.** *Let $\psi = Q : \varphi$ be a DQBF and $\psi' = Q' : \varphi'$ a DQBF derived from $\psi$ by universally expanding some variables in $\psi$. If there exists a resolution Z-path from $C_1 \in \varphi'$ to $C_2 \in \varphi'$, then the connecting sequence $y_1, \ldots, y_n$ (see Definition 7) does not include a pair $(y_i, y_{i+1})$ where $y_i$ and $y_{i+1}$ are two copies of the same existential variable in $\psi$.*

*Proof.* Assume that the resolution $Z$-path has a connecting sequence with a pair $(y_i, y_{i+1})$ where $y_i$ and $y_{i+1}$ are two copies of the same existential variable in $\psi$. According to the definition of resolution $Z$-paths, $y_i \neq y_{i+1}$, i.e., $y_i$ and $y_{i+1}$ are two *different* copies of the same existential variable in $\psi$. Then there exists a clause $C \in \varphi'$ with $y_i, y_{i+1} \in \text{var}(C)$, $y_i \neq y_{i+1}$. This contradicts the definition of universal expansion (see Definition 3), since clauses in a universal expansion can contain at most one copy of the same original variable. □

Now we come to the proof of Theorem 3. After the proof, its construction is illustrated by Example 1.

*Proof.* Let $\psi = \forall x_1 \ldots \forall x_n \exists y_1 (D_{y_1}) \ldots \exists y_m (D_{y_m}) : \varphi$ be a DQBF. W. l. o. g. assume that $x_1 \in D_{y_1}$ and $y_1$ is not independent of $x_1$. We have to prove that $(x_1, y_1) \in \text{rqdep}^{\text{rp}}(\psi)$.

Since, in $\psi$, $y_1$ is not independent of $x_1$, $\psi$ has to be satisfiable and $\psi \ominus (x_1, y_1)$ is unsatisfiable. In $\psi$, we universally expand on the remaining universal variables $x_2, \ldots, x_n$. Let $\boldsymbol{y}_{\text{indep}}$ be the copies of existential variables $y_i$ ($i \in \{2, \ldots, m\}$) with $x_1 \notin D_{y_i}$, $\boldsymbol{y}_{\text{dep}}$ be the copies of existential variables $y_i$ ($i \in \{2, \ldots, m\}$) with $x_1 \in D_{y_i}$, and $y_1^1, \ldots, y_1^k$ the copies of $y_1$ with $k = 2^{|D_{y_1}|-1}$. Thus, universal expansion results in the following **QBF** $\psi'$:

$$\psi' = \exists \boldsymbol{y}_{\text{indep}} \forall x_1 \exists y_1^1 \ldots \exists y_1^k \exists \boldsymbol{y}_{\text{dep}} : \varphi'.$$

Now we start to move copies $y_1^i$ from the right of $x_1$ to the left of $x_1$. Since $\psi \ominus (x_1, y_1)$ is unsatisfiable, $\psi'$ will get unsatisfiable when *all* $y_1^1 \ldots y_1^k$ have been moved to the left of $x_1$. So there exists a maximal subset $\boldsymbol{y}_1^{\text{mov}} \subsetneq \{y_1^1, \ldots, y_1^k\}$ (with $\boldsymbol{y}_1^{\text{stay}} = \{y_1^1, \ldots y_1^k\} \setminus \boldsymbol{y}_1^{\text{mov}}$ containing the remaining variables from $\{y_1^1 \ldots y_1^k\}$) with the property that

$$\psi'' = \exists \boldsymbol{y}_{\text{indep}} \exists \boldsymbol{y}_1^{\text{mov}} \forall x_1 \exists \boldsymbol{y}_1^{\text{stay}} \exists \boldsymbol{y}_{\text{dep}} : \varphi'$$

is satisfied and, in $\psi''$, each variable from $\boldsymbol{y}_1^{\text{stay}}$ is *not* independent of $x_1$. That means that moving an arbitrary variable from $\boldsymbol{y}_1^{\text{stay}}$ to the left of $x_1$ will turn $\psi''$ from satisfiable to unsatisfiable. The set $\boldsymbol{y}_1^{\text{stay}}$ is not empty, since otherwise $y_1$ would be independent of $x_1$ in $\psi$. Choose an arbitrary existential variable $y_1^j$ in $\boldsymbol{y}_1^{\text{stay}}$. Since $\psi''$ is a QBF where $y_1^j$ is not independent of $x_1$, we have $(x_1, y_1^j) \in \text{qdep}^{\text{rp}}(\psi'')$ due to Theorem 1, i.e., $\exists C_1, C_2, C_3, C_4 \in \varphi'$ with $x_1 \in C_1$, $\neg x_1 \in C_2$, $y_1^j \in C_3$, $\neg y_1^j \in C_4$, and – w. l. o. g. – $C_1 \xleftrightarrow[\text{rp}]{Z_{x_1} \setminus \{y_1^j\}} C_3$ and $C_2 \xleftrightarrow[\text{rp}]{Z_{x_1} \setminus \{y_1^j\}} C_4$ with $Z_{x_1} = \boldsymbol{y}_1^{\text{stay}} \cup \boldsymbol{y}_{\text{dep}}$.

Now we make use of a simple property of the process of universal expansion in order to turn the two constructed resolution $(Z_{x_1} \setminus \{y_1^j\})$-paths for $\psi''$ into suitable resolution paths for the original DQBF $\psi$ proving that $(x_1, y_1) \in \text{rqdep}^{\text{rp}}(\psi)$: Each clause $C$ in a universal expansion can be mapped back to a clause $C^{\text{orig}}$ of the original formula 'it results from' by (1) adding universal literals which have been removed by the universal expansion, since they are unsatisfied in the copy of the clause at hand, and by (2) replacing copies of existential literals $\ell$ by the existential literals $\ell^{\mathrm{orig}}$ in the original formula. Now consider one of the constructed resolution $(Z_{x_1} \setminus \{y_1^j\})$-paths $D_1, \ldots, D_n$ with $D_1 = C_1$ and $D_n = C_3$ (or $D_1 = C_2$ and $D_n = C_4$). For all $1 \leq i < n$ there is a literal $\ell_i$ with $\mathrm{var}(\ell_i)$ from $Z_{x_1} \setminus \{y_1^j\} = \boldsymbol{y_{\mathrm{dep}}} \cup (\boldsymbol{y_1^{\mathrm{stay}}} \setminus \{y_1^j\})$, $\ell_i \in C_i$, $\neg \ell_i \in C_{i+1}$. It is easy to see that $D_1^{\mathrm{orig}}, \ldots, D_n^{\mathrm{orig}}$ is a resolution $Z'_{x_1}$-path for $\psi$ with $Z'_{x_1} = \{y \in V_\psi^\exists \mid x_1 \in D_y\}$: For all $1 \leq i < n$ there is a literal $\ell_i^{\mathrm{orig}}$ with $\mathrm{var}(\ell_i^{\mathrm{orig}}) \in Z'_{x_1}$, $\ell_i^{\mathrm{orig}} \in C_i$, $\neg \ell_i^{\mathrm{orig}} \in C_{i+1}$. (Note that $\ell_i^{\mathrm{orig}}$ may also be $y_1$ or $\neg y_1$, since the connecting sequence of $D_1, \ldots, D_n$ may contain a copy of $y_1$ which is different from $y_1^j$.) Due to Lemma 1, $\mathrm{var}(\ell_i)$ and $\mathrm{var}(\ell_{i+1})$ are copies of *different* existential variables and thus $\mathrm{var}(\ell_i^{\mathrm{orig}}) \neq \mathrm{var}(\ell_{i+1}^{\mathrm{orig}})$ for all $1 \leq i < n$. Altogether we have proven that there are two resolution $Z'_{x_1}$-paths for $\psi$ with $Z'_{x_1} = \{y \in V_\psi^\exists \mid x_1 \in D_y\}$ leading from $C_1^{\mathrm{orig}} \in \varphi$ with $x_1 \in C_1^{\mathrm{orig}}$ to $C_3^{\mathrm{orig}} \in \varphi$ with $y_1 \in C_3^{\mathrm{orig}}$ and from $C_2^{\mathrm{orig}} \in \varphi$ with $\neg x_1 \in C_2^{\mathrm{orig}}$ to $C_4^{\mathrm{orig}} \in \varphi$ with $\neg y_1 \in C_4^{\mathrm{orig}}$. Thus $(x_1, y_1) \in \mathrm{rqdep}^{\mathrm{rp}}(\psi)$.[2] $\square$

*Example 1.* We illustrate the construction of the proof for Theorem 3 by means of an example. Here we use the same example as in the proof of Theorem 2, i.e.,

$$D_1 = (x_1 \vee x_2 \vee \neg y_1), \quad D_2 = (x_2 \vee y_1 \vee y_2)$$
$$D_3 = (\neg x_2 \vee y_1 \vee \neg y_2), \quad D_4 = (\neg x_1 \vee \neg x_2 \vee \neg y_1),$$
$$\varphi = D_1 \wedge D_2 \wedge D_3 \wedge D_4,$$

and the DQBF $\psi = \forall x_1 \forall x_2 \exists y_1(x_1, x_2) \exists y_2(x_1) : \varphi$. For $\psi$, $y_1$ is *not* independent of $x_1$. In $\psi$, we universally expand on the universal variables different from $x_1$, i.e., we expand on $x_2$. This results in the DQBF

$$\psi' = \forall x_1 \exists y_1^0(x_1) \exists y_1^1(x_1) \exists y_2(x_1) :$$
$$\left[ (x_1 \vee \neg y_1^0) \wedge (y_1^0 \vee y_2) \right] \wedge \left[ (y_1^1 \vee \neg y_2) \wedge (\neg x_1 \vee \neg y_1^1) \right].$$

Since only one universal variable is left, the DQBF is a QBF and can be written (in QBF notation) as

$$\psi' = \forall x_1 \exists y_1^0 \exists y_1^1 \exists y_2 : \left[ (x_1 \vee \neg y_1^0) \wedge (y_1^0 \vee y_2) \right] \wedge \left[ (y_1^1 \vee \neg y_2) \wedge (\neg x_1 \vee \neg y_1^1) \right].$$

There are no existential variables which do not have $x_1$ in their dependency set, i.e., in the notions used in the proof, $\boldsymbol{y_{\mathrm{indep}}} = \emptyset$, $\boldsymbol{y_{\mathrm{dep}}} = \{y_2\}$, and there are two copies ($y_1^0$ and $y_1^1$) of $y_1$. Moving $y_1^0$ or $y_1^1$ to the left of $\forall x_1$ turns the QBF from satisfiable to unsatisfiable, thus, in the notions used in the proof, $\boldsymbol{y_1^{\mathrm{mov}}} = \emptyset$ and $\boldsymbol{y_1^{\mathrm{stay}}} = \{y_1^0, y_1^1\}$, and each variable from $\boldsymbol{y_1^{\mathrm{stay}}}$ is not independent of $x_1$. Choose $y_1^0 \in \boldsymbol{y_1^{\mathrm{stay}}}$. Due to Theorem 1, $(x_1, y_1^0) \in \mathrm{qdep}^{\mathrm{rp}}(\psi')$. This can be seen by choosing $C_1 = C_4 = (x_1 \vee \neg y_1^0)$, $C_2 = (\neg x_1 \vee \neg y_1^0)$, and $C_3 = (y_1^0 \vee y_2)$.

---

[2] The construction does *not* lead to $(Z'_{x_1} \setminus \{y_1\})$-paths and thus $(x_1, y_1) \in \mathrm{qdep}^{\mathrm{rp}}(\psi)$ cannot be proven (which is not surprising due to Theorem 2).

(1) $(x_1 \vee \neg y_1^0)$ is a (trivial) resolution $\{y_1^1, y_2\}$-path with empty connecting sequence showing $C_1 \xleftarrow{\{y_1^1, y_2\}}{}_{\mathrm{rp}} C_4$ and

(2) $(\neg x_1 \vee \neg y_1^1), (y_1^1 \vee \neg y_2), (y_1^0 \vee y_2)$ is a resolution $\{y_1^1, y_2\}$-path with connecting sequence $y_1^1$, $y_2$ showing $C_2 \xleftarrow{\{y_1^1, y_2\}}{}_{\mathrm{rp}} C_3$.

Mapping the clauses of the universal expansion $\psi'$ back to the corresponding original clauses from $\psi$ turns (1) the first path into $(x_1 \vee x_2 \vee \neg y_1)$ (again with empty connecting sequence) and (2) the second path into $(\neg x_1 \vee \neg x_2 \vee \neg y_1)$, $(\neg x_2 \vee y_1 \vee \neg y_2)$, $(x_2 \vee y_1 \vee y_2)$ with connecting sequence $y_1, y_2$.

These two paths are resolution $Z_{x_1}$-paths for $\psi$ with $Z_{x_1} = \{z \in V_\psi^\exists \mid x_1 \in D_z\} = \{y_1, y_2\}$. They prove $(x_1, y_1) \in \mathrm{rqdep}^{\mathrm{rp}}(\psi)$.

## 3.1  Monotonicity of Dependency Schemes

Monotonicity of dependency schemes has been considered before for QBF [23, 26]. In general, dependency schemes are not monotone [23, 26]. Monotone dependency schemes have the advantage that removing pseudo-dependencies identified by that scheme has a unique fixed point, i.e., the order of removal has no influence on the final result.

**Definition 9 (Monotone).**  *Let* ds *be a dependency scheme. It is called monotone if for all DQBFs* $\psi$, *variables* $x^* \in V_\psi^\forall$ *and* $y^* \in V_\psi^\exists$ *such that* $(x^*, y^*) \notin \mathrm{ds}(\psi)$ *the condition* $\mathrm{ds}(\psi \ominus (x^*, y^*)) \subseteq \mathrm{ds}(\psi)$ *is satisfied.*

That means, a dependency scheme is monotone if removing a pseudo-dependency from a DQBF never turns a pseudo-dependency into a proper one.

All dependency schemes which are sound for DQBF in the sense that they can be used to remove pseudo-dependencies have the nice property of monotonicity:

**Theorem 4.** *All dependency schemes in* $\Delta^{\mathrm{dqbf}}$ *are monotone.*

Monotonicity simply follows from the definition of 'connected' (Definition 6) and 'resolution path connected' (Definition 7): Removing a pseudo-dependency $(x^*, y^*)$ means removing $x^*$ from the dependency set of $y^*$. Let $\mathrm{ds} \in \Delta^{\mathrm{dqbf}}$. For $(x^{**}, y^{**}) \in \mathrm{ds}(\psi \ominus (x^*, y^*))$, the existence of certain (resolution) $Z_{x^{**}}$-paths from clauses containing $x^{**}$ or $\neg x^{**}$ to clauses containing $y^{**}$ or $\neg y^{**}$ is needed with $Z_{x^{**}} = \{y \in V_\psi^\exists \mid x^{**} \in D_y\}$. Since the dependency sets of $\psi \ominus (x^*, y^*)$ can only be smaller than or equal to the dependency sets of $\psi$, the corresponding paths are valid for proving $(x^{**}, y^{**}) \in \mathrm{ds}(\psi)$ as well.

## 3.2  Adding Dependencies

For DQBFs, it may not only be beneficial to remove dependencies from the dependency set of an existential variable, but also to add dependencies in order to make the formula more QBF-like. Consider, for instance, the formula

$\forall x_1 \forall x_2 \exists y_1(x_1) \exists y_2(x_2) : \varphi$ with some matrix $\varphi$. It can either be turned into a QBF by removing one of the dependencies or by adding one dependency.

Dependency schemes can be used to add pseudo-dependencies. The intuition is that one may add a dependency $(x^*, y^*)$ according to a dependency scheme ds if ds allows to remove the dependency afterwards.

**Lemma 2.** *Let* ds $\in \Delta^{\mathrm{dqbf}}$ *be a dependency scheme,* $\psi$ *a DQBF, and* $(x^*, y^*) \in V_\psi^\forall \times V_\psi^\exists$ *such that* $x^* \notin D_{y^*}$. *If* $(x^*, y^*) \notin \mathrm{ds}(\psi \oplus (x^*, y^*))$, *then* $\psi$ *and* $\psi \oplus (x^*, y^*)$ *are equisatisfiable.*

*Proof.* If $(x^*, y^*) \notin \mathrm{ds}(\psi \oplus (x^*, y^*))$, then $\psi \oplus (x^*, y^*)$ and $\big(\psi \oplus (x^*, y^*)\big) \ominus (x^*, y^*) = \psi$ are equisatisfiable.    $\square$

We can show that even *adding* dependencies has a unique fixed point. The reason for this is not as simple as for removing pseudo-dependencies in Sect. 3.1 and needs a slightly refined analysis. It relies on the following lemma which says that the order in which dependencies are added does not matter.

**Lemma 3.** *Let* ds $\in \Delta^{\mathrm{dqbf}}$ *be a dependency scheme. Additionally, for* $i = 1, 2$, *let* $x_i \in V_\psi^\forall$, $y_i \in V_\psi^\exists$ *such that* $x_i \notin D_{y_i}$ *and* $(x_i, y_i) \notin \mathrm{ds}(\psi \oplus (x_i, y_i))$. *Then* $(x_2, y_2) \notin \mathrm{ds}(\psi \oplus (x_1, y_1) \oplus (x_2, y_2))$.

*Proof.* W.l.o.g. we assume that ds = sdep and define the following abbreviations:

$$\psi_1 := \psi \oplus (x_1, y_1), \quad \psi_2 := \psi \oplus (x_2, y_2), \quad \psi_{12} := \psi \oplus (x_1, y_1) \oplus (x_2, y_2).$$

Now we distinguish the following two cases:

- **Case 1:** $x_1 \neq x_2$.
  Assume that $(x_2, y_2) \in \mathrm{sdep}(\psi_{12})$. That means, there are clauses $C, C' \in \varphi$ such that $x_2 \in \mathrm{var}(C)$, $y_2 \in \mathrm{var}(C')$ and $C \xleftrightarrow{Z_{\psi_{12}}^{x_2}} C'$ where $Z_{\psi_{12}}^{x_2} = \{y \in V_\psi^\exists \mid x_2 \in D_y^{\psi_{12}}\}$. Since $x_2 \in D_y^{\psi_{12}}$ iff $x_2 \in D_y^{\psi_2}$, we have $C \xleftrightarrow{Z_{\psi_2}^{x_2}} C'$ with $Z_{\psi_2}^{x_2} = \{y \in V_\psi^\exists \mid x_2 \in D_y^{\psi_2}\}$. This in turn means that $(x_2, y_2) \in \mathrm{sdep}(\psi_2)$, which contradicts the assumption of the lemma.
- **Case 2:** $x_1 = x_2$.
  Let $x := x_1 = x_2$. We have $(x, y_i) \notin \mathrm{sdep}(\psi_i)$ for $i = 1, 2$. Assume that $(x, y_2) \in \mathrm{sdep}(\psi_{12})$.
  $(x, y_2) \in \mathrm{sdep}(\psi_{12})$ means there are clauses $C, C' \in \varphi$ such that $x \in \mathrm{var}(C)$, $y_2 \in \mathrm{var}(C')$ and $C \xleftrightarrow{Z_{\psi_{12}}^{x}} C'$, $Z_{\psi_{12}}^{x} = \{y \in V_\psi^\exists \mid x \in D_y^{\psi_{12}}\}$. Say the $Z_{\psi_{12}}^{x}$-path proving $C \xleftrightarrow{Z_{\psi_{12}}^{x}} C'$ is $C = C^{(1)}, C^{(2)}, \ldots, C^{(n)} = C'$ with the connecting sequence $y^{(1)}, y^{(2)}, \ldots, y^{(n-1)}$ and assume w.l.o.g. that there is no $1 \leq i < n$ with $y^{(i)} = y_2$ (otherwise we simply shorten the path). As $(x, y_2) \notin \mathrm{sdep}(\psi_2)$, the connecting sequence has to contain $y_1$, i.e., $y^{(i)} = y_1$ for some $1 \leq i < n$. That means, $C \xleftrightarrow{Z_{\psi_1}^{x}} C^{(i)}$ with $x \in \mathrm{var}(C)$ and $y_1 \in \mathrm{var}(C^{(i)})$, $Z_{\psi_1}^{x} = \{y \in V_\psi^\exists \mid x \in D_y^{\psi_1}\}$. Therefore $(x, y_1) \in \mathrm{sdep}(\psi_1)$, which is a contradiction.

The same argumentation can be applied for all dependency schemes in $\Delta^{\mathrm{dqbf}}$, replacing "connected" by "resolution path connected" where appropriate.    □

**Corollary 1.** *Adding as many dependencies as possible according to* ds $\in \Delta^{\mathrm{dqbf}}$ *leads to a unique result, irrespective of the order.*

### 3.3    Manipulation of Dependencies Using Functional Definitions

CNFs (especially those derived from circuit representations) may contain so-called functional definitions. These are clauses which are logically equivalent to a formula $y \equiv f(v_1, \ldots, v_k)$ with an existential variable $y$, an arbitrary boolean function $f$ and (universal or existential) variables $v_i$ $(1 \leq i \leq k)$. Here $y$ is called the *defined variable*, $f$ is the *defn* of $y$, and the clauses corresponding to the relationship $y \equiv f(v_1, \ldots, v_k)$ are also called the *defining clauses*.

As already mentioned in [28], the detection of functional definitions provides a method for manipulating dependency sets which is completely orthogonal to the dependency schemes presented so far: If the CNF contains defining clauses for $y \equiv f(v_1, \ldots, v_k)$ and $\bigcup_{i=1}^{k} \mathrm{dep}(v_i) \subsetneq D_y$,[3] then $D_y$ can be replaced by $\bigcup_{i=1}^{k} \mathrm{dep}(v_i)$, by $V_\psi^\forall$, or by any set in between without changing the truth value of the DQBF, since any assignment to the universal variables from $\bigcup_{i=1}^{k} \mathrm{dep}(v_i)$ already fixes the only value of $y$ which is able to satisfy the defining clauses.

The manipulation of dependencies using functional definitions can be combined with the manipulation of dependencies using dependency schemes:

If the goal is to remove as many dependencies as possible, then we first remove dependencies based on functional definitions. Then, the dependency sets are already reduced when we start computing dependency schemes. Since this reduces the number of existing (resolution) $Z$-paths, this leads to smaller dependency schemes and thus to the detection of more pseudo-dependencies.

If the goal is to add as many dependencies as possible, we proceed the other way round: We make use of dependency schemes first and then add dependencies due to functional definitions. If we started by adding dependencies due to functional definitions, then the potential to add dependencies due to dependency schemes would be reduced with the same argumentation as above.

## 4    Experiments

We have implemented all of the presented dependency schemes that are sound for DQBF as well as the detection of functional definitions.[4] For each dependency scheme ds, we used an implementation which needs linear run-time in the size of the graphs representing $\overset{Z}{\longleftrightarrow} / \overset{Z}{\underset{\mathrm{rp}}{\longleftrightarrow}}$ for computing whether $(x, y) \in \mathrm{ds}(\psi)$ as in [25]. Our tool supports both adding and removing pseudo-dependencies. We want to compare the effectiveness of the different dependency schemes.

---

[3] For notational convenience, we use here $\mathrm{dep}(v_i) = D_{v_i}$ if $v_i$ is existential and $\mathrm{dep}(v_i) = \{v_i\}$ if $v_i$ is universal.

[4] Our tool and all benchmarks we used are available at https://projects.informatik. uni-freiburg.de/projects/dqbf.

**Table 1.** Effectiveness of the different dependency schemes (f. d. = extraction of functional definitions)

| Benchmark | f. d.? | Original | sdep | sdep$^{rp}$ | ssdep | ssdep$^{rp}$ | rtdep | rtdep$^{rp}$ | rqdep | rqdep$^{rp}$ |
|---|---|---|---|---|---|---|---|---|---|---|
| *adding dependencies* | | | | | | | | | | |
| pec_adder_n_bit_9_7 | no | 19751 | 123628 | 123628 | 123628 | 123628 | 123628 | 123628 | 123628 | 123628 |
| | yes | 113805 | 124038 | 124038 | 124038 | 124038 | 124038 | 124038 | 124038 | 124038 |
| term1_0.50_1.00_3_1 | no | 48678 | 48730 | 48730 | 48745 | 48745 | 48730 | 48866 | 48745 | 48869 |
| | yes | 48768 | 48730 | 48730 | 48745 | 48745 | 48730 | 48866 | 48745 | 48869 |
| C432_0.20_1.00_5_3 | no | 32560 | 32649 | 32960 | 32649 | 32965 | 32649 | 32996 | 32649 | 33001 |
| | yes | 32560 | 32649 | 32960 | 32649 | 32965 | 32649 | 32996 | 32649 | 33001 |
| *removing dependencies* | | | | | | | | | | |
| genbuf15f14unrealy | no | 198412 | 198412 | 198412 | 198412 | 198412 | 198412 | 198412 | 198412 | 198412 |
| | yes | 60873 | 60873 | 60873 | 60873 | 60873 | 60873 | 60873 | 60873 | 60873 |
| term1_0.50_1.00_3_1 | no | 48678 | 32570 | 32506 | 29683 | 29623 | 32570 | 32234 | 29683 | 29375 |
| | yes | 41061 | 27543 | 27479 | 25142 | 25082 | 27543 | 27207 | 25142 | 24834 |
| C432_0.20_1.00_5_3 | no | 32560 | 28397 | 22915 | 28389 | 22883 | 28397 | 21362 | 28389 | 21346 |
| | yes | 32264 | 28109 | 22627 | 28101 | 22595 | 28109 | 21074 | 28101 | 21058 |

All experiments were run on one Intel Xeon E5-2650v2 CPU core at 2.60 GHz clock frequency and 64 GB of main memory under Ubuntu Linux as operating system. As benchmarks we used 4811 DQBF instances from different sources. They encompass the 4381 instances already used in [28]: DQBFs resulting from equivalence checking of incomplete circuits [13,15,17] and controller synthesis problems [6]. We have added 34 instances which were obtained from converting SAT instances into DQBFs that depend only on a logarithmic number of variables [3]. Additionally we used 396 instances resulting from partial equivalence checking problems [17,24].

We applied detection of functional definitions and then the reflexive quadrangle resolution path dependency scheme rqdep$^{rp}$ to remove as many dependencies as possible. The same was done for adding dependencies, using first rqdep$^{rp}$ and then detection of functional definitions. For all instances, our tool terminated within fractions of a second and used less than 50 MB of main memory.

Figure 2 shows the results. We compare the original formula with the formula both after the removal (blue crosses) and the addition (green stars) of pseudo-dependencies. For a fixed instance, a mark on the diago-

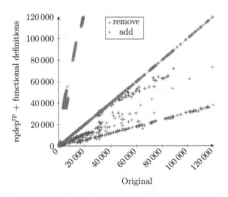

**Fig. 2.** Effectiveness of dependency manipulation using rqdep$^{rp}$ and detection of functional definitions.(Color figure online)

nal means that the dependency set could not be modified, above that it could be increased and below that it could be reduced. We can observe that almost all

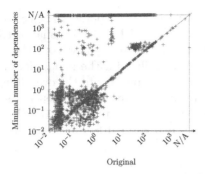 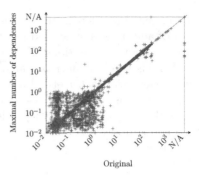

**Fig. 3.** Running times of HQS [18] compared to the original benchmark after removing (left) and adding (right) dependencies using rqdep$^{rp}$ and detecting functional definitions. "N/A" means that the benchmark could not be solved within 3600 s computation time and 8 GB of memory.

instances could be modified; in some cases, the size of the maximal dependency set is ten times the size of the minimal one.

Table 1 shows the numbers of dependencies for some selected instances before and after modifying the dependency sets using the different dependency schemes. We can observe that they are not equally powerful and that the detection of functional definitions can amplify the effectiveness of the dependency schemes. Since the computation times for rqdep$^{rp}$ are in the same range as the computation times for the other dependency schemes, we propose to use rqdep$^{rp}$ together with the detection of functional definitions.

Our work demonstrates the potential of adding/removing dependencies and lays the groundwork for exploiting this flexibility later on. Figure 3 demonstrates that trivial solutions (just adding or removing dependencies) for exploiting the flexibility do not help much. It shows the solution times of HQS [18] on all 4811 instances after applying rqdep$^{rp}$ and detection of functional definitions compared to the original benchmark. The left image is for removing dependencies, the right one for adding dependencies. We can see that manipulating the dependency sets can have a huge effect on the solution times. They can both increase and decrease. This is not really surprising, since both adding and removing dependencies may increase or decrease the 'distance' of the DQBF from an equisatisfiable QBF. Some of the benchmarks even became solvable or unsolvable within the resource limits of 3600 s computation time and 8 GB of memory. These results show that a clever way for exploiting the flexibility is definitely needed to solve more instances in less time.

## 5   Conclusion

We have presented a complete characterization of those dependency schemes which can be generalized from QBF to DQBF and those which cannot. The generalizations are suitable to remove and add dependencies in DQBFs. Both adding and removing dependencies lead to a unique fixed point, irrespective of the order of adding/removing and can be combined with an orthogonal method based on functional definitions. First experimental results show that the presented methods give an enormous amount of flexibility for the manipulation of dependency sets and each method has its own contributions to the overall flexibility. A central task for future research is to develop appropriate heuristics for exploiting the flexibility in the dependency sets in order to turn a DQBF into a QBF at lower costs. The groundwork for such a method has been laid by the presented paper.

**Acknowledgments.** We thank the anonymous reviewers for their really helpful comments.

# References

1. Ashar, P., Ganai, M.K., Gupta, A., Ivancic, F., Yang, Z.: Efficient SAT-based bounded model checking for software verification. In: Margaria, T., Steffen, B., Philippou, A., Reitenspieß, M. (eds.) International Symposium on Leveraging Applications of Formal Methods (ISoLA). Technical Report, vol. TR-2004-6, pp. 157–164, Department of Computer Science, University of Cyprus, Paphos, Cyprus, October 2004
2. Balabanov, V., Chiang, H.K., Jiang, J.R.: Henkin quantifiers and boolean formulae: a certification perspective of DQBF. Theor. Comput. Sci. **523**, 86–100 (2014)
3. Balabanov, V., Jiang, J.H.R.: Reducing satisfiability and reachability to DQBF, September 2015. Talk at the International Workshop on Quantified Boolean Formulas (QBF)
4. Biere, A.: Resolve and expand. In: H. Hoos, H., Mitchell, D.G. (eds.) SAT 2004. LNCS, vol. 3542, pp. 59–70. Springer, Heidelberg (2005)
5. Biere, A., Cimatti, A., Clarke, E.M., Strichman, O., Zhu, Y.: Bounded model checking. Adv. Comput. **58**, 117–148 (2003)
6. Bloem, R., Könighofer, R., Seidl, M.: SAT-based synthesis methods for safety specs. In: McMillan, K.L., Rival, X. (eds.) VMCAI 2014. LNCS, vol. 8318, pp. 1–20. Springer, Heidelberg (2014)
7. Bubeck, U.: Model-based transformations for quantified boolean formulas. Ph.D. thesis, University of Paderborn (2010)
8. Bubeck, U., Büning, H.K.: Dependency quantified horn formulas: models and complexity. In: Biere, A., Gomes, C.P. (eds.) SAT 2006. LNCS, vol. 4121, pp. 198–211. Springer, Heidelberg (2006)
9. Bubeck, U., Kleine Büning, H.: Bounded universal expansion for preprocessing QBF. In: Marques-Silva, J., Sakallah, K.A. (eds.) SAT 2007. LNCS, vol. 4501, pp. 244–257. Springer, Heidelberg (2007)
10. Clarke, E.M., Biere, A., Raimi, R., Zhu, Y.: Bounded model checking using satisfiability solving. Formal Methods Syst. Design **19**(1), 7–34 (2001)
11. Czutro, A., Polian, I., Lewis, M.D.T., Engelke, P., Reddy, S.M., Becker, B.: Thread-parallel integrated test pattern generator utilizing satisfiability analysis. Int. J. Parallel Prog. **38**(3–4), 185–202 (2010)
12. Eggersglüß, S., Drechsler, R.: A highly fault-efficient SAT-based ATPG flow. IEEE Des. Test Comput. **29**(4), 63–70 (2012)

13. Finkbeiner, B., Tentrup, L.: Fast DQBF refutation. In: Sinz, C., Egly, U. (eds.) SAT 2014. LNCS, vol. 8561, pp. 243–251. Springer, Heidelberg (2014)
14. Fröhlich, A., Kovásznai, G., Biere, A.: A DPLL algorithm for solving DQBF. In: International Workshop on Pragmatics of SAT (POS), Trento, Italy (2012)
15. Fröhlich, A., Kovásznai, G., Biere, A., Veith, H.: iDQ: instantiation-based DQBF solving. In: Berre, D.L. (ed.) International Workshop on Pragmatics of SAT (POS). EPiC Series, vol. 27, pp. 103–116. EasyChair, Vienna, Austria. http://www.easychair.org/publications/?page=2037484173
16. Van Gelder, A.: Variable independence and resolution paths for quantified boolean formulas. In: Lee, J. (ed.) CP 2011. LNCS, vol. 6876, pp. 789–803. Springer, Heidelberg (2011)
17. Gitina, K., Reimer, S., Sauer, M., Wimmer, R., Scholl, C., Becker, B.: Equivalence checking of partial designs using dependency quantified Boolean formulae. In: Proceedings of ICCD, pp. 396–403. IEEE CS, Asheville, October 2013
18. Gitina, K., Wimmer, R., Reimer, S., Sauer, M., Scholl, C., Becker, B.: Solving DQBF through quantifier elimination. In: Proceedings of DATE. IEEE, Grenoble, March 2015
19. Lonsing, F., Biere, A.: Efficiently representing existential dependency sets for expansion-based QBF solvers. ENTCS **251**, 83–95 (2009)
20. Lonsing, F., Biere, A.: Integrating dependency schemes in search-based QBF solvers. In: Strichman, O., Szeider, S. (eds.) SAT 2010. LNCS, vol. 6175, pp. 158–171. Springer, Heidelberg (2010)
21. Peterson, G., Reif, J., Azhar, S.: Lower bounds for multiplayer non-cooperative games of incomplete information. Comput. Math. Appl. **41**(7–8), 957–992 (2001)
22. Rintanen, J., Heljanko, K., Niemelä, I.: Planning as satisfiability: parallel plans and algorithms for plan search. Artif. Intell. **170**(12–13), 1031–1080 (2006)
23. Samer, M., Szeider, S.: Backdoor sets of quantified Boolean formulas. J. Autom. Reasoning **42**(1), 77–97 (2009)
24. Scholl, C., Becker, B.: Checking equivalence for partial implementations. In: ACM/IEEE Design Automation Conference (DAC), pp. 238–243. ACM Press, Las Vegas, June 2001
25. Slivovsky, F., Szeider, S.: Computing resolution-path dependencies in linear time. In: Cimatti, A., Sebastiani, R. (eds.) SAT 2012. LNCS, vol. 7317, pp. 58–71. Springer, Heidelberg (2012)
26. Slivovsky, F., Szeider, S.: Quantifier reordering for QBF. J. Autom. Reasoning **56**(4), 459–477 (2016)
27. Slivovsky, F., Szeider, S.: Soundness of Q-resolution with dependency schemes. Theor. Comput. Sci. **612**, 83–101 (2016)
28. Wimmer, R., Gitina, K., Nist, J., Scholl, C., Becker, B.: Preprocessing for DQBF. In: Heule, M., et al. (eds.) SAT 2015. LNCS, vol. 9340, pp. 173–190. Springer, Heidelberg (2015). doi:10.1007/978-3-319-24318-4_13
29. Wimmer, R., Gitina, K., Nist, J., Scholl, C., Becker, B.: Preprocessing for DQBF. Reports of SFB/TR 14 AVACS 110. http://www.avacs.org

# Lifting QBF Resolution Calculi to DQBF

Olaf Beyersdorff[1], Leroy Chew[1(✉)], Renate A. Schmidt[2], and Martin Suda[2]

[1] School of Computing, University of Leeds, Leeds, UK
{o.beyersdorff,mm12lnc}@leeds.ac.uk
[2] School of Computer Science, University of Manchester, Manchester, UK
{Renate.Schmidt,martin.suda}@manchester.ac.uk

**Abstract.** We examine existing resolution systems for quantified Boolean formulas (QBF) and answer the question which of these calculi can be lifted to the more powerful Dependency QBFs (DQBF). An interesting picture emerges: While for QBF we have the strict chain of proof systems Q-Res < IR-calc < IRM-calc, the situation is quite different in DQBF. The obvious adaptations of Q-Res and likewise universal resolution are too weak: they are not complete. The obvious adaptation of IR-calc has the right strength: it is sound and complete. IRM-calc is too strong: it is not sound any more, and the same applies to long-distance resolution. Conceptually, we use the relation of DQBF to effectively propositional logic (EPR) and explain our new DQBF calculus based on IR-calc as a subsystem of first-order resolution.

## 1 Introduction

The logic of dependency quantified Boolean formulas (DQBF) [23] generalises the notion of quantified Boolean formulas (QBF) that allow Boolean quantifiers over a propositional problem. DQBF is a relaxation of QBF in that the quantifier order is no longer necessarily linear and the dependencies of the quantifiers are completely specified. This is achieved using Henkin quantifiers [16], usually put into a Skolem form. DQBF is NEXPTIME-complete [1], compared to the PSPACE-completeness of QBF [28]. Thus, unless the classes are equal, many problems that are difficult to express in QBF can be succinctly represented in DQBF.

Recent developments in QBF proof complexity [5–11,17–19,27] have increased our theoretical understanding of QBF proof systems and proof systems in general and have shown that there is an intrinsic link between proof complexity and SAT-, QBF-, and DQBF-solving. Lower bounds in resolution-based proof systems give lower bounds to CDCL-style algorithms. In propositional logic there is only one resolution system (although many subsystems have been studied [24,29]), but in QBF, resolution can be adapted in different ways to get sound and complete calculi of varying strengths [7,15,19,30].

The first and best-studied QBF resolution system is Q-Res introduced in [21]. For Q-Res there are two main enhanced versions: QU-Res [30], which allows resolution on universal variables, and LD-Q-Res [15], which introduces a process of merging positive and negative universal literals under certain conditions. These two concepts were combined into a single calculus LQU⁺-Res [5].

© Springer International Publishing Switzerland 2016
N. Creignou and D. Le Berre (Eds.): SAT 2016, LNCS 9710, pp. 490–499, 2016.
DOI: 10.1007/978-3-319-40970-2_30

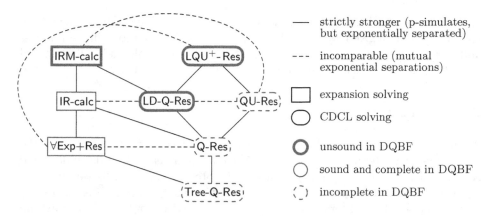

**Fig. 1.** The simulation order of QBF resolution systems [8] and soundness/completeness of their lifted DQBF versions (Color figure online)

While these calculi model CDCL solving, another group of resolution systems were developed with the goal to express ideas from expansion solving. The first and most basic of these systems is ∀Exp+Res [19], which also uses resolution, but is conceptually very different from Q-Res. In [7] two further proof systems IR-calc and IRM-calc are introduced, which unify the CDCL- and expansion-based approaches in the sense that IR-calc simulates both Q-Res and ∀Exp+Res. The system IRM-calc enhances IR-calc and additionally simulates LD-Q-Res. The relative strength of these QBF resolution systems is illustrated in Fig. 1.

The aim of this paper is to clarify which of these QBF resolution systems can be lifted to DQBF. This is motivated both by the theoretical quest to understand which QBF resolution paradigms are robust enough to work in the more powerful DQBF setting, as well as from the practical perspective, where recent advances in DQBF solving [12–14,31] prompt the question of how to model and analyse these solvers proof-theoretically.

Our starting point is the work of Balabanov, Chiang, and Jiang [3], who show that Q-Res can be naturally adapted to a sound calculus for DQBF, but they show it is not strong enough and lacks completeness. Using an idea from [5] we extend their result to QU-Res, thus showing that the lifted version of this system to DQBF is not complete either. We present an example showing that the lifted version of LD-Q-Res is not sound, and this transfers to the DQBF analogues of the stronger systems LQU⁺-Res and IRM-calc.

While this rules out most of the existing QBF resolution calculi already—and in fact all CDCL-based systems (cf. Fig. 1)—we show that IR-calc, lifted in a natural way to a DQBF calculus D-IR-calc, is indeed sound and complete for DQBF; and this holds as well for the lifted version of the weaker expansion system ∀Exp+Res.

Conceptually, our soundness and completeness arguments use the known correspondence of QBF and DQBF to first-order logic [25], and in particular to the fragment EPR, also known as the Bernays-Schönfinkel class, the universal fragment of first-order logic without function symbols of non-zero arity. Similarly to DQBF, EPR is NEXPTIME-complete [22]. In addition to providing soundness and completeness this explains the semantics of both IR-calc and D-IR-calc and identifies these systems as special cases of first-order resolution.

## 2    Preliminaries

A *literal* is a Boolean variable or its negation. If $l$ is a literal, $\neg l$ denotes the complementary literal, i.e., $\neg\neg x = x$. A *clause* is a set of literals understood as their disjunction. The empty clause is denoted by $\bot$, which is semantically equivalent to false. A formula in *Conjunctive Normal Form* (CNF) is a conjunction of clauses. For a literal $l = x$ or $l = \neg x$, we write var($l$) for $x$ and extend this notation to var($C$) for a clause $C$.

A *Dependency Quantified Boolean Formula (DQBF)* $\phi$ in prenex Skolem form consists of a quantifier prefix $\Pi$ and a propositional matrix $\psi$. QBF $\phi$ can also be written as $\Pi \cdot \psi$. Here we mainly study DQBFs where $\psi$ is in CNF. The propositional variables of $\psi$ are partitioned into sets $Y$ and $X$. We define $Y$ as the set of universal variables and $X$ the set of existential variables. For every $y \in Y$, $\Pi$ contains the quantifier $\forall y$. For every $x \in X$ there is a predefined subset $Y_x \subseteq Y$ and $\Pi$ contains the quantifier $\exists x(Y_x)$.

The semantics of DQBF is defined in terms of Skolem functions. A Skolem function $f_x : \{0,1\}^{Y_x} \to \{0,1\}$ describes the evaluation of an existential variable $x$ under the possible assignments to its dependencies $Y_x$. Given a set $F$ of Skolem functions, where $F = \{f_x \mid x \in X\}$ for all the existential variables and an assignment $\alpha : Y \to \{0,1\}$ for the universal variables, the *extension* of $\alpha$ by $F$ is defined as $\alpha_F(x) = f_x(\alpha \restriction Y_x)$ for $x \in X$ and $\alpha_F(y) = \alpha(y)$ for $y \in Y$. A DQBF $\phi$ is *true* if there exist Skolem functions $F = \{f_x \mid x \in X\}$ for the existential variables such that for every assignment $\alpha : Y \to \{0,1\}$ to the universal variables the matrix $\psi$ propositionally evaluates to 1 under the extension $\alpha_F$ of $\alpha$ by $F$.

In QBF, the quantifier prefix is a sequence of standard quantifiers of the form $\exists x$ and $\forall y$. To see that this is a special case of DQBF, we use the sequence from left to right to assign to every variable in the prefix a unique index ind : $X \cup Y \to \mathbb{N}$, and make every existential variable $x$ depend on all the preceding universal variables by setting $Y_x = \{y \in Y \mid \text{ind}(y) < \text{ind}(x)\}$.

We now give a brief overview of the main existing resolution-based calculi for QBF. We start by describing the proof systems modelling *CDCL-based QBF solving*; their rules are summarized in Fig. 2. The most basic and important system is *Q-resolution (Q-Res)* by Kleine Büning et al. [21]. It is a resolution-like calculus that operates on QBFs in prenex form with CNF matrix. In addition to the axioms, Q-Res comprises the resolution rule S∃R and universal reduction ∀-Red (cf. Fig. 2).

*Long-distance resolution (LD-Q-Res)* appears originally in the work of Zhang and Malik [32] and was formalized into a calculus by Balabanov and Jiang [4]. It merges complementary literals of a universal variable $u$ into the special literal $u^*$.

$$\frac{}{C} \text{ (Axiom)} \qquad \frac{D \cup \{u\}}{D} \text{ (∀-Red)} \qquad \frac{D \cup \{u^*\}}{D} \text{ (∀-Red}^*\text{)}$$

$C$ is a clause in the matrix. Literal $u$ is universal and $\mathrm{ind}(u) \geq \mathrm{ind}(l)$ for all $l \in D$.

$$\frac{C_1 \cup U_1 \cup \{x\} \qquad C_2 \cup U_2 \cup \{\neg x\}}{C_1 \cup C_2 \cup U} \text{ (Res)}$$

We consider four instantiations of the Res-rule:

**S∃R:** $x$ is existential. If $z \in C_1$, then $\neg z \notin C_2$. $U_1 = U_2 = U = \emptyset$.
**S∀R:** $x$ is universal. Otherwise same conditions as S∃R.
**L∃R:** $x$ is existential. If $l_1 \in C_1, l_2 \in C_2$, $\mathrm{var}(l_1) = \mathrm{var}(l_2) = z$ then $l_1 = l_2 \neq z^*$. $U_1, U_2$ contain only universal literals with $\mathrm{var}(U_1) = \mathrm{var}(U_2)$. $\mathrm{ind}(x) < \mathrm{ind}(u)$ for each $u \in \mathrm{var}(U_1)$. If $w_1 \in U_1, w_2 \in U_2$, $\mathrm{var}(w_1) = \mathrm{var}(w_2) = u$ then $w_1 = \neg w_2$, $w_1 = u^*$ or $w_2 = u^*$. $U = \{u^* \mid u \in \mathrm{var}(U_1)\}$.
**L∀R:** $x$ is universal. Otherwise same conditions as L∃R.

**Fig. 2.** The rules of CDCL-based proof systems

These special literals prohibit certain resolution steps. In particular, different literals of a universal variable $u$ may be merged only if $\mathrm{ind}(x) < \mathrm{ind}(u)$, where $x$ is the resolved variable. LD-Q-Res uses the rules L∃R, ∀-Red and ∀-Red$^*$.

*QU-resolution (QU-Res)* [30] removes the restriction from Q-Res that the resolved variable must be an existential variable and allows resolution of universal variables. The rules of QU-Res are S∃R, S∀R and ∀-Red. *LQU$^+$-Res* [5] extends LD-Q-Res by allowing short and long distance resolved literals to be universal; however, the resolved literal is never a merged literal $z^*$. LQU$^+$-Res uses the rules L∃R, L∀R, ∀-Red and ∀-Red$^*$.

The second type of calculi models *expansion-based QBF solving*. These calculi are based on *instantiation* of universal variables: ∀Exp+Res [20], IR-calc, and IRM-calc [7]. All these calculi operate on clauses that comprise only existential variables from the original QBF, which are additionally *annotated* by a substitution to some universal variables, e.g. $\neg x^{0/u_1 1/u_2}$. For any annotated literal $l^\sigma$, the substitution $\sigma$ must not make assignments to variables at a higher quantification index than that of $l$, i.e., if $u \in \mathrm{dom}(\sigma)$, then $u$ is universal and $\mathrm{ind}(u) < \mathrm{ind}(l)$.

To preserve this invariant we use the following definition. Fix a DQBF $\Pi \cdot \psi$. Let $\tau$ be a partial assignment of the universal variables $Y$ to $\{0, 1\}$ and let $x$ be an existential variable. $\mathrm{restrict}_x(\tau)$ is the assignment where $\mathrm{dom}(\mathrm{restrict}_x(\tau)) = \mathrm{dom}(\tau) \cap Y_x$ and $\mathrm{restrict}_x(\tau)(u) = \tau(u)$.

The simplest of the instantiation-based calculi we consider is ∀Exp+Res from [19] (cf. also [7,8]). The system IR-calc extends ∀Exp+Res by enabling partial assignments in annotations. To do so, we utilize the auxiliary operation of *instantiation*. We define $\mathrm{inst}_\tau(C)$ to be the clause containing all the literals

$$\frac{}{\left\{ x^{\mathrm{restrict}_x(\tau)} \mid x \in C, x \text{ is existential} \right\}} \text{ (Axiom)}$$

$C$ is a non-tautological clause from the matrix. $\tau = \{0/u \mid u \text{ is universal in } C\}$, where the notation $0/u$ for literals $u$ is shorthand for $0/x$ if $u = x$ and $1/x$ if $u = \neg x$.

$$\frac{\{x^{\tau}\} \cup C_1 \qquad \{\neg x^{\tau}\} \cup C_2}{C_1 \cup C_2} \text{ (Resolution)} \qquad\qquad \frac{C}{\mathrm{inst}_{\tau}(C)} \text{ (Instantiation)}$$

$\tau$ is an assignment to universal variables with $\mathrm{rng}(\tau) \subseteq \{0, 1\}$.

**Fig. 3.** The rules of IR-calc [7] and of D-IR-calc (Sect. 4)

$l^{\mathrm{restrict}_{\mathrm{var}(l)}(\sigma)}$, where $l^{\xi} \in C$ and $\mathrm{dom}(\sigma) = \mathrm{dom}(\xi) \cup \mathrm{dom}(\tau)$ and $\sigma(u) = \xi(u)$ if $u \in \mathrm{dom}(\xi)$ and $\sigma(u) = \tau(u)$ otherwise.

The calculus IRM-calc from [7] further extends IR-calc by enabling annotations containing $*$, similarly as in LD-Q-Res.

# 3  Problems with Lifting QBF Calculi to DQBF

There is no unique method for lifting calculi from QBF to DQBF. However, we can consider 'natural' generalisations of these calculi, where we interpret index conditions as dependency conditions. This means that when a proof system requires for an existential variable $x$ and a universal variable $y$ with $\mathrm{ind}(y) < \mathrm{ind}(x)$, this should be interpreted as $y \in Y_x$. Analogously $\mathrm{ind}(x) < \mathrm{ind}(y)$ should be interpreted as $y \notin Y_x$. This approach was followed when taking Q-Resolution to D-Q-Resolution in Theorem 7 of [3]. Balabanov et al. showed there that D-Q-Resolution is not complete for DQBF using some specific formula. This formula is easily shown to be false, but no steps are possible in D-Q-Resolution, hence D-Q-Resolution is not complete [3]. Consider now the following modification of that formula where the universal variables are doubled:

$$\forall x_1 \forall x_1' \forall x_2 \forall x_2' \exists y_1(x_1, x_1') \exists y_2(x_2, x_2') \tag{1}$$

$$\{y_1, y_2, x_1, x_1'\} \qquad\qquad \{\neg y_1, \neg y_2, x_1, x_1'\}$$
$$\{y_1, y_2, \neg x_1, \neg x_1', \neg x_2, \neg x_2'\} \quad \{\neg y_1, \neg y_2, \neg x_1, \neg x_1', \neg x_2, \neg x_2'\}$$
$$\{y_1, \neg y_2, \neg x_1, \neg x_1', x_2, x_2'\} \quad \{\neg y_1, y_2, \neg x_1, \neg x_1', x_2, x_2'\}.$$

The falsity of (1) follows from the fact that its hypothetical Skolem model would immediately yield a Skolem model for the original formula using assignments with $x_1 = x_1'$, $x_2 = x_2'$. But there is no such model because the original formula is false. However, since we have doubled the universal literals we cannot perform any generalised QU-Res steps to begin a refutation. This technique of doubling literals was first used in [5].

Now we look at another portion of the calculi from Fig. 1, namely the calculi that utilise merging. As a specific example we consider LD-Q-Res and show that it is not sound when lifted to DQBF in the natural way.

To do this we look at the condition of (L∃R) given in Fig. 2. Here instead of requiring $\text{ind}(x) < \text{ind}(u)$ as a condition for $u$ becoming merged, we require $u \notin Y_x$. This is unsound as we show by the following DQBF:

$$\forall u \forall v \exists x(u) \exists y(v) \exists z(u,v). \quad \begin{matrix} \{x, v, z\} & \{\neg x, \neg v, z\} \\ \{y, u, \neg z\} & \{\neg y, \neg u, \neg z\} \end{matrix}$$

Its truth is witnessed by the Skolem functions $x(u) = u$, $y(v) = \neg v$, and $z(u,v) = (u \wedge v) \vee (\neg u \wedge \neg v)$. However, the lifted version of LD-Q-Res admits a refutation:

$$\frac{\dfrac{\{x, v, z\} \qquad \{\neg x, \neg v, z\}}{\{v^*, z\}} \qquad \dfrac{\{y, u, \neg z\} \qquad \{\neg y, \neg u, \neg z\}}{\{u^*, \neg z\}}}{\dfrac{\{u^*, v^*\}}{\dfrac{\{u^*\}}{\bot}}}$$

This shows that LD-Q-Res is unsound for DQBF. Likewise, since IRM-calc, LQU-Res and LQU$^+$-Res step-wise simulate LD-Q-Res, this proof would also be available, showing that these are all unsound calculi in the DQBF setting.

# 4    A Sound and Complete Proof System for DQBF

In this section we introduce the D-IR-calc refutation system and prove its soundness and completeness for DQBF. The calculus takes inspiration from IR-calc, a system for QBF [7], which in turn is inspired by first-order translations of QBF. One such translation is to the EPR fragment, i.e., the universal fragment of first-order logic without function symbols of non-zero arity (this means we only allow constants). We broaden this translation to DQBF and then bring this back down to D-IR-calc in a similar way as in IR-calc.

We adapt annotated literals $l^\tau$ to DQBF, such that $l$ is an existential literal and $\tau$ is an annotation which is a partial assignment to universal variables in $Y_x$. In QBF, $Y_x$ contains all universal variables with an index lower than $x$, and this is exactly the maximal range of the potential annotation to $x$ literals. Thus our definition of annotated literals generalises those used in IR-calc.

The definitions of restrict and inst were defined for QBF, but with dependency already in mind. With these definitions at hand we can now define the new calculus D-IR-calc. Its rules are exactly the same as the ones for IR-calc stated in Fig. 3, but applied to DQBF.

Before analysing D-IR-calc further we present the translation of DQBF into EPR. We use an adaptation of the translation described for QBF [25], which becomes straightforward in the light of the DQBF semantics based on Skolem functions. The key observation is that for the intended two valued Boolean domain the Skolem functions can be represented by predicates.

To translate a DQBF $\Pi \cdot \psi$ we introduce on the first-order side (1) a predicate symbol $p$ of arity one and two constant symbols 0 and 1 to describe the Boolean domain, (2) for every existential variable $x \in X$ a predicate symbol $x$ of arity $|Y_x|$, and (3) for every universal variable $y \in Y$ a first-order variable $y$.

Now we can define a translation mapping $t_\Pi$. It translates each occurrence of an existential variable $x$ with dependencies $Y_x = \{y_1, \dots, y_k\}$ to the atom $t_\Pi(x) = x(y_1, \dots, y_k)$ (we assume an arbitrary but fixed order on the dependencies which dictates their placement as arguments) and each occurrence of a universal variable $y$ to the atom $t_\Pi(y) = p(y)$. The mapping is then homomorphically extended to formulas: $t_\Pi(\neg\psi) = \neg t_\Pi(\psi)$, $t_\Pi(\psi_1 \vee \psi_2) = t_\Pi(\psi_1) \vee t_\Pi(\psi_2)$, and $t_\Pi(\psi_1 \wedge \psi_2) = t_\Pi(\psi_1) \wedge t_\Pi(\psi_2)$. This means a CNF matrix $\psi$ is mapped to a corresponding first-order CNF $t_\Pi(\psi)$. As customary, the first-order variables of $t_\Pi(\psi)$ are assumed to be implicitly universally quantified at the top level.

**Lemma 1.** *A DQBF $\Pi \cdot \psi$ is true if and only if $t_\Pi(\psi) \wedge p(1) \wedge \neg p(0)$ is satisfiable.*

*Proof (Idea).* When the DQBF $\Pi \cdot \psi$ is true, this is witnessed by the existence of Skolem functions $F = \{f_x \mid x \in X\}$. On the other hand, if $t_\Pi(\psi) \wedge p(1) \wedge \neg p(0)$ is satisfiable then we can by Herbrand's theorem assume it has a Herbrand model $H$ over the base $\{0,1\}$. We can naturally translate between one and the other by setting $f_x(\boldsymbol{v}) = 1$ iff $x(\boldsymbol{v}) \in H$ for every $x \in X$ and $\boldsymbol{v} \in \{0,1\}^{|Y_x|}$. The lemma then follows by structural induction over $\psi$. □

For the purpose of analysing D-IR-calc, the mapping $t_\Pi$ is further extended to annotated literals: $t_\Pi(x^\tau) = t_\Pi(x)\tau$ for an existential variable $x$. Here we continue to slightly abuse notation and treat $\tau$, an annotation in the propositional context, as a first-order substitution over the corresponding translated variables in the first-order context (recall point (3) above).

We aim to show soundness and completeness of D-IR-calc by relating it via the above translation to a first-order resolution calculus FO-res. This calculus consists of (1) a lazy grounding rule: given a clause $C$ and a substitution $\sigma$ derive the instance $C\sigma$, and (2) the resolution rule: given two clauses $C \cup \{l\}$ and $D \cup \{\neg l\}$, where $l$ is a first-order literal, derive $C \cup D$. Note that similarly to propositional clauses, we understand first-order clauses as *sets* of literals. Thus we do not need any explicit factoring rule. Also note that we require the resolved literals of the two premises of the resolution rule to be equal (up to the polarity). Standard first-order resolution, which involves *unification* of the resolved literals, can be simulated in FO-res by combining the instantiation and the resolution rule. It is clear that FO-res is sound and complete for first-order logic.

Our argument for the soundness of D-IR-calc is now the following. Given $\pi = (L_1, L_2, \dots, L_\ell)$, a D-IR-calc derivation of the empty clause $L_\ell = \bot$ from DQBF $\Pi \cdot \psi$, we show by induction that $t_\Pi(L_n)$ is derivable from $\Psi = t_\Pi(\psi) \wedge p(1) \wedge \neg p(0)$ by FO-res for every $n \leq \ell$. Because $t_\Pi(\bot) = \bot$ is unsatisfiable, so must $\Psi$ be, by soundness of FO-res and therefore $\Pi \cdot \psi$ is false by Lemma 1.

We need to consider the three cases by which a clause is derived in D-IR-calc. First, it is easy to verify that D-IR-calc instantiation by an annotation $\tau$ corresponds to FO-res instantiation by $\tau$ as a substitution, i.e., $t_\Pi(\text{inst}_\tau(C)) = t_\Pi(C)\tau$. Also the D-IR-calc and FO-res resolution rules correspond one to one in an obvious way. Thus the most interesting case concerns the Axiom rule.

Intuitively, the Axiom rule of D-IR-calc removes universal variables from a clause while recording their past presence (and polarity) within the applied annotation $\tau$. We simulate this step in FO-res by first instantiating the translated clause by $\tau$ and then resolving the obtained clause with the unit $p(1)$ and/or $\neg p(0)$. Here is an example for a DQBF prefix $\Pi = \forall u \, \forall v \, \forall w \, \exists x(u,v) \, \exists y(v,w)$:

$$(\text{D-IR-calc}) \ \frac{\{x, y, \neg u, v\}}{\{x^{1/u, 0/v}, y^{0/v}\}} \qquad (\text{FO-res}) \ \frac{\{x(u,v), y(v,w), \neg p(u), p(v)\}}{\{x(1,0), y(0,w)\}}$$

**Theorem 2.** *D-IR-calc is sound.*

We now show completeness. Let $\Pi \cdot \psi$ be a false DQBF and let us consider $\mathcal{G}(t_\Pi(\psi))$, the set of all ground instances of clauses in $t_\Pi(\psi)$. Here, by a ground instance of a clause $C$ we mean the clause $C\sigma$ for some substitution $\sigma : \text{var}(C) \rightarrow \{0, 1\}$. By the combination of Lemma 1 and Herbrand's theorem, $\mathcal{G}(t_\Pi(\psi)) \wedge p(1) \wedge \neg p(0)$ is unsatisfiable and thus it has a FO-res refutation. Moreover, by completeness of *ordered* resolution [2], we can assume that (1) the refutation does not contain clauses subsumed by $p(1)$ or $\neg p(0)$, and (2) any clause containing the predicate $p$ is resolved on a literal containing $p$. From this it is easy to see that any leaf in the refutation gives rise (in zero, one or two resolution steps with $p(1)$ or $\neg p(0)$) to a clause $D = t_\Pi(C)$ where $C$ can be obtained by D-IR-calc Axiom from a $C_0 \in \psi$. The rest of the refutation consists of FO-res resolution steps which can be simulated by D-IR-calc. Thus we obtain the following.

**Theorem 3.** *D-IR-calc is refutationally complete for DQBF.*

Although one can lift the above argument with ordered resolution to show that the set $\{t_\Pi(C) \mid C \text{ follows by Axiom from some } C_0 \in \psi\}$ is unsatisfiable for any false DQBF $\Pi \cdot \psi$, we have shown how to simulate *ground* FO-res steps by D-IR-calc. That is because a lifted FO-res derivation may contain instantiation steps which *rename variables apart* for which a subsequent resolvent cannot be represented in D-IR-calc. An example is the resolvent $\{y(v), z(v')\}$ of clauses $\{x(u), y(v)\}$ and $\{\neg x(u), z(v')\}$ which is obviously stronger than the clause $\{y(v), z(v)\}$. However, only the latter has a counterpart in D-IR-calc.

We also remark that in a similar way we can also lift to DQBF the QBF calculus $\forall$Exp+Res from [19]. It is easily verified that the simulation of $\forall$Exp+Res by IR-calc shown in [7] directly transfers from QBF to DQBF. Hence Theorem 2 immediately implies the soundness of $\forall$Exp+Res lifted to DQBF. Moreover, because all ground instances are also available in $\forall$Exp+Res lifted to DQBF, this system is also complete as can be shown by repeating the argument of Theorem 3.

**Acknowledgments.** This research was supported by grant no. 48138 from the John Templeton Foundation, EPSRC grant EP/L024233/1, and a Doctoral Training Grant from the EPSRC (2nd author).

Martin Suda was supported by the EPSRC grant EP/K032674/1 and the ERC Starting Grant 2014 SYMCAR 639270.

# References

1. Azhar, S., Peterson, G., Reif, J.: Lower bounds for multiplayer non-cooperative games of incomplete information. J. Comput. Math. Appl. **41**, 957–992 (2001)
2. Bachmair, L., Ganzinger, H.: Resolution theorem proving. In: Robinson, J.A., Voronkov, A. (eds.) Handbook of Automated Reasoning, pp. 19–99. Elsevier and MIT Press (2001)
3. Balabanov, V., Chiang, H.J.K., Jiang, J.H.R.: Henkin quantifiers and boolean formulae: a certification perspective of DQBF. Theor. Comput. Sci. **523**, 86–100 (2014)
4. Balabanov, V., Jiang, J.H.R.: Unified QBF certification and its applications. Formal Methods Syst. Des. **41**(1), 45–65 (2012)
5. Balabanov, V., Widl, M., Jiang, J.-H.R.: QBF resolution systems and their proof complexities. In: Sinz, C., Egly, U. (eds.) SAT 2014. LNCS, vol. 8561, pp. 154–169. Springer, Heidelberg (2014)
6. Beyersdorff, O., Bonacina, I., Chew, L.: Lower bounds: from circuits to QBF proof systems. In: Proceedings of ACM Conference on Innovations in Theoretical Computer Science (ITCS 2016), pp. 249–260. ACM (2016)
7. Beyersdorff, O., Chew, L., Janota, M.: On unification of QBF resolution-based calculi. In: Csuhaj-Varjú, E., Dietzfelbinger, M., Ésik, Z. (eds.) MFCS 2014, Part II. LNCS, vol. 8635, pp. 81–93. Springer, Heidelberg (2014)
8. Beyersdorff, O., Chew, L., Janota, M.: Proof complexity of resolution-based QBF calculi. In: Proceedings of Symposium on Theoretical Aspects of Computer Science, pp. 76–89. LIPIcs series (2015)
9. Beyersdorff, O., Chew, L., Mahajan, M., Shukla, A.: Feasible interpolation for QBF resolution calculi. In: Halldórsson, M.M., Iwama, K., Kobayashi, N., Speckmann, B. (eds.) ICALP 2015. LNCS, vol. 9134, pp. 180–192. Springer, Heidelberg (2015)
10. Beyersdorff, O., Chew, L., Mahajan, M., Shukla, A.: Are short proofs narrow? QBF resolution is not simple. In: Proceedings of Symposium on Theoretical Aspects of Computer Science (STACS 2016) (2016)
11. Egly, U.: On sequent systems and resolution for QBFs. In: Cimatti, A., Sebastiani, R. (eds.) SAT 2012. LNCS, vol. 7317, pp. 100–113. Springer, Heidelberg (2012)
12. Finkbeiner, B., Tentrup, L.: Fast DQBF refutation. In: Sinz, C., Egly, U. (eds.) [26], pp. 243–251
13. Fröhlich, A., Kovásznai, G., Biere, A., Veith, H.: iDQ: Instantiation-based DQBF solving. In: Sinz, C., Egly, U. (eds.) [26], pp. 103–116
14. Gitina, K., Wimmer, R., Reimer, S., Sauer, M., Scholl, C., Becker, B.: Solving DQBF through quantifier elimination. In: Nebel, W., Atienza, D. (eds.) Proceedings of the 2015 Design, Automation & Test in Europe Conference & Exhibition, DATE 2015, Grenoble, France, March 9–13, 2015. pp. 1617–1622. ACM (2015)
15. Giunchiglia, E., Marin, P., Narizzano, M.: Reasoning with quantified boolean formulas. In: Biere, A., Heule, M., van Maaren, H., Walsh, T. (eds.) Handbook of Satisfiability, Frontiers in Artificial Intelligence and Applications, vol. 185, pp. 761–780. IOS Press (2009)
16. Henkin, L.: Some remarks on infinitely long formulas. J. Symbolic Logic, pp. 167–183 (1961). Pergamon Press
17. Heule, M.J., Seidl, M., Biere, A.: Efficient extraction of skolem functions from QRAT proofs. In: Formal Methods in Computer-Aided Design (FMCAD), 2014, pp. 107–114. IEEE (2014)

18. Heule, M.J.H., Seidl, M., Biere, A.: A unified proof system for QBF preprocessing. In: Demri, S., Kapur, D., Weidenbach, C. (eds.) IJCAR 2014. LNCS, vol. 8562, pp. 91–106. Springer, Heidelberg (2014)
19. Janota, M., Marques-Silva, J.: Expansion-based QBF solving versus Q-resolution. Theor. Comput. Sci. **577**, 25–42 (2015)
20. Janota, M., Marques-Silva, J.: On propositional QBF expansions and Q-resolution. In: Järvisalo, M., Van Gelder, A. (eds.) SAT 2013. LNCS, vol. 7962, pp. 67–82. Springer, Heidelberg (2013)
21. Büning, K.H., Karpinski, M., Flögel, A.: Resolution for quantified Boolean formulas. Inf. Comput. **117**(1), 12–18 (1995)
22. Lewis, H.R.: Complexity results for classes of quantificational formulas. J. Comput. Syst. Sci. **21**(3), 317–353 (1980)
23. Peterson, G.L., Reif, J.: Multiple-person alternation. In: 20th Annual Symposium on Foundations of Computer Science, 1979, pp. 348–363, October 1979
24. Segerlind, N.: The complexity of propositional proofs. Bull. Symbolic Logic **13**(4), 417–481 (2007)
25. Seidl, M., Lonsing, F., Biere, A.: qbf2epr: a tool for generating EPR formulas from QBF. In: Fontaine, P., Schmidt, R.A., Schulz, S. (eds.) PAAR-2012. Third Workshop on Practical Aspects of Automated Reasoning. EPiC Series in Computing, vol. 21, pp. 139–148. EasyChair (2013)
26. Sinz, C., Egly, U. (eds.): SAT 2014. LNCS, vol. 8561. Springer, Heidelberg (2014)
27. Slivovsky, F., Szeider, S.: Variable dependencies and Q-resolution. In: Sinz, C., Egly, U. (eds.) [26], pp. 269–284
28. Stockmeyer, L.J., Meyer, A.R.: Word problems requiring exponential time: preliminary report. In: Aho, A.V., Borodin, A., Constable, R.L., Floyd, R.W., Harrison, M.A., Karp, R.M., Strong, H.R. (eds.) Proceedings of the 5th Annual ACM Symposium on Theory of Computing, April 30 – May 2, 1973, Austin, Texas, USA, pp. 1–9. ACM (1973)
29. Urquhart, A.: The complexity of propositional proofs. Bull. Symbolic Logic **1**, 425–467 (1995)
30. Van Gelder, A.: Contributions to the theory of practical quantified boolean formula solving. In: Milano, M. (ed.) CP 2012. LNCS, vol. 7514, pp. 647–663. Springer, Heidelberg (2012)
31. Wimmer, R., Gitina, K., Nist, J., Scholl, C., Becker, B.: Preprocessing for DQBF. In: Heule, M., et al. (eds.) SAT 2015. LNCS, vol. 9340, pp. 173–190. Springer, Heidelberg (2015). doi:10.1007/978-3-319-24318-4_13
32. Zhang, L., Malik, S.: Conflict driven learning in a quantified Boolean satisfiability solver. In: ICCAD, pp. 442–449 (2002)

# Long Distance Q-Resolution with Dependency Schemes

Tomáš Peitl, Friedrich Slivovsky, and Stefan Szeider$^{(\boxtimes)}$

Algorithms and Complexity Group, TU Wien, Vienna, Austria
{peitl,fslivovsky,sz}@ac.tuwien.ac.at

**Abstract.** Resolution proof systems for quantified Boolean formulas (QBFs) provide a formal model for studying the limitations of state-of-the-art search-based QBF solvers, which use these systems to generate proofs. In this paper, we define a new proof system that combines two such proof systems: Q-resolution with generalized universal reduction according to a dependency scheme and long distance Q-resolution. We show that the resulting proof system is sound for the reflexive resolution-path dependency scheme—in fact, we prove that it admits strategy extraction in polynomial time. As a special case, we obtain soundness and polynomial-time strategy extraction for long distance Q-resolution with universal reduction according to the standard dependency scheme. We report on experiments with a configuration of DepQBF that generates proofs in this system.

## 1 Introduction

Quantified Boolean Formulas (QBFs) offer succinct encodings for problems from domains such as formal verification, synthesis, and planning [5,10,13,27,33,37]. Although the combination of (more verbose) propositional encodings with SAT solvers is still the state-of-the-art approach to many of these problems, QBF solvers are gaining ground. An arsenal of new techniques has been introduced over the past few years [8,9,11,18,19,21,23,26,29,30,32], and these advances in solver technology have been accompanied by the development of a better understanding of the underlying QBF proof systems and their limitations [4,6, 7,15,22,24,36].

Search-based solvers based on the QDPLL algorithm [12] represent one of the principal state-of-the-art approaches in QBF solving. Akin to modern SAT solvers, these solvers rely on successive variable assignments in combination with fast constraint propagation and learning. Unlike SAT solvers, however, search-based QBF solvers are constrained by the variable dependencies induced by the quantifier prefix[1]: while SAT solvers can assign variables in any order, search-based QBF solvers can only assign variables from the leftmost quantifier block

---

This research was partially supported by FWF grants P27721 and W1255-N23.
[1] We consider QBFs in prenex normal form.

N. Creignou and D. Le Berre (Eds.): SAT 2016, LNCS 9710, pp. 500–518, 2016.
DOI: 10.1007/978-3-319-40970-2_31

that contains unassigned variables, since the assignment of a variable further to the right might depend on the variable assignment to this block. In the most extreme case, this forces solvers into a fixed order of variable assignments, rendering decision variable heuristics ineffective.

The search-based solver DepQBF uses dependency schemes to partially bypass this restriction [8,28]. Dependency schemes can sometimes identify pairs of variables as independent, allowing the solver to assign them in any order. This gives decision heuristics more freedom and results in increased performance [8].

While this provides a strong motivation to use dependency schemes, their integration with QDPLL poses challenges of its own. Soundness of the proof system underlying QDPLL with the standard dependency scheme as implemented in DepQBF was shown only recently [36], and combining other state-of-the-art techniques with dependency schemes is often highly nontrivial. In this paper, we focus on two such issues:

(a) Long distance Q-resolution permits the derivation of tautological clauses in certain cases [2,39,40]. This system can be used in constraint learning as an alternative to Q-resolution, leading to fewer backtracks during search and, sometimes, reduced runtime [16]. In addition, clause learning based on long distance Q-resolution is substantially easier to implement. Currently, however, DepQBF does not permit learning based on long distance Q-resolution in conjunction with dependency schemes, as the resulting proof system is not known to be sound.

(b) For applications in verification and synthesis, it is not enough for solvers to decide whether an input QBF is true or false—they also have to generate a certificate. Such certificates can be efficiently constructed from Q-resolution [2] and even long distance Q-resolution proofs [3]. However, it is not clear whether this is possible for proofs generated by DepQBF with the standard dependency scheme, and proof generation with the standard dependency scheme is disabled by default.

We address (a) by showing that long distance Q-resolution combined with universal reduction according to the reflexive resolution-path dependency scheme [36] is sound. In fact, we prove that this proof system allows for certificate extraction in polynomial time, thus resolving (b). In particular, these results hold for the standard dependency scheme, which is weaker than the reflexive resolution-path dependency scheme.

Our proof of this result relies on an interpretation of Q-resolution refutations as winning strategies for the universal player in the evaluation game [20]. Defining LDQ(D) as the proof system consisting of long distance Q-resolution with universal reduction according to a dependency scheme D, we identify a natural property of a dependency scheme D that not only allows for the interpretation of an LDQ(D)-refutation as a winning strategy for the universal player, but even implies certificate extraction in time $O(|\mathcal{P}| \cdot n)$ from an LDQ(D)-refutation $\mathcal{P}$ of a QBF with $n$ variables. We then show that the reflexive resolution path dependency scheme in fact has this property.

One of our motivations for studying the combination of long distance Q-resolution and dependency schemes is that all of the required logic is already implemented in DepQBF—it is simply that, by default, these features cannot be enabled at the same time because it is unclear whether the resulting solver configuration is sound. To complement our theoretical results, we bypassed these checks and performed experiments with DepQBF and constraint learning based on LDQ(D) with the standard dependency scheme. Our experiments show that performance with this type of learning is on par with and—in some cases—even surpasses the performance of DepQBF with other configurations of constraint learning.

Due to space constraints, several proofs had to be omitted.

## 2   Preliminaries

*Formulas and Assignments.* A *literal* is a negated or unnegated variable. If $x$ is a variable, we write $\bar{x} = \neg x$ and $\overline{\neg x} = x$, and let $var(x) = var(\neg x) = x$. If $X$ is a set of literals, we write $\overline{X}$ for the set $\{\, \bar{x} : x \in X \,\}$. A *clause* is a finite disjunction of literals. We call a clause *tautological* if it contains the same variable negated as well as unnegated. A *CNF formula* is a finite conjunction of non-tautological clauses. Whenever convenient, we treat clauses as sets of literals, and CNF formulas as sets of sets of literals. We write $var(C)$ for the set of variables occuring (negated or unnegated) in a clause $C$, that is, $var(C) = \{\, var(\ell) : \ell \in C \,\}$. Moreover, we let $var(\varphi) = \bigcup_{C \in \varphi} var(C)$ denote the set of variables occurring in a CNF formula $\varphi$.

A *truth assignment* (or simply *assignment*) to a set $X$ of variables is a mapping $\tau : X \to \{0,1\}$. We write $[X]$ for the set of truth assignments to $X$, and extend $\tau : X \to \{0,1\}$ to literals by letting $\tau(\neg x) = 1 - \tau(x)$ for $x \in X$. Let $\tau : X \to \{0,1\}$ be a truth assignment. The restriction $C[\tau]$ of a clause $C$ by $\tau$ is defined as follows: if there is a literal $\ell \in C \cap (X \cup \overline{X})$ such that $\tau(\ell) = 1$ then $C[\tau] = 1$. Otherwise, $C[\tau] = C \setminus (X \cup \overline{X})$. The restriction $\varphi[\tau]$ of a CNF formula $\varphi$ by the assignment $\tau$ is defined $\varphi[\tau] = \{\, C[\tau] : C[\tau] \neq 1 \,\}$.

*PCNF Formulas.* A *PCNF formula* is denoted by $\Phi = \mathcal{Q}.\varphi$, where $\varphi$ is a CNF formula and $\mathcal{Q} = Q_1 X_1 \ldots Q_n X_n$ is a sequence such that $Q_i \in \{\forall, \exists\}$, $Q_i \neq Q_{i+1}$ for $1 \leq i < n$, and the $X_i$ are pairwise disjoint sets of variables. We call $\varphi$ the *matrix* of $\Phi$ and $\mathcal{Q}$ the *(quantifier) prefix* of $\Phi$, and refer to the $X_i$ as *quantifier blocks*. We require that $var(\varphi) = X_1 \cup \cdots \cup X_n$ and write $var(\Phi) = var(\varphi)$. We define a partial order $<_\Phi$ on $var(\varphi)$ as $x <_\Phi y \Leftrightarrow x \in X_i, y \in X_j, i < j$. We extend $<_\Phi$ to a relation on literals in the obvious way and drop the subscript whenever $\Phi$ is understood. For $x \in var(\Phi)$ we let $R_\Phi(x) = \{\, y \in var(\Phi) : x <_\Phi y \,\}$ and $L_\Phi(x) = \{\, y \in var(\Phi) : y <_\Phi x \,\}$ denote the sets of variables *to the right* and *to the left* of $x$ in $\Phi$, respectively. Relative to the PCNF formula $\Phi$, variable $x$ is called *existential* (*universal*) if $x \in X_i$ and $Q_i = \exists$ ($Q_i = \forall$). The set of existential (universal) variables occurring in $\Phi$ is denoted $var_\exists(\Phi)$ ($var_\forall(\Phi)$). The *size* of a PCNF formula $\Phi = \mathcal{Q}.\varphi$ is defined as $|\Phi| = \sum_{C \in \varphi} |C|$. If $\tau$ is an assignment, then $\Phi[\tau]$ denotes the PCNF formula $\mathcal{Q}'.\varphi[\tau]$, where $\mathcal{Q}'$ is the quantifier prefix obtained from $\mathcal{Q}$ by deleting variables that do not occur in $\varphi[\tau]$. *True* and *false* PCNF formulas are defined in the usual way.

*Countermodels.* Let $\Phi = \mathcal{Q}.\varphi$ be a PCNF formula. A *countermodel* of $\Phi$ is an indexed family $\{f_u\}_{u \in var_\forall(\Phi)}$ of functions $f_u : [L_\Phi(u)] \to \{0,1\}$ such that $\varphi[\tau] = \{\emptyset\}$ for every assignment $\tau : var(\Phi) \to \{0,1\}$ satisfying $\tau(u) = f_u(\tau|_{L_\Phi(u)})$ for $u \in var_\forall(\Phi)$.

**Proposition 1. (Folklore)** *A PCNF formula is false if, and only if, it has a countermodel.*

# 3   Dependency Schemes and LDQ(D)-Resolution

In this section, we introduce the proof system LDQ(D), which combines Q(D)-resolution [36] with long-distance Q-resolution [2]. Q-resolution is a generalization of propositional resolution to PCNF formulas [25]. Q-resolution is of practical interest due to its relation to search based QBF solvers that implement the QDPLL algorithm [12]: the trace of a QDPLL solver generated for a false PCNF formula corresponds to a Q-resolution refutation [17]. QDPLL generalizes the well-known DPLL procedure [14] from SAT to QSAT. In a nutshell, DPLL searches for a satisfying assignment of an input formula by propagating unit clauses and assigning pure literals until the formula cannot be simplified any further, at which point it picks an unassigned variable and branches on the assignment of this variable. Although any of the remaining variables can be chosen for assignment, the order of assignment can have significant effects on the runtime, and modern SAT solvers derived from the DPLL algorithm use sophisticated heuristics to determine what variable to assign next [31].

In QDPLL, the quantifier prefix imposes constraints on the order of variable assignments: a variable may be assigned only if it occurs in the leftmost quantifier block with unassigned variables. Often, this is more restrictive than necessary. For instance, variables from disjoint subformulas may be assigned in any order. Intuitively, a variable can be assigned as long as it *does not depend* on any unassigned variable. This is the intuition underlying a generalization of QDPLL implemented in the solver DepQBF [8,28]. DepQBF uses a *dependency scheme* [34] to compute an overapproximation of variable dependencies. Dependency schemes are mappings that associate every PCNF formula with a binary relation on its variables that refines the order of variables in the quantifier prefix.[2]

**Definition 1 (Dependency Scheme).** A *dependency scheme* is a mapping $D$ that associates each PCNF formula $\Phi$ with a relation $D_\Phi \subseteq \{(x,y) : x <_\Phi y\}$ called the *dependency relation* of $\Phi$ with respect to $D$.

---

[2] The original definition of dependency schemes [34] is more restrictive than the one given here, but the additional requirements are irrelevant for the purposes of this paper.

The mapping which simply returns the prefix ordering of an input formula can be thought of as a baseline dependency scheme:

**Definition 2 (Trivial Dependency Scheme).** The *trivial dependency scheme* $D^{trv}$ associates each PCNF formula $\Phi$ with the relation $D^{trv}_{\Phi} = \{(x,y) : x <_{\Phi} y\}$.

DepQBF uses a dependency relation to determine the order in which variables can be assigned: if $y$ is a variable and there is no unassigned variable $x$ such that $(x, y)$ is in the dependency relation, then $y$ is considered ready for assignment. DepQBF also uses the dependency relation to generalize the $\forall$-reduction rule used in clause learning [8]. As a result of its use of dependency schemes, DepQBF generates proofs in a generalization of Q-resolution called Q(D)-resolution [36], a proof system that takes a dependency scheme D as a parameter.

Dependency schemes can be partially ordered based on their dependency relations: if the dependency relation computed by a dependency scheme $D_1$ is a subset of the dependency relation computed by a dependency scheme $D_2$, then $D_1$ is *more general* than $D_2$. The more general a dependency scheme, the more freedom DepQBF has in choosing decision variables. Currently, (aside from the trivial dependency scheme) DepQBF supports the so-called *standard dependency scheme* [34].[3] We will work with the more general *reflexive resolution-path dependency scheme* [36], a variant of the resolution-path dependency scheme [35,38]. This dependency scheme computes an overapproximation of variable dependencies based on whether two variables are connected by a (pair of) resolution path(s).

**Definition 3 (Resolution Path).** Let $\Phi = Q.\varphi$ be a PCNF formula and let $X$ be a set of variables. A *resolution path* (from $\ell_1$ to $\ell_{2k}$) via $X$ (in $\Phi$) is a sequence $\ell_1, \ldots, \ell_{2k}$ of literals satisfying the following properties:

1. For all $i \in [k]$, there is a $C_i \in \varphi$ such that $\ell_{2i-1}, \ell_{2i} \in C_i$.
2. For all $i \in [k]$, $var(\ell_{2i-1}) \neq var(\ell_{2i})$.
3. For all $i \in [k-1]$, $\{\ell_{2i}, \ell_{2i+1}\} \subseteq X \cup \overline{X}$.
4. For all $i \in [k-1]$, $\overline{\ell_{2i}} = \ell_{2i+1}$.

If $\pi = \ell_1, \ldots, \ell_{2k}$ is a resolution path in $\Phi$ via $X$, we say that $\ell_1$ and $\ell_{2k}$ are *connected in $\Phi$* (with respect to $X$). For every $i \in \{1, \ldots, k\}$ we say that $\pi$ *goes through* $var(\ell_{2i})$.

One can think of a resolution path as a potential chain of implications: if each clause $C_i$ contains exactly two literals, then assigning $\ell_1$ to 0 requires setting $\ell_{2k}$ to 1. If, in addition, there is such a path from $\overline{\ell_1}$ to $\overline{\ell_{2k}}$, then $\ell_1$ and $\ell_{2k}$ have to be assigned the same value. Accordingly, the resolution path dependency scheme identifies variables connected by a pair of resolution paths as potentially dependent on each other.

---

[3] Strictly speaking, it uses a refined version of the standard dependency scheme [28, p. 49].

$$\frac{}{C} \text{ (input clause)} \qquad \frac{C_1 \vee e \qquad \neg e \vee C_2}{C_1 \vee C_2} \text{ (resolution)}$$

An *input clause* $C \in \varphi$ can be used as an axiom. From two clauses $C_1 \vee e$ and $\neg e \vee C_1$, where $e$ is an existential variable, the *(long-distance) resolution* rule can derive the clause $C_1 \vee C_2$, provided that $e < var(\ell)$ for every universal literal $\ell$ such that $\ell \in C_1$ and $\bar{\ell} \in C_2$.

$$\frac{C}{C \setminus \{u, \neg u\}} \text{ ($\forall$-reduction)}$$

The $\forall$-*reduction* rule derives the clause $C \setminus \{u, \neg u\}$ from $C$, where $u \in var(C)$ is a universal variable such that $(u, e) \notin D_\Phi$ for every existential variable $e \in var(C)$.

**Fig. 1.** Derivation rules of LDQ(D)-resolution for a PCNF formula $\Phi = \mathcal{Q}.\varphi$.

**Definition 4 (Dependency pair).** Let $\Phi$ be a PCNF formula and $x, y \in var(\Phi)$. We say $\{x, y\}$ is a *resolution-path dependency pair* of $\Phi$ with respect to $X \subseteq var_\exists(\Phi)$ if at least one of the following conditions holds:

- $x$ and $y$, as well as $\neg x$ and $\neg y$, are connected in $\Phi$ with respect to $X$.
- $x$ and $\neg y$, as well as $\neg x$ and $y$, are connected in $\Phi$ with respect to $X$.

**Definition 5.** The *reflexive resolution-path dependency scheme* is the mapping $D^{rrs}$ that assigns to each PCNF formula $\Phi = \mathcal{Q}.\varphi$ the relation $D_\Phi^{rrs} = \{\, x <_\Phi y : \{x, y\}$ is a resolution-path dependency pair in $\Phi$ with respect to $R_\Phi(x) \setminus var_\forall(\Phi) \,\}$.

Both Q-resolution and Q(D)-resolution only allow for the derivation of non-tautological clauses, that is, clauses that do not contain a literal negated as well as unnegated. *Long-distance Q-resolution* is a variant of Q-resolution that admits tautological clauses in certain cases [2]. Variants of QDPLL that allow for learnt clauses to be tautological [39,40] have been shown to generate proofs in long-distance Q-resolution [16].

We combine long-distance resolution with Q(D)-resolution to obtain long-distance Q(D)-resolution (LDQ(D)-resolution). The derivation rules of LDQ(D)-resolution are shown in Fig. 1. Here, as in the rest of the paper, D denotes an arbitrary dependency scheme.

A derivation in a proof system consists of repeated applications of the derivation rules to derive a clause from the clauses of an input formula. Here, derivations will be represented by node-labeled directed acyclic graphs (DAGs). More specifically, we require these DAGs to have a unique sink (that is, a node without outgoing edges) and each of their nodes to have at most two incoming edges. We further assume an ordering on the in-neighbors (or parents) of every node with two incoming edges—that is, each node has a "first" and a "second" in-neighbor. Referring to such DAGs as *proof DAGs*, we define the following two operations to represent resolution and $\forall$-reduction:

1. If $\ell$ is a literal and $\mathcal{P}_1$ and $\mathcal{P}_2$ are proof DAGs with distinct sinks $v_1$ and $v_2$, then $\mathcal{P}_1 \odot_\ell \mathcal{P}_2$ is the proof DAG consisting of the union of $\mathcal{P}_1$ and $\mathcal{P}_2$ along with a new sink $v$ that has two incoming edges, the first one from $v_1$ and the second one from $v_2$. Moreover, if $C_1$ is the label of $v_1$ in $\mathcal{P}_1$ and $C_2$ is the label of $v_2$ in $\mathcal{P}_2$, then $v$ is labeled with the clause $(C_1 \setminus \{\ell\}) \cup (C_2 \setminus \{\bar{\ell}\})$.
2. If $u$ is a variable and $\mathcal{P}$ is a proof DAG with a sink $w$ labeled with $C$, then $\mathcal{P} - u$ denotes the proof DAG obtained from $\mathcal{P}$ by adding an edge from $w$ to a new node $v$ such that $v$ is labeled with $C \setminus \{u, \neg u\}$.

**Definition 6. (Derivation).** An *LDQ(D)-resolution derivation* (or *LDQ(D)-derivation*) of a clause $C$ from a PCNF formula $\Phi = \mathcal{Q}.\varphi$ is a proof DAG $\mathcal{P}$ satisfying the following properties. Source nodes are labeled with input clauses from $\varphi$. If a node with label $C$ has parents labeled $C_1$ and $C_2$ then $C$ can be derived from $C_1$ and $C_2$ by (long-distance) resolution. If a node labeled with a clause $C$ has a single parent with label $C'$ then $C$ can be derived from $C'$ by $\forall$-reduction with respect to the dependency scheme D. We refer to these nodes as *input nodes*, *resolution nodes*, and $\forall$ *-reduction nodes*, respectively.

Let $\mathcal{P}$ be an LDQ(D)-derivation from a PCNF formula $\Phi$. The (clause) label of the sink node is called the *conclusion* of $\mathcal{P}$, denoted $Cl(\mathcal{P})$. If the conclusion of $\mathcal{P}$ is the empty clause then we refer to $\mathcal{P}$ as an *LDQ(D)-refutation* of $\Phi$. For a node $v$ of $\mathcal{P}$, the *subderivation* (of $\mathcal{P}$) rooted at $v$ is the proof DAG induced by $v$ and its ancestors in $\mathcal{P}$. It is straightforward to verify that the resulting proof DAG is again an LDQ(D)-derivation from $\Phi$. For convenience, we will identify (sub)derivations with their sinks. The *size* of $\mathcal{P}$, denoted $|\mathcal{P}|$, is the total number of literal occurrences in clause labels of $\mathcal{P}$.

## 4   Soundness of and Strategy Extraction for LDQ(D$^{\mathrm{rrs}}$)

A PCNF formula can be associated with an evaluation game played between an existential and a universal player. These players take turns assigning quantifier blocks in the order of the prefix. The existential player wins if the matrix evaluates to 1 under the resulting variable assignment, while the universal player wins if the matrix evaluates to 0. One can show that the formula is true (false) if and only if the existential (universal) player has a winning strategy in this game, and this winning strategy is a (counter)model.

Goultiaeva, Van Gelder and Bacchus [20] proved that a Q-resolution refutation can be used to compute winning moves for the universal player in the evaluation game. The idea is that universal maintains a "restriction" of the refutation by the assignment constructed in the evaluation game, which is a refutation of the restricted formula.

For assignments made by the existential player, the universal player only needs to consider each instance of resolution whose pivot variable is assigned: one of the premises is not satisfied and can be used to (re)construct a refutation.

When it is universal's turn, the quantifier block for which she needs to pick an assignment is leftmost in the restricted formula. This means that $\forall$-reduction of these variables is blocked by any of the remaining existential variables and can only be applied to a purely universal clause. In a Q-resolution refutation, these variables must therefore be reduced at the very end, and because Q-resolution does not permit tautological clauses, only one polarity of each universal variable from the leftmost block can appear in a refutation. It follows that universal can maintain a Q-resolution refutation by assigning variables from the leftmost block in such a way as to map the associated literals to 0, effectively deleting them from the remaining Q-resolution refutation.

In this manner, the universal player can maintain a refutation until the end of the game, when all variables have been assigned. At that point, a refutation can consist only of the empty clause, which means that the assignment chosen by the two players falsifies a clause of the original matrix and universal has won the game.

Egly, Lonsing, and Widl [16] observed that this argument goes through even in the case of long-distance Q-resolution, since a clause containing both $u$ and $\neg u$ for a universal variable $u$ can only be derived by resolving on an existential variable to the left of $u$, but no such existential variable exists if $u$ is from the leftmost block.

In this section, we will prove that this argument can be generalized to $LDQ(D^{rrs})$-refutations. We illustrate this correspondence with an example:

*Example 1.* Consider the PCNF formula

$$\Phi = \exists x \, \forall u \, \exists e, y \quad (x \vee u \vee \overline{y}) \wedge (\overline{x} \vee \overline{u} \vee \overline{y}) \wedge (x \vee y) \wedge (\overline{x} \vee e) \wedge (\overline{u} \vee y) \wedge (\overline{y} \vee e)$$

Figure 2 shows an $LDQ(D^{rrs})$-refutation of $\Phi$. The only universal variable is $u$, so a strategy for the universal player in the evaluation game associated with $\Phi$ has to determine an assignment to $u$ given an assignment to $x$, the only (existential) variable preceding $u$. The figure illustrates how to compute the assignment to $u$ for the two possible assignments $\tau : \{x\} \rightarrow \{0, 1\}$ from the restriction of the refutation by $\tau$. In both cases, only one polarity of $u$ occurs in the restricted refutation and therefore it is easy for universal to determine the correct assignment. Notice that in one of the cases, a generalized $\forall$-reduction node remains present in the restriction—this shows that we cannot limit ourselves to looking at the final reduction step in the proof when looking for the variables to assign (as is the case with ordinary Q-resolution refutations, cf. [20]).

In all of the above cases, the key property that allows universal to maintain a refutation is that universal variables from a leftmost quantifier block may appear in at most one polarity. We will show that, indeed, this property is sufficient for soundness of $LDQ(D)$ when combined with a natural monotonicity property of dependency schemes.

**Definition 7.** Let D be a dependency scheme. We say that D is *monotone*, if for every PCNF formula $\Phi$ and every assignment $\tau$ to a subset of $var_\Phi$, $D_{\Phi[\tau]} \subseteq D_\Phi$. We say that D is *simple*, if for every PCNF formula $\Phi = \forall X \, \mathcal{Q}.\varphi$, every $LDQ(D)$-derivation $\mathcal{P}$ from $\Phi$, and every universal variable $u \in X$, $u$ or $\overline{u}$ does not appear in $\mathcal{P}$. We say that D is *normal* if it is both monotone and simple.

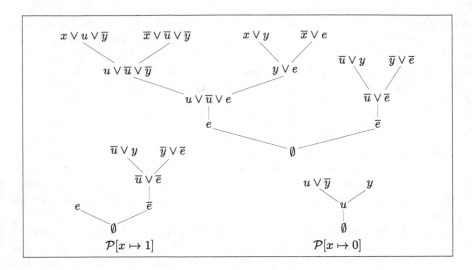

**Fig. 2.** An LDQ(D$^{\mathrm{rrs}}$)-refutation of the formula $\Phi$ from Example 1 and two restrictions.

As in the case of Q-resolution, universal's move for a particular quantifier block can be computed from the assignment corresponding to the previous moves and the refutation in polynomial time. Since every polynomial-time algorithm can be implemented by a family of polynomially-sized circuits, and because these circuits can even be computed in polynomial time [1, p. 109], it follows that LDQ(D) admits polynomial-time strategy extraction when D is normal. While the strategy extraction algorithm based on these general considerations is unlikely to be efficient, the algorithm for computing winning moves for universal is simple enough to be amenable to efficient simulation by a Boolean circuit. In Sect. 4.1, we give a direct construction that leads to the following result.

**Theorem 1.** *Let D be a normal dependency scheme. Then, there is an algorithm that computes a countermodel of a PCNF formula $\Phi$ with n variables from an LDQ(D)-refutation $\mathcal{P}$ of $\Phi$ in time $O(|\mathcal{P}| \cdot n)$.*

As an application of this general result, we will prove that the reflexive resolution-path dependency scheme is normal in Sect. 4.2.

**Theorem 2.** $D^{\mathrm{rrs}}$ *is normal.*

**Corollary 1.** *There is an algorithm that computes a countermodel of a PCNF formula $\Phi$ with n variables from an LDQ(D$^{rrs}$)-refutation $\mathcal{P}$ of $\Phi$ in time $O(|\mathcal{P}| \cdot n)$.*

This result immediately carries over to the less general standard dependency scheme:

**Corollary 2.** *There is an algorithm that computes a countermodel of a PCNF formula $\Phi$ with n variables from an LDQ(D$^{std}$)-refutation $\mathcal{P}$ of $\Phi$ in time $O(|\mathcal{P}| \cdot n)$.*

In combination with Proposition 1, these results imply soundness of both proof systems.

**Corollary 3.** *The systems LDQ($D^{std}$) and LDQ($D^{rrs}$) are sound.*

## 4.1    Certificate Extraction for Normal Dependency Schemes

We begin by formally defining the "restriction" of an LDQ(D)-derivation by an assignment, which is a straightforward generalization of this operation for Q-resolution derivations [20].[4] The result of restricting a derivation is either a derivation or the object $\top$, which can be interpreted as representing the tautological clause containing every literal. Accordingly, we stipulate that $\ell \in \top$ for every literal $\ell$.

**Definition 8. (Restriction).** Let $\Phi$ be a PCNF formula and let $\mathcal{P}$ be an LDQ(D)-derivation from $\Phi$. Further, let $X \subseteq var(\Phi)$ and let $\tau : X \to \{0,1\}$ be a truth assignment. The *restriction of $\mathcal{P}$ by $\tau$*, in symbols $\mathcal{P}[\tau]$, is defined as follows.

1. If $\mathcal{P}$ is an input node there are two cases. If $Cl(\mathcal{P})[\tau] = 1$ then $\mathcal{P}[\tau] = \top$. Otherwise, $\mathcal{P}[\tau]$ is the proof DAG consisting of a single node labeled with $Cl(\mathcal{P})[\tau]$.
2. If $\mathcal{P} = \mathcal{P}_1 \odot_\ell \mathcal{P}_2$, that is, if $\mathcal{P}$ is a resolution node, we distinguish four cases:
   (a) If $\ell \notin Cl(\mathcal{P}_1[\tau])$ then $\mathcal{P}[\tau] = \mathcal{P}_1[\tau]$.
   (b) If $\ell \in Cl(\mathcal{P}_1[\tau])$ and $\bar{\ell} \notin Cl(\mathcal{P}_2[\tau])$ then $\mathcal{P}[\tau] = \mathcal{P}_2[\tau]$.
   (c) If $\ell \in Cl(\mathcal{P}_1[\tau])$, $\bar{\ell} \in Cl(\mathcal{P}_2[\tau])$, and $\mathcal{P}_1[\tau] = \top$ or $\mathcal{P}_2[\tau] = \top$, we let $\mathcal{P}[\tau] = \top$.
   (d) If $\ell \in Cl(\mathcal{P}_1[\tau])$, $\bar{\ell} \in Cl(\mathcal{P}_2[\tau])$, $\mathcal{P}_1[\tau] \neq \top$, and $\mathcal{P}_2[\tau] \neq \top$, we define $\mathcal{P}[\tau] = \mathcal{P}_1[\tau] \odot_\ell \mathcal{P}_2[\tau]$.
3. If $\mathcal{P} = \mathcal{P}' - u$, that is, if $\mathcal{P}$ is a $\forall$-reduction node, we distinguish three cases:
   (a) If $\mathcal{P}'[\tau] = \top$ then $\mathcal{P}[\tau] = \top$.
   (b) If $\mathcal{P}'[\tau] \neq \top$ and $u \notin var(Cl(\mathcal{P}'[\tau]))$ then $\mathcal{P}[\tau] = \mathcal{P}'[\tau]$.
   (c) If $\mathcal{P}'[\tau] \neq \top$ and $u \in var(Cl(\mathcal{P}'[\tau]))$ then $\mathcal{P}[\tau] = \mathcal{P}'[\tau] - u$.

If D is a monotone dependency scheme, LDQ(D)-refutations are preserved under restriction by an existential assignment (cf. [20, Lemma4]). This is stated in the following lemma, which can be proved by a straightforward induction on the structure of the LDQ(D)-derivation.

**Lemma 1.** *Let D be a monotone dependency scheme, let $\mathcal{P}$ be an LDQ(D)-derivation from a PCNF formula $\Phi$, let $E \subseteq var_\exists(\Phi)$, and let $\tau : E \to \{0,1\}$ be an assignment. If $\mathcal{P}[\tau] = \top$ then $Cl(\mathcal{P})[\tau] = 1$. Otherwise, $\mathcal{P}[\tau]$ is an LDQ(D)-derivation from $\Phi[\tau]$ such that $Cl(\mathcal{P}[\tau]) \subseteq Cl(\mathcal{P})[\tau]$.*

---

[4] Our definition slightly differs from the original for the resolution rule: if restriction removes the pivot variable from both premises, we simply pick the (restriction of the) first premise as the result (rather than the clause containing fewer literals). This simplifies the certificate extraction argument given below.

Above, we argued that the universal player can use an LDQ(D)-refutation for a normal dependency scheme D in order to compute winning moves in the evaluation game associated with a PCNF formula and that this can be used to compute a countermodel of the formula in polynomial time. We now prove this directly, by showing how to construct a circuit implementing a countermodel from an LDQ(D)-refutation.

We begin by describing auxiliary circuits simulating the restriction operation. Let $\Phi = Q_1 X_1 \ldots Q_k X_k.\varphi$ be a PCNF formula and let $\mathcal{P}$ be a refutation of $\Phi$. For each quantifier block $X_i$, each subderivation $\mathcal{S}$ of $\mathcal{P}$, and each literal $\ell$, we will construct circuits $\text{TOP}_{\mathcal{S}}^i$ and $\text{CONTAINS}_{\mathcal{S},\ell}^i$ with inputs from $X = \bigcup_{j<i} X_j$ such that, for every assignment $\sigma : X \to \{0,1\}$,

$$\text{TOP}_{\mathcal{S}}^i[\sigma] = 1 \iff \mathcal{S}[\sigma] = \top \tag{1}$$

$$\text{CONTAINS}_{\mathcal{S},\ell}^i[\sigma] = 1 \iff \ell \in Cl(\mathcal{S}[\sigma]) \tag{2}$$

We first describe our construction and then prove that it satisfies the above properties in Lemma 2. Let $\mathcal{S}$ be an input node. We let

$$\text{TOP}_{\mathcal{S}}^1 := \bigvee \{ Cl(\mathcal{S}) \cap (X_1 \cup \overline{X_1}) \},$$

and define $\text{TOP}_{\mathcal{S}}^i$ for $1 < i \leq k$ as

$$\text{TOP}_{\mathcal{S}}^i := \text{TOP}_{\mathcal{S}}^{i-1} \vee \bigvee \{ Cl(\mathcal{S}) \cap (X_i \cup \overline{X_i}) \}.$$

Moreover, for $1 \leq i \leq k$ we define $\text{CONTAINS}_{\mathcal{S},\ell}^i$ as

$$\text{CONTAINS}_{\mathcal{S},\ell}^i = \begin{cases} 1 & \text{if } \ell \in Cl(\mathcal{S}) \setminus (X \cup \overline{X}), \\ \text{TOP}_{\mathcal{S}}^i & \text{otherwise.} \end{cases}$$

For non-input nodes, we proceed as follows. If $\mathcal{S} = \mathcal{S}_1 \odot_q \mathcal{S}_2$, we define $\text{TOP}_{\mathcal{S}}^i$ as

$$\text{TOP}_{\mathcal{S}}^i = (\text{CONTAINS}_{\mathcal{S}_1,q}^i \wedge \text{TOP}_{\mathcal{S}_2}^i) \vee (\text{CONTAINS}_{\mathcal{S}_2,\overline{q}}^i \wedge \text{TOP}_{\mathcal{S}_1}^i),$$

and if $\mathcal{S} = \mathcal{S}' - u$, we let

$$\text{TOP}_{\mathcal{S}}^i := \text{TOP}_{\mathcal{S}'}^i.$$

For the $\text{CONTAINS}_{\mathcal{S},\ell}^i$ circuit, we distinguish two cases. Let $\ell$ be a literal and $\mathcal{S}$ a derivation. If $\ell \notin Cl(\mathcal{S})$ we simply let

$$\text{CONTAINS}_{\mathcal{S},\ell}^i := \text{TOP}_{\mathcal{S}}^i.$$

Otherwise, if $\ell \in Cl(\mathcal{S})$, we have to consider two cases. First, if $\mathcal{S} = \mathcal{S}_1 \odot_q \mathcal{S}_2$, we let

$$\text{CONTAINS}_{\mathcal{S},\ell}^i = \text{TOP}_{\mathcal{S}}^i \vee$$
$$(\neg\text{CONTAINS}_{\mathcal{S}_1,q}^i \wedge \text{CONTAINS}_{\mathcal{S}_1,\ell}^i) \vee$$
$$(\text{CONTAINS}_{\mathcal{S}_1,q}^i \wedge \neg\text{CONTAINS}_{\mathcal{S}_2,\bar{q}}^i \wedge \text{CONTAINS}_{\mathcal{S}_2,\ell}^i) \vee$$
$$(\text{CONTAINS}_{\mathcal{S}_1,q}^i \wedge \text{CONTAINS}_{\mathcal{S}_2,\bar{q}}^i \wedge (\text{CONTAINS}_{\mathcal{S}_1,\ell}^i \vee \text{CONTAINS}_{\mathcal{S}_2,\ell}^i)).$$

Second, if $\mathcal{S} = \mathcal{S}' - u$, then

$$\text{CONTAINS}_{\mathcal{S},\ell}^i := \text{CONTAINS}_{\mathcal{S}',\ell}^i.$$

To implement the winning strategy for universal sketched above, we further construct circuits $\text{POLARITY}_{\mathcal{S},u}$ for each node $\mathcal{S}$ of $\mathcal{P}$ and each universal variable $u \in var_\forall(\Phi)$, such that, for each assignment $\tau : L_\Phi(u) \rightarrow \{0,1\}$,

$$\text{POLARITY}_{\mathcal{S},u}[\tau] = 1 \iff u \text{ occurs in } \mathcal{S}[\tau]. \tag{3}$$

Let $u \in X_i$ be a universal variable from the $i$th quantifier block. If $\mathcal{S}$ is an input node, we simply define

$$\text{POLARITY}_{\mathcal{S},u} := \text{CONTAINS}_{\mathcal{S},u}^i,$$

and if $\mathcal{S} = \mathcal{S}' - u$ is a $\forall$-reduction node, we let

$$\text{POLARITY}_{\mathcal{S},u} := \text{POLARITY}_{\mathcal{S}',u}.$$

If $\mathcal{S} = \mathcal{S}_1 \odot_q \mathcal{S}_2$, then

$$\text{POLARITY}_{\mathcal{S},u} := (\neg\text{CONTAINS}_{\mathcal{S}_1,q}^i \wedge \text{POLARITY}_{\mathcal{S}_1,u}) \vee$$
$$(\text{CONTAINS}_{\mathcal{S}_1,q}^i \wedge \neg\text{CONTAINS}_{\mathcal{S}_2,\bar{q}}^i \wedge \text{POLARITY}_{\mathcal{S}_2,u}) \vee$$
$$(\text{CONTAINS}_{\mathcal{S}_1,q}^i \wedge \text{CONTAINS}_{\mathcal{S}_2,\bar{q}}^i \wedge$$
$$(\text{POLARITY}_{\mathcal{S}_1,u} \vee \text{POLARITY}_{\mathcal{S}_2,u})).$$

**Lemma 2.** *Let $\Phi = Q_1 X_1 \ldots Q_k X_k . \varphi$ be a PCNF formula and let $\mathcal{P}$ be an LDQ(D)-derivation from $\Phi$. For each $1 \leq i \leq k$, each literal $\ell$, every $u \in var_\forall(\Phi) \cap X_i$, and every truth assignment $\sigma : \bigcup_{j=1}^{i-1} X_j \rightarrow \{0,1\}$, $\text{TOP}_{\mathcal{P}}^i$ satisfies (1), $\text{CONTAINS}_{\mathcal{P},\ell}^i$ satisfies (2), and $\text{POLARITY}_{\mathcal{P},u}$ satisfies (3).*

These auxiliary circuits can be efficiently constructed in a top-down manner, from the input nodes to the conclusion. By a careful analysis, we obtain the following:

**Lemma 3.** *There is an algorithm that, given a PCNF formula $\Phi$ and an LDQ(D)-derivation $\mathcal{P}$ from $\Phi$, computes the circuits $\text{POLARITY}_{\mathcal{P},u}$ for every universal variable $u$ in time $O(|\mathcal{P}| \cdot n)$, where $n = |var(\Phi)|$.*

Using Lemma 1, we can spell out the argument sketched at the beginning of this section and prove that for normal dependency schemes D, the universal player can maintain an LDQ(D)-refutation throughout the evaluation game by successively restricting an initial LDQ(D)-refutation by both players' moves and assigning universal variables from the leftmost remaining block $X$ so as to falsify the (unique) literals from $X$ remaining the refutation. Lemma 2 tells us that the POLARITY circuits can be used to implement this strategy, which leads to the following lemma.

**Lemma 4.** *Let $D$ be a normal dependency scheme, let $\mathcal{P}$ be an LDQ(D)-refutation of a PCNF formula $\Phi$. Then the family $\{f_u\}_{u \in var_\forall(\Phi)}$ of functions $f_u = \neg \text{POLARITY}_{\mathcal{P},u}$ is a countermodel of $\Phi$.*

Theorem 1 immediately follows from from Lemmas 3 and 4.

## 4.2   The Reflexive Resolution-Path Dependency Scheme is Normal

In order to prove Theorem 2 and show that $D^{rrs}$ is normal, we will need some insight into the relationship between resolution paths and LDQ(D)-derivations. We begin by observing that no clause derived by LDQ(D) from a PCNF formula $\forall X Q.\varphi$ can contain both $u$ and $\neg u$ if $u \in X$, as such a clause would have to be derived by resolving on a variable to the left of $u$.

**Lemma 5.** *Let $\mathcal{P}$ be an LDQ(D)-derivation from a PCNF formula $\Phi = \forall X Q.\varphi$ and let $u \in X$. Then $Cl(\mathcal{P})$ cannot contain both $u$ and $\neg u$.*

Let $\mathcal{P}$ be an LDQ(D)-refutation of a PCNF formula $\Phi$ and let $u$ be a universal variable such that both $u$ and $\neg u$ appear in $\mathcal{P}$. Since $\mathcal{P}$ is a refutation, both $u$ and $\neg u$ have to be eventually removed by $\forall$-reduction. By the preceding lemma, $u$ and $\neg u$ cannot occur together in a clause. Now, if $C$ is a purely universal clause appearing in $\mathcal{P}$, then $C$ must be followed by a sequence of $\forall$-reduction steps that lead to the empty clause. Since $\mathcal{P}$ contains only a single node labeled with the empty clause, this means that each purely universal clause appearing in $\mathcal{P}$ must be part of a single final sequence of $\forall$-reduction steps concluding the refutation, so that $C \subseteq C'$ or $C' \subseteq C$ holds for every pair $C, C'$ of such clauses. It follows that we cannot have a purely universal clause $C$ with $u \in C$ and a purely universal clause $C'$ with $\neg u \in C'$ such that $C$ and $C'$ both occur in $\mathcal{P}$. So at least one of $u$ and $\neg u$ must be reduced from a clause which still contains an existential variable. We will prove that $(u, e) \in D^{rrs}_\Phi$ for some such variable $e$.

To show this, we establish a connection between resolution paths and the structure of LDQ(D)-refutations: resolution paths can be used to determine whether two literals may occur together in a clause derived by LDQ(D). The following statement can be proved by a simple induction on the structure of $\mathcal{P}$ (cf. [35, 38]).

**Lemma 6.** *Let $\mathcal{P}$ be an LDQ(D)-derivation from a PCNF formula $\Phi$ and let $\ell, \ell'$ be literals in $Cl(\mathcal{P})$. Then $\ell$ and $\ell'$ are connected in $\Phi$ with respect to the set of variables appearing as pivots in $\mathcal{P}$.*

If a clause $C'$ appears in the derivation of a clause $C$ and $C \nsubseteq C'$, then $C$ cannot be derived from $C'$ by $\forall$-reduction alone, and some existential variable $e \in var(C')$ must be resolved on in the derivation of $C$. It follows that there have to be resolution paths connecting literals in $C'$ with literals in $C \setminus C'$ that go through an existential variable appearing in $C'$. This is made precise in the following lemma.

**Lemma 7.** *Let $\mathcal{P}$ be an LDQ(D)-derivation from a PCNF formula $\Phi$, let $\mathcal{P}'$ be a node of $\mathcal{P}$, and let $\ell' \in Cl(\mathcal{P}')$. For each literal $\ell \in Cl(\mathcal{P}) \setminus Cl(\mathcal{P}')$, there is an existential variable $e \in var(Cl(\mathcal{P}'))$ and a resolution path from $\ell'$ to $\ell$ in $\Phi$ via the set of pivots in $\mathcal{P}$ that goes through $e$.*

Together, the two preceding lemmas can be used to show that an LDQ(D)-refutation that contains both $u$ and $\neg u$ from an outermost universal quantifier block induces a resolution path from $u$ to $\neg u$ through an existential variable $e$ such that $u$ and $e$ or $\neg u$ and $e$ occur together in a clause.

**Lemma 8.** *Let $\mathcal{P}$ be an LDQ(D)-refutation of a PCNF formula $\Phi = \forall X\, Q.\varphi$ and let $u \in X$ be a universal variable such that $\mathcal{P}$ contains nodes $\mathcal{P}_1$ and $\mathcal{P}_2$ with $u \in Cl(\mathcal{P}_1)$ and $\neg u \in Cl(\mathcal{P}_2)$. Then there is an existential variable $e \in var(Cl(\mathcal{P}_1) \cup Cl(\mathcal{P}_2))$ and a resolution path in $\Phi$ from $u$ to $\neg u$ via variables resolved on in $\mathcal{P}$ that goes through $e$.*

*Proof (of Theorem 2).* We first prove that $D^{rrs}$ is simple. Let $\mathcal{P}$ be an LDQ($D^{rrs}$)-refutation of a PCNF formula $\Phi = \forall X\, Q.\varphi$ and let $u \in X$. Suppose that $u$ and $\neg u$ both appear in $\mathcal{P}$. Since $\mathcal{P}$ is a refutation, there have to be $\forall$-reduction nodes $\mathcal{P}_1 = \mathcal{P}_1' - u$ such that $u \in Cl(\mathcal{P}_1')$ and $\mathcal{P}_2 = \mathcal{P}_2' - u$ such that $\neg u \in Cl(\mathcal{P}_2')$. By Lemma 8, there has to be a resolution path $\pi$ from $u$ to $\neg u$ via variables resolved on in $\mathcal{P}$ such that $\pi$ goes through an existential variable $e \in var(Cl(\mathcal{P}_1) \cup Cl(\mathcal{P}_2))$. We can think of $\pi$ as the concatenation of two resolution paths $\pi_1$ and $\pi_2$, where $\pi_1$ is a resolution path from $u$ to $\ell$ and $\pi_2$ is a resolution path from $\bar{\ell}$ to $\neg u$, where $\ell$ is a literal with $var(\ell) = e$. That is, $\{u, e\}$ is a resolution-path dependency pair of $\Phi$ with respect to variables resolved on in $\mathcal{P}$. Since $u$ is leftmost in the quantifier prefix, $\mathcal{P}$ only resolves on existential variables to the right of $u$ and thus $(u, e) \in D_\Phi^{rrs}$. But $(u, e) \notin D_\Phi^{rrs}$ for each existential variable $e \in var(Cl(\mathcal{P}_1') \cup Cl(\mathcal{P}_2'))$ by definition of LDQ($D^{rrs}$), a contradiction. To see that $D^{rrs}$ is monotone, note that a resolution path in any restriction $\Phi[\tau]$ of a PCNF formula $\Phi$ is also a resolution path in $\Phi$.  □

## 5    Experiments

To gauge the potential of clause learning based on LDQ($D^{std}$), we ran experiments with the search-based solver DepQBF.[5] By default, DepQBF supports proof generation only in combination with the trivial dependency scheme—in

---

[5] http://lonsing.github.io/depqbf/

that case, it generates Q-resolution or long distance Q-resolution proofs (depending on whether long distance resolution is enabled). However, by uncommenting a few lines in the source code, proof generation can also be enabled with the standard dependency scheme, and this option can even be combined with long distance resolution. For false formulas, the resulting proofs are $Q(D^{std})$-resolution or $LDQ(D^{std})$-resolution refutations.

We compared the performance of DepQBF in four configurations,[6] each using a different proof system for constraint learning:

1. Long distance Q-resolution with $\forall/\exists$-reduction according to $D^{std}$ (LDQD).
2. Long distance Q-resolution with ordinary $\forall/\exists$-reduction (LDQ).
3. Q-resolution with $\forall/\exists$-reduction according to $D^{std}$ (QD).
4. Ordinary Q-resolution (Q).

These experiments were performed on a cluster with Intel Xeon E5649 processors at 2.53 GHz running 64-bit Linux. We set time and memory limits of 900 seconds and 4 GB, respectively.

Instances were taken from two tracks of the QBF Gallery 2014: the *application* track consisting of 6 instance families and a total of 735 formulas, and the *QBFLib* track consisting of 276 formulas.

For our first set of experiments, we disabled dynamic QBCE (Quantified Blocked Clause Elimination), a technique introduced with version 5.0 of DepQBF [29]. We further used bloqqer[7] with default settings as a preprocessor. Since $LDQ(D^{std})$ generalizes both long distance Q-resolution and $Q(D^{std})$-resolution, we were expecting a performance increase with $LDQ(D^{std})$-learning compared to learning based on the other proof systems. However, all four configurations showed virtually identical performance on both the application and QBFlib benchmark sets in terms of total runtime and instances solved within the time limit.

To get a more detailed picture, we broke down the results for the application track by instance family, limiting ourselves to instances that were solved by at least one configuration. The barplot in Fig. 3 shows that there are considerable differences in performance between solver configurations for individual instances families, with each solver configuration being outperformed by another configuration on at least one family.

For our second set of experiments, we turned on dynamic QBCE. This led to a significant performance increase both in terms of number of instances solved within the time limit and total runtime for both benchmark sets, a result that is consistent with the findings in [29]. However, as far as the performance of $LDQ(D^{std})$-learning is concerned, the application and QBFlib tracks differed significantly for this experiment.

While $LDQ(D^{std})$-learning fared *worst* among the configurations both with respect to instances solved and total runtime on the application track, it was

---

[6] As a sanity check, we verified that all configurations that were able to solve a particular instance returned the same result.

[7] http://fmv.jku.at/bloqqer/

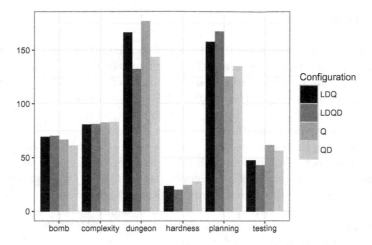

**Fig. 3.** Average runtime in seconds (y-axis) for instances from the application track for each instance family (x-axis), by solver configuration (with preprocessing, but without dynamic QBCE). Here, we only considered instances that were solved by at least one configuration.

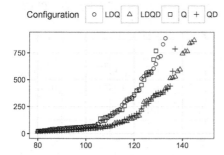

**Fig. 4.** Solved instances from the QBFLib track (x-axis) sorted by runtime (y-axis), by solver configuration (with preprocessing and dynamic QBCE).

**Table 1.** Number of solved instances, solved true instances, solved false instances, and total runtime in seconds (including timeouts) for the application track, with QBCE (but without preprocessing)

| Configuration | Solved | True | False | Time |
|---|---|---|---|---|
| LDQD | 440 | 223 | 217 | 287012 |
| LDQ | 435 | 223 | 212 | 291574 |
| QD | 440 | 225 | 215 | 291661 |
| Q | 437 | 221 | 216 | 337141 |

the *best* configuration for the QBFlib track in both respects. As Fig. 4 shows, using the standard dependency scheme was beneficial both with and without long distance resolution for the QBFlib instances.

For our final set of experiments, we left dynamic QBCE enabled but *disabled* preprocessing for the application track, as this was shown to lead to a performance *increase* in the case of learning with ordinary Q-resolution [29]. As expected, this resulted in a performance increase across the board (see Table 1). Moreover, LDQ($D^{std}$)-learning was the best configuration in terms of instances solved (on par with Q($D^{std}$)-resolution) as well as in terms of overall runtime.

## 6   Discussion

The experiments in Sect. 5 show that DepQBF can benefit from learning based on $LDQ(D^{std})$. This benefit is essentially "for free", in that it does not require any changes to the implementation, but soundness of the resulting solver configuration is not immediate. The results of Sect. 4 contribute to a soundness proof, but they remain partial in two respects: first, soundness of $LDQ(D^{std})$ only implies that we can trust the solver when it outputs "false". To prove that "true" answers can be trusted as well, one has to show soundness of quantified term resolution when combined with the standard dependency scheme and long distance resolution. Alternatively, one could use $LDQ(D^{std})$ for clause learning only, in combination with ordinary long distance Q-resolution for cube learning. Second, we observed synergies only when dynamic QBCE was activated, and it remains to show that clause learning based on $LDQ(D^{std})$ is sound in combination with this technique.

We take Theorem 1 as proof that, in principle, efficient certificate extraction from $LDQ(D^{rrs})$-refutations is possible. For practical purposes, the time bound of $O(|\mathcal{P}| \cdot n)$ is not good enough. For $LDQ(D^{std})$, a modified extraction algorithm achieves a runtime of $O(|\mathcal{P}| \cdot k)$, where $k$ is the number of quantifier alternations of the input formula. A proof-of-concept implementation currently does not scale to proofs larger than a few megabytes, but we are confident that further improvements will lead to an efficient enough algorithm for practical use.

**Acknowledgments.** We would like to thank Florian Lonsing for helpful discussions and for pointing out how to modify DepQBF so that it generates $LDQ(D^{std})$ proofs.

## References

1. Arora, S., Barak, B.: Computational Complexity - A Modern Approach. Cambridge University Press, New York (2009)
2. Balabanov, V., Jiang, J.R.: Unified QBF certification and its applications. Formal Methods Syst. Des. **41**(1), 45–65 (2012)
3. Balabanov, V., Jiang, J. R., Janota, M., Widl, M.: Efficient extraction of QBF (counter)models from long-distance resolution proofs. In: Bonet, B., Koenig, S. (eds.), Proceedings of the Twenty-Ninth AAAI Conference on Artificial Intelligence, January 25–30, Austin, Texas, USA, pp. 3694–3701. AAAI Press (2015)
4. Balabanov, V., Widl, M., Jiang, J.H.R.: QBF resolution systems and their proof complexities. In: Sinz, C., Egly, U. (eds.) SAT 2014. LNCS, vol. 8561, pp. 154–169. Springer, Heidelberg (2014)
5. Benedetti, M., Mangassarian, H.: QBF-based formal verification: Experience and perspectives. J. Satisfiability, Boolean Model. Comput. **5**(1–4), 133–191 (2008)
6. Beyersdorff, O., Chew, L., Janota, M.: Proof complexity of resolution-based QBF calculi. In: Mayr, E.W., Ollinger, N. (ed.), 32nd International Symposium on Theoretical Aspects of Computer Science, STACS 2015, March 4–7, 2015, Garching, Germany, vol. 30 of LIPIcs, pp. 76–89. Schloss Dagstuhl - Leibniz-Zentrum fuer Informatik (2015)

7. Beyersdorff, O., Chew, L., Mahajan, M., Shukla, A.: Feasible interpolation for QBF resolution calculi. In: Halldórsson, M.M., Iwama, K., Kobayashi, N., Speckmann, B. (eds.) ICALP 2015. LNCS, vol. 9134, pp. 180–192. Springer, Heidelberg (2015)
8. Biere, A., Lonsing, F.: Integrating dependency schemes in search-based QBF solvers. In: Strichman, O., Szeider, S. (eds.) SAT 2010. LNCS, vol. 6175, pp. 158–171. Springer, Heidelberg (2010)
9. Biere, A., Lonsing, F., Seidl, M.: Blocked clause elimination for QBF. In: Bjørner, N., Sofronie-Stokkermans, V. (eds.) CADE 2011. LNCS, vol. 6803, pp. 101–115. Springer, Heidelberg (2011)
10. Bloem, R., Könighofer, R., Seidl, M.: SAT-based synthesis methods for safety specs. In: McMillan, K.L., Rival, X. (eds.) VMCAI 2014. LNCS, vol. 8318, pp. 1–20. Springer, Heidelberg (2014)
11. Bubeck, U.: Model-based transformations for quantified Boolean formulas. Ph.D. thesis, University of Paderborn (2010)
12. Cadoli, M., Schaerf, M., Giovanardi, A., Giovanardi, M.: An algorithm to evaluate Quantified Boolean Formulae and its experimental evaluation. J. Autom. Reasoning **28**(2), 101–142 (2002)
13. Cashmore, M., Fox, M., Giunchiglia, E.: Partially grounded planning as Quantified Boolean Formula. In: Borrajo, D., Kambhampati, S., Oddi, A., Fratini, S. (eds.), 23rd International Conference on Automated Planning and Scheduling, ICApPS. AAAI (2013)
14. Davis, M., Logemann, G., Loveland, D.: A machine program for theorem-proving. Commun. ACM **5**, 394–397 (1962)
15. Egly, U.: On sequent systems and resolution for QBFs. In: Cimatti, A., Sebastiani, R. (eds.) SAT 2012. LNCS, vol. 7317, pp. 100–113. Springer, Heidelberg (2012)
16. Egly, U., Lonsing, F., Widl, M.: Long-distance resolution: Proof generation and strategy extraction in search-based QBF solving. In: McMillan, K., Middeldorp, A., Voronkov, A. (eds.) LPAR-19 2013. LNCS, vol. 8312, pp. 291–308. Springer, Heidelberg (2013)
17. Giunchiglia, E., Narizzano, M., Tacchella, A.: Clause/term resolution and learning in the evaluation of Quantified Boolean Formulas. J. Artif. Intell. Res. **26**, 371–416 (2006)
18. Goultiaeva, A., Bacchus, F.: Exploiting QBF duality on a circuit representation. In: Fox, M., Poole, D. (eds.), Proceedings of the Twenty-Fourth AAAI Conference on Artificial Intelligence, AAAI . AAAI Press (2010)
19. Goultiaeva, A., Seidl, M., Biere, A.: Bridging the gap between dual propagation and CNF-based QBF solving. In: Macii, E. (ed.) Design. Automation and Test in Europe. EDA Consortium San Jose, pp. 811–814. ACM DL, CA, USA (2013)
20. Goultiaeva, A., Van Gelder, A., Bacchus, F.: A uniform approach for generating proofs and strategies for both true and false QBF formulas. In Walsh, T. (ed), Proceedings of IJCAI, pp. 546–553. IJCAI/AAAI (2011)
21. Heule, M., Seidl, M., Biere, A.: A unified proof system for QBF preprocessing. In: Demri, S., Kapur, D., Weidenbach, C. (eds.) IJCAR 2014. LNCS, vol. 8562, pp. 91–106. Springer, Heidelberg (2014)
22. Janota, M., Chew, L., Beyersdorff, O.: On unification of QBF resolution-based calculi. In: Csuhaj-Varjú, E., Dietzfelbinger, M., Ésik, Z. (eds.) MFCS 2014, Part II. LNCS, vol. 8635, pp. 81–93. Springer, Heidelberg (2014)
23. Janota, M., Klieber, W., Marques-Silva, J., Clarke, E.: Solving QBF with counterexample guided refinement. In: Cimatti, A., Sebastiani, R. (eds.) SAT 2012. LNCS, vol. 7317, pp. 114–128. Springer, Heidelberg (2012)

24. Janota, M., Marques-Silva, J.: On propositional QBF expansions and Q-Resolution. In: Järvisalo, M., Van Gelder, A. (eds.) SAT 2013. LNCS, vol. 7962, pp. 67–82. Springer, Heidelberg (2013)
25. Kleine Büning, H., Karpinski, M., Flögel, A.: Resolution for quantified Boolean formulas. Inf. Comput. **117**(1), 12–18 (1995)
26. Klieber, W., Sapra, S., Gao, S., Clarke, E.: A non-prenex, non-clausal QBF solver with game-state learning. In: Strichman, O., Szeider, S. (eds.) SAT 2010. LNCS, vol. 6175, pp. 128–142. Springer, Heidelberg (2010)
27. Kronegger, M., Pfandler, A., Pichler, R.: Conformant planning as benchmark for QBF-solvers. In: International Workshop on Quantified Boolean Formulas - QBF (2013). http://fmv.jku.at/qbf2013/
28. Lonsing, F.: Dependency Schemes and Search-Based QBF : Theory and Practice. Ph.D. thesis, Johannes Kepler University, Linz, Austria, Apr 2012
29. Lonsing, F., Bacchus, F., Biere, A., Egly, U., Seidl, M.: Enhancing search-based QBF solving by dynamic blocked clause elimination. In: Davis, M. (ed.) LPAR-20 2015. LNCS, vol. 9450, pp. 418–433. Springer, Heidelberg (2015). doi:10.1007/978-3-662-48899-7_29
30. Lonsing, F., Egly, U., Van Gelder, A.: Efficient clause learning for quantified boolean formulas via QBF pseudo unit propagation. In: Järvisalo, M., Van Gelder, A. (eds.) SAT 2013. LNCS, vol. 7962, pp. 100–115. Springer, Heidelberg (2013)
31. Marques-Silva, J.: The impact of branching heuristics in propositional satisfiability algorithms. In: Barahona, P., Alferes, J.J. (eds.) EPIA 1999. LNCS (LNAI), vol. 1695, pp. 62–74. Springer, Heidelberg (1999)
32. Niemetz, A., Preiner, M., Lonsing, F., Seidl, M., Biere, A.: Resolution-based certificate extraction for QBF. In: Cimatti, A., Sebastiani, R. (eds.) SAT 2012. LNCS, vol. 7317, pp. 430–435. Springer, Heidelberg (2012)
33. Rintanen, J.: Asymptotically optimal encodings of conformant planning in QBF. In: 22nd AAAI Conference on Artificial Intelligence, pp. 1045–1050. AAAI (2007)
34. Samer, M., Szeider, S.: Backdoor sets of quantified Boolean formulas. J. Autom. Reasoning **42**(1), 77–97 (2009)
35. Slivovsky, F., Szeider, S.: Computing resolution-path dependencies in linear time. In: Cimatti, A., Sebastiani, R. (eds.) SAT 2012. LNCS, vol. 7317, pp. 58–71. Springer, Heidelberg (2012)
36. Slivovsky, F., Szeider, S.: Soundness of Q-resolution with dependency schemes. Theor. Comput. Sci. **612**, 83–101 (2016)
37. Staber, S., Bloem, R.: Fault localization and correction with QBF. In: Marques-Silva, J., Sakallah, K.A. (eds.) SAT 2007. LNCS, vol. 4501, pp. 355–368. Springer, Heidelberg (2007)
38. Van Gelder, A.: Variable independence and resolution paths for quantified boolean formulas. In: Lee, J. (ed.) CP 2011. LNCS, vol. 6876, pp. 789–803. Springer, Heidelberg (2011)
39. Zhang, L., Malik, S.: Conflict driven learning in a quantified Boolean satisfiability solver. In: Pileggi, L.T. Kuehlmann, A. (eds.), Proceedings of the IEEE/ACM International Conference on Computer-aided Design, ICCAD, San Jose, California, USA, November 10–14, pp. 442–449. ACM/IEEE Computer Society (2002)
40. Zhang, L., Malik, S.: Towards a symmetric treatment of satisfaction and conflicts in quantified boolean formula evaluation. In: Van Hentenryck, P. (ed.) CP 2002. LNCS, vol. 2470, pp. 200–215. Springer, Heidelberg (2002)

**Tools**

# BEACON: An Efficient SAT-Based Tool for Debugging $\mathcal{EL}^+$ Ontologies

M. Fareed Arif[1], Carlos Mencía[2(✉)], Alexey Ignatiev[3,6], Norbert Manthey[4], Rafael Peñaloza[5], and Joao Marques-Silva[3]

[1] University College Dublin, Dublin, Ireland
muhammad.arif.1@ucdconnect.ie
[2] University of Oviedo, Oviedo, Spain
cmencia@gmail.com
[3] University of Lisbon, Lisbon, Portugal
{aignatiev,jpms}@ciencias.ulisboa.pt
[4] TU Dresden, Dresden, Germany
norbert.manthey@tu-dresden.de
[5] Free University of Bozen-Bolzano, Bolzano, Italy
rafael.penaloza@unibz.it
[6] ISDCT SB RAS, Irkutsk, Russia

**Abstract.** Description Logics (DLs) are knowledge representation and reasoning formalisms used in many settings. Among them, the $\mathcal{EL}$ family of DLs stands out due to the availability of polynomial-time inference algorithms and its ability to represent knowledge from domains such as medical informatics. However, the construction of an ontology is an error-prone process which often leads to unintended inferences. This paper presents the BEACON tool for debugging $\mathcal{EL}^+$ ontologies. BEACON builds on earlier work relating minimal justifications (MinAs) of $\mathcal{EL}^+$ ontologies and MUSes of a Horn formula, and integrates state-of-the-art algorithms for solving different function problems in the SAT domain.

## 1 Introduction

The importance of Description Logics (DLs) cannot be overstated, and impact a growing number of fields. The $\mathcal{EL}$-family of tractable DLs in particular has been used to build large ontologies from the life sciences [34,35]. Ontology development is an error-prone task, with potentially critical consequences in the life sciences; thus it is important to develop automated tools to help debugging large ontologies. Axiom pinpointing refers to the task of finding the precise axioms in an ontology that cause a (potentially unwanted) consequence to follow [25]. Recent years have witnessed remarkable improvements in axiom pinpointing technologies, including for the case of the $\mathcal{EL}$ family of DLs [1,2,5–7,20,21,30,31]. Among these, the use of SAT-based methods [1,2,30] was shown to outperform other alternative approaches very significantly. This is achieved by reducing the problem to a propositional Horn formula, which is then analyzed with a dedicated decision engine for Horn formulae.

© Springer International Publishing Switzerland 2016
N. Creignou and D. Le Berre (Eds.): SAT 2016, LNCS 9710, pp. 521–530, 2016.
DOI: 10.1007/978-3-319-40970-2_32

This paper describes a tool, BEACON, that builds on recent work on efficient enumeration of Minimal Unsatisfiable Subsets (MUSes) of group Horn formulae, which finds immediate application in axiom pinpointing of $\mathcal{EL}$ ontologies [2]. In contrast to earlier work [2], which used EL+SAT [30,31] as front-end, this paper proposes an integrated tool to perform analysis on ontologies, offering a number of new features.

The rest of the paper is organized as follows: Sect. 2 introduces some preliminaries. Section 3 describes the organization of BEACON. Experimental results are given in Sect. 4. Finally, Sect. 5 concludes the paper.

## 2    Preliminaries

### 2.1    The Lightweight Description Logic $\mathcal{EL}^+$

$\mathcal{EL}^+$ [4] is a light-weight DL that has been successfully used to build large ontologies, most notably from the bio-medical domains. As with all DLs, the main elements in $\mathcal{EL}^+$ are concepts. $\mathcal{EL}^+$ *concepts* are built from two disjoint sets $\mathsf{N_C}$ and $\mathsf{N_R}$ of *concept names* and *role names* through the grammar rule $C ::= A \mid \top \mid C \sqcap C \mid \exists r.C$, where $A \in \mathsf{N_C}$ and $r \in \mathsf{N_R}$. The knowledge of the domain is stored in a *TBox* (ontology), which is a finite set of *general concept inclusions* (GCIs) $C \sqsubseteq D$, where $C$ and $D$ are $\mathcal{EL}^+$ concepts, and *role inclusions* (RIs) $r_1 \circ \cdots \circ r_n \sqsubseteq s$, where $n \geq 1$ and $r_i, s \in \mathsf{N_R}$. We will often use the term *axiom* to refer to both GCIs and RIs. As an example, `Appendix` $\sqsubseteq \exists \texttt{partOf}.\texttt{Intestine}$ represents a GCI.

The semantics of this logic is based on *interpretations*, which are pairs of the form $\mathcal{I} = (\Delta^{\mathcal{I}}, \cdot^{\mathcal{I}})$ where $\Delta^{\mathcal{I}}$ is a non-empty set called the *domain* and $\cdot^{\mathcal{I}}$ is the *interpretation function* that maps every $A \in \mathsf{N_C}$ to a set $A^{\mathcal{I}} \subseteq \Delta^{\mathcal{I}}$ and every $r \in \mathsf{N_R}$ to a binary relation $r^{\mathcal{I}} \subseteq \Delta^{\mathcal{I}} \times \Delta^{\mathcal{I}}$. The interpretation $\mathcal{I}$ *satisfies* the GCI $C \sqsubseteq D$ iff $C^{\mathcal{I}} \subseteq D^{\mathcal{I}}$; it *satisfies* the RI $r_1 \circ \cdots \circ r_n \sqsubseteq s$ iff $r_1^{\mathcal{I}} \circ \cdots \circ r_n^{\mathcal{I}} \subseteq s^{\mathcal{I}}$, with $\circ$ denoting composition of binary relations. $\mathcal{I}$ is a *model* of $\mathcal{T}$ iff $\mathcal{I}$ satisfies all its GCIs and RIs.

The main reasoning problem in $\mathcal{EL}^+$ is to decide subsumption between concepts. A concept $C$ is *subsumed* by $D$ w.r.t. $\mathcal{T}$ (denoted $C \sqsubseteq_{\mathcal{T}} D$) if for every model $\mathcal{I}$ of $\mathcal{T}$ it holds that $C^{\mathcal{I}} \subseteq D^{\mathcal{I}}$. *Classification* refers to the task of deciding all the subsumption relations between concept names appearing in $\mathcal{T}$. Rather than merely deciding whether a subsumption relation follows from a TBox, we are interested in understanding the causes of this consequence, and repairing it if necessary.

**Definition 1 (MinA, diagnosis).** *A* MinA *for* $C \sqsubseteq D$ *w.r.t. the TBox* $\mathcal{T}$ *is a minimal subset (w.r.t. set inclusion)* $\mathcal{M} \subseteq \mathcal{T}$ *such that* $C \sqsubseteq_{\mathcal{M}} D$. *A diagnosis for* $C \sqsubseteq D$ *w.r.t.* $\mathcal{T}$ *is a minimal subset (w.r.t. set inclusion)* $\mathcal{D} \subseteq \mathcal{T}$ *such that* $C \not\sqsubseteq_{\mathcal{T} \setminus \mathcal{D}} D$.

MinAs and diagnoses are closely related by minimal hitting set duality [19,29].

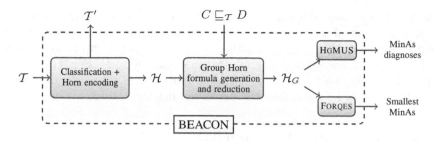

**Fig. 1.** BEACON organization

*Example 2.* Consider the TBox $\mathcal{T}_{\mathrm{exa}} = \{A \sqsubseteq \exists r.A, A \sqsubseteq Y, \exists r.Y \sqsubseteq B, Y \sqsubseteq B\}$. There are two MinAs for $A \sqsubseteq B$ w.r.t. $\mathcal{T}_{\mathrm{exa}}$, namely $\mathcal{M}_1 = \{A \sqsubseteq Y, Y \sqsubseteq B\}$, and $\mathcal{M}_2 = \{A \sqsubseteq \exists r.A, A \sqsubseteq Y, \exists r.Y \sqsubseteq B\}$. The diagnoses for this subsumption relation are $\{A \sqsubseteq Y\}$, $\{A \sqsubseteq \exists r.A, Y \sqsubseteq B\}$, and $\{\exists r.Y \sqsubseteq B, Y \sqsubseteq B\}$.

## 2.2 Propositional Satisfiability

We assume familiarity with propositional logic [9]. A CNF formula $\mathcal{F}$ is defined over a set of Boolean variables $X$ as a finite conjunction of clauses, where a clause is a finite disjunction of literals and a literal is a variable or its negation. A truth assignment is a mapping $\mu\colon X \rightarrow \{0,1\}$. If $\mu$ satisfies $\mathcal{F}$, $\mu$ is referred to as a *model* of $\mathcal{F}$. Horn formulae are those composed of clauses with at most one positive literal. Satisfiability of Horn formulae is decidable in polynomial time [12,15,24]. Given an unsatisfiable formula $\mathcal{F}$, the following subsets are of interest [19,22]:

**Definition 3 (MUS, MCS).** $\mathcal{M} \subseteq \mathcal{F}$ *is a* Minimally Unsatisfiable Subset *(MUS) of* $\mathcal{F}$ *iff* $\mathcal{M}$ *is unsatisfiable and* $\forall c \in \mathcal{M}, \mathcal{M} \setminus \{c\}$ *is satisfiable.* $\mathcal{C} \subseteq \mathcal{F}$ *is a* Minimal Correction Subset *(MCS) iff* $\mathcal{F} \setminus \mathcal{C}$ *is satisfiable and* $\forall c \in \mathcal{C}, \mathcal{F} \setminus (\mathcal{C} \setminus \{c\})$ *is unsatisfiable.*

MUSes and MCSes are related by hitting set duality [8,10,28,33]. Besides, these concepts have been extended to formulae where clauses are partitioned into *groups* [19].

**Definition 4 (Group-MUS).** *Given an explicitly partitioned unsatisfiable CNF formula* $\mathcal{F} = \mathcal{G}_0 \cup ... \cup \mathcal{G}_k$, *a group-MUS of* $\mathcal{F}$ *is a set of groups* $\mathcal{G} \subseteq \{\mathcal{G}_1, ..., \mathcal{G}_k\}$, *such that* $\mathcal{G}_0 \cup \mathcal{G}$ *is unsatisfiable, and for every* $\mathcal{G}_i \in \mathcal{G}, \mathcal{G}_0 \cup (\mathcal{G} \setminus \mathcal{G}_i)$ *is satisfiable.*

## 3 The BEACON Tool

The main problem BEACON is aimed at is the enumeration of the MinAs and diagnoses for a given subsumption relation w.r.t. an $\mathcal{EL}^+$ TBox $\mathcal{T}$. BEACON consists of three main components: The first one *classifies* $T$ and encodes this process into a set of Horn clauses. Given a subsumption to be analyzed, the second component creates and simplifies an unsatisfiable group Horn formula. Finally, the third one computes group-MUSes and group-MCSes, corresponding to MinAs and diagnoses resp. Figure 1 depicts the main organization of BEACON. Each of its components is explained below.

## 3.1　Classification and Horn Encoding

During the classification of $T$, a Horn formula $\mathcal{H}$ is created according to the method introduced in EL$^+$SAT [30,31]. To this end, each axiom $a_i \in T$ is initially assigned a unique selector variable $s_{[a_i]}$. The classification of $T$ is done in two phases [4,6].

First, $T$ is normalized so that each of its axioms are of the form $(i)$ $(A_1 \sqcap ... \sqcap A_k) \sqsubseteq B$ $(k \geq 1)$, $(ii)$ $A \sqsubseteq \exists r.B$, $(iii)$ $\exists r.A \sqsubseteq B$, or $(iv)$ $r_1 \circ ... \circ r_n \sqsubseteq s$ $(n \geq 1)$, where $A, A_i, B \in N_C$ and $r, r_i, s \in N_R$. This process results in a TBox $T_N$ where each axiom $a_i \in T$ is substituted by a set of axioms in normal form $\{a_{i1}, ..., a_{im_i}\}$. At this point, the clauses $s_{[a_i]} \to s_{[a_{ik}]}$, with $1 \leq k \leq m_i$, are added to $\mathcal{H}$.

Second, $T_N$ is saturated through the exhaustive application of the *completion rules* shown in Table 1, resulting in the extended TBox $T'$. Each of the rows in Table 1 constitute a completion rule. Their application is sound and complete for inferring subsumptions [4]. Whenever a rule $r$ can be applied (with antecedents $\mathsf{ant}(r)$) leading to inferring an axiom $a_i$, the Horn clause $(\bigwedge_{\{a_j \in \mathsf{ant}(r)\}} s_{[a_j]}) \to s_{[a_i]}$ is added to $\mathcal{H}$.

As a result, $\mathcal{H}$ will eventually encode all possible derivations of completion rules inferring any axiom such that $X \sqsubseteq_T Y$, with $X, Y \in N_C$.

## 3.2　Generation of Group Horn Formulae

After classifying $T$, some axioms $C \sqsubseteq D$ may be included in $T'$ for which a justification or diagnosis may be required. Each of these *queries* will result in a group Horn formula defined as: $\mathcal{H}_G = \{\mathcal{G}_0, \mathcal{G}_1, ..., \mathcal{G}_{|T|}\}$, where $\mathcal{G}_0 = \mathcal{H} \cup \{(\neg s_{[C \sqsubseteq D]})\}$ and for each axiom $a_i$ $(i > 0)$ in the original TBox $T$, group $\mathcal{G}_i = \{(s_{[a_i]})\}$ is defined with a single unit clause. $\mathcal{H}_G$ is unsatisfiable and, as shown in [1,2], its

Table 1. $\mathcal{EL}^+$ completion rules

| Preconditions | | Inferred axiom |
|---|---|---|
| $A \sqsubseteq A_i, 1 \leq i \leq n$ | $A_1 \sqcap .... \sqcap A_n \sqsubseteq B$ | $A \sqsubseteq B$ |
| $A \sqsubseteq A_1$ | $A_1 \sqsubseteq \exists r.B$ | $A \sqsubseteq \exists r.B$ |
| $A \sqsubseteq \exists r.B,\ B \sqsubseteq B_1$ | $\exists r.B_1 \sqsubseteq B_2$ | $A \sqsubseteq B_2$ |
| $A_{i-1} \sqsubseteq \exists r_i.A_i, 1 \leq i \leq n$ | $r_1 \circ ... \circ r_n \sqsubseteq r$ | $A_0 \sqsubseteq \exists r.A_n$ |

---

**Algorithm 1.** EMUS [26] / MARCO [18]

---

**Input:** $\mathcal{F}$ a CNF formula
**Output:** Reports the set of MUSes (and MCSes) of $\mathcal{F}$

```
1   ⟨I, Q⟩ ← ⟨{p_i | c_i ∈ F}, ∅⟩                    // Variable p_i picks clause c_i
2   while true do
3   │   (st, P) ← MaximalModel(Q)
4   │   if not st then return
5   │   F' ← {c_i | p_i ∈ P}                          // Pick selected clauses
6   │   if not SAT(F') then
7   │   │   M ← ComputeMUS(F')
8   │   │   ReportMUS(M)
9   │   │   b ← {¬p_i | c_i ∈ M}                      // Negative clause blocking the MUS
10  │   else
11  │   │   ReportMCS(F \ F')
12  │   │   b ← {p_i | p_i ∈ I \ P}                   // Positive clause blocking the MCS
13  └   Q ← Q ∪ {b}
```

---

group-MUSes correspond to the MinAs for $C \sqsubseteq_{\mathcal{T}} D$. Equivalently, due to the hitting set duality for MinAs/diagnoses, which also holds for MUSes/MCSes, group-MCSes of $\mathcal{H}_G$ correspond to diagnoses for $C \sqsubseteq_{\mathcal{T}} D$.

BEACON simplifies $\mathcal{H}_G$ with the techniques introduced in [30,31], which often reduce the formulas to a great extent.

### 3.3  Computation of Group-MUSes/Group-MCSes

For enumerating group-MUSes and group-MCSes of the formula $\mathcal{H}_G$ defined above, BEACON integrates the state-of-the-art HGMUS enumerator [2]. HGMUS exploits hitting set dualization between (group) MCSes and (group) MUSes and, hence, it shares ideas also explored in MaxHS [11], EMUS/MARCO [17,26], among others. As shown in Algorithm 1, these methods rely on a two (SAT) solvers approach. Formula $\mathcal{Q}$ is defined over a set of selector variables corresponding to clauses in $\mathcal{F}$, and it is used to enumerate subsets of $\mathcal{F}$. Iteratively, the algorithm computes a maximal model $P$ of $\mathcal{Q}$ and tests whether the subformula $\mathcal{F}' \subseteq \mathcal{F}$ containing the clauses associated to $P$ is satisfiable. If it is, $\mathcal{F} \setminus \mathcal{F}'$ is an MCS of $\mathcal{F}$. Otherwise, $\mathcal{F}'$ is reduced to an MUS. MCSes and MUSes are blocked adding clauses to $\mathcal{Q}$.

HGMUS shares the main organization of Algorithm 1, with $\mathcal{F} = \mathcal{G}_0$ and $\mathcal{Q}$ defined over selector variables for groups $\mathcal{G}_i$ of $\mathcal{H}_G$, with $i > 0$. It also includes some specific features. First, it uses the Horn satisfiability algorithhm LTUR [24]. Besides, it integrates a dedicated insertion-based MUS extractor as well as an efficient algorithm for computing maximal models based on a reduction to computing MCSes [23].

### 3.4  BEACON's Additional Specific Features

Besides computing MinAs/diagnoses, BEACON offers additional functionalities.

**Diagnosing Multiple Subsumption Relations at a Time.** After classifying $\mathcal{T}$, there could be several *unintended* subsumption relations $C_i \sqsubseteq_{\mathcal{T}} D_i$ that need to be removed. BEACON allows for diagnosing this multiple unintended inferences at the same time. By adding the unit clauses $(\neg s_{[C_i \sqsubseteq D_i]})$ to $\mathcal{G}_0$ in $\mathcal{H}_G$, each computed group-MCS corresponds to a diagnosis that would eliminate *all* the indicated subsumption relations.

**Computing Smallest MinAs.** Alternatively to enumerating all the possible MinAs, one may want to compute only those of the minimum possible size. To enable this functionality, BEACON integrates a state-of-the-art solver for the smallest MUS problem (SMUS) called FORQES [14]. The decision version of the SMUS problem is known to be $\Sigma_2^P$-complete (e.g. see [13,16]). As HGMUS, FORQES is based on the hitting set dualization between (group) MUSes and (group) MCSes. The tool iteratively computes *minimum* hitting sets of a set of MCSes of a formula detected so far. While these minimum hitting sets are satisfiable, they are grown into an MSS, whose complement is an MCS which is added to the set of MCSes. The process terminates when an unsatisfiable minimum hitting set is identified, representing a smallest MUS of the formula.

## 4  Experimental Results

This section reports a summary of results that illustrates the performance of BEACON[1] w.r.t. other $\mathcal{EL}^+$ axiom pinpointing tools in the literature. It also provides information on its capability of computing diagnoses and enumerating smallest MinAs.

The experiments were run on a Linux cluster (2 Ghz) with a limit of 3600 s and 4Gbyte, considering 500 subsumption relations from five well-known $\mathcal{EL}^+$ bio-medical ontologies: GALEN [27] (FULL-GALEN and NOT-GALEN), Gene [3], NCI [32] and SNOMED-CT [34]. The experiments use Horn formulae encoded by EL⁺SAT [30,31] applying the reduction techniques that BEACON incorporates by default. These formulae are fed to BEACON's engines, namely HGMUS and FORQES.

The results reported focus on HGMUS and FORQES. Due to lack of space, running times for classifying the ontologies and formula reduction are not reported. Classification is done in polynomial time once for each ontology, so it is amortized among all queries for the ontology. Formula reduction usually takes very short time. Detailed results are available with the distribution of BEACON, including an analysis on the size of the Horn formulae and the reductions achieved.

---

[1] Available at http://logos.ucd.ie/web/doku.php?id=beacon-tool.

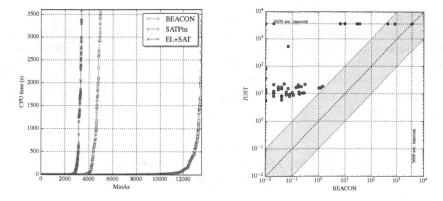

**Fig. 2.** Plots comparing BEACON to EL$^+$SAT, SATPin and JUST

**Axiom Pinpointing.** BEACON shows significant improvements over the existing tools EL$^+$SAT [30,31], SATPin [21], EL2MCS [1], CEL [5] and JUST [20]. BEACON often achieves remarkable reductions in the running times, and exhibits a clear superiority in enumerating MinAs for 19 very hard instances that cannot be solved by a time limit of 3600 s. This is illustrated in Fig. 2. The cactus plot shows the number of MinAs reported over time. BEACON computes much more MinAs faster than other tools. The scatter plot compares BEACON with JUST regarding the running times on a subset of the instances JUST can cope with. BEACON shows a significant performance gap. Similar results have been observed for EL2MCS and CEL [2].

**Computing Diagnoses.** For all solved instances (481 out of 500), BEACON enumerates all diagnoses, where its number ranges from 2 to 565409. Interestingly, for the 19 aborted instances, the number of reported diagnoses ranges from 1011164 to 1972324. These numbers illustrate the efficiency of BEACON at computing diagnoses, and explain the difficulty of these aborted instances. Of the other tools, only EL2MCS reports diagnoses, which, for hard instances, computes around 33 % fewer diagnoses.

**Computing Smallest MinAs.** The last experiments consider the 19 instances for which BEACON is unable to enumerate all MinAs. Notably, BEACON is very efficient at computing the smallest MinAs using FORQES. In all cases, each set of smallest MinAs is computed in negligible time (less than 0.1 s). The sizes of the smallest MinAs range from 5 to 13 axioms, and their number ranges from 1 to 7.

## 5   Conclusions

This paper describes BEACON, an axiom pinpointing tool for the $\mathcal{EL}$-family of DLs. BEACON integrates HGMUS [2], a group MUS enumerator for propositional Horn formulae, with a dedicated front-end, interfacing a target ontology, and generating group Horn formulae for HGMUS. Besides enumerating MinAs (and associated diagnoses), BEACON enables the simultaneous diagnosis of multiple inferences, and the computation of the smallest MinA (or smallest MUS [14]). The experimental results indicate that the computation of the smallest MinA is very efficient in practice, in addition to the already known top performance of HGMUS.

**Acknowledgement.** This work was funded in part by SFI grant BEACON (09/IN.1/-I2618), by DFG grant DFG HO 1294/11-1, and by Spanish grant TIN2013-46511-C2-2-P. The contribution of the researchers associated with the SFI grant BEACON is also acknowledged.

# References

1. Arif, M.F., Mencía, C., Marques-Silva, J.: Efficient axiom pinpointing with EL2MCS. In: Hölldobler, S., Krötzsch, M., Peñaloza, R., Rudolph, S., Edelkamp, S., Edelkamp, S. (eds.) KI 2015. LNCS, vol. 9324, pp. 225–233. Springer, Heidelberg (2015). doi:10.1007/978-3-319-24489-1_17
2. Arif, M.F., Mencía, C., Marques-Silva, J.: Efficient MUS enumeration of Horn formulae with applications to axiom pinpointing. In: Heule, M., Weaver, S. (eds.) SAT 2015. LNCS, vol. 9340, pp. 324–342. Springer, Heidelberg (2015). doi:10.1007/978-3-319-24318-4_24
3. Ashburner, M., Ball, C.A., Blake, J.A., Botstein, D., Butler, H., Cherry, J.M., Davis, A.P., Dolinski, K., Dwight, S.S., Eppig, J.T., et al.: Gene ontology: tool for the unification of biology. Nat. Genet. **25**(1), 25–29 (2000)
4. Baader, F., Brandt, S., Lutz, C.: Pushing the $\mathcal{EL}$ envelope. In: IJCAI, pp. 364–369 (2005)
5. Baader, F., Lutz, C., Suntisrivaraporn, B.: CEL — a polynomial-time reasoner for life science ontologies. In: Furbach, U., Shankar, N. (eds.) IJCAR 2006. LNCS (LNAI), vol. 4130, pp. 287–291. Springer, Heidelberg (2006)
6. Baader, F., Peñaloza, R., Suntisrivaraporn, B.: Pinpointing in the description logic $\mathcal{EL}^+$. In: KI, pp. 52–67 (2007)
7. Baader, F., Suntisrivaraporn, B.: Debugging SNOMED CT using axiom pinpointing in the description logic $\mathcal{EL}^+$. In: KR-MED (2008)
8. Bailey, J., Stuckey, P.J.: Discovery of minimal unsatisfiable subsets of constraints using hitting set dualization. In: Hermenegildo, M.V., Cabeza, D. (eds.) PADL 2004. LNCS, vol. 3350, pp. 174–186. Springer, Heidelberg (2005)
9. Biere, A., Heule, M., van Maaren, H., Walsh, T. (eds.): Handbook of Satisfiability. Frontiers in Artificial Intelligence and Applications, vol. 185. IOS Press, Amsterdam (2009)
10. Birnbaum, E., Lozinskii, E.L.: Consistent subsets of inconsistent systems: structure and behaviour. J. Exp. Theor. Artif. Intell. **15**(1), 25–46 (2003)
11. Davies, J., Bacchus, F.: Solving MAXSAT by solving a sequence of simpler SAT instances. In: Lee, J. (ed.) CP 2011. LNCS, vol. 6876, pp. 225–239. Springer, Heidelberg (2011)
12. Dowling, W.F., Gallier, J.H.: Linear-time algorithms for testing the satisfiability of propositional Horn formulae. J. Log. Program. **1**(3), 267–284 (1984)

13. Gupta, A.: Learning Abstractions for Model Checking. Ph.D. thesis, Carnegie Mellon University, June 2006
14. Ignatiev, A., Previti, A., Liffiton, M., Marques-Silva, J.: Smallest MUS extraction with minimal hitting set dualization. In: Pesant, G. (ed.) CP 2015. LNCS, vol. 9255, pp. 173–182. Springer, Heidelberg (2015)
15. Itai, A., Makowsky, J.A.: Unification as a complexity measure for logic programming. J. Log. Program. **4**(2), 105–117 (1987)
16. Liberatore, P.: Redundancy in logic I: CNF propositional formulae. Artif. Intell. **163**(2), 203–232 (2005)
17. Liffiton, M.H., Malik, A.: Enumerating infeasibility: finding multiple MUSes quickly. In: Gomes, C., Sellmann, M. (eds.) CPAIOR 2013. LNCS, vol. 7874, pp. 160–175. Springer, Heidelberg (2013)
18. Liffiton, M.H., Previti, A., Malik, A., Marques-Silva, J.: Fast, flexible MUs enumeration. Constraints (2015). Online version: http://link.springer.com/article/10.1007/s10601-015-9183-0
19. Liffiton, M.H., Sakallah, K.A.: Algorithms for computing minimal unsatisfiable subsets of constraints. J. Autom. Reasoning **40**(1), 1–33 (2008)
20. Ludwig, M.: Just: a tool for computing justifications w.r.t. ELH ontologies. In: ORE (2014)
21. Manthey, N., Peñaloza, R.: Exploiting SAT technology for axiom pinpointing. Technical report LTCS 15–05, Chair of Automata Theory, Institute of Theoretical Computer Science, Technische Universität Dresden, April 2015. https://ddll.inf.tu-dresden.de/web/Techreport3010
22. Marques-Silva, J., Heras, F., Janota, M., Previti, A., Belov, A.: On computing minimal correction subsets. In: IJCAI, pp. 615–622 (2013)
23. Mencía, C., Previti, A., Marques-Silva, J.: Literal-based MCS extraction. In: IJCAI, pp. 1973–1979 (2015)
24. Minoux, M.: LTUR: A simplified linear-time unit resolution algorithm for Horn formulae and computer implementation. Inf. Process. Lett. **29**(1), 1–12 (1988)
25. Peñaoza, R.: Axiom pinpointing in description logics and beyond. Ph.D. thesis, Dresden University of Technology (2009)
26. Previti, A., Marques-Silva, J.: Partial MUS enumeration. In: AAAI, pp. 818–825 (2013)
27. Rector, A.L., Horrocks, I.R.: Experience building a large, re-usable medical ontology using a description logic with transitivity and concept inclusions. In: Workshop on Ontological Engineering, pp. 414–418 (1997)
28. Reiter, R.: A theory of diagnosis from first principles. Artif. Intell. **32**(1), 57–95 (1987)
29. Schlobach, S., Cornet, R.: Non-standard reasoning services for the debugging of description logic terminologies. In: IJCAI, pp. 355–362 (2003)
30. Sebastiani, R., Vescovi, M.: Axiom pinpointing in lightweight description logics via Horn-SAT encoding and conflict analysis. In: Schmidt, R.A. (ed.) CADE-22. LNCS, vol. 5663, pp. 84–99. Springer, Heidelberg (2009)
31. Sebastiani, R., Vescovi, M.: Axiom pinpointing in large $\mathcal{EL}^+$ ontologies via SAT and SMT techniques. Technical report DISI-15-010, DISI, University of Trento, Italy, Under Journal Submission, April 2015. http://disi.unitn.it/~rseba/elsat/elsat_techrep.pdf
32. Sioutos, N., de Coronado, S., Haber, M.W., Hartel, F.W., Shaiu, W., Wright, L.W.: NCI thesaurus: A semantic model integrating cancer-related clinical and molecular information. J. Biomed. Inform. **40**(1), 30–43 (2007)

33. Slaney, J.: Set-theoretic duality: A fundamental feature of combinatorial optimisation. In: ECAI, pp. 843–848 (2014)
34. Spackman, K.A., Campbell, K.E., Côté, R.A.: SNOMED RT: a reference terminology for health care. In: AMIA (1997)
35. Stefan, S., Ronald, C., Spackman, K.A.: Consolidating SNOMED CT's ontological commitment. Appl. Ontol. **6**, 111 (2011)

# HordeQBF: A Modular and Massively Parallel QBF Solver

Tomáš Balyo[1] and Florian Lonsing[2](✉)

[1] Karlsruhe Institute of Technology (KIT), Karlsruhe, Germany
[2] Knowledge-Based Systems Group, Vienna University of Technology,
Vienna, Austria
florian.lonsing@tuwien.ac.at

**Abstract.** The recently developed massively parallel satisfiability (SAT) solver HordeSAT was designed in a modular way to allow the integration of any sequential CDCL-based SAT solver in its core. We integrated the QCDCL-based quantified Boolean formula (QBF) solver DepQBF in HordeSAT to obtain a massively parallel QBF solver—HordeQBF. In this paper we describe the details of this integration and report on results of the experimental evaluation of HordeQBF's performance. HordeQBF achieves superlinear average and median speedup on the hard application instances of the 2014 QBF Gallery.

## 1 Introduction

HordeSAT [3] is a modular massively parallel SAT solver which allows the integration of any sequential CDCL-based SAT solver in its core. This enables the transfer of advancements in CDCL SAT solving to a parallel setting. Experiments showed that HordeSAT can achieve superlinear average speedup on hard benchmarks.

The logic of quantified Boolean formulas (QBFs) extends SAT by explicit quantification of propositional variables. Problems in complexity classes beyond NP, particularly PSPACE-complete problems in domains like, e.g., formal verification, reactive synthesis, or planning, can naturally be encoded as QBFs.

QBF solvers based on QCDCL, the QBF-specific variant of CDCL, apply techniques similar to CDCL SAT solvers. Thanks to this fact, it is possible to replace the SAT solver in the core of HordeSAT by *any* QCDCL QBF solver. Thereby, it is not necessary to change the framework of HordeSAT which controls the sharing of learned information and the execution of the core solver instances.

We integrated the latest public version 5.0 of the QCDCL-based solver DepQBF [18] in HordeSAT to obtain the massively parallel QBF solver Horde-QBF. We present the implementation of HordeQBF, which is not tailored

T. Balyo—Supported by DFG project SA 933/11-1.

F. Lonsing—Supported by the Austrian Science Fund (FWF) under grant S11409-N23.

N. Creignou and D. Le Berre (Eds.): SAT 2016, LNCS 9710, pp. 531–538, 2016.
DOI: 10.1007/978-3-319-40970-2_33

towards the use of DepQBF as a core solver, and evaluate its scalability on a computer cluster with 1024 processor cores. Experiments using the application benchmarks of the 2014 QBF Gallery show that HordeQBF achieves superlinear average and median speedup for hard instances.

## 2   Preliminaries

We consider closed QBFs $\psi := \Pi.\phi$ in *prenex CNF (PCNF)* consisting of a quantifier-free CNF $\phi$ over a set $V$ of variables and a *quantifier prefix* $\Pi := Q_1 v_1 \ldots Q_n v_n$ in which $Q_i \in \{\exists, \forall\}$ and $v_i \in V$. QBF solving with clause and cube learning (*QCDCL*) [9,15,27], also called *constraint learning*, is a generalization of *conflict-driven clause learning (CDCL)* for SAT. The variables in a PCNF $\psi$ are assigned by *decision making*, *unit propagation*, and *pure literal detection*. Assignments by decision making have to follow the prefix ordering from left to right. If a clause is falsified under the current assignment $A$, then a *learned clause $C$* is derived from $\psi$ by *Q-resolution* [14] and added conjunctively to $\psi$. If all clauses are satisfied under $A$, then a *learned cube* is constructed from $A$ and added disjunctively to $\psi$. Learned cubes may also be derived by *term resolution* [9], a variant of Q-resolution applied to previously learned cubes. After a new clause or cube has been learned, assignments are retracted during *backtracking*. QCDCL terminates if and only if the *empty clause (resp. cube)* is derived during learning, indicating that $\psi$ is unsatisfiable (resp. satisfiable).

## 3   Related Work

Approaches to parallel QBF solving are based on shared and distributed memory architectures. PQSolve [7] is an early parallel DPLL [5] solver without knowledge sharing. It comes with a dynamic master/slave framework implemented using the message passing interface (MPI) [10]. Search space is partitioned among master and slaves by variable assignments. QMiraXT [16] is a multi-threaded QCDCL solver with search space partitioning. PAQuBE [17] is an MPI-based parallel variant of the QCDCL solver QuBE [8]. Clause and cube sharing in PAQuBE can be adapted dynamically at run time. Search space is partitioned like in the SAT solver PSATO based on *guiding paths* [25]. The MPI-based solver MPIDepQBF [13] implements a master/worker architecture without knowledge sharing. A worker consists of an instance of the QCDCL solver DepQBF [19]. The master balances the workload by generating subproblems defined by variable assignments (assumptions), which are solved by the workers. Parallel solving approaches have also been presented for quantified CSPs [24] and non-PCNF QBFs [21].

HordeQBF is a parallel portfolio solver with clause and cube sharing. Whereas sequential portfolio solvers like AQME [23] include different QBF solvers, HordeQBF integrates instances of the same QCDCL solver (i.e., DepQBF). Unlike MPIDepQBF, HordeQBF does not rely on search space partitioning. Instead, the parallel instances of DepQBF are diversified by different parameter settings.

## 4    The HordeSAT Parallelization Framework

HordeSAT is a portfolio SAT solver with clause sharing [3]. It can be viewed as a multithreaded program running several instances of a sequential SAT solver and communicating via MPI with other instances of the same program.

The parallelization framework has three main tasks: to ensure that the core solvers are diversified, to handle the clause exchange, and to stop all the solvers when one of them has solved the problem. To communicate with the core solvers it uses an API which is described in detail in the HordeSAT paper [3]. Since the HordeQBF interface is identical, we only briefly list the most relevant methods:
**void diversify(int rank, int size):** This method tells the core solver to diversify its settings. The specifics of diversification are left to the solver. The description for DepQBF is given in the following section.
**void addLearnedClause(vector<int> clause):** This method is used to import learned clauses (and cubes) received from other solvers of the portfolio.
**void setLearnedClauseCallback(LCCallback\* callback):** This method sets a callback class that will process the clauses (and cubes) shared by this solver.

## 5    QBF Solver Integration

In parallel QCDCL-based QBF solving, learned cubes may be shared among the solver instances in addition to learned clauses. Although HordeSAT does not provide API functions dedicated to cube sharing, its available API readily supports it. We describe the integration of the QCDCL-based QBF solver DepQBF[1] in HordeQBF, which applies to any QCDCL-based QBF solver.

We rely on version 5.0 of DepQBF which comes with a dynamic variant of *blocked clause elimination (QBCE)* [18] for advanced cube learning. QBCE allows to eliminate redundant clauses from a PCNF [11]. Dynamic QBCE is applied frequently during the solving process. If all clauses in the PCNF are satisfied under the current assignment or removed by QBCE, then a cube is learned.

DepQBF features a sophisticated analysis of variable dependencies in a PCNF [19,20] to relax the linear ordering of variables in the prefix. For the experiments in this paper, however, we disabled dependency analysis for both HordeQBF and the sequential variant of DepQBF since the use of dependency information causes run time overhead (during clause/cube learning) in addition to overhead already caused by dynamic blocked clause elimination (QBCE) [18].

We modified DepQBF as follows to integrate it in HordeQBF. Learned constraints are exported to the master process right after they have been learned. The master does not distinguish between learned clauses and cubes but treats them as sorted lists of literals. We add special marker literals to learned clauses

---

[1] http://lonsing.github.io/depqbf/.

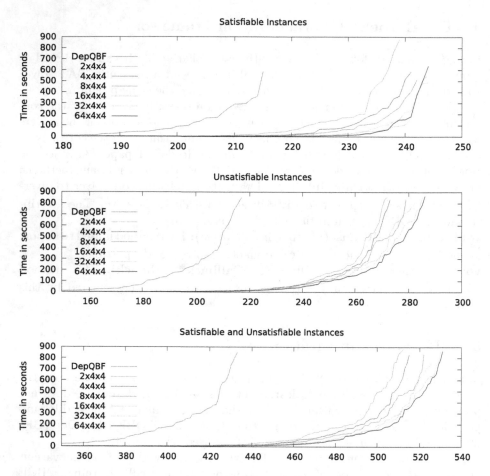

**Fig. 1.** Cactus plots for the benchmarks solved under 900 s by DepQBF and various configurations of HordeQBF. The left regions of the plots (containing easy instances) are omitted.

and cubes to distinguish between them at the time when the master provides the workers with sets of shared learned constraints.

In DepQBF we check whether shared constraints are available for import after a restart has been carried out. To this end, we modified the restart policy of DepQBF to always backtrack to decision level zero. This is different from the original restart policy of DepQBF [19], where the solver backtracks to higher decision levels depending on the current assignment. After a restart, available shared constraints are imported, the watched data structures are updated, and QCDCL continues by propagating unit literals resulting from imported constraints.

Every instance of DepQBF receives a random seed from the master and diversifies the solving process as follows. The values of variables in the *assignment*

**Table 1.** The speedup of HordeQBF configurations relative to DepQBF. The second column is the number of instances solved by HordeQBF, the third is the number of instances solved by both DepQBF (in 50000 s) and the HordeQBF (in 900 s). The following six columns contain the average, total, and median speedups for either all the instances solved by HordeQBF or only big instances (solved after 10×#cores seconds by DepQBF). The last column is the parallel efficiency (median speedup/#cores).

| Core | Parallel | Both | Speedup all | | | Speedup big | | | |
|---|---|---|---|---|---|---|---|---|---|
| Solvers | Solved | Solved | Avg. | Tot. | Med. | Avg. | Tot. | Med. | Eff. |
| 2×4×4 | 513 | 483 | 622 | 107.30 | 0.82 | 3328 | 127.36 | 303.26 | 9.48 |
| 4×4×4 | 516 | 484 | 667 | 137.36 | 0.92 | 3893 | 176.27 | 458.34 | 7.16 |
| 8×4×4 | 523 | 492 | 748 | 128.35 | 0.96 | 4655 | 175.26 | 553.53 | 4.32 |
| 16×4×4 | 527 | 493 | 754 | 140.37 | 0.96 | 5154 | 236.18 | 1449.28 | 5.66 |
| 32×4×4 | 531 | 496 | 780 | 132.41 | 0.96 | 6282 | 269.87 | 2461.84 | 4.81 |
| 64×4×4 | 532 | 496 | 762 | 141.99 | 0.89 | 6702 | 307.29 | 2557.54 | 2.49 |

*cache* [22] are initialized at random. In general, decision variables are assigned to the cached value (if any). The assignment cache is updated with values assigned by unit propagation and pure literal detection. As an effect of random initialization, the first value assigned to a decision variable is always a random value. Parameters of *variable activity scaling* are set at random. DepQBF implements variable activities similar to MiniSAT [6]. Additionally, the amount (percentage) of learned constraints that are removed periodically is initialized at random. DepQBF stores learned clauses and cubes in separate lists with certain capacities. If a list has been filled during learning then less frequently used constraints are removed and the capacity of the list is increased. DepQBF implements a *nested restart scheme* similar to PicoSAT [4], the parameters of which are randomly selected. Variants of *dynamic QBCE* [18] are enabled at random, including switching off dynamic QBCE at all, or applying QBCE only as a preprocessing or inprocessing step. Finally, applications of *long-distance resolution* [1,26], an extension of traditional Q-resolution [14] used to derive learned constraints, are toggled at random.

# 6   Experimental Evaluation

To examine our portfolio-based parallel QBF solver HordeQBF we performed experiments using all the 735 benchmark problems from the application track of the 2014 QBF Gallery [12]. We compared HordeQBF with DepQBF, which is the QBF solver in the core of HordeQBF.

The experiments were run on a cluster with nodes having two octa-core 2.6 GHz Intel Xeon E5-2670 processors (Sandy Bridge) and 64 GB of main memory. Each node has 16 cores and we used 64 nodes which amounts in the total of 1024 cores. The nodes communicate using an InfiniBand 4X QDR Interconnect and use the SUSE Linux Enterprise Server 11 (x86_64) (patch level 3) operating

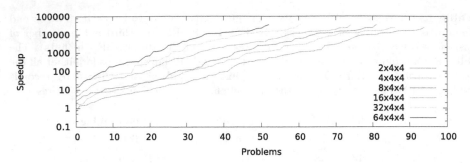

**Fig. 2.** Distribution of speedups on the "big instances" (solved after 10×#cores seconds by DepQBF – the data corresponding to Columns 7–9 of Table 1).

system. HordeQBF was compiled using the icpc compiler version 15.0.2. The complete source code and detailed experimental results are available at http://baldur.iti.kit.edu/hordesat/.

We ran experiments using 2, 4, ..., 64 cluster nodes. On each node we ran four processes with four threads each, which amounts to 16 core solver (DepQBF) instances per node. The results are summarized in Fig. 1 using cactus plots. We can observe that increasing the number of cores is beneficial for both SAT and UNSAT instances since the number of solved instances steadily increases and runtimes are reduced.

However, it is not easy to see from a cactus plot whether the additional performance is a reasonable return on the invested hardware resources. Therefore we include Table 1 in order to quantify the overall scalability of HordeQBF. We compute speedups for all the instances solved by the parallel solver. We ran DepQBF with a time limit $T = 50\,000\,s$ and for the instances it did not solve we use the runtime of $T$ in speedup calculation. The parallel configurations have a time limit of $900\,s$. Columns 4, 5, and 6 of Table 1 show the average, total (sum of sequential runtimes divided by the sum of parallel runtimes) and median speedup values respectively. While the average and total speedup values are high, the median speedup is below one.

Nevertheless, these figures treat HordeQBF unfairly since the majority of the benchmarks is easy (solvable under a minute by DepQBF) and it makes no sense to use large computer clusters to solve them. In parallel computing, it is usual to analyze the performance on many processors using *weak scaling* where one increases the amount of work involved in the considered instances proportionally to the number of processors. Therefore in columns 7–9 we restrict ourselves to "big instances" – where DepQBF needs at least 10×(the number of cores used by HordeQBF) seconds to solve them. The average, total and median speedup values get significantly larger and in fact we obtain highly superlinear average and median speedups. Figure 2 shows the distribution of speedups for these instances, it also reveals how many instances (x-axis) qualify as "big instances".

# 7   Conclusion

We showed that QBF solving can be successfully parallelized using the same techniques as for massively parallel SAT solving. Our parallel QBF solver HordeQBF achieved superlinear total and median speedups for hard instances, i.e., instances where parallelization makes sense.

As future work it would be interesting to consider further variants of Q-resolution systems [2] (apart from traditional [14] and long-distance resolution [1,26]) as a means of diversification in HordeQBF, which would amount to a combination of QBF proof systems with different power. Further, it may be promising to equip HordeQBF with search space partitioning as in MPI-DepQBF [13].

# References

1. Balabanov, V., Jiang, J.R.: Unified QBF certification and its applications. Formal Meth. Syst. Des. **41**(1), 45–65 (2012)
2. Balabanov, V., Widl, M., Jiang, J.H.R.: QBF resolution systems and their proof complexities. In: Sinz, C., Egly, U. (eds.) SAT 2014. LNCS, vol. 8561, pp. 154–169. Springer, Heidelberg (2014)
3. Balyo, T., Sanders, P., Sinz, C.: HordeSat: A massively parallel portfolio SAT solver. In: Heule, M. (ed.) SAT 2015. LNCS, vol. 9340, pp. 156–172. Springer, Heidelberg (2015)
4. Biere, A.: PicoSAT essentials. JSAT **4**(2–4), 75–97 (2008)
5. Cadoli, M., Giovanardi, A., Schaerf, M.: An algorithm to evaluate quantified boolean formulae. In: AAAI, pp. 262–267. AAAI Press / The MIT Press (1998)
6. Eén, N., Sörensson, N.: An extensible SAT-solver. In: Giunchiglia, E., Tacchella, A. (eds.) SAT 2003. LNCS, vol. 2919, pp. 502–518. Springer, Heidelberg (2004)
7. Feldmann, R., Monien, B., Schamberger, S.: A distributed algorithm to evaluate quantified boolean formulae. In: AAAI, pp. 285–290. AAAI Press / The MIT Press (2000)
8. Giunchiglia, E., Marin, P., Narizzano, M.: QuBE7.0. JSAT **7**(2–3), 83–88 (2010)
9. Giunchiglia, E., Narizzano, M., Tacchella, A.: Clause/Term resolution and learning in the evaluation of quantified boolean formulas. JAIR **26**, 371–416 (2006)
10. Gropp, W., Lusk, E., Doss, N., Skjellum, A.: A high-performance, portable implementation of the MPI message passing interface standard. Parallel Comput. **22**(6), 789–828 (1996)
11. Heule, M., Järvisalo, M., Lonsing, F., Seidl, M., Biere, A.: Clause elimination for SAT and QSAT. J. Artif. Intell. Res. (JAIR) **53**, 127–168 (2015)
12. Janota, M., Jordan, C., Klieber, W., Lonsing, F., Seidl, M., Van Gelder, A.: The QBF Gallery 2014: The QBF competition at the FLoC olympic games. JSAT **9**, 187–206 (2016)
13. Jordan, C., Kaiser, L., Lonsing, F., Seidl, M.: MPIDepQBF: Towards parallel QBF solving without knowledge sharing. In: Sinz, C., Egly, U. (eds.) SAT 2014. LNCS, vol. 8561, pp. 430–437. Springer, Heidelberg (2014)
14. Kleine Büning, H., Karpinski, M., Flögel, A.: A resolution for quantified boolean formulas. Inf. Comput. **117**(1), 12–18 (1995)

15. Letz, R.: Lemma and model caching in decision procedures for quantified boolean formulas. In: Egly, U., Fermüller, C. (eds.) TABLEAUX 2002. LNCS (LNAI), vol. 2381, pp. 160–175. Springer, Heidelberg (2002)

16. Lewis, M., Schubert, T., Becker, B.: QMiraXT - a multithreaded QBF solver. In: Methoden und Beschreibungssprachen zur Modellierung und Verifikation von Schaltungen und Systemen (MBMV) (2009)

17. Lewis, M., Schubert, T., Becker, B., Marin, P., Narizzano, M., Giunchiglia, E.: Parallel QBF solving with advanced knowledge sharing. Fundamenta Informaticae **107**(2–3), 139–166 (2011)

18. Lonsing, F., Bacchus, F., Biere, A., Egly, U., Seidl, M.: Enhancing search-based QBF solving by dynamic blocked clause elimination. In: Davis, M. (ed.) LPAR-20 2015. LNCS, vol. 9450, pp. 418–433. Springer, Heidelberg (2015)

19. Lonsing, F., Biere, A.: DepQBF: A dependency-aware QBF solver. JSAT **7**(2–3), 71–76 (2010)

20. Lonsing, F., Biere, A.: Integrating dependency schemes in search-based QBF solvers. In: Strichman, O., Szeider, S. (eds.) SAT 2010. LNCS, vol. 6175, pp. 158–171. Springer, Heidelberg (2010)

21. Mota, B.D., Nicolas, P., Stéphan, I.: A new parallel architecture for QBF tools. In: Proceedings of the International Conferference on High Performance Computing and Simulation (HPCS 2010), pp. 324–330. IEEE (2010)

22. Pipatsrisawat, K., Darwiche, A.: A lightweight component caching scheme for satisfiability solvers. In: Marques-Silva, J., Sakallah, K.A. (eds.) SAT 2007. LNCS, vol. 4501, pp. 294–299. Springer, Heidelberg (2007)

23. Pulina, L., Tacchella, A.: AQME'10. JSAT **7**(2–3), 65–70 (2010)

24. Vautard, J., Lallouet, A., Hamadi, Y.: A parallel solving algorithm for quantified constraints problems. In: ICTAI, pp. 271–274. IEEE Computer Society (2010)

25. Zhang, H., Bonacina, M.P., Hsiang, J.: PSATO: A distributed propositional prover and its application to quasigroup problems. J. Symb. Comput. **21**(4), 543–560 (1996)

26. Zhang, L., Malik, S.: Conflict driven learning in a quantified boolean satisfiability solver. In: ICCAD, pp. 442–449. ACM / IEEE Computer Society (2002)

27. Zhang, L., Malik, S.: Towards a symmetric treatment of satisfaction and conflicts in quantified boolean formula evaluation. In: Van Hentenryck, P. (ed.) CP 2002. LNCS, vol. 2470, pp. 200–215. Springer, Heidelberg (2002)

# LMHS: A SAT-IP Hybrid MaxSAT Solver

Paul Saikko, Jeremias Berg, and Matti Järvisalo[(✉)]

Helsinki Institute for Information Technology HIIT,
Department of Computer Science, University of Helsinki, Helsinki, Finland
matti.jarvisalo@cs.helsinki.fi

**Abstract.** We describe LMHS, an open-source weighted partial maximum satisfiability (MaxSAT) solver. LMHS is a hybrid SAT-IP MaxSAT solver that implements the implicit hitting set approach to MaxSAT. On top of the main algorithm, LMHS offers integrated preprocessing, solution enumeration, an incremental API, and the use of a choice of SAT and IP solvers. We describe the main features of LMHS, and give empirical results on the influence of preprocessing and the choice of the underlying SAT and IP solvers on the performance of LMHS.

## 1 Introduction

LMHS is a weighted partial maximum satisfiability (MaxSAT) solver. Weighted partial MaxSAT is a common generalization of maximum satisfiability that allows some clauses to be designated as mandatory and assigns weights to clauses that may be left unsatisfied. LMHS implements the so-called implicit hitting set approach [16,18] for weighted partial MaxSAT, and can be viewed as an independent from-scratch re-implementation of the MaxHS solver [8–10]. On top of the main algorithm, LMHS integrates MaxSAT preprocessing [3–6] into the solver, and offers solution enumeration, an incremental API, as well as the use of a choice of SAT and IP solvers. The solver entered the 2015 MaxSAT Evaluation [2], where it solved the most problems (among non-portfolio solvers) in the crafted and industrial weighed partial MaxSAT categories. This paper gives an overview of key features of the LMHS solver as well as the effects of preprocessing and the choice of the SAT and IP solvers on its performance.

## 2 Overview of LMHS

LMHS implements an instantiation of an implicit hitting set algorithm [16] for weighted partial MaxSAT, following MaxHS [8–10]. Given an unsatisfiable CNF formula $\mathcal{F}$, the MaxSAT problem is to identify a minimum (minimum-cost

---

Work funded by Academy of Finland, grants 251170 COIN, 276412, and 284591; and Doctoral School in Computer Science DoCS and Research Funds of the University of Helsinki.

N. Creignou and D. Le Berre (Eds.): SAT 2016, LNCS 9710, pp. 539–546, 2016.
DOI: 10.1007/978-3-319-40970-2_34

**Algorithm 1.** The MaxHS implicit hitting set algorithm for MaxSAT [8].

1: **function** MAXHS($\mathcal{F}_h, \mathcal{F}_s, c$)
2:     $\mathcal{K} \leftarrow \emptyset$, $H \leftarrow \emptyset$
3:     $(sat, \kappa, \tau) \leftarrow$ SOLVESAT($\mathcal{F}_h \cup \mathcal{F}_s$)
4:     **while not** $sat$ **do**
5:         $\mathcal{K} \leftarrow \mathcal{K} \cup \{\kappa\}$
6:         $H \leftarrow$ SOLVEMCHS($\mathcal{K}, c$)
7:         $(sat, \kappa, \tau) \leftarrow$ SOLVESAT($\mathcal{F}_h \cup (\mathcal{F}_s \backslash H)$)
8:     **return** $\tau$

for weighted problems) set $H$ of (soft) clauses such that $\mathcal{F} \backslash H$ is satisfiable. A connection to implicit hitting set problems comes from the unsatisfiable subsets of clauses, or cores $\kappa$, or the formula. The set of clauses $H$ must hit a clause from each core $\kappa$, so an optimal MaxSAT solution can be obtained by computing a minimum-cost hitting set (MCHS) over the set of all cores. When the set of all cores is not known, this becomes an implicit hitting set problem.

The implicit hitting set approach for weighted partial MaxSAT, given hard clauses $\mathcal{F}_h$, soft clauses $\mathcal{F}_s$, and cost function $c : \mathcal{F}_s \to \mathbb{R}^+$, is described in more detail in Algorithm 1 and Fig. 1. It uses both a SAT and an IP solver. During the solving process it accumulates a set $\mathcal{K}$ of cores and stores a MCHS of $\mathcal{K}$ in $H$. Starting with $H = \emptyset$, the algorithm tests the satisfiability of $\mathcal{F} \backslash H$ using the SAT solver. If satisfiable, the variable assignment $\tau$ returned by the SAT solver is optimal. If unsatisfiable, a new core $\kappa$ of $\mathcal{F} \backslash H$ is obtained from the SAT solver and added to $\mathcal{K}$. Finally, the IP solver is used to update $H$ to a new hitting set by computing a MCHS of $\mathcal{K}$.

In practice, every soft clause $C_i \in \mathcal{F}_s$ is augmented with a unique auxiliary variable $a_i$ so that if $a_i = 1$, then $C_i$ need not be satisfied, i.e., creating the clause $C_i \vee a_i$. Arbitrary sets of soft clauses can then be excluded from the formula by assuming $a_i = 1$ for the corresponding auxiliary variables in a SAT solver call. To obtain a MCHS of $\mathcal{K}$, the IP solver minimizes $\sum_{C_i \in \mathcal{F}_s} a_i \cdot c(C_i)$ subject to the constraint $\sum_{C_i \in \kappa} a_i \geq 1$ for each core $\kappa \in \mathcal{K}$, enforcing that each core in $\mathcal{K}$ is hit.

Besides the features elaborated on in Sect. 3, some design choices differentiate LMHS from the MaxHS solver of Davies and Bacchus (http://maxhs. org). LMHS never enforces the equivalence $\neg a_i \leftrightarrow C_i$ of auxiliary variables and clauses explicitly in CNF. Instead, the value of each $a_i$ is fixed via the assumptions interface for every SAT solver call, which ensure that $\neg a_i \leftrightarrow C_i$ implicitly holds. In terms of heuristic optimizations, by default LMHS finds a maximal disjoint set of cores on each iteration and uses a greedy hitting set algorithm in place of an IP solver call whenever possible. At each iteration, the greedy hitting set algorithm is used in place of an IP solver call as long as this results in an unsatisfiable formula (i.e., a core is produced). When the greedy method does not yield a core, the IP solver is used to compute a minimum-cost hitting set. The 2015 MaxSAT Evaluation versions LMHS-I and LMHS-C differ slightly in this regard: LMHS-I did not use the greedy algorithm, while LMHS-C finds a set

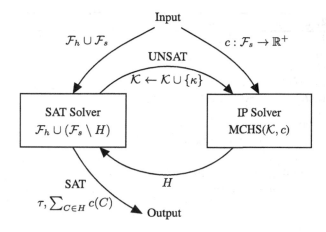

**Fig. 1.** Information flow in the implicit hitting set approach to MaxSAT

of possibly overlapping cores at each iteration. Furthermore, as a consequence of the integration of SAT-based preprocessing, auxiliary variables may not be limited to a single unique $a_i$ per soft clause.

## 3 Features

Here we give an overview of the main features offered by LMHS on top of the main algorithm it implements.

**Integrated Preprocessing.** The use of SAT preprocessing techniques for MaxSAT [4] is integrated into LMHS using the Coprocessor 2.0 SAT preprocessor [17]. Many SAT preprocessing techniques, such as bounded variable elimination [11], are not sound for MaxSAT on their own [4]. However, they can be made sound by introducing a layer of auxiliary variables (labels) and forbidding their removal during preprocessing [3,4]. Concretely, a new variable $l_i$ is introduced for each soft clause $C_i$ prior to applying preprocessing. The original soft clause replaced by a hard clause $(C_i \vee l_i)$ with the restriction that the variable $l_i$ may not be eliminated from the formula. After preprocessing, soft clauses $(\neg l_i)$ with the weights of the original clauses $C_i$ are added to the formula.

Efficient integration of SAT-based preprocessing in LMHS is enabled by the observation that the MaxHS algorithm is sound even in cases where an assumption variable is shared between clauses or a clause contains more than one assumption variable [5]. This allows LMHS to re-use the variables $l_i$ introduced by preprocessing in place of the auxiliary variables $a_i$, avoiding the introduction of a new layer of assumption variables for SAT-based core extraction within the main algorithm.

Following [6], we further avoid the addition of unnecessary auxiliary variables in LMHS by detecting variables in the original instance which can be reused already in the preprocessing phase. Any literal $l \in \{x, \neg x\}$ which occurs only

---

**Algorithm 2.** Enumeration of optimal solutions.

```
1: function ENUMERATE(𝓕_h, 𝓕_s, c)
2:     𝓚 ← ∅, H ← ∅
3:     while true do
4:         while true do
5:             (sat, κ, τ) ← SOLVESAT(𝓕_h ∪ (𝓕_s\H))
6:             if not sat then
7:                 𝓚 ← 𝓚 ∪ {κ}
8:                 H ← SOLVEMCHS(𝓚, c)
9:             else break
10:        if opt is undefined then opt ← cost(τ)
11:        if cost(τ) > opt then break
12:        else
13:            yield τ
14:            𝓕_h ← 𝓕_h ∪ {⋁_{τ(x)=1} ¬x ∨ ⋁_{τ(x)=0} x}
```

---

in a single unit soft clause $(\neg l)$ and some hard clauses $(C_1 \vee l), \ldots, (C_n \vee l)$ of the input instance is detected by simple pattern matching and re-used by the preprocessor and thereafter by the main algorithm. Such variables are introduced by, e.g., a straightforward encoding of group constraints [13].

**Solution Enumeration.** LMHS offers command-line options for enumerating MaxSAT solutions. The solver can enumerate the $k$ best solutions or all optimal solutions. Enumeration can be based on variable assignments or satisfied clauses. In the latter case, only solutions which satisfy a unique set of soft clauses are considered. Solution enumeration in LMHS is implemented as Algorithm 2. The MaxHS algorithm is enclosed within the loop on Line 3. When the first solution is found, Line 10 records its cost as the optimal cost. On subsequent optimal solutions, Line 14 adds a single clause which forbids the latest obtained optimal solution. The termination condition on Line 11 is met when all optimal solutions have been found.

To enumerate unique solutions in terms of satisfied clauses, the refinement of $\mathcal{F}$ on Line 14 is replaced by adding the constraint $\sum_{C_i \in H} a_i - \sum_{C_i \in \mathcal{F}_s \setminus H} a_i < |H|$ to the hitting set IP, followed by a re-computation of the hitting set. A fixed number of best solutions can be found by modifying the condition of Line 11 to only break after enough solutions have been found. An application of the solution enumeration interface—and the incremental API described next—is presented in [19], where LMHS is used for MaxSAT to deriving cutting planes in an IP-based approach to learning optimal Bayesian network structures.

**Incremental API.** LMHS also implements a more general type of incrementality. Through a C or C++ API, the working formula can be incrementally extended with arbitrary clauses and the solver subsequently incrementally used for finding optimal solutions to the altered formula without starting search from scratch. In terms of Algorithm 2, operations performed through the API in effect replace Line 14. An overview of the interface follows.

- `reset` Resets the internal state of LMHS, allowing a new instance to be started.
- `initialize` Initializes LMHS and its components. Three variants of this method are offered. An instance can be initialized from a file, from clauses in memory, or as an empty instance to be built using the API.
- `getNewVariable` Requests a new variable from the internal SAT solver.
- `addHardClause` Adds a hard clause to the working MaxSAT instance.
- `addSoftClause` Adds a soft clause to the working MaxSAT instance. This automatically internally creates a blocking (auxiliary) variable for the clause. This variable is returned by the function in case the user wishes to make use of it. As a rule, the blocking variable created will always have a larger index number than the last variable created with `getNewVariable`.
- `addCoreConstraint` If a subset of soft clauses is known to be unsatisfiable, it can be explicitly added to the set of cores, expressed using the blocking variables of the soft clauses.
- `forbidLastModel` Internally creates a SAT constraint forbidding the previously found variable assignment.
- `forbidLastSolution` Internally creates an IP constraint forbidding the previously found set of satisfied soft clauses.
- `getOptimalSolution` Optimally solves the current MaxSAT instance.

**Choice of SAT and IP Solvers.** A lightweight interface between LMHS and its SAT solver component allows for flexibility in the choice of solver. Any solver which provides an assumption-based incremental interface can be integrated into LMHS by implementing a small interface class and making minor modifications to the build process. Interfaces for two such solvers, MiniSat 2.2 [12] and the inprocessing [15] SAT solver Lingeling [7], are included in the current release of LMHS. Similarly, LMHS was also designed to allow for the use of different IP solvers. Currently LMHS includes interfaces to the state-of-the-art commercial IP solver IBM CPLEX [14] and the open-source non-commercial IP solver SCIP [1].

**Input Format.** In addition to adhering to the DIMACS WCNF input format for weighted partial MaxSAT, LMHS also supports the use of floats (without preprocessing) in the input WCNF, i.e., MaxSAT with cost functions associating real-valued non-negative weights to clauses. Within the solver, the costs are handled by the IP solver.

## 4    Performance Overview

This section examines some interesting aspects of the performance of LMHS. We evaluate the solver on the complete set of 2209 crafted and industrial partial and weighted partial benchmarks of the 2015 MaxSAT Evaluation [2]. The experiments were run on machines with 32-GB memory and Intel Xeon E5540 processors under Ubuntu Linux 12.04. A per-instance time limit of 1800 seconds (30 min) was enforced. Figure 2 is a plot of the number of instances

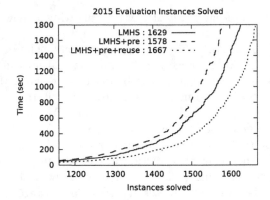

**Fig. 2.** Effect of integrated preprocessing on LMHS on 2015 MaxSAT evaluation instances.

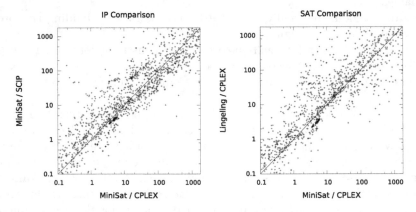

**Fig. 3.** A comparison of SAT (right) and IP (left) solver components within LMHS.

solved at different per-instance timeouts, showing the impact of integrating pre-processing into the solver by reusing auxiliary variables. It shows an interest-ing effect, in that the ordinary application of MaxSAT preprocessing to the instance (LMHS+pre) degrades solver performance compared to no preprocess-ing (LMHS), but the tighter integration of preprocessing by reusing variables (LMHS+pre+reuse) produces a clear improvement. While the reasons for this effect are not entirely clear at present, we suspect it to be at least in part due to the larger search space resulting from the added auxiliary variables. The extra layer of variables can also be detrimental in terms of potential additional con-straints available for the IP solver; see [5, Example 1].

Figure 3 compares the use of the SCIP 3.0.1 IP solver to CPLEX 12.6.0 (left) and the MiniSat 2.2 solver to Lingeling (right) as the MIP and SAT components, respectively, in LMHS. The combination of CPLEX and MiniSat was mainly used during the development of LMHS, so these components can be expected to perform better in the default configuration. Figure 3 plots per-instance runtimes for solved instances, and clearly shows better results with CPLEX and MiniSat. However, although SCIP and Lingeling results in worse performace overall compared to CPLEX and MiniSat, there is significant number of instances which they enable solving faster. This suggests that choosing a combination of a SAT solver and an IP solver on a per-instance basis could result in improved performance. Additionally, more in-depth analysis of which instances are best suited to each solver component could yield interesting further insights.

## 5   Availability and Conclusions

LMHS is competitive with the current state-of-the-art in MaxSAT solvers, recently having reached top positions in the 2015 MaxSAT Evaluation. LMHS integrates MaxSAT preprocessing into the solver. LMHS was designed to be flexible in allowing for integrating different SAT and IP solvers. The solver is open source and released under the MIT license at http://www.cs.helsinki.fi/group/coreo/lmhs/.

## References

1. Achterberg, T.: SCIP: solving constraint integer programs. Math. Program. Comput. **1**(1), 1–41 (2009)
2. Argelich, J., Li, C.M., Manyá, F., Planes, J.: Max-SAT 2015: Tenth Max-SAT Evaluation (2015). http://www.maxsat.udl.cat/15/
3. Belov, A., Järvisalo, M., Marques-Silva, J.: Formula preprocessing in MUS extraction. In: Piterman, N., Smolka, S.A. (eds.) TACAS 2013 (ETAPS 2013). LNCS, vol. 7795, pp. 108–123. Springer, Heidelberg (2013)
4. Belov, A., Morgado, A., Marques-Silva, J.: SAT-based preprocessing for MaxSAT. In: McMillan, K., Middeldorp, A., Voronkov, A. (eds.) LPAR-19 2013. LNCS, vol. 8312, pp. 96–111. Springer, Heidelberg (2013)
5. Berg, J., Saikko, P., Järvisalo, M.: Improving the effectiveness of SAT-based preprocessing for MaxSAT. In: Proceedings of IJCAI, pp. 239–245. AAAI Press (2015)
6. Berg, J., Saikko, P., Järvisalo, M.: Re-using auxiliary variables for MaxSAT preprocessing. In: Proceedings of ICTAI, pp. 813–820. IEEE (2015)
7. Biere, A.: Yet another local search solver and Lingeling and friends entering the SAT competition 2014. In: Proceedings of SAT Competition 2014, vol. B-2014-2, pp. 39–40. Department of Computer Science Series of Publications B, University of Helsinki (2014)
8. Davies, J., Bacchus, F.: Solving MAXSAT by solving a sequence of simpler SAT instances. In: Lee, J. (ed.) CP 2011. LNCS, vol. 6876, pp. 225–239. Springer, Heidelberg (2011)
9. Davies, J., Bacchus, F.: Exploiting the power of MIP solvers in MAXSAT. In: Järvisalo, M., Van Gelder, A. (eds.) SAT 2013. LNCS, vol. 7962, pp. 166–181. Springer, Heidelberg (2013)
10. Davies, J., Bacchus, F.: Postponing optimization to speed up MAXSAT solving. In: Schulte, C. (ed.) CP 2013. LNCS, vol. 8124, pp. 247–262. Springer, Heidelberg (2013)

11. Eén, N., Biere, A.: Effective preprocessing in SAT through variable and clause elimination. In: Bacchus, F., Walsh, T. (eds.) SAT 2005. LNCS, vol. 3569, pp. 61–75. Springer, Heidelberg (2005)
12. Eén, N., Sörensson, N.: An extensible SAT-solver. In: Giunchiglia, E., Tacchella, A. (eds.) SAT 2003. LNCS, vol. 2919, pp. 502–518. Springer, Heidelberg (2004)
13. Heras, F., Morgado, A., Marques-Silva, J.: MaxSAT-based encodings for group MaxSAT. AI Commun. **28**(2), 195–214 (2015)
14. IBM ILOG: CPLEX optimizer 12.6.0 (2014). http://www-01.ibm.com/software/commerce/optimization/cplex-optimizer/
15. Järvisalo, M., Heule, M.J.H., Biere, A.: Inprocessing rules. In: Gramlich, B., Miller, D., Sattler, U. (eds.) IJCAR 2012. LNCS, vol. 7364, pp. 355–370. Springer, Heidelberg (2012)
16. Karp, R.M.: Implicit hitting set problems and multi-genome alignment. In: Amir, A., Parida, L. (eds.) CPM 2010. LNCS, vol. 6129, pp. 151–151. Springer, Heidelberg (2010)
17. Manthey, N.: Coprocessor 2.0 – a flexible CNF simplifier. In: Cimatti, A., Sebastiani, R. (eds.) SAT 2012. LNCS, vol. 7317, pp. 436–441. Springer, Heidelberg (2012)
18. Moreno-Centeno, E., Karp, R.M.: The implicit hitting set approach to solve combinatorial optimization problems with an application to multigenome alignment. Oper. Res. **61**(2), 453–468 (2013)
19. Saikko, P., Malone, B., Järvisalo, M.: MaxSAT-based cutting planes for learning graphical models. In: Michel, L. (ed.) CPAIOR 2015. LNCS, vol. 9075, pp. 347–356. Springer, Heidelberg (2015)

# OpenSMT2: An SMT Solver for Multi-core and Cloud Computing

Antti E.J. Hyvärinen[(✉)], Matteo Marescotti, Leonardo Alt,
and Natasha Sharygina

Università della Svizzera italiana, Lugano, Switzerland
antti.hyvarinen@gmail.com

**Abstract.** This paper describes a major revision of the OpenSMT solver developed since 2008. The version 2 significantly improves its predecessor by providing a design that supports extensions, several critical bug fixes and performance improvements. The distinguishing feature of the new version is the support for a wide range of parallelization algorithms both on multi-core and cloud-computing environments. Presently the solver implements the quantifier free theories of uninterpreted functions and equalities and linear real arithmetics, and is released under the MIT license.

## 1 Introduction

SMT solvers constitute an attractive approach for constraint programming that combines the efficiency of the solvers for propositional formulas with the expressiveness of higher-order logics. While the underlying principle of SMT solvers is simple, the state-of-the-art SMT solvers are wonderfully complex software that, while offering superior performance, are challenging to approach for developers new to the code. In this paper we present the OpenSMT2 SMT solver. The new release puts a particular emphasis to easy approachability by being compact but still supporting the important quantifier-free theories of uninterpreted functions and equalities (QF_UF) and linear real arithmetics (QF_LRA). The solver is available at http://verify.inf.usi.ch/opensmt, is open source under the relatively liberal MIT license, and uses the MiniSat 2.0 SAT solver as the search engine. Compared to the previous version [2] the major improvements are the complete re-design of the data structure used for representing terms to allow extensibility to new theories; a modular framework for building expressions in different logics matched with a similarly modular framework for logic solvers; and several critical bug fixes. We have also improved the performance of the solver in particular on the instances of QF_LRA. The system supports scalability through parallel SMT solving on both multi-core and cloud computing environments with impressive increase in performance for cloud computing.

## 2 OpenSMT2

This section gives an overview of the implementation of OpenSMT2. For a more detailed and generic description the reader is referred to [3]. Figure 1 describes

© Springer International Publishing Switzerland 2016
N. Creignou and D. Le Berre (Eds.): SAT 2016, LNCS 9710, pp. 547–553, 2016.
DOI: 10.1007/978-3-319-40970-2_35

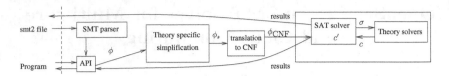

**Fig. 1.** Overview of the OpenSMT2 architecture.

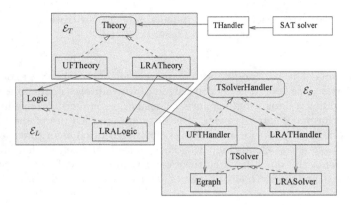

**Fig. 2.** The architecture of the theory framework in OpenSMT2. The element $\mathcal{E}_L$ contains the logic implementation, $\mathcal{E}_S$ contains the solver implementation, and $\mathcal{E}_T$ contains the theory elements that glue the logic to the solvers.

the functionality of OpenSMT2 on a high-level. The solver supports reading an smt-lib file and interacting through an application-program-interface. The problem is then converted into an SMT formula $\phi$ and simplified into $\phi_s$ using both simplifications working on the propositional level, where for example nested conjunctions are flattened and Boolean constraints removed, as well as on the theory level, where for instance asserted equalities are used to compute variable substitutions. These simplifications are fairly standard in SMT solvers. The resulting formula is translated into conjunctive normal form $\phi_{\mathrm{CNF}}$. The formula $\phi_{\mathrm{CNF}}$ is fed to the SAT solver which initializes the theory solvers based on the variables seen by the solver. The SAT solver then provides the theory solvers with assignments $\sigma$ satisfying $\phi_{\mathrm{CNF}}$. If a theory solver deems $\sigma$ unsatisfiable it returns a clause $c$ that prevents the SAT solver from reproducing similar inconsistent assignments. The clauses are simplified to *learned clauses* $c'$ using resolution guided by the conflict graph [9] and the CNF formula is updated to $\phi_{\mathrm{CNF}} := \phi_{\mathrm{CNF}} \wedge c'$. The process finishes when either $\phi_{\mathrm{CNF}}$ becomes unsatisfiable or the SAT solver finds a theory-consistent truth assignment $\sigma$.

Figure 2 shows an abstraction the framework OpenSMT2 uses for implementing theories together with two concrete examples, QF_UF [3] and QF_LRA [4]. The figure follows a UML-style representation where the boxes with rounded corners represent abstract classes that cannot be instantiated, and sharp-cornered boxes represent concrete classes. The dashed arrows point to base classes while

**Table 1.** Abstract methods that must be overridden in the classes described in Fig. 2 to implement new theories.

| Method | Description |
|---|---|
| **Theory** | |
| *simplify* | Entry point for theory specific simplifications. |
| **Logic** | |
| *mkConst* | Create logic-specific constants. |
| *isUFEquality* | Check whether a given equality is uninterpreted. |
| *isTheoryEquality* | Check whether a given equality is from a theory. |
| *insertTerm* | Insert a theory term. |
| *retrieveSubstitutions* | Get the substitutions based on the logic. |
| **TSolverHandler** | |
| *assertLit_special* | Assert literals in the simplification phase. |
| **TSolver** | |
| *assertLit* | Assert a theory literal. |
| *pushBacktrackPoint, popBacktrackPoint* | Incrementally add and remove asserted theory literals. |
| *check* | Check theory consistency of the asserted literals. |
| *getValue* | obtain a value of a theory term once a model has been found. |
| *computeModel* | compute a concrete model for the theory terms once the theory solver finds a model consistent. |
| *getConflict* | return a compact explanation of the theory-inconsistency in the form of theory literals. |
| *getDeduction* | get theory literals implied under the current assignment. |
| *declareTerm* | inform the theory solvers about a theory literal. |

solid arrows point to instances held by a particular class. The framework implements the interactions in Fig. 1 related to the API, theory specific simplifications, and the theory solvers. The architecture is divided into three elements: the Logic element $\mathcal{E}_L$ implementing the logical language, the Solver element $\mathcal{E}_S$ that implements the solver, and the Theory element $\mathcal{E}_T$ which combines the logic and the solver. Extending the solver with new theories is done by introducing classes for the new theory solver and theory solver handler to $\mathcal{E}_S$, the new logic to $\mathcal{E}_L$ and the new theory to $\mathcal{E}_T$. Table 1 provides a brief overview of the most important methods that need to be implemented for the classes in new theories.

Figure 3 compares the solver performance to other solvers and the previous version OpenSMT1 in the QF_UF and QF_LRA categories of smt-lib. The solver is competitive in particular in the logic QF_UF compared to other solvers, and is a clear improvement over the previous version in QF_LRA.

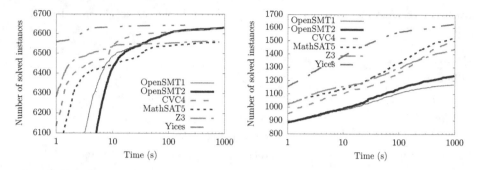

**Fig. 3.** Number of solved instances in a given timeout for OpenSMT1, OpenSMT2, and certain other solvers for the logics QF_UF (*left*) and QF_LRA (*right*) (Color figure online).

## 3  The Parallel Solvers

This section describes two parallel SMT solvers based on OpenSMT2. Section 3.1 details an implementation designed to run on a distributed cloud computing environment, and Sect. 3.2 describes a thread-based parallel SMT solver. The implementation of both parallel solvers is based on the *safe partitioning* algorithm [7,8] where an input formula $\phi_{\mathrm{CNF}}$ is partitioned into $\phi_{\mathrm{CNF}}^1 \cdots \phi_{\mathrm{CNF}}^n$ that are pairwise unsatisfiable ($\phi_{\mathrm{CNF}}^i \wedge \phi_{\mathrm{CNF}}^j$ is unsatisfiable whenever $i \neq j$) and whose disjunction is equisatisfiable with $\phi_{\mathrm{CNF}}$ ($\bigvee_{i=1}^n \phi^i \equiv \phi_{\mathrm{CNF}}$). Each partition is then solved with a set $S^i$ of SMT solvers each using different random seeds. When a partition $\phi_{\mathrm{CNF}}^i$ is shown satisfiable the parallel solver terminates showing also $\phi_{\mathrm{CNF}}$ satisfiable, whereas if $\phi_{\mathrm{CNF}}^i$ is shown unsatisfiable for all $i$, also $\phi_{\mathrm{CNF}}$ is unsatisfiable. In case $\phi_{\mathrm{CNF}}^i$ is shown unsatisfiable, the parallel implementations reallocate the solvers $S^i$ evenly for solving the yet unsolved partitions. We also consider an important special case of safe partitioning where $n = 1$ called the *portfolio* approach [6].

### 3.1  OpenSMT2 for Cloud Computing

This tool consists of a framework using OpenSMT2 to provide an SMT solver designed to run on distributed cloud computing environments. The design follows a client-server model based on native TCP/IP message passing. The server receives input instances in the smt-lib format from the user, handles the connection with the clients, and acts as a front-end to the user. Both client disappearance and asynchronous connection of new clients at run-time are handled transparently by the server making the system more user-friendly and maintaining the soundness of the result also in case of disappearing clients. The clients are OpenSMT2 solvers whose task is to solve the instance $\phi_{\mathrm{CNF}}$ received from the server. Depending on the configuration of the server the system runs either in the safe partitioning or portfolio mode. Figure 4 gives an overview of the framework.

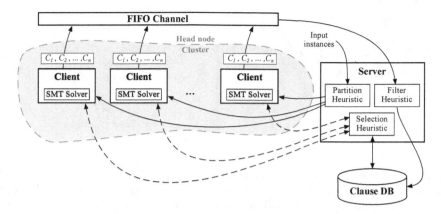

**Fig. 4.** The distributed SMT solver framework with clause sharing

The cloud computing implementation supports learned clause sharing among the clients $S^i$ working on the same partition $\phi^i_{\mathrm{CNF}}$. During the solving task, each client periodically sends new learned clauses through a FIFO channel which acts as a light push mechanism to the server. The clauses are stored to the Clause DB (see Fig. 4) where the clients periodically query new clauses. Heuristics for filtering promising clauses are used both when storing clauses to the Clause DB and when answering the client queries. In the current implementation the heuristics prefer short clauses over longer clauses. The connection required to update the clients is bidirectional since it is not possible to foresee when a client is ready to accept new clauses. The bidirectional connections are shown with dashed lines with double arrows.

In order to partition and share clauses the system must ensure that the internal clausal representation of each instance is the same in every client. The smt-lib format does not guarantee this since small changes in the input formula might result in optimizations that will dramatically change the formula $\phi_{\mathrm{CNF}}$. Instead OpenSMT2 uses a custom binary format storing its state. This format is used both for data transfers between each client and the server, and for initializing the solvers in the multi-threaded implementation.

The Clause DB and the FIFO queue are implemented with the in-memory database REDIS[1]. We chose REDIS since it supports both the *publisher / subscriber* messaging paradigm used as FIFO channel for clauses exchange, and the *hash set* feature which is useful to store clauses and handling sets operations used by both the filter and the selection heuristics. The use of the cloud implementation is explained in the source code repository.

Figure 5 reports an experimental evaluation of the cloud computing version on randomly selected, hard instances from QF_UF (*left*) and QF_LRA (*right*). The server is executed with six different configurations: partitioning the input instance into one (portfolio), two and eight partitions and spreading them among

---

[1] http://redis.io.

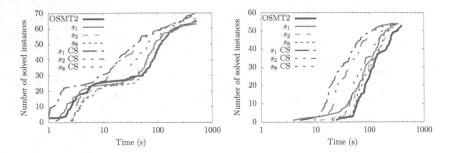

**Fig. 5.** OpenSMT2 cloud version comparison between partitioning in 1,2 and 8 partitions with and without clause sharing on QF_UF (*left*) and QF_LRA (*right*). $s_n$ stands for partitioning into $n$, and CS stands for using clause sharing and filtering clauses that contain more than 5 literals (Color figure online).

the 64 solvers in the cloud, with and without clause sharing. As a reference we also report the corresponding result with OpenSMT2. Clause sharing with 64 solvers results in a significant two-fold speed-up compared to not using clause sharing, and a three-fold speed-up compared to the sequential solver. The safe partitioning approach works better on QF_UF than on QF_LRA, suggesting that the role of the partitioning heuristic in QF_LRA might be more critical.

### 3.2  Multi-threaded OpenSMT2

OpenSMT2 supports also multi-threaded solving. The client dispatching in this case is similar to the cloud-computing version. A main thread partitions the input instance, creates the requested number of POSIX threads, and starts clients on the threads for solving the partitions. The communication between the main thread and the solver threads are handled with the efficient POSIX pipes.

The multi-threading feature of OpenSMT2 can be activated by setting the -p and -t arguments to, respectively, the number of partitions and solving threads. For example solving `instance.smt2` by partitioning the instance into two and using four threads is done by executing `opensmt -p2 -t4 instance.smt2`.

## 4  Conclusions

This work presents the SMT solver OpenSMT2, its architecture, two parallel variants, and a brief performance evaluation. The solver is used as the back-end in model-checking tools eVolCheck [5], FunFrog [11], and PeRIPLO [1,10]. We are currently working on improving the solver performance and its capabilities in theory interpolation.

**Acknowledgements.** This work was financially supported by Swiss National Science Foundation (SNSF) project number 153402.

# References

1. Alt, L., Fedyukovich, G., Hyvärinen, A.E.J., Sharygina, N.: A proof-sensitive approach for small propositional interpolants. In: Gurfinkel, A., Seshia, S.A. (eds.) VSTTE 2015. LNCS, vol. 9593, pp. 1–18. Springer, Heidelberg (2016). doi:10.1007/978-3-319-29613-5_1

2. Bruttomesso, R., Pek, E., Sharygina, N., Tsitovich, A.: The openSMT solver. In: Esparza, J., Majumdar, R. (eds.) TACAS 2010. LNCS, vol. 6015, pp. 150–153. Springer, Heidelberg (2010)

3. Detlefs, D., Nelson, G., Saxe, J.B.: Simplify: a theorem prover for program checking. J. ACM **52**(3), 365–473 (2005)

4. Dutertre, B., de Moura, L.: A fast linear-arithmetic solver for DPLL(T). In: Ball, T., Jones, R.B. (eds.) CAV 2006. LNCS, vol. 4144, pp. 81–94. Springer, Heidelberg (2006)

5. Fedyukovich, G., Sery, O., Sharygina, N.: eVolCheck: incremental upgrade checker for C. In: Piterman, N., Smolka, S.A. (eds.) TACAS 2013 (ETAPS 2013). LNCS, vol. 7795, pp. 292–307. Springer, Heidelberg (2013)

6. Huberman, B.A., Lukose, R.M., Hogg, T.: An economics approach to hard computational problems. Science **275**(5296), 51–54 (1997)

7. Hyvärinen, A.E.J., Marescotti, M., Sharygina, N.: Search-space partitioning for parallelizing SMT solvers. In: Heule, M., Weaver, S. (eds.) SAT 2015. LNCS, vol. 9340, pp. 369–386. Springer, Heidelberg (2015). doi:10.1007/978-3-319-24318-4_27

8. Hyvärinen, A.E.J., Junttila, T., Niemelä, I.: Partitioning search spaces of a randomized search. In: Cucchiara, R., Serra, R. (eds.) AI*IA 2009. LNCS, vol. 5883, pp. 243–252. Springer, Heidelberg (2009)

9. Marques-Silva, J.P., Sakallah, K.A.: GRASP: a search algorithm for propositional satisfiability. IEEE Trans. Comput. **48**(5), 506–521 (1999)

10. Rollini, S.F., Alt, L., Fedyukovich, G., Hyvärinen, A.E.J., Sharygina, N.: PeRIPLO: a framework for producing effective interpolants in SAT-based software verification. In: McMillan, K., Middeldorp, A., Voronkov, A. (eds.) LPAR-19 2013. LNCS, vol. 8312, pp. 683–693. Springer, Heidelberg (2013)

11. Sery, O., Fedyukovich, G., Sharygina, N.: FunFrog: bounded model checking with interpolation-based function summarization. In: Chakraborty, S., Mukund, M. (eds.) ATVA 2012. LNCS, vol. 7561, pp. 203–207. Springer, Heidelberg (2012)

# SpyBug: Automated Bug Detection in the Configuration Space of SAT Solvers

Norbert Manthey[1] and Marius Lindauer[2(⊠)]

[1] Technische Universität Dresden, Dresden, Germany
norbert.manthey@tu-dresden.de
[2] University of Freiburg, Freiburg, Germany
lindauer@cs.uni-freiburg.de

**Abstract.** Automated configuration is used to improve the performance of a SAT solver. Increasing the space of possible parameter configurations leverages the power of configuration but also leads to harder maintainable code and to more undiscovered bugs. We present the tool SpyBug that finds erroneous minimal parameter configurations of SAT solvers and their parameter specification to help developers to identify and narrow down bugs in their solvers. The importance of SpyBug is shown by the bugs we found for four well-known SAT solvers that won prices in international competitions.

## 1 Introduction

Recently, algorithm configuration [12] has shown to considerably improve the performance of SAT solvers by tuning their parameters, e.g. in the Configurable SAT Solver Challenge (CSSC; [13]). As a consequence, SAT solvers became highly flexible and configurable such that their performance can be optimized for broad range of different SAT formulas. One well-known example is the solver *Lingeling* [4] that exposes 323 parameters to the user and gave rise to $10^{1341}$ possible parameter configurations in the CSSC'14. In combination with its already strong default performance, *Lingeling* won the industrial track of the CSSC'14.

This flexibility of solvers comes with a price: as it is infeasible to assess the performance of all possible parameter settings, it is also infeasible to verify the correctness of all these settings. The high number of parameters of SAT solvers motivated this work, as we expect to find untested (possibly buggy) configurations when searching long enough. Since the components of a SAT solver highly interact with each other and due to their increasing size and complexity, it is hard to apply traditional methods aiming at formally proving SAT solver to be bug free (see Sect. 2). Therefore, we propose a new tool, called SpyBug (see Sect. 3) that searches in the space of possible parameter settings and instances

---

N. Manthey—Supported by the DFG grant HO 1294/11-1.

M. Lindauer—Supported by the DFG under Emmy Noether grant HU 1900/2-1.

N. Creignou and D. Le Berre (Eds.): SAT 2016, LNCS 9710, pp. 554–561, 2016.
DOI: 10.1007/978-3-319-40970-2_36

to find erroneous solver runs that unexpectedly terminated or returned a wrong proof. Since SAT solvers have many parameters and in most cases only a few parameters actually trigger the bug (often 2–10 out of hundreds of parameters), we furthermore use a heuristic to minimize the parameter configuration by removing all parameters that are not necessary for the bug. This automatic procedure helps the developer to narrow down the reason of the bug. In Sect. 4, we demonstrate the effectiveness of SpyBug on the participants of the industrial track of the CSSC 2014 and show that in 4 out of the 6 participants we were able to find bugs which can be often explained by less than 10 parameters. With a few modifications, SpyBug could also be applied to other problem domains, e.g. mixed integer programming (MIP), maximum satisfiability (MaxSAT), quantified Boolean formula (QBF) or SAT modulo theory (SMT).

## 2   Related Work: Bug Detection in Software

Bugs in software are found in several ways. On one hand, *formal methods* can be used to verify the software with respect to a specification, or to formally prove that certain error states cannot be reached. Typically recognizable states are assertions in the code, numerical overflows, or passing memory bounds. Similar bugs are found with *symbolic execution*, or by *fuzz testing*. Finally, we can use *unit tests* to test single components of a software system. A survey on formal methods for software verification can be found in [20].

Depending on the size of the software system under test, certain methods are not feasible due to their computational complexity. Unit tests are possible as soon as there are single components in a system. Here, for example the solvers data structures could be tested. Fuzz testing executes the software binary many times and checks each run for errors. For SAT solvers, there exists the fuzz testing suite [1,7], whose latter versions also support passing parameters to the used SAT solver and minimizing faulty combinations. However, the required feature to parse parameters from the formulas is not supported by many solvers. The approach of SpyBug treats the software under examination as a black box. Instead of the binary, the API of a solver library can be tested [7].

Formal methods have been applied to write verified SAT solvers with clause learning: F. Marić [16] specified a SAT solver in the *Isabelle* interactive theorem prover and then a solver implemented in *Haskell* is created. The SAT solver *versat* [17] is created and verified by using the language *Guru* and translating the code to *C*. In both projects, many techniques of recent systems have not been implemented, and there is a performance gap to modern systems.

For symbolic execution and proving the reachability of erroneous states it is hard to add assertions that check whether a solver state is valid. Therefore, we currently focus on improved fuzz testing, as we want to check the correctness of a highly configurable system without modifying the tool itself. With fuzz testing, we only require that there exists a formal specification of the parameters of the solver, and a standard input format and output format.

Hutter et al. [10] implemented a similar approach as used in SpyBug to identify bugs in two MIP solvers. They modified the algorithm configurator

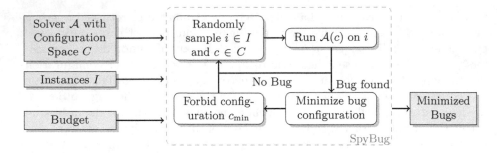

**Fig. 1.** Workflow of SpyBug

*ParamILS* [12] to record erroneous configurations and to minimize them. The main difference to SpyBug is that *ParamILS* will find bugs mostly in well-performing areas of the configuration space since its goal is not to find bugs but to optimize the performance of a solver. As *ParamILS* aims at optimization, it is more complicated to set up *ParamILS* compared to setting up SpyBug.

## 3   SpyBug's Framework

The tool SpyBug is supposed to help developers by finding bugs in large configuration spaces. We require only a parameter specification, the solver binary and a list of formulas to perform testing.[1] In the following sections, we explain the internals of SpyBug and present a use case. The tool is open-source under GPLv2 and available at http://www.ml4aad.org/spybug/.

### 3.1   Workflow

Since the idea of SpyBug is to search in the space of parameter configurations, it is built upon a simplified subset of the already well-known interfaces of the state-of-the-art algorithm configurators *ParamILS* [12] and *SMAC* [11] that was also used in the Configurable SAT Solver Challenge [13]. Hence, developers already using algorithm configuration can use SpyBug directly out-of-the-box. As illustrated in Fig. 1, the input of SpyBug is a solver with its configuration space, a set of instances to run the solver on, and a budget how long SpyBug runs (i.e., a wall clock time budget or a maximal number of solver runs). In the end, SpyBug returns a list of all minimized found erroneous configurations of the given solver.

The main workflow of SpyBug consists of a loop:

1. It samples an instance and a valid configuration uniformly at random.
2. It runs the solver with the configuration on the instance with some resource limits (i.e., the used runtime cutoff or memory limit). The solver is wrapped by the so-called *generic wrapper*, which was used in the CSSC and which

---

[1] Such a collection could be generated with fuzzing tools [7].

is responsible to determine whether it was a valid run or not. In a valid run, the solver either returned SATISFIABLE with a corresponding model or UNSATISFIABLE – to verify the output, we check the returned model – or it violated the resource limits. Hence, segmentation faults or failed assertions are also considered as bugs.

3. If it was a valid run, we continue with Step 1. Otherwise, we first minimize the found bug and then forbid the minimized configuration (or any super set of this configuration) to be sampled again. If the minimized configuration is empty (i.e., the bug is also triggered by the default configuration), we remove the instance from the instance set because we assume that the solver will always fail to run on this instance.

The *generic wrapper* judges a run based on the output of the solver, and the report of the *runsolver* tool [19]. By replacing the SAT-specific component of this script, SpyBug can easily be adapted to other types of solvers, for example MaxSAT solvers, Pseudo-Boolean solvers, QBF solver or SMT solvers.

## 3.2 Bug Minimization

The minimization of a configuration that triggered a bug is important to narrow down the reason for this bug. In the end, we are interested in the changed error-prone parameters with respect to the default configuration of a solver since the default configuration is often well tested and well studied by the developers. We therefore apply two iterations of a backward elimination, i.e., we iterate over all changed parameters and flip them to the default if the bug remains. The second iteration is done in reversed order of the parameters because a single traversal of the parameters could keep many non-relevant parameters. In preliminary tests, many further non-relevant parameters were removed with the help of a reverse traversal. To not spend too much time on the minimization, we do not use a full minimization procedure.

In this minimization process, we have to consider that the parameter configuration space (pcs) of a solver could be structured. According to the used pcs-format[2], the configuration space can consist of forbidden parameter combinations and conditional parameter clauses. For the first, we simply have to ensure that we do not violate these forbidden combinations. In contrast, conditional parameter clauses define when a parameter is active (e.g., a parameter of a heuristic is only active if this heuristic is indeed used). If we flip a parameter value we check which parameters are inactive and which ones need to be activated. Since a parameter will be set to its default value when it gets activated, the number of parameters to activate does not increase by flipping parameters during minimization.

In order to not find the same faulty configuration twice, the obtained combination of parameters is disallowed for all further solver calls. We furthermore remove all continuous parameters from this parameter configuration, as for such

---

[2] http://www.cs.ubc.ca/labs/beta/Projects/SMAC/v2.08.00/manual.pdf.

**Table 1.** Overview of participants of the industrial track of the Configurable SAT Solver Challenge 2014. For illustration, the number of configurations refers to a discretized version of the configuration space where each parameter has at most 8 possible values.

| Solver | # Parameters | # Configurations | Reference |
|---|---|---|---|
| *Minisat-HACK-999ED* | 10 | $8 \cdot 10^5$ | [18] |
| *Cryptominisat* | 36 | $3 \cdot 10^{24}$ | [21] |
| *Clasp-3.0.4-p8* | 75 | $1 \cdot 10^{49}$ | [9] |
| *Riss-4.27* | 214 | $5 \cdot 10^{86}$ | [15] |
| *SparrowToRiss* | 222 | $1 \cdot 10^{112}$ | [2] |
| *Lingeling* | 323 | $2 \cdot 10^{1341}$ | [5] |

a kind of parameter there might be too many other values that would lead to the same fault in the solver.

### 3.3    An Exemplary Use Case

We present a small use case for the solver *Riss-4.27* [15] (RISS) with its 214 parameters (specified in the file RISS.PCS). We use the solver and parameter specification that were submitted to the CSSC 2014 [13]. As a benchmark set, we use *Circuit Fuzz* formulas [7].

When starting SpyBug for the submitted version of *Riss-4.27* with the above setup, the following lines will be printed.

```
found a bug in -agil-r no -probe yes [...] on fuzz_100_20562.cnf
[...]
Minimal bug config: -cp3_strength no [...] -inprocess yes
```

The example highlights the major properties of SpyBug: when it finds a bug, it reports the full parameters that have been passed to the solvers wrapper script.[3] Furthermore, the formula that triggered the bug is printed (SAT_dat.k50.cnf). Next, the minimized list of parameters is printed. For the given example, the parameter -agil-r is not set any longer in the minimized set, as the bug is reproducible with the default value for this parameter. This combination is reported to the user, as shown in the example above.

## 4    Case Study: Configurable SAT Solver Challenge

To demonstrate the usefulness of SpyBug, we evaluated the 6 solvers that have been submitted to industrial track of the CSSC 2014 [13]. We independently analyze the solvers on the three benchmark families, i.e., hardware verification

---

[3] When reproducing the buggy configuration with the native binary, the parameters that might be added to the solver call in the wrapper script must be considered.

**Table 2.** Number of found bugs and their minimized average sizes ($\pm$ standard deviation) in the configuration space after 4 independent runs of SpyBug for at most 2 days or 10.000 randomly sampled configuration-instance pairs.

| | # Bugs | | | ∅ Bugs | | |
|---|---|---|---|---|---|---|
| | *IBM* | *CF* | *BMC* | *IBM* | *CF* | *BMC* |
| *Minisat-HACK-999ED* | 0 | 0 | 0 | – | – | – |
| *Cryptominisat* | 2 | 0 | 7 | $2 \pm 0$ | – | $3 \pm 2$ |
| *Clasp-3.0.4-p8* | 0 | 0 | 0 | – | – | – |
| *Riss-4.27* | 3 | 5 | 2 | $5 \pm 1$ | $13 \pm 6$ | $6 \pm 4$ |
| *SparrowToRiss* | 0 | 0 | 1 | – | – | $7 \pm 0$ |
| *Lingeling* | 0 | 2 | 0 | – | $8 \pm 2$ | – |

(*IBM* [22]), circuit fuzz (*CF* [7]), and bounded model checking (*BMC* [6]). Similar to automated configuration, we repeat the analysis for the combination of a solver and a benchmark family four times. The execution of a solver on a formula is limited to 300 CPU seconds and 3 GB RAM. We stop the analysis as soon as we tested 10 000 random configurations, or when reaching 2 days wall clock time. We communicated the found erroneous configurations to the solver developers.

The experiment has been executed on a cluster with 64 GB RAM that is shared among two Intel Xeon E5-2650v2 8-core CPUs with 20 MB L2 cache; running Ubuntu 14.04 LTS 64 bit. Table 1 presents the 6 solvers of the analysis and their number of parameters. While larger configuration spaces might yield better configurations, verifying the absence of bugs becomes harder.

Next, we present a summary of the bugs we found during the analysis in Table 2. For *Minisat-HACK-999ED* and *Clasp-3.0.4-p8*, we did not find a bug during our experiments. *Lingeling* has two misbehaving configurations for the *CF* family and *SparrowToRiss* shows bugs for the *BMC* family. For *Cryptominisat* and *Riss-4.27*, more than 5 bugs could be revealed on different families. The table furthermore presents how many parameters have to be changed from the default configuration to reach the faulty configuration. We note that SpyBug minimized the erroneous configurations of *Lingeling* with 323 parameters down to 8 parameters on average which drastically reduces the possible reasons of the responsible bugs. SpyBug got the best narrowing of responsible parameters for *Cryptominisat* for which only 2 or 3 parameters on average were necessary.

## 5 Conclusion

State-of-the-art SAT solvers contain many different techniques such as simplification during search. Combining only a few of these techniques can already result in unsound sequential systems [14]. Implementing competitive SAT solvers is a difficult task, especially if the system should be tuned for different applications. We presented the tool SpyBug, which orthogonally to existing tools finds bugs in the configuration space of the SAT solver. One specific use-case of SpyBug is to reveal bugs in solvers, before tuning them with many expensive resources. We reported bugs for four SAT solvers that won first prices in international competitions in 2014. For the future, we plan to integrate SpyBug into the framework of *SpySMAC* [8], i.e., an automatic framework to tune and analyse of SAT solvers.

# References

1. Artho, C., Biere, A., Seidl, M.: Model-based testing for verification back-ends. In: Veanes, M., Viganò, L. (eds.) TAP 2013. LNCS, vol. 7942, pp. 39–55. Springer, Heidelberg (2013)
2. Balint, A., Manthey, N.: SparrowToRiss. In: Belov et al. [3], pp. 77–78
3. Belov, A., Diepold, D., Heule, M., Järvisalo, M. (eds.): Proceedings of SAT Competition 2014: Solver and Benchmark Descriptions, Department of Computer Science Series of Publications B, vol. B-2014-2. University of Helsinki (2014)
4. Biere, A.: Lingeling, plingeling and treengeling entering the sat competition 2013. In: Proceedings of SAT Competition 2013, pp. 51–52 (2013)
5. Biere, A.: Yet another local search solver and Lingeling and friends entering the SAT competition. In: Belov et al. [3], pp. 39–40 (2014)
6. Biere, A., Cimatti, A., Claessen, K.L., Jussila, T., McMillan, K., Somenzi, F.: Benchmarks from the 2008 hardware model checking competition (HWMCC 2008) (2008). http://fmv.jku.at/hwmcc08/benchmarks.html
7. Brummayer, R., Lonsing, F., Biere, A.: Automated testing and debugging of SAT and QBF solvers. In: Strichman, O., Szeider, S. (eds.) SAT 2010. LNCS, vol. 6175, pp. 44–57. Springer, Heidelberg (2010)
8. Falkner, S., Lindauer, M., Hutter, F.: SpySMAC: automated configuration and performance analysis of SAT solvers. In: Heule, M., Weaver, S. (eds.) SAT 2015. LNCS, vol. 9340, pp. 215–222. Springer, Heidelberg (2015). doi:10.1007/978-3-319-24318-4_16
9. Gebser, M., Kaufmann, B., Schaub, T.: Conflict-driven answer set solving: from theory to practice. Artif. Intell. **187–188**, 52–89 (2012)
10. Hutter, F., Hoos, H.H., Leyton-Brown, K.: Automated configuration of mixed integer programming solvers. In: Lodi, A., Milano, M., Toth, P. (eds.) CPAIOR 2010. LNCS, vol. 6140, pp. 186–202. Springer, Heidelberg (2010)
11. Hutter, F., Hoos, H.H., Leyton-Brown, K.: Sequential model-based optimization for general algorithm configuration. In: Coello, C.A.C. (ed.) LION 2011. LNCS, vol. 6683, pp. 507–523. Springer, Heidelberg (2011)
12. Hutter, F., Hoos, H., Leyton-Brown, K., Stützle, T.: ParamILS: an automatic algorithm configuration framework. JAIR **36**, 267–306 (2009)
13. Hutter, F., Lindauer, M., Balint, A., Bayless, S., Hoos, H.H., Leyton-Brown, K.: The configurable SAT solver challenge. CoRR (2015). http://arxiv.org/abs/1505.01221
14. Järvisalo, M., Heule, M.J.H., Biere, A.: Inprocessing rules. In: Gramlich, B., Miller, D., Sattler, U. (eds.) IJCAR 2012. LNCS, vol. 7364, pp. 355–370. Springer, Heidelberg (2012)
15. Manthey, N.: Riss 4.27. In: Belov et al. [3], pp. 65–67
16. Maric, F.: Formal verification of a modern SAT solver by shallow embedding into isabelle/hol. Theor. Comput. Sci. **411**(50), 4333–4356 (2010)

17. Oe, D., Stump, A., Oliver, C., Clancy, K.: versat: a verified modern SAT solver. In: Kuncak, V., Rybalchenko, A. (eds.) Verification, Model Checking, and Abstract Interpretation. LNCS, vol. 7148, pp. 363–378. Springer, Heidelberg (2012)
18. Oh, C.: MiniSat HACK 999ED, MiniSat HACK 1430ED and SWDiA5BY. In: Belov et al.[3], p. 46
19. Roussel, O.: Controlling a solver execution with the runsolver tool. J. Satisfiability Boolean Model. Comput. **7**(4), 139–144 (2011)
20. Silva, V., Kroening, D., Weissenbacher, G.: A survey of automated techniques for formal software verification. IEEE Trans. Comput. Aided Des. Integr. Circuits Syst. **27**(7), 1165–1178 (2008)
21. Soos, M.: CryptoMiniSat v4. In: Belov et al. [3], p. 23
22. Zarpas, E.: Benchmarking SAT solvers for bounded model checking. In: Bacchus, F., Walsh, T. (eds.) SAT 2005. LNCS, vol. 3569, pp. 340–354. Springer, Heidelberg (2005)

# Author Index

Printed in the United States
By Bookmasters